Difficult to Diagnose Rare Diffuse Lung Disease

Difficult to Diagnose Rare Diffuse Lung Disease

Edited by

Alexander Averyanov

ELSEVIER

ACADEMIC PRESS

An imprint of Elsevier

Academic Press is an imprint of Elsevier
125 London Wall, London EC2Y 5AS, United Kingdom
525 B Street, Suite 1650, San Diego, CA 92101, United States
50 Hampshire Street, 5th Floor, Cambridge, MA 02139, United States
The Boulevard, Langford Lane, Kidlington, Oxford OX5 1GB, United Kingdom

Library of Congress Cataloging-in-Publication Data
A catalog record for this book is available from the Library of Congress

British Library Cataloguing-in-Publication Data
A catalogue record for this book is available from the British Library

ISBN: 978-0-12-815375-8

For information on all Academic Press publications
visit our website at https://www.elsevier.com/books-and-journals

Publisher: Stacy Masucci
Acquisition Editor: Katie Chan
Editorial Project Manager: Barbara Makinster
Production Project Manager: Punithavathy Govindaradjane
Cover Designer: Christian Bilbow

Typeset by SPi Global, India

Contents

Contributors

Numbers in parentheses indicate the pages on which the authors' contributions begin.

Alexander Averyanov (1, 29, 141, 165, 209, 221, 239, 265, 321, 361, 393), Clinical Department, Pulmonology Research Institute under FMBA of Russia; Pulmonary Division, Federal Research Clinical Center under FMBA of Russia, Moscow, Russia

Olesya Danilevskaya (1, 29, 141, 165, 239, 265, 321), Endoscopy Department, Pulmonology Research Institute under FMBA of Russia, Moscow, Russia

Evgeniya Kogan (29, 141, 165, 209, 221, 239, 265, 321, 361, 393), Anatomic Pathology Department, Sechenov University, Moscow, Russia

Victor Lesnyak (1, 29, 141, 165, 209, 221, 239, 265, 321, 361, 393), Radiology Department, Federal Research Clinical Center under FMBA of Russia, Moscow, Russia

Oleg Pikin (361), Thoracic Surgery Department, Hertzen Research Institute of Oncology, Moscow, Russia

Igor E. Stepanyan (29, 209, 221, 239, 321), Central TB Research Institute, Moscow, Russia

Preface

The diagnosis of diffuse parenchymal lung diseases (DPLD) is a very difficult task in respiratory medicine. On one hand, this is owing to the rare occurrence of such diseases and the lack of doctors with sufficient experience for a confident diagnosis. On the other hand, many diffuse lung diseases, which are often referred to as interstitial, have very similar clinical and radiological signs that make them difficult to differentiate. Moreover, different diseases can have the same morphological pattern or can be simultaneously represented by several mixed histopathologic patterns in one patient. All these aspects often do not allow physicians to consider the morphological conclusion as a decisive argument for final diagnosis. To improve the quality of diagnosis of DPLD, a multidisciplinary discussion (MDD) has been recently proposed, with participation of a pulmonologist, radiologist, morphologist, and sometimes rheumatologist and thoracic surgeon, in patients with controversial diagnosis or disagreements in the interpretation of the identified respiratory anomalies.

The idea of this monograph is to attempt to create a unified guide for physicians, which not only contains general information about rare DPLD, its histopathology, and high-resolution computed tomography (HRCT) features but also contains a detailed analysis of similar diseases that should be differentiated. We prepared a book that describes the theoretical basis for diagnosis and treatment of rare DPLD and practical issues that appear in the process of an MDD for choosing one or another examination method and interprets the results obtained. Based on our own long-term experience, which includes hundreds of difficult diagnostic cases, we discuss the diseases with similar clinical, radiological, and histopathologic signs and highlight the unique features that allow one to get closer to the correct diagnosis. Each chapter contains a large visual material with morphological characteristics, HRCT data, and alveoloscopic images of a number of diseases obtained during probe-based confocal laser endomicroscopy of distal airways. Due to a large number of images with comments, the monograph also carries the features of an atlas using HRCT for DPLD.

According to the definition of the European Respiratory Society, rare lung diseases are those that have a prevalence of <1 case per 2000 population, and their total list includes more than 6000 disorders, approximately 80% of which are genetically determined. We are aware of the fact that not all diseases that are among the orphan or rare DPLD are discussed in this monograph. We primarily chose those conditions that are most difficult to diagnose. Parts of these diseases are devoted to separate chapters; the rest are considered in the framework of the discussion of differential diagnosis and can be found with the index at the end of the book. However, for the majority of DPLD, such as idiopathic interstitial pneumonia, interstitial lung disease associated with connective tissue diseases, diffuse cystic lung lesions, airspace-predominant diseases, and hypersensitive pneumonitis, the material is given in detail with lining up of the differential rows. We hope that this work, prepared by a team of pulmonologists, a morphologist, a radiologist, a bronchologist, and a thoracic surgeon, will be useful and interesting for physicians of these specialties and for the course of MDD in patients with difficult diagnosis of DPLD.

On behalf of the author team,
Editor
Alexander Averyanov
Moscow, Russia

Acknowledgments

Authors are grateful to the team of the pulmonary department of the Federal Research Clinical Center of the Federal Medical and Biological Agency of Russia for their help in preparing the practical material and clinical cases for this monograph. We would like to emphasize the merits of Andrey Vorobyov in the preparation of CT scans for this publication. We are extremely grateful to Professor Oleg Epstein and the company Materia Medica Holding for sponsoring the translation of the manuscript. We remember with gratitude the role of the Eco-Vector LLC and personally Artem Bogachev in the language and copy editing support of this publication. We wish to gratefully acknowledge American Thoracic Society, American College of Rheumatology, European Respiratory Society and Fleischner Society granted their materials to the book.

We would also like to thank our families for their patience, which allowed the authors to fully immerse themselves in the preparation of this book.

Abbreviations

AA	amyloid protein A
AAH	atypical adenomatous hyperplasia
ABPA	allergic bronchopulmonary aspergillosis
ACCP	antibodies to cyclic citrullinated peptide
ACPA	anticyclic citrullinated peptide antibody
ACR	American College of Rheumatology
AE ILD	acute exacerbations of interstitial lung diseases
AE	acute exacerbation
AEP	acute eosinophilic pneumonia
AFOP	acute fibrinous organizing pneumonia
AH	alveolar hemorrhage
AHP	acute forms of HP
AIDS	acquired immunodeficiency syndrome
AILD	amiodarone-induced lung disease
AIP	acute interstitial pneumonia
AL	light-chain amyloidosis
ALP	acute lupus pneumonia
ALT	alanine aminotransferase
AM	amyloidosis
AML	alveolar microlithiasis
ANA	antinuclear antibodies
ANCA	antineutrophil cytoplasmic antibodies
anti-*MDA5*	anti-melanoma differentiation-associated gene 5
AP	alveolar proteinosis
ARDS	acute respiratory distress syndrome
AST	aspartate aminotransferase
ATS	American Thoracic Society
ATTR	amyloidosis transthyretin
AVM	arteriovenous malformation
AZA	azathioprine
BAC	bronchioloalveolar carcinoma
BAL	bronchoalveolar lavage
BG	bronchocentric granulomatosis
BHDS	Birt-Hogg-Dubé syndrome
BIILD	bleomycin-induced interstitial lung disease
BO	bronchiolitis obliterans
BOOP	bronchiolitis obliterans organizing pneumonia
BP	bacterial pneumonia
CEP	chronic eosinophilic pneumonia
CF	cyclophosphamide
CK	cytokeratin
CM	cystic metastases
COP	cryptogenic organizing pneumonia
COPD	chronic obstructive pulmonary disease
CPAM	congenital pulmonary airway malformation
CPK	creatine phosphokinase
CRP	C-reactive protein
CS	corticosteroids
CT	computed tomography
CTD	connective tissue disease
CYC	cyclophosphamide
DAD	diffuse alveolar damage
DAH	diffuse alveolar hemorrhage
DIEP	drug-induced eosinophilic pneumonia
DIILD	drug-induced interstitial lung disease
DIP	desquamative interstitial pneumonia
DIP	drug-induced pneumonitis
DIPD	drug-induced pulmonary diseases
DKC1	dyskerin pseudouridine synthase 1
DLCO	diffusing capacity for carbon monoxide
DLH	diffuse lymphoid hyperplasia
DNA	deoxyribonucleic acid
DOL	diffuse ossification of the lungs
DPDs	diffuse pulmonary diseases
DPL	diffuse pulmonary lymphangiomatosis
DPOD	dendriform parenchymal osteodystrophy
ECMO	extracorporeal membrane oxygenation
EGPA	eosinophilic granulomatosis with polyangiitis
EHS	exuberant honeycombing sign
ELD	eosinophilic lung disease
EP	eosinophilic pneumonia
EPD	eosinophilic pulmonary disease
ERS	European Respiratory Society
ESR	erythrocyte sedimentation rate
EULAR	European League Against Rheumatism
FB	follicular bronchiolitis
FCS	"four corners" sign
FDA	Food and Drug Administration
FDG	fluorodeoxyglucose
FEV1	forced expiratory volume in one second
FFS	five-factor score
FLCN	folliculin
FVC	forced vital capacity
GBM	glomerular basement membrane
GCS	glucocorticosteroids
GER	gastroesophageal reflux
GERD	gastroesophageal reflux disease
GGO	ground-glass opacity
GM-CSF	granulocyte-macrophage colony-stimulating factor
GPA	granulomatosis with polyangiitis
H&E	hematoxylin and eosin
HC	honeycombing
HES	hypereosinophilic syndrome
hITLN-1	human intelectin-1

HP hypersensitivity pneumonitis
HRCT high-resolution computed tomography
HU Hounsfield units
ICEP idiopathic chronic eosinophilic pneumonia
ID ill-defined
IGCS inhaled glucocorticosteroids
IIP idiopathic interstitial pneumonia
IL interleukin
ILD interstitial lung disease
ILD-MP interstitial lung disease associated with idiopathic autoimmune myopathies
ILD-PM/DM interstitial lung disease associated with polymyositis/dermatomyositis
ILD-SLE interstitial lung disease associated with systemic lupus erythematosus
ILD-SS interstitial lung disease associated with systemic scleroderma
IPA invasive pulmonary aspergillosis
IPAF interstitial pneumonia with autoimmune features
IPF idiopathic pulmonary fibrosis
IPHS idiopathic pulmonary hemosiderosis
IPPFE idiopathic pleuroparenchymal fibroelastosis
LAM lymphangioleiomyomatosis
LAP lymphadenopathy
LC lymphangitic carcinomatosis
LCB lung cryobiopsy
LCDD light-chain deposition disease
LCH Langerhans cell histiocytosis
LCM lymphangitic carcinomatosis
LDH lactate dehydrogenase
LIA lepidic invasive adenocarcinoma
LIP lymphoid (lymphocytic) interstitial pneumonia
LP lipoid pneumonia
LTOT long-term oxygen therapy
MALT mucosa-associated lymphoid tissue
MC metastatic calcification
MCL metastatic calcification in the lungs
MCLD multiple cystic lung disease
MDD multidisciplinary discussion
MILD methotrexate-induced lung disease
MIP methotrexate-induced pneumonitis
MM mycophenolate mofetil
MMP matrix metalloproteinase
MPA microscopic polyangiitis
MPC metastatic pulmonary calcification
MRI magnetic resonance imaging
MT methotrexate
MTOR mammalian target of rapamycin
ND no data
NLH nodular lymphoid hyperplasia
Non-ENT other than ear, nose, and throat disease
NPPO nodular parenchymal pulmonary osteodystrophy
NSAID nonsteroidal antiinflammatory drugs
NSCLC non-small-cell lung cancer
NSIP nonspecific interstitial pneumonia
OB obliterative (constrictive) bronchiolitis
OP organizing pneumonia
PAC pulmonary adenocarcinoma
PAH pulmonary arterial hypertension
PAP pulmonary alveolar proteinosis
PC pneumoconiosis
pCLE probe-based confocal laser endomicroscopy
PCR polymerase chain reaction
PE, PEX plasma exchange
PE pulmonary edema
PE pulmonary embolism
PEC perivascular epithelioid cells
PET positron emission tomography
PH pulmonary hypertension
PI pulmonary infections
PM pulmonary metastases
PM/DM polymyositis/dermatomyositis
PP *pneumocystis jirovecii* pneumonia
PPFE pleuroparenchymal fibroelastosis
RA rheumatoid arthritis
RAF VII recombinant activated factor VII
RA-ILD rheumatoid arthritis associated with interstitial lung disease
RB respiratory bronchiolitis
RB-ILD respiratory bronchiolitis–associated interstitial lung disease
RTEL1 regulator of telomere elongation helicase 1
RTX rituximab
SAA serum amyloid A protein
SAP serum amyloid P component
SBE subacute bacterial endocarditis
Scl-70 antibody to topoisomerase I
SCT stem cell transplantation
SES straight-edge sign
SLE systemic lupus erythematosus
SP-A surfactant protein A
SP-D surfactant protein D
SPE simple pulmonary eosinophilia
SR sarcoidosis
SS Sjögren syndrome
SSC systemic sclerosis
SSC-ILD systemic sclerosis associated with interstitial lung disease
ST septal thickening
TB tuberculosis
TBB transbronchial biopsy
TBLB transbronchial forceps lung biopsy
TBLC transbronchial lung cryobiopsy
TBOH tracheobronchopathia osteochondroplastica
TBW thickening of the bronchial walls
TERT telomerase reverse transcriptase
TGF-β transforming growth factor beta
TILS thickening of the intralobular interstitium or intralobular septa
TIS thickening of interlobular septa
TS tuberous sclerosis
TSC tuberous sclerosis complex
TTF thyroid transcription factor
UIP usual interstitial pneumonia
VAT video-assisted thoracoscopy
VATS video-assisted thoracoscopic surgery
VEGF vascular endothelial growth factor
WD well defined
WLL whole lung lavage

Chapter 1

HRCT and alveoloscopic assessment of diffuse lung disease

Alexander Averyanov[a,b], Victor Lesnyak[c], Olesya Danilevskaya[d]

[a]*Clinical Department, Pulmonology Research Institute under FMBA of Russia, Moscow, Russia,* [b]*Pulmonary Division, Federal Research Clinical Center under FMBA of Russia, Moscow, Russia,* [c]*Radiology Department, Federal Research Clinical Center under FMBA of Russia, Moscow, Russia,* [d]*Endoscopy Department, Pulmonology Research Institute under FMBA of Russia, Moscow, Russia*

A modern approach to the diagnosis of diffuse pulmonary diseases (DPDs) is based on a multidisciplinary discussion (MDD) among clinicians, pathologists, and radiologists. A study by Grenier et al. including patients with infiltrating DPDs revealed that the diagnostic accuracy with only clinical data was 29%, whereas that based on data from high-resolution computed tomography (HRCT) was 36%; the diagnostic accuracy reached 80% if all data, both clinical and radiographic, were considered [1]. The obligatory morphological evaluation for cases with difficult pulmonological diagnosis, which was the long prevailing approach, appears to have become obsolescent. Burg et al. compared the diagnoses for interstitial lung diseases based on the evaluation of biopsy material or MDD and found that the histological diagnosis based on MDD was revised in 30% of the cases [2]. Many rare pulmonary diseases have very clear and specific HRCT signs, and there is often no need for invasive diagnostic approaches to obtain histological material. For example, in cases of idiopathic pulmonary fibrosis, lymphangioleiomyomatosis, alveolar microlithiasis, and alveolar proteinosis as well as others, chest HRCT can be a definitive diagnostic method; in other cases, it can narrow the range of possible diagnoses so that additional laboratory tests or patient history can conclude the diagnostic quest satisfactorily. Thus knowledge of the basic interpretation of HRCT signs of DPDs is a prerequisite for the professional competence of a chest physician. In this chapter, we will overview the main HRCT signs of DPDs, which will be discussed in more detail in subsequent chapters that describe the radiological picture of individual DPDs. In addition, we provide basic information on probe-based confocal laser endomicroscopy (pCLE) of the lower respiratory tract, a relatively new method, which, similar to HRCT, is aimed at structural assessment of the lung tissue; however, unlike HRCT, pCLE does not utilize X-ray radiation but analyzes reflected laser beam from intraacinar structures and inclusions that have taking autofluorescence properties [3]. In this monograph, we also present our own data and conclusions regarding the prospects for the use of this technology based on our accumulating experience in pCLE for patients with DPDs.

The lung is a natural contrasting organ for X-ray techniques, and most pathological processes in the lung parenchyma are associated with increased or reduced attenuation of the pulmonary parenchyma. HRCT manifestations of DPDs are very diverse, which, however, can be categorized into several large patterns that can dominate in heterogeneous diseases (Table 1.1). In the previously cited study by Grenier et al. based on the analysis of computer tomography scans in 208 patients with interstitial pulmonary diseases, signs such as ground-glass opacity (GGO), reticular abnormalities, micronodules (<5 mm), and consolidation were the most common HRCT findings. Very rarely, only one HRCT pattern is present in DPDs; a combination of two or more patterns (Table 1.2) is more common, often forming specific radiological phenomena that are found only in certain diseases (e.g., the halo sign and crazy-paving sign). The specific distribution of pulmonary abnormalities is also important for the likelihood of an accurate diagnosis (Table 1.3).

Ground-glass opacity (GGO) is defined as a hazy increase in lung tissue attenuation with preservation of the bronchial and vascular margins (Figs. 1.1–1.2) [4]. GGO is one of the most common HRCT signs that characterize diffuse parenchymal lung diseases and may reflect both acute and chronic lung diseases [5]. It should be noted that a moderate

Difficult to Diagnose Rare Diffuse Lung Disease. https://doi.org/10.1016/B978-0-12-815375-8.00001-7

TABLE 1.1 Most common HRCT patterns in DPD

Pattern	Disease/s in which the pattern can dominate
Septal pattern	AP, AM, AH, DPL, LC, LP, PE, PM
Air traps	FB, HP, LIP, PP, RB-ILD
Honeycombing	CTD-ILD, IPF, IPPF, HP, NSIP, PC
Reticular abnormalities	CTD-ILD, DI-ILD, IPF, IPPF, HP, NSIP, SR
Cystic pattern	AM, BHDS, LAM, LCDD, LCH, LIP, TS
Ground-glass opacity	AP, COP, CTD-ILD, DI-ILD, DIP, DPL, EGPA, HP, IPF, AE-ILD, LIP, LP, NSIP, PAC, PE, PH, PP, RB-ILD
Consolidation	AEP, AE-ILD, AIP, AM, ARDS, COP, DAD, DI-ILD, GPA, LP, PA, PAC, PI, PE, SR
Nodular pattern	AML, FB, HP, GPA, LCH, LIP, PC, PI, PM, SR

AEP, acute eosinophilic pneumonia; *AE-ILD*, acute exacerbation of interstitial lung disease; *AIP*, acute interstitial pneumonia; *AM*, amyloidosis; *AML*, alveolar microlithiasis; *AH*, alveolar hemorrhage; *AP*, alveolar proteinosis; *BHDS*, Birt-Hogg-Dube syndrome; *CTD*, connective tissue disease; *COP*, cryptogenic organizing pneumonia; *DAD*, diffuse alveolar damage; *DI-ILD*, drug-induced interstitial lung disease; *DIP*, desquamative interstitial pneumonia; *DPL*, diffuse pulmonary lymphangiomatosis; *EGPA*, eosinophilic granulomatosis with polyangiitis; *ELD*, eosinophilic lung disease; *FB*, follicular bronchiolitis; *GPA*, granulomatosis with polyangiitis; *HP*, hypersensitivity pneumonitis; *IPF*, idiopathic pulmonary fibrosis; *IPPF*, idiopathic pleuroparenchymal fibroelastosis; *LAM*, lymphangioleiomyomatosis; *LC*, lymphangitic carcinomatosis; *LCH*, Langerhans cell histiocytosis; *LIP*, lymphoid interstitial pneumonia; *LCDD*, light-chain deposition disease; *LP*, lipoid pneumonia; *PAC*, pulmonary adenocarcinoma; *PI*, pulmonary infection; *PC*, pneumoconiosis; *PE*, pulmonary edema; *PM*, pulmonary metastasis; *PP*, *Pneumocystis* pneumonia; *SS*, Sjögren syndrome; *NSIP*, nonspecific interstitial pneumonia; *SR*, sarcoidosis; *RB-ILD*, respiratory bronchiolitis-associated interstitial lung disease; *TS*, tuberous sclerosis; *TB*, pulmonary tuberculosis.

decrease in the opacity of the pulmonary parenchyma can be observed in healthy people at expiration scanning, which, unlike GGO, usually manifests locally or in a mosaic distribution; the physiological decrease in lung density associated with respiration phase is diffuse and symmetrical. Sometimes, limited areas of increased attenuation, similar to GGO, can be visualized in the basal lung segments, which is associated with a gravitational effect of the blood flow.

The morphological substrate of GGO is thickening of the intralobular and interlobular interstitia, mainly inflammatory in origin, with partial filling of the alveolar spaces with cellular elements or fluid (exudate or transudate) as well as other substances such as protein-lipid complexes [6]. If the alveoli are filled entirely with a substance, the GGO can transform into a consolidation zone (Fig. 1.3). GGO can also be observed due to increased lung tissue attenuation due to other pathological processes including subatelectasis and pulmonary fibrosis [7] (Table 1.4).

GGO without fibrosis, traction bronchiectasis, or lung tissue distortion is often a sign of reversible and treatable pulmonary disease [8]. Importantly the vessel diameter in areas of GGO is almost unchanged on HRCT scans.

GGO should be differentiated from changes in the attenuation of lung tissue that occurs with the so-called mosaic attenuation or mosaic perfusion phenomenon [9], which, also known as false GGO, is characterized by a combination of high- and low-density zones with a patchwork or mosaic distribution in the volume of lung parenchyma. In fact, mosaic perfusion is a common term as it can occur by two different pathological mechanisms: (a) redistribution of blood flow from areas of hyperinflation to healthy areas, followed by an increase in lung parenchymal density in that area, and (b) hemodynamic disorders (e.g., pulmonary embolism and pulmonary hypertension) in the pulmonary artery system with blood flow to adjacent, unchanged anatomical units [10]. Mosaic perfusion is best detected on inspiratory scans. Reduction of vessel size in lucent lung regions bordering areas of GGO is an important feature distinguishing mosaic perfusion from true GGO (Fig. 1.4). Expiratory HRCT is used to clarify the etiology underlying mosaic perfusion, which reveals that areas of air trapping do not change the lucency, unlike zones of oligemia that is associated with increased attenuation on expiratory scans [11].

Lung tissue consolidation. Airspace consolidation sign on HRCT is defined as a high-attenuation area without visible vascular components in the lung tissue that results from the replacement of air in the alveoli with pathological contents. The morphological substrate of consolidation may be transudate (pulmonary edema), blood (pulmonary embolism and pulmonary hemorrhage), exudate with cellular contents (bacterial pneumonia and diffuse alveolar damage), cells (e.g., tumor, organizing pneumonia, and eosinophilic pneumonia), and substance accumulation (protein-lipid complexes with alveolar proteinosis and lipoid pneumonia) [12].

On HRCT scans, air-filled bronchi are frequently visible within the area of consolidation to create the air bronchogram sign (Fig. 1.5). Absence of bronchial lumens reflects filling of the bronchi by pathological content such as mucus, tumor,

TABLE 1.2 DPD with a typical combination of HRCT signs

	Ground-glass opacity	Consolidation	Cysts/cavities	Septal thickening	Intralobular septal thickening	Honeycombing	Ill-defined nodules	Well-defined nodules	Air traps
Ground-glass opacity		AH, AIP, COP, DAD, ELD, GPA, PAC, PE, PI	DIP, HP, LIP, PP, SS	AH, AM, AP, DPL, LC, LP, PAC, PE	AH, AP, HP, LIP, NSIP, PP	CTD, HP, LIP, NSIP	AH, DIP, EGPA, FB, HP, MPC, PE, PI, PP, RB-ILD	GPA, PM, SR	EGPA, FB, HP, PP
Consolidation	AH, AIP, COP, DAD, ELD, GPA, PAC, PE, PI		AM, GPA, LCDD, PAC, PI, TB	AM, LCDD PAC, PC, SR	AH, AP, PAC, PE	PC, SR	AH, COP, PAC, PE	GPA, PAC, PC, SR	EGPA, PP
Cysts/cavities	DIP, HP, LIP, PP, SS	AM, GPA, LCDD, PAC, PI, TB		AM, LCDD	AM, LIP	HP, LIP	LCH, PM, TB	LCH, PM, TB, TS	HP, PP
Septal thickening	AH, AM, AP, DPL, LC, LP, PAC, PE	AM, AH, PE, SR	AM, LCDD		HP, IPF, IPPF, NSIP	IPF, IPPF, NSIP, PC, SR	AH, LP, PAC, PE	PAC, PC, SR	PP
Intralobular septal thickening	AH, AP, HP, NSIP, PP	AH, AP, PAC, PE	AM, LIP	HP, IPF, IPPF, NSIP		HP, IPF, IPPF, NSIP	LIP	LCH	LCH, PP
Honeycombing	CTD, HP, LIP, NSIP	PC, SR	HP, LIP	IPF, IPPF, NSIP, PAC, SR	HP, IPF, IPPF, NSIP		HP	PC, SR	HP, IPF
Ill-defined nodules	AH, DIP, EGPA, FB, HP, MPC, PE, PI, PP, RB-ILD	AH, COP, PAC, PE	LCH, PM, TB	PAC, PE, AH, LP	LIP	HP		LCH, LIP, PC	EGPA, FB, HP PP
Well-defined nodules	GP, PM, SR	GPA, PAC, PC, SR	LCH, PM, TS, TB	PAC, PC, SR	LCH	PC, SR	LCH, LIP, PC		LCH
Air traps	FB, HP, PP	PP	HP, PP	PP	PP, LCH	HP, IPF	EGPA, FB, HP, PP	LCH	

AIP, acute interstitial pneumonia; *AM*, amyloidosis; *AH*, alveolar hemorrhage; *AP*, alveolar proteinosis; *COP*, cryptogenic organizing pneumonia; *CTD*, connective tissue disease; *DAD*, diffuse alveolar damage; *DIP*, desquamative interstitial pneumonia; *DPL*, diffuse pulmonary lymphangiomatosis; *EGPA*, eosinophilic granulomatosis with polyangiitis; *ELD*, eosinophilic lung disease; *FB*, follicular bronchiolitis; *GPA*, granulomatosis with polyangiitis; *HP*, hypersensitivity pneumonitis; *IPF*, idiopathic pulmonary fibrosis; *IPPF*, idiopathic pleuroparenchymal fibroelastosis; *LC*, lymphangitic carcinomatosis; *LCH*, Langerhans cell histiocytosis; *LIP*, lymphoid interstitial pneumonia; *LCDD*, light-chain deposition disease; *LP*, lipoid pneumonia; *MPC*, metastatic pulmonary calcification; *PAC*, pulmonary adenocarcinoma; *PI*, pulmonary infections; *PC*, pneumoconiosis; *PE*, pulmonary edema; *PM*, pulmonary metastasis; *SS*, Sjögren syndrome; *NSIP*, nonspecific interstitial pneumonia; *SR*, sarcoidosis; *PP*, *Pneumocystis* pneumonia; *RB-ILD*, respiratory bronchiolitis–associated interstitial lung disease; *TS*, tuberous sclerosis; *TB*, pulmonary tuberculosis.

TABLE 1.3 The most typical distribution patterns of abnormalities in DPDs

Distribution	Disease
Parachilar	AH, DPL, PE, PP
Perilymphatic	AM, LC, LCH, PC, SR
Basal, subpleural	AML, DIP, IPF, LIP, NSIP
Peribronchovascular	AP, COP, SR
Upper and middle zones	CEP, EGPA, IPPF, MPC, PP, RB-ILS, SR, TB
Random	COP, GPA, LAM, PAC

AM, amyloidosis; *AML*, alveolar microlithiasis; *AH*, alveolar hemorrhage; *AP*, alveolar proteinosis; *CEP*, chronic eosinophilic pneumonia; *COP*, cryptogenic organizing pneumonia; *DIP*, desquamative interstitial pneumonia; *DPL*, diffuse pulmonary lymphangiomatosis; *EGPA*, eosinophilic granulomatosis with polyangiitis; *GPA*, granulomatosis with polyangiitis; *HP*, hypersensitivity pneumonitis; *IPF*, idiopathic pulmonary fibrosis; *IPPF*, idiopathic pleuroparenchymal fibroelastosis; *LC*, lymphangitic carcinomatosis; *LCH*, Langerhans cell histiocytosis; *LIP*, lymphoid interstitial pneumonia; *MPC*, metastatic pulmonary calcification; *NSIP*, nonspecific interstitial pneumonia; *PI*, pulmonary infection; *PC*, pneumoconiosis; *PE*, pulmonary edema; *PM*, pulmonary metastasis; *PAC*, pulmonary adenocarcinoma; *SR*, sarcoidosis; *PP*, *pneumocystis jirovecii* pneumonia; *RB-ILD*, respiratory bronchiolitis–associated interstitial lung disease; *TB*, pulmonary tuberculosis.

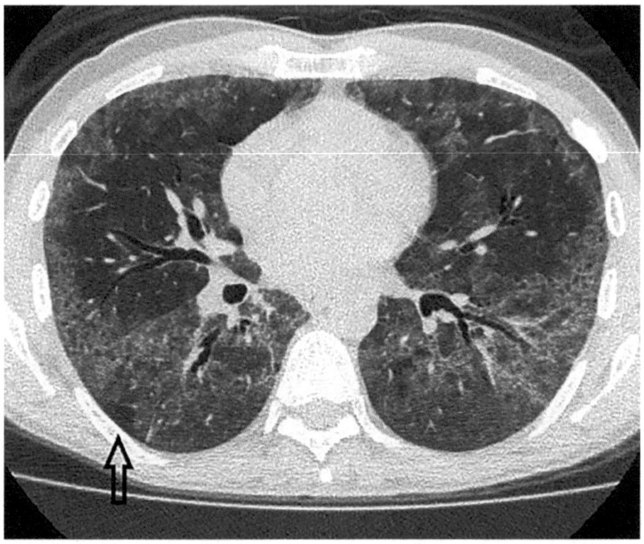

FIG. 1.1 *Pneumocystis jirovecii* pneumonia in a patient with acquired immunodeficiency syndrome. Extensive bilateral areas of ground-glass opacity associated with intralobular thickening. Air trap *(arrow)* is visualized within the GGO. Note the clear outline from intact pulmonary parenchyma.

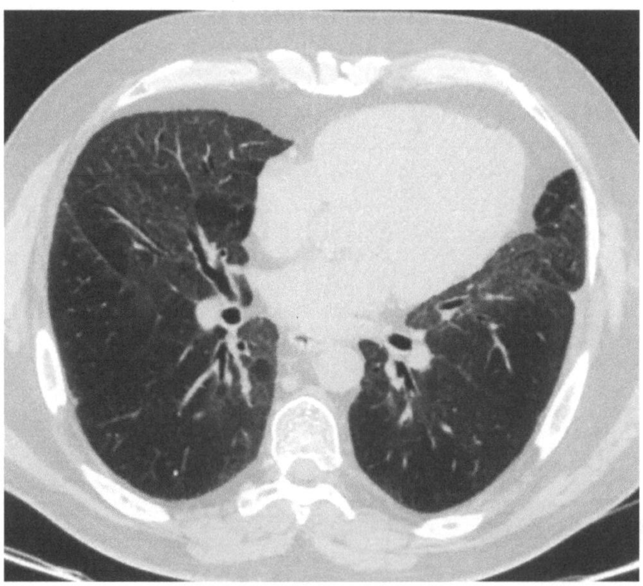

FIG. 1.2 Valsartan-induced pneumonitis. Areas of ground-glass opacity in the anterior segments. Vessels and bronchi are clearly visualized inside.

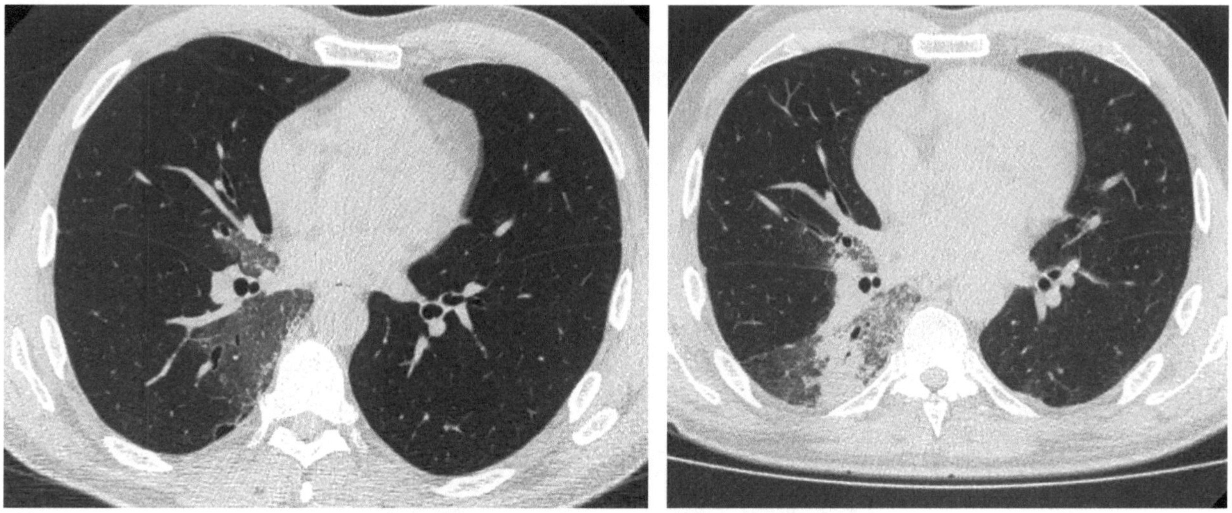

(A) (B)

FIG. 1.3 Dynamics of the high-resolution computed tomography pattern of bacterial pneumonia in a 55-year-old male. Corresponding HRCT scans obtained in 1-day intervals. Area of ground-glass opacity in the lower lobe of the right lung, with a clear border from the unaffected parenchyma (A). Expansion of the ground-glass opacity, with the appearance the consolidation zones inside (B).

TABLE 1.4 Main causes of GGO

Course	Disease	Additional HRCT signs
Acute	Pulmonary edema	Diffuse or patchy GGO, a relatively parahilar symmetrical distribution (i.e., butterfly sign), does not reach visceral pleura
	Alveolar hemorrhage	Patchy GGO with parahilar distribution, thickening of the interlobular septa, often in combination with consolidation
	ARDS, diffuse alveolar damage, acute interstitial pneumonia	Diffuse bilateral massive consolidation zones
	Bacterial pneumonia	Asymmetrical, often unilateral consolidation areas
Subacute	*Pneumocystis* pneumonia	Multiple nodules and areas of GGO in early stage, predominantly in the upper lobes
	Fungal pneumonia	Foci of consolidation, often associated with destructive cavities
	Hypersensitivity pneumonitis	Air trapping, centrilobular nodules
	Organizing pneumonia (cryptogenic organizing pneumonia)	Consolidation areas surrounded by GGO (i.e., halo sign), subpleural and peribronchovascular distribution, mild reticular changes
	Desquamative interstitial pneumonia	Patchy or diffuse areas of GGO predominantly basal localization, centrilobular nodules
	Chronic eosinophilic pneumonia	Consolidation areas in the upper lobes, migration of the affected areas, atoll sign
	Mycobacteriosis	Consolidation and bronchiectasis usually present, distribution predominantly in the upper lobes
	Acute exacerbation of interstitial lung disease	Diffuse or patchy GGO with or without consolidation which superimposed on the features of underlying disease
Chronic	Nonspecific interstitial pneumonia	Posterior basal distribution, moderate honeycombing, reticular changes
	Idiopathic pulmonary fibrosis	Honeycombing, pronounced reticular changes, subpleural and basal localization
	Vasculitis	Foci of consolidation with cavities (granulomatosis with polyangiitis), centrilobular nodules (eosinophilic granulomatosis with polyangiitis)
	Lung sarcoidosis	Increased intrathoracic lymphatic nodes, multiple small nodules with perilymphatic distribution
	Adenocarcinoma with lepidic growth (former bronchioloalveolar carcinoma)	Zones of consolidation, pseudocavity, the air bubble sign
	Alveolar proteinosis	Areas of "geographic" distribution of GGO associated with interlobular septal thickening

ARDS, acute respiratory distress syndrome; *GGO*, ground-glass opacity.

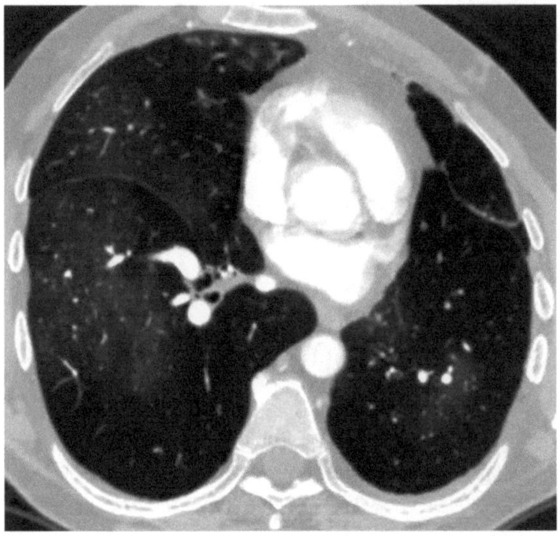

FIG. 1.4 Mosaic perfusion in a patient with constrictive bronchiolitis. Vast areas of increased attenuation bordering hyperlucent regions containing vessels with decreased size.

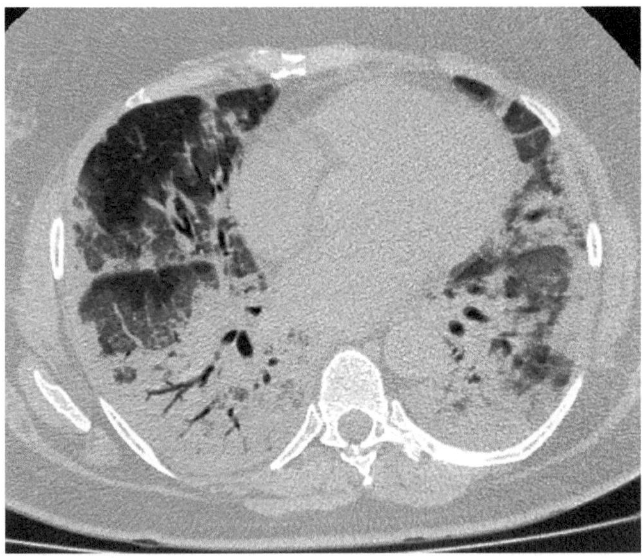

FIG. 1.5 Pneumonia caused by H1N1 influenza. Bilateral massive consolidation zones turning into areas of ground-glass opacity area. On the right, bronchial lumens are clearly visible (the air bronchogram sign).

and blood. Densitometric indicators of consolidation may closely resemble those that are typical for soft tissues. The consolidation zone structure may be homogeneous or heterogeneous due to the presence of cavities or areas of partial alveolar damage (Fig. 1.6) [13]. Consolidation zones often have fuzzy contours similar to GGO, due to the irregular involvement of the peripheral alveoli in the pathological process.

Consolidation, unlike atelectasis, is usually not accompanied by a change in the volume of the affected anatomical pulmonary unit [12].

Given the diversity of pathological processes underlying this HRCT finding, the range of diseases manifesting with the consolidation of lung tissue is quite wide and heterogeneous (Table 1.1). The prevalence and localization of the abnormalities are important for the differential diagnosis of the consolidation sign. Diffuse distribution is most typical for diseases with diffuse alveolar or vascular damage including acute interstitial pneumonia, acute eosinophilic pneumonia, severe infectious diseases, pulmonary edema, and diffuse alveolar hemorrhage. Subpleural and peribronchovascular distributions are typical for organized pneumonia, sarcoidosis, adenocarcinoma with lepidic growth, and alveolar proteinosis (Fig. 1.7).

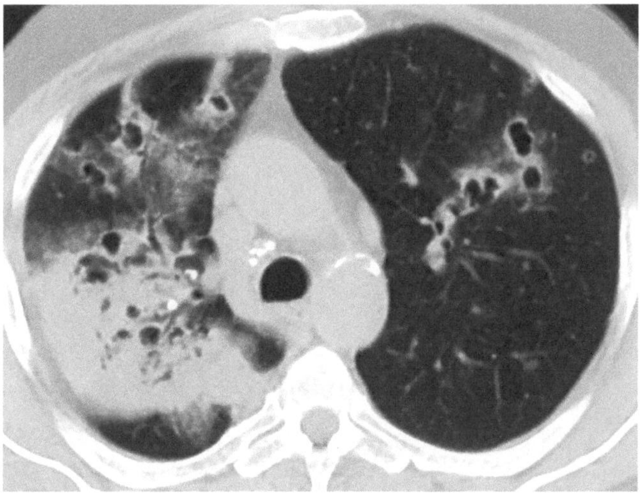

FIG. 1.6 Mucinous invasive adenocarcinoma. Massive consolidation zone in the upper right lobe. Multiple cavities are visualized inside. Heterogeneous cavities associated with ground-glass opacity are also defined in the anterior segments. Several small, randomly located, poorly differentiated nodules in the left lung.

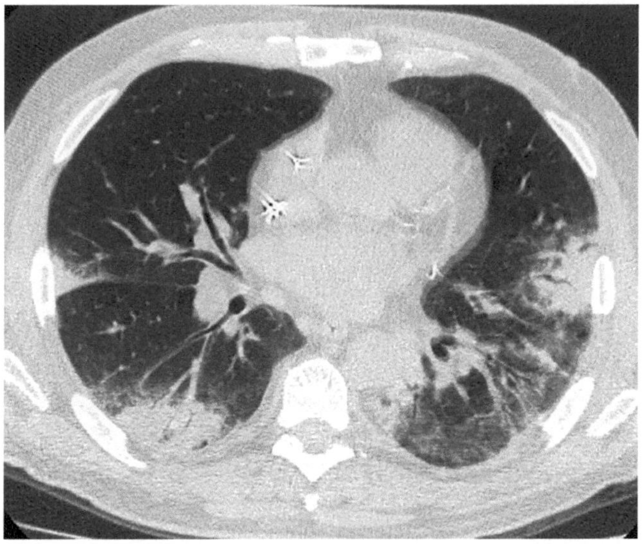

FIG. 1.7 Cryptogenic organizing pneumonia. Bilateral subpleural consolidation zones surrounded by ground-glass opacity (halo sign).

The combination of GGO and consolidation can lead to the appearance of special radiological presentation; the halo (nimbus) sign, named after its resemblance to the luminous ring around the sun, is the most common. The halo sign is characterized by the presence of a GGO shadow around the area of consolidation (Figs. 1.7 and 1.8). The underlying etiology of the halo sign can be tumor infiltration of the lung stroma and alveolar walls, edema and inflammatory reaction, and hemorrhagic filling of alveolar spaces that violates the permeability of the alveolar-capillary membranes. The halo sign can be observed with organized pneumonia, infectious lesions, granulomatosis with polyangiitis, and primary and secondary lung tumors [14].

Another specific manifestation of the combination of GGO with consolidation is the reversed halo, or the atoll, sign. In contrast to the halo sign, the atoll sign manifests as a dense annular consolidation zone around the area of GGO or unaffected parenchyma, resembling a coral reef or a crescent. The atoll sign is a characteristic of organizing pneumonia (secondary or cryptogenic) and eosinophilic pneumonia and, as a rule, is characterized by the presence of multiple sites (Fig. 1.9) [15].

Linear and reticular opacities in the lung tissue result from thickening of the pulmonary interstitium due to the presence of fluid, cellular infiltration, extracellular material (amyloid), and/or fibrous tissue or expansion of the local lymphatic vessels [16]. As a rule, such processes are diffuse and constitute the pathological basis of several DPDs. Therefore the HRCT

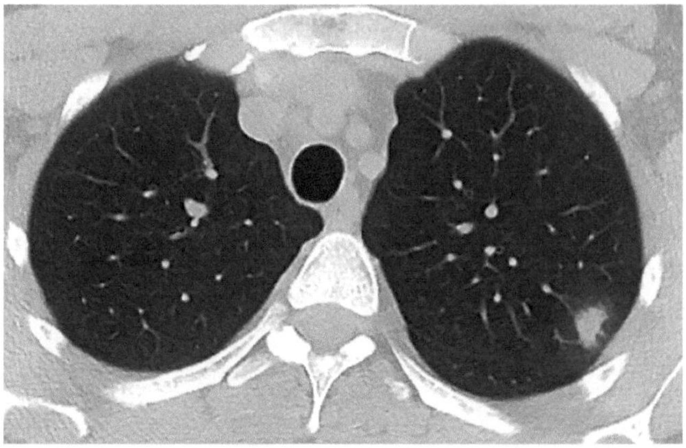

FIG. 1.8 Atypical presentation of community-acquired pneumonia in a 22-year-old female. A single focus of consolidation in the upper lobe of the left lung, surrounded by a rim of ground-glass opacity, resembling adenocarcinoma, which disappeared completely after a 5-day course of antibacterial therapy.

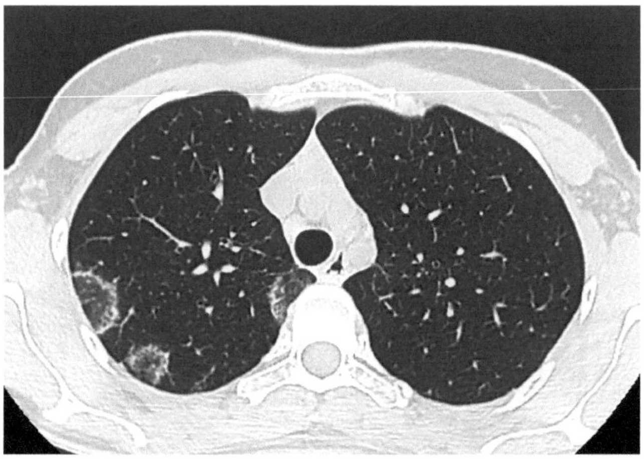

FIG. 1.9 Chronic eosinophilic pneumonia. Foci of ground-glass opacity surrounded by consolidation rings, that is, the reversed halo sign.

manifestations of changes in the pulmonary interstitium are quite diverse and depend on the underlying morphological substrate. There are three main types of reticular changes: thickening of the interlobular septa, thickening of the intralobular septa, and honeycombing [16].

Thickening of the interlobular septa resembles a mesh pattern on HRCT: the segments, approximately 1–2 cm in length, form polygonal shapes corresponding to the secondary pulmonary lobule (Fig. 1.10). The thickening may be smooth, nodular, or irregular. The dominance of the thickening of the interlobular septa over other reticular abnormalities is termed the septal pattern, which is a characteristic of several rare DPDs (Table 1.5).

The uniform thickening of the interlobular septa in areas of GGO is termed the crazy-paving sign (Fig. 1.11), which was initially described in patients with alveolar proteinosis and later identified in many other diseases (see Chapter 4.1, Table 4.1.2).

On HRCT scans, nodular thickening of the interlobular septa is as unevenly thickened lines resembling beads or rosaries. This sign is often accompanied with an increase in lymph nodes and central and peribronchovascular interstitium. Nodular thickening of the interlobular septa is typical for sarcoidosis, amyloidosis, lymphogenous spread of tumors, lymphoproliferative diseases and vascular lymphatic abnormalities, silicosis, and other pneumoconioses (Fig. 1.12). Irregular thickening of the interlobular interstitium is often associated with interstitial intralobular fibrosis and occurs in idiopathic pulmonary fibrosis, in late-stage nonspecific interstitial pneumonia, sarcoidosis, hypersensitivity pneumonitis, and pneumoconiosis (Fig. 1.10).

By HRCT, thickening of the intralobular interstitium or intralobular septa (TILS) is observed as thin lines that are several millimeters long, which are located more often in the peripheral part of the lung parenchyma, resembling a fine web [17]. TILS that is not accompanied by GGO can be an early sign of interstitial fibrosis, whereas intralobular thickening

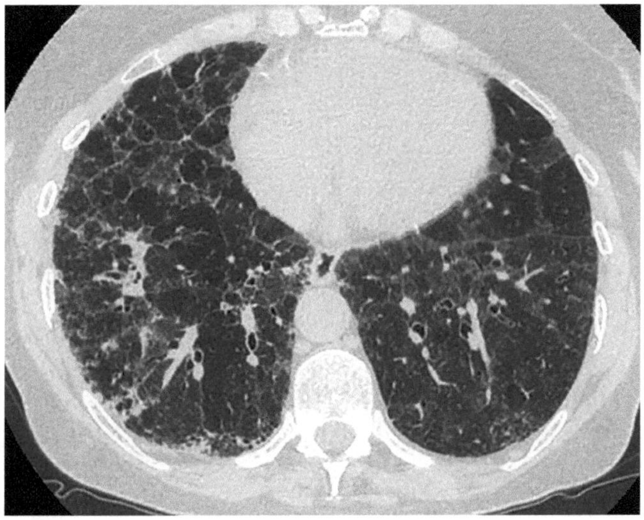

FIG. 1.10 Chronic hypersensitivity pneumonitis. Irregular thickening of the interlobular septa, primarily in the right middle lobe. Diffuse zones of ground-glass opacity associated with air traps. The initial manifestations of honeycombing.

TABLE 1.5 Specific aspects of the septal pattern in DPDs

Smooth	Irregular	Nodular
Pulmonary edema	Post-ARDS pulmonary fibrosis	Lymphangitic carcinomatosis
Alveolar hemorrhage	Radiation fibrosis	Sarcoidosis
Alveolar proteinosis	Idiopathic pulmonary fibrosis	Pneumoconiosis
Lipoid pneumonia	Chronic hypersensitivity pneumonitis	Amyloidosis
Pulmonary postcapillary hypertension	Fibrous nonspecific interstitial pneumonia	Light-chain deposition disease
Lymphangiosis carcinomatosa	Idiopathic pleuroparenchymal fibroelastosis	Alveolar microlithiasis
Diffuse pulmonary lymphangiomatosis	Sarcoidosis	
Pneumocystis pneumonia	Pneumoconiosis	
Adenocarcinoma (formerly bronchioloalveolar carcinoma)		
Amyloidosis		
Light-chain deposition disease		
Niemann-Pick disease		
Erdheim-Chester disease		

ARDS, acute respiratory distress syndrome; *DPD*, diffuse pulmonary disease.

associated with areas of GGO indicates concurrent interstitial inflammation, which can be a manifestation of infection (*Pneumocystis jirovecii*, viral pneumonia) or noninfectious interstitial lesions of the lungs (Fig. 1.13). Fibrotic processes are indicated by the appearance of bronchiectasis in areas of TILS and adjacent transseptal irregular lines usually extending from the visceral pleura deep into the pulmonary parenchyma, with a thickness of 1–3 mm and a length of up to 5 cm. Most often, these are associated with the formation of linear fibrosis and often observed in combination with pleural thickening [1] (Fig. 1.14).

Honeycombing is defined as clustered cystic airspaces, typically with a comparable diameter of 3–10 mm, which are usually subpleural and have well-defined walls [18]. On HRCT scans, honeycombing resembles air cavities that are round or oval in shape, with quite distinct and relatively thick walls (1–3 mm), which are localized typically in peripheral and

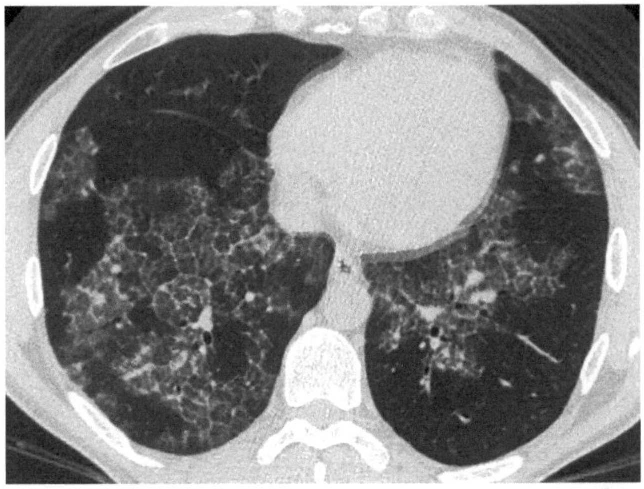

FIG. 1.11 Alveolar proteinosis. Bilateral patchy areas of ground-glass opacity associated with thickened interlobular septa, that is, the crazy-paving sign.

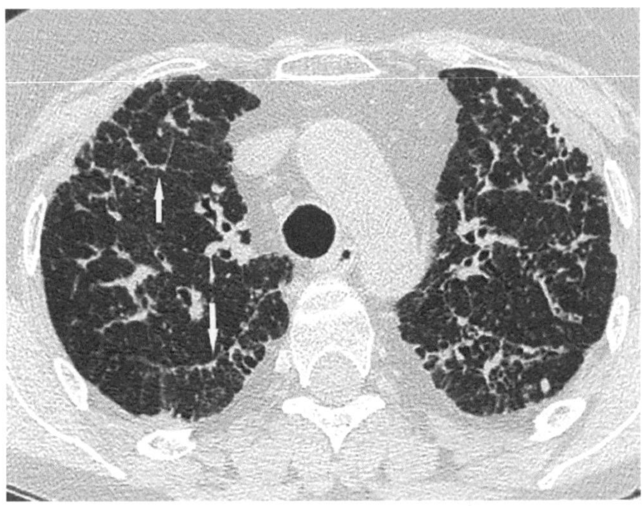

FIG. 1.12 Chronic pulmonary sarcoidosis. Significant reticular changes. Irregular thickening of the interlobular septa with nodular patches *(arrows)*.

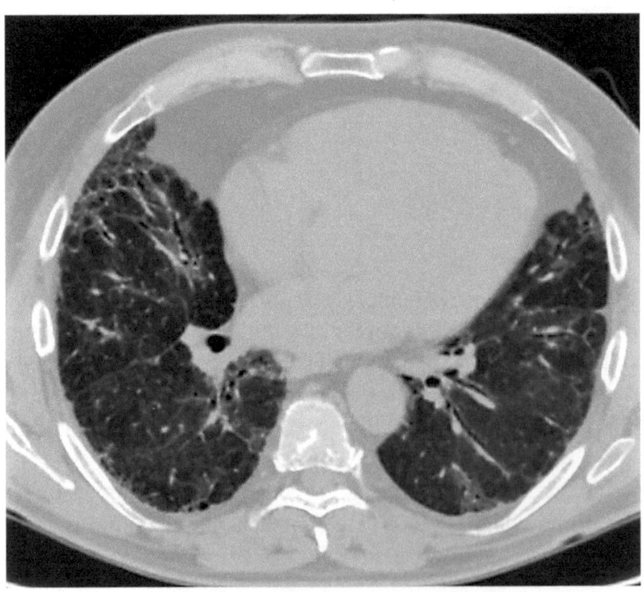

FIG. 1.13 Interstitial lung disease associated with systemic sclerosis. Diffuse ground-glass opacity. Reticular abnormalities (thickened intralobular septa in subpleural zones and isolated thickened interlobular septa) and traction bronchiectasis.

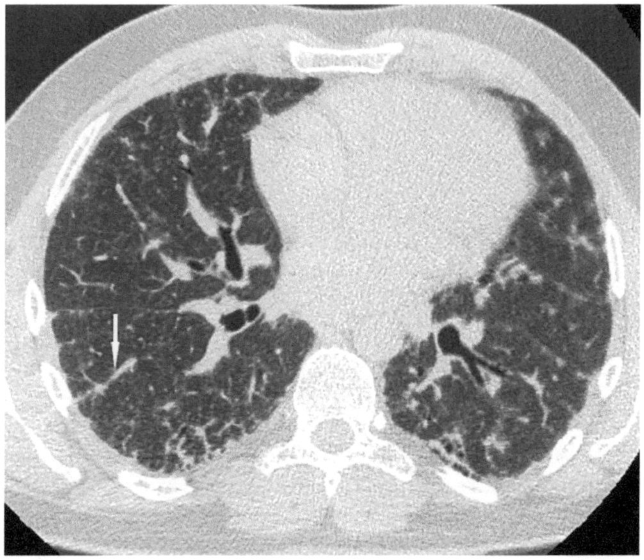

FIG. 1.14 Idiopathic pulmonary fibrosis. Pronounced subpleural reticular changes and manifestations of honeycombing. Interseptal line *(arrow)*.

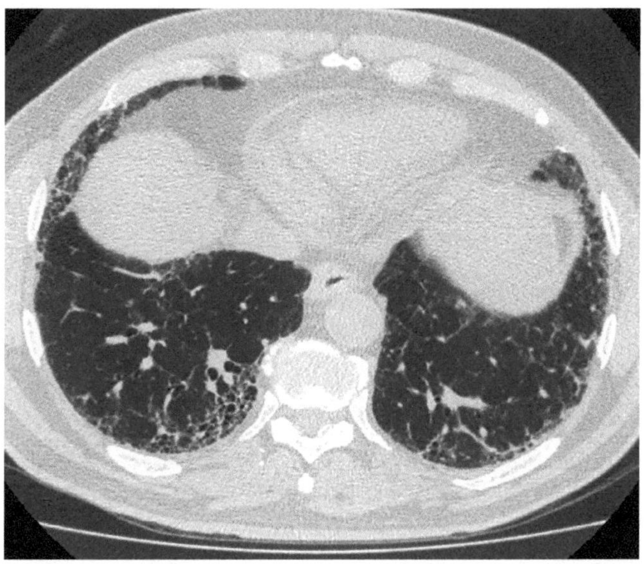

FIG. 1.15 Idiopathic pulmonary fibrosis. Irregular reticular abnormalities and subpleural honeycombing, most pronounced in the right lung.

subpleural fields of the lungs (Fig. 1.15) [19]. Histologically the HRCT areas of honeycombing correspond to respiratory-lined cysts (dilated cysts partially lined with respiratory epithelium and fibroblast foci) and peripheral bronchiolectasis (subpleural cysts lined by respiratory epithelium with a discrete smooth muscle layer, often in continuity with a proximal bronchus, extending to within 1 cm of the visceral pleura) [20]. The presence of honeycombing indicates advanced-stage fibrotic lung diseases; however, in patients with idiopathic pulmonary fibrosis, a fragmented honeycombing pattern can be observed at relatively early stages of the disease, hindering interpretation. According to Watadani et al. the interobserver agreement on the assessment of honeycombing differed among clinicians, expert radiologists, and board-certified chest radiologists (κ value, 0.4–0.7) [19]. An important distinguishing feature of honeycombing is the multirow arrangement of small cysts (Fig. 1.16), which is distinct from paraseptal emphysema that is also located subpleurally, albeit in one layer, which most often observed in the upper lobes and usually (but not always) larger than 1 cm in diameter (Fig. 1.17) [16]. Nevertheless, some experts believe that single-row cysts, especially at the onset of the disease, can be an HRCT sign of honeycombing [18]. Such minimal changes propose the mandatory presence of at least three subpleural cysts, 3–10 mm in diameter, with visible walls arranged in a row or cluster (Fig. 1.18) [16]. Honeycombing is the most important component

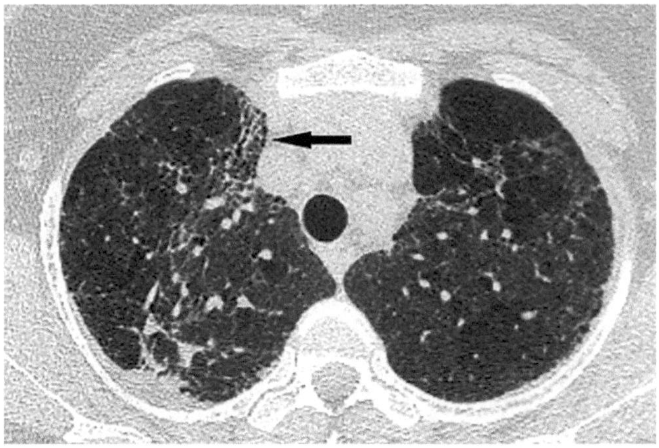

FIG. 1.16 Chronic hypersensitivity pneumonitis after prolonged contact with birds. Pronounced reticular changes, pleural layering, subpleural honeycombing, and air traps. Unlike paraseptal emphysema, the cysts are small in size and arranged in several layers *(arrow)*.

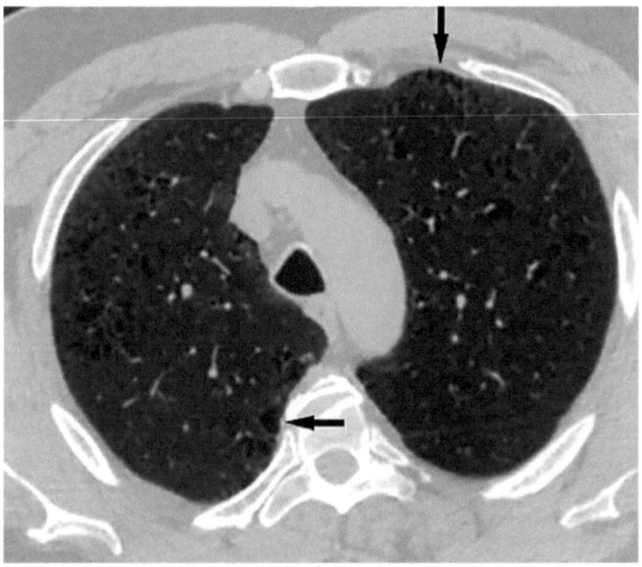

FIG. 1.17 Paraseptal emphysema. Multiple small cavities in subpleural zones *(arrows)*. Unlike honeycombing, emphysematous clarification is located in one layer, accompanied by centriacinar emphysema in the absence of reticular abnormalities.

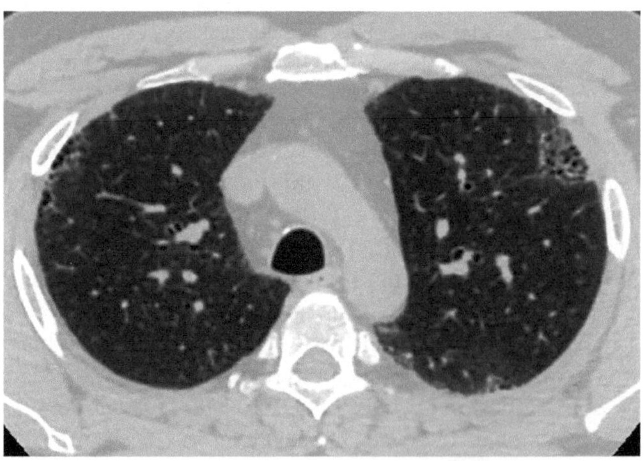

FIG. 1.18 Honeycombing of a worker in wood manufacturing. On the left, multirow subpleural honeycombs are seen; on the right, there is one layer of small cysts.

of the radiological manifestations and a hallmark of usual interstitial pneumonia [21]. Therefore, diseases that can manifest with honeycombing include those that assume the development of UIP: idiopathic pulmonary fibrosis, nonspecific interstitial pneumonia, connective tissue diseases, chronic hypersensitivity pneumonitis, and drug-induced pulmonary fibrosis, among others [22].

The UIP pattern is discussed in more detail in Chapter 2.1 (Idiopathic pulmonary fibrosis). In addition to paraseptal emphysema, honeycombing sometimes must be differentiated from diffuse cystic lung diseases (lymphangioleiomyomatosis and Langerhans cell histiocytosis), especially in cases where cystic changes are grouped subpleurally (Fig. 1.19).

Pulmonary cysts are round airspaces with thin walls (usually <2 mm) located inside the pulmonary parenchyma [4]. Pulmonary cysts can occur in numerous DPDs; however, the cystic pattern dominates over the remaining radiological findings in a substantially small subset of disorders (Table 1.1). Cystic lung lesions are discussed in detail in Chapter 9. Pulmonary cysts are not necessarily a disease manifestation; the so-called incidental pulmonary cysts are also found in people without lung diseases [23]. In the Framingham Heart Study, among 2633 participants who underwent chest HRCT, at least one random cyst was found in 7.6% of the subjects, whereas more than five cysts were found in 0.9% of the cohort; the cysts were significantly associated with older age, low body mass index, and calcification of the coronary arteries [24]. Of note, incidental cysts were not found in patients under the age of 40 years, indicating their relation to age [24].

The valve mechanism is the most often considered among the proposed pathogenic pathways of cyst formation, which involves obstruction at the level of bronchioles with blocking of the expiratory flow, as well as ischemic dilatation of the distal airways and high protease activity in the acinus zone [25]. The same mechanisms are also considered in the formation of pulmonary emphysema with which pulmonary cysts most often have to be differentiated. Cysts on HRCT may be similar to centrilobular pulmonary emphysema that is sometimes observed in a round shape (Fig. 1.20). However, the absence of a visible wall and the frequent detection of the intralobular artery in the center of the focal zone of lucency usually aid in differentiating the two conditions [26]. Conversely, it is significantly more difficult and sometimes impossible to clearly distinguish a bulla from a cyst. The bullae are larger than 1 cm in diameter, with <1-mm thick walls, usually located subpleurally, and often associated with paraseptal, centriacinar emphysema, and other signs of chronic obstructive pulmonary disease [27]. Differentiating cysts from cystic bronchiectasis is usually not difficult. The latter are usually grouped in separate segments resembling bunches of grapes, have an adjacent vessel, and are accompanied by signs of infectious bronchiolitis [28].

Air traps are a radiographic finding that reflects air retention with stretching and hyperinflation of secondary lobules due to the obstruction of adductor bronchioles [29]. The most common causes of air traps are diseases of the small bronchi (chronic obstructive pulmonary disease, severe asthma, noncystic fibrosis bronchiectasis, and bronchiolitis) and interstitial lung diseases involving respiratory bronchioles (e.g., hypersensitivity pneumonitis, sarcoidosis, and idiopathic pulmonary fibrosis); however, air traps can also occur in healthy adults (Figs. 1.1, 1.10, and 1.21) [30,31].

Areas of decreased attenuation usually have clear anatomical boundaries. The presence of air traps is confirmed by expiratory scanning: The density of intact areas of lung tissue is physiologically reduced, leading to an increase in attenuation, whereas the air traps remain unchanged compared with the inspiratory scans (Fig. 1.22) [32]. The most frequent signs

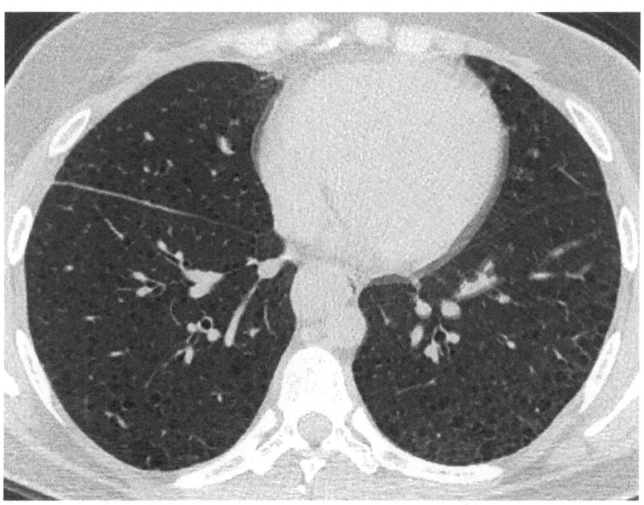

FIG. 1.19 Lymphangioleiomyomatosis. The subpleural location of the cysts resembles honeycombing. However, unlike honeycombing, the cystic walls are thinner, and cysts are distributed not only under the pleura but also in the middle of the pulmonary parenchyma; there are no other signs of fibrosis (reticular abnormalities or bronchiectasis).

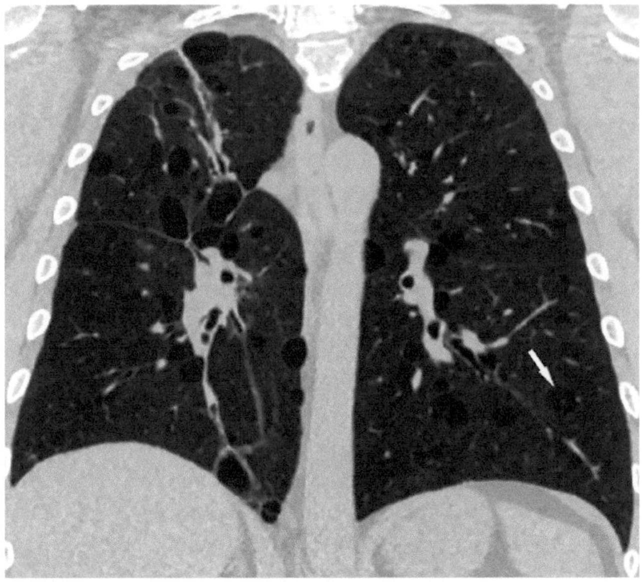

FIG. 1.20 Bullous and centriacinar emphysema (confirmed with video-assisted thoracoscopic surgical biopsy) in a 54-year-old heavy smoker male that resembles cystic lung disease (coronal reconstruction). Bilateral cyst-like hyperlucent abnormalities, some of which have a visible thin wall. In some places, intralobular vessels are visible *(arrow)*. In the right lung, fibrous bands are formed after right-sided pneumonia.

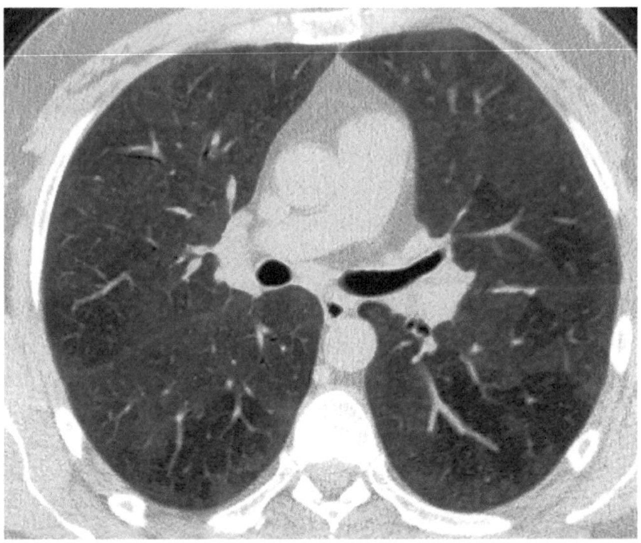

FIG. 1.21 Subacute hypersensitivity pneumonitis in a brewery worker. Bilateral diffuse areas of ground-glass opacity associated with irregular air trapping.

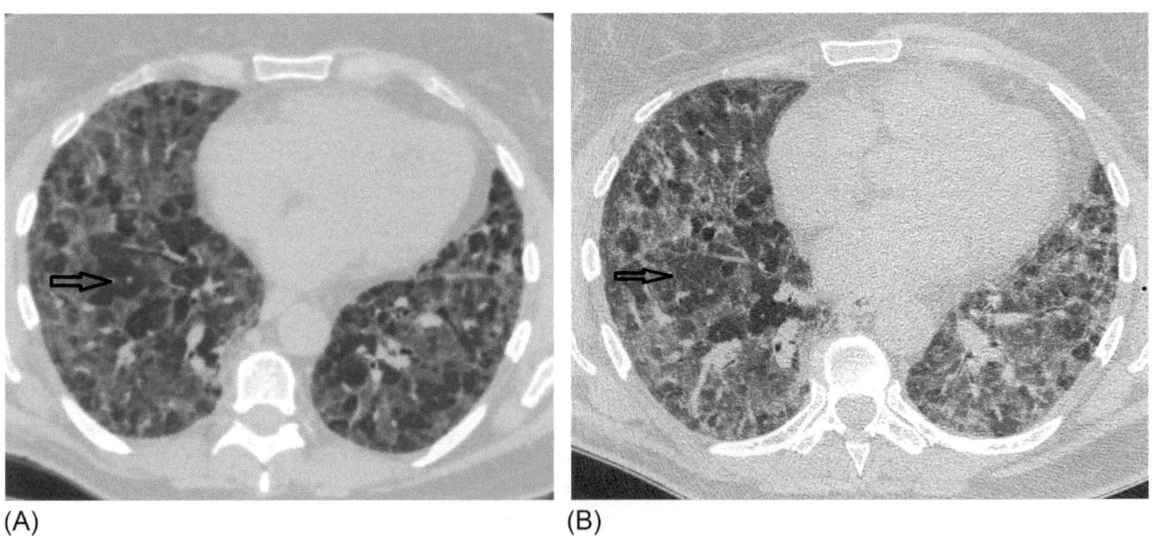

FIG. 1.22 Hypersensitivity pneumonitis. Inspiratory (A) and expiratory (B) scans. After expiration true air traps are visualized as remaining areas of decreased attenuation, whereas non-affected zones that looked like air traps, have become more dense (*arrows*).

associated with the presence of air traps on HRCT are bronchiectasis, the tree-in-bud sign, reticular changes, and honeycombing as well as GOO and architectural distortion [30].

Nodules are defined as round or irregular opacities that are well or poorly defined, which measure up to 3 cm in diameter [4]. Micronodules have a size of <3 mm, whereas nodes more than 3 cm in diameter are considered as masses [4]. Characteristics (i.e., contours, density, and distribution) of nodules are the most important differential diagnostic signs of many pulmonary diseases. Well-defined nodules with a higher density are more common in granulomatosis, malignancies, and diseases with hematogenous or lymphogenous spread, whereas ill-defined nodules often have an inflammatory origin and lower density close to that of a GGO (Table 1.2).

The distribution of nodules relative to the secondary pulmonary lobule is an important aspect in their differential diagnosis. *Perilymphatic distribution* suggests localization of the nodules along the lymphatic vessels and bronchovascular bundles, in the interlobular septa, adjacent to the pleura, including the pleurae of the interlobar fissures, and in the intralobular septa. This distribution is typical for sarcoidosis, silicosis, and other types of pneumoconiosis, lymphoproliferative diseases, and lymphogenous metastasis (Fig. 1.23) [33]. *Centrilobular nodules* occur near intralobular arteries and bronchioles and, due to their similarity to secondary lobules in size, are often evenly distributed [34]. Their size typically ranges from a few millimeters to 1 cm, and their density ranges from ground-glass to solid opacities [33]. The classic distinguishing feature of centrilobular nodules is the location that is no closer than 5–10 mm from the visceral pleura [35]; however, in patients with severe diffuse lesions, the nodules may be located close to the pleura (Fig. 1.24). Some studies classify centrilobular nodules into two distinct patterns: those associated with the tree-in-bud sign and those that are ill-defined [36].

The tree-in-bud sign appears with obstruction of the peripheral bronchioles (usually by inflammatory secretion) and manifests as typical Y-shaped structures with nodules at the ends (Fig. 1.25). In most cases, this sign indicates an infectious origin for the nodular lesions, although it can rarely be a manifestation of aspiration, juvenile laryngotracheobronchial papillomatosis, or obliterating bronchiolitis [37]. Poorly defined centrilobular nodules usually appear in patients with diseases affecting the intralobular structures such as vessels, walls of the bronchioles, and alveolar spaces, which most often are DPDs (Table 1.1). With random distribution the nodules are located within the lung tissue without a link to the anatomical lung structures; this nodular distribution corresponds to hematogenous spread of the pathological processes such as that seen in tuberculosis, metastases, and fungal infections (Fig. 1.26) [38]. Nodules can be detected in pleural surfaces and interlobular interstitium and near vascular branches; in typical cases the process is bilateral and symmetrical, involving both lungs.

In conclusion the analysis of all HRCT findings in patients with DPDs can significantly approximate the diagnosis; however, comprehensive analysis of all clinical, radiological, laboratory, and often morphological data in an MDD is usually necessary for a reliable definitive diagnosis.

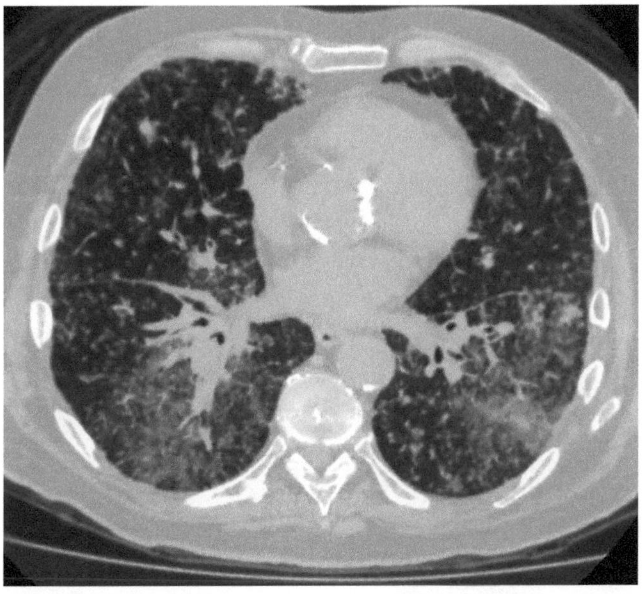

FIG. 1.23 Gastric sarcoma with lymphogenic metastases to the lungs. Multiple nodules of various sizes maximally distributed along the visceral pleura, around the bronchi, and vessels. Areas of ground-glass opacity in the posterior segments.

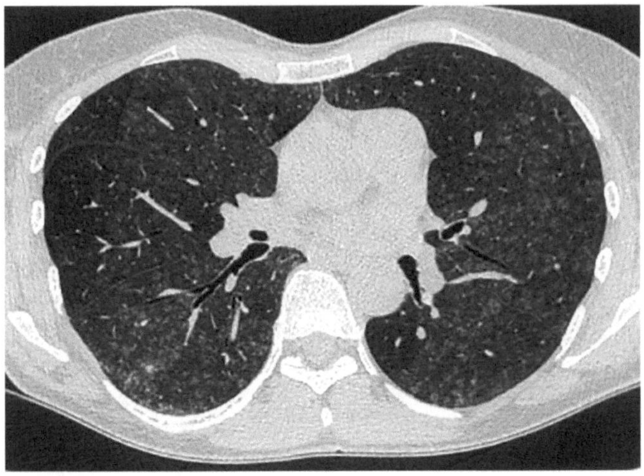

FIG. 1.24 Subacute hypersensitivity pneumonitis after contact with birds. Diffuse, innumerous, poorly differentiated centrilobular nodules, some of which are merging with the visceral pleura. Lobular air trapping.

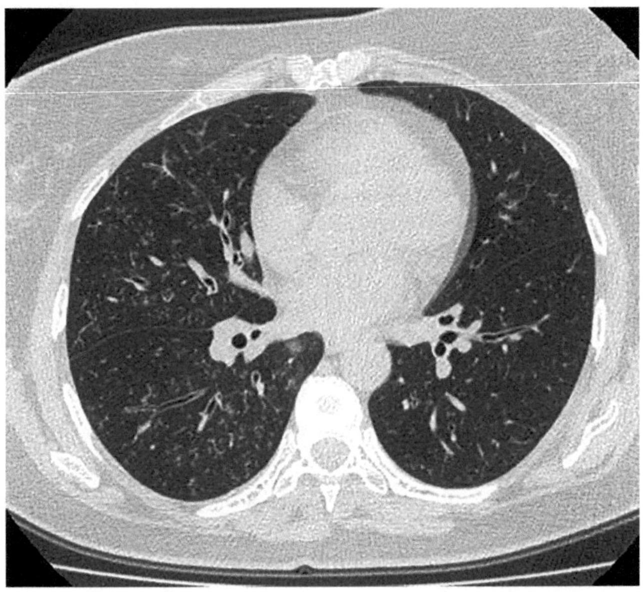

FIG. 1.25 Infectious bronchiolitis in a patient with right-sided non-cystic fibrosis bronchiectasis. Multiple small nodules associated with the subpleural tree-in-bud sign.

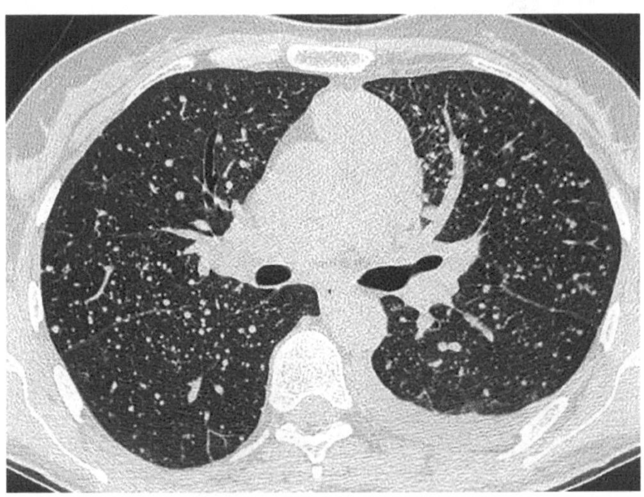

FIG. 1.26 Disseminated pulmonary tuberculosis. Diffuse, random distribution of well-defined nodules. Left-sided pleural effusion.

Probe-based confocal laser endomicroscopy of distal airways (pCLE)

pCLE is a relatively new minimally invasive diagnostic approach that allows real time in vivo evaluation of the lung tissue structure based on to their fluorescent response. The degree of resolution capacity on an order of several micrometers provides visualization of the elastic framework, blood vessels, and individual cellular and noncellular structures of the terminal bronchioles and alveoli [39]. The fiber-optic probe of the endomicroscope, while in contact with the tissue under examination, excites the fluorescence of tissue by laser radiation passing through the probe; the optical signal, that is, the image, of the fluorescent tissue response through the same optical probe returns to the recording spectral optical system.

Based on our experience with pCLE of distal airways in patients with certain DPDs, we herein summarize our findings and specific endomicroscopic patterns associated with specific lesions. Distal airway pCLE or alveoloscopy was performed during bronchoscopy after completion of the endoscopic examination. The Cellvizio diagnostic system (Mauna Kea Technologies, France), with an AlveoFlex miniprobe (1.4mm in diameter) and laser radiation with a wavelength of 488nm, was used. The resolution capacity of the miniprobe was 3.5μm, the diameter of the optical field was 600μm, and the depth of investigation was 0–50μm, with 12 images per second. Additional fluorophores were not used in alveoloscopy, since the method was based on the natural autofluorescence of the structures of the bronchi and the pulmonary parenchyma.

The miniprobe was inserted through the working channel of the videobronchoscope (EB-530T; Fujinon, Japan) and advanced to the distal regions until a weak resistance from the lung tissue was achieved, and the alveolar sacs and ducts were visualized. All available segmental and subsegmental bronchi were examined sequentially both sides, except RB1 and LB1+2 in some cases due to technical unattainability. A dynamic monochrome microscopic picture, which was observed on the monitor screen, was recorded in film format and reproduced for analysis.

Terminal bronchioles with a diameter of 1–2mm change into respiratory bronchioles, the diameter of which is no more than 0.6mm [40]. The diameter of the AlveoFlex miniprobe is more than two times the diameter of the terminal bronchiole. In our experience the alveolar structures could be evaluated by penetration of the bronchiole walls in majority of the patients undergoing this procedure (approximately 90%), by slightly increasing the pressure on the miniprobe while advancing forward along the instrumental channel, which was reported in other studies [41], with the exception of those patients with pronounced atrophy or individual anatomical features (Fig. 1.27).

In single cases a miniprobe could be inserted directly into the alveoli through the respiratory bronchiole, without damaging the wall of the terminal bronchiole; in such vases the shape and, if possible, dimensions of the alveolar mouth were also recorded (Fig. 1.28).

Terminal bronchioles are visualized based on the transverse enhancement of fluorescence, which is due to the presence of smooth muscle fibers, a specific anatomical feature of this part of the respiratory system (Fig. 1.29A) [42]. Based on our survey of 10 healthy individuals, the alveoli of an individual who never smoked were visualized as a clear, round structure with a septal thickness of 10.27±0.54μm. There were no secretions or any cellular structures in the lumen of the alveoli (Fig. 1.29B).

In smokers without any lung diseases, the structure of the alveoli was similar to that of nonsmokers; however, some (score of 1–3 on a scale from 0 to 6) brightly fluorescent macrophages, 15–30μm in diameter, were noted in the lumen of the alveoli (Fig. 1.30); the observed autofluorescence of macrophages is due to inclusions of tobacco tar [41]. For a semi-quantitative assessment of the presence of alveolar macrophages in the alveoli and respiratory bronchioles, we developed a 6-point scale, where a score of 0 was used for no cells, a score of 1 reflected single cells in the visual field (with additional

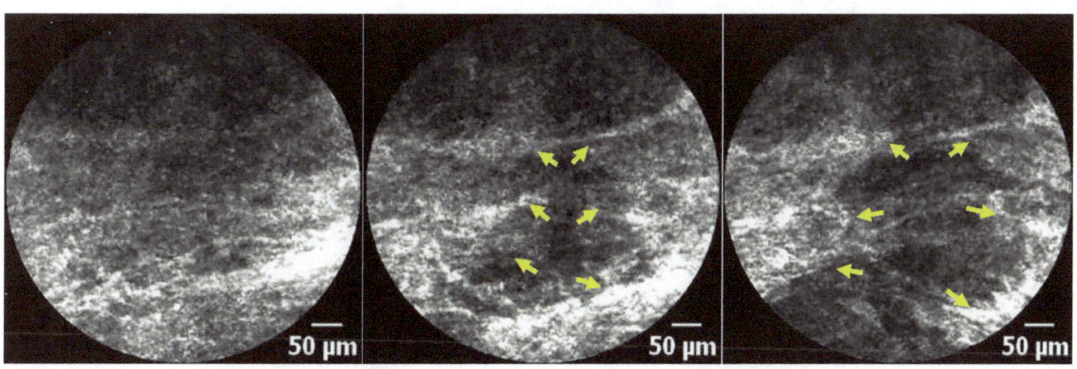

FIG. 1.27 Technique for visualization of alveolar structures in pCLE. Penetration of the terminal bronchiole wall is characterized by the rupture of elastic fibers with a slight increase in the pressure on the miniprobe. *Yellow arrows* indicate the ends of the elastic fibers, the distance between which increases progressively as the miniprobe moves forward.

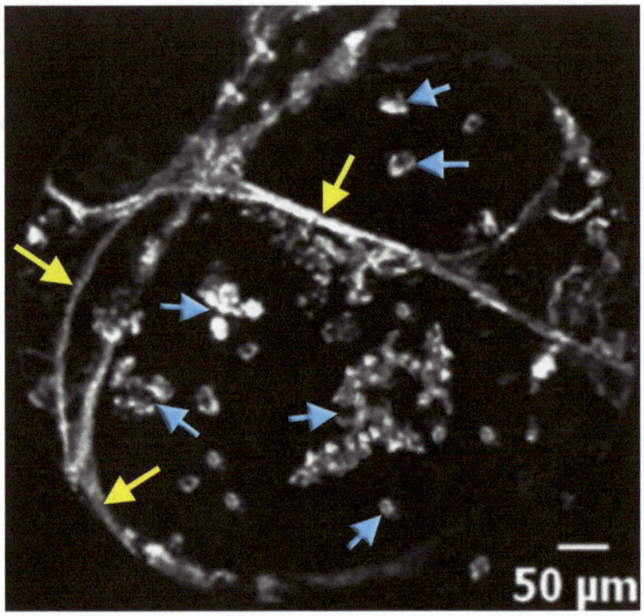

FIG. 1.28 Visualization of the alveolar mouth. The miniprobe is inserted directly through the terminal bronchiole without penetration. *Yellow arrows* indicate the elastic fibers of the alveolar septa; *blue arrows* show intraluminal macrophages located individually or in groups.

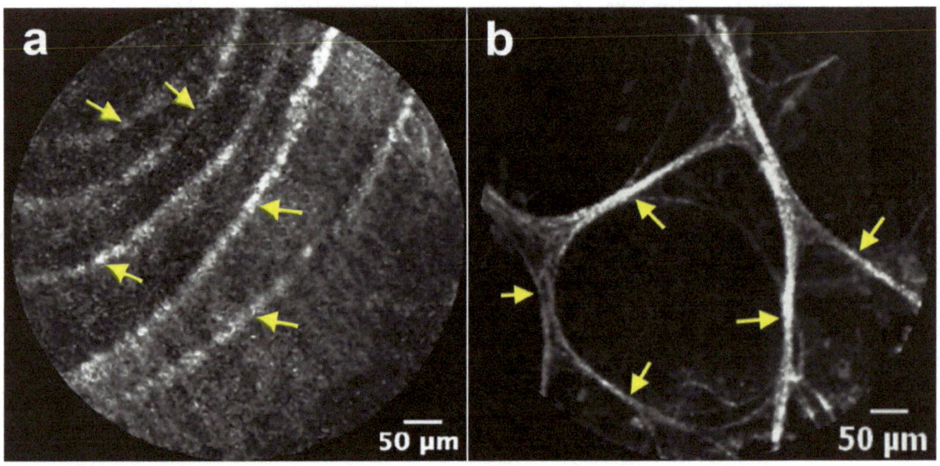

FIG. 1.29 Normal endomicroscopic pattern of the distal respiratory tract in a nonsmoker. (A) A terminal bronchiole with the enhanced cross pattern of elastic fibers *(arrows)* due to the presence of smooth muscle cells. (B) Alveolar ducts. The elastic fibers of the alveolar septa are indicated with *arrows*.

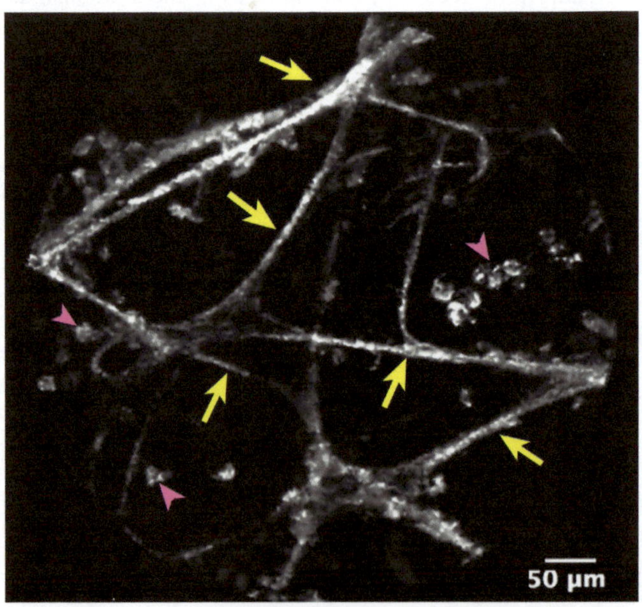

FIG. 1.30 Normal alveoloscopic pattern in a smoker. The alveolar structure is not violated; the cavities are rounded. Alveolar macrophages *(pink arrowheads)* are identified in the lumen. Interalveolar septa are indicated with *yellow arrows*.

measurement of the cell size), a score of 2 was used to note that the cells occupied less than half of the visual field, a score of 3 indicated that the cells occupied approximately half of the visual field, a score of 4 reflected that the cells occupied more than half of the visual field, and a score of 5 indicated that the cells covered the entire visual field (Fig. 1.31).

In addition to the scale described earlier, we developed a 6-point scale for estimating the total number of floating intra-alveolar substances, whether they were exclusively noncellular complexes or comingled with macrophages, in which a score of 0 meant the absence of these complexes and a score of 5 indicated maximum severity (see Fig. 4.1.11 in Chapter 4) [43].

The average length of a complete pCLE per patient was 11.1 ± 5.3 min. The alveoloscopy data were registered in a video recording mode (MKT format), with 15–30 videos and 100–135 graphic images collected for each patient.

The findings revealed that pCLE was informative if it achieved the visualization of the alveolar structures and that only then pCLE of distal airways could be considered as alveoloscopy.

After pCLE, endomicroscopic images were analyzed and evaluated retrospectively using the Cellvizio Viewer software (version 1.6.0; Mauna Kea Technologies). The parameters available for measurement were evaluated in absolute values, and other signs inherent in specific diseases were evaluated semiquantitatively based on their presence in fields of view that were examined and defined as percentages. By pCLE the following parameters can be quantified: thickness of the interalveolar septa; size of alveolar macrophages; diameter of microvessels; size of fluorescent intra-alveolar complexes; and the fluorescence intensity of the interalveolar septa, alveolar macrophages, and fluorescent intra-alveolar complexes. Semiquantitatively the following can be assessed by pCLE: signs of decreased alveolar airiness, the presence of fluorescent viscous fluid and/or secretion in the alveolar lumen, the presence of giant alveolar macrophages and fluorescent intra-alveolar complexes, and signs of replacement of alveolar structures with connective tissue elements.

The cumulative results of the examination of patients with DPDs using pCLE are summarized in Table 1.6. More detailed descriptions are presented in respective chapters.

Briefly, pCLE revealed the following changes in the distal respiratory tract of patients with interstitial lung disease: an increase or decrease in the thickness of the interalveolar septa and the size of the alveoli (Fig. 1.32); damage to the integrity of the elastic fibers within the interalveolar septa (Fig. 1.33); increase or decrease in the level of autofluorescence of

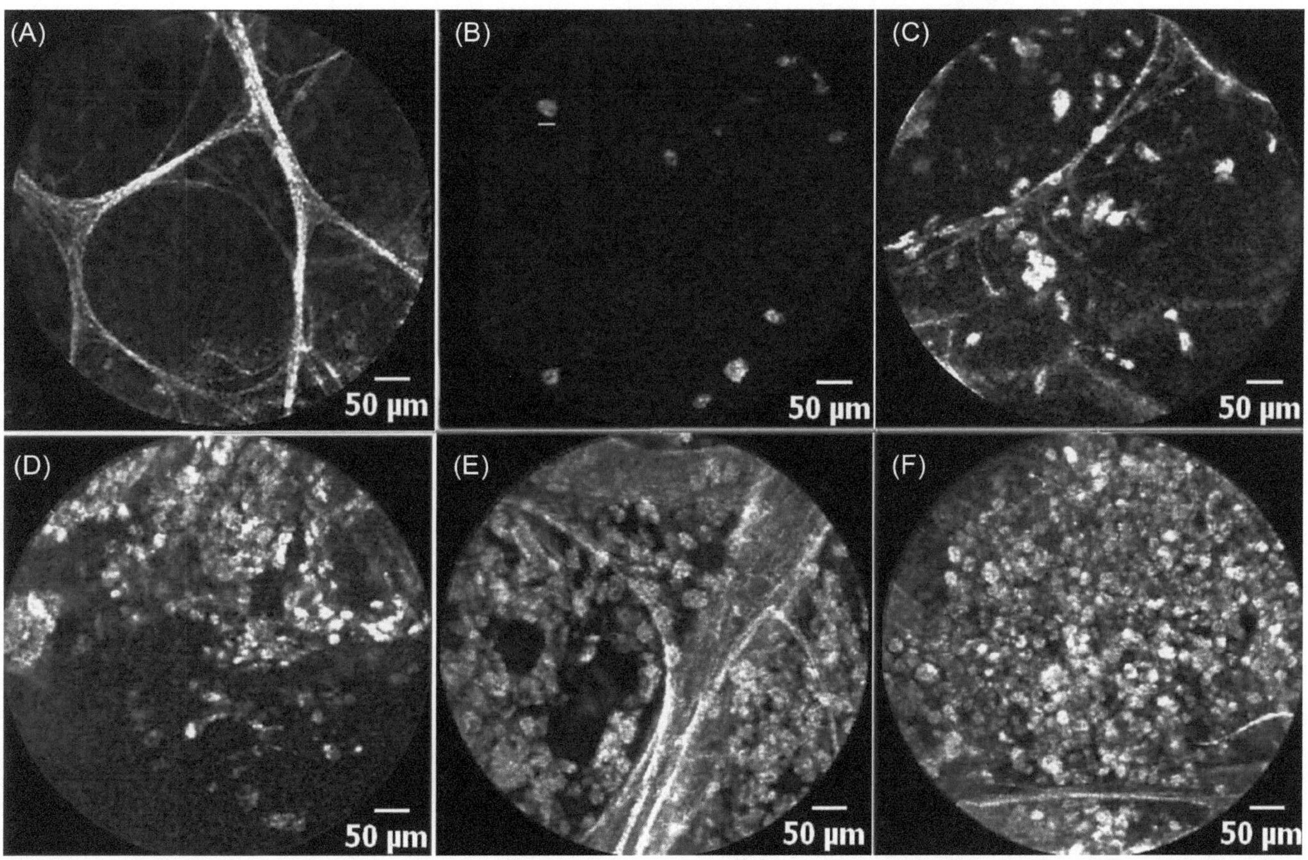

FIG. 1.31 Estimation of the number of alveolar macrophages with alveoloscopy: (A) 0 points, (B) 1 point, (C) 2 points, (D) 3 points, (E) 4 points, and (F) 5 points.

TABLE 1.6 Summary of the characteristic pCLE patterns in certain interstitial lung diseases

	NSIP	IPF	ILD-SS	LCH	LAM	HP	EGPA	AP	LP	IPA
Alveolar diameter (μm)	249.6 ± 33.6	270.9 ± 35.8	269.0 ± 30.0	359.9 ± 63.3	400.2 ± 53.1	269.7 ± 38.0	285.3 ± 24.9	250.8 ± 16.8	259.4 ± 24.7	313.8 ± 58.2
Alveolar elastic fiber thickness (μm)	12.4 ± 1.8	15.7 ± 2.7	15.4 ± 3.1	8.1 ± 2.9	8.6 ± 3.5	12.6 ± 1.8	12.8 ± 2.6	10.9 ± 1.0	11.2 ± 1.3	12.2 ± 1.8
Size of alveolar macrophages (μm)	21.5 ± 1.7	–	21.7 ± 0.3	25.5 ± 2.9	–	22.8 ± 1.9	20.0 ± 1.3	25.6 ± 3.5	22.6 ± 2.4	24.6 ± 2.3
Microvessel diameter (μm)	120.3 ± 38.0	125.9 ± 35.7	118.8 ± 32.7	123.0 ± 43.0	130.7 ± 37.1	141.2 ± 46.8	123.3 ± 40.6	94.8 ± 20.8	108.6 ± 32.1	129.5 ± 43.8
Size of FIC (μm)	–	–	–	215.6 ± 25.3	–	–	–	81.3 ± 21.6	55.4 ± 16.3	–
Fluorescence intensity of intra-alveolar septa (AU)	63.9 ± 8.9	70.2 ± 11.3	54.9 ± 8.3	54.2 ± 9.8	32.1 ± 5.0	57.2 ± 8.2	58.1 ± 7.2	60.3 ± 10.3	49.6 ± 4.9	55.7 ± 8.1
Fluorescence intensity of alveolar macrophages (AU)	79.5 ± 11.6	–	54.3 ± 9.6	96.4 ± 6.1	–	83.2 ± 7.2	52.0 ± 5.2	88.3 ± 20.5	73.5 ± 11.4	78.8 ± 11.2
Fluorescence intensity of fluorescent intra-alveolar complexes (AU)	–	–	–	85.9 ± 21.7	–	–	–	113.7 ± 25.6	78.7 ± 25.1	–
Dystelectasis	47/127[a] (37.0%)	–	5/42 (11.9%)	2/35 (5.7%)	12/33 (36.4%)	8/35 (22.9%)	5/15 (33.3%)	3/121 (2.5%)	1/23 (4.3%)	9/19 (47.4%)
Thickening of the elastic fiber of intra-alveolar septa >12 μm	34/127 (26.8%)	74/117 (63.2%)	28/42 (66.7%)	5/35 (14.3%)	6/33 (18.2%)	11/35 (31.4%)	8/15 (53.3%)	23/121 (19%)	3/23 (13.0%)	6/19 (31.6%)
Thickening of the elastic fiber of intra-alveolar septa >20 μm	17/127 (13.4%)	45/117 (38.5%)	15/42 (35.7%)	2/35 (5.7%)	2/33 (6%)	2/35 (5.7%)	4/15 (26.7%)	15/121 (12.4%)	–	4/19 (21.1%)
Thinning of the elastic fiber of intra-alveolar septa	–	–	3/42 (7.1%)	16/35 (45.7%)	16/33 (48.5%)	8/35 (22.9%)	2/15 (13.3%)	2/121 (1.7%)	2/23 (8.7%)	4/19 (21.1%)
Significant increase in alveoli diameter	–	–	–	7/35 (20.0%)	10/33 (30.3%)	2/35 (5.7%)	1/15 (6.7%)	–	–	4/19 (21.1%)

Fragmentation of the elastic fiber of intra-alveolar septa	6/127 (4.7%)	–	–	–	–	–	–	4/121 (3.3%)	–	–
Partial fibrosis per field of vision	–	40/117 (34.2%)	10/42 (23.8%)	–	–	7/35 (20.0%)	–	3/121 (2.5%)	–	–
Total fibrosis per field of vision	–	14/117 (12.0%)	4/42 (9.5%)	–	–	2/35 (5.7%)	–	–	–	–
Alveolar macrophages >40 µm in diameter	12/127 (9.4%)	–	4/42 (9.5%)	25/35 (71.4%)	0%	13/35 (37.1%)	–	58/121 (47.9%)	15/23 (65.2%)	11/19 (57.9%)
Fluorescent particles 3–7.5 µm	29/127 (22.8%)	16/117 (13.7%)	6/42 (14.3%)	4/35 (11.4%)	9/33 (27.3%)	9/35 (25.7%)	7/15 (46.7%)	–	–	–
Mucous secretion in alveoli (with air bubbles)	33/127 (26%)	26/117 (22.2%)	–	–	–	–	–	–	–	–
Fluorescent liquid in alveoli	–	0%	3/42 (7.1%)	3/35 (8.6%)	18/33 (54.5%)	10/35 (28.5%)	5/15 (33.3%)	91/121 (75.2%)	17/23 (73.9%)	16/19 (84.2%)
Normal pCLE imaging	18/127 (14.2%)	23/117 (19.6%)	11/42 (26.2%)	2/35 (5.7%)	11/33 (33.3%)	7/35 (20.0%)	3/15 (20%)	24/121 (19.8%)	1/23 (4.3%)	4/19 (21.1%)

[a] *Here and elsewhere in the table, number of zones with the presence of the sign/total number of the explored bronchopulmonary zones.*

AP, alveolar proteinosis; *EGPA*, eosinophilic granulomatosis with polyangiitis; *FIC*, fluorescent intra-alveolar complexes; *HP*, hypersensitivity pneumonitis; *ILD-SS*, interstitial lung disease associated with systemic scleroderma; *IPA*, invasive pulmonary aspergillosis; *IPF*, idiopathic pulmonary fibrosis; *LAM*, lymphangioleiomyomatosis; *LCH*, Langerhans cell histiocytosis; *LP*, lipoid pneumonia; *NSIP*, nonspecific interstitial pneumonia.

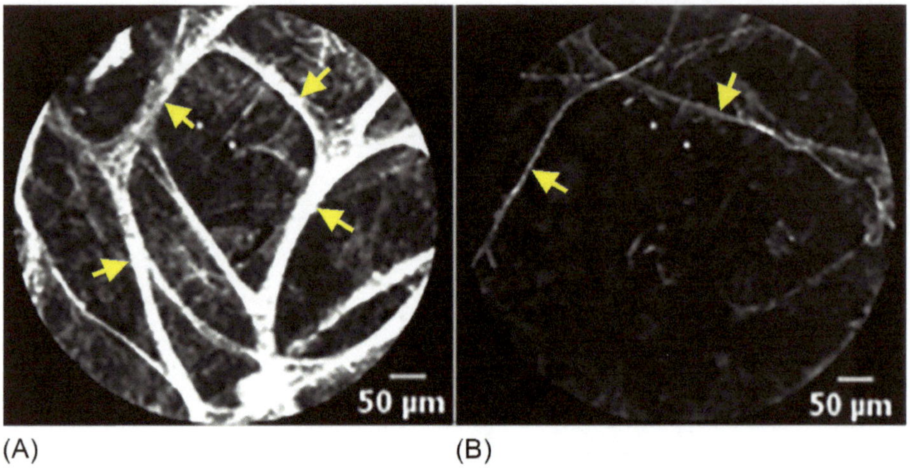

FIG. 1.32 Thickness of the elastic fibers *(yellow arrows)* of alveoli during pCLE: (A) thickened with high autofluorescence (a patient with chronic bronchitis) and (B) thinned and flaccid-appearing with low autofluorescence (a patient with lymphangioleiomyomatosis).

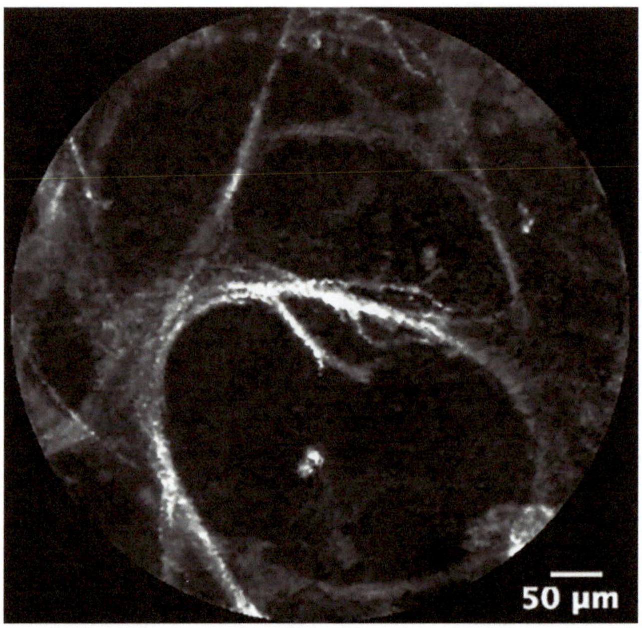

FIG. 1.33 pCLE in a patient with COPD. In the middle of the pCLE image, fragmented elastic fibers of the interalveolar septum are visible.

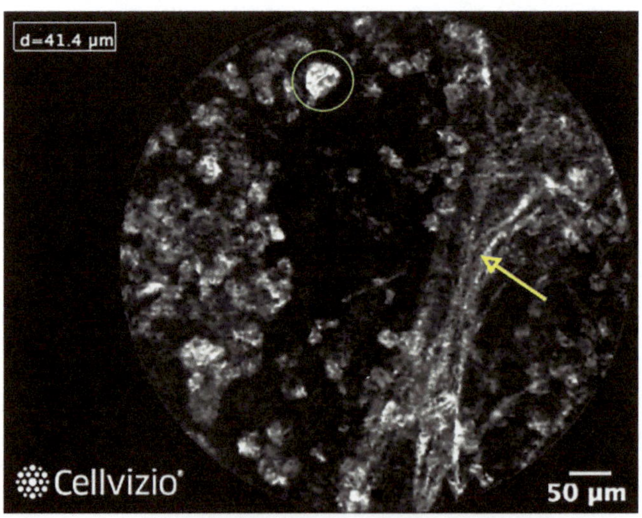

FIG. 1.34 Alveoloscopy in a patient with nonspecific interstitial pneumonia. The ellipse denotes a giant alveolar macrophage with a higher fluorescence intensity than those of other visualized cells.

the interalveolar septa and alveolar macrophages (Figs. 1.32 and 1.34); secretion, giant alveolar macrophages, fluorescent intra-alveolar complexes, and bipalmate structures in the lumen of the alveoli (Fig. 1.35); dystelectasis with an increase in the total number of alveoli within the field of view, where the alveoli have lost their round shape (Fig. 1.36); indirect signs of fibrotic changes in the lung parenchyma observed as the lack of alveolar differentiation, with their replacement with intensely fluorescent fibrillar and nodular structures (Fig. 1.37); signs of complete destruction of the elastic framework of the alveoli with visualization of blurred elements including fibers, nodules, and cellular mucus-containing structures (Fig. 1.38); partial visualization of microcysts and cysts (Fig. 1.39); higher frequency of vascular imaging with an increase in the fluorescence intensity of the vascular walls (Fig. 1.40); and intensely fluorescent rounded elements (3–7.5 μm in diameter) in the lumen of the terminal bronchioles and alveoli (Fig. 1.41).

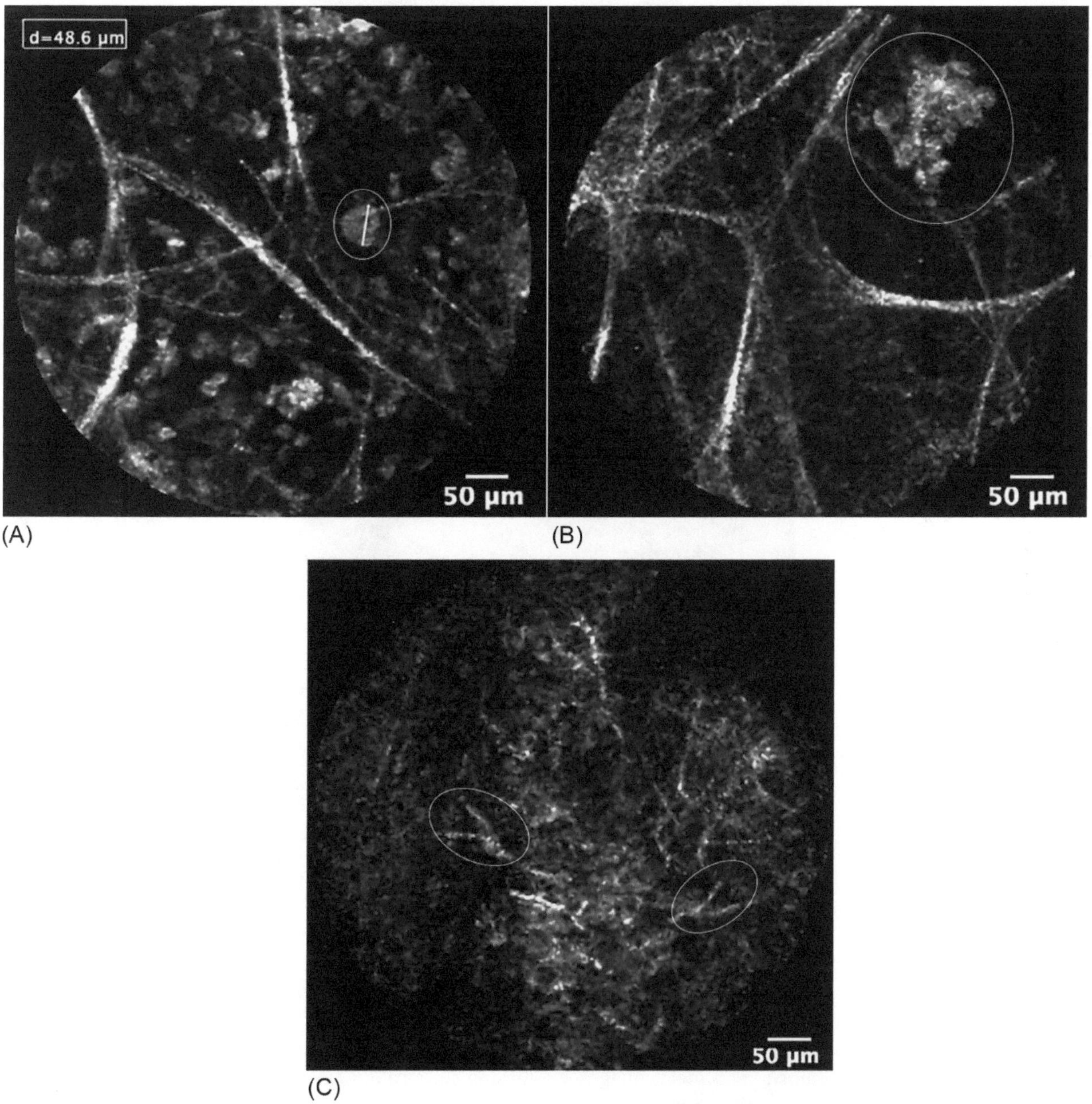

FIG. 1.35 During alveoloscopy in a patient with pulmonary proteinosis, a large number of macrophages (four points) are visualized with normal alveolar structure in the background (A); some of the macrophages are giant-sized, whereas others are comingled with amorphous mass in the alveolar lumen (B). (C) The normal alveolar structure is lost completely. Dichotomous branching structures, representing *Aspergillus niger* mycelium, are visualized in a patient with invasive pulmonary aspergillosis.

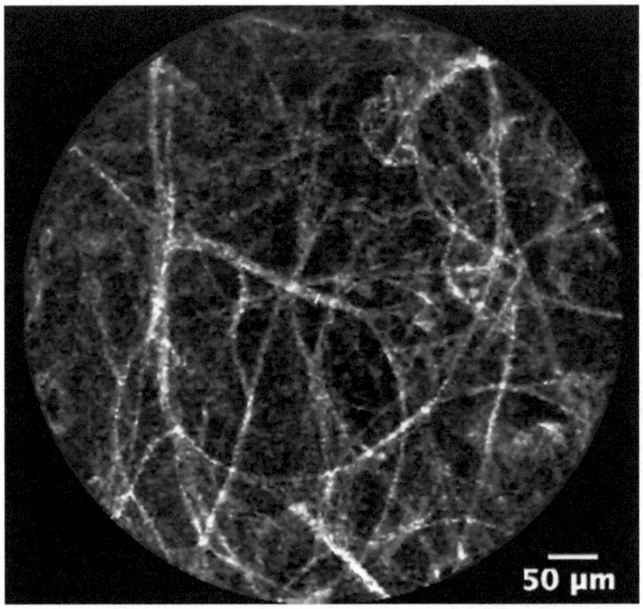

FIG. 1.36 Signs of a decrease in airiness of the lung parenchyma by pCLE imaging in a patient with bacterial pneumonia. The total number of alveolar structures in the field of vision is increased.

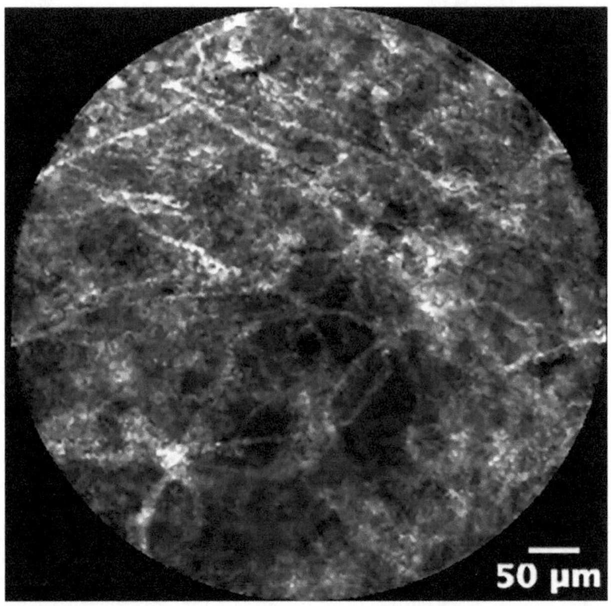

FIG. 1.37 Endomicroscopic signs of fibrotic changes of the lung parenchyma in a patient with idiopathic pulmonary fibrosis. Alveoli are not differentiated. Fibrillar and nodular structures with bright autofluorescence are visualized.

The most specific pCLE patterns were established in interstitial lung diseases with predominant intra-alveolar damage (alveolar proteinosis and exogenous lipoid pneumonia). Floating complexes, 55–350 µm in size, with a high level of autofluorescence were detected in the alveolar lumen in 74.4%–87.0% of the bronchopulmonary fields, which were more pronounced with alveolar proteinosis, and macrophages containing lipoprotein (alveolar proteinosis) (Figs. 1.35B and 4.1.12–4.1.13) or lipid (lipoid pneumonia) components (Fig. 1.42). In addition, the resolution capacity of pCLE is greater than HRCT in patients with alveolar proteinosis [43] (Fig. 4.1.13). In this regard, pCLE can be used for primary diagnosis and monitoring of the disease course in patients with accumulation of protein-lipid substances in the alveolar lumen and in those with unclear diagnosis despite clinical respiratory symptoms even in the absence of any HRCT changes.

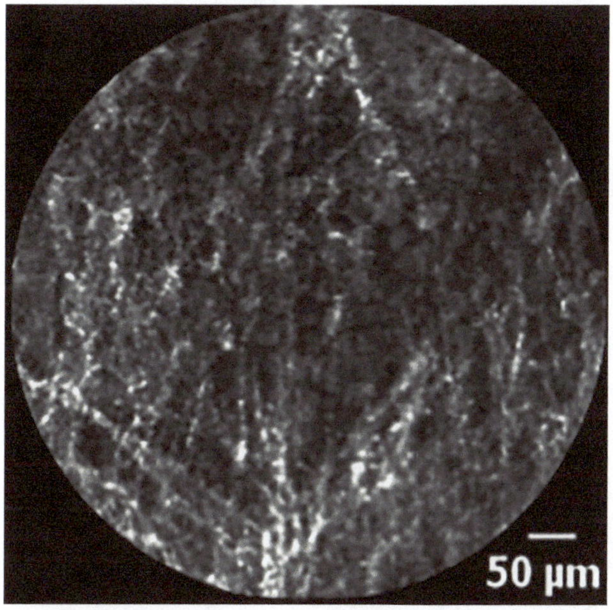

FIG. 1.38 pCLE of distal airways in a patient with peripheral lung adenocarcinoma. The alveolar structure is destroyed completely. The pathological substrate of the observed individual fibers cannot be determined.

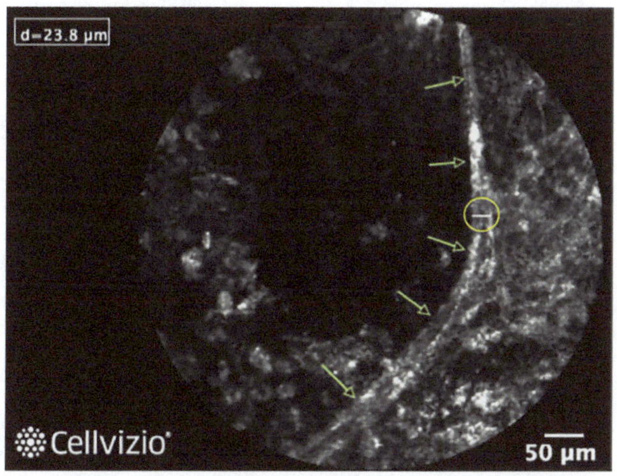

FIG. 1.39 Alveoloscopy in a patient with Langerhans cell histiocytosis. A part of the cystic structure, the dimensions of which exceed the size of the field of vision, is visualized.

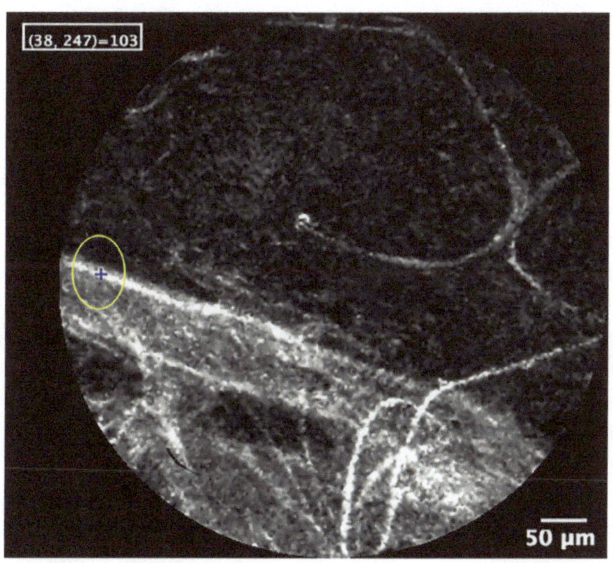

FIG. 1.40 Increased autofluorescence of vessel walls during alveoloscopy in a patient with eosinophilic granulomatosis with polyangiitis.

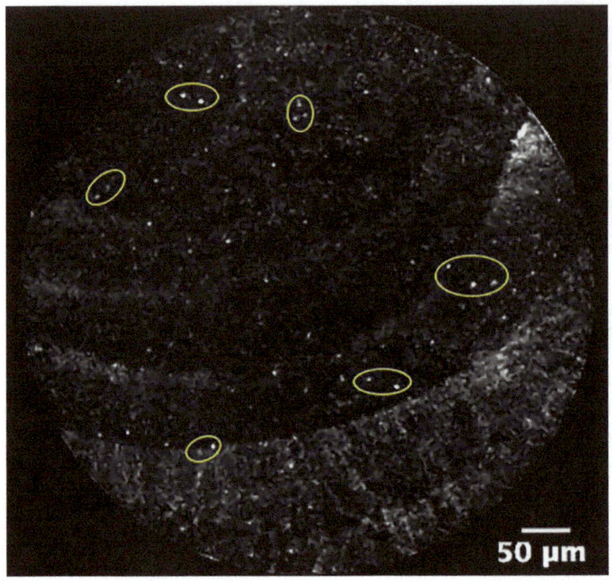

FIG. 1.41 Brightly fluorescent, small, rounded elements in the lumen of the terminal bronchiole during pCLE in a patient with chronic bronchitis.

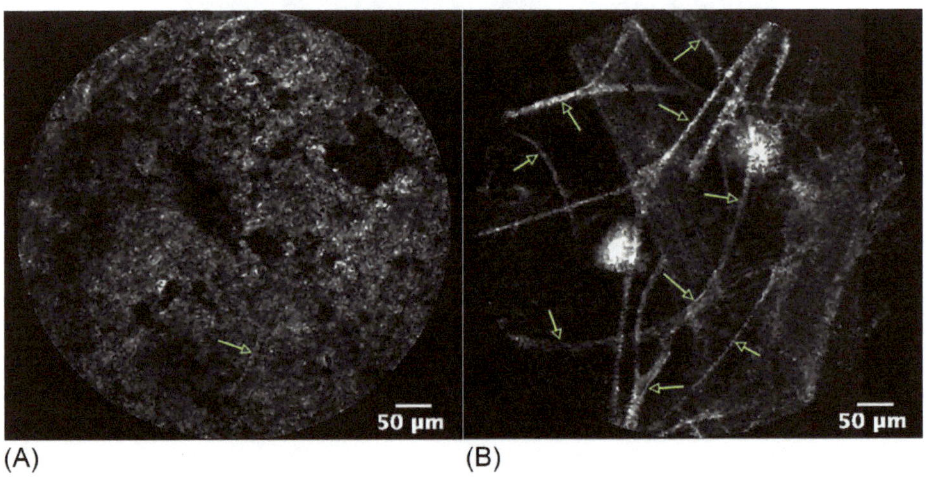

FIG. 1.42 Alveoloscopy in a patient with exogenous lipoid pneumonia. The field of vision is completely filled (5 points) with amorphous lipid masses and cellular elements (A) or roundish formations with high irregular fluorescent activity with diameters 61.9 and 84.7 μm are determined in the alveolar lumen (B). Interalveolar septa are indicated with *arrows*.

The most specific characteristics of the endomicroscopic pattern have been also established in a patient with invasive pulmonary aspergillosis in the form of fluorescent bipalmate fibrillar structures in both the central and the distal airways [44] (Fig. 1.35C).

Less specific changes in pCLE patterns were established with other interstitial lung diseases. In idiopathic pulmonary fibrosis (Fig. 2.1.20) and interstitial lung disease associated with systemic scleroderma (Fig. 8.2.24), the normal alveolar structure was partially or completely replaced by connective tissue elements with pronounced autofluorescent activity (Fig. 1.37). Patients with pulmonary Langerhans cell histiocytosis (Fig. 9.1.16) and lymphangioleiomyomatosis (Fig. 9.2.18) exhibited increased alveolar size to 350–450 μm, with thinning of the interalveolar septa to <9 μm, and partial visualization of the alveolar structures with a diameter exceeding the size of the visual field and a wall thickness of 45–65 μm in 20%–30% of the fields examined. Alveolar macrophages, which were not observed in the alveolar lumen in lymphangioleiomyomatosis, were visualized in the lumen (2–4 points on a 6-point scale) of patients with pulmonary Langerhans cell histiocytosis, with the presence of giant macrophages in 71.4% of the positive areas. In hypersensitivity pneumonitis, in addition to the overall moderate thickening of the interalveolar septa, there were distinct fields of view with interalveolar septal walls that were thickened by more than six times and alveoli with increased diameter (450–515 μm); giant macrophages were found in 37.1% of the regions (Fig. 3.19). Eosinophilic granulomatosis with polyangiitis was characterized by the visualization of a large number of vessels with pronounced fluorescence intensity of the vessel walls, which occurred in an amount of at least 2–3 within the same examined area in 80% of the cases (Fig. 7.1.9). In nonspecific interstitial pneumonia, endomicroscopic pattern did not have distinct specific characteristics (Fig. 2.2.10).

References

[1] Grenier P, Chevret S, Beigelman C, Brauner MW, Chastang C, Valeyre D. Chronic diffuse infiltrative lung disease: determination of the diagnostic value of clinical data, chest radiography, and CT with Bayesian analysis. Radiology 1994;191(2):383–90.

[2] Burge PS, Reynolds J, Trotter S, Burge GA, Walters G. Histologist's original opinion compared with multidisciplinary team in determining diagnosis in interstitial lung disease. Thorax 2017;72(3):280–1.

[3] Thiberville L, Bourg-Heckly G, Peltier E, Cave C. Per-endoscopic alveolar imaging using fluorescent confocal fibered microscopy. Eur Respir J 2006;28(Suppl 50):155s–6s.

[4] Hansell DM, Bankier AA, MacMahon H, McLoud TC, Müller NL, Remy J. Fleischner society: glossary of terms for thoracicimaging. Radiology 2008;246(3):697–722.

[5] Remy-Jardin M, Giraud F, Remy J, Copin MC, Gosselin B, Duhamel A. Importance of ground glass attenuation in chronic diffuse infiltrative lung disease: pathologic-CT correlation. Radiology 1993;189(3):693–8.

[6] Engeler CE, Tashijian JH, Trenkner SW, Walsh JW. Ground-glass opacity of the lung parenchyma: guide to analysis with high resolution CT. AJR Am J Roentgenol 1993;160(2):249–51.

[7] Lee JS, Im J, Ahn JM, Kim YM, Han MC. Fibrosing alveolitis: prognostic implication of ground glass attenuation at high-resolution CT. Radiology 1992;184(2):451–4.

[8] Zompatori M, Palmarini D, Canini R, Battista G, Sassi C. Ground-glass opacity: interpretation of high resolution CT findings. Radiol Med 2003;106(5–6):425–42.

[9] Stern EJ, Müller NL, Swensen SJ, Hartman TE. CT mosaic pattern of lung attenuation: etiologies and terminology. J Thorac Imaging 1995;10(4):294–7.

[10] Agrawal A, Agrawal A, Bansal V, Pandit M. A systematic approach to interpretation of heterogeneous lung attenuation on computed tomography of the chest. Lung India 2013;30(4):327–34.

[11] Ridge CA, Bankier AA, Eisenberg RL. Mosaic attenuation. Am J Roentgenol 2011;197(6):W970–7.

[12] Lee KS, Kim EA. High-resolution CT of alveolar filling disorders. Radiol Clin North Am 2001;39(6):1211–30.

[13] Johkoh T, Ikezoe J, Kohno N, Takeuchi N, Ichikado K, Arisawa J, et al. Usefulness of high-resolution CT for differential diagnosis of multi-focal pulmonary consolidation. Radiat Med 1996;14(3):139–46.

[14] Alves GR, Marchiori E, Irion K, Nin CS, Watte G, Pasqualotto AC, et al. The halo sign: HRCT findings in 85 patients. J Bras Pneumol 2016;42(6):435–9.

[15] Maturu VN, Agarwal R. Reversed halo sign: a systematic review. Respir Care 2014;59(9):1440–9.

[16] Webb WR, Müller NL, Naidich DP. High-resolution CT of the lung. 5th ed. Philaderphia, PA: Lippincot Williams and Wilkins; 2015. p. 74–104.

[17] Webb WR, Stein MG, Finkbeiner WE, Im JG, Lynch D, Gamsu G. Normal and diseased isolated lungs: high-resolution CT. Radiology 1988;166 (1 Pt 1):81–7.

[18] Johkoh T, Sakai F, Noma S, Akira M, Fujimoto K, Watadani T, Sugiyama Y. Honeycombing on CT; its definition, pathologic correlation, and future direction of its diagnosis. Eur J Radiol 2014;83(1):27–31.

[19] Watadani T, Sakai F, Johkoh T, Noma S, Akira M, Fujimoto K, et al. Interobserver variability in the CT assessment of honeycombing in the lungs. Radiology 2013;266(3):936–44.

[20] Staats P, Kligerman S, Todd N, Tavora F, Xu L, Burke A. A comparative study of honeycombing on high resolution computed tomography with histologic lung remodeling in explants with usual interstitial pneumonia. Pathol Res Pract 2015;211(1):55–61.

[21] Nishimura K, Kitaichi M, Izumi T, Nagai S, Kanaoka M, Itoh H. Usual interstitial pneumonia: histologic correlation with high-resolution CT. Radiology 1992;182(2):337–42.

[22] Raghu G, Remy-Jardin M, Myers JL, Richeldi L, Ryerson CJ, Lederer DJ, et al. Diagnosis of idiopathic pulmonary fibrosis. An official ATS/ERS/JRS/ALAT clinical practice guideline. Am J Respir Crit Care Med 2018;198(5):e44–68.

[23] Winter DH, Manzini M, Salge JM, Busse A, Jaluul O, Jacob Filho W, et al. Aging of the lungs in asymptomatic lifelong nonsmokers: findings on HRCT. Lung 2015;193(2):283–90.

[24] Araki T, Nishino M, Gao W, Dupuis J, Putman RK, Washko GR, et al. Pulmonary cysts identified on chest CT: are they part of aging change or of clinical significance? Thorax 2015;70(12):1156–62.

[25] Park S, Lee EJ. Diagnosis and treatment of cystic lung disease. Korean J Intern Med 2017;32(2):229–38.

[26] Beddy P, Babar J, Devaraj A. A practical approach to cystic lung disease on HRCT. Insights Imaging 2011;2(1):1–7.

[27] Raoof S, Bondalapati P, Vydyula R, Ryu JH, Gupta N, Raoof S, et al. Cystic lung diseases: algorithmic approach. Chest 2016;150(4):945–65.

[28] Kumar NA, Nguyen B, Maki D. Bronchiectasis: current clinical and imaging concepts. Semin Roentgenol 2001;36(1):41–50.

[29] Lucidarme O, Grenier PA, Cadi M, Mourey-Gerosa I, Benali K, Cluzel P. Evaluation of air trapping at CT: comparison of continuous-versus suspended-expiration CT techniques. Radiology 2000;216(3):768–77.

[30] Miller Jr WT, Chatzkel J, Hewitt MG. Expiratory air trapping on thoracic computed tomography. A diagnostic subclassification. Ann Am Thorac Soc 2014;11(6):874–81.

[31] Mets OM, van Hulst RA, Jacobs C, van Ginneken B, de Jong PA. Normal range of emphysema and air trapping on CT in young men. AJR Am J Roentgenol 2012;199(2):336–40.

[32] Webb WR, Stern EJ, Kanth N, Gamsu G. Dynamic pulmonary CT: findings in healthy adult men. Radiology 1993;186(1):117–24.

[33] Webb WR, Müller NL, Naidich DP. High-resolution CT of the lung. 5th ed. Philaderphia, PA: Lippincot Williams and Wilkins; 2015106–39.

[34] Gruden JF, Webb WR, Warnock M. Centrilobular opacities in the lung on high-resolution CT: diagnostic considerations and pathologic correlation. AJR Am J Roentgenol 1994;162(3):569–74.

[35] Murata K, Itoh H, Todo G, Kanaoka M, Noma S, Itoh T, et al. Centrilobular lesions of the lung: demonstration by high-resolution CT and pathologic correlation. Radiology 1986;161(3):641–5.

[36] Okada F, Ando Y, Yoshitake S, Ono A, Tanoue S, Matsumoto S, et al. Clinical/pathologic correlations in 553 patients with primary centrilobular findings on high-resolution CT scan of the thorax. Chest 2007;132(6):1939–48.

[37] Collins J, Blankenbaker D, Stern EJ. CT patterns of bronchiolar disease: what is "tree-in-bud"? AJR Am J Roentgenol 1998;171(2):365–70.

[38] Nishino M, Itoh H, Hatabu H. A practical approach to high-resolution CT of diffuse lung disease. Eur J Radiol 2014;83(1):6–19.

[39] Thiberville L, Moreno-Swirc S, Vercauteren T, Peltier E, Cave C, Bourg Heckly G. In vivo imaging of the bronchial wall microstructure using fibered confocal fluorescence microscopy. Am J Respir Crit Care Med 2007;175(1):22–31.

[40] Toshima M, Ohtani Y, Ohtani O. Three-dimensional architecture of elastin and collagen fiber networks in the human and rat lung. Arch Histol Cytol 2004;67(1):31–40.

[41] Thiberville L, Salaun M, Lachkar S, Dominique S, Moreno-Swirc S, Vever-Bizet C, et al. Human in vivo fluorescence microimaging of the alveolar ducts and sacs during bronchoscopy. Eur Respir J 2009;33(5):974–85.

[42] Wellikoff AS, Holladay RC, Downie GH, Chaudoir CS, Brandi L, Turbat-Herrera EA. Comparison of in vivo probe-based confocal laser endomicroscopy with histopathology in lung cancer: a move toward optical biopsy. Respirology 2015;20(6):967–74.

[43] Danilevskaya O, Averyanov A, Lesnyak V, Chernyaev A, Sorokina A. Confocal laser endomicroscopy for diagnosis and monitoring of pulmonary alveolar proteinosis. J Bronchology Interv Pulmonol 2015;22(1):33–40.

[44] Danilevskaya O, Averyanov A, Klimko N, Lesnyak V, Sorokina A, Sazonov D, et al. A case of diagnostics of invasive pulmonary aspergillosis using in vivo probe-based confocal laser endomicroscopy of central and distal airways. Med Mycol Case Rep 2014;5:35–9.

Chapter 2

Idiopathic interstitial pneumonias

Alexander Averyanov[a,b], Evgeniya Kogan[c], Victor Lesnyak[d], Igor E. Stepanyan[e], Olesya Danilevskaya[f]

[a]Clinical Department, Pulmonology Research Institute under FMBA of Russia, Moscow, Russia, [b]Pulmonary Division, Federal Research Clinical Center under FMBA of Russia, Moscow, Russia, [c]Anatomic Pathology Department, Sechenov University, Moscow, Russia, [d]Radiology Department, Federal Research Clinical Center under FMBA of Russia, Moscow, Russia, [e]Central TB Research Institute, Moscow, Russia, [f]Endoscopy Department, Pulmonology Research Institute under FMBA of Russia, Moscow, Russia

Idiopathic interstitial pneumonia (IIP) includes a heterogeneous group of nonneoplastic diseases of unknown origin, which develop in response to damage to the lung parenchyma and present as various combinations of inflammation and interstitial fibrosis, often spreading to the distal lower respiratory tract and blood vessels [1].

The history of IIP as a distinct entity begins in 1892, when W. Osler [2] first described diffuse fibrosis in the lung tissue, which he termed "pulmonary cirrhosis" in his famous monograph that withstood many reprints and was translated into six languages. In 1935 American physicians L. Hamman and A. Rich presented a case of a special variant of severe, rapidly progressing lung lesions with interstitial fibrosis and severe respiratory failure, which they called "fulminant diffuse interstitial fibrosis of the lungs" [3].

For several decades the term "Hamman-Rich disease" referred to both acutely and chronically occurring pulmonary lesions with interstitial fibrosis. In his letter to the editor of the *British Medical Journal* in 1964, J. Scadding proposed the term "fibrosing alveolitis," which combined the acute forms described by Hamman and Rich with the more common chronic presentations [4]. Further evolution of views on IIP is associated with two American pathologists A. Liebow and C. Carrington, who published classification of interstitial pneumonias in 1969, which later became the basis of the modern clinical and pathological separation [5]. American pathologists A.L. Katzenstein, R.F. Fiorelli, and J.L. Myers also made significant contribution to the development of the modern IIP classification [6,7].

The consensus developed by experts from the American Thoracic Society and the European Respiratory Society in 2002 classified IIP into seven forms: usual interstitial pneumonia (UIP)/idiopathic pulmonary fibrosis (IPF), nonspecific interstitial pneumonia (NSIP), cryptogenic organizing pneumonia (COP), acute interstitial pneumonia (AIP), lymphocytic interstitial pneumonia (LIP), desquamative interstitial pneumonia (DIP), and respiratory bronchiolitis associated with interstitial lung disease (RB-ILD) [8]. This classification was revised in 2013 [1] to include idiopathic pleuroparenchymal fibroelastosis (IPPFE), which, together with LIP, was attributed to rare forms of IIP, with the remaining IIPs belonging to the major forms. The major forms were grouped into three categories: (1) chronic fibrosing interstitial pneumonia (IPF and NSIP), (2) IIP associated with smoking (RB-ILD and DIP), and (3) acute/subacute forms (COP and AIP).

Each IIP variant has its own clinical, morphological, and radiological characteristics (Table 2.1); however, IIP in up to 15% of the patients cannot be definitely attributed to any of the classified forms due to the presence of morphologically heterogeneous forms, such as NSIP and COP or AIP and LIP in different lung regions of the same patient [9].

Difficult to Diagnose Rare Diffuse Lung Disease. https://doi.org/10.1016/B978-0-12-815375-8.00002-9

TABLE 2.1 Clinical and radiological characteristics of idiopathic interstitial pneumonias

Feature	IPF[a]	NSIP	COP	LIP	AIP	DIP	RB-ILD	IPPFE
Anamnesis	Older than 50 years, more often smoker males, history of GERD	40–50 Years, more often nonsmoking females, frequent familial association with connective tissue diseases	40–50 Years old, most often nonsmoking females	40–50 Years old, most often females	Any age and anamnesis	30–40 Years old, 90% smokers	30–40 Years old, 100% smokers	Any sex and age
Disease onset	Slow	Slow	Acute or subacute	Slow	Acute	Slow	Slow	Slow
GGO	+	+++	+++	+++	+++	+++	++	+
Consolidation	–[a]	+	+++	+	+++	–	–	++
Honeycombing	+++	++	–	+	+	+	–	++
Traction bronchiectasis	+++	++	–	+	+	–	–	++
Thickening of intralobular, interlobular septa	+++	+++	+	++	+	+	+	+++
Intralobular nodules	–	++	++	++	–	++	+++	
Reversed halo sign	–	–	++	–	–	–	–	–
Cysts	+	+	–	++	–	+	–	++
Isolated lesions of the upper lobes	–	–	+	–	–	+	++	+++
Subpleural sparing	–	++	–	+	–	+	+	–
Response to steroids	–[a]	++	+++	++	+	+++	+++	–[a]
Possibility of full recovery	–	–	++	–	+	++	++	–

[a] *Not in acute exacerbation.*
+++, typical sign; ++, frequent sign; +, possible sign; –, atypical sign.
AIP, acute interstitial pneumonia; *COP*, cryptogenic organized pneumonia; *DIP*, desquamative interstitial pneumonia; *IPF*, idiopathic pulmonary fibrosis; *IPPFE*, idiopathic pleuroparenchymal fibroelastosis; *GERD*, gastroesophageal reflux disease; *LIP*, lymphocytic interstitial pneumonia; *NSIP*, nonspecific interstitial pneumonia; *RB-ILD*, respiratory bronchiolitis associated with interstitial lung disease.

References

[1] Travis WD, Costabel U, Hansell DM, King Jr TE, Lynch DA, Nicholson AG, et al. An official American Thoracic Society/European Respiratory Society statement: update of the international multidisciplinary classification of the idiopathic interstitial pneumonias. Am J Respir Crit Care Med 2013;188(6):733–48.

[2] Osler W. The principles and practice of medicine. 1st ed. New York: Appleton; 1892. p. 529.

[3] Hamman L, Rich AR. Fulminating diffuse interstitial fibrosis of the lungs. Trans Am Clin Climatol Assoc 1935;51:154–63.

[4] Scadding JD. Fibrosing alveolitis. Br Med J 1964;2(5410):686.

[5] Liebow AA, Carrington CB. The interstitial pneumonias. In: Simon M, Potchen EJ, LeMay M, editors. Frontiers of pulmonary radiology. 1st ed. New York: Grune & Stratton; 1969. p. 102–41.

[6] Katzenstein AL, Fiorelli RF. Nonspecific interstitial pneumonia/fibrosis. Histologic features and clinical significance. Am J Surg Pathol 1994;18(2):136–47.

[7] Katzenstein AL, Myers JL. Idiopathic pulmonary fibrosis. Clinical relevance of pathologic classification. Am J Respir Crit Care Med 1998;157 (4 Pt 1):1301–15.

[8] American Thoracic Society; European Respiratory Society. American Thoracic Society/European Respiratory Society international multidisciplinary consensus classification of the idiopathic interstitial pneumonias. Am J Respir Crit Care Med 2002;165(2):277–304.

[9] Skolnik K, Ryerson CJ. Unclassifiable interstitial lung disease: a review. Respirology 2016;21(1):51–6.

Chapter 2.1

Idiopathic pulmonary fibrosis

IPF is a specific form of chronic progressive fibrosing interstitial pneumonia of unknown origin that occurs mainly in the elderly, is limited to the lungs, and is associated with the histological and/or radiological pattern of UIP [1].

IPF is the most common form of IIP, comprising between 50% and 60% of all IIP cases [2]. Epidemiological studies in the United States and Canada reveal that the incidence of IPF ranges from 14 to 42.7 per 100,000 population, with a widespread increase in the incidence in the last 20 years [3–7]. Given the increase in life expectancy, general expert opinion predicts a further increase in the incidence and prevalence of IPF. Conversely, approximately 40,000 new IPF cases are diagnosed annually in the European Union countries [8].

Risk factors

The etiology of IPF is unknown; however, disease-associated risk factors are well studied.

Genetic risk factors

Three to five percent of the patients have a family history of IPF [1]. Genetically determined IPF is suggested to manifest at an earlier age than the sporadic forms, exhibit a subacute onset, follow a rapidly progressive course, and occur in combination with dyskeratosis, aplastic anemia, and liver damage [9,10].

Mutations that determine the risk of IPF can be divided into two categories: those associated with the metabolism of surfactant or mucin proteins and changes in the biology of telomeres. Among the surfactant genes whose polymorphisms are associated with the development of IPF, the most studied are *SFTPC* that encodes surfactant protein C, *SFTPA2* that encodes surfactant protein A2, *ABCA3* that encodes ATP-binding cassette subfamily A member 3 that is involved in surfactant homeostasis, and *MUC5B* that is localized in chromosome 11 and responsible for the synthesis of mucin 5B [11]. Moreover, nucleotide polymorphisms of MUC5B are detected in both familial and sporadic cases of IPF [12].

Telomeres are the end sections of chromosomes that are shortened during cell division, and the rate of telomere shortening is a determinant of aging. One hypothesis on the development of IPF proposes premature aging of the epithelium, confirmed by the presence of shortened telomeres in blood cells and respiratory tract epithelium of patients with IPF [13]. In addition, telomere shortening in peripheral blood cells was shown to be a negative predictor of survival in patients with IPF [14]. Changes in telomere regulators including telomerase reverse transcriptase (*TERT*), dyskerin pseudouridine synthase (*DKC1*), and regulator of telomere elongation helicase 1 (*RTEL1*) are associated with the development of IPF, and *TERT*, the most frequently mutated gene in familial pulmonary fibrosis, is rarely found in sporadic forms [15,16].

Telomere shortening appears to be a significant factor for the early onset and poor prognosis of IPF. A study by Planas-Cerezales et al. showed that 66% of patients with shortened telomeres were younger than 60 years, died or required lung transplantation, and exhibited the most severe impairment of pulmonary function at diagnosis [17].

Furthermore a human leukocyte antigen locus containing genes with significantly different expression levels was detected in patients with fibrosing interstitial pneumonia in comparison with healthy individuals, suggesting autoimmune mechanisms may be at play during IPF development [18].

Smoking

Smoking is one of the main risk factors for IPF. Most often the disease develops in patients with a smoking history of more than 20 pack-year [19]. At the same time the dose-dependent effect of smoking on the development of IPF has been proved. Each pack-year increases the likelihood of IPF by 3% for those who continue to smoke and by 2% for those who quit smoking [20]. Perhaps the predominance of males among IPF patients reflects the higher prevalence of smoking in the male population. The mechanisms that determine the effect of smoking on the development of IPF have not been established. However, direct toxic effects of tobacco smoke components on epithelium; free radical products induced by inhalation of tobacco smoke; and activation of neutrophils with increased protease activity, similar to those that occur in chronic obstructive pulmonary disease (COPD), are postulated to be important in IPF pathogenesis [21].

Oxidative stress due to smoking is an inducer of *MUC5B* gene expression, which, as mentioned earlier, is associated with the development of pulmonary fibrosis [22]. Recently the epigenetic effects of smoking and the impact of pollutants on DNA methylation processes, which are also observed in lung cancer, have been described. Specifically, hypermethylation of the *THY-1* DNA region leads to a decrease in the synthesis of the corresponding glycoprotein by fibroblasts and transformation of fibroblasts into myofibroblasts [23]. Despite the strong correlation between tobacco smoking and the development of IPF, the disease is not classified as an IIP associated with smoking [24]. Data on the role of smoking in the course and prognosis of IPF are contradictory. Some studies report increased frequency of acute IPF exacerbations and lower survival rates among patients with no history of smoking in comparison with current and ex-smokers [25,26]. Conversely, cardiovascular diseases, COPD, and lung cancer, which are more frequent in smokers with IPF, are associated with a worse prognosis [27].

Air pollutants

The relationship between the development of IPF and exposure to air pollutants, especially industrial, has been studied in less detail than that for other risk factors. However, chronic inhalation of metal, wood, silicon-containing particles, and organic dust was shown to be associated with increased risk of local lung fibrosis and IPF; this risk is higher in males [28]. In a case-control study of a Swedish registry including 137 patients with IPF, Ekström et al. established that 68% of the patients had more than 10 years of history of contact with various dust particles in addition to a history of smoking [20]. It is unclear, however, whether patients with chronic hypersensitivity pneumonitis, which is almost indistinguishable from IPF in late stages, were included in the IPF group in that study.

There is evidence supporting that some of the features of IPF are associated with dust factors. In a large Korean study, the disease onset was earlier, and the prognosis was worse in patients with IPF who had contact with industrial dust compared with other patients [29]. However, in general, the role of air pollutants in IPF development is likely to be underestimated. Careful assessment of the professional history and the air environment of patients with IPF is necessary to exclude air pollutants as potential aggravating factors for IPF.

Old age

IPF is a disease of the elderly, with a mean patient age of 66 years, and is rarely diagnosed until the age of 50 years [1]. A large-scale epidemiological study in the United States found that the incidence and prevalence rates of IPF were 93.7 and 494.5 per 100,000 population, respectively, among those aged over 65 years, which are an order of magnitude higher than those in the younger age group (under 65 years) [30]. As mentioned earlier, accelerated aging of lung tissue is one of the hypotheses for IPF, with a number of the mechanisms that underlie aging shown to be involved in patients with IPF [31]. First, genomic instability due to the accumulation of DNA damage is much greater in patients with IPF compared with age-matched healthy controls [32]. Epigenetic mechanism, DNA hypermethylation of epithelial and stem cells in particular, is also associated with accelerated aging and uncontrolled fibrosis [33]. The second proved mechanism of accelerated aging leading to defects in cell division and full regeneration is telomere shortening [16]. Accelerated physiological aging in IPF is observed as an increase in the number of epithelial cells expressing markers of aging, senescence-associated β-galactosidase, and cyclin-dependent kinase inhibitor [34]. Within the framework of disturbance of the cellular homeostasis, mitochondrial dysfunction and degradation of cellular proteins due to disruption in the activity of proteolytic systems are active topics of discussion [35,36]. Finally, depletion of the resources for resident stem cells, that is, stem cell exhaustion, which is the source of lung regeneration, also reflects accelerated aging at play in IPF [37].

Gastroesophageal reflux

A significant number of studies have shown a close correlation between IPF and gastroesophageal reflux (GER). Daily monitoring of esophagus pH reveals an increase in acidity in 87%–94% of patients with IPF, which is significantly higher than in the general population, whereas GER is present in the absence of classical clinical manifestations such as heartburn and belching in 54%–75% of patients with IPF [38,39]. Conversely the number of GER episodes was found to be directly associated with the severity of fibrosis by high-resolution computed tomography (HRCT) [40]. In addition, the incidence of nonacid reflux is higher in patients with IPF, as confirmed by the detection of pepsin in bronchoalveolar lavage (BAL) fluid, which is significantly higher during the exacerbation of IPF [40]. Microaspiration of gastric juice is assumed to be a contributing factor for the initiation of lung damage in IPF. However, a reverse effect—the development of reflux due to changes in intrathoracic pressure following a decrease in the elasticity of the lungs—cannot be ruled out. The undisputable association of GER with IPF led to the inclusion of antireflux drugs in the empirical regimen of IPF in 2015, albeit with an emphasis on the limited evidence for such recommendations [41].

Indeed, several studies revealed that antacid drugs conferred no benefit for IPF. Raghu et al. [39] showed that the administration of proton pump inhibitors did not normalize the acidity of the esophagus in 63% of patients. In a large INPULSIS trial, antacid therapy was found not to have a beneficial effect on the disease course or the dynamics of pulmonary function in the entire study cohort [42]. Kreuter et al. assessed the effect of proton pump inhibitors and H_2 histamine blockers on lung function in patients receiving pirfenidone and concluded that their inclusion in the treatment regimen did not improve the outcomes but increased the risk of respiratory infections in patients with a forced vital capacity (FVC) <70% [43].

Infection

The percentages of carriers of herpesviruses (especially Epstein-Barr virus), hepatitis C virus, and cytomegalovirus are higher in patients with IPF than in the general population [44,45]. It is possible that these same viruses and the influenza virus may also play a role in the induction of IPF acute exacerbations [46].

Bacterial infections are also clear contributors to the development and the changes in the disease course of IPF. BAL studies showed that bacterial colonization was significantly higher in patients with IPF than in healthy individuals, including both smokers and nonsmokers, and in patients with moderate COPD. The presence of above-threshold levels of *Streptococcus* and *Staphylococcus* DNA in BAL fluid was associated with decreased survival of patients with IPF [47].

In the COMET study the most frequent bacterial pathogens present in the BAL fluid were *Prevotella*, *Veillonella*, and *Cronobacter* sp., and colonization with *Streptococcus* and *Staphylococcus* was associated with more rapid progression of the disease [48]. Administration of co-trimoxazole at a dose of 960mg twice a day for 12 months in IPF patients led to a significant decrease in mortality and improvement in the quality of life in the antibiotic therapy group, albeit not improving functional parameters [49].

Acute exacerbation of IPF is associated with up to fourfold increase in bacterial burden and a high level of neutrophils in the BAL fluid [50].

Further study of the microbial community in patients with IPF may reveal new directions for the treatment of IPF and prevention of acute exacerbations [51].

Pathogenesis

The current understanding of the pathogenesis underlying IPF suggests chronic microinjury to the respiratory epithelial cells by various factors including tobacco smoke, pollutants, acid, viruses, and bacteria, with impairment in physiological repair, epithelial-mesenchymal transition, proliferation of fibroblasts, and myofibroblasts that are resistant to apoptosis, uncontrolled production of collagen, and fibronectin that deposit in the extracellular matrix [52]. Pathological immune response and atypical remodeling of the lung tissue appear to be due to genetic factors [53]. A number of biologically active molecules play a role in fibrogenesis during IPF and include transforming growth factor β1 (TGF-β1); platelet-derived growth factor; interleukins 1β, 13, and 17; and chemokine ligands 2, 12, 17, and 18 [54].

Unlike other forms of IIP, interstitial inflammation is not the pathogenic component of IPF, which is the reason underlying the predominance of signs of fibrosis over other changes.

Morphology

The IPF morphology is UIP, which is characterized by predominance of irregular lesions in the respiratory interstitium with subsequent development of pathological repair and fibrosis [55]. An extreme presentation of IPF is honeycombing lung, which is a combination of interstitial fibrosis and cystic transformation of terminal and respiratory bronchioles and alveoli with the development of lung cancer in 13%–14% of the cases [56,57]. Fibrous changes are accompanied by block of the air-blood barrier and the development of pulmonary hypertension (pH), right ventricular hypertrophy, and pulmonary heart.

Macroscopic examination reveals that the most pronounced changes are in the subpleural regions and the lower lobes of both lungs. Visually the lungs are unevenly airy and engorged with blood, with a rubber density on palpation; a cellular pattern is visible on the cut surface, and pleura exhibits a poorly paved cobblestone appearance (Fig. 2.1.1).

Key histopathologic findings of UIP are patchy fibrosis zones that alternate with unaffected or slightly altered lung parenchyma, honeycombing with architectural distortion of the parenchyma and subpleural distribution, and fibroblastic foci [58]. These abnormalities, along with the absence of changes that are not characteristic of UIP, including granulomas, organized pneumonia, and significant inertial inflammation, confirm definitive morphological diagnosis [58]. The diagnosis of UIP can also be valid in the presence of other signs without honeycombing [58].

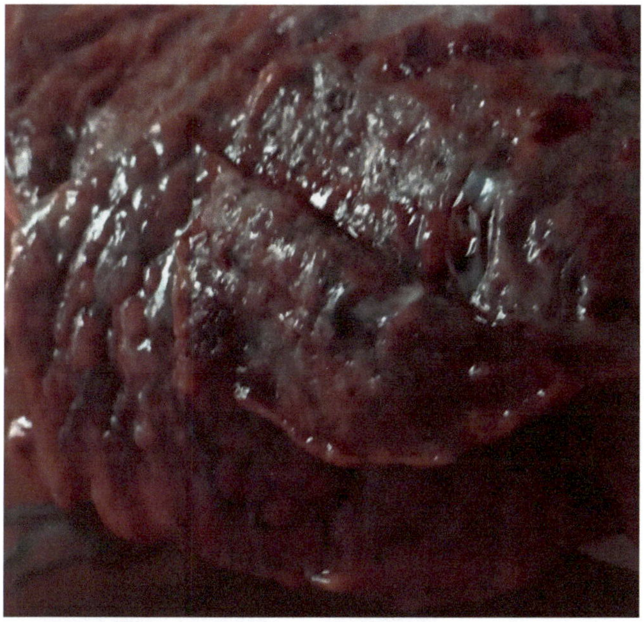

FIG. 2.1.1 Idiopathic pulmonary fibrosis (IPF). Lung view: cysts formation (honeycomb lung) in the subpleural areas presented as irregular pleural tuberosity.

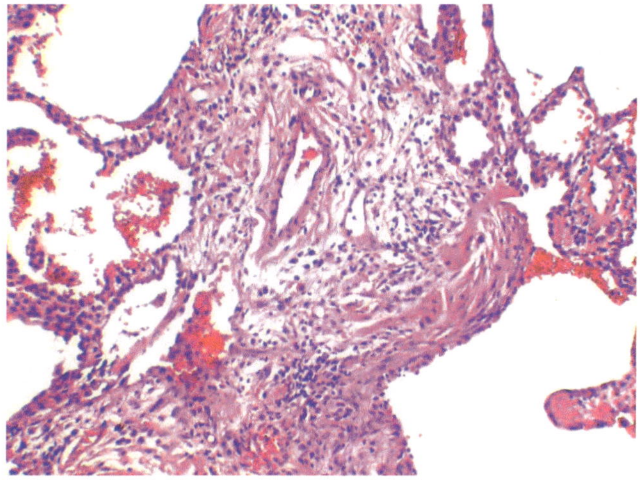

FIG. 2.1.2 IPF. Usual interstitial pneumonia. The bronchiolar-alveolar junction with proliferation of myofibroblasts and the formation of myofibroblastic focus. Hematoxylin and eosin (H&E) staining, 200×.

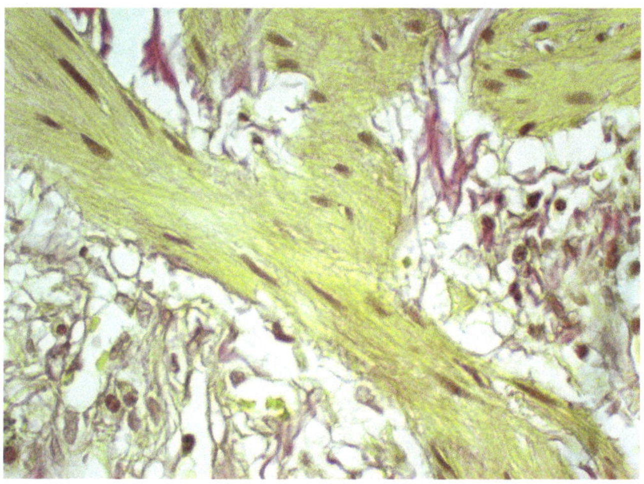

FIG. 2.1.3 IPF. The bronchiolar-alveolar junction with proliferation of myofibroblasts and the formation of myofibroblastic focus not stained with picrofucsin. Van Gieson staining, 400×.

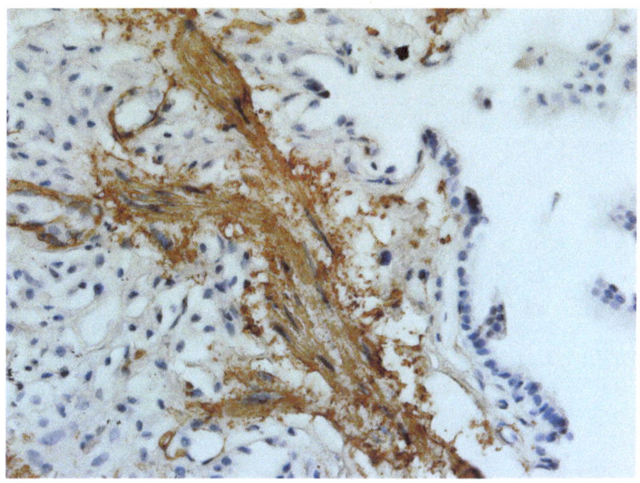

FIG. 2.1.4 IPF. Usual interstitial pneumonia with proliferation of myofibroblasts and the formation of myofibroblastic focus. Smooth muscle actin in myofibroblasts within the myofibroblastic focus of the zone of bronchiolar-alveolar junction. Immunoperoxidase reaction, 200×.

Microscopic manifestations of early-stage UIP include signs of pathological repair with early formation of interstitial fibrosis, possibly in combination with minimal interstitial exudative and exudative-productive inflammation (Figs. 2.1.2–2.1.11). UIP differs from other ILDs in that there is more severe damage, pathological repair, and sclerosis of the interstitium of the respiratory regions of the lungs, accompanied with the formation of fibroblastic and myofibroblastic foci in the interstitium. Fibrotic zones among the unchanged pulmonary parenchyma are a characteristic pattern. The longer the disease is, the smaller the unaffected areas. Pathological repair develops in bronchiolar-alveolar junction zones, that is, the junction of the alveoli and bronchioles where destruction of the epithelial lining and the basement membrane are initially replaced immature mesenchymal cells exhibiting a myofibroblastic phenotype (Fig. 2.1.5).

Subsequently, in the same areas, fibroblastic and myofibroblastic foci arise, and epithelialization due to mesenchymal-epithelial transition of immature mesenchymal cells is observed. In parallel, proliferation of the epithelial component with formation of adenomatous structures on the background of fibrosis can develop. The bronchiolar-alveolar junction zones

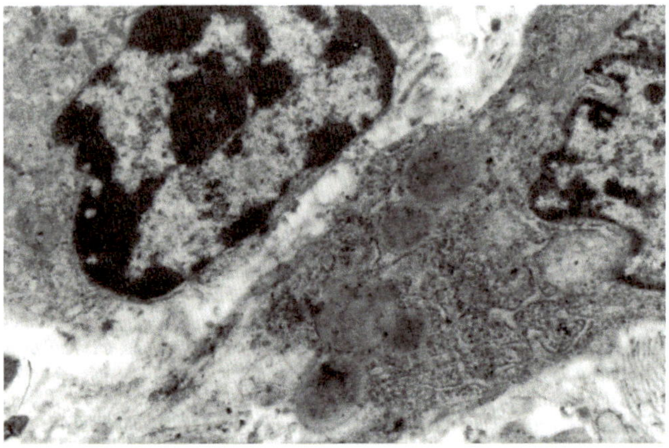

FIG. 2.1.5 IPF. Usual interstitial pneumonia. Myofibroblasts of the myofibroblastic focus with bundles of myofilaments and lipid droplets in the cytoplasm. Electron microphotograph, 18,000×.

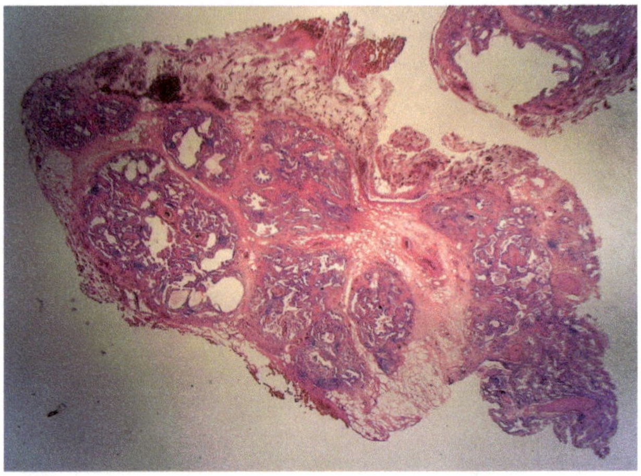

FIG. 2.1.6 IPF. The honeycomb lung. Mosaic changes due to the alternation of the affected and intact parts of the lung. H&E staining, 100×.

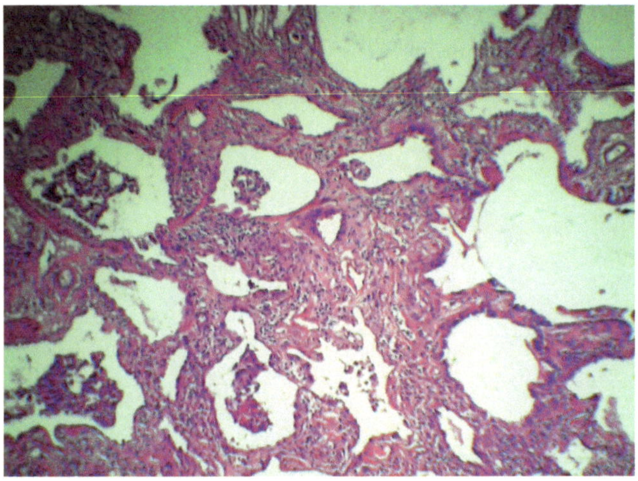

FIG. 2.1.7 IPF. Usual interstitial pneumonia. Honeycomb lung. General view with thickening and sclerosis of the alveolar septa and formation of cystic structures. H&E staining, 100×.

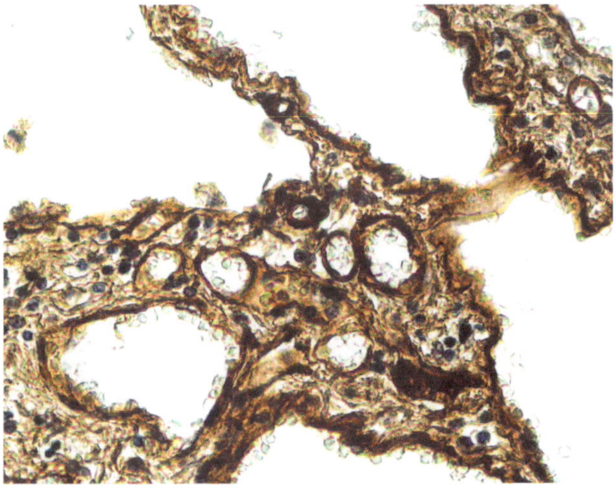

FIG. 2.1.8 IPF. Usual interstitial pneumonia. Increased levels of type 4 collagen in the basement membranes of the alveolar capillaries and epithelial structures in the honeycomb lung. Immunoperoxidase reaction, 200×.

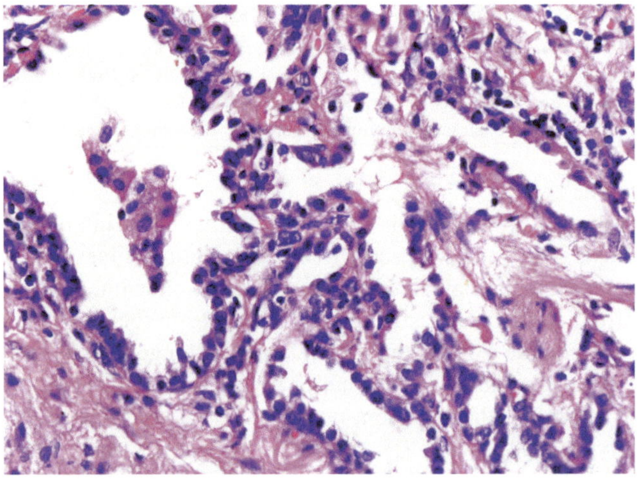

FIG. 2.1.9 IPF. Usual interstitial pneumonia. Adenomatosis in the honeycombing area with squamous metaplasia and epithelial atypia. H&E staining, 200×.

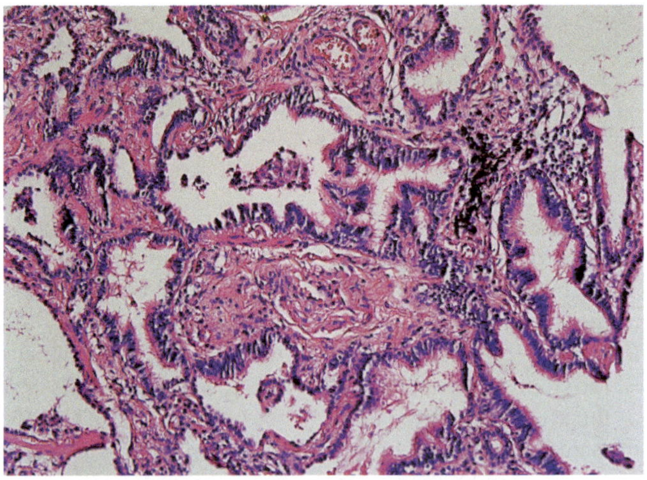

FIG. 2.1.10 IPF. Usual interstitial pneumonia. Adenomatosis in the honeycombing area with ciliary and cylindrical epithelium. H&E staining, 200×.

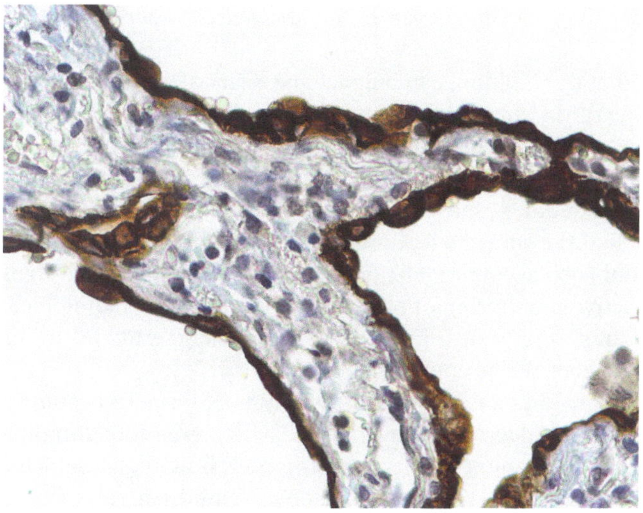

FIG. 2.1.11 IPF. Usual interstitial pneumonia. Cytokeratin-19 in cells of adenomatous structures in the honeycombing area. Immunoperoxidase reaction, 200×.

are established as the most important stem cell niches in the peripheral regions of the lung [59]. We showed that persistent and/or significantly pronounced damage to the bronchiolar-alveolar junction zones, that is, the stem cell niches, led to impaired regeneration of the lung tissue, followed by fibrosis and adenomatosis [60].

Microscopic examination in the later stages of UIP reveals marked sclerosis of the respiratory interstitium of the lungs and microcystic reorganization of the lung tissue (Figs. 2.1.7 and 2.1.8). The lung parenchyma is replaced by connective tissue, honeycombing lined by adenomatous hyperplastic alveolar epithelium. Type 1 pneumocytes are replaced by type 2 hyperplastic pneumocytes exhibiting signs of atypia, and focal adenomatous hyperplasia, atypical adenomatous hyperplasia, and foci of squamous metaplasia and epithelial dysplasia are observed (Figs. 2.1.9–2.1.11). The air-blood barrier is blocked and ceases to function due to both pronounced interstitial fibrosis of the alveolar septa and dysregenerative changes in the epithelial lining.

Pulmonary interstitium expands due to sclerosis and reduplication of epithelial and endothelial basement membranes; all types of collagens accumulate within a sharp increase in the specific gravity of degradation-resistant type IV and V collagens. A large number of collagen fibers, activated fibroblasts, fibrocytes, and inflammatory cells, which are dominated by lymphoid elements and histiocytes, are introduced into the air-blood barrier zone. Interstitial fibroblasts with signs of collagen synthetic activity are detected not only in the area around the air-blood barrier between the thickened basement membranes, that is, interposition, but also in the alveolar lumen, which can lead to PH [61].

The pulmonary epithelium in the later stages of UIP also undergoes remodeling. Type I pneumocytes in large areas are replaced by type 2 pneumocytes with immature osmiophil multilamellar bodies, often with signs of immaturity and cell atypism [60]. In connection with impaired production of surfactant and obliteration of the bronchioles, focal microatelectases develop. Damage to the epithelium of the transitional bronchiolar-alveolar zone may explain the development of early and severe dysregenerative changes in the epithelium, including precancerous changes and lung cancer.

Differential diagnosis of the histological pattern of UIP

Ideally the diagnosis of IPF requires concurrence of clinical, morphological, and radiological data with the identification of a definitive UIP pattern. In real clinical settings, clinicians have deal with mixed or insufficient data from diagnostic tests to reach a definitive diagnosis. Moreover, histopathologic evidence of UIP is not synonymous with that of IPF. UIP can also occur in a number of other diseases such as rheumatoid arthritis, dermatomyositis/polymyositis, systemic sclerosis, systemic lupus erythematosus, hypersensitivity pneumonitis, asbestosis, and drug-induced pneumonitis [61]. However, the morphological features of UIP in these clinical presentations may be distinct. For example, UIP lesions in hypersensitivity pneumonitis tend to exhibit a bronchocentric location, and peribronchiolar granulomas and lymphocytic bronchiolitis are often encountered. Conversely, asbestos bodies are observed in UIP cases associated with asbestosis. The appearance of follicular bronchiolitis or lymphoid hyperplasia with germinal centers with the clinical presentation of UIP argues in favor of the Sjögren syndrome or drug-induced interstitial lung disease (DI-ILD) [62]. In systemic sclerosis an overlap is often observed between UIP and NSIP with a significantly smaller number of myofibroblastic foci and more limited honeycombing [63].

Collection of representative biopsy material allows the differentiation of UIP from the fibrous variant of idiopathic NSIP, which is characterized by the homogeneity of fibrotic changes and the presence of varying severity of interstitial inflammation [64].

Clinical presentation and course

The main symptoms of IPF are slow progressive dyspnea and unproductive cough, which are present in most patients. In later stages, cough often becomes productive. The characteristic Velcro crackles are often present in zones corresponding to honeycombing. The presence of these sounds in patients older than 70 years, even in the absence of other symptoms, requires HRCT to check for IPF [65].

In 40%–75% of patients, deformation of the nail phalanges, that is, clubbing, can outpace the onset of other symptoms [66]. In the early stages, IPF can be almost asymptomatic. In one of the substudies of the large-scale Farmington trial on the prevalence of cardiac pathology, no one did have a pulmonary diagnosis and minimally had symptoms of pulmonary pathology approximately 2% of the patients over 50 years of age in the general population who underwent chest computed tomography (CT) and found to have interstitial abnormalities that could be interpreted as definite fibrosis [67]. Further observation of the study cohort revealed that approximately 75% of the subjects had progression in both the radiological signs of fibrosis and the reduction in pulmonary function [68]. For such asymptomatic patients with minimal HRCT signs of definite or possible UIP, the term "preIPF" is proposed, and active monitoring of physical, functional, and radiological status and determining the expression of MUC5B as an important predictor of IPF development are suggested [69]. The average time from the emergence of first symptoms to the diagnosis of IPF is approximately 2 years [26].

Symptoms of IPF in predominantly elderly patients are often accompanied by manifestations of numerous comorbidities such as COPD, coronary heart disease, diabetes mellitus, and deep vein thrombosis, which are more common than in the general population [70]. These facts must be taken into account when assessing the clinical status of the patient, since dyspnea can be a manifestation of not only IPF but also heart failure, pulmonary embolism, and PH.

Although the well-established opinion states that IPF is a progressive disease with a permanent decline in pulmonary function, albeit at different rates, studies of antifibrotic medications, including pirfenidone and nintedanib, in large cohorts demonstrated the possibility that a stable status can be achieved without any drugs or other interventions [71,72]. Conventionally, IPF clinical course can be categorized into three types based on the rate of FVC decline: slow deterioration (<10% decline in FVC over 6 months or no changes in disease indicators for many months), intermittent deterioration with episodes of exacerbations, and rapid deterioration (more than 10% decline in FVC for 6 months) [2,73]. In the latter two classes, the life expectancy of patients does not exceed 2 years from the time of diagnosis [2].

Acute exacerbation of IPF

Acute exacerbation (AE) of IPF has been a focus of active debate since 2007. AE of IPF is associated with mortality and hospitalization in approximately 40%–60% and 35% of patients, respectively [74,75]. In the 2011 consensus statement for IPF by the ATS/ERS/JRS/ALAT, the frequency of AEs among all patients with IPF was 5%–10% per year [1]. A metaanalysis of clinical studies on IPF revealed an exacerbation frequency of 41 per 1000 patient-years [76]. However, given that patients with severely impaired respiratory function and those with serious comorbidities were excluded from these studies, this number may be considerably higher in clinical practice.

The definition and criteria for AEs of IPF, proposed in 2007 [77], were revised in 2016 [78] (Table 2.1.1).

Although the definition of an AE of IPF and the criteria for its diagnosis hint at an idiopathic nature, AEs are known to be triggered by viral infections, surgical lung biopsy, aspiration, and BAL [79,80]. Moreover, it is clinically and morphologically impossible to differentiate diffuse alveolar damage of exogenous and endogenous origin [81]. In this regard the definition and diagnostic criteria for the AEs of IPF were criticized and revised, the definition of "unknown origin" was excluded from its definition, and the current interpretation suggests an AE of IPF as an acute clinically significant deterioration in respiratory symptoms characterized by the appearance of new common alveolar lesions [78].

Diagnosis of the AEs of IPF initially involves the exclusion of other diseases that can lead to increased dyspnea in a patient with IPF, such as pulmonary embolism, pneumothorax, pleural effusion, infection of distal airways, decompensated heart failure, drug damage, and aspiration [78]. All these etiologies should be considered in the differential diagnosis, since they require distinct approaches in patient management in most cases. Furthermore, infection, aspiration, and drug toxicity can act as not only separate, new diseases but also triggers of exacerbation, along with surgical or bronchological intervention. Clinical and radiographic differentiation of these conditions such as drug-induced pneumonitis or bacterial pneumonia from an IPF exacerbation can be quite challenging, since most of the symptoms and changes in laboratory values can occur in both conditions.

Risk factors for IPF exacerbations are reductions in initial functional parameters including FVC, carbon monoxide diffusion capacity in the lungs (DLCO), and 6-min walking distance as well as PH, low arterial oxygenation, and a rapid FVC decline [78]. In other words the more severe the disease, the higher the risk of AE. Less significant risk factors for AEs IPF are younger age, concomitant coronary heart disease, and a higher body mass index [82,83].

As stated earlier, during an AE of IPF, there are almost always morphological signs of diffuse alveolar damage, which manifest as ground-glass opacity (GGO) and consolidation on HRCT; these findings are not characteristic for UIP with a stable IPF course [84].

TABLE 2.1.1 Definition and criteria for the diagnosis of acute exacerbations of IPF

	2007	2016
Definition	Acute, clinically significant deterioration of unknown origin in patients with IPF	Acute clinically significant deterioration of respiratory symptoms, characterized by the emergence of new widespread areas of alveolar damage
Criteria for diagnosis	Previous or probable diagnosis of IPF	Previous or probable diagnosis of IPF
	Unexplained gain or development of shortness of breath within the last 30 days	Acute exacerbation or development of dyspnea over a course of usually less than 1 month
	Appearance of new bilateral GGO zones and/or consolidations of the background of reticular changes and honeycombing by HRCT, corresponding to the CT pattern of UIP	Appearance of new bilateral GGO zones and/or consolidations on the background of CT pattern of UIP by HRCT
	The lack of evidence for infection of distal airways based on BAL or endotracheal aspirate examination	
	Exclusion of alternative causes, such as heart failure, pulmonary embolism, and acute lung injury of known etiology (sepsis, trauma, drug damage, aspiration, surgery, stem cell transplantation, and massive blood transfusion)	Worsening course that cannot be fully explained by heart failure or fluid overload

BAL, bronchoalveolar lavage; *CT*, computed tomography; *GGO*, ground-glass opacity; *HRCT*, high-resolution computed tomography; *IPF*, idiopathic pulmonary fibrosis; *UIP*, usual interstitial pneumonia.
Extract from Collard HR, Moore BB, Flaherty KR, Brown KK, Kaner RJ, King TE Jr, et al. Acute exacerbations of idiopathic pulmonary fibrosis. Am J Respir Crit Care Med 2007;176(7):636–43; Collard HR, Ryerson CJ, Corte TJ, Jenkins G, Kondoh Y, Lederer DJ, et al. Acute exacerbation of idiopathic pulmonary fibrosis: an International Working Group Report. Am J Respir Crit Care Med 2016;194(3):265–75.

Laboratory tests and biomarkers

Among laboratory studies used for the definitive diagnosis of IPF, serological tests to detect connective tissue diseases (CTDs) are the most important and should be performed in all patients with interstitial lung lesions [58]. Recommended routine tests include C-reactive protein, erythrocyte sedimentation rate, antinuclear antibodies, rheumatoid factor, myositis panel, and anticyclic citrullinated peptide antibody level. Other tests should be performed depending on the clinical symptoms and disease history and may include antibodies to Scl-70 (topoisomerase-1), centromere, RNA polymerase III, U1-RNP, Th/To, PM-Scl, U3-RNP (fibrillarin), and Ku for suspicious systemic sclerosis; antibodies to SSA/Ro and SSB/La for suspicious the Sjögren syndrome; and antibodies to creatinine phosphokinase, myoglobin, and aldolase for suspicious myopathy, as well as others including antibodies to Jo-1, MDA5, Mi-2, NXP2, and TIF1-g [58].

A frequently observed laboratory phenomenon of IPF is elevated antineutrophil cytoplasmic antibodies (ANCA) titers in 4%–35% of the patients. Moreover, only 7%–23% of the patients develop ANCA-associated vasculitis in the future [85]. The pathophysiology of IPF with increased ANCA titers is not well understood. These patients also often exhibit increased serum levels of rheumatoid factor, antinuclear antibodies, anti-Scl-70, and others; therefore IPF in these cases can be classified as interstitial pneumonia with autoimmune features [86]. However, an isolated increase in ANCA without associated autoimmune serum markers does not by itself allow classification of these patients as interstitial pneumonia with autoimmune features according to the majority of the experts. For these patients the term ANCA-associated pulmonary fibrosis can be used [85].

The clinical and radiological signs of ANCA-positive and ANCA-negative IPF are almost indistinguishable [87]. Nevertheless, higher ANCA titers are associated with lower survival rates in patients with IPF [88]. Smoking cessation can lead to normalization of ANCA titers and decrease the risk of ANCA-associated vasculitis [89].

Currently, there is no one diagnostic biomarker with sufficient specificity and sensitivity that can be recommended for use in routine practice for definitive or differential diagnosis of IPF [58]. However, several studies demonstrated that increased levels of surfactant protein (SP)-A, SP-D, matrix metalloproteinase (MMP)-7, MMP-28, and osteopontin in patients with IPF were significantly different than those observed in other fibrosing ILDs [90–92]. Several potential biomarkers (KL-6, periostin, SP-A, and SP-D) reflect the severity of the disease, correlate with the degree of functional impairment, and are relatively useful for assessing the prognosis of IPF [93,94].

The search for biomarkers from various sources is one of the diagnostic directions in the IPF field. Recently, several micro-RNAs (miR-142-3p, miR-33a-5p, and let-7d-5p) isolated from the sputum of patients distinguished IPF from hypersensitivity pneumonitis with high confidence when used as a panel [95].

Functional diagnostics

Several methods such as regular spirometry, body plethysmography, lung diffusion capacity, and the 6-min walking test are used to assess the functional status of the patients and monitor the disease course [96,97]. The characteristic spirometric signs for IPF and other restrictive lung diseases are a decrease in FVC with a preserved or slightly reduced forced expiratory volume in one second (FEV1), which leads to an increase in the FEV1/FVC ratio. Additionally, total lung capacity and DLCO are usually reduced [98]. The 6-min walking distance decreases in proportion to the severity of the disease and is usually accompanied by desaturation [97]. The gas levels in arterial blood reflect a disruption in gas exchange in the lungs. A decrease in partial arterial CO_2 pressure occurs only during physical exertion initially, which is observed even at rest in advance-stage disease [99].

Hypercapnia is detected in the terminal stage of IPF. According to several studies, decreases in the aforementioned functional parameters, along with older age and male gender, are predictors of disease progression and worse survival. Ley et al. developed a multivariate index to assess the severity and predict the course of IPD, which they termed gender, age, and physiology (GAP), which includes two functional predictors: FVC and DLCO. According to the GAP index, an FVC level <50% of predicted value and a DLCO <35% of predicted value are thresholds for the most severe functional compromise [100].

FVC and DLCO are also surrogate markers for estimating the rate of progression in IPF. An annual (according to other data for 6 months) FVC decline of >10% and/or an annual (6 months) DLCO decline of >15% indicate that the disease is in a rapidly progressive course with a high risk of death [2,101].

The standard interval for monitoring pulmonary function in IPF is 3–4 months; however, the final decision on the frequency of monitoring rests on the clinician, based on clinical data and the rate of disease progression.

High-resolution computed tomography

Together with histopathologic evaluation, HRCT is the most important, and often the decisive, tool for the diagnosis of IPF with which surgical lung biopsy can be avoided in many patients. Chest X-ray has insufficient resolution to be useful for initial assessment of the lung parenchyma and disease monitoring and usually reveals bilateral reticular changes with a basal-subpleural distribution.

The typical HRCT presentation of IPF is the characteristic UIP pattern that includes the following features (Figs. 2.1.12 and 2.1.13) [102,103]:

- Reticular abnormalities: intralobular septal thickening, irregular interlobular septal thickening, and linear opacities
- Traction bronchiectasis and bronchiolectasis
- Honeycombing
- Bilateral, basal, and subpleural distribution of abnormalities
- Little GGO in the area of reticular changes

Honeycombing is a classic sign of UIP, but is not always present, found in 24%–95% of cases depending on the disease stage, according to different studies [104,105].

At the onset of disease, there are no signs of honeycombing, and reticular changes may be minimal; therefore unambiguous interpretation is difficult (Fig. 2.1.14), especially since clinical symptoms are often absent in these patients. However, reticular signs are the earliest and easily attributable to UIP; they are detected in almost all patients. As the disease progresses, severity of the reticular changes increases, and bronchiectases are visualized inside the reticular areas, which then merge with small cysts that eventually form the honeycombing pattern.

GGO zones, albeit not conflicting with the pattern of UIP, do not extend beyond reticular abnormalities and are usually adjacent to the sites of honeycombing (Fig. 2.1.15). The predominance of GGOs over the reticular and honeycombing zones and their spread beyond their limits during periods of no AE indicates the dominance of interstitial inflammation over fibrosis, which renders the diagnosis of IPF less likely but does not exclude it [102]. However, architectural distortion of the lung parenchyma and volume loss in the basal regions can lead to the appearance of widespread GGOs in areas of fibrosis that, in that case, reflects not interstitial inflammation but a decrease in opacity and increase in attenuation of the lung parenchyma (Fig. 2.1.16).

In IPF, intrathoracic lymph nodes can become enlarged, but their size rarely exceeds 1.5 cm [106]. A nonspecific, albeit quite frequent, finding in patients with IPF is pulmonary emphysema (Fig. 2.1.17). According to Cottin et al., emphysema

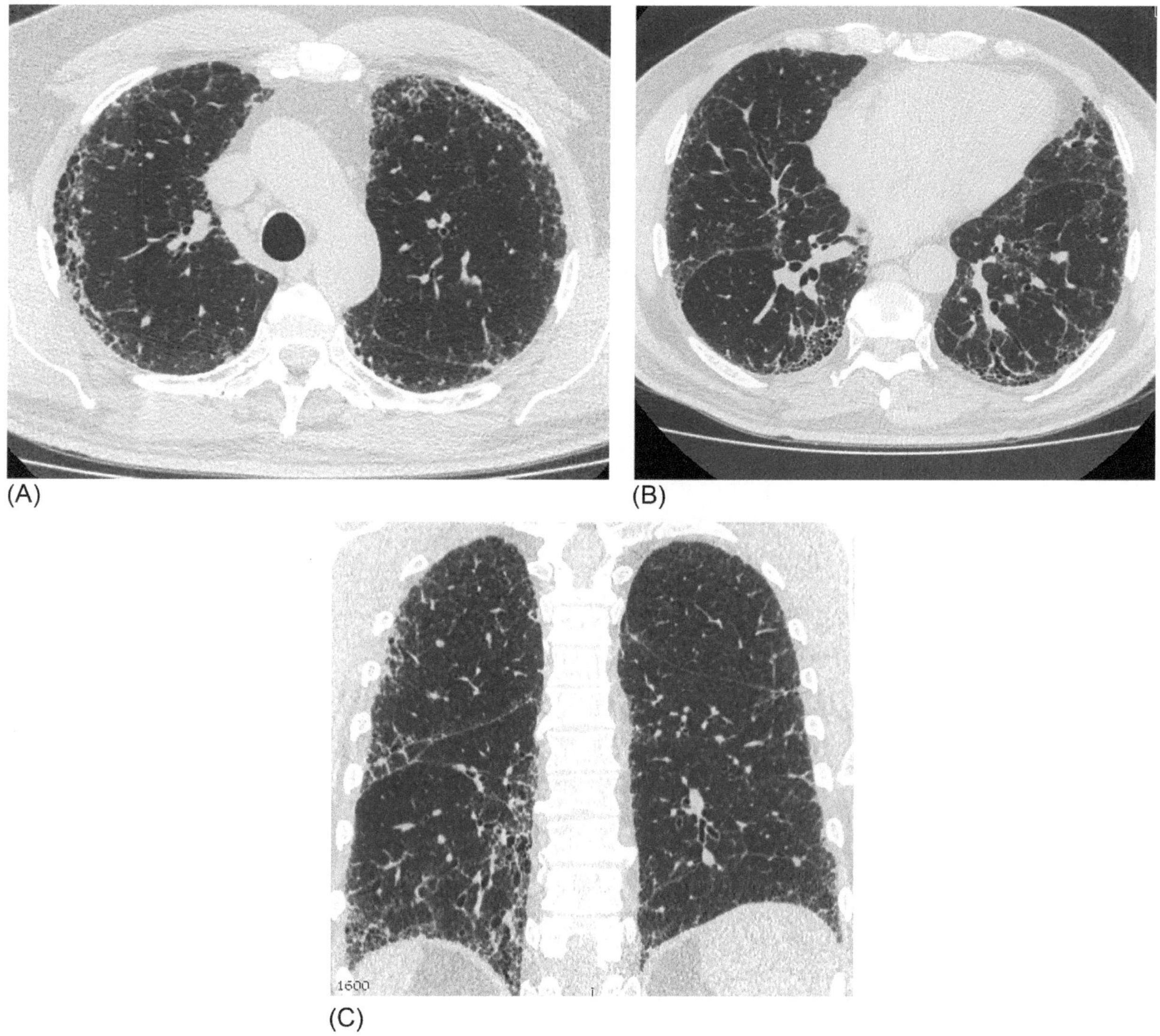

FIG. 2.1.12 IPF. Subpleural zones of honeycombing. Irregular intralobular and interlobular septal thickening. Outside the affected areas, lung tissue appears healthy (A and B). Abnormalities are expressed predominantly subpleurally in the basal and paramediastinal areas (C), corresponding to the pattern of usual interstitial pneumonia.

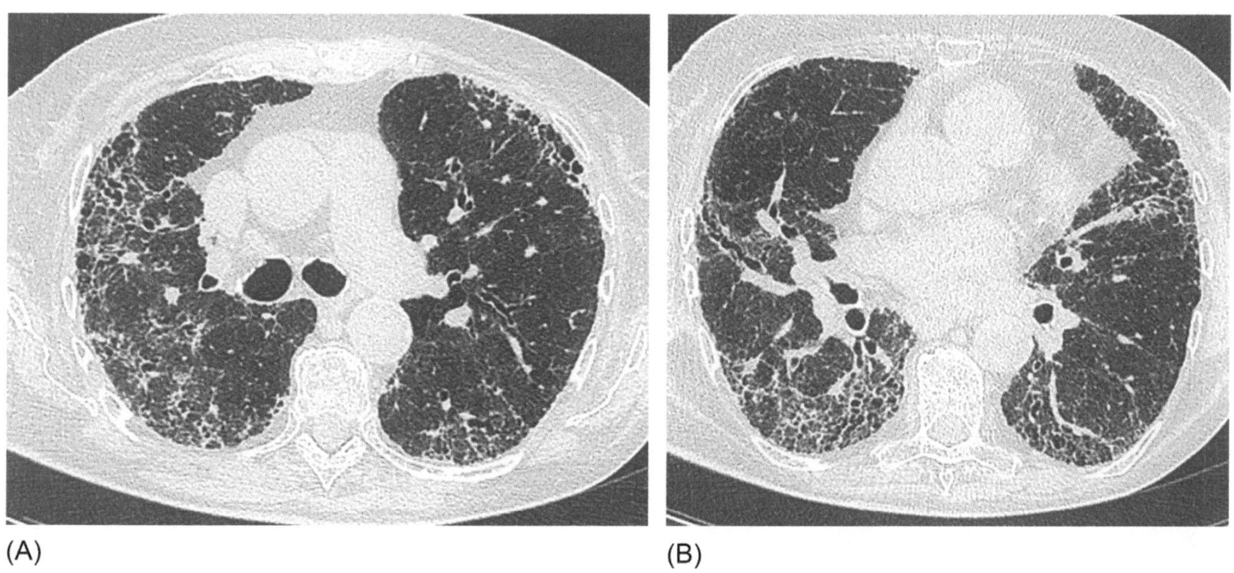

FIG. 2.1.13 IPF. Pronounced reticular changes, honeycombing, and traction bronchiectasis. Changes are predominantly subpleural, corresponding to the pattern of usual interstitial pneumonia (A and B).

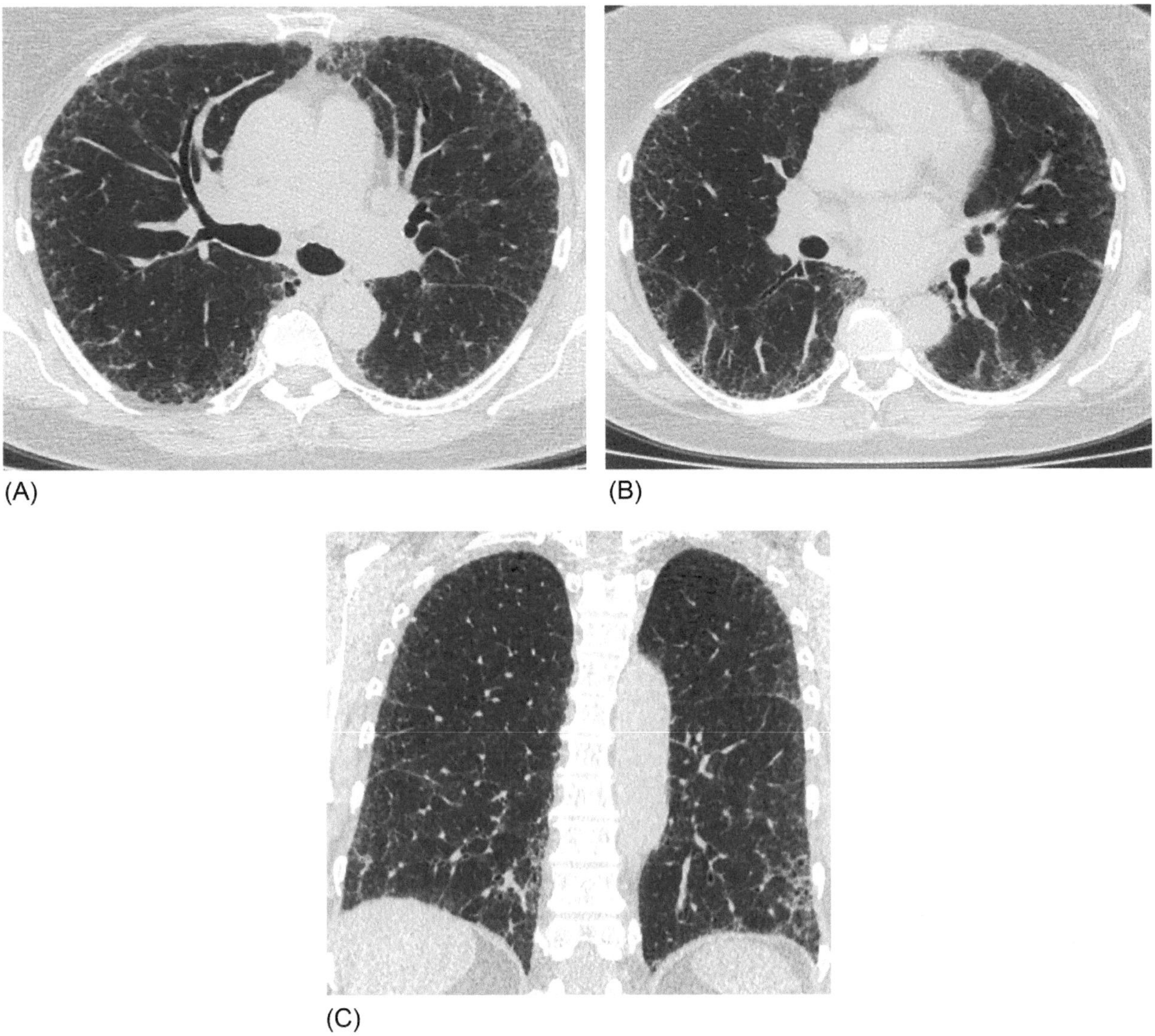

FIG. 2.1.14 IPF, confirmed by video-assisted thoracoscopic lung biopsy. Early usual interstitial pneumonia pattern. Thickened intralobular septa, linear fibrosis, and single-row small cysts with subpleural distribution. Light ground-glass opacity (GGO) within fibrous changes (A and B). Abnormalities are distributed predominantly in the basal and subpleural segments (C).

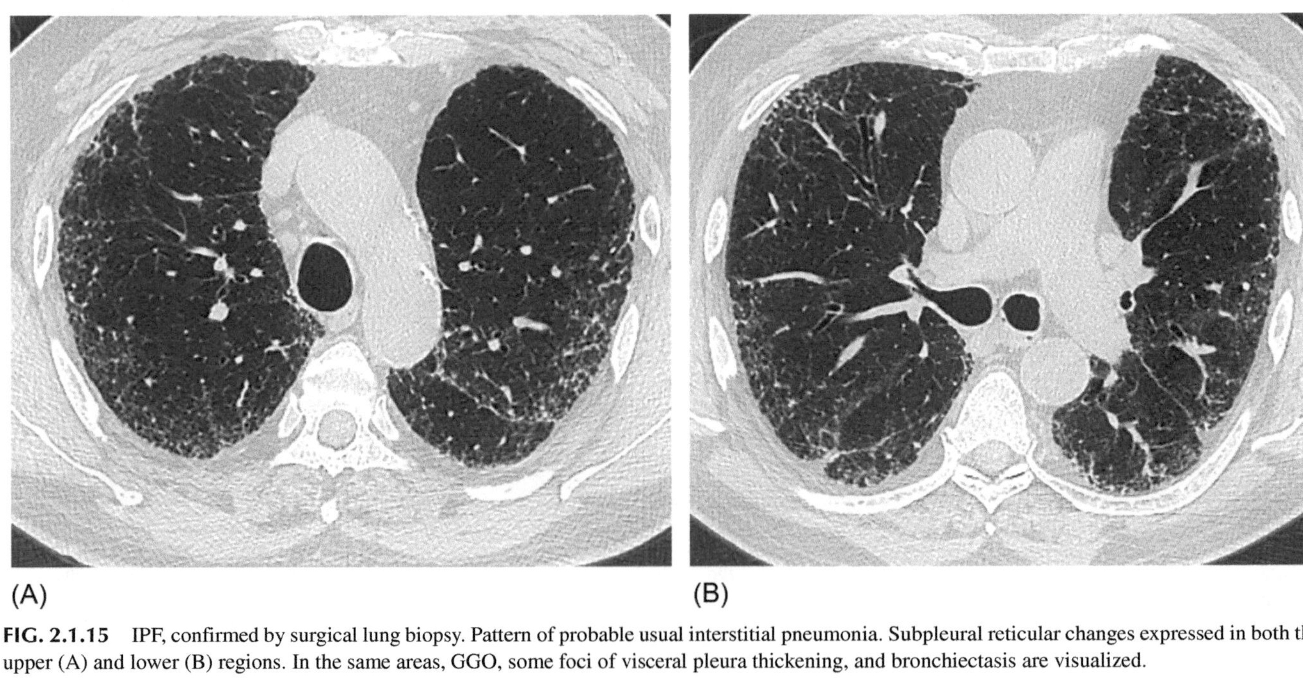

FIG. 2.1.15 IPF, confirmed by surgical lung biopsy. Pattern of probable usual interstitial pneumonia. Subpleural reticular changes expressed in both the upper (A) and lower (B) regions. In the same areas, GGO, some foci of visceral pleura thickening, and bronchiectasis are visualized.

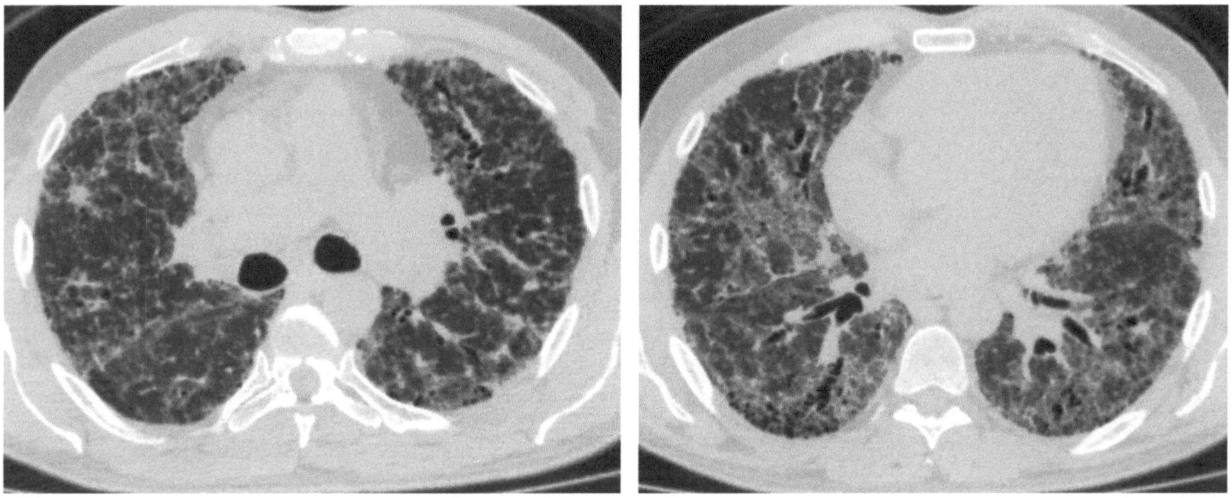

(A) (B)

FIG. 2.1.16 IPF, advanced stage. Diffuse reticular changes and traction bronchiectasis (A and B). In the lower zones (with the exception of the middle lobe), there are diffuse GGOs, subpleural honeycombing, and traction bronchiectasis.

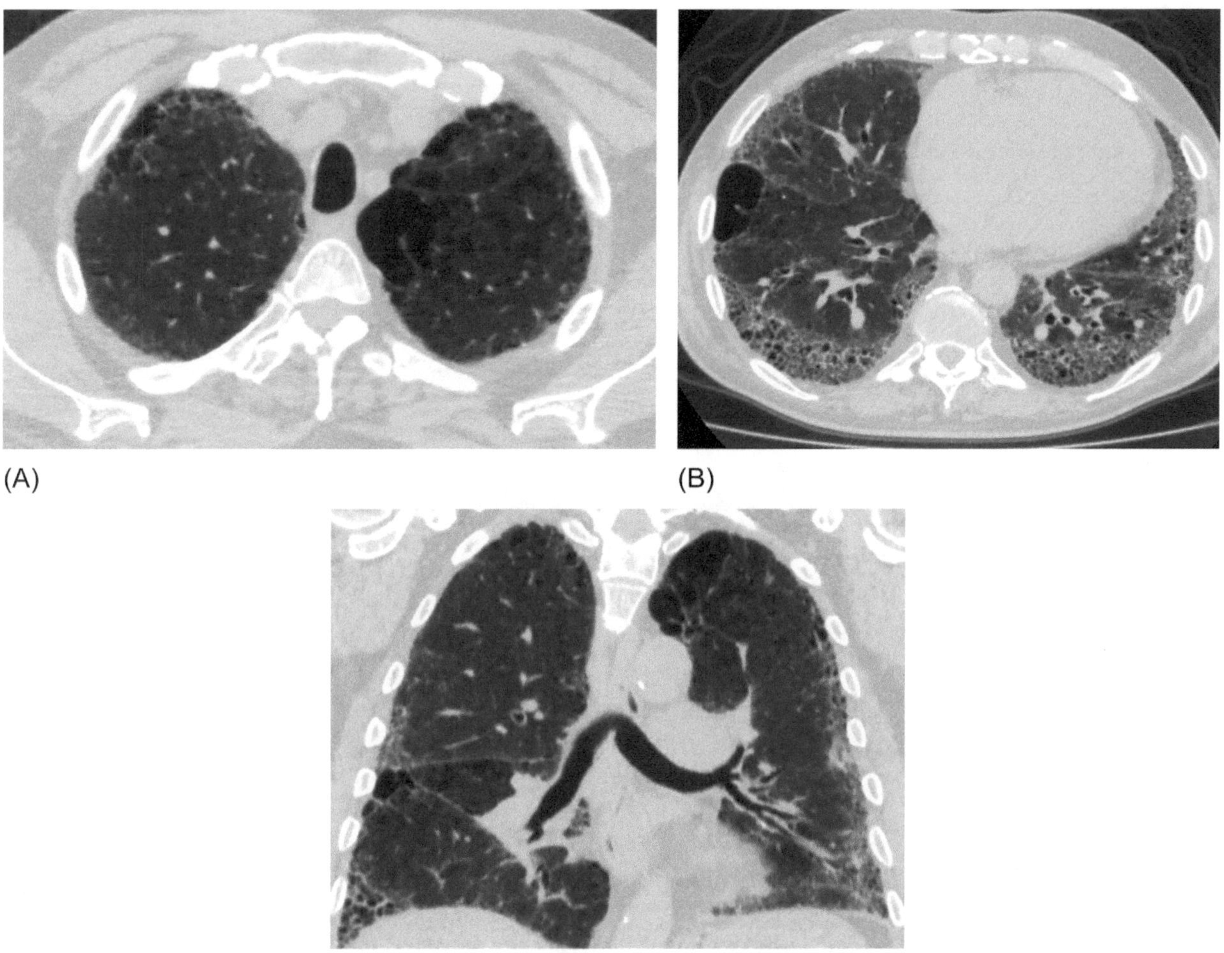

(A) (B) (C)

FIG. 2.1.17 IPF. Bullous and paraseptal emphysema of the upper lobes (A). Subpleural honeycombing, traction bronchiectasis in the lower lobes, and single bulla in the right lung (B). Coronal reconstruction shows the distribution of emphysema in the upper lobes and fibrosis mainly in the lower zones (C).

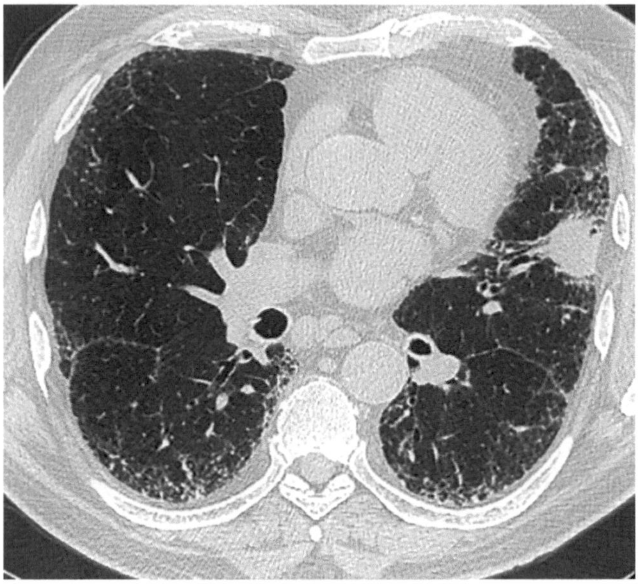

FIG. 2.1.18 Squamous cell carcinoma of the lingular segments of the left lung associated with signs of usual interstitial pneumonia.

is usually localized in the upper parts of the lungs, both centrilobular and paraseptal forms are present in majority of the cases, 54% of patients also have bullae, and 47% of patients have signs of PH [107]. The appearance of air traps in the upper lobes is found in no more than 4% of patients with IPF and usually argues in favor of other diagnoses, primarily hypersensitivity pneumonitis; however, air traps associated with IPF are detected more often in lower levels, not different than that observed in other fibrotic lung diseases [108].

HRCT signs such as nodules, infiltrates, and multiple cysts in the lung parenchyma; changes that are one-sided or include the upper lobes or peribronchovascular region; and subpleural sparing are not typical for UIP [109]. Their presence in a patient with IPF means either an alternative diagnosis or the appearance of an additional new disease, including lung cancer, pneumonia, and pulmonary embolism (Fig. 2.1.18).

The radiological diagnosis of UIP is especially difficult in periods of AE. Nonetheless, along with the classical HRCT presentation, there are signs of acute alveolar damage represented by massive bilateral GGO zones, which often overlap reticular changes and honeycombing (Fig. 2.1.19). Consolidation areas may also be observed. In an analysis of 64 AE episodes of IPF, Akira et al. concluded that the distribution of GGOs in an AE can be divided into three categories. The most frequent, the peripheral pattern, occurs in more than half of the cases and is characterized by a subpleural spread, resembling a sickle. The second most frequent is the diffuse variant, which is characterized by bilateral lesions that capture both the central and peripheral zones. Finally the rarest, multifocal pattern presents with patchy areas of GGO in different parts of the lungs [84].

Bronchoalveolar lavage

Cytological analysis of BAL fluid in IPF has characteristic findings among which the most commonly reported is an increase in the fraction of neutrophils (6%–22%) and eosinophils (2.4%–7.5%) and a decrease in alveolar macrophages (46%–83%) [58]. Individual patients may experience lymphocytosis and an increase in the CD4/CD8 ratio; however, BAL analysis is recognized primarily for its utility in the differential diagnosis of lung diseases with specific BAL findings such as hypersensitivity pneumonitis, sarcoidosis, and pulmonary eosinophilia [58,110].

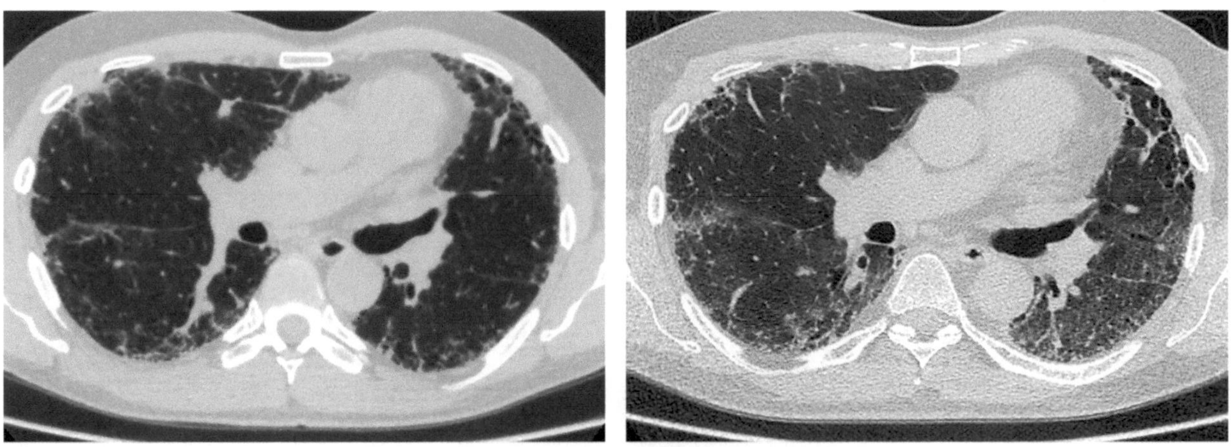

(A) (B)

FIG. 2.1.19 IPF. Subpleural reticular changes and honeycombing on the left (A). The same patient after 3 months. Acute exacerbation of IPF. Diffuse GGOs in the posterior zones. Worsening of reticular abnormalities (B).

Probe-based confocal laser endomicroscopy of distal airways

Our unpublished study examined 117 bronchopulmonary areas in six patients with morphologically proved IPF using probe-based confocal laser endomicroscopy (pCLE) of distal airways, and 843 informative images were obtained.

A typical endomicroscopic presentation in IPF is characterized by the lack of visualization of normal alveolar structures in the affected pulmonary segments due to the replacement of the alveolar sacs with connective tissue elements, which are defined as brightly randomly located fluorescent fibrils and nodules that seal the alveoli; this was found in 40 out of a total of 117 areas that were examined (Fig. 2.1.20A). In parts within the same bronchopulmonary segment, normal alveoli with preserved structures alternated with alveoli with completely (Fig. 2.1.20A) or partially (Fig. 2.1.20B) sealed components of fibrous tissue and pronounced fluorescence. In 9 out of the 117 regions, it was not possible to differentiate any structures with normal alveolar pCLE pattern. None of the patients at the time of the examination smoked, and there were no cellular elements in the alveolar lumen. Additionally, mucus was present in the alveolar lumen (Fig. 2.1.20C). In 16 of the 117 areas that were examined, a moderate amount of rounded, small (3–5 μm in diameter), brightly fluorescent particles was found in the terminal bronchioles (Fig. 2.1.20D). The thickness of the interalveolar septa field could vary from normal to two to three times the normal within the same visual field in patients with IPF (Fig. 2.1.20E).

The role of transbronchial lung biopsy (TBB) is currently limited to patients who cannot undergo video-assisted thoracoscopic lung biopsy (VATS) due to the disease severity or a high risk of complications [111].

The small size of the species and their deformation with forceps TBB often do not allow differentiation of IPF from NSIP and HP using TBB [112]. The main value of this diagnostic approach is not the evidence it provides for IPF but its utility in exclusion of other diseases such as sarcoidosis and other pulmonary granulomatous diseases; therefore decision to perform TBB should be reached individually [58]. Cryobiopsy is a relatively new, minimally invasive biopsy method for ILDs, allows the collection of lung fragments approximately five times larger than those obtained with forceps biopsy, and is not subject to artifacts [113]. This method increases the likelihood of correct histological diagnosis of ILDs by up to 80% [58]. A metaanalysis of 14 studies on cryobiopsy in ILDs, including IPF, showed a significantly greater diagnostic value of cryobiopsy in comparison with TBB, albeit with a rather high rate of bleeding (23.7%) [114].

Due to the heterogeneity in results, primarily due to complications and the lack of a standardized protocol, cryobiopsy has not yet been incorporated into routine practice, and its use should be limited to specialized pulmonary clinics with extensive experience in cryobiopsy [115,116].

Currently, VATS remains the gold standard for obtaining complete histological material to reach a diagnosis of IPF [70]. However, as an invasive surgical intervention, VATS cannot always be performed in patients with severe respiratory failure, who have contraindications for surgery or simply refuse the procedure. In addition, for unknown reasons, IPF exacerbates after VATS in a small number of patients [117].

Open surgical lung biopsy is associated with a significant risk of complications compared with VATS in patients with ILDs (30-day mortality, 4.3% vs 2.1%; nonfatal complications, 18.1% vs 9.6%) and can be considered as a diagnostic tool only if its potential utility is considered to be superior to the associated surgical risks [118]. Biopsy specimens should be collected from several lobes, and sampling from areas that are affected by honeycombing the most should be avoided, since the diagnostic benefits of biopsy material from areas of maximum fibrosis are low [119].

Regardless of the method, lung biopsy should not be prescribed in patients with an HRCT pattern of UIP [58].

Diagnosis

The current diagnostic approach of IPF, presented in the recommendations of the Fleischner Society and the 2018 practical guidelines for the diagnosis of IPF by the ATS/ERS/JRS/ALAT [58,109], is based on assessment of the likelihood that a patient has an UIP pattern according to the findings of HRCT and biopsy material, if other possible diseases are excluded (Fig. 2.1.21). First the proposed diagnostic algorithm provides a thorough evaluation of the patient's history for characteristic clinical manifestations to exclude other ILDs due to exogenous factors and CTDs. For the same purpose, laboratory tests are required to exclude CTDs (see section "Laboratory tests and biomarkers").

If an exogenous factor or a competing serum-positive disease is excluded, the next step is to analyze the HRCT data together with signs that are characteristic for UIP to various degrees (see Table 2.1.2). In cases with predominantly subpleural-basal localization of the honeycombing in the absence of not relevant signs of UIP, the diagnosis of IPF is valid without histological confirmation (Figs. 2.1.12 and 2.1.13). In remaining patients with CT findings that don't clarify UIP

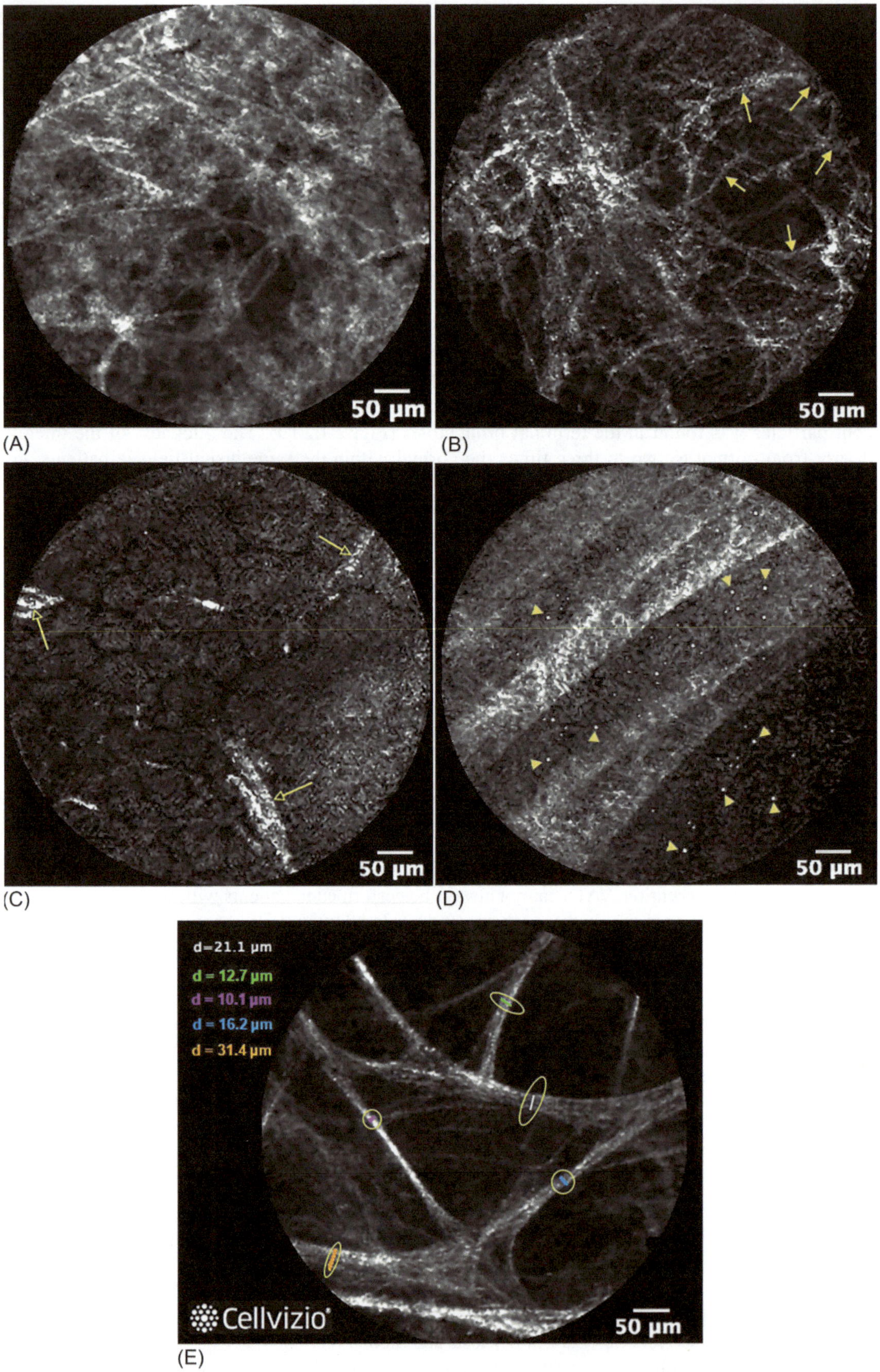

FIG. 2.1.20 PCLE images of distal airways in patients with IPF. (A) Brightly fluorescent, chaotically located connective tissue elements are seen in most segments of the lung on both sides; the walls of the alveoli are not differentiated. (B) The left part of the field of vision is almost completely occupied by connective tissue and brightly fluorescent elements without differentiation of the walls of the alveoli. In the right side of the field of vision, *yellow arrows* show some clearly visible interalveolar septa. C) *Arrows* show the ends of the elastic fibers of the interalveolar septa, fragmented because of pressure by a miniprobe. In the lumen, there is mucous secretion with air bubbles. (D) Characteristic cross striation, typical for the walls of terminal bronchioles. Rounded corpuscles with high autofluorescence stand out from the rest of the structures (*yellow arrow tips* indicate the largest corpuscules). (E) Interalveolar septa thickened to varying degrees. Measurements are marked with corresponding color ellipses.

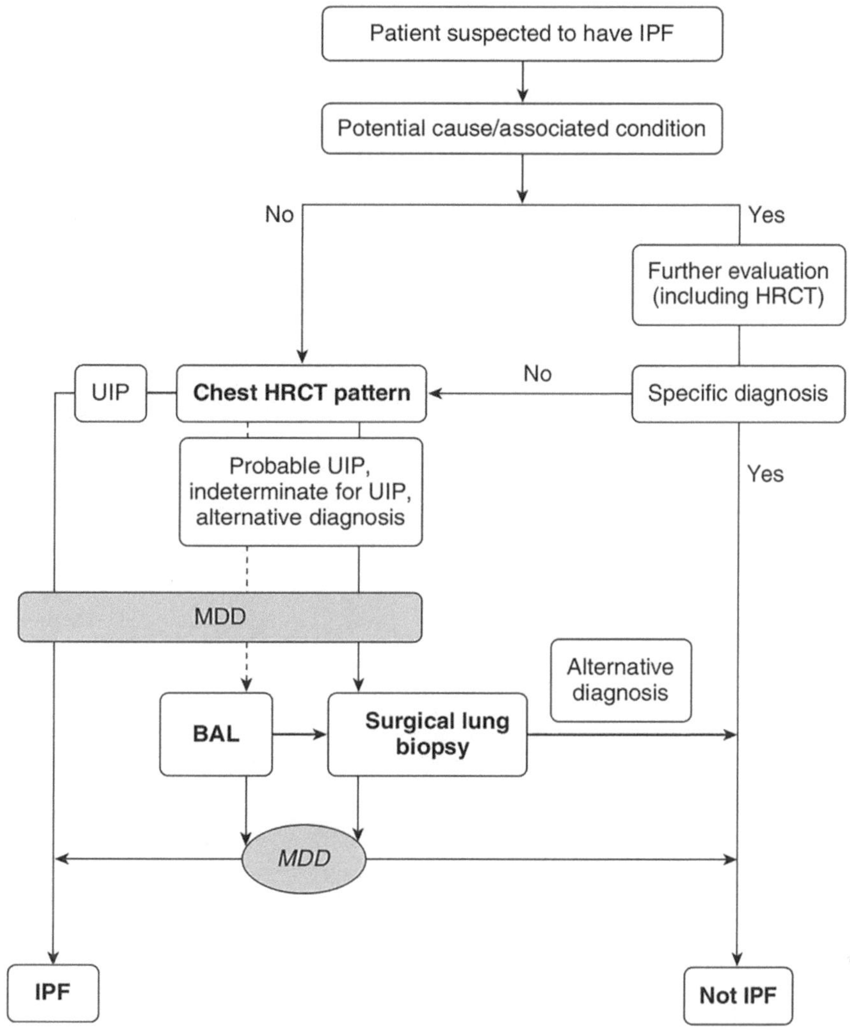

FIG. 2.1.21 Diagnostic algorithm for idiopathic pulmonary fibrosis (IPF). *MDD*, multidisciplinary discussion; *UIP*, usual interstitial pneumonia. *(Extracted from Raghu G, Remy-Jardin M, Myers JL, Richeldi L, Ryerson CJ, Lederer DJ, et al. Diagnosis of idiopathic pulmonary fibrosis. An official ATS/ERS/JRS/ALAT clinical practice guideline. Am J Respir Crit Care Med 2018;198(5):e44–e68 with the permission of ATS.)*

pattern, the next steps in diagnosis (BAL in cases with suspicious pulmonary infection or a disease with a specific BAL features or surgical lung biopsy) require multidisciplinary discussion (MDD) (Fig. 2.1.21).

If there is a risk of complication with surgical biopsy, that is, DLCO <25% of predicted, severe hypoxemia, pH, and oxygen dependence ≥2 L/min [120], the next step can be a decision to conduct cryobiopsy in a specialized clinic or TBB if some diseases are suspected (sarcoidosis and Langerhans cell histiocytosis). Following histological examination a repeated MDD can confirm or excludes the final diagnosis of IPF. MDD is always required at differential diagnosis of similar to IPF diseases.

Differential diagnosis

The range of disorders that should be included in the differential diagnosis of IPF is quite wide and includes diseases with HRCT and morphological features that resemble the UIP pattern, primarily honeycombing and/or reticular abnormalities. Among these disorders are NSIP, chronic HP, CTDs, drug-induced pulmonary fibrosis, pneumoconiosis, and radiation pneumonitis (Table 2.1.3). Less commonly, conditions such as the fibrous stage of sarcoidosis, fibrosis after diffuse alveolar damage, and LIP as well as certain cystic lung diseases and paraseptal emphysema, which can sometimes simulate a honeycombing, may be mistakenly diagnosed as an UIP.

TABLE 2.1.2 High-resolution computed tomography scan patterns

UIP	Probable UIP	Indeterminate for UIP	Alternative diagnosis
Subpleural and basal predominant; distribution is often heterogeneous[a] Honeycombing with or without peripheral traction bronchiectasis or bronchiolectasis	Subpleural and basal predominant; distribution is often heterogeneous Reticular pattern with peripheral traction bronchiectasis or bronchiolectasis May have mild GGO	Subpleural and basal predominant Subtle reticulation; may have mild GGO or distortion (early UIP pattern) CT features and/ or distribution of lung fibrosis that do not suggest any specific etiology (truly indeterminate for UIP)	Findings suggestive of another diagnosis, including *CT features[b] of cysts* Marked mosaic attenuation Predominant GGO Profuse micronodules Centrilobular nodules Nodules Consolidation *Predominant distribution*: Peribronchovascular Perilymphatic Upper or mid-lung *Others*: Pleural plaques (consider asbestosis) Dilated esophagus (consider CTD) Distal clavicular erosions (consider RA) Extensive lymph node enlargement (consider other etiologies) Pleural effusions, pleural thickening (consider CTD/drugs)

[a] *Variants of distribution: occasionally diffuse and may be asymmetrical.*
[b] *Superimposed CT features: mild GGO, reticular pattern, and pulmonary ossification.*
CT, computed tomography; *CTD*, connective tissue disease; *GGO*, ground-glass opacities; *RA*, rheumatoid arthritis; *UIP*, usual interstitial pneumonia.
Extracted from Raghu G, Remy-Jardin M, Myers JL, Richeldi L, Ryerson CJ, Lederer DJ, et al. Diagnosis of idiopathic pulmonary fibrosis. An official ATS/ERS/JRS/ALAT clinical practice guideline. Am J Respir Crit Care Med 2018;198(5):e44–e68, with the permission of ATS.

Regarding NSIP, signs of interstitial inflammation predominate over fibrosis, especially in the early stages (Fig. 2.1.22) and can be radiographically observed as dominance of the GGO. At the late stage of NSIP, as with IPF, reticular changes, traction bronchiectasis, and honeycombing are also seen [121]. In patients with NSIP, there may be foci of consolidation, which do not occur in UIP [122].

With the progression of NSIP, the severity of pulmonary fibrosis increases, and the HRCT pattern of NSIP becomes similar to that of UIP (Fig. 2.1.23) [123]. Two radiological phenomena are noted to accompany NSIP frequently and not occur in IPF: subpleural sparing [124] (Fig. 2.1.24) and the straight-edge sign [125]. The latter reflects a reasonably clean interface between the unaffected pulmonary parenchyma and the affected zones, which are localized to the basal regions and do not extend along the visceral pleura to the overlying parts of the lungs (Fig. 2.1.25) [125]. It should be noted that NSIP is often a pulmonary manifestation of a CTD; therefore thorough examination for complete history, additional symptoms, and laboratory tests aimed at CTD markers can aid in achieving the correct diagnosis.

The chronic course of hypersensitivity pneumonitis, especially without adequate treatment, leads to the development of interstitial pulmonary fibrosis, which resembles not only the HRCT findings but also the histological pattern of UIP [111]. The combination of reticular signs, honeycombing with a moderately expressed GGO, is possible in the advanced stages of HP (Fig. 2.1.26). Distinctive features of the latter are the predominance of GGOs over reticular and honeycombing changes, frequent presence of foci of consolidation and centrilobular nodules, a more homogenous distribution of pathological areas with the upper lung lobes, and subpleural sparing [124]. Silva et al. propose that the predominance of sites with honeycombing in the upper lobes is a characteristic feature of HP [30]. Conversely, air traps are equally common (45%) in zones of pathological changes (lower and middle zones) in IPF and HP; however, the appearance of air trappings in upper lobes is not typical in IPF and can be used in differential diagnosis as a characteristic feature of HP (Fig. 2.1.26) [108].

CTDs can affect the lungs in the form of UIP or NSIP. This is especially true for rheumatoid arthritis (RA) and systemic sclerosis (SSC) [126]. In patients with RA the lung lesion is usually preceded by a long articular history; in SSC the pulmonary manifestations may dominate and outpace other symptoms. Therefore, in patients with an HRCT picture of UIP or NSIP, careful history-taking and physical and other evaluations are very important to identify signs typical of SSC including the Raynaud phenomenon, arthralgia, dysphagia, sclerodactyly, amimia, mummified skin, pH, and esophageal dilatation (Fig. 2.1.27).

TABLE 2.1.3 The differential signs of IPF and other fibrotic lung diseases

	IPF	NSIP	Chronic HP	SSC-ILD	RA-ILD	DI-ILD
Anamnesis	Smoking, pollutants, GER	Nonsmokers	Nonsmokers; contact with allergens from birds, fungi, organic dust	The Raynaud syndrome	Polyarthritis	Medication (bleomycin, amiodarone, and biologic drugs)
Laboratory tests	No specific data	No specific data	No specific data	Increases in CRP, antinuclear antibodies, and anti-Scl-70 (topoisomerase-1), anticentromere, and anti-RNA polymerase III antibodies	Increases in CRP, rheumatoid factor, anticyclic citrullinated peptide	Increased CRP
BAL	Neutrophilia Mild eosinophilia	Moderate neutrophilia Moderate eosinophilia	Lymphocytosis >50%	Mild neutrophilia, mild eosinophilia	Moderate neutrophilia, moderate lymphocytosis	Lymphocytosis >30%, increased $CD4^+/CD8^+$ ratio
Reticular abnormalities	+++	++	++	+++	++	++
GGO	+	+++	+++	++	++	+++
Honeycombing	+++	++	+	++	++	+
Straight-edge	+	++	−	++	++	−
Consolidation	−	+	+	+	+	++
Air traps in nonfibrous areas	−	−	+++	−	−	+
Subpleural sparring	−	++	++	+	+	++
Abnormalities distribution	Subpleural, posterior-basal divisions	Basal segments	Middle and lower zones Possible predominance of changes in the upper lobes	Basal segments	Basal and subpleural zones	No preference

+++, Typical sign; ++, often sign; +, possible sign; −, atypical sign.
IPF, idiopathic pulmonary fibrosis; *NSIP*, nonspecific interstitial pneumonia; *HP*, hypersensitivity pneumonitis; *SSC-ILD*, systemic scleroderma associated with interstitial lung disease; *RA-ILD*, rheumatoid arthritis associated with interstitial lung disease; *DI-ILD*, drug-induced interstitial lung disease.

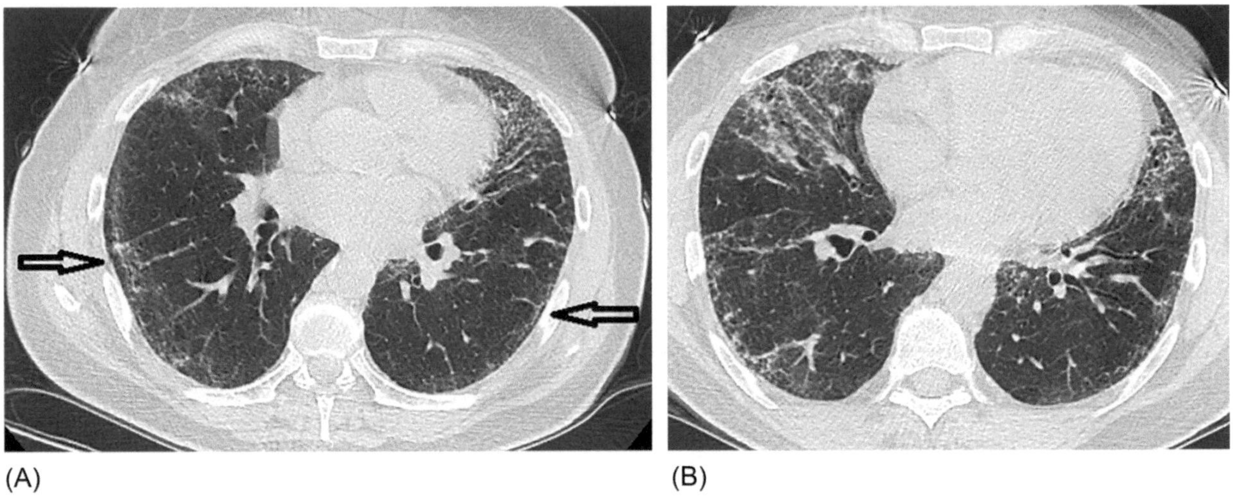

FIG. 2.1.22 Nonspecific interstitial pneumonia. Significant reticular changes associated with GGO in the anterior segments (A and B). Subpleural sparing *(arrows)* (A).

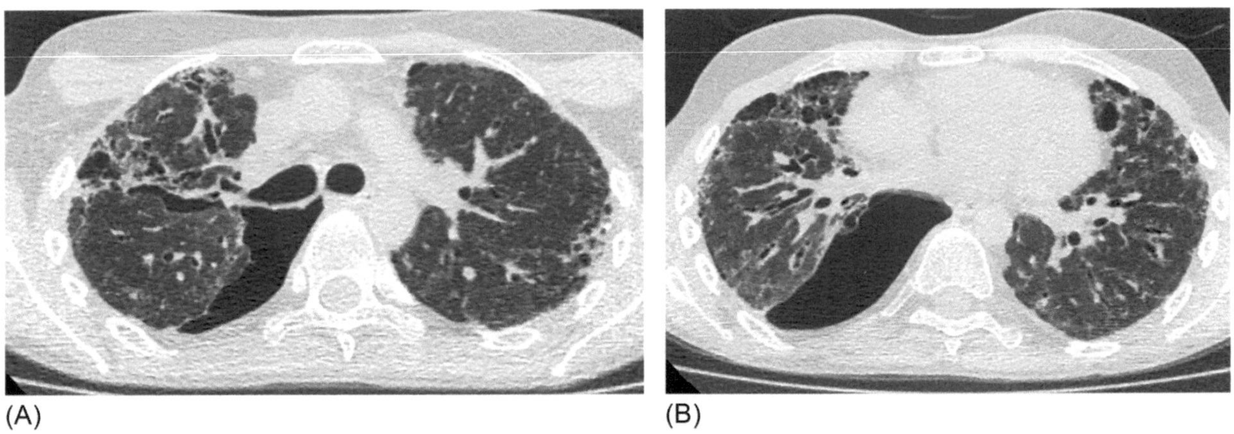

FIG. 2.1.23 Nonspecific interstitial pneumonia, confirmed by lung biopsy in late stage, in a 23-year-old patient. Subpleural honeycombing, traction bronchiectasis, and moderate reticular changes. Right-sided limited pneumothorax (A and B).

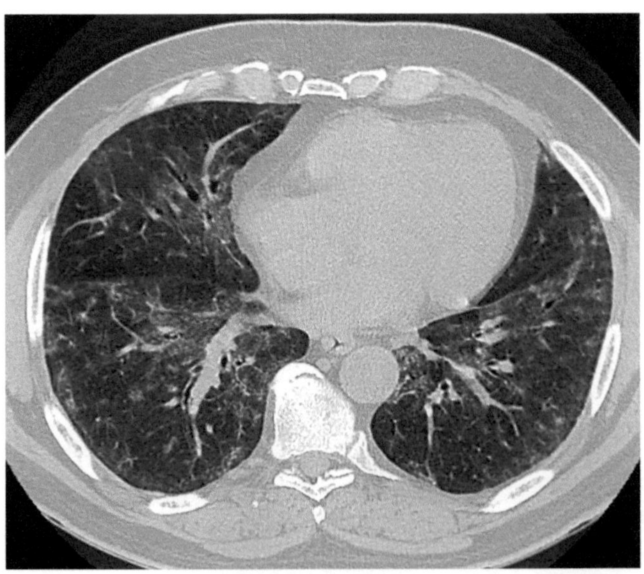

FIG. 2.1.24 Nonspecific interstitial pneumonia, confirmed by lung biopsy. Diffuse bilateral areas of GGO and moderate reticular changes. Subpleural sparing.

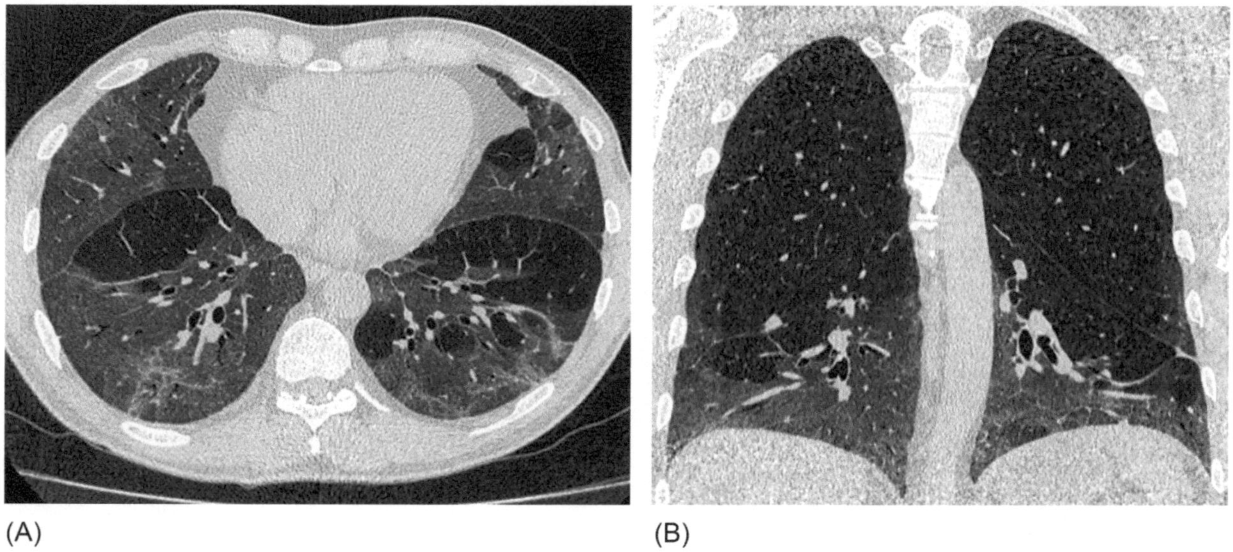

FIG. 2.1.25 Nonspecific interstitial pneumonia. Diffuse zones of GGO with fragmented reticular signs. Healthy and affected areas are separated by straight edges (A). Coronal slice shows localization of the lesion only in the basal segments (B).

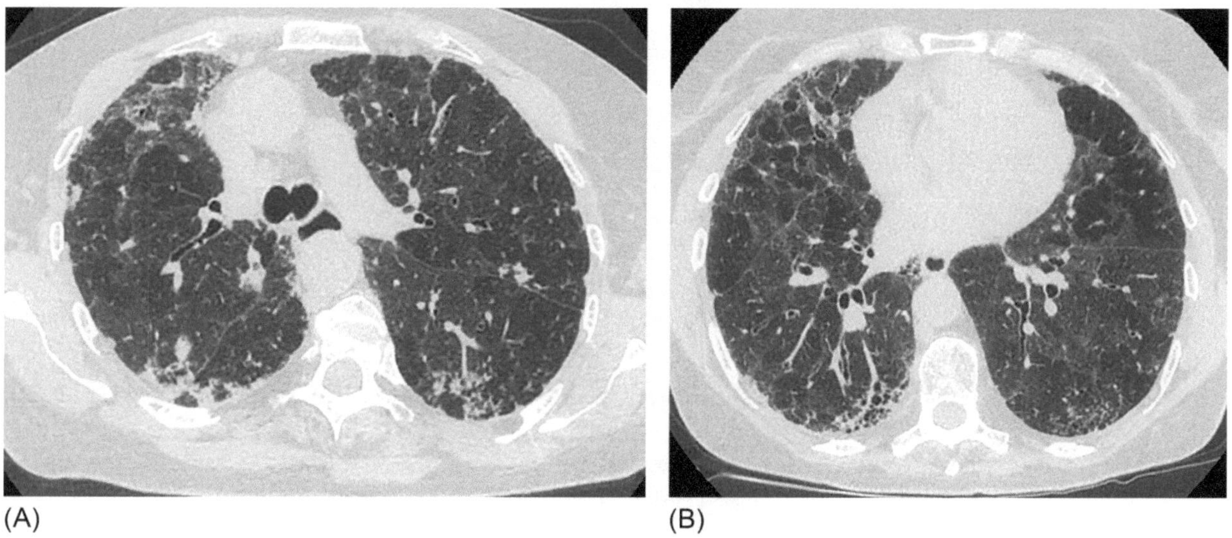

FIG. 2.1.26 Chronic hypersensitivity pneumonitis caused by prolonged contact with birds. Diffuse reticular changes. Honeycombing areas, air traps in both the upper (A) and lower zones (B).

In most cases it is extremely difficult to distinguish UIP in IPF from that in CTDs radiologically; nevertheless, Chung et al. [127] showed that radiological findings such as exuberant honeycombing and the straight-edge sign were present in 22%–25% of patients with CTD (Figs. 2.1.27 and 2.1.28), whereas their frequencies did not exceed 6% in patients with IPF. Exuberant honeycombing reflects the formation of multiple honeycomb-like cyst formations that make up more than 70% of the fibrous areas in the lung tissue [127]. Another CT sign to distinguish IPF from SSC is the so-called four-corner sign, the presence of UIP signs (honeycombing and reticular abnormalities) in symmetrical zones: the anterior corners corresponding to the anterolateral mid-upper lobes from the top of the aortic arch to the carina and the posterior corners corresponding to the posterosuperior lower lobes defined as an area between the carina and the inferior pulmonary veins [128]. Although this sign has a low sensitivity (16.4%), its specificity is 100% for ILD associated with systemic sclerosis (Fig. 2.1.29) [128].

Lymphoid interstitial pneumonia (LIP) is characterized by combination of signs of interstitial inflammation with cystic transformation of the lungs; however, the cysts are usually larger in size and more rounded than those observed in UIP. However, the progression of LIP may lead to a picture that is similar to IPF, and the differential diagnosis with UIP becomes relevant (Fig. 2.1.30) [42].

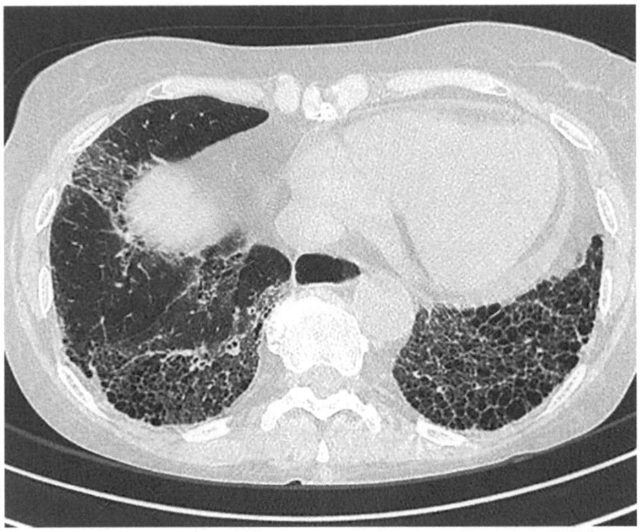

FIG. 2.1.27 Systemic sclerosis associated with interstitial lung disease. Exuberant honeycombing in the lower lobes, which are clearly separated from the intact parenchyma on the right. The dilated esophagus is visualized.

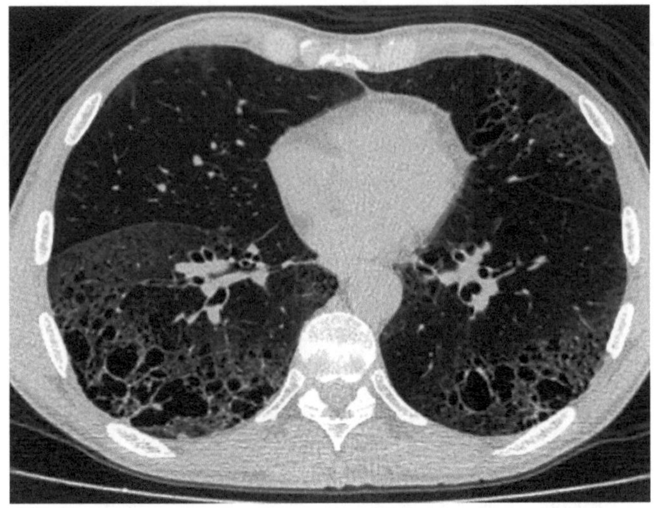

FIG. 2.1.28 Systemic sclerosis. Diffuse areas of GGO associated with excessive cystic transformation; unlike the honeycombing in IPF, cysts have not only subpleural but also parenchymal distribution. The affected areas are clearly distinguished from healthy tissue.

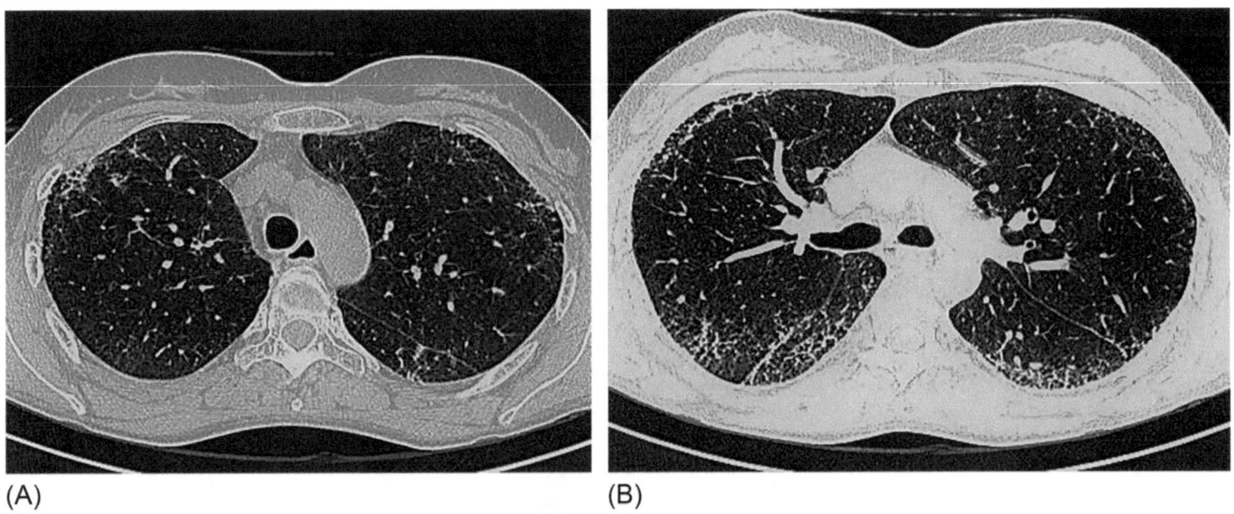

FIG. 2.1.29 Interstitial lung disease associated with systemic sclerosis. The four-corner sign. Symmetrical subpleural honeycombing in the anterior segments of the upper lobes (A) and reticular abnormalities in the posterior segments of the lower lobes (B).

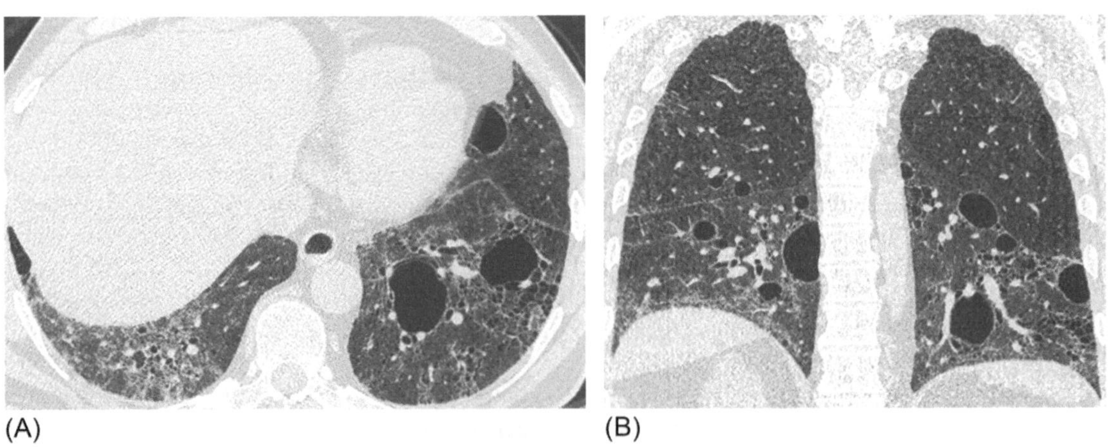

FIG. 2.1.30 Lymphocytic interstitial pneumonia. Bilateral reticular changes, honeycombing, and GGO in the lower sections (A and B). In contrast to usual interstitial pneumonia, thin-walled cysts of different sizes are additionally visualized to be predominantly subpleural.

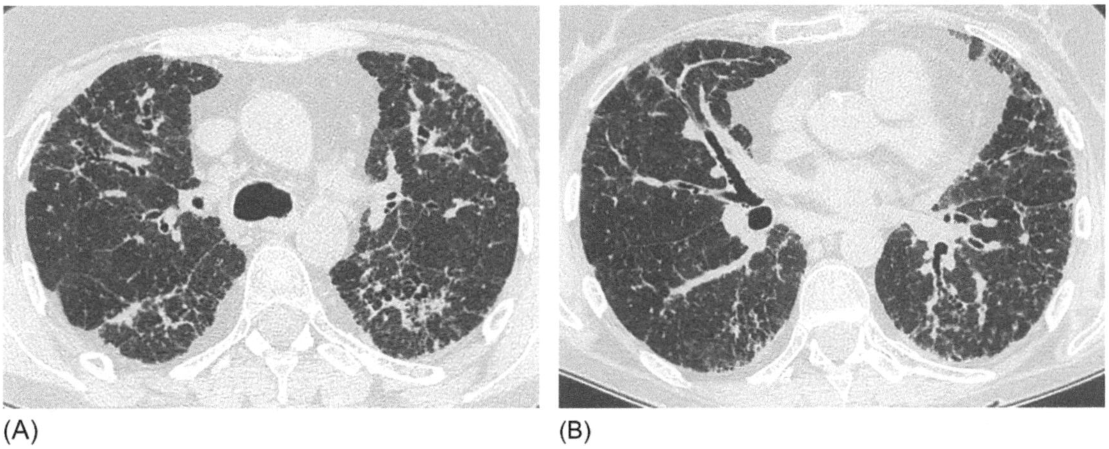

FIG. 2.1.31 Sarcoidosis, fibrous stage. Pronounced reticular changes and traction bronchiectasis. Fibrous cords located peribronchovascularly and along the interlobar pleura. Multiple small nodules of different density (A and B). Irregular thickened visceral pleura (A).

The fibrous stage of sarcoidosis of the lungs and pneumoconiosis are characterized by the appearance of changes in the lungs by chest HRCT, which resemble UIP in some cases. However, these changes are preceded by a long history of pulmonary abnormalities; furthermore the presence of nodular formations, areas of consolidation, and intrathoracic lymphadenopathy aid in differentiating sarcoidosis from IPF in almost all cases (Fig. 2.1.31).

The HRCT changes in drug-induced interstitial lung disease (DI-ILD) caused by the administration of certain medications, such as amiodarone, bleomycin, methotrexate, nitrofurans, and new biologics, often proceed as NSIP but sometimes can resemble the UIP pattern (Fig. 2.1.32).

Careful collection of the patient history with clarification of the facts associated with long-term use of certain drugs aid in linking fibrotic changes in the lungs with a drug-induced reaction. Lymphocytosis in the BAL fluid, usually exceeding 30%, is an additional argument in favor of DI-ILD.

Unilateral fibrosis of the lungs, a rare condition, is usually associated with radiation pneumonitis, impaired blood flow in the corresponding lung area, or drug-induced fibrosis (Fig. 2.1.33). Presence of one-sided honeycombing argues against an IPF diagnosis.

The pandemic of severe forms of the H1N1 flu in the 2009–10 and 2015–16 seasons resulted in interstitial pulmonary fibrosis in a subset of surviving patients who developed diffuse alveolar damage, followed by fibrosis of the lungs that resembled UIP by radiological characteristics (Fig. 2.1.34). However, unlike that observed in UIP, honeycombing in these cases is significantly less prominent than reticular signs, and pulmonary abnormalities are stable, with no significant disease progression.

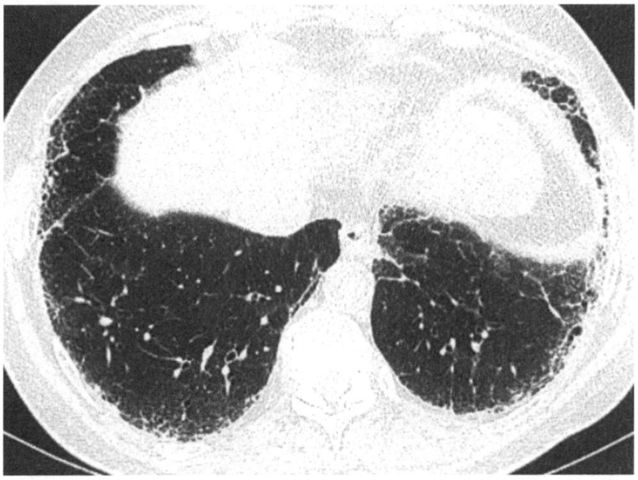

FIG. 2.1.32 Usual interstitial pneumonia caused by uncontrolled nitrofurantoin intake. Reticular changes, the beginning of the formation of the honeycombing in the basal-subpleural zones. *(Case courtesy of Prof. S.N. Avdeev, Sechenov First Moscow State Medical University, Moscow, Russia.)*

Complications of IPF

A number of complications that change the clinical and radiological signs may develop in patients with IPF. One of the most frequent complications is PH, noted in almost half of the patients [129,130]. There are several mechanisms underlying PH in IPF: prolonged angiospasm due to chronic hypoxemia with irreversible remodeling of the vascular wall, reduction of the capillary bed in zones of honeycombing, and endothelial dysfunction with increased expression of vasoconstrictors, primarily endothelin-1 [131]. Nevertheless, many studies reported no significant correlations between IPF severity, functional parameters, and PH severity [130,132]. Generally, clinical studies on the treatment of PH in patients with IPF have not effective, and drugs such as ambrisentan, bosentan, and sildenafil are currently not recommended for use [41].

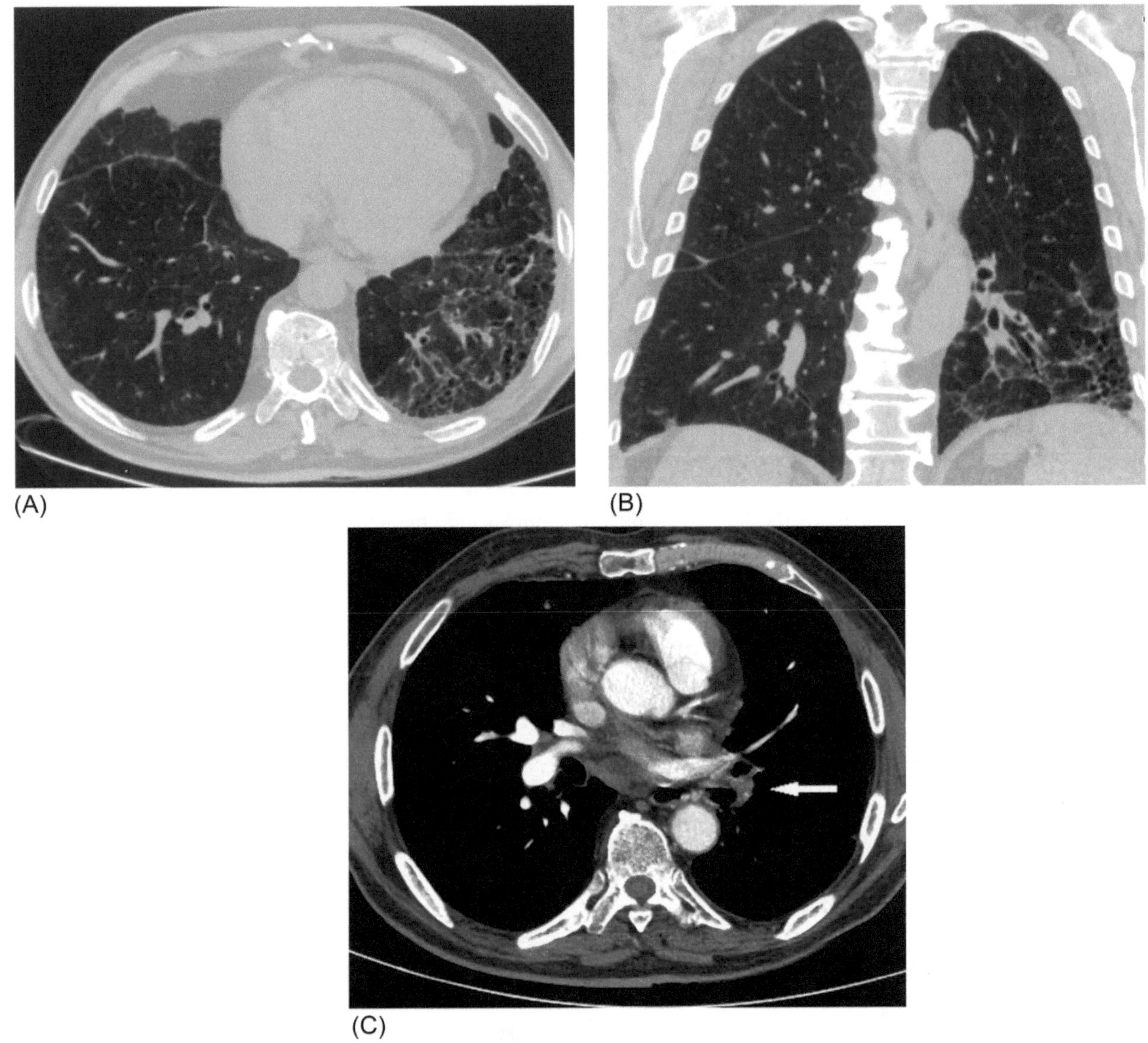

FIG. 2.1.33 Unilateral pulmonary fibrosis in the left lower lobe. Honeycombing and traction bronchiectasis. Initially regarded as an atypical form of IPF (A and B). Visualization using contrast reveals thrombotic occlusion of the left inferior pulmonary artery *(arrow)* (C).

Moreover, increased respiratory failure on the background of antihypertensive treatment is frequently observed, in particular among patients treated with sildenafil, due to the deterioration of the ventilation-perfusion relationship and enhancement of intrapulmonary shunting [133]. In fact the only way to impact PH in patients with IPF is prolonged oxygen therapy in cases of hypoxemia. Another possible cause of PH in IPF is recurrent pulmonary embolism (PE), which is predisposed by the use of corticosteroids, sedentary lifestyle, old age, and chronic hypoxemia, all of which contribute to the development of deep venous thrombosis [134]. Progression of dyspnea and appearance of new areas of reduced opacities in a patient with IPF are most often regarded as an AE of IPF, whereas acute PE often manifests with similar signs [135]. In cases of a rapid increase in respiratory failure in patients with IPF, contrast HRCT is recommended to rule out PE (Fig. 2.1.35). The question regarding the utility of anticoagulant treatment for the prevention of thrombosis and PE in patients with IPF remains open. A study of 624 patients with IPF who received oral anticoagulants or placebo demonstrated that the mortality rate was higher in the anticoagulant therapy group [136], although the first study evaluating the effects of warfarin during IPF exacerbation reached opposite conclusions [137].

Secondary immunodeficiency contributes to the development of opportunistic pulmonary infections, with *Pneumocystis jirovecii* pneumonia, aspergillosis, and pulmonary tuberculosis in descending frequency [138]. Clinically, these situations are similar, with predominance of increased dyspnea and cough; however, *Pneumocystis jirovecii* pneumonia is characterized by high fever and more pronounced increases in erythrocyte sedimentation rate and C-reactive protein. Foremost among the diagnostic tests in such patients is detection of *P. jirovecii* in the respiratory tract secretion by bacterioscopy or analysis of sputum or BAL fluid by polymerase chain reaction.

The development of invasive pulmonary aspergillosis in IPF contributes not only to immunodeficiency but also to the presence of cavitary lesions, that is, bronchiectasis and cysts.

Cavitation associated with upper lobe consolidation and intracavitary fungus ball are characteristic HRCT signs of pulmonary aspergillosis. Some studies describe a relatively high incidence of aspergillosis (20%) in ILD [139].

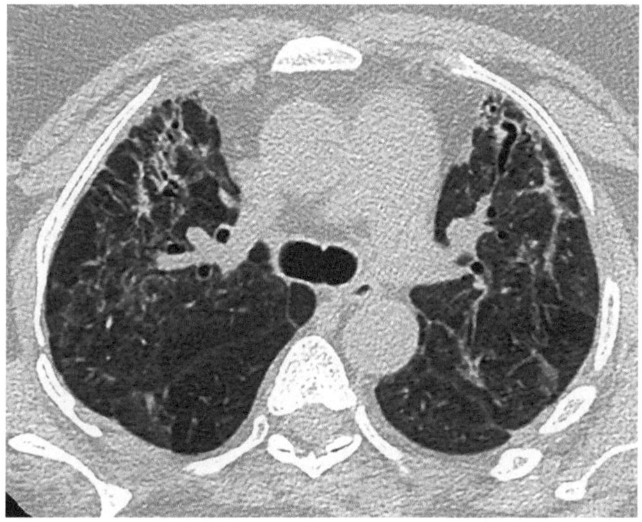

FIG. 2.1.34 Pulmonary fibrosis that developed after diffuse alveolar damage with H1N1 influenza infection. Linear subpleural fibrosis, traction bronchiectasis and bronchiolectases, and irregular reticular changes.

Pulmonary tuberculosis is significantly more common in patients with IPF than in the general population. According to Chung et al., tuberculosis on the background of IPF presents as a peripheral single solid infiltrate without cavitation on HRCT in 67% of the cases, which hinders its differential diagnosis from lung tumor [140].

Lung cancer is another pulmonary disease associated with IPF. In patients with IPF, there is a fivefold increase in the risk of lung cancer, due to common risk factors (smoking and old age) and chronic epithelial damage and epithelial metaplasia in areas of fibrotic changes [141]. Additionally, overlap of tumorigenesis and fibrogenesis pathways promoted by TGF-β1 may also play a role in the increased frequency of lung cancer in IPF [142].

The tumor in most patients with lung cancer and IPF is peripheral (Fig. 2.1.18) and occurs in the area of honeycombing in 82% of the cases according to Sakai et al. [143].

Finally, pneumothorax and pneumomediastinum are other possible complications of IPF [144]. They can be asymptomatic since most cases have a limited distribution due to pleural adhesions.

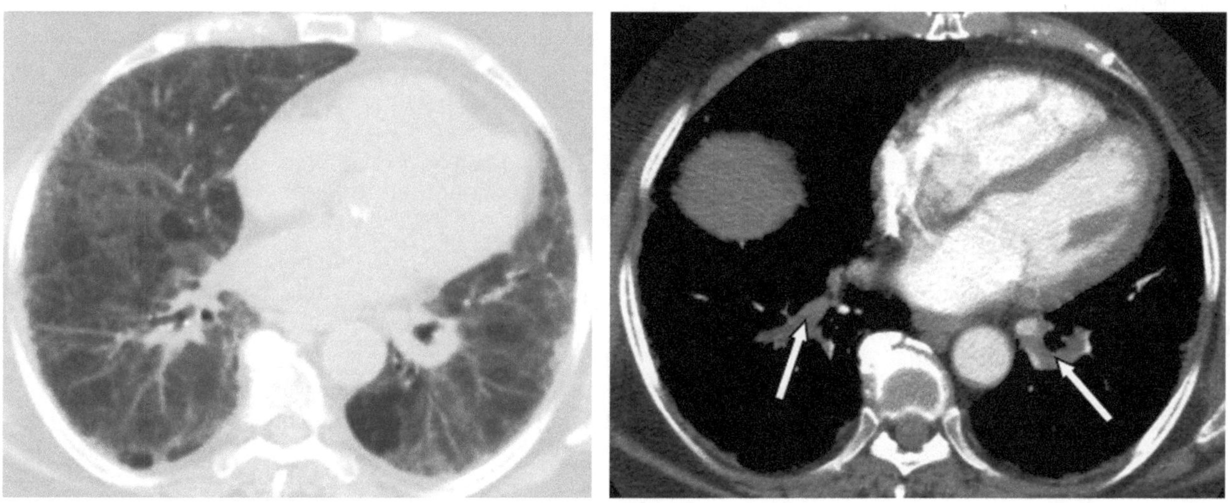

(A) (B)

FIG. 2.1.35 Bilateral pulmonary embolism in a patient with IPF. Bilateral areas of GGO, associated and reticular changes, single subpleural cysts (A). At contrast study, bilateral filling defects in the inferior branches of the pulmonary artery are visualized *(arrows)* (B).

Treatment

Therapeutic approaches to patients with IPF have fundamentally changed in recent years. For many years, systemic steroids were used; although they were effective in certain patients in the short term, generally deteriorating overall IPF survival, they also lead to complications; the severity of which was often comparable with the severity of the underlying disease [145].

Currently, systemic corticosteroids are limited to pulse therapy (0.5–1 g methylprednisolone per day) during AE of IPF, after exclusion of infection [146]. Triple therapy with systemic corticosteroids, azathioprine, and acetylcysteine, which was widely used previously in patients with IPF, was found to be ineffective and harmful due to the increase in risk of hospitalization and mortality compared with the placebo by Raghu et al. [147].

In recent years, there have been significant breakthroughs in the treatment approaches for IPF, thanks to the emergence of two antifibrotic drugs that have passed all phases of clinical trials and were approved for use in most countries. The first, pirfenidone, was synthesized in 1976 and originally patented as an antiinflammatory agent [148]. The antifibrotic effect of pirfenidone was discovered in 1992 in an experimental study, and the results of the first clinical study were published in 1999 [149]. The main mechanism underlying the antifibrotic action of pirfenidone involves the suppression of proliferation of fibroblasts and their transformation to myofibroblasts and suppression of collagen synthesis by blocking the effects of TGFβ-1, the main mediator of fibroblast activity [150,151]. In addition, pirfenidone was demonstrated to suppress the expression of proinflammatory cytokines, primarily tumor necrosis factor α, and to exert an antioxidant effect in experimental models [152,153]. In European Union countries, pirfenidone as for IPF treatment was approved in 2011 after the publication of the results of the CAPACITY trials, including the CAPACITY 1 and CAPACITY 2 studies, which, albeit different in design, demonstrated an overall dose efficiency at 2403 mg/day in comparison with placebo, with decreases in the rate of FVC reduction and the proportion of patients with a rapid decline of >10% in FVC and a 36% reduction in the risk of death and progression of IPF (CAPACITY 2) and an increase in the distance covered in the 6-min walking test (CAPACITY 1) [154]. In the United States, pirfenidone was approved after the ASCEND study including 555 patients with IPF according to the 2011 ATS/ERS criteria, which showed a significant reduction in the number of patients with a rapid decline in FVC and a reduction in the rate of decrease in the distance covered in the 6-min walking test [71]. Side effects of pirfenidone include gastrointestinal symptoms such as nausea, vomiting, loss of appetite, and increased liver enzymes as well as photosensitivity, skin rash, and insomnia [155].

The second antifibrotic drug is nintedanib, a tyrosine kinase inhibitor, primarily developed as an antitumor agent [156]. The antifibrotic action of nintedanib is based on the suppression of endothelial, vascular, fibroblastic, and platelet growth factors such as vascular endothelial growth factor, fibroblast growth factor, and platelet-derived growth factor. As a result the activity of fibroblasts and their transition into myofibroblasts are suppressed, and the production of extracellular matrix and, ultimately, fibrogenesis is blocked [157].

Three clinical trials for nintedanib in patients with IPF, TOMORROW, INPULSIS I, and INPULSIS II showed that, at a dose of 300 mg/day compared with placebo, nintedanib reduced the rate of decrease in FVC annually by more than twofolds, reduced the number of acute exacerbations, increased the time until the first exacerbation, and improved the quality of life [72,73]. Among the side effects, diarrhea (61%) and nausea (23%) were the most frequent, which nevertheless allowed therapy to continue in 95% of the patients [72]. In 2014 nintedanib was approved for use in IPF in the European Union and the United States, and both nintedanib and pirfenidone were officially recommended for the treatment of IPF in by a consensus of experts with the ATS/ERS/JRS/ALAT 2015 [41].

Over the years, antifibrotic therapy was evaluated for efficacy and safety in numerous studies. Metaanalyses of the pooled data suggest that both nintedanib and pirfenidone can reduce mortality from IPF [158,159]. Earlier studies indicate the possibility of combined use of the two antifibrotic drugs. While the frequency of side effects increased in parallel with combination use, key parameters of lung function improved simultaneously [160,161]. In observational studies, both antifibrotic drugs (named as disease-modifying therapy [162]) were also evaluated, revealing that the side effects of pirfenidone (nausea and fatigue) and nintedanib (diarrhea, nausea, and loss of appetite) with long-term administration were comparable with those reported during the initial 52-week study in general [163,164]. Both drugs demonstrate similar efficacy and safety profiles; currently, there are no consensus recommendations for choosing a specific drug to initiate starting therapy [119].

Despite the evident progress in potential approaches in IPF therapy, a number of practical questions remain unresolved. The unresolved question of antifibrotic treatment for IPF as a whole will determine the scientific directions that are being implemented in the framework of ongoing clinical studies.

Nonpharmacological treatment of IPF

Maintenance of patients with IPF should not be limited to drug treatments. As with most other serious chronic diseases, a multifaceted approach is necessary to correct various physical and psychological aspects of disease in one way or another and to improve the quality of life.

Long-term oxygen therapy (LTOT) should be considered as a treatment of choice in patients with hypoxemia according to the 2011 ATS/ERS recommendations. However, a clear threshold for implementation of LTOT has not been identified; nevertheless, based on the practices for managing COPD, LTOT can be recommended in patients with hypoxemia with a partial arterial O_2 pressure of <55 mmHg or a blood O_2 saturation of <88% by pulse oximetry. LTOT generally improves the quality of life in patients with IPF [165].

Rehabilitation activities should include measures to maintain physical activity and psychological support. Physical rehabilitation in the form of regular exercise with weight on skeletal muscles in the presence of adequate oxygenation can improve the functional parameters of patients, including increases in FVC and the 6-min walking distance [166].

Psychological assistance to patients with IPF is limited to detailed information provided to patients about the disease course, including the possibility of relative stabilization with adequate therapy and rehabilitation and a delicate preparation of the patients for their final stage in life, as is often used in patients with lung cancer. In a number of countries, hospice-type palliative care centers have been established for patients with IPF to alleviate the physical and psychological suffering of patients and their relatives in the final stage of the disease.

Today, *lung transplantation* is the only method that allows a radical change in the quality of life of patients and significantly increases their life expectancy. However, the outcomes of unilateral and bilateral lung transplantation in patients with IPF are worse than those in patients with other terminal lung diseases. Among patients with IPF, 1-year survival rates are 81% and 67% for bilateral and unilateral transplantation, respectively, whereas 5-year survival rates are 55% and 34% for bilateral and unilateral transplantation, respectively [167].

Indications for lung transplantation in IPF are the following [168]: histologically or radiologically confirmed UIP plus one of the following criteria: DLCO ≤39%; ≥10% decrease in FVC within 6 months; reduction in blood O_2 saturation to <88% during the 6-min walking test; and honeycombing on HRCT, with a fibrosis index of more than 2.

Contraindications for placement on a transplantation waiting are far more. Absolute contraindications are the following: history of malignant neoplasm during the last 2 years; decompensated function of another vital organ (the heart, liver, and/or kidneys); significant chest deformation; documented inability or refusal to follow medical prescriptions; intractable psychiatric or psychological disorders; inability to receive social assistance; and use of alcohol, drugs, or tobacco products in the last 6 months. Relative contraindications are also numerous and include an age over 65 years. Although the authors of the recommendations emphasize that transplantation can be performed in older patients, the survival rate of these patients is far worse [169].

Prevention of acute exacerbations

Patients with IPF should be vaccinated against influenza and pneumococcus, as these infections can be fatal in patients with limited body reserves [1]. Long courses of macrolides are widely used for the prevention of exacerbations of bronchiectasis, cystic fibrosis, and COPD. The first study reporting a two-third reduction in the hospitalization rate and a possible reduction in the number of AE with the use of 250-mg azithromycin thrice weekly in patients with IPF patients was published by Macaluso et al. in 2018 [170]. The authors emphasized that in addition to the prophylactic effect, azithromycin exhibited good tolerability and safety. However, the mechanism underlying these effects of azithromycin remains unclear [170], although macrolides' ability to stimulate immune responses may partially explain these outcomes.

Management of comorbidities in patients with IPF is important. The most common cardiovascular diseases encountered in patients with IPF require monitoring and regular therapy [171,172]. Regarding antacid therapy for gastroesophageal reflux disease, which, in accordance with the 2015 consensus statement, should be considered in patients with IPF; however, the INPULSIS substudy evaluating the efficacy of antacid therapy did not confirm the results from previous studies showing that the decline in functional parameters was attenuated and that the survival rate was increased in patients with IPF who were using antacids [42].

Prognosis

The prognosis of IPF is less optimistic than most other forms, except for acute interstitial pneumonia and idiopathic interstitial pneumonia. Five-year survival does not usually exceed 30%, which is significantly less than that for most forms of cancer except for lung and pancreas cancer [111]. Only approximately 15% of patients live more than 10 years from the onset of the first IPF symptoms [26,173].

Among the various approaches used to determine prognosis in patients with IPF are estimation of the initial value and rate of decline in FVC, distance of 6-min walking test, fibrosis index by HRCT, and indicators of quality of life, all assessed according to different scales [100,174–176]. A relatively simple method is the GAP scale, which is based on the score depending on gender, age, FVC, and DLCO values [175]. Jacob et al. proposed the vascular structure index, which is based on HRCT fibrosis score and respiratory function, which had a high predictive value for survival of patients with IPF [176].

Most of the studies on prognosis, course of disease, and survival in IPF were conducted before the era of antifibrotic drugs, and there is reason to believe that progress will be made in the prognosis of IPF patients in the coming years.

References

[1] Raghu G, Collard HR, Egan JJ, Martinez FJ, Behr J, Brown KK, et al. ATS/ERS/JRS/ALAT Committee on Idiopathic Pulmonary Fibrosis. An official ATS/ERS/JRS/ALAT statement: idiopathic pulmonary fibrosis: evidence-based guide-lines for diagnosis and management. Am J Respir Crit Care Med 2011;183(6):788–824.

[2] Kim DS, Collard HR, King Jr TE. Classification and natural history of the idiopathic interstitial pneumonias. Proc Am Thorac Soc 2006;3(4):285–92.

[3] Raghu G, Weycker D, Edelsberg J, Bradford WZ, Oster G. Incidence and prevalence of idiopathic pulmonary fibrosis. Am J Respir Crit Care Med 2006;174(7):810–6.

[4] Harari S, Caminati A, Madotto F, Conti S, Cesana G. Epidemiology, survival, incidence and prevalence of idiopathic pulmonary fibrosis in the USA and Canada. Eur Respir J 2017;49(1). pii:1601504.

[5] Hutchinson J, Fogarty A, Hubbard R, McKeever T. Global incidence and mortality of idiopathic pulmonary fibrosis: a systematic review. Eur Respir J 2015;46(3):795–806.

[6] Raghu G, Chen S-Y, Hou Q, Yeh W-S, Collard HR. Incidence and prevalence of idiopathic pulmonary fibrosis in US adults 18–64 years old. Eur Respir J 2016;48(1):179–86.

[7] Hopkins RB, Burke N, Fell C, Dion G, Kolb M. Epidemiology and survival of idiopathic pulmonary fibrosis from national data in Canada. Eur Respir J 2016;48(1):187–95.

[8] Navaratnam V, Fleming KM, West J, Smith CJ, Jenkins RG, Fogarty A, et al. The rising incidence of idiopathic pulmonary fibrosis in the UK. Thorax 2011;66(6):462–7.

[9] Hoffman H, Rice C, Skordalakes E. Structural analysis reveals the deleterious effects of telomerase mutations in bone marrow failure syndromes. J Biol Chem 2017;292(11):4593–601.

[10] Thomas PB, Hanae H, Kim V, Antoine F. Telomerase-related monogenic lung fibrosis presenting with subacute onset: a case report and review of literature. Acta Clin Belg 2018;19:1–6.

[11] Roy MG, Livraghi-Butrico A, Fletcher AA, et al. Muc5b is required for airway defence. Nature 2014;505(7483):412–6.

[12] Seibold MA, Wise AL, Speer MC, Steele MP, Brown KK, Loyd JE, et al. A common MUC5B promoter polymorphism and pulmonary fibrosis. N Engl J Med 2011;364:1503–12.

[13] Selman M, Buendía-Roldán I, Pardo A. Aging and pulmonary fibrosis. Rev Investig Clin 2016;68(2):75–83.

[14] Stuart BD, Lee JS, Kozlitina J, Noth I, Devine MS, Glazer CS, et al. Effect of telomere length on survival in patients with idiopathic pulmonary fibrosis: an observational cohort study with independent validation. Lancet Respir Med 2014;2(7):557–65.

[15] Calado RT. Telomeres in lung diseases. Prog Mol Biol Transl Sci 2014;125:173–83.

[16] Hoffman TW, van Moorsel CHM, Borie R, Crestani B. Pulmonary phenotypes associated with genetic variation in telomere-related genes. Curr Opin Pulm Med 2018;24(3):269–80.

[17] Planas-Cerezales L, Arias-Salgado EG, Buendia-Roldán I, Montes-Worboys A, López CE, Vicens-Zygmunt V, et al. Predictive factors and prognostic effect of telomere shortening in pulmonary fibrosis. Respirology 2018; https://doi.org/10.1111/resp.13423. [Epub ahead of print].

[18] Fingerlin TE, Zhang W, Yang IV, Ainsworth HC, Russell PH, Blumhagen RZ, et al. Genome-wide imputation study identifies novel HLA locus for pulmonary fibrosis and potential role for auto-immunity in fibrotic idiopathic interstitial pneumonia. BMC Genet 2016;17(1):74.

[19] Baumgartner KB, Samet JM, Stidley CA, Colby TV, Waldron JA. Cigarette smoking: a risk factor for idiopathic pulmonary fibrosis. Am J Respir Crit Care Med 1997;155(1):242–8.

[20] Ekström M, Gustafson T, Boman K, Nilsson K, Tornling G, Murgia N, et al. Effects of smoking, gender and occupational exposure on the risk of severe pulmonary fibrosis: a population-based case-control study. BMJ Open 2014;4(1):e004018.

[21] Oh C, Murray LA, Molfino NA. Smoking and idiopathic pulmonary fibrosis. Pulm Med 2012;2012:808260.

[22] Lee MG, Lee YH. A meta-analysis examining the association between the MUC5B rs35705950 T/G polymorphism and susceptibility to idiopathic pulmonary fibrosis. Inflamm Res 2015;64(6):463–70.

[23] Sanders YY, Pardo A, Selman M, Nuovo GJ, Tollefsbol TO, Siegal GP, et al. Thy-1 promoter hypermethylation: a novel epigenetic pathogenic mechanism in pulmonary fibrosis. Am J Respir Cell Mol Biol 2008;39(5):610–8.

[24] Travis WD, Costabel U, Hansell DM, King Jr TE, Lynch DA, Nicholson AG, et al. An official American Thoracic Society/European Respiratory Society statement: update of the international multidisciplinary classification of the idiopathic interstitial pneumonias. Am J Respir Crit Care Med 2013;188(6):733–48.

[25] Kishaba T, Nagano H, Nei Y, Yamashiro S. Clinical characteristics of idiopathic pulmonary fibrosis patients according to their smoking status. J Thorac Dis 2016;8(6):1112–20.

[26] King Jr TE, Tooze JA, Schwarz MI, Brown KR, Cherniack RM. Predicting survival in idiopathic pulmonary fibrosis: scoring system and survival model. Am J Respir Crit Care Med 2001;164(7):1171–81.

[27] Kärkkäinen M, Kettunen H-P, Nurmi H, Selander T, Purokivi M, Kaarteenaho R. Effect of smoking and comorbidities on survival in idiopathic pulmonary fibrosis. Respir Res 2017;18(1):160.

[28] Gustafson T, Dahlman-Höglund A, Nilsson K, Ström K, Tornling G, Torén K. Occupational exposure and severe pulmonary fibrosis. Respir Med 2007;101(10):2207–12.

[29] Lee SH, Kim DS, Kim YW, Chung MP, Uh ST, Park CS. Association between occupational dust exposure and prognosis of idiopathic pulmonary fibrosis: a Korean national survey. Chest 2015;147(2):465–74.

[30] Raghu G, Chen SY, Yeh WS, Maroni B, Li Q, Lee YC, et al. Idiopathic pulmonary fibrosis in US Medicare beneficiaries aged 65 years and older: incidence, prevalence, and survival, 2001–11. Lancet Respir Med 2014;2(7):566–72.

[31] López-Otín C, Blasco MA, Partridge L, Serrano M, Kroemer G. The hallmarks of aging. Cell 2013;153(6):1194–217.

[32] Moskalev AA, Shaposhnikov MV, Plyusnina EN, Zhavoronkov A, Budovsky A, Yanai H, et al. The role of DNA damage and repair in aging through the prism of Koch-like criteria. Ageing Res Rev 2013;12(2):661–84.

[33] Pollina EA, Brunet A. Epigenetic regulation of aging stem cells. Oncogene 2011;30(28):3105–26.

[34] Álvarez D, Cárdenes N, Sellarés J, Bueno M, Corey C, Hanumanthu VS, et al. IPF lung fibroblasts have a senescent phenotype. Am J Physiol Lung Cell Mol Physiol 2017;313(6):L1164–73.

[35] Schuliga M, Pechkovsky D, Read J, Waters D, Blokland K, Reid A, et al. Mitochondrial dysfunction contributes to the senescent phenotype of IPF lung fibroblasts. J Cell Mol Med 2018;22(12):5847–61.

[36] Bueno M, Lai YC, Romero Y, Brands J, St Croix CM, Kamga C, et al. PINK1 deficiency impairs mitochondrial homeostasis and promotes lung fibrosis. J Clin Invest 2015;125(2):521–38.

[37] Chakkalakal JV, Jones KM, Basson MA, Brack AS. The aged niche disrupts muscle stem cell quiescence. Nature 2012;490(7420):355–60.

[38] Tobin RW, Pope 2nd CE, Pellegrini CA, Emond MJ, Sillery J, Raghu G. Increased prevalence of gastroesophageal reflux in patients with idiopathic pulmonary fibrosis. Am J Respir Crit Care Med 1998;158(6):1804–8.

[39] Raghu G, Freudenberger TD, Yang S, Curtis JR, Spada C, Hayes J, et al. High prevalence of abnormal acid gastro-oesophageal reflux in idiopathic pulmonary fibrosis. Eur Respir J 2006;27(1):136–42.

[40] Savarino E, Carbone R, Marabotto E, Furnari M, Sconfienza L, Ghio M, et al. Gastro-oesophageal reflux and gastric aspiration in idiopathic pulmonary fibrosis patients. Eur Respir J 2013;42(5):1322–31.

[41] Raghu G, Rochwerg B, Zhang Y, Garcia CA, Azuma A, Behr J, et al. An official ATS/ERS/JRS/ALAT clinical practice guideline: treatment of idiopathic pulmonary fibrosis. An update of the 2011 clinical practice guideline. Am J Respir Crit Care Med 2015;192(2):e3–19.

[42] Costabel U, Behr J, Crestani B, Stansen W, Schlenker-Herceg R, Stowasser S, et al. Anti-acid therapy in idiopathic pulmonary fibrosis: insights from the INPULSIS® trials. Respir Res 2018;19(1):167.

[43] Kreuter M, Wuyts W, Renzoni E, Koschel D, Maher TM, Kolb M, et al. Antacid therapy and disease outcomes in idiopathic pulmonary fibrosis: a pooled analysis. Lancet Respir Med 2016;4(5):381–9.

[44] Arase Y, Ikeda K, Tsubota A, Saitoh S, Suzuki Y, Kobayashi M, et al. Usefulness of serum KL-6 for early diagnosis of idiopathic pulmonary fibrosis in patients with hepatitis C virus. Hepatol Res 2003;27(2):89–94.

[45] Tang YW, Johnson JE, Browning PJ, Cruz-Gervis RA, Davis A, Graham BS, et al. Herpesvirus DNA is consistently detected in lungs of patients with idiopathic pulmonary fibrosis. J Clin Microbiol 2003;41(6):2633–40.

[46] Vannella KM, Moore BB. Viruses as co-factors for the initiation or exacerbation of lung fibrosis. Fibrogenesis Tissue Repair 2008;1(1):2.

[47] Molyneaux PL, Cox MJ, Willis-Owen SAG, Mallia P, Russell KE, Russell AM, et al. The role of bacteria in the pathogenesis and progression of idiopathic pulmonary fibrosis. Am J Respir Crit Care Med 2014;190(8):906–13.

[48] Han MLK, Zhou Y, Murray S, Tayob N, Noth I, Lama VN, et al. Lung microbiome and disease progression in idiopathic pulmonary fibrosis: an analysis of the COMET study. Lancet Respir Med 2014;2(7):548–56.

[49] Shulgina L, Cahn AP, Chilvers ER, Parfrey H, Clark AB, Wilson EC, et al. Treating idiopathic pulmonary fibrosis with the addition of co-trimoxazole: a randomised controlled trial. Thorax 2013;68(2):155–62.

[50] Molyneaux PL, Cox MJ, Wells AU, Kim HC, Ji W, Cookson WOC, et al. Changes in the respiratory microbiome during acute exacerbations of idiopathic pulmonary fibrosis. Respir Res 2017;18(1):29.

[51] Fastrès A, Felice F, Roels E, Moermans C, Corhay J-L, Bureau F, et al. The lung microbiome in idiopathic pulmonary fibrosis: a promising approach for targeted therapies. Int J Mol Sci 2017;18(12). pii E2735.

[52] Fernandez IE, Eickelberg O. New cellular and molecular mechanisms of lung injury and fibrosis in idiopathic pulmonary fibrosis. Lancet 2012;380(9842):680–8.

[53] Loomis-King H, Flaherty KR, Moore BB. Pathogenesis, current treatments and future directions for idiopathic pulmonary fibrosis. Curr Opin Pharmacol 2013;13(3):377–85.

[54] Kolahian S, Fernandez IE, Eickelberg O, Hartl D. Immune mechanisms in pulmonary fibrosis. Am J Respir Cell Mol Biol 2016;55(3):309–22.

[55] Larsen BT, Colby TV. Update for pathologists on idiopathic interstitial pneumonias. Arch Pathol Lab Med 2012;136(10):1234–41.

[56] Meyer EC, Liebow AA. Relationship of interstitial pneumonia honeycombing and atypical epithelial proliferation to cancer of the lung. Cancer 1965;18:322–51.

[57] Aubry MC, Myers JL, Douglas WW, Tazelaar HD, Washington Stephens TL, Hartman TE, et al. Primary pulmonary carcinoma in patients with idiopathic pulmonary fibrosis. Mayo Clin Proc 2002;77(8):763–70.

[58] Raghu G, Remy-Jardin M, Myers JL, Richeldi L, Ryerson CJ, Lederer DJ, et al. Diagnosis of idiopathic pulmonary fibrosis. An official ATS/ERS/JRS/ALAT clinical practice guideline. Am J Respir Crit Care Med 2018;198(5):e44–68.

[59] Giangreco A, Reynolds SD, Stripp BR. Terminal bronchioles harbor a unique airway stem cell population that localizes to the bronchoalveolar duct junction. Am J Pathol 2002;161(1):173–82.

[60] Demura SA, Kogan EA, Paukov VS. Morphology and molecular basis of damage to the niche of respiratory acinus stem cells in idiopathic interstitial pneumonia. Pathology Arch 2014;76(6):28–36.

[61] Nicholson AG, Rice AJ. Interstitial lung diseases. In: Hasleton P, Flieder DB, editors. Spencer's pathology of the lung. 6th ed. New York: Cambridge Medicine; 2016. p. 366–408.

[62] Wuyts WA, Cavazza A, Rossi G, Bonella F, Sverzellati N, Spagnolo P. Differential diagnosis of usual interstitial pneumonia: when is it truly idiopathic? Eur Respir Rev 2014;23(133):308–19.

[63] Song JW, Do KH, Kim MY, Jang SJ, Colby TV, Kim DS. Pathologic and radiologic differences between idiopathic and collagen vascular disease-related usual interstitial pneumonia. Chest 2009;136(1):23–30.

[64] Travis WD, Matsui K, Moss J, Ferrans VJ. Idiopathic nonspecific interstitial pneumonia: prognostic significance of cellular and fibrosing patterns: survival comparison with usual interstitial pneumonia and desquamative interstitial pneumonia. Am J Surg Pathol 2000;24(1):19–33.

[65] Cordier JF, Cottin V. Neglected evidence in idiopathic pulmonary fibrosis: from history to earlier diagnosis. Eur Respir J 2013;42(4):916–23.

[66] Kanematsu T, Kitaichi M, Nishimura K, Nagai S, Izumi T. Clubbing of the fingers and smooth-muscle proliferation in fibrotic changes in the lung in patients with idiopathic pulmonary fibrosis. Chest 1994;105(2):339–42.

[67] Hunninghake GM, Hatabu H, Okajima Y, Gao W, Dupuis J, Latourelle JC, et al. MUC5B promoter polymorphism and interstitial lung abnormalities. N Engl J Med 2013;368:2192–200.

[68] Araki T, Putman RK, Hatabu H, Gao W, Dupuis J, Latourelle JC, et al. Development and progression of interstitial lung abnormalities in the Framingham Heart Study. Am J Respir Crit Care Med 2016;194:1514–22.

[69] Schwartz DA. Idiopathic pulmonary fibrosis is a genetic disease involving mucus and the peripheral airways. Ann Am Thorac Soc 2018;15(Suppl 3):S192–7.

[70] Lynch 3rd JP, Huynh RH, Fishbein MC, Saggar R, Belperio JA, Weigt SS. Idiopathic pulmonary fibrosis: epidemiology, clinical features, prognosis, and management. Semin Respir Crit Care Med 2016;37(3):331–57.

[71] King T, Bradford W, Castro-Bernardini S, Fagan EA, Glaspole I, Glassberg MK, et al. A phase 3 trial of pirfenidone in patients with idiopathic pulmonary fibrosis. N Engl J Med 2014;370(22):2083–92.

[72] Richeldi L, du Bois RM, Raghu G, Azuma A, Brown KK, Costabel U, et al. Efficacy and safety of nintedanib in idiopathic pulmonary fibrosis. New England J Med 2014;370(22):2071–82.

[73] Naira G, Matelaa A, Kurbanov D, Raghu G. Newer developments in idiopathic pulmonary fibrosis in the era of anti-fibrotic medications. Expert Rev Respir Med 2016;10(6):699–711.

[74] Song JW, Hong SB, Lim CM, Koh Y, Kim DS. Acute exacerbation of idiopathic pulmonary fibrosis: incidence, risk factors and outcome. Eur Respir J 2011;37(2):356–63.

[75] Natsuizaka M, Chiba H, Kuronuma K, et al. Epidemiologic survey of Japanese patients with idiopathic pulmonary fibrosis and investigation of ethnic differences. Am J Respir Crit Care Med 2014;190(7):773–9.

[76] Atkins CP, Loke YK, Wilson AM. Outcomes in idiopathic pulmonary fibrosis: a meta-analysis from placebo controlled trials. Respir Med 2014;108(2):376–87.

[77] Collard HR, Moore BB, Flaherty KR, Brown KK, Kaner RJ, King Jr TE, et al. Acute exacerbations of idiopathic pulmonary fibrosis. Am J Respir Crit Care Med 2007;176(7):636–43.

[78] Collard HR, Ryerson CJ, Corte TJ, Jenkins G, Kondoh Y, Lederer DJ, et al. Acute exacerbation of idiopathic pulmonary fibrosis: an International Working Group Report. Am J Respir Crit Care Med 2016;194(3):265–75.

[79] Matthay MA, Ware LB, Zimmerman GA. The acute respiratory distress syndrome. J Clin Invest 2012;122(8):2731–40.

[80] Ghatol A, Ruhl AP, Danoff SK. Exacerbations in idiopathic pulmonary fibrosis triggered by pulmonary and nonpulmonary surgery: a case series and comprehensive review of the literature. Lung 2012;190:373–80.

[81] Ryerson CJ, Cottin V, Brown KK, Collard HR. Acute exacerbation of idiopathic pulmonary fibrosis: shifting the paradigm. Eur Respir J 2015;46(2):512–20.

[82] Collard HR, Yow E, Richeldi L, Anstrom KJ, Glazer C, IPFnet Investigators. Suspected acute exacerbation of idiopathic pulmonary fibrosis as an outcome measure in clinical trials. Respir Res 2013;14:73.

[83] Kondoh Y, Taniguchi H, Katsuta T, Kataoka K, Kimura T, Nishiyama O, et al. Risk factors of acute exacerbation of idiopathic pulmonary fibrosis. Sarcoidosis Vasc Diffuse Lung Dis 2010;27(2):103–10.

[84] Akira M, Kozuka T, Yamamoto S, Sakatani M. Computed tomography findings in acute exacerbation of idiopathic pulmonary fibrosis. Am J Respir Crit Care Med 2008;178(4):372–8.

[85] Borie R, Crestani B. Antineutrophil cytoplasmic antibody-associated lung fibrosis. Semin Respir Crit Care Med 2018;39(4):465–70.

[86] Fischer A, Antoniou KM, Brown KK, Cadranel J, Corte TJ, du Bois RM, et al. An official European Respiratory Society/American Thoracic Society research statement: interstitial pneumonia with autoimmune features. Eur Respir J 2015;46(4):976–87.

[87] Hosoda C, Baba T, Hagiwara E, Ito H, Matsuo N, Kitamura H, et al. Clinical features of usual interstitial pneumonia with anti-neutrophil cytoplasmic antibody in comparison with idiopathic pulmonary fibrosis. Respirology 2016;21(5):920–6.

[88] Nozu T, Kondo M, Suzuki K, Tamaoki J, Nagai A. A comparison of the clinical features of ANCA-positive and ANCA-negative idiopathic pulmonary fibrosis patients. Respiration 2009;77(4):407–15.

[89] Hozumi H, Oyama Y, Yasui H, Suzuki Y, Kono M, Karayama M, et al. Clinical significance of myeloperoxidase-anti-neutrophil cytoplasmic antibody in idiopathic interstitial pneumonias. PLoS One 2018;13(6):e0199659.

[90] Kennedy B, Branagan P, Moloney F, Haroon M, O'Connell OJ, O'Connor TM, et al. Biomarkers to identify ILD and predict lung function decline in scleroderma lung disease or idiopathic pulmonary fibrosis. Sarcoidosis Vasc Diffuse Lung Dis 2015;32(3):228–36.

[91] Maldonado M, Buendía-Roldán I, Vicens-Zygmunt V, Planas L, Molina-Molina M, Selman M, et al. Identification of MMP28 as a biomarker for the differential diagnosis of idiopathic pulmonary fibrosis. PLoS One 2018;13(9):e0203779.

[92] White ES, Xia M, Murray S, Dyal R, Flaherty CM, Flaherty KR, et al. Plasma surfactant protein-D, matrix metalloproteinase-7, and osteopontin index distinguishes idiopathic pulmonary fibrosis from other idiopathic interstitial PNEUMONIAS. Am J Respir Crit Care Med 2016;194(10):1242–51.

[93] Jiang Y, Luo Q, Han Q, Huang J, Ou Y, Chen M, et al. Sequential changes of serum KL-6 predict the progression of interstitial lung disease. J Thorac Dis 2018;10(8):4705–14.

[94] Chiba H, Otsuka M, Takahashi H. Significance of molecular biomarkers in idiopathic pulmonary fibrosis: a mini review. Respir Investig 2018;56(5):384–91.

[95] Njock MS, Guiot J, Henket MA, Nivelles O, Thiry M, Dequiedt F, et al. Sputum exosomes: promising biomarkers for idiopathic pulmonary fibrosis. Thorax 2018. pii: thoraxjnl-2018-211897.

[96] Erbes R, Schaberg T, Loddenkemper R. Lung function tests in patients with idiopathic pulmonary fibrosis. Are they helpful for predicting outcome? Chest 1997;111(1):51–7.

[97] Hallstrand TS, Boitano LJ, Johnson WC, Spada CA, Hayes JG, Raghu G. The timed walk test as a measure of severity and survival in idiopathic pulmonary fibrosis. Eur Respir J 2005;25(1):96–103.

[98] Schwartz DA, Helmers RA, Galuin JR, Van Fossen DS, Frees KL, Dayton CS, et al. Determinants of survival in idiopathic pulmonary fibrosis. Am J Respir Crit Care Med 1994;149(2 Pt 1):450–4.

[99] Miki K, Maekura R, Hiraga T, Okuda Y, Okamoto T, Hirotani A, et al. Impairments and prognostic factors for survival in patients with idiopathic pulmonary fibrosis. Respir Med 2003;97(5):482–90.

[100] Ley B, Ryerson CJ, Vittinghoff E, Ryu JH, Tomassetti S, Lee JS, et al. A multidimensional index and staging system for idiopathic pulmonary fibrosis. Ann Intern Med 2012;156(10):684–91.

[101] Latsi PI, du Bois RM, Nicholson AG, Colby TV, Bisirtzoglou D, Nikolakopoulou A, et al. Fibrotic idiopathic interstitial pneumonia: the prognostic value of longitudinal functional trends. Am J Respir Crit Care Med 2003;168(5):531–7.

[102] Lynch DA, Godwin JD, Safrin S, Starko KM, Hormel P, Brown KK, et al. High-resolution computed tomography in idiopathic pulmonary fibrosis: diagnosis and prognosis. Am J Respir Crit Care Med 2005;172(4):488–93.

[103] Martin MD, Chung JH, Kanne JP. Idiopathic pulmonary fibrosis. J Thorac Imaging 2016;31(3):127–39.

[104] Nishimura K, Kitaichi M, Izumi T, Nagai S, Kanaoka M, Itoh H. Usual interstitial pneumonia: histologic correlation with high-resolution CT. Radiology 1992;182(2):337–42.

[105] Rabeyrin M, Thivolet F, Ferretti GR, Chalabreysse L, Jankowski A, Cottin V, et al. Usual interstitial pneumonia end-stage features from explants with radiologic and pathological correlations. Ann Diagn Pathol 2015;19(4):269–76.

[106] Souza CA, Muller NL, Lee KS, Johkoh T, Mitsuhiro H, Chong S. Idiopathic interstitial pneumonias: prevalence of mediastinal lymph node enlargement in 206 patients. AJR Am J Roentgenol 2006;186(4):995–9.

[107] Cottin V, Nunes H, Brillet PY, Delaval P, Devouassoux G, Tillie-Leblond I, et alGroupe d'Etude et de Recherche sur les Maladies Orphelines Pulmonaires (GERM O P). Combined pulmonary fibrosis and emphysema: a distinct underrecognised entity. Eur Respir J 2005;26(4):586–93.

[108] Hochhegger B, Sanches FD, Altmayer SPL, Pacini GS, Zanon M, Guedes ÁDCB, et al. Air trapping in usual interstitial pneumonia pattern at CT: prevalence and prognosis. Sci Rep 2018;8(1):17267.

[109] Lynch DA, Sverzellati N, Travis WD, Brown KK, Colby TV, Galvin JR, et al. Diagnostic criteria for idiopathic pulmonary fibrosis: a Fleischner Society white paper. Lancet Respir Med 2018;6(2):138–53.

[110] Domagała-Kulawik J, Skirecki T, Maskey-Warzechowska M, Grubek-Jaworska H, Chazan R. Bronchoalveolar lavage total cell count in interstitial lung diseases—does it matter? Inflammation 2012;35(3):803–9.

[111] du Bois RM. An earlier and more confident diagnosis of idiopathic pulmonary fibrosis. Eur Respir Rev 2012;21(124):141–6.

[112] Kendall DM, Gal AA. Interpretation of tissue artifacts in transbronchial lung biopsy specimens. Ann Diagn Pathol 2003;7(1):20–4.

[113] Poletti V, Casoni GL, Gurioli C, Ryu JH, Tomassetti S. Lung cryobiopsies: a paradigm shift in diagnostic bronchoscopy? Respirology 2014;19(5):645–54.

[114] Dhooria S, Sehgal I, Aggarwal A, Behera D, Agarwal R. Diagnostic yield and safety of cryoprobe transbronchial lung biopsy in diffuse parenchymal lung diseases: systematic review and meta-analysis. Respir Care 2016;61(5):700–12.

[115] DiBardino DM, Haas AR, Lanfranco AR, Litzky LA, Sterman D, Bessich JL. High complication rate after introduction of transbronchial cryobiopsy into clinical practice at an academic medical center. Ann Am Thorac Soc 2017;14(6):851–7.

[116] Ganganah O, Guo SL, Chiniah M, Li YS. Efficacy and safety of cryobiopsy versus forceps biopsy for interstitial lung diseases and lung tumours: a systematic review and meta-analysis. Respirology 2016;21(5):834–41.

[117] Kondoh Y, Taniguchi H, Kitaichi M, Yokoi T, Johkoh T, Oishi T, et al. Acute exacerbation of interstitial pneumonia following surgical lung biopsy. Respir Med 2006;100(10):1753–9.

[118] Nguyen W, Meyer KC. Surgical lung biopsy for the diagnosis of interstitial lung disease: a review of the literature and recommendations for optimizing safety and efficacy. Sarcoidosis Vasc Diffuse Lung Dis 2013;30(1):3–16.

[119] Lederer DJ, Martinez FJ. Idiopathic pulmonary fibrosis. N Engl J Med 2018;378(19):1811–23.

[120] Hutchinson JP, Fogarty AW, McKeever TM, Hubbard RB. In-hospital mortality after surgical lung biopsy for interstitial lung disease in the United States: 2000 to 2011. Am J Respir Crit Care Med 2016;193(10):1161–7.

[121] MacDonald SL, Rubens MB, Hansell DM, Copley SJ, Desai SR, du Bois RM, et al. Nonspecific interstitial pneumonia and usual interstitial pneumonia: comparative appearances at and diagnostic accuracy of thin-section CT. Radiology 2001;221(3):600–5.

[122] Screaton NJ, Hiorns MP, Lee KS, Franquet T, Johkoh T, Fujimoto K, et al. Serial high resolution CT in non-specific interstitial pneumonia: prognostic value of the initial pattern. Clin Radiol 2005;60(1):96–104.

[123] Silva CI, Müller NL, Hansell DM, Lee KS, Nicholson AG, Wells AU. Nonspecific interstitial pneumonia and idiopathic pulmonary fibrosis: change in pattern and distribution of disease over time. Radiology 2008;247(1):251–9.

[124] Silva CI, Müller NL, Lynch DA, Curran-Everett D, Brown KK, Lee KS, et al. Chronic hypersensitivity pneumonitis: differentiation from idiopathic pulmonary fibrosis and nonspecific interstitial pneumonia by using thin-section CT. Radiology 2008;246(1):288–97.

[125] Zhan X, Koelsch T, Montner SM, Zhu A, Vij R, Swigris JJ, et al. Differentiating usual interstitial pneumonia from nonspecific interstitial pneumonia using high-resolution computed tomography: the "straight-edge sign". J Thorac Imaging 2018;33(4):266–70.

[126] Ahuja J, Arora D, Kanne JP, Henry TS, Godwin JD. Imaging of pulmonary manifestations of connective tissue diseases. Radiol Clin N Am 2016;54(6):1015–31.

[127] Chung JH, Cox CW, Montner SM, Adegunsoye A, Oldham JM, Husain AN, et al. CT features of the usual interstitial pneumonia pattern: differentiating connective tissue disease-associated interstitial lung disease from idiopathic pulmonary fibrosis. AJR Am J Roentgenol 2018;210(2):307–13.

[128] Walkoff L, White DB, Chung JH, Asante D, Cox CW. The four corners sign: a specific imaging feature in differentiating systemic sclerosis-related interstitial lung disease from idiopathic pulmonary fibrosis. J Thorac Imaging 2018;33(3):197–203.

[129] Nadrous HF, Pellikka PA, Krowka MJ, Swanson KL, Chaowalit N, Decker PA, et al. Pulmonary hypertension in patients with idiopathic pulmonary fibrosis. Chest 2005;128(4):2393–9.

[130] Shorr AF, Wainright JL, Cors CS, Lettieri CJ, Nathan SD. Pulmonary hypertension in patients with pulmonary fibrosis awaiting lung transplant. Eur Respir J 2007;30(4):715–21.

[131] Patel NM, Lederer DJ, Borczuk AC, Kawut SM. Pulmonary hypertension in idiopathic pulmonary fibrosis. Chest 2007;132(3):998–1006.

[132] Hamada K, Nagai S, Tanaka S, Handa T, Shigematsu M, Nagao T, et al. Significance of pulmonary arterial pressure and diffusion capacity of the lung as prognosticator in patients with idiopathic pulmonary fibrosis. Chest 2007;131(3):650–6.

[133] Ghofrani HA, Wiedemann R, Rose F, Schermuly RT, Olschewski H, Weissmann N, et al. Sildenafil for treatment of lung fibrosis and pulmonary hypertension: a randomised controlled trial. Lancet 2002;360(9337):895–900.

[134] Hubbard RB, Smith C, Le Jeune I, Gribbin J, Fogarty AW. The association between idiopathic pulmonary fibrosis and vascular disease: a population-based study. Am J Respir Crit Care Med 2008;178(12):1257–61.

[135] Oikonomou A, Prassopoulos P. Mimics in chest disease: interstitial opacities. Insights Imaging 2013;4(1):9–27.

[136] Kreuter M, Wijsenbeek MS, Vasakova M, Spagnolo P, Kolb M, Costabel U, et al. Unfavourable effects of medically indicated oral anticoagulants on survival in idiopathic pulmonary fibrosis. Eur Respir J 2016;47(6):1776–84.

[137] Kubo H, Nakayama K, Yanai M, Suzuki T, Yamaya M, Watanabe M, et al. Anticoagulant therapy for idiopathic pulmonary fibrosis. Chest 2005;128(3):1475–82.

[138] Lloyd CR, Walsh SL, Hansell DM. High-resolution CT of complications of idiopathic fibrotic lung disease. Br J Radiol 2011;84(1003):581–92.

[139] Saraceno JL, Phelps DT, Ferro TJ, Futerfas R, Schwartz DB. Chronic necrotizing pulmonary aspergillosis: approach to management. Chest 1997;112(2):541–8.

[140] Chung MJ, Goo JM, Im JG. Pulmonary tuberculosis in patients with idiopathic pulmonary fibrosis. Eur J Radiol 2004;52(2):175–9.

[141] Raghu G, Nyberg F, Morgan G. The epidemiology of interstitial lung disease and its association with lung cancer. Br J Cancer 2004;91 (Suppl 2):S3–10.

[142] Saito A, Horie M, Micke P, Nagase T. The role of TGF-β signaling in lung cancer associated with idiopathic pulmonary fibrosis. Int J Mol Sci 2018;19(11). pii: E3611.

[143] Sakai S, Ono M, Nishio T, Kawarada Y, Nagashima A, Toyoshima S. Lung cancer associated with diffuse pulmonary fibrosis: CT-pathologic correlation. J Thorac Imaging 2003;18(2):67–71.

[144] Franquet T, Giménez A, Torrubia S, Sabaté JM, Rodriguez-Arias JM. Spontaneous pneumothorax and pneumomediastinum in IPF. Eur Radiol 2000;10(1):108–13.

[145] Collard HR, King Jr TE. Treatment of idiopathic pulmonary fibrosis: the rise and fall of corticosteroids. Am J Med 2001;110(4):326–8.

[146] Murray and Nadel's textbook of respiratory medicine. 6th ed. USA: Elsevier Inc; 2016. p. 1144.

[147] Raghu G, Anstrom KJ, King Jr TE, Lasky JA, Martinez FJ. Prednisone, azathioprine, and N-acetylcysteine for pulmonary fibrosis. N Engl J Med 2012;366(21):1968–77.

[148] Inventor. 5-Methyl-1-phenyl-2-(1H)-pyridone compositions and methods of use; 1976. US patent 3,974,281.

[149] Raghu G, Johnson WC, Lockhart D, Mageto Y. Treatment of idiopathic pulmonary fibrosis with a new antifibrotic agent, pirfenidone: results of a prospective, open-label Phase II study. Am J Respir Crit Care Med 1999;159(4 Pt 1):1061–9.

[150] Nakayama S, Mukae H, Sakamoto N, Kakugawa T, Yoshioka S, Soda H, et al. Pirfenidone inhibits the expression of HSP47 in TGF-beta1-stimulated human lung fibroblasts. Life Sci 2008;82(3–4):210–7.

[151] Conte E, Gili E, Fagone E, Fruciano M, Iemmolo M, Vancheri C, et al. Effect of pirfenidone on proliferation, TGF-β-induced myofibroblast differentiation and fibrogenic activity of primary human lung fibroblasts. Eur J Pharm Sci 2014;58:13–9.

[152] Nakazato H, Oku H, Yamane S, Tsuruta Y, Suzuki R. A novel anti-fibrotic agent pirfenidone suppresses tumor necrosis factor-alpha at the translational level. Eur J Pharmacol 2002;446(1–3):177–85.

[153] Misra HP, Rabideau C. Pirfenidone inhibits NADPH-dependent microsomal lipid peroxidation and scavenges hydroxyl radicals. Mol Cell Biochem 2000;204(1–2):119–26.

[154] Noble PW, Albera C, Bradford WZ, Costabel U, Glassberg MK, Kardatzke D, et al. CAPACITY Study Group. Pirfenidone in patients with idiopathic pulmonary fibrosis (CAPACITY): two randomised trials. Lancet 2011;377(9779):1760–9.

[155] Costabel U, Bendstrup E, Cottin V, Dewint P, Egan JJ, Ferguson J, et al. Pirfenidone in idiopathic pulmonary fibrosis: expert panel discussion on the management of drug-related adverse events. Adv Ther 2014;31(4):375–91.

[156] Ledermann JA, Hackshaw A, Kaye S, Jayson G, Gabra H, McNeish I, et al. Randomized phase II placebo-controlled trial of maintenance therapy using the oral triple angiokinase inhibitor BIBF 1120 after chemotherapy for relapsed ovarian cancer. J Clin Oncol 2011;29(28):3798–804.

[157] Wollin L, Maillet I, Quesniaux V, Holweg A, Ryffel B. Antifibrotic and anti-inflammatory activity of the tyrosine kinase inhibitor nintedanib in experimental models of lung fibrosis. J Pharmacol Exp Ther 2014;349(2):209–20.
[158] Nathan SD, Albera C, Bradford WZ, Costabel U, Glaspole I, Glassberg MK, et al. Effect of pirfenidone on mortality: pooled analyses and meta-analyses of clinical trials in idiopathic pulmonary fibrosis. Lancet Respir Med 2017;5(1):33–41.
[159] Richeldi L, Cottin V, du Bois RM, Selman M, Kimura T, Bailes Z, et al. Nintedanib in patients with idiopathic pulmonary fibrosis: combined evidence from the TOMORROW and INPULSIS trials. Respir Med 2016;113:74–9.
[160] Vancheri C, Kreuter M, Richeldi L, Ryerson CJ, Valeyre D, Grutters JC, et al. Nintedanib with add-on pirfenidone in idiopathic pulmonary fibrosis: results of the INJOURNEY Trial. Am J Respir Crit Care Med 2018;197(3):356–63.
[161] Flaherty KR, Fell CD, Huggins JT, Nunes H, Sussman R, Valenzuela C, et al. Safety of nintedanib added to pirfenidone treatment for idiopathic pulmonary fibrosis. Eur Respir J 2018;52:2. pii: 1800230.
[162] Richeldi L, Collard HR, Jones MG. Idiopathic pulmonary fibrosis. Lancet 2017;389(10082):1941–52.
[163] Cottin V, Koschel D, Günther A, Albera C, Azuma A, Sköld CM, et al. Long-term safety of pirfenidone: results of the prospective, observational PASSPORT study. ERJ Open Res 2018;4(4). pii: 00084-2018.
[164] Fletcher SV, Jones MG, Renzoni EA, Parfrey H, Hoyles RK, Spinks K, et al. Safety and tolerability of nintedanib for the treatment of idiopathic pulmonary fibrosis in routine UK clinical practice. ERJ Open Res 2018;4(4). pii: 00049-2018.
[165] Visca D, Mori L, Tsipouri V, Fleming S, Firouzi A, Bonini M, et al. Effect of ambulatory oxygen on quality of life for patients with fibrotic lung disease (AmbOx): a prospective, open-label, mixed-method, crossover randomised controlled trial. Lancet Respir Med 2018;6(10):759–70.
[166] Troy LK, Young IH, Lau EM, Corte TJ. Exercise pathophysiology and the role of oxygen therapy in idiopathic interstitial pneumonia. Respirology 2016;21(6):1005–14.
[167] Mason DP, Brizzio ME, Alster JM, McNeill AM, Murthy SC, Budev MM, et al. Lung transplantation for idiopathic pulmonary fibrosis. Ann Thorac Surg 2007;84(4):1121–8.
[168] Weill D, Benden C, Corris PA, Dark JH, Davis RD, Keshavjee S, et al. A consensus document for the selection of lung transplant candidates: 2014—an update from the Pulmonary Transplantation Council of the International Society for Heart and Lung Transplantation. J Heart Lung Transplant 2015;34(1):1–15.
[169] Orens JB, Estenne M, Arcasoy S, Conte JV, Corris P, Egan JJ, et al. International guidelines for the selection of lung transplant candidates: 2006 update—a consensus report from the Pulmonary Scientific Council of the International Society for Heart and Lung Transplantation. J Heart Lung Transplant 2006;25(7):745–55.
[170] Macaluso C, Furcada JM, Alzaher O, Chaube R, Chua F, Wells AU, Al e. The potential impact of azithromycin in idiopathic pulmonary fibrosis. Eur Respir J 2018. pii: 1800628.
[171] Fulton BG, Ryerson CJ. Managing comorbidities in idiopathic pulmonary fibrosis. Int J Gen Med 2015;8:309–18.
[172] King CS, Nathan SD. Idiopathic pulmonary fibrosis: effects and optimal management of comorbidities. Lancet Respir Med 2017;5(1):72–84.
[173] Bando M, Sugiyama Y, Azuma A, Ebina M, Taniguchi H, Taguchi Y, et al. A prospective survey of idiopathic interstitial pneumonias in a web registry in Japan. Respir Investig 2015;53(2):51–9.
[174] du Bois RM. 6-minute walk distance as a predictor of outcome in idiopathic pulmonary fibrosis. Eur Respir J 2014;43(6):1823–4.
[175] Lee SH, Kim SY, Kim DS, Kim YW, Chung MP, Uh ST, et al. Predicting survival of patients with idiopathic pulmonary fibrosis using GAP score: a nationwide cohort study. Respir Res 2016;17(1):131.
[176] Jacob J, Bartholmai BJ, Rajagopalan S, van Moorsel CHM, van Es HW, van Beek FT, et al. Predicting outcomes in idiopathic pulmonary fibrosis using automated computed tomographic analysis. Am J Respir Crit Care Med 2018;198(6):767–76.

Chapter 2.2

Nonspecific interstitial pneumonia

Nonspecific interstitial pneumonia (NSIP) is the second most common variant of interstitial pneumonia after usual interstitial pneumonia (UIP), occurring in 14%–35% of biopsy cases [1]. NSIP may be an independent disease of unknown origin; however, it is more often a manifestation of connective tissue diseases (CTDs) and drug-induced pulmonary lesions, and NSIP may be associated with hypersensitivity pneumonitis (HP) or diffuse alveolar damage with subsequent development of interstitial fibrosis [2,3]. Primary diagnosis of idiopathic NSIP does not rule out further development of CTD as reflected in a study by Cono et al., where heterogeneous CTDs developed within 5 years of follow-up in 17.1% of 35 patients with idiopathic NSIP [4].

NSIP can be a presentation of not only CTD but also a recently introduced, new class of interstitial pneumonia, interstitial pneumonia with autoimmune features (IPAF). Yamakawa et al. [5] analyzed 159 patients with a morphologically proved NSIP and found that 40.8% of the patients have signs of CTDs, 36.9% could be classified as IPAF, and only 22.3% fell into the idiopathic NSIP category.

Katzenstein and Fiorelli identified three NSIP subtypes: cellular with predominance of inflammation, fibrous with predominance of fibrosis, and mixed with comparable ratios of both inflammatory and fibrotic processes [2]. The cellular variant of NSIP usually responds better to treatment and has a more favorable prognosis than fibrous NSIP; it can occur at any age, but most cases arise in the second half of life, usually affecting females (60%–70%) and those who never smoked (70%) [6].

Morphology

NSIP is one of the most challenging diseases for diagnosis due to the lack of specific features. It is characterized by signs of uniform interstitial inflammation and fibrosis with a tendency to subpleural and paraseptal location [2]. Morphologically the cellular subtype of NSIP is characterized by interstitial inflammation with the involvement of macrophages, $CD3^+$ and $CD20^+$ lymphocytes, stromal edema, and marked changes in microvessels, including fibrinoid necrosis and the development of destructive and productive vasculitis [3,7] (Figs. 2.2.1–2.2.4). Foci of organizing pneumonia may be noted, although they do not dominate the overall pattern [2]. Pathological changes are usually uniform and can include both diffuse and patchy distribution. Unlike UIP the zone of the bronchiolar-alveolar junction is not affected. In the mixed subtype of NSIP, the morphological signs of productive interstitial inflammation and interstitial fibrosis are detected simultaneously. In patients with the fibrous subtype, paucicellular interstitial fibrosis dominates the morphological pattern. In contrast to UIP, fibrosis is represented by uniform collagen deposition in alveolar septa and peribronchiolar interstitium, but interlobular septa and visceral pleura may also be involved [8]. Myofibroblastic foci and honeycombing are rare but possible findings in NSIP: in a study by Li et al., honeycombing and accumulation of myofibroblasts were observed in 16.7% of patients with NSIP, whereas their incidence was 93%–100% in those with UIP [9].

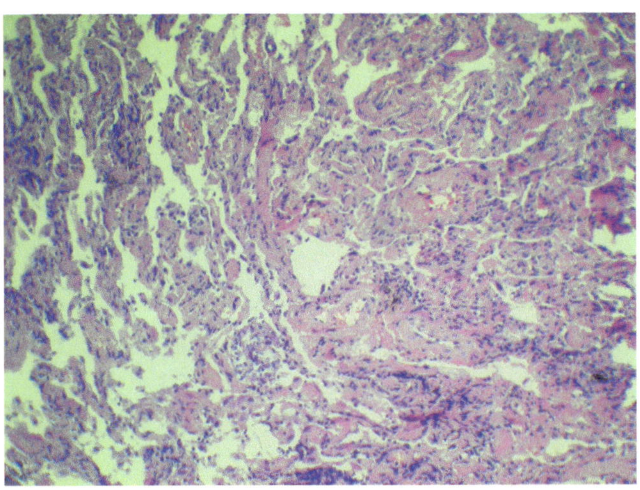

FIG. 2.2.1 Nonspecific interstitial pneumonia. Relative uniformity of edema, lymphohistiocytic infiltration, and sclerosis of the pulmonary interstitium. Hematoxylin and eosin staining, 100×.

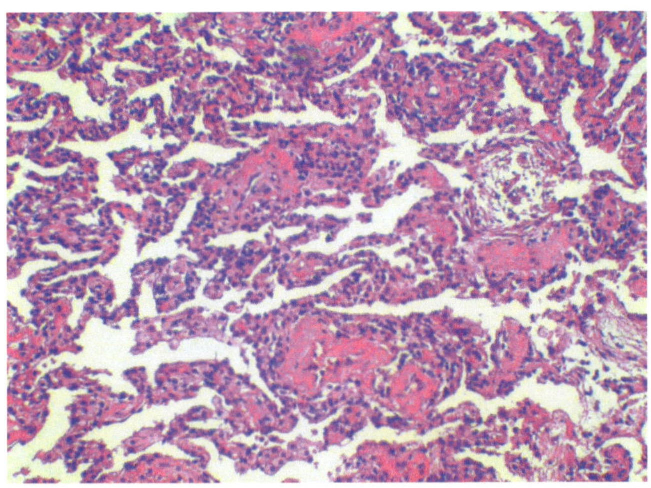

FIG. 2.2.2 Nonspecific interstitial pneumonia. Edema, lymphohistiocytic infiltration, and sclerosis of the pulmonary interstitium and small vessels. Hematoxylin and eosin staining, 200×.

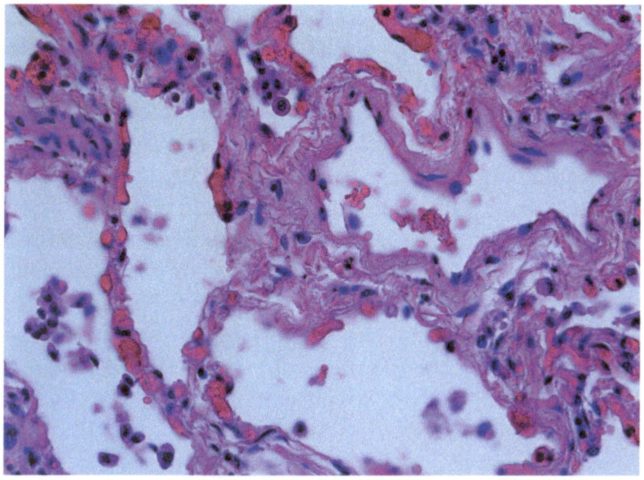

FIG. 2.2.3 Nonspecific interstitial pneumonia. Diffuse sclerosis, edema, and lymphohistiocytic infiltration of the pulmonary interstitium and small vessels. Hematoxylin and eosin staining, 400×.

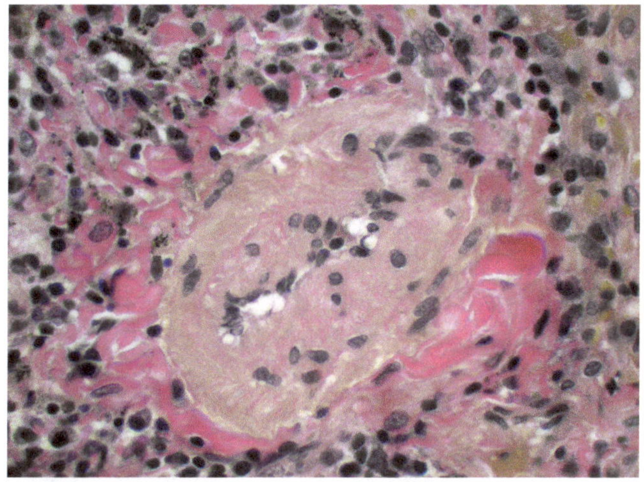

FIG. 2.2.4 Nonspecific interstitial pneumonia. Small vessel sclerosis, edema, and lymphohistiocytic infiltration of the pulmonary interstitium. Van Gieson staining, 400×.

Differential diagnosis. The diagnosis of NSIP is established after excluding all other variants of diffuse lung diseases, based not only on morphological features but also on history and clinical, radiological, and laboratory signs [7]. NSIP-type morphological changes often represent the pulmonary involvement of a number of diseases such as systemic sclerosis (SSC), rheumatoid arthritis (RA), systemic lupus erythematosus, and dermatomyositis/polymyositis and can be a manifestation of HP, asbestosis, and drug-induced pneumonitis [10,11]. Differential diagnosis is further complicated due to its possible overlap with UIP, organizing pneumonia, or HP in the same patient [12,13]. It can be very difficult, and sometimes impossible, to differentiate NSIP and HP, and only the presence of peribronchiolar granulomas (up to 80% of cases) weight the scales on behalf of the latter [8,11].

The pulmonary changes in NSIP can resemble lymphocytic interstitial pneumonia [7]; however, the latter, in addition to lymphocytic interstitial infiltration and lymphocytic aggregates that are characteristics of both diseases, usually has associated granulomas and lymphocytic infiltration, most pronounced in the bronchiolar walls, in half of the patients [14].

Clinical presentation

Clinical presentation of idiopathic NSIP has no special signs that aid in distinguishing significantly from other chronic forms of IIP. Dyspnea and unproductive cough occur in 70%–96% and 50%–87% of patients, respectively; however, some systemic manifestations are also often present, including weight loss (18%–25%), fever (22%–33%), arthralgia (14%–17%), Raynaud phenomenon (8%–10%), myalgia (7%), skin rash (5%–40%), and arthritis (3%) [6,15]. These symptoms can be noticed in the history of patients before the development of NSIP. Auscultatory data, that is, velcro crackles, depend on the severity of fibrotic changes and are registered in 53%–64% of patients [15]. Clubbing, which is characteristic of idiopathic pulmonary fibrosis (IPF), is rare (2%–10%) in patients with NSIP [4,15]. There are no significant differences in the majority of clinical symptoms between secondary and idiopathic NSIP [15].

High-resolution computed tomography

The most common radiological signs of NSIP include ground-glass opacity (GGO), irregular linear fibrosis, traction bronchiectasis, thickening of the interlobular septa, and decrease in the volume of the lower lobes (Figs. 2.2.5–2.2.7) [16,17]. While NSIP is traditionally viewed to exhibit GGO, found in 66%–100% of the cases, that dominates over all other signs [16,18], a study by Travis et al. reported that GGO occurred in only 44% of patients with NSIP, whereas bronchiectasis, reticular pattern, and lobar volume loss were found in 82%, 96%, and 77% of the patients, respectively [6]. The characteristic of NSIP is the most extensive distribution of the pathological findings in the lower and peripheral pulmonary regions [16–19]. Honeycombing zones are generally not typical for NSIP and occur in 5%–30% of patients according to different studies, while their prevalence does not exceed 10% of the total lung surface [16,18]. At advanced stages of the disease, a HRCT pattern of UIP can be observed (Fig. 2.2.8). Relative subpleural sparing is one of the frequent signs of NSIP, occurring in 20%–64% of cases [19]; this sign has relatively high specificity for NSIP and helps to differentiate from UIP [20]. Another specific sign of NSIP is the recently described straight-edge sign (SES), that is, reticulation and GGO isolated

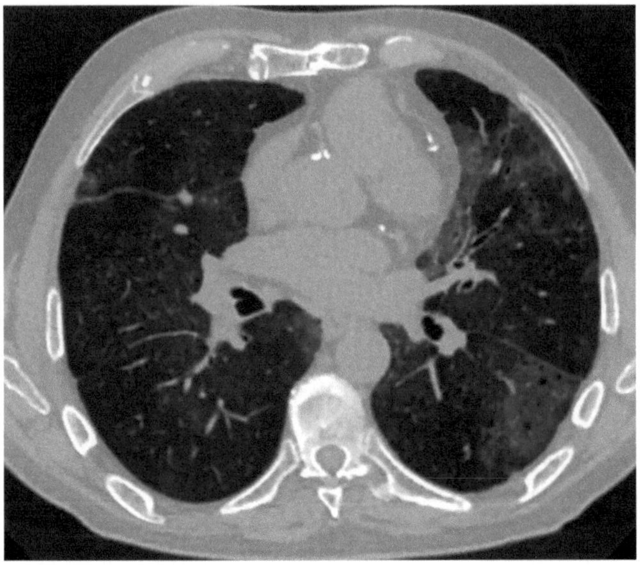

FIG. 2.2.5 Nonspecific interstitial pneumonia. Bilateral patchy areas of ground-glass opacity and moderate reticular changes. Dilated bronchi are visible inside the abnormalities.

to the lung bases with sharp demarcation from unaffected lung tissue, without substantial extension along the lateral margins of the lungs on coronal images (Fig. 2.2.9) [21]. In a study by Zhan et al. [21], the SES was detected in 14 of 15 patients and was unilateral in 26.7% of the cases (bilateral in the remaining 67.7%) with a histologically proved NSIP.

HRCT generally reflects the histological pattern of NSIP. The inflammatory, that is, cellular, subtype is characterized by predominance of GGO and absence of honeycombing (Figs. 2.2.5, 2.2.7, and 2.2.9). Fibrous (Fig. 2.2.8) and mixed (Fig. 2.2.6) NSIP subtypes suggest more varied features, with all four primary radiological signs and rarely honeycombing, are simultaneously present, albeit with varying degrees of severity [22]. A possible HRCT sign of NSIP is foci of consolidation (Fig. 2.2.6), which may indicate the simultaneous presence of organizing pneumonia, which overlaps with NSIP in two-thirds of the patients [13].

As with IPF the course of NSIP may be accompanied by periods of deterioration of clinical symptoms, usually interpreted as acute exacerbation (AE) of NSIP. The exact causes underlying the AE are not conclusively established, although infectious factors are considered as the most likely culprits. AE should be differentiated from other destabilizing events such as pulmonary artery thromboembolism, pneumothorax, and acute cardiac failure. HRCT obtained at the time of AE of NSIP reveal expansion of GGO and appearance of new consolidation areas [23].

Enlarged lymph nodes of the mediastinum, which are typical for NSIP, occur in other IIPs as well. Souza et al. reported that, among 206 patients with IIP, intrathoracic lymphadenopathy was found in 81%, 71%, and 66% of patients with NSIP, respiratory bronchiolitis associated with interstitial lung disease, and IPF, respectively [24].

Bronchoscopy and bronchoalveolar lavage

Bronchoscopy and bronchoalveolar lavage (BAL) are more useful for the differential diagnosis of NSIP from pulmonary infections and diseases that have a specific cellular pattern that can be detected by analysis of BAL fluid. In general,

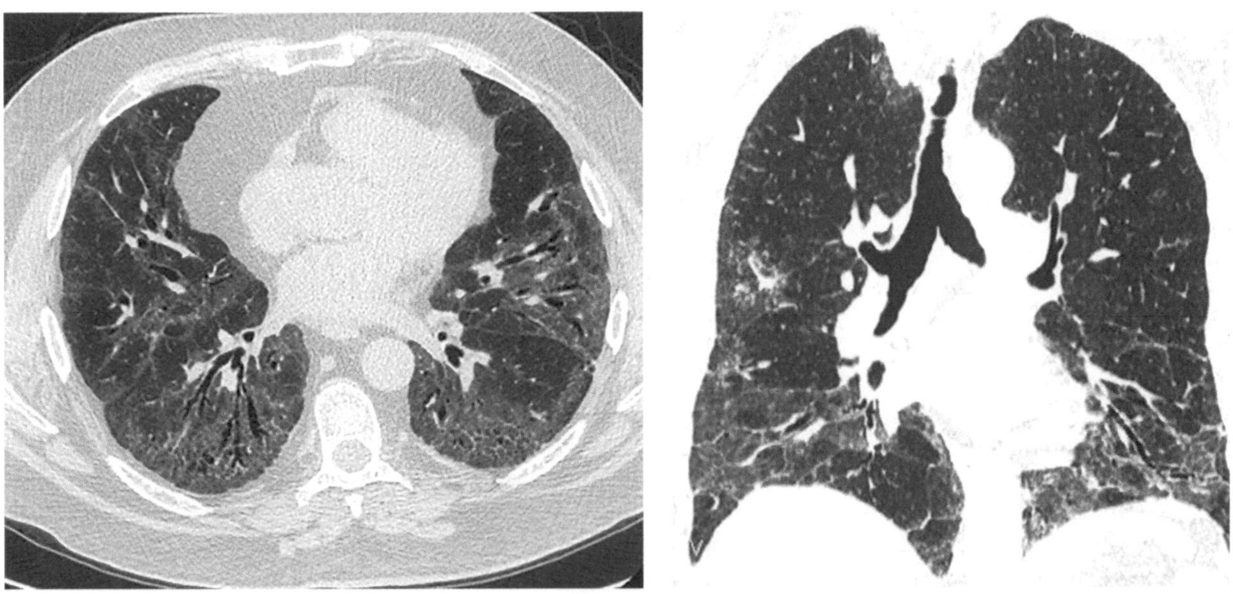

(A) (B)

FIG. 2.2.6 Nonspecific interstitial pneumonia. Bilateral diffuse areas of ground-glass opacity associated with reticular abnormalities and traction bronchiectasis. Subpleural sparing (A). Coronal reconstruction reveals the maximum distribution of changes in the lower parts of the lungs with architectural distortion of the parenchyma and a decrease in the volume of the lower lobes. Separate foci of consolidation are visualized, which reflect the pattern of organizing pneumonia (B).

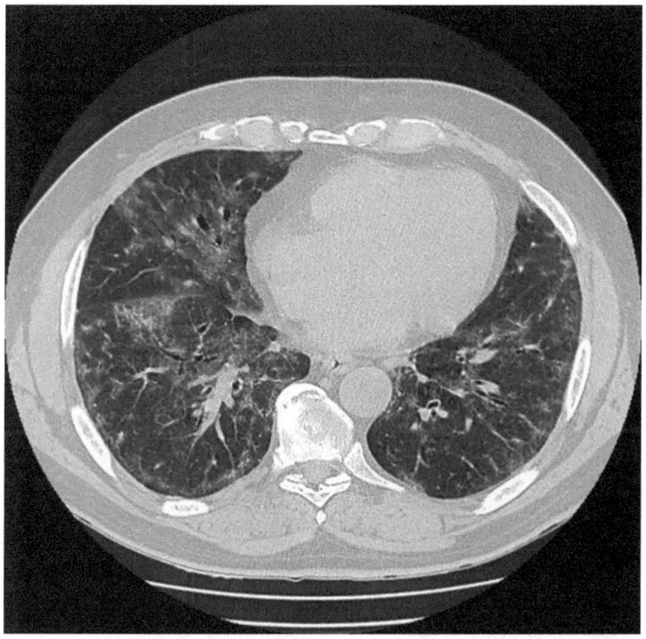

FIG. 2.2.7 Nonspecific interstitial pneumonia. Bilateral diffuse-patchy areas of ground-glass opacity associated with reticular changes. Bilateral subpleural sparing.

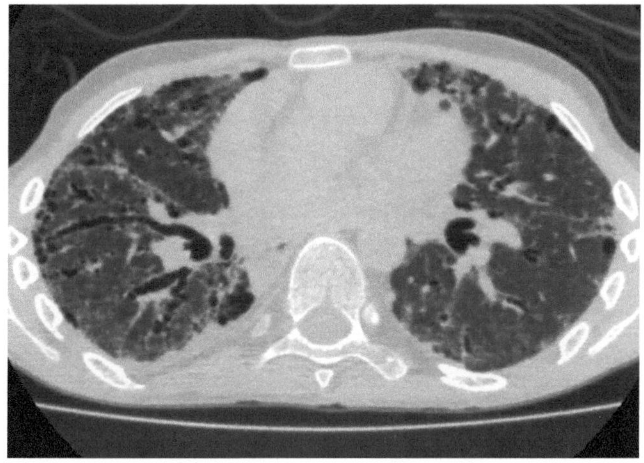

FIG. 2.2.8 Nonspecific interstitial pneumonia, late stage, confirmed by video-assisted thoracoscopic biopsy in a 23-year-old patient. Subpleural honeycombing, traction bronchiectasis, reticular abnormalities, and local pleural thickenings.

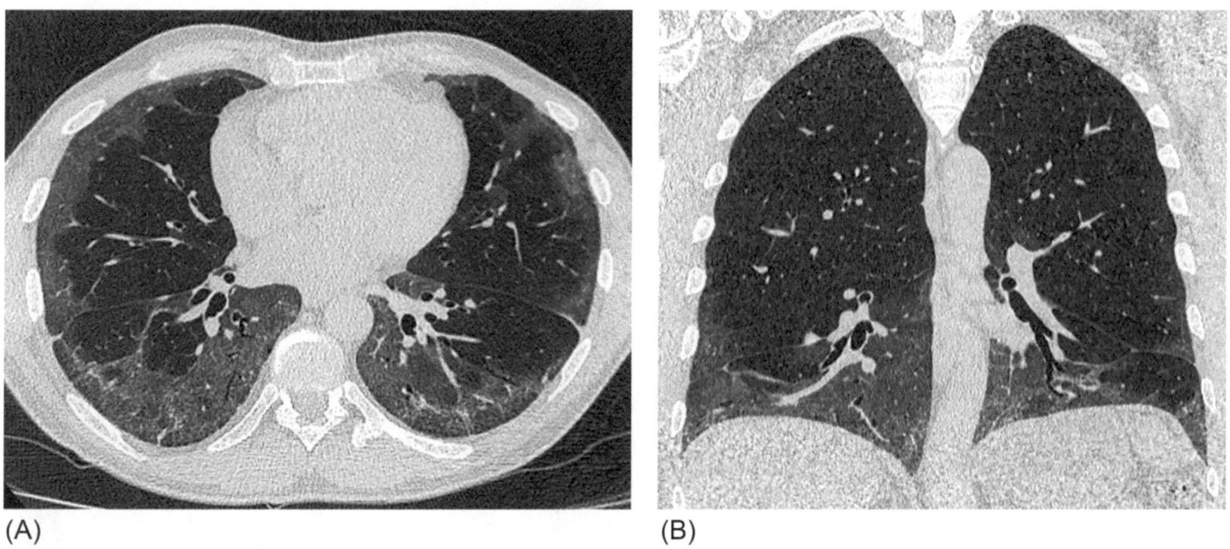

FIG. 2.2.9 Nonspecific interstitial pneumonia. Diffuse subpleural areas of ground-glass opacity associated with mild reticular signs clearly demarcated from healthy tissue. Bilateral subpleural sparing (A). Coronal reconstruction reveals the distribution of abnormalities in the basal segments with distinct demarcation from the nonaffected areas (straight-edge sign) (B).

changes in BAL fluid in NSIP include an increase in the percentage of lymphocytes, ranging from 19% to 75%; moderate neutrophilia; and mild eosinophilia [25–27]. However, in the fibrous variant of NSIP, lymphocytosis is usually mild, and changes in the BAL cellular profile closely resemble that of UIP [27]. Transbronchial lung biopsy is of limited importance for a sufficient volume of histological material for definitive diagnosis: Its diagnostic value is 16%–33%, and video-assisted thoracoscopic biopsy remains the gold standard for obtaining tissue samples for NSIP [28].

Probe-based confocal laser endomicroscopy of distal airways

We studied eight patients with NSIP by probe-based confocal laser endomicroscopy (pCLE) and examined 127 broncho-pulmonary zones to capture 621 informative images. The endomicroscopic pattern of NSIP did not reveal specific features. Because most of the patients were smokers, there were alveolar macrophages in the alveolar lumen, although their numbers exceeded those of smokers without pathological changes in the lungs. Even with a completely normal alveolar structure with unchanged interalveolar septa, the number of alveolar macrophages reached 2–5 points, in accordance with the scale presented in the Fig. 1.31 (Fig. 2.2.10A and B). Approximately in the same amount, macrophages were encountered with thickening of the interalveolar septa (Fig. 2.2.10C). In addition to cells of normal size, single giant macrophages exceeding

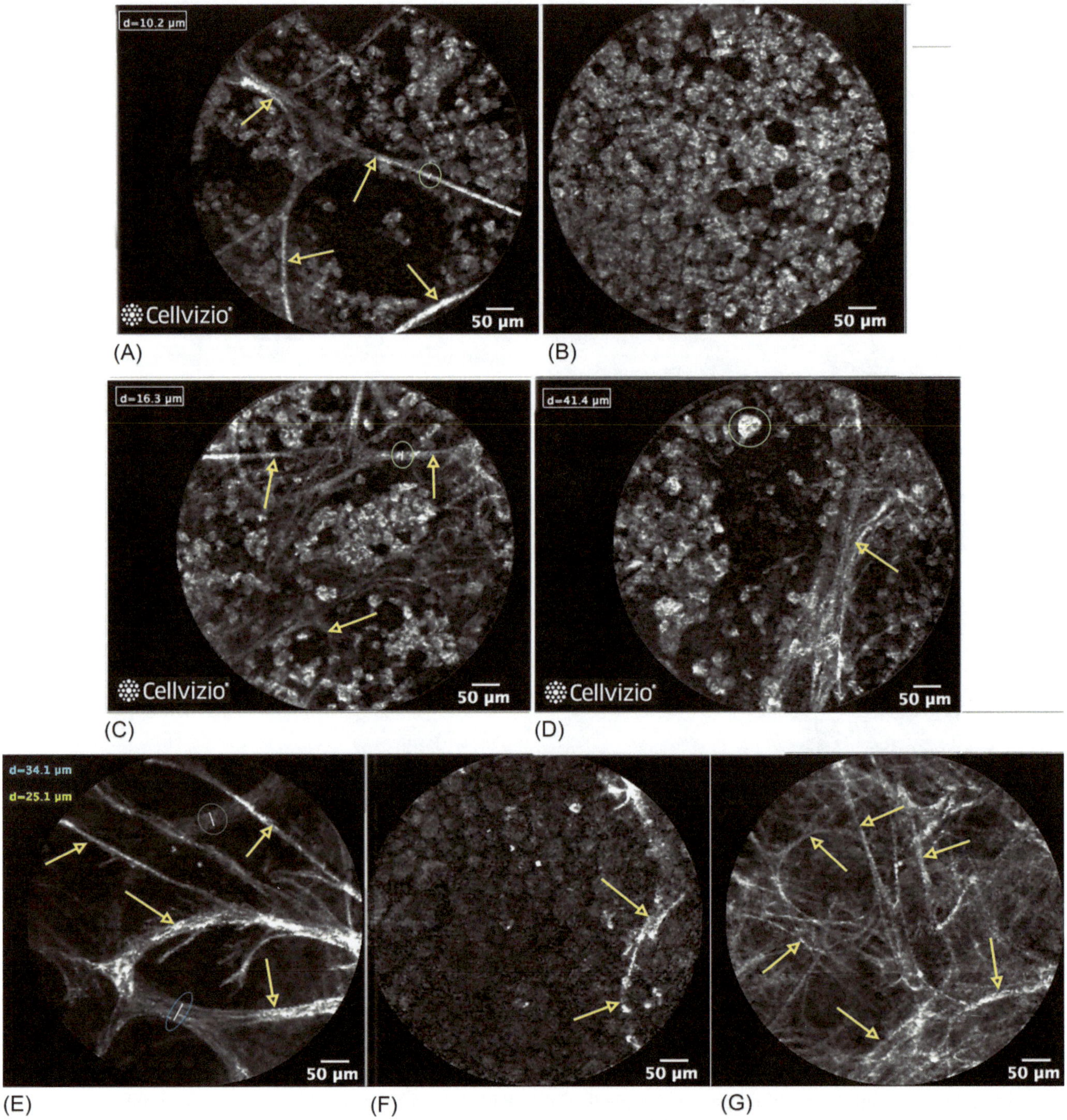

FIG. 2.2.10 Confocal laser endomicroscopy patterns in patients with nonspecific interstitial pneumonia (cellular form, A–C; fibrous form, D–F; and mixed form, D–G). (A) Alveoli with normal structure, alveolar septa are not thickened; alveolar macrophages of 3–4 points are present in the lumen. (B) The entire field of vision is occupied by intraluminal alveolar macrophages, and single air bubbles *(black circles)* are visible. (C) Alveolar shape is preserved, interalveolar septa are thickened up to 16.3 µm, and alveolar macrophages of 4–5 points are visualized. (D) Partial visualization of the alveolar structures and alveolar macrophages of 3 points are present; a giant intraluminal cell, 41.4 µm in diameter, is indicated with an ellipse. (E) Airiness of the alveoli is normal. Almost all visible septa are thickened to a significant extent (up to 34.1 µm). Some elastic fibers of the interalveolar septa are fragmented, their ends protruding loosely into the lumen. (F) A large volume of secretion with air bubbles is visualized in the alveolar lumen. In the lower right corner, single interalveolar septum is visualized. (G) Note the decrease in alveolar airiness with an increase in visualization of the elastic fibers of the alveolar walls within the same field of vision. Individual alveoli have a normal shape.

40 μm in size were found in 12 of the 127 areas (Fig. 2.2.10D). There were significantly thickened interalveolar septa with fragmentation of elastic fibers in each field of vision (Fig. 2.2.10E). Along with alveolar macrophages a moderately fluorescent liquid secretion was present in the alveolar lumen of 43 of the 127 surveyed areas (Fig. 2.2.10F). In contrast the intraluminal fluorescent cellular elements were significantly less prominent in nonsmoking patients (1–3 points). Additionally, dystelectasis was observed in 47 of the 127 regions (Fig. 2.2.10G).

Differential diagnosis of idiopathic NSIP

The absence of specific clinical and HRCT patterns renders the diagnosis of NSIP dependent on biopsy results. Unfortunately, even the presence of a apparent histological pattern does not always lead to a final clinical diagnosis of idiopathic NSIP. In these cases a multidisciplinary discussion is important, according to the results of which the consensus diagnosis of NSIP can be established two to three times more often than that based on the opinion of the pathologist [29,30]. By definition the term nonspecific interstitial pneumonia implies several clinical conditions with similar morphological characteristics. Thus idiopathic NSIP is more likely a diagnosis of exclusion. CTDs, drug-induced pulmonary lesions, and HP can exhibit the histological and radiological patterns of NSIP. In these cases, detailed history and identification of additional clinical symptoms are important (Table 2.2.1). SSC is characterized by the presence of Raynaud phenomenon that often manifests many years ahead of the pulmonary parenchymal lesions. Visual signs of SSC include mummification of the skin, which is especially noticeable on the hands and face, and sclerodactyly. In 78% of the SSC cases, pulmonary damage has the histological pattern of NSIP [31]. However, pulmonary manifestations of SSC may be ahead of the lesions of other organs, which hinder its diagnosis.

HRCT findings of SSC with the NSIP pattern include those manifestations of NSIP that have already been described earlier. An important additional sign that often appears on chest CT scan is dilatation of the esophagus due to sclerosis of

TABLE 2.2.1 Differential diagnosis of NSIP

	NSIP	IPF	Subacute HP	CTD	DIPD
Anamnesis	Frequent systemic manifestations, onset of the disease is at ages 40–50 years	Onset of disease after 50 years, long-term smoking	Exposure with a potential allergen (e.g., birds and fungi)	Onset of pulmonary lesions are preceded by symptoms in other organs	Intake of bleomycin, amiodarone, methotrexate, sulfasalazine, and other drugs
Clinical presentation	More often in nonsmoking females	More often in males, clubbing, velcro crackles in most	Increased dyspnea, cough, fever after contact with an allergen	Raynaud phenomenon, polyarthritis, dry mouth and dry eyes, sclerodactyly, dry thinned skin	Emergence of respiratory symptoms usually coincides with the intake of the drug
Laboratory tests	No abnormalities or moderately increased rheumatoid factor and antinuclear antibodies	No abnormalities	Moderately increased CRP and ESR	High titers of antinuclear antibodies, rheumatoid factor; moderate increase in ESR, CRP	Moderate leukocytosis in the blood; increased ESR, CRP
BAL	Moderate lymphocytosis, moderate neutrophilia	Moderate neutrophilia, eosinophilia	High lymphocytosis more than 50%	Moderate lymphocytosis, moderate neutrophilia	Lymphocytosis more than 30%
CT signs	Ground-glass opacity, moderate reticular changes, subpleural sparing, straight-edge sign. Predominantly basal localization	Honeycombing with preferential subpleural and basal localization. Pronounced reticular changes	Ground-glass opacity, lobular air traps. Intralobular nodules, subpleural sparing Uniform distribution with the capture of the upper lobes	Identical to NSIP. Dilatation of the esophagus, dilatation of the pulmonary artery (with SSC), straight-edge sign. Four-corners sign	Identical to NSIP. Frequent areas of consolidation as a manifestation of OP. May localize to upper lobes

CRP, C-reactive protein; *CTD*, connective tissue disorder; *CT*, computed tomography; *DIPD*, drug-induced pulmonary disease; *ESR*, erythrocyte sedimentation rate; *HP*, hypersensitivity pneumonitis; *IPF*, idiopathic pulmonary fibrosis; *NSIP*, nonspecific interstitial pneumonia; *OP*, organizing pneumonia; *SSC*, systemic scleroderma.

the mediastinal periesophageal tissue (Fig. 2.2.11). This sign can be detected earlier than the lung parenchymal damage [32]. Another radiographic marker of SSC is dilatation of the pulmonary artery, which indicates severe pulmonary hypertension and develops in approximately 20% of patients with SSC [33]. In the majority of the patients with SSC, high levels of antinuclear antibodies are found in the blood, including antibodies against topoisomerase (anti-Scl-70), anticentromere antibodies, and anti-RNA polymerase III, which are important diagnostic criteria [34].

NSIP is the second most common morphological pattern of interstitial lung disease associated with RA [35]. The diagnosis of RA is not usually very difficult in patients with a history of erosive polyarthritis; however, if the disease manifests with pulmonary lesions or serositis such as pleural effusion and pericarditis, differential diagnosis may be challenging. Therefore examination of the characteristic serological markers of RA, namely, rheumatoid factor and specific anticyclic citrullinated peptide antibodies, and markers of other CTDs must be performed in the first stage of diagnostic workup for all patients with interstitial lung disease [36].

Dermatomyositis/polymyositis is a heterogeneous group of inflammatory myopathies, in which NSIP is observed in the affected lungs. In addition to interstitial lung disease, the clinical pattern usually has signs of muscle weakness, systemic inflammation, and increased levels of serum muscle enzymes and antisynthetase antibodies, primarily anti-histidyl-tRNA synthetase (anti-Jo1) [37], which should also be investigated in patients with interstitial pulmonary disease with an obscure etiology, since the lungs are the primary target of the autoimmune process in 20% of the cases with dermatomyositis/polymyositis [38].

Interstitial lung disease in the form of NSIP is possible in systemic lupus erythematosus, the Sjögren syndrome, and a mixed CTD. Initial examination of the patient, evaluation for additional symptoms of the disease except those of pulmonary nature, and levels of serum immunologic markers (rheumatoid factor, antinuclear antibodies, and antidouble stranded DNA antibody) often aid in the inclusion of these diseases in differential diagnosis at the first visit. The more detailed differential diagnostic workup for CTDs is described in Chapter 8.

Increased antibody titers specific for certain CTDs in combination with interstitial lung disease do not automatically lead to a CTD diagnosis. For example, in a study of 25 patients with idiopathic NSIP, Huo et al. [26] found that 76% of the patients had increased antinuclear antibody titers. For patients who do not fulfill the criteria for a CTD but exhibit clinical, serological, and morphological signs of an autoimmune process, the term IPAF has been proposed [39]. Most often, IPAF is associated with HRCT and morphological patterns of UIP or NSIP, both of which have a more favorable course and a better prognosis for survival than idiopathic forms [40].

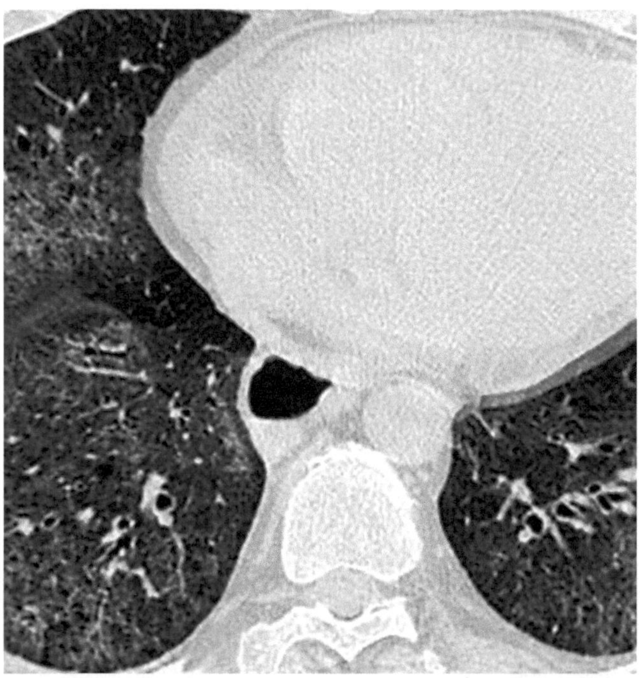

FIG. 2.2.11 Dilatation of the esophagus in a patient with interstitial lung disease associated with systemic sclerosis. Marked bilateral reticular abnormalities, bronchiolectases, and subpleural sparing indicating the nonspecific interstitial pneumonia pattern.

With the exception of SSC, the HRCT pattern of NSIP associated with a CTD or IPAF does not have specific characteristics that differ from those of idiopathic NSIP. Chronic and subacute forms of HP can nearly completely replicate the histological pattern of NSIP [11]. Radiological differentiation of these cases is very difficult. Silva et al. [20], based on a comparative analysis of the CT data in 66 patients, considered that the appearance of lobular air traps associated with reduction of blood vessels (Fig. 2.2.12), centrilobular nodules, and the lack of dominance of the lower lobe localization were more typical for HP.

A thorough history regarding the potential etiologic factors of HP surely enables to tilt the balance toward the correct diagnosis; however, even in cases with an obvious diagnosis of HP, the patients are not always able to identify the trigger allergen. For such cases the term cryptogenic HP has been proposed [41]. An important diagnostic test to facilitate the differential diagnosis of HP from IIP is the cytological analysis of BAL fluid. The presence of more than 50% lymphocytes in the cell sediment indicates the diagnosis if HP [25]. However, this sign is not sufficiently sensitive and specific; a high level of up to 75% lymphocytosis in the BAL fluid is also possible in patients with NSIP [26].

NSIP is one of the most common histological variants of drug-induced pulmonary disease (DIPD). Bleomycin, cyclophosphamide, methotrexate, and amiodarone are the most common causes of drug-induced NSIP. The apparent chronological association of drug administration with the development of interstitial pulmonary abnormalities usually facilitates the correct definitive diagnosis. However, cases of drug retention for several years after the completion of treatment were also described. For example, NSIP was observed 15–17 years after the last course of chemotherapy in patients receiving cytostatic carmustine for brain tumors [42]. There are no clear signs that can differentiate drug-induced NSIP from idiopathic NSIP, with the exception of possible preferential localization of the abnormalities in the upper lobes in DIPD (Fig. 2.2.13), which, however, does not always occur [43].

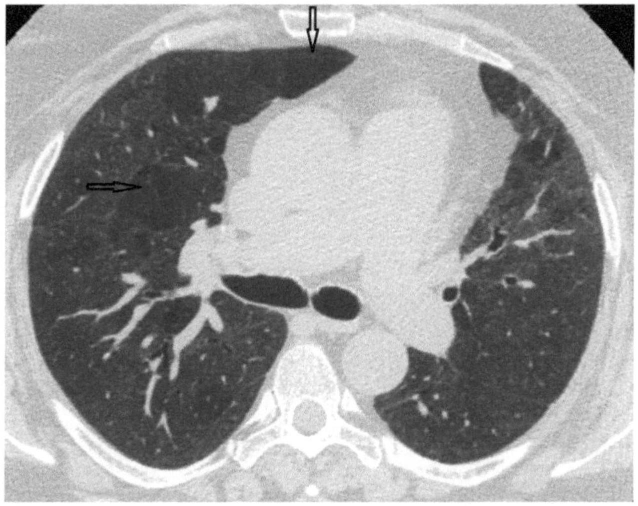

FIG. 2.2.12 Subacute hypersensitivity pneumonitis resulting from contact with birds. Air traps *(arrows)* associated with bilateral diffuse areas of ground-glass opacity.

The most challenging is the radiological differential diagnosis of the fibrous variant of NSIP and IPF. Before the results of large-scale IPF studies became available, erroneous diagnosis of NSIP was very common. Thus Johkoh et al. [44] evaluated the compliance of radiological and morphological diagnoses in 129 patients with IIP in 1999 and found that two independent radiologists reached correct diagnosis in 71%, 79%, and 63% of the patients with IPF, obliterating bronchiolitis with OP, and desquamative interstitial pneumonia (DIP), respectively, whereas NSIP was interpreted correctly only in 9% of the patients. Several studies have since shown that the principal differences between IPF and NSIP on HRCT included the severity of honeycombing (typical for IPF) and the presence of GGO (typical for NSIP) (Fig. 2.2.14) [6,18]. However, during the acute exacerbation of IPF, GGO is observed on CT scans, which renders the radiological pattern of IPF to closely resemble that of NSIP (Fig. 2.2.15). Other common signs of NSIP, which are rarely noted in IPF, are subpleural sparing and SES (Fig. 2.2.9) [20]. Bilateral SES, which was found only in 3.3% and 6.6% of patients with UIP and HP, respectively, was noted in 33% of patients with NSIP [21].

Differentiation of NSIP from DIP is also difficult. The latter is characterized almost exclusively in active smokers, majority of whom are males, while NSIP develops in nonsmoking females mostly [45]. Inflammatory changes in blood such as an elevated erythrocyte sedimentation rate above 50 mm/h and an increased C-reactive protein level, as well as an elevated lactate dehydrogenase level are more common in DIP than in NSIP [46].

Brown macrophages and moderate eosinophilia (approximately 18%) and neurophilia (approximately 11%) with increased total cell count are commonly found in the BAL fluid of patients with DIP [46], whereas moderate lymphocytosis and/or neutrophilia are characteristics of NSIP. By HRCT, bilateral areas of GGO that tend to localize in the peripheral and

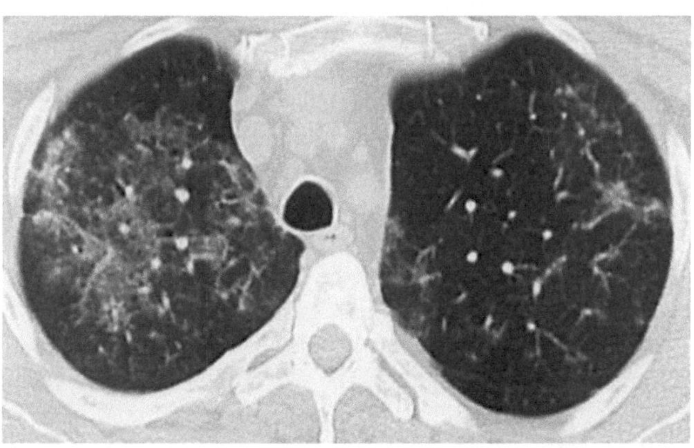

FIG. 2.2.13 Bleomycin-induced pneumonitis. Bilateral diffuse areas of ground-glass opacity associated with pronounced reticular changes. High-resolution computed tomography pattern of nonspecific interstitial pneumonia.

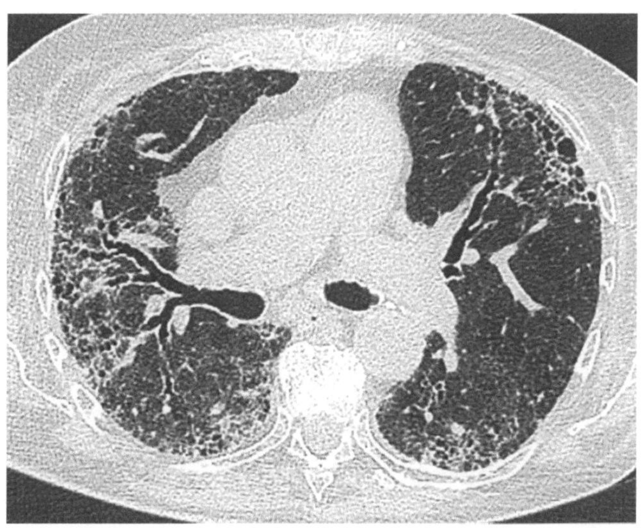

FIG. 2.2.14 Idiopathic pulmonary fibrosis. Bilateral symmetrical subpleural areas of honeycombing, severe reticular signs, and traction bronchiectasis. Pattern of usual interstitial pneumonia.

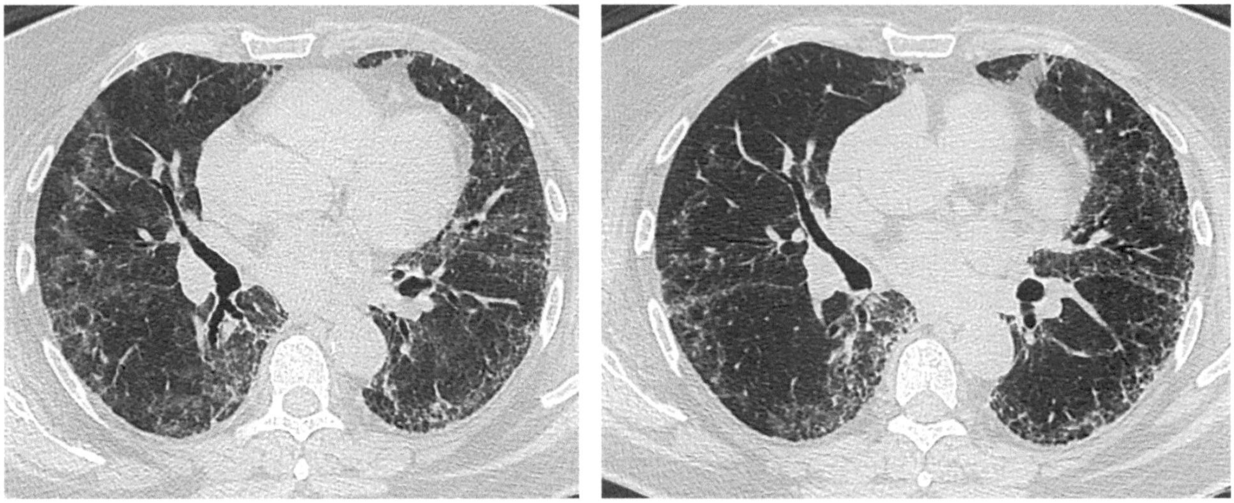

FIG. 2.2.15 Idiopathic pulmonary fibrosis. Acute exacerbation: bilateral patchy areas of ground-glass opacity, pronounced subpleural reticular abnormalities. High-resolution computed tomography pattern is similar to that of nonspecific interstitial pneumonia (A). The same patient 2 weeks later. Significant reduction in ground-glass opacity, more clear visualization of honeycombing in the paravertebral subpleural areas. High-resolution computed tomography findings correspond to the pattern of probable usual interstitial pneumonia (B).

basal regions of the lungs are usually observed in both NSIP and DIP. However, in patients with DIP, significant changes may also be localized in the upper regions of the lungs, which is not a characteristic of NSIP [47]. In general a mosaic pattern is typical for GGO with DIP, which is very similar to that of air traps in patients with HP (Fig. 2.2.16). As a rule, reticular changes are poorly expressed, but thickened bronchi resulting for long-term smoking are found frequently. DIP usually responds very quickly to treatment with systemic steroids and is often resolved without leaving any trace. NSIP and DIP may have a completely identical HRCT pattern, which does not aid in their radiological differentiation [47].

Treatment and prognosis

Treatment for NSIP includes systemic corticosteroids and immunosuppressants, administered separately or in combination. Standard doses are 0.5–0.75 mg/kg weight per day for prednisone, 2 mg/kg/day for cyclophosphamide maximum 2 mg/kg/day for azathioprine (no more than 150 mg/day) and 2000 mg daily for mycophenolate. When combined with immunosuppressive drugs, the dose of prednisone is usually reduced to 0.25 mg/kg/day [48]. In a study by Lee et al. [49], 86% of patients with NSIP responded to steroid therapy with an average dose of 0.54 mg/kg/day. Additionally, they reported that the factors that influenced poor response to the therapy included positive antinuclear antibody titers and the presence of additional diseases such as CTDs, acute leukemia, and hepatitis B infection. Further analysis of the factors that influenced the relapsing course of NSIP revealed that relapse was observed in patients who received 0.5 mg/kg/day as the initial prednisone dose and an average duration of treatment was 4.7 months, while nonrelapsing patients received a starting prednisone dose of 0.6 mg/kg for an average of 7.7 months with no steroid dose reduction [49]. The development of NSIP in patients with a CTD requires, as a rule, the administration of immunosuppressive therapy (i.e., cyclophosphamide alone or in combination with prednisone) while combination treatment is not substantially better than monotherapy [50]. A study evaluating the efficacy of the potentially useful antifibrosis drugs pirfenidone and nintedanib for other fibrotic intestinal lung diseases except IPF is in its final phase [51].

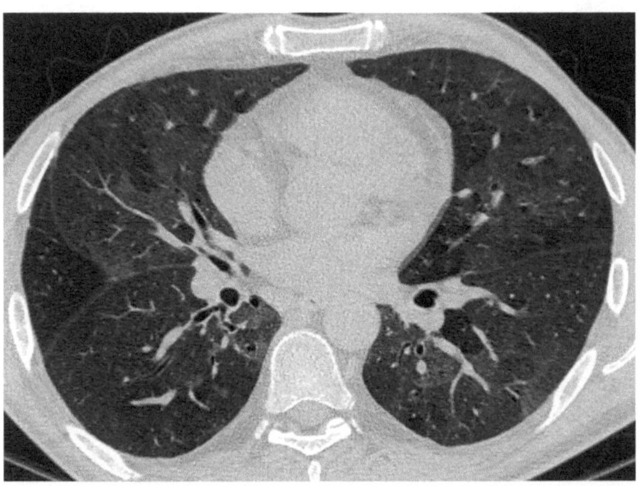

FIG. 2.2.16 Desquamative interstitial pneumonia. Diffuse bilateral areas of ground-glass opacity in the lower segments, minimally expressed reticular abnormalities. Clearly demarcated areas of the unaffected parenchyma, resembling air traps, are seen.

The prognosis of NSIP is more favorable than that of IPF. Under the influence of steroid therapy, there is often a decrease or disappearance of GGO with increased reticular abnormalities; however, the rate of fibrosis is significantly lower in NSIP than IPF [52]. Akira et al. evaluated the dynamics of CT patterns over 72 months in 50 patients with NSIP and found that nearly all patients had a decrease in areas of GGO and foci of consolidation, whereas areas of fibrosis and the number of bronchiectasis expanded simultaneously; furthermore the formation of honeycombing was observed in 34% of the patients [53]. Wide areas of GGO without traction bronchiectasis on HRCT and the presence of subpleural sparing are factors of a favorable prognosis, whereas reticular abnormalities worsen the course and prognosis [54,55].

The development of pulmonary hypertension (PH) is also a negative factor in NSIP prognosis. In a study by King et al., PH, confirmed by right-heart catheterization, was detected in 31% of the patients with NSIP, and the median transplant-free survival in the PH group was 17.6 months, which was 47.9 months in patients with NSIP without PH [56]. Travis et al. reported that the average 5- and 10-year survival rates among NSIP patients were 82.3% and 73.2%, respectively [6]. However, these rates are significantly worse for fibrous NSIP, with 5- and 10-year survival rates of 64.9% and 37.1%, respectively [5].

References

[1] Visscher DW, Myers JL. Histologic spectrum of idiopathic interstitial pneumonias. Proc Am Thorac Soc 2006;3(4):322–9.

[2] Katzenstein AL, Fiorelli RF. Nonspecific interstitial pneumonia/fibrosis: histologic features and clinical significance. Am J Surg Pathol 1994;18(2):136–47.

[3] Flaherty K, Travis W, Colby T, Toews GB, Kazerooni EA, Gross BH, et al. Histopathologic variability in usual and nonspecific interstitial pneumonias. Am J Respir Crit Care Med 2001;164(9):1722–7.

[4] Kono M, Nakamura Y, Yoshimura K, Enomoto Y, Oyama Y, Hozumi H, et al. Nonspecific interstitial pneumonia preceding diagnosis of collagen vascular disease. Respir Med 2016;117:40–7.

[5] Yamakawa H, Kitamura H, Takemura T, Ikeda S, Sekine A, Baba T, et al. Prognostic factors and disease behaviour of pathologically proven fibrotic non-specific interstitial pneumonia. Respirology 2018;23(11):1032–40.

[6] Travis WD, Hunninghake G, King Jr TE, Lynch DA, Colby TV, Galvin JR, et al. Idiopathic nonspecific interstitial pneumonia: report of an American Thoracic Society project. Am J Respir Crit Care Med 2008;177:1338–47.

[7] Travis WD, Matsui K, Moss J, Ferrans VJ. Idiopathic nonspecific interstitial pneumonia: prognostic significance of cellular and fibrosing patterns: survival comparison with usual interstitial pneumonia and desquamative interstitial pneumonia. Am J Surg Pathol 2000;24(1):19–33.

[8] Myers JL. Reprint of: Nonspecific interstitial pneumonia: pathologic features and clinical implications. Semin Diagn Pathol 2018;35(5):334–8.

[9] Li X, Chen C, Xu J, Liu J, Yi X, Sun X, et al. Nonspecific interstitial pneumonia and usual interstitial pneumonia: comparison of the clinicopathologic features and prognosis. J Thorac Dis 2014;6(10):1476–81.

[10] Nicholson AG, Rice AJ. Interstitial lung diseases. In: Hasleton P, Flieder DB, editors. Spencer's pathology of the lung. 6th ed. New York: Cambridge University Press; 2016. p. 366–408.

[11] Vourlekis JS, Schwarz MI, Cool CD, Tuder RM, King TE, Brown KK. Nonspecific interstitial pneumonitis as the sole histologic expression of hypersensitivity pneumonitis. Am J Med 2002;112(6):490–3.

[12] Monaghan H, Wells A, Colby TV, du Bois RM, Hansell DM, Nicholson AG. Prognostic implications of histologic patterns in multiple surgical lung biopsies from patients with idiopathic interstitial pneumonias. Chest 2004;125(2):522–6.

[13] American Thoracic Society/European Respiratory Society. American Thoracic Society/European Respiratory Society international multidisciplinary consensus classification of the idiopathic interstitial pneumonias. Am J Respir Crit Care Med 2002;165(2):277–304.

[14] Nicholson AG, Wotherspoon AC, Diss TC, Hansell DM, Du Bois R, Sheppard MN, et al. Reactive pulmonary lymphoid disorders. Histopathology 1995;26(5):405–12.

[15] Xu W, Xiao Y, Liu H, Qin M, Zheng W, Shi J. Nonspecific interstitial pneumonia: clinical associations and outcomes. BMC Pulm Med 2014;14:175.

[16] Johkoh T, Muller NL, Colby TV, Ichikado K, Taniguchi H, Kondoh Y, et al. Nonspecific interstitial pneumonia: correlation between thin-section CT findings and pathologic subgroups in 55 patients. Radiology 2002;225(1):199–204.

[17] Hartman TE, Swensen SJ, Hansell DM, Colby TV, Myers JL, Tazelaar HD, et al. Nonspecific interstitial pneumonia: variable appearance at high-resolution chest CT. Radiology 2000;217(3):701–5.

[18] MacDonald SL, Rubens MB, Hansell DM, Copley SJ, Desai SR, du Bois RM, et al. Nonspecific interstitial pneumonia and usual interstitial pneumonia: comparative appearances at and diagnostic accuracy of thin-section CT. Radiology 2001;221(3):600–5.

[19] Webb WR, Müller NL, Naidich DP. High-resolution CT of the lung. 5th ed. Philadelphia, PA: Lippincot Williams and Wilkins; 2015220–31.

[20] Silva CI, Muller NL, Lynch DA, Curran-Everett D, Brown KK, Lee KS, et al. Chronic hypersensitivity pneumonitis: differentiation from idiopathic pulmonary fibrosis and nonspecific interstitial pneumonia by using thin-section CT. Radiology 2008;246(1):288–97.

[21] Zhan X, Koelsch T, Montner SM, Zhu A, Vij R, Swigris JJ, et al. Differentiating usual interstitial pneumonia from nonspecific interstitial pneumonia using high-resolution computed tomography: the "straight-edge sign". J Thorac Imaging 2018;33(4):266–70.

[22] Tsubamoto M, Muller NL, Johkoh T, Ichikado K, Taniguchi H, Kondoh Y, et al. Pathologic subgroups of nonspecific interstitial pneumonia: differential diagnosis from other idiopathic interstitial pneumonias on high-resolution computed tomography. J Comput Assist Tomogr 2005;29(6):793–800.

[23] Silva CI, Muller NL, Fujimoto K, Kato S, Ichikado K, Taniguchi H, et al. Acute exacerbation of chronic interstitial pneumonia: high-resolution computed tomography and pathologic findings. J Thorac Imaging 2007;22(3):221–9.

[24] Souza CA, Muller NL, Lee KS, Johkoh T, Mitsuhiro H, Chong S. Idiopathic interstitial pneumonias: prevalence of mediastinal lymph node enlargement in 206 patients. AJR Am J Roentgenol 2006;186(4):995–9.

[25] Meyer KC, Raghu G, Baughman RP, Brown KK, Costabel U, du Bois RM, et al. An official American Thoracic Society clinical practice guideline: the clinical utility of bronchoalveolar lavage cellular analysis in interstitial lung disease. Am J Respir Crit Care Med 2012;185(9):1004–14.

[26] Huo Z, Li J, Li S, Zhang H, Jin Z, Pang J, et al. Organizing pneumonia components in nonspecific interstitial pneumonia (NSIP): a clinicopathological study of 33 NSIP cases. Histopathology 2016;68(3):347–55.

[27] Veeraraghavan S, Latsi PI, Wells AU, Pantelidis P, Nicholson AG, Colby TV, et al. BAL findings in idiopathic nonspecific interstitial pneumonia and usual interstitial pneumonia. Eur Respir J 2003;22(2):239–44.

[28] Tomassetti S, Ryu JH, Piciucchi S, Chilosi M, Poletti V. Nonspecific interstitial pneumonia: what is the optimal approach to management? Semin Respir Crit Care Med 2016;37(3):378–94.

[29] Jo HE, Glaspole IN, Levin KC, McCormack SR, Mahar AM, Cooper WA, et al. Clinical impact of the interstitial lung disease multidisciplinary service. Respirology 2016;21(8):1438–44.

[30] Burge PS, Reynolds J, Trotter S, Burge GA, Walters G. Histologist's original opinion compared with multidisciplinary team in determining diagnosis in interstitial lung disease. Thorax 2017;72(3):280–1.

[31] Tansey D, Wells AU, Colby TV, Ip S, Nikolakoupolou A, du Bois RM, et al. Variations in histological patterns of interstitial pneumonia between connective tissue disorders and their relationship to prognosis. Histopathology 2004;44(6):585–96.

[32] Desai SR, Veeraraghavan S, Hansell DM, Nikolakopolou A, Goh NS, Nicholson AG, et al. CT features of lung disease in patients with systemic sclerosis: comparison with idiopathic pulmonary fibrosis and nonspecific interstitial pneumonia. Radiology 2004;232(2):560–7.

[33] Devaraj A, Wells AU, Meister MG, Corte TJ, Wort SJ, Hansell DM. Detection of pulmonary hypertension with multidetector CT and echocardiography alone and in combination. Radiology 2010;254(2):609–16.

[34] van den Hoogen KD, Fransen J, Johnson SR, Baron M, Tyndall A, et al. 2013 Classification criteria for systemic sclerosis: an American college of rheumatology/European league against rheumatism collaborative initiative. Ann Rheum Dis 2013;72(11):1747–55.

[35] Kim EJ, Collard HR, King Jr TE. Rheumatoid arthritis-associated interstitial lung disease: the relevance of histopathologic and radiographic pattern. Chest 2009;136(5):1397–405.

[36] Raghu G, Remy-Jardin M, Myers JL, Richeldi L, Ryerson CJ, Lederer DJ, et al. Diagnosis of idiopathic pulmonary fibrosis. An official ATS/ERS/JRS/ALAT clinical practice guideline. Am J Respir Crit Care Med 2018;198(5):e44–68.

[37] Troyanov Y, Targoff IN, Tremblay J, Goulet J, Raymond Y, Senécal J. Novel classification of idiopathic inflammatory myopathies based on overlap syndrome features and autoantibodies: analysis of 100 French Canadian patients. Medicine (Baltimore) 2005;84(4):231–49.

[38] Vij R, Strek M. Diagnosis and treatment of connective tissue disease-associated interstitial lung disease. Chest 2013;143(3):814–24.

[39] Fischer A, Antoniou KM, Brown KK, Cadranel J, Corte TJ, du Bois RM, et al. An official European Respiratory Society/American Thoracic Society research statement: interstitial pneumonia with autoimmune features. Eur Respir J 2015;46(4):976–87.

[40] Yoshimura K, Kono M, Enomoto Y, Nishimoto K, Oyama Y, Yasui H, et al. Distinctive characteristics and prognostic significance of interstitial pneumonia with autoimmune features in patients with chronic fibrosing interstitial pneumonia. Respir Med 2018;137:167–75.

[41] Vasakova M, Morell F, Walsh S, Leslie K, Raghu G. Hypersensitivity pneumonitis: perspectives in diagnosis and management. Am J Respir Crit Care Med 2017;196(6):680–9.

[42] O'Driscoll BR, Kalra S, Gattamaneni HR, Woodcock AA. Late carmustine lung fibrosis. Age at treatment may influence severity and survival. Chest 1995;107(5):1355–7.

[43] Silva CI, Muller NL. Drug-induced lung diseases: most common reaction patterns and corresponding high-resolution CT manifestations. Semin Ultrasound CT MR 2006;27(2):111–6.

[44] Johkoh T, Muller NL, Cartier Y, Kavanagh PV, Hartman TE, Akira M, et al. Idiopathic interstitial pneumonias: diagnostic accuracy of thin-section CT in 129 patients. Radiology 1999;211:555–60.

[45] Vassalo R, Ryu JH. Smoking-related interstitial lung diseases. Clin Chest Med 2012;33(1):165–78.

[46] Kawabata Y, Takemura T, Hebisawa A, Sugita Y, Ogura T, Nagai S, et al. Desquamative interstitial pneumonia may progress to lung fibrosis as characterized radiologically. Respirology 2012;17(8):1214–21.

[47] Lynch DA, Travis WD, Müller NL, Galvin JR, Hansell DM, Grenier PA, et al. Idiopathic interstitial pneumonias: CT features. Radiology 2005;236(1):10–21.

[48] Flaherty KR, Toews GB, Lynch JPIII, Kazerooni EA, Gross BH, Strawderman RL, et al. Steroids in idiopathic pulmonary fibrosis: a prospective assessment of adverse reactions, response to therapy, and survival. Am J Med 2001;110(4):278–82.

[49] Lee JY, Jin S-M, Lee BJ, Chung DH, Jang BG, Park HS, et al. Treatment response and long term follow-up results of nonspecific interstitial pneumonia. J Korean Med Sci 2012;27(6):661–7.

[50] Domiciano DS, Bonfá E, Borges CT, Kairalla RA, Capelozzi VL, Parra E, et al. A long-term prospective randomized controlled study of non-specific interstitial pneumonia (NSIP) treatment in scleroderma. Clin Rheumatol 2011;30(2):223–9.

[51] Richeldi L, Varone F, Bergna M, de Andrade J, Falk J, Hallowell R, et al. Pharmacological management of progressive-fibrosing interstitial lung diseases: a review of the current evidence. Eur Respir Rev 2018;27(150).

[52] Lee HY, Lee KS, Jeong JY, Hwang JH, Kim HJ, Chung MP, et al. High-resolution CT findings in fibrotic idiopathic interstitial pneumonias with little honeycombing: serial changes and prognostic implications. AJR Am J Roentgenol 2012;199(5):982–9.
[53] Akira M, Inoue Y, Arai T, Okuma T, Kawata Y. Long-term follow-up high-resolution CT findings in non-specific interstitial pneumonia. Thorax 2011;66(1):61–5.
[54] Hozumi H, Nakamura Y, Johkoh T, et al. Nonspecific interstitial pneumonia: prognostic significance of high-resolution computed tomography in 59 patients. J Comput Assist Tomogr 2011;35(5):583–9.
[55] Sumikawa H, Johkoh T, Fujimoto K, Arakawa H, Colby TV, Fukuoka J, et al. Pathologically proved nonspecific interstitial pneumonia: CT pattern analysis as compared with usual interstitial pneumonia CT pattern. Radiology 2014;272(2):549–56.
[56] King CS, Brown AW, Shlobin OA, Weir N, Libre M, Franco-Palacios D, et al. Prevalence and impact of WHO group 3 pulmonary hypertension in advanced idiopathic nonspecific interstitial pneumonia. Eur Respir J 2018;52(1). pii: 1800545.

Chapter 2.3

Cryptogenic organizing pneumonia

Cryptogenic organizing pneumonia (COP) is one of the clinical and histological forms of idiopathic interstitial pneumonia (IIP) that belongs to a group of entities with acute or subacute course, the morphological substrate of which is bronchiolitis obliterans with organizing pneumonia (BOOP) [1]. BOOP is an inflammatory reaction of the terminal regions of the respiratory tract, characterized by the organization of exudative and fibroblastic proliferative processes as ingrowth of the granulation tissue into respiratory bronchioles, alveolar ducts, and alveoli [2].

Pathophysiologically, BOOP is a nonspecific inflammatory proliferative response of the distal respiratory tract to various systemic diseases and exogenous factors. The most common causes of BOOP include infection, drugs (e.g., amiodarone, sulfanilamides, methotrexate, and biologics), radiation therapy, connective tissue diseases (CTDs), and organ transplantation (e.g., the bone marrow, lungs, heart, liver, and kidneys) [3]. BOOP in patients with no apparent etiology is termed COP; therefore COP is a subtype of BOOP. Currently the term BOOP is recommended to describe the histological changes in lung tissue lesions, whereas etiologic terms such as drug-induced pneumonitis and radiation pneumonitis or COP in cases with unknown etiology should be used for clinical diagnosis [4].

The precise epidemiology of COP is unclear due to diagnostic challenges. A large-scale study in Iceland revealed that the general prevalence of BOOP was 1.97 cases per 100,000 populations and that the prevalence of COP was 1.1 cases per 100, 000 populations [5]. Males and females are affected to a similar extent, whereas the patient age ranges from 12 to 93 years, with an average age of 50 years according to various studies. Majority of the patients do not have a history of smoking or otherwise ceased smoking a long time before developing COP [6,7].

Morphology

The histopathology of BOOP is characterized by a combination of patchy intraluminal organizing fibrosis in the distal parts of the respiratory tract such as respiratory bronchioles, alveolar ducts, and alveoli with retention of the lung architectonics and chronic moderate interstitial inflammation (Figs. 2.3.1–2.3.6).

Typical findings during the early disease stage are peribronchiolar lymphocytic infiltration, accumulation of inflammatory exudate in bronchiolar and alveolar lumen, and basal epithelial cell hyperplasia. Organization of the inflammatory exudate involves fibroblasts, which acquire the phenotypic features of myofibroblasts with increased overproduction of collagen that is deposited as granulation tissue in the alveolar lumen [3].

The hallmark of BOOP in the later stage is budding of the granulation tissue within distal airspaces that dominates the overall pattern, with the absence of signs common in other diseases such as granulomas, prominent neutrophilic or eosinophilic infiltration, hyaline membranes, and giant cells [8,9].

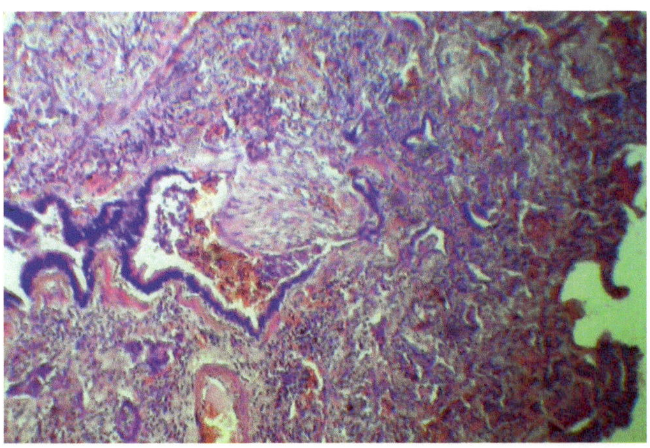

FIG. 2.3.1 BOOP. General appearance. Hematoxylin and eosin staining, 100×.

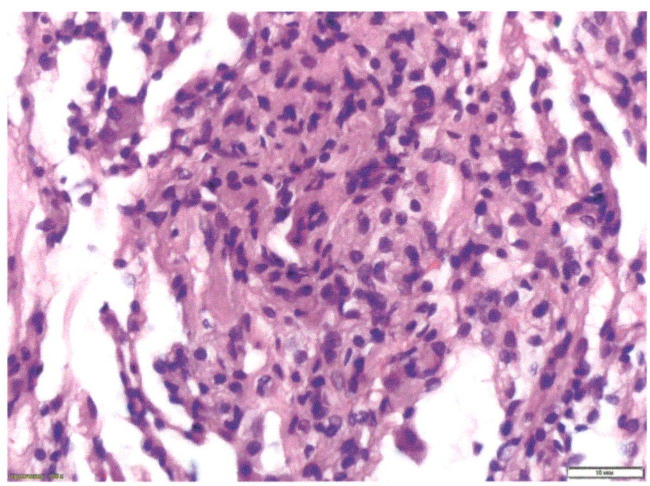

FIG. 2.3.2 Focus of BOOP with an organized exudate in alveoli—area of OP. Hematoxylin and eosin staining, 400×.

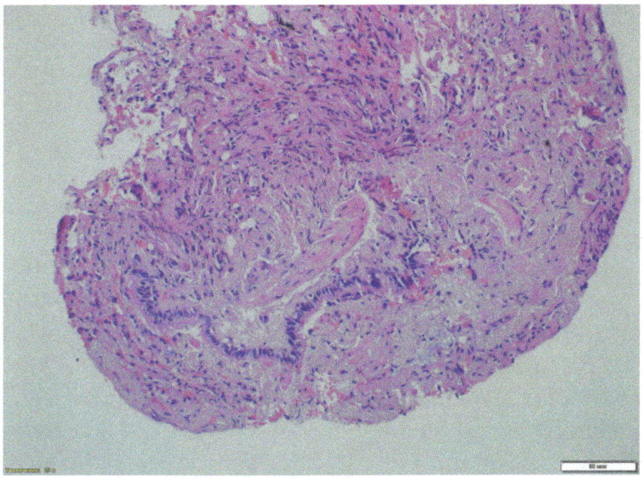

FIG. 2.3.3 Focus of bronchiolitis obliterans with an organized exudate in the bronchiole lumen observed as a granulation tissue polyp. Hematoxylin and eosin staining, 100×.

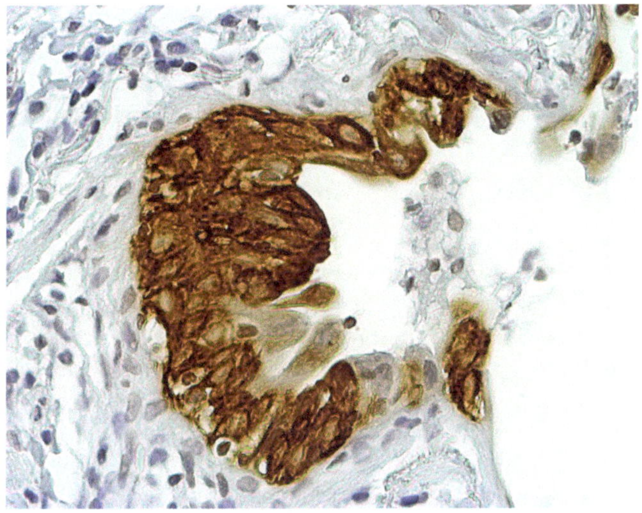

FIG. 2.3.4 BOOP at the stage of sclerotic changes. Focus of bronchiolitis obliterans with dysregeneratory changes in the epithelium and accumulation of cytokeratin-7. Immunoperoxidase reaction, 600×.

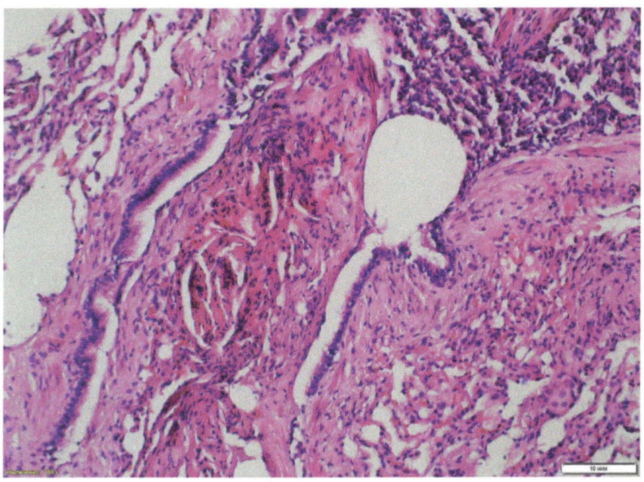

FIG. 2.3.5 BOOP at the stage of sclerotic changes. Focus of bronchiolitis obliterans with dysregeneratory changes and alveolar macrophages in the lumen. Hematoxylin and eosin staining, 200×.

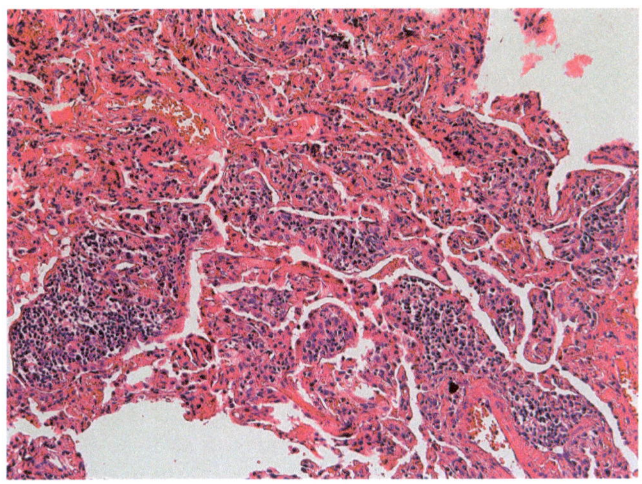

FIG. 2.3.6 BOOP. The focus of OP with sclerotic changes. Hematoxylin and eosin staining, 100×.

Transforming growth factor beta, fibroblast growth factor beta, cytokeratin-19, and cytokeratin-5 are actively expressed by epithelial cells in areas with atypical basal cell hyperplasia (Fig. 2.3.5).

Differential diagnosis of BOOP should include slowly resolving and organizing infectious pneumonia. In these cases the prevalence of lesions should be considered for definitive diagnosis. In patients with unilateral nodular or lobular lesions, the definitive diagnosis is unresolved organizing pneumonia of infectious origin. Foci of BOOP can be detected in nonspecific interstitial pneumonia, hypersensitivity pneumonitis (HP), fibrosis after diffuse alveolar damage, granulomatosis with polyangiitis, and chronic eosinophilic pneumonia (CEP); however, these foci do not dominate the overall histological pattern [8,10]. In many cases a multidisciplinary discussion in combination with the evaluation of clinical, laboratory, and radiological data is required for the correct interpretation of the histopathology.

Clinical presentation

The clinical presentation of COP is very similar to that of bacterial pneumonia. As a rule, COP has an acute or subacute onset and is accompanied by fever, fatigue, nonproductive cough, weight loss, sweating, and chest pain as well as and crackles in areas of infiltration in up to 80% of the patients [6,8,11]. COP is also characterized by moderate dyspnea starting in early days of the disease. Antibiotics are usually not effective, or their intake leads to a partial but not complete improvement. In about half of the patients, COP may resemble influenza infection [12]. Cases of rapidly progressing COP that leads to acute respiratory failure that requires respiratory support are also described. These cases must be differentiated from diffuse alveolar damage due to influenza, opportunistic infections, and acute interstitial pneumonia [13].

Diagnosis

Traditional laboratory data are not specific and include blood leukocytosis, increased C-reactive protein level, and erythrocyte sedimentation rate [6,14]. Among pulmonary functional tests a decrease in the diffusing capacity of the lung for carbon monoxide is the most diagnostically valuable parameter. Moderate deterioration in DLCO and forced vital capacity is typical for COP [15,16].

Examination of the bronchoalveolar lavage (BAL) fluid is useful for definitive diagnosis, especially in cases where morphological verification of the diagnosis is impossible. The diagnostic criteria of BAL for COP with a predictive value of 85% or above are the following [17]: lymphocyte count > 25%, $CD4^+/CD8^+$ T-cell ratio <0.9, foamy macrophages >20%, neutrophils >5%, and eosinophils 2%–25%. The definitive diagnosis of COP requires the presence of the first two criteria plus any two of the remaining three criteria [17]. Cytological, microbiological, and polymerase chain reaction analyses of the BAL fluid are also important to rule out infectious or neoplastic causes that present with a similar pattern by high-resolution computed tomography (HRCT).

Transbronchial lung biopsy, despite the small size of the samples obtained, is often sufficient for morphological diagnosis of BOOP, especially in patients with typical radiological patterns and relevant clinical presentation, with a reported sensitivity of 64% and specificity of 86% [17].

A higher diagnostic accuracy of 88% based on histological specimens can be achieved using a computed tomography (CT)-guided transthoracic needle biopsy [18].

In cases with a solitary consolidation area or an atypical HRCT pattern of COP, biopsy by video-assisted thoracoscopic surgery remains the gold standard.

High-resolution computed tomography

For COP, there are several characteristic signs detected by HRCT [19–23]:

1. Patchy areas of consolidation in different sizes and shapes, including band-like consolidation, in 80%–90% of the patients.
2. Bilateral abnormalities. Solitary infiltrates of COP are described; however, these are generally very rare.
3. Subpleural and/or peribronchovascular distribution of consolidation areas.
4. Simultaneously with consolidation, there are patchy areas of ground-glass opacity (GGO), which often exhibit a perilobular distribution around the interlobular septa, in up to 80% of patients.
5. Small nodules, often centrilobular, in 30%–50% of patients.
6. Large nodular abnormalities in up to 20% of patients.
7. Reversed halo sign (atoll sign), in up to 20% of patients.
8. Thickening of the bronchial walls and expansion of the bronchial lumen.
9. Changes in localization and attenuation of the lesions, which can migrate from one place to another, if steroid therapy is not administered. Of note, this characteristic is also typical for CEP.

In general the first four signs in various combinations are found in most patients, which render the radiological diagnosis of COP one of the most predictable among other interstitial pulmonary diseases (Figs. 2.3.7–2.3.9). In a study by Johkoh et al. [24], COP was correctly diagnosed by independent radiologists in 79% of the patients, whereas the correct diagnosis was achieved for idiopathic pulmonary fibrosis, desquamative interstitial pneumonia, and nonspecific interstitial pneumonia in 71%, 63%, and 9% of the patients with histologically verified disease, respectively.

Interestingly, consolidation areas in COP are typical for patients with a normal immune status, whereas these areas were noted only 40% of the patients among those with immune compromise, according to Lee et al. [25].

Focal and nodular abnormalities of COP detected on HRCT are in fact the same foci of granulation tissue within the alveoli and the bronchioles, which are rounder in shape that is therefore treated as an independent radiological phenomenon.

The reversed halo sign was first described in COP and considered as an exclusive finding for COP for some time [26]. This sign represents the characteristic pattern of the central focus of normal tissue or GGO that is surrounded by a dense ring of consolidation, which creates the peculiar appearance of a coral reef (Fig. 2.3.8) [27]. Later studies revealed that this sign had a low specificity for COP and was very common in many other diseases of infectious and noninfectious nature including bacterial pneumonia, tuberculosis, fungal lesions, pulmonary embolism, and primary and metastatic tumors of the lungs [28].

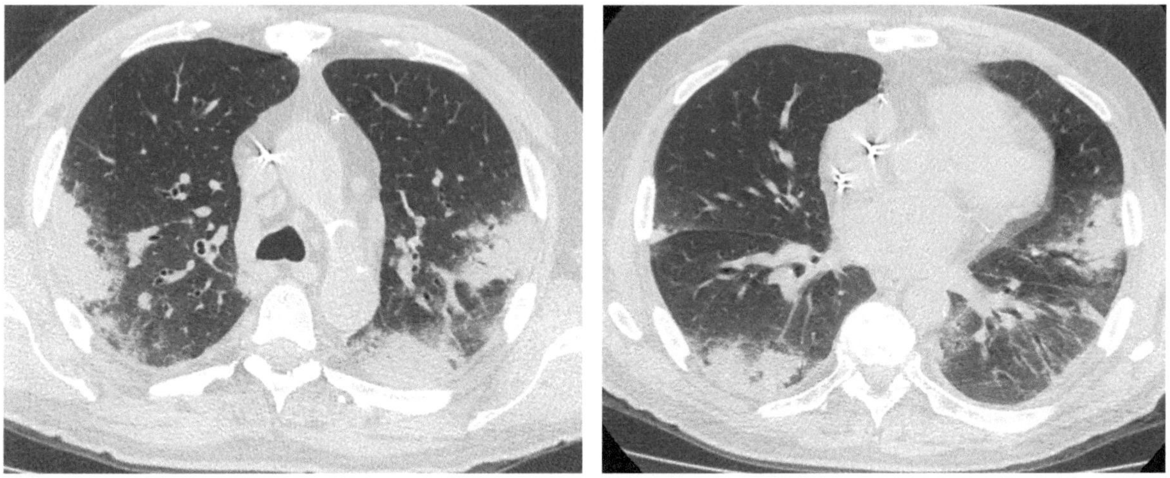

(A) (B)

FIG. 2.3.7 COP. Bilateral subpleural areas of consolidation and GGO. Changes are observed in both the upper (A) and lower (B) fields.

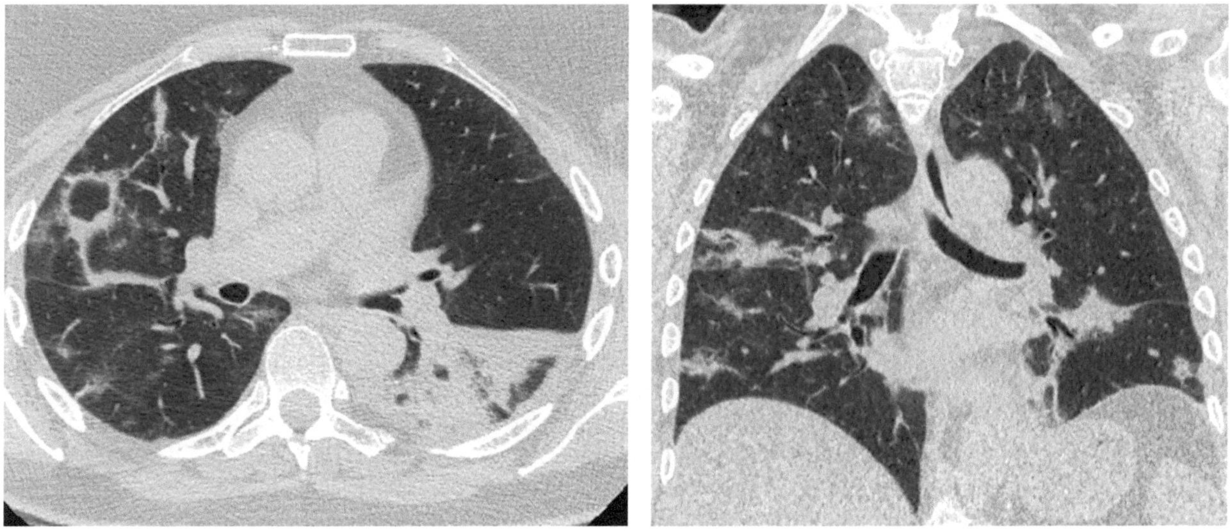

(A) (B)

FIG. 2.3.8 COP. On the left, a massive heterogeneous area of consolidation is visualized; on the right, there is a pronounced reversed halo sign, moderately pronounced reticular changes, and nodules (A). On coronal reconstruction, there is a noticeable, spotted peribronchovascular and subpleural arrangement of areas of consolidation, multiple nodules, and isolated thickened interlobular septa (B).

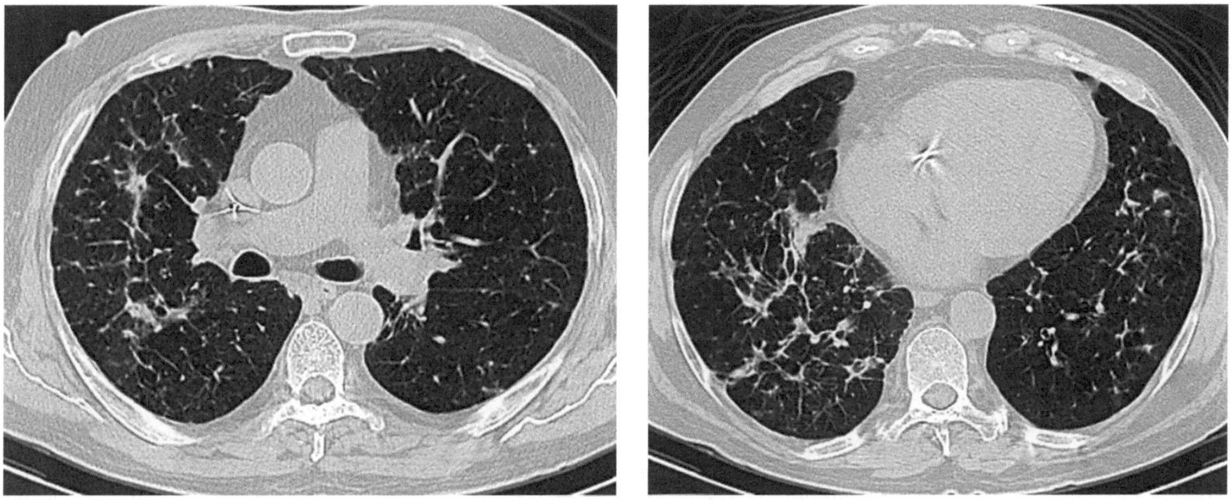

(A) (B)

FIG. 2.3.9 COP. Note the predominance of reticular abnormalities. Foci of consolidation with peribronchovascular distribution, irregular thickening of the interlobular septa, and multiple nodules of various sizes (A and B).

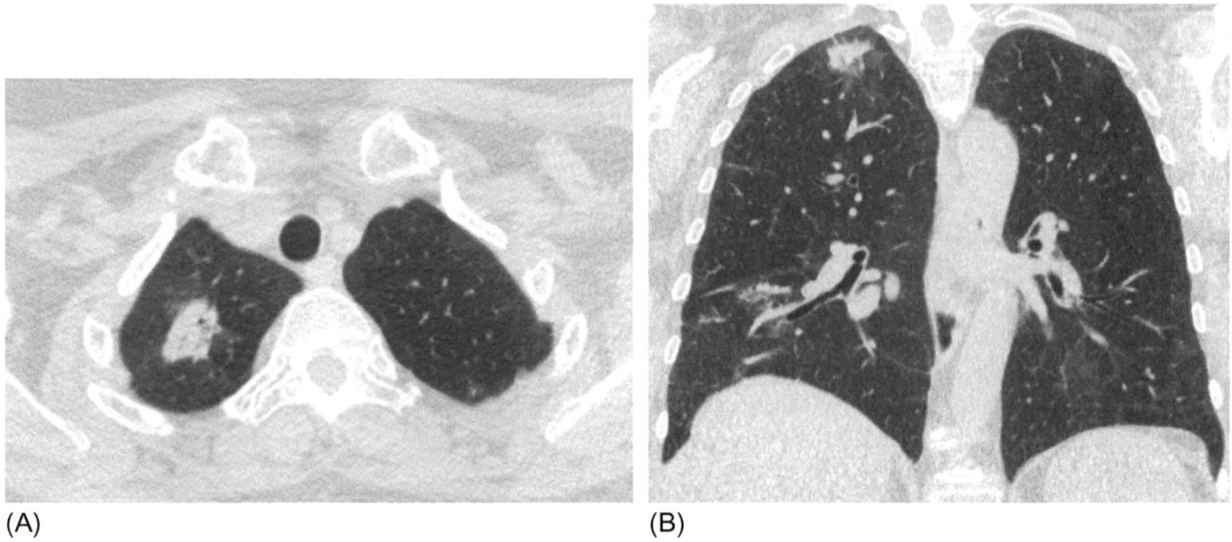

FIG. 2.3.10 COP, confirmed by video-assisted thoracoscopic surgical biopsy. The area of consolidation is seen as an irregular spiculated shape (A and B). Areas of GGO in the lower parts of the lungs are visible on the coronal scan (B).

Rarer but possible findings of COP are the crazy-paving phenomenon, linear opacities (7%–29%), minor pleural effusion (10%–30%), and intrathoracic lymphadenopathy (20%–40%) [18,20,25]. The presence of honeycombing rules out the diagnosis of COP [29].

Cordier and Cottin [30] suggest that COP can be characterized by three distinct pulmonary lesions: (i) bilateral subpleural and/or peribronchovascular areas of consolidation in combination with patchy GGO (typical pattern); (ii) a single local area of consolidation (focal pattern), often localized at the periphery in the upper lobe with minimal clinical manifestations, which leads to suspicion of a tumor and often leads to surgical intervention (Fig. 2.3.10); and (iii) diffuse infiltrative interstitial process with predominance of GGO with minimal consolidation areas, which resembles other types of IIP [30].

COP may overlap with other forms of IIP, such as nonspecific interstitial pneumonia, desquamative interstitial pneumonia, and idiopathic pulmonary fibrosis, in patients with several simultaneous histological variants and HRCT signs of interstitial pulmonary lesions [4]. Moreover the HRCT pattern of COP changes during treatment. Chung et al. observing 93 patients with histologically confirmed COP for a median of 7 months found that the size and number of consolidation areas, GGOs, and nodules/mass were significantly reduced after steroid therapy (Fig. 2.3.11) but that the reticular changes increased, which led to the resemblance of abnormalities on HRCT to fibrous nonspecific interstitial pneumonia [31].

Secondary OP caused by CTDs or radiation- or drug-induced pneumonitis does not have specific HRCT characteristics that distinguish them from COP (Fig. 2.3.12).

Differential diagnosis

Despite the fact that experienced specialists can reach a fairly confident COP diagnosis, it is however significantly delayed, usually 6–13 weeks after the disease onset [12]; the clinical presentation of COP is not very specific, and the presence of consolidation areas in the lungs, as the most frequent HRCT sign, is usually interpreted toward the more common diagnosis of bacterial pneumonia that is unresolved or slowly resolving. During the first days of the disease, reaching a definitive diagnosis of COP without an attempt for antibiotic prescription is extremely difficult in the presence of an acute inflammatory reaction with respiratory symptoms and lung infiltrates. However, if a properly selected antibiotic does not provide the expected result, the physician should consider possible opportunistic or antibiotic-resistant infection or common diseases mimicking bacterial pneumonia, such as pulmonary embolism, tumor, mycoses, and mycobacterioses, as well as diseases manifesting with OP including radiation- and drug-induced OP, CTDs, posttransplantation OP, and COP.

Serum procalcitonin level is an important diagnostic test to distinguish bacterial and noninfectious inflammation during the first days of the disease. A low serum procalcitonin level aids in ruling out an active bacterial infection. However, due to its low specificity, an elevated serum procalcitonin level cannot exclude a COP diagnosis. The lack of improvement with the antibiotic therapy is also important supporting the COP diagnosis, although in our practice we met several patients with histologically confirmed COP who exhibited some clinical, but not radiological, improvement including reduced fever and fatigue after antibiotic treatment.

Reduction in the DLCO to below 60% of the predicted value is another finding that supports interstitial pulmonary lesions. Conversely, almost all characteristic HRCT signs of COP can occur at different stages of bacterial pneumonia, especially in multifocal pneumonia (Fig. 2.3.13). Pneumonia caused by opportunistic infections often exhibits a pattern of interstitial pneumonia and

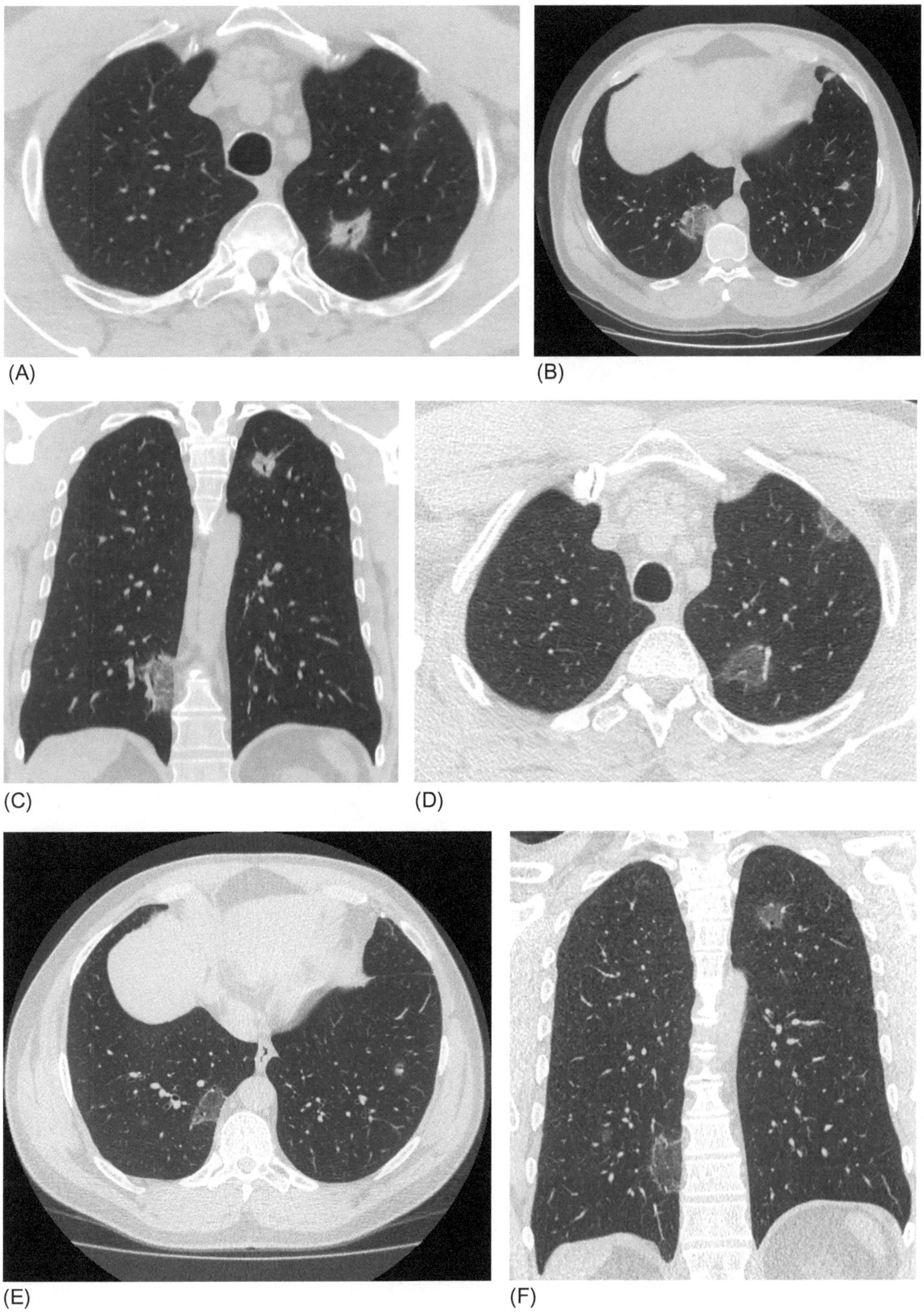

FIG. 2.3.11 Spontaneous dynamics of high-resolution computed tomographic abnormalities in a patient with COP confirmed by biopsy. On the initial scans, irregularly shaped consolidation foci are located in subpleura in the left upper lobe and round areas of GGO in the right lower lobe (A, B, and C). One month later, in the left upper lobe, areas of GGO are observed at sites of the consolidation foci; the density of areas with GGO in the lower right lobe is decreased (except for the outer boundary), which led to the appearance of the atoll sign (D, E, and F).

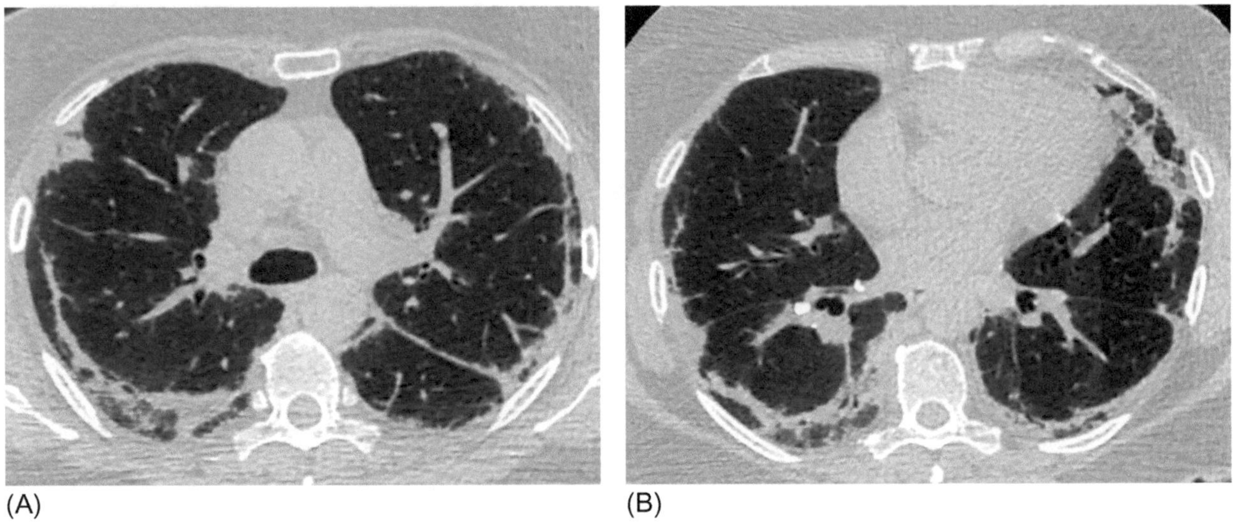

FIG. 2.3.12 Statin-induced OP. Symmetrical massive subpleural and linear consolidation areas in both the upper (A) and lower (B) lobes. *(Case courtesy of Prof. S.N. Avdeev, Sechenov First Moscow State Medical University, Moscow, Russia.)*

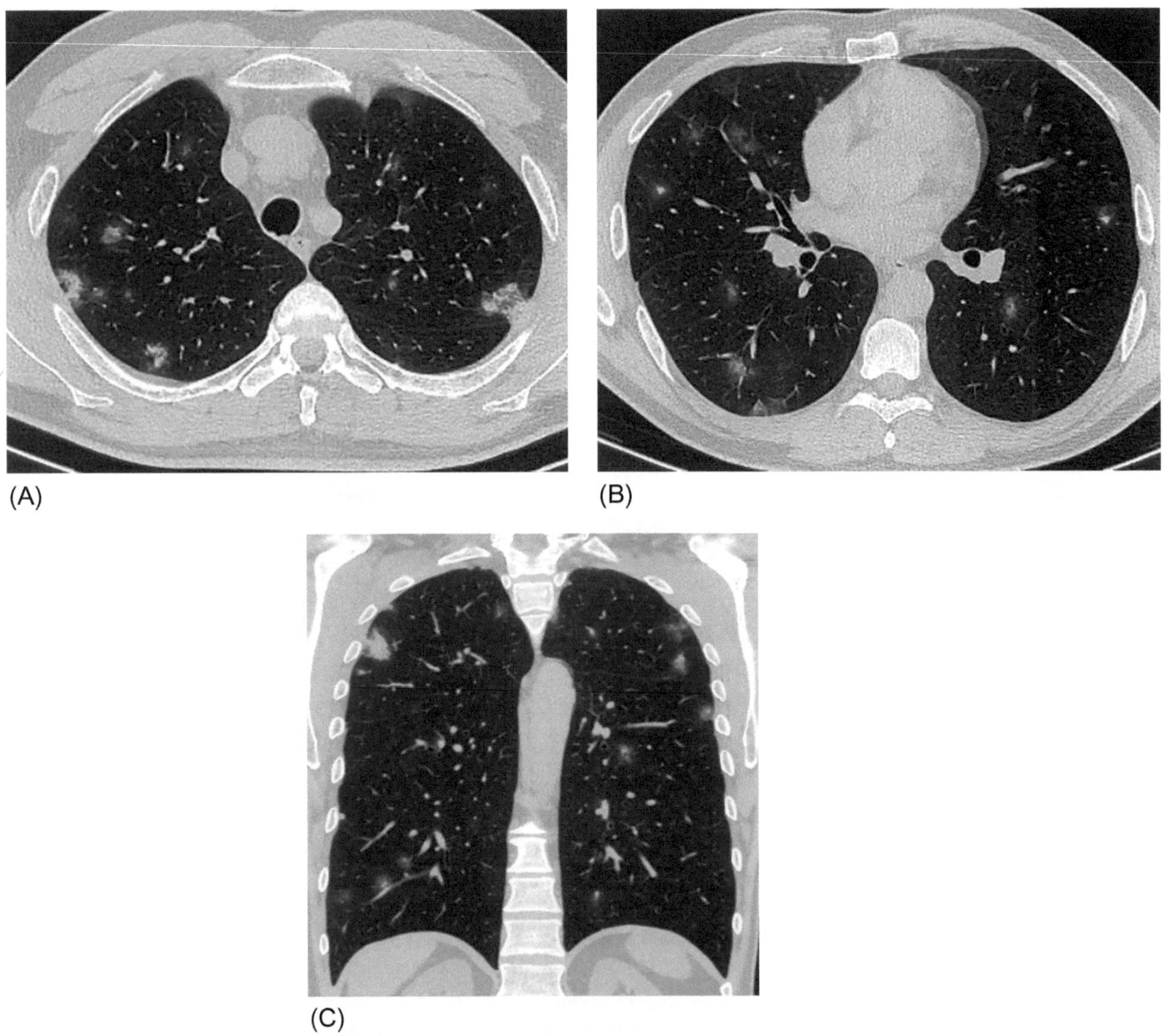

FIG. 2.3.13 Multifocal bacterial pneumonia, completely resolved after antibiotic therapy. Multiple ground-glass opacities and consolidation areas scattered randomly across all fields of the lungs. In addition to the subpleural and peribronchovascular distribution, intraparenchymal lesions are also visible (A, B, and C).

HRCT pattern resembling that of COP. However, the dominant HRCT sign for pneumonia caused by *Pneumocystis jirovecii* and *Cytomegalovirus* is GGO, whereas consolidation foci are less common (Figs. 2.3.14 and 2.3.15).

Other rarer but possible causes of infectious and inflammatory processes of the lung, such as mycobacterioses, tuberculosis, and invasive mycoses, should also be considered. With these etiologies the disease process has a subacute onset, with less pronounced symptoms than those observed with bacterial pneumonia; there is also a lack of response to conventional antibiotics. The HRCT pattern can also be very similar to that of COP, except for the cavitation of infiltrates, which is a characteristic of both tuberculosis and mycotic lesions; however, this sign is not always observed (Fig. 2.3.16). The presence of bronchiectasis is typical for nontuberculous mycobacteriosis, as well as the primarily upper lobe localization of the infiltrates (Fig. 2.3.17). The definitive diagnosis in such cases relies on the microbiological and polymerase chain reaction analysis of the BAL fluid to isolate the potential pathogens.

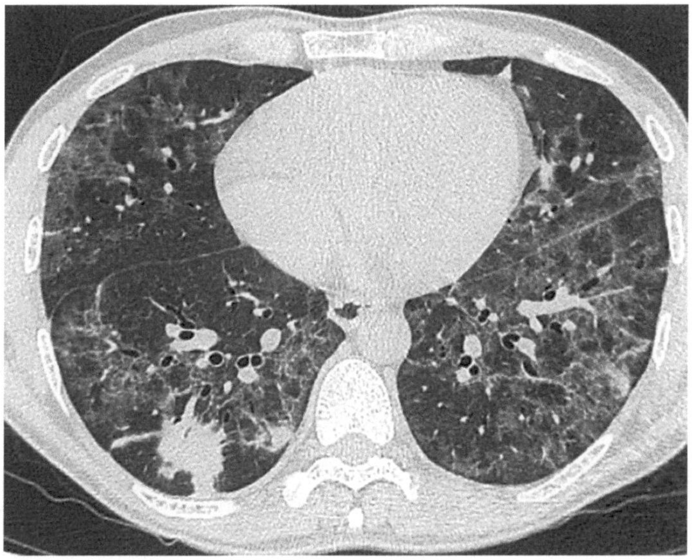

FIG. 2.3.14 *Pneumocystis jirovecii* pneumonia in a patient with acquired immunodeficiency syndrome. Diffuse areas of GGO and subpleural foci of consolidation in the lower right lobe.

In the absence of sufficient tools for verification of the diagnosis, steroid therapy may be prescribed in patients with suspicious COP after the exclusion of an infectious cause [30]. With steroid treatment, clinical changes occur very fast and include a decrease in fever, cough, dyspnea, and C-reactive protein level. Usually a 5-day course of therapy is sufficient to verify the efficiency of systemic corticosteroids. However, other interstitial pulmonary lesions can also respond quickly to steroid therapy, in particular HP, which should also be differentiated from COP. The clinical, laboratory, and functional signs of COP can be very similar to those of acute and subacute HP (Table 2.3.1). If there is no clear indication of a responsible allergen, the next step in the differential diagnosis should be the analysis of BAL fluid. In patients with HP who have not been treated with systemic steroids, BAL lymphocytosis (>50%) is characteristic [32], which ranges from 25% to 50% in patients with COP.

In general, all characteristic HRCT signs of COP can be observed in subacute HP; however, consolidation areas are a possible but not typical finding that is detected in only up to 20% of patients with HP (Fig. 2.3.18). Moreover, air traps, which are usually found in HP, are not a sign of COP [33]. By some, consolidation is not considered an independent sign of HP but additional lung tissue reactivity in the form of BOOP, which can be observed in the same patient within the histological pattern of HP [34].

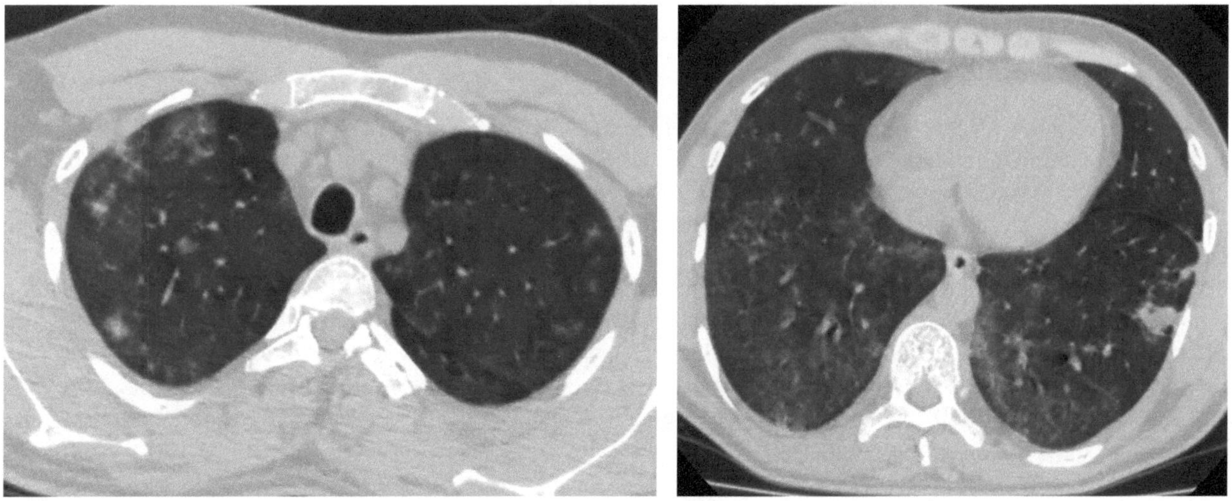

(A) (B)

FIG. 2.3.15 Cytomegalovirus pneumonia in a patient with secondary immunodeficiency. Diffuse areas of GGO associated with subpleural sparing. Moderate reticular abnormalities and subpleural foci of consolidation in both the upper (A) and lower (B) lobes.

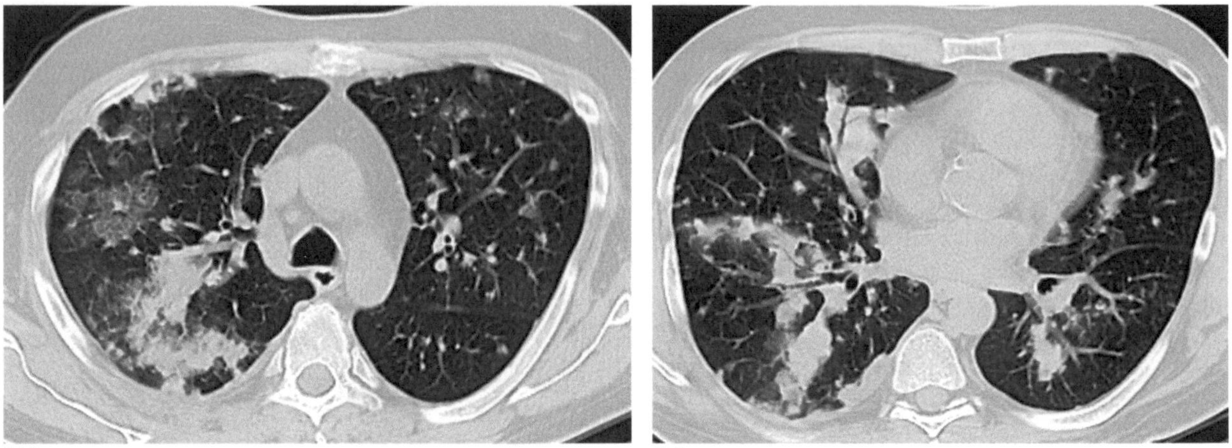

(A) (B)

FIG. 2.3.16 Invasive aspergillosis. Bilateral, primarily peribronchovascular, areas of consolidation and GGO, multiple randomly distributed nodules.

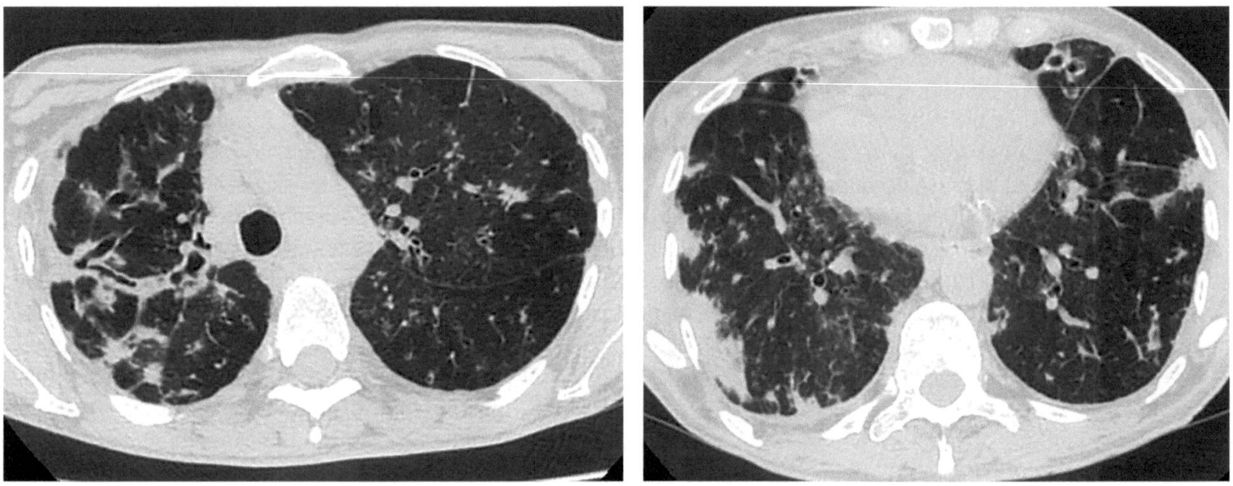

(A) (B)

FIG. 2.3.17 Mycobacteriosis caused by *Mycobacterium abscessus*. Bilateral, mostly subpleural, foci of consolidation, GGO, and multiple nodules. Cylindrical bronchiectasis, mainly in the upper right lobe (A) and lingular lobes of the left lung (B).

Our experience indicates the necessity to consider recurrent pulmonary embolism in the differential diagnosis of COP. Pulmonary embolism of small branches of the pulmonary artery with the development of peripheral infarctions, followed by an inflammatory reaction, may mimic COP (Fig. 2.3.19). Foci of consolidation with subpleural localization combined with the emergence of dyspnea and moderate fever, and a lack of response to antibiotic therapy may closely resemble the clinical presentation of COP.

It is important to note that the absence of signs of pulmonary artery filling defect on CT angiography and the absence of pulmonary hypertension cannot completely rule out peripheral PE that occurred several days prior. For pulmonary infarctions, perilobular areas of opacity, centrilobular nodules, and movement of the affected areas are not typical.

The radiological pattern of pulmonary lymphoma, in which consolidation areas alternate with areas of GGO, is similar to that of COP (Fig. 2.3.20); fever, nonproductive cough, dyspnea, and failure of antibiotic treatment are also usually present. Establishing lymphoma diagnosis without morphological and immunohistochemical verification is almost impossible.

Eosinophilic pulmonary diseases, that is, CEP and eosinophilic granulomatosis with polyangiitis, have CT findings that are very similar to those of COP, rendering their radiological distinction almost impossible (Fig. 2.3.21) [35]. However, Mehrian et al. found that upper lobe localization and prevalence of areas of GGO over consolidation

TABLE 2.3.1 Differential diagnosis of COP

	COP	Bacterial pneumonia	PE	GPA	Subacute HP
Contact with potential allergen	–	–	–	–	+++
Fever	++	+++	+	++	+
Thoracic pain	+	+	++	+	+
Dyspnea	++	+	++	+	+++
Hemoptysis	–	+	+++	++	–
Risk factors of deep venous thrombosis	–	–	+++	–	–
CRP	++	+++	+	++	++
ANCA	–	–	–	+++	–
Procalcitonin (ng/mL)	<0.5	>0.5	<0.5	<0.5	<0.5
Increased D-dimer level	–	+	+++	+	–
DLCO reduction	++	–	+	+	+++
BAL	Foamy macrophages, eosinophilia 2%–25%, lymphocytosis >25%	Neutrophilia	Erythrocytosis	Lymphocytosis in the low-activity period; $CD4^+/CD8^+ > 3$, neutrophilia in highly active phase	Lymphocytosis >50%
Subpleural distribution	+++	++	+++	+	+
Peribronchovascular distribution	+++	+	–	+++	–
Bilateral localization	+++	+	+	++	+++
GGO	+++	++	+	+	+++
Intralobular nodules	++	++	–	+	++
Reversed halo sign	++	+	++	+	–
Spontaneous migration of consolidation areas	++	+	–	–	+

–, Nonrelevant sign; +, possible sign; ++, frequent sign; +++, typical sign.
ANCA, antineutrophilic cytoplasmic antibodies; *BAL*, bronchoalveolar lavage; *COP*, cryptogenic organizing pneumonia; *CRP*, C-reactive protein; *DLCO*, diffusing capacity of the lung for carbon monoxide; *GGO*, ground-glass opacity; *GPA*, granulomatosis with polyangiitis; *HP*, hypersensitivity pneumonitis; *PE*, pulmonary embolism.

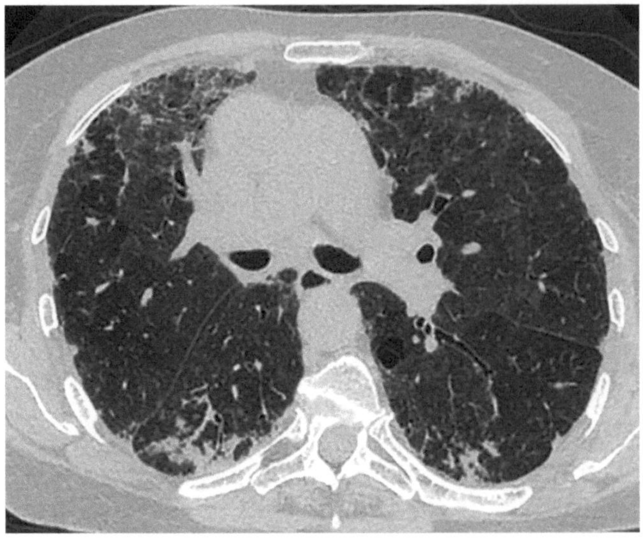

FIG. 2.3.18 Hypersensitivity pneumonitis. Bilateral areas of GGO comprising the foci of consolidation in the posterior segments. Significant reticular abnormalities and air traps are also visible.

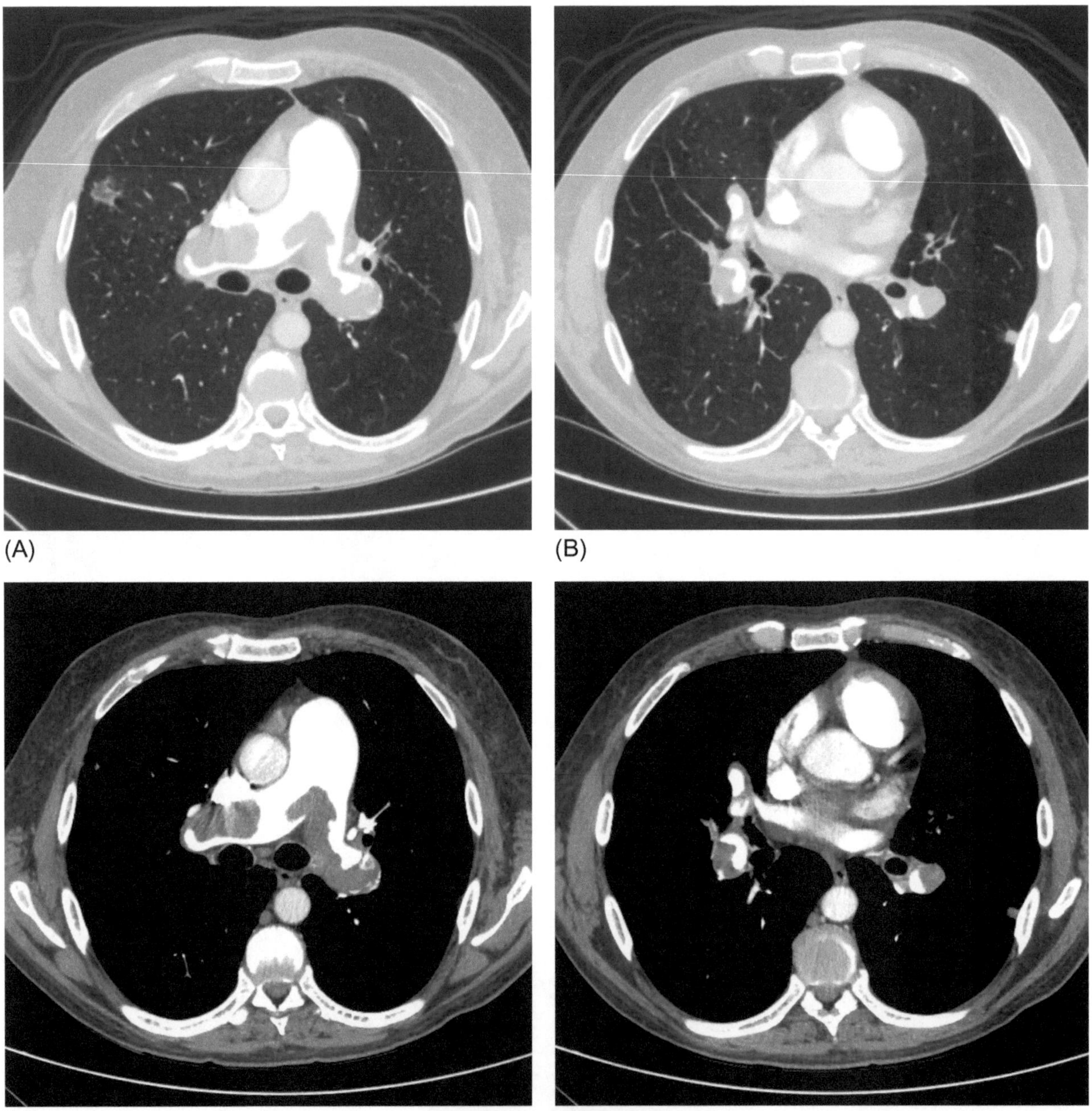

(A) (B) (C) (D)

FIG. 2.3.19 Multiple pulmonary embolism. Subpleural area of GGO in the right lung (A) and consolidation in the left lung (B) correspond to the thrombus in the right (C) and left (D) pulmonary arteries on a contrast-enhanced spiral HRCT.

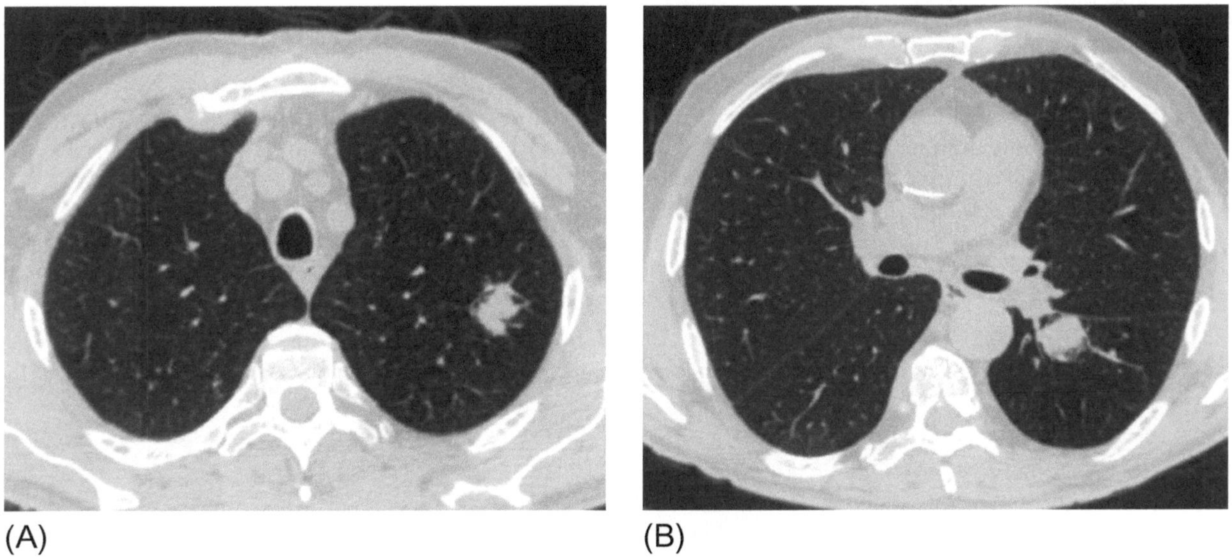

FIG. 2.3.20 B-cell lymphoma. Focus of consolidation in the upper lobe (A) and lower lobe (B) of the left lung.

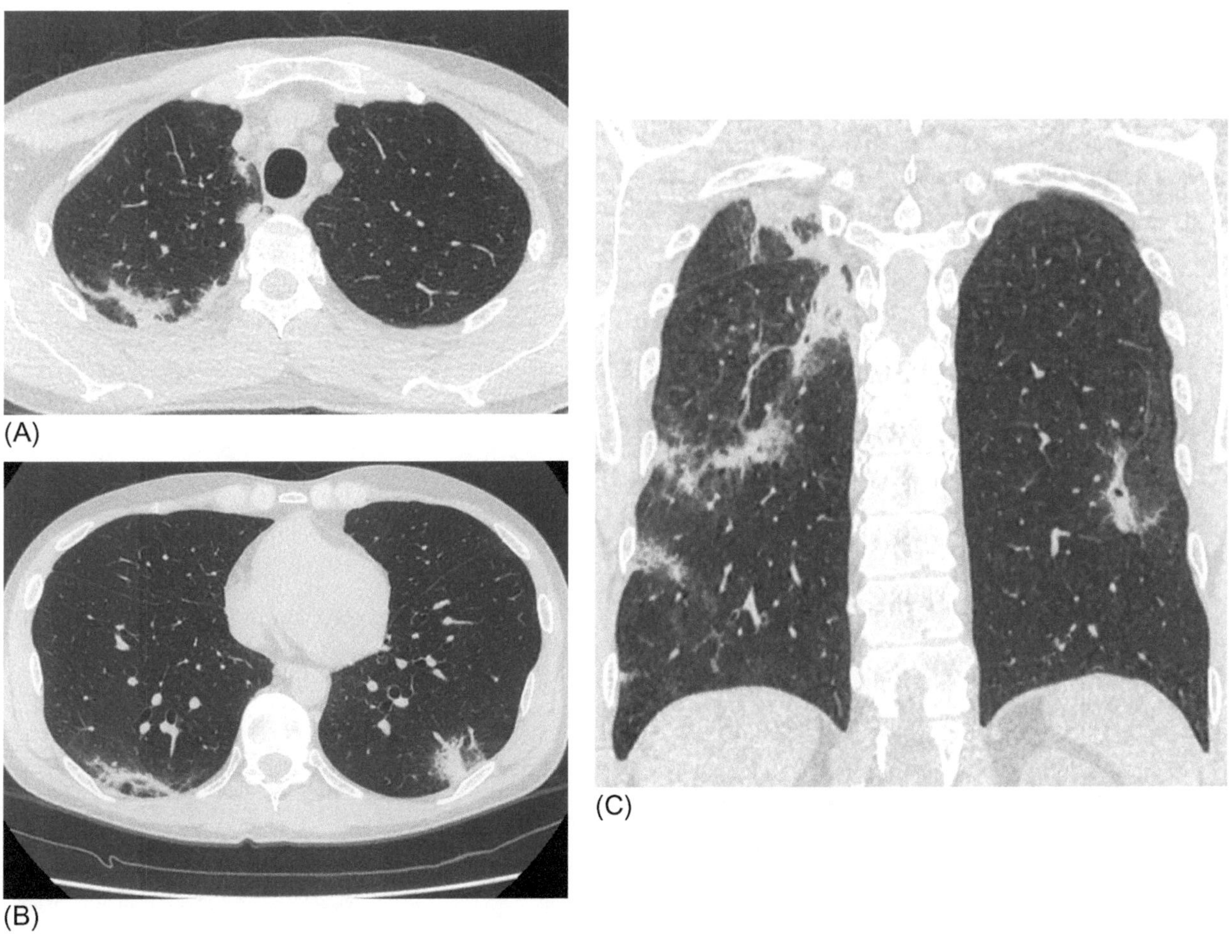

FIG. 2.3.21 Chronic eosinophilic pneumonia. Bilateral areas of irregular consolidation and GGO with subpleural and peribronchovascular distribution. Atoll sign is visible in the right lung (A and B). Abnormalities indistinguishable from COP (A–C).

were significantly more common in CEP and that the reversed halo sign and the presence of nodules less than 1 cm in diameter in combination with masses were more typical for COP [36]. The differential diagnosis is facilitated based on the fact that eosinophilic granulomatosis with polyangiitis and CEP are always accompanied by a high level of eosinophilia in peripheral blood, a frequent clinical pattern of asthma, and a high level of eosinophilia in BAL fluid. Granulomatosis with polyangiitis (GPA) presenting with pulmonary lesions can have a histological pattern of BOOP or induce secondary BOOP [37–39]. The presence of multiorgan involvement with lesions of the upper respiratory tract and kidneys and high titers of antineutrophil cytoplasmic antibodies (ANCA) usually facilitates to differentiate GPA from COP. However, if vasculitis manifests only with pulmonary lesions, GPA can be very similar clinically and radiologically to COP (Fig. 2.3.22); therefore all patients with consolidations in the lungs are recommended to be evaluated for to examine ANCA levels to rule out early-stage GPA. Another disease that must be considered in the differential diagnosis of COP is mucinous adenocarcinoma. Since the clinical and radiological patterns (Fig. 2.3.23) and the laboratory and functional parameters of the two diseases can be very similar, the cytological or histological verification of tumor is decisive.

Acute forms of noninfectious lung lesions, which are accompanied with respiratory failure and radiologically similar to both COP and acute interstitial pneumonia, are considered to be a special pattern of acute fibrinous organizing pneumonia (AFOP), in which, in addition to granulation tissue, many fibrin balls are noted in the alveolar lumen [40]. Clinically and morphologically, AFOP is an intermediate diagnosis between COP and diffuse alveolar damage; however, in contrast to the latter, AFOP is not accompanied by the formation of hyaline membranes [40,41]. Similar to COP, AFOP usually responds well to systemic steroids and is sometimes considered as a special form of OP [42]. Radiologically, AFOP manifests as patchy or diffuse areas of consolidation and GGO (Fig. 2.3.24).

Differential diagnosis of focal (solitary) COP is the most challenging. A study from the Mayo Clinic in Rochester, the United States [43], conducted a retrospective analysis of 26 patients with morphologically confirmed focal COP of over 8 years. Prior to surgery, all patients underwent CT angiography, which confirmed contrast accumulation in lesions typical for a tumor, or positron emission tomography, which demonstrated active uptake of [18F]-fluorodeoxyglucose by the suspicious focus. The average size of the consolidation areas was 1.9 cm (0.6–6.75 cm). The majority of the patients (62%) had no clinical symptoms of a respiratory illness, whereas the rest had different complaints, with cough as the most frequent. The authors concluded that the diagnosis of focal COP was extremely difficult without biopsy of the affected tissue (Fig. 2.3.10) [43]. Interestingly, after the surgical resection of the lesion, only one patient experienced a relapse within 15 months, and none of the patients received steroid therapy [43].

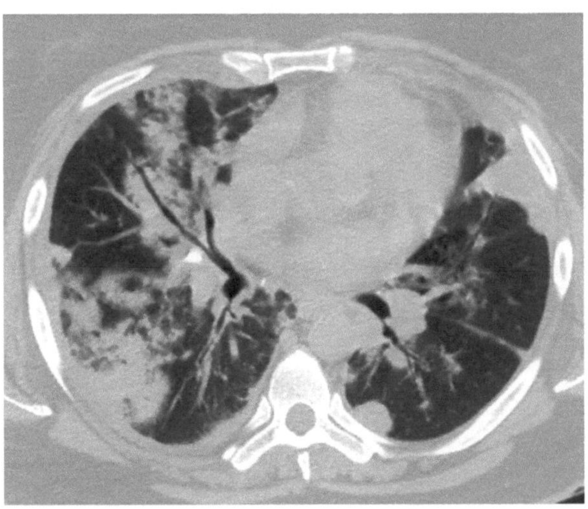

FIG. 2.3.22 Atypical presentation of granulomatosis with polyangiitis. Bilateral peribronchovascular and subpleural areas of consolidation without visible cavitation, radiologically indistinguishable from OP.

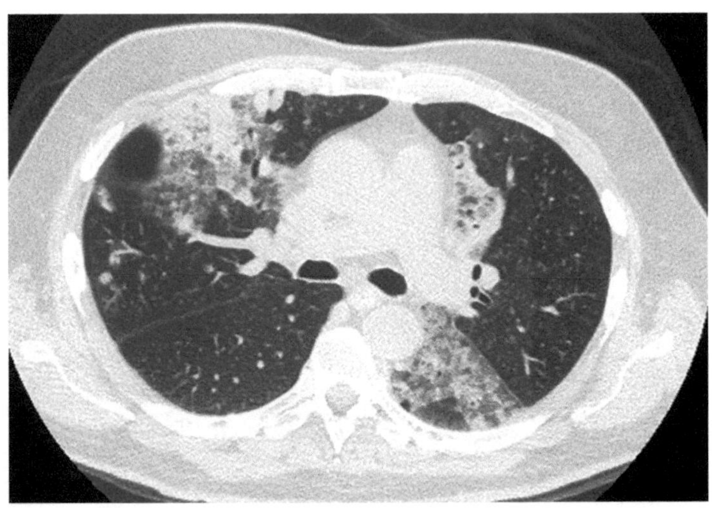

FIG. 2.3.23 Invasive mucinous adenocarcinoma. Bilateral subpleural consolidation areas and adjacent nodules are visualized. Lumens of small bronchi are detectable within the affected areas, creating the air bubble sign.

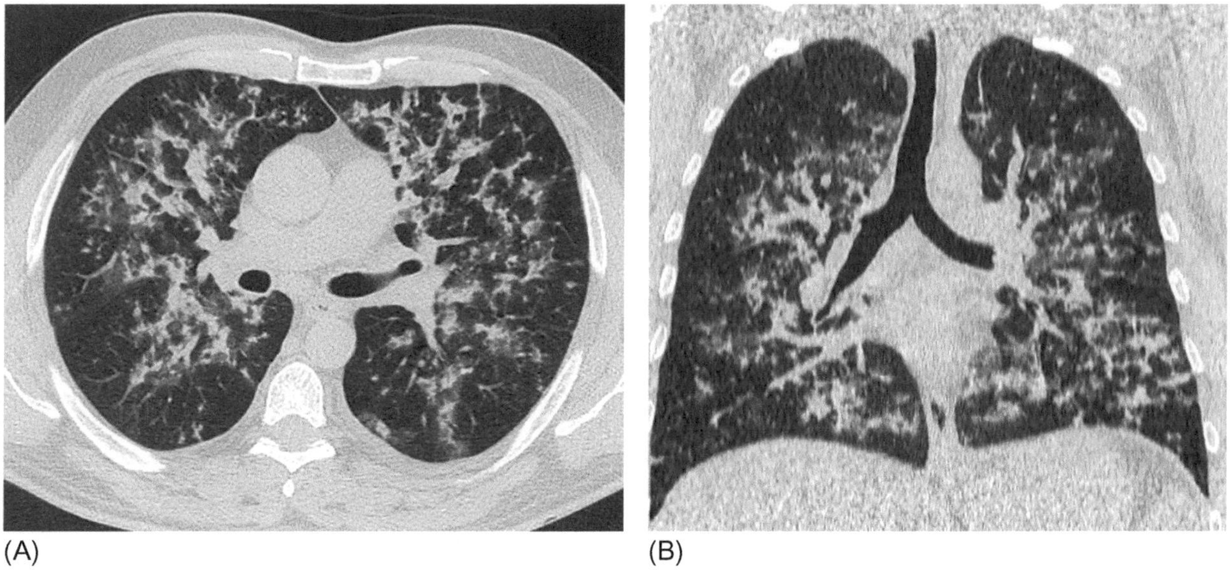

FIG. 2.3.24 AFOP, confirmed by video-assisted thoracoscopic surgical biopsy. Bilateral symmetrical patchy areas of consolidation and GGO with peribronchovascular distribution (A and B).

Treatment and prognosis

The standard treatment for COP is administration of systemic corticosteroids, with a starting dose of prednisone at 0.75 mg/kg body weight per day, followed by a dose reduction every 4–6 weeks, up to complete withdrawal within 6 months [12]. Patients with severe COP who develop severe respiratory failure require administration of pulse doses of methylprednisolone [3,44]. Despite the usually good response to steroid therapy, 30%–50% of the patients relapse with a milder form of COP after the withdrawal of corticosteroids. In relapsed cases, lower-dose prednisone (20 mg/kg) for 12 weeks with a subsequent decrease and withdrawal by the sixth month is recommended [45]. The presence of fibrin balls in biopsy samples, which is a sign AFOP; the presence of lesions in three or more lung lobes; smoking; old age; and late initiation of treatment are risk factors for the recurrence of COP [46–48].

Among other therapeutic approaches for COP, long-term treatment with macrolide antibiotics with antiinflammatory properties (azithromycin, clarithromycin, or erythromycin) is considered [49,50].

Indications for macrolides include patients with COP that does not respond to treatment with steroids or those with complications due to steroid therapy [51]. An observational study by Radzikowska et al. [11], which enrolled 40 patients with COP who received either steroids or clarithromycin (500 mg, twice daily for 3 months), found that the efficiency of the macrolide was maximum in patients with a baseline predicted forced vital capacity above 80% and a predicted FEV1 above 70%. However, a consensus on the use of macrolides in COP has not yet been achieved [3].

Treatment with rituximab can be an alternative treatment for steroid-resistant forms of COP. In a series of four patients with a biopsy-confirmed COP not responding to steroids, the administration of rituximab resulted in complete radiological and clinical recovery in one patient and led to improvement, enabling a reduction in the steroid dose in the remaining patients [52]. Cases of patients with secondary OP successfully treated with rituximab were also described [53].

The prognosis for COP is favorable, with a 5-year survival of approximately 90% [45].

References

[1] Travis WD, Costabel U, Hansell DM, King Jr TE, Lynch DA, Nicholson AG, et al. An official American Thoracic Society/European Respiratory Society statement: update of the international multidisciplinary classification of the idiopathic interstitial pneumonias. Am J Respir Crit Care Med 2013;188(6):733–48.

[2] Epler GR. Bronchiolitis obliterans organizing pneumonia: definition and clinical features. Chest 1992;102(1 Suppl):2S–6S.

[3] Cordier JF, Cottin V, Lazor R, Thivolet-Béjui F. Many faces of bronchiolitis and organizing pneumonia. Semin Respir Crit Care Med 2016;37(3):421–40.

[4] American Thoracic Society/European Respiratory Society. International multidisciplinary consensus classification of the idiopathic interstitial pneumonias. Am J Respir Crit Care Med 2002;165(2):277–304.

[5] Gundmundsson G, Sveinsson D, Isaksson HJ, Jonsson S, Frodadottir H, Aspelund T. Epidemiology of organizing pneumonia in Iceland. Thorax 2006;61(9):805–8.

[6] Cazzato S, Zompatori M, Baruzzi G, Schiattone ML, Burzi M, Rossi A, et al. Bronchiolitis obliterans organizing pneumonia: an Italian experience. Respir Med 2000;94(7):702–8.

[7] Oymak FS, Demirbas HM, Mavili E, Akgun H, Gulmez I, Demir R, et al. Bronchiolitis obliterans organizing pneumonia. Clinical and roentgenological features in 26 cases. Respiration 2005;72(3):254–62.

[8] King Jr TE. Organizing pneumonia. In: Interstitial lung disease. 5th ed. Shelton, CT: People's Medical Publishing House; 2011. p. 981–90.

[9] Colby TV. Pathologic aspects of bronchiolitis obliterans organizing pneumonia. Chest 1992;102(1 Suppl):38S–43S.

[10] Nicholson AG, Rice AJ. Interstitial lung diseases. In: Ph H, Flieder DB, editors. Spencer's pathology of the lung. 6th ed. Cambridge Medicine; 2016. p. 366–408.

[11] Radzikowska E, Wiatr E, Langfort R, Bestry I, Skoczylas A, Szczepulska-Wójcik E, et al. Cryptogenic organizing pneumonia. Results of treatment with clarithromycin versus corticosteroids. Observational study. PLoS One 2017;12(9):e0184739.

[12] Cordier JF. Cryptogenic organising pneumonia. Eur Respir J 2006;28(2):422–46.

[13] Epler GR. Bronchiolitis obliterans organizing pneumonia, 25 years: a variety of causes, but what are the treatment options? Expert Rev Respir Med 2011;5(3):353–61.

[14] Cordier JF. Cryptogenic organizing pneumonitis. Bronchiolitis obliterans organizing pneumonia. Clin Chest Med 1993;14(4):677–92.

[15] Epler GR, Colby TV, McLoud TC, Carrington CB, Gaensler EA. Bronchiolitis obliterans organizing pneumonia. N Engl J Med 1985;312(3):152–8.

[16] King TE, Mortenson RL. Cryptogenic organizing pneumonitis: the North American experience. Chest 1992;102(1 Suppl):8S–13S.

[17] Poletti V, Cazzato S, Minicuci N, Zompatori M, Burzi M, Schiattone ML. The diagnostic value of bronchoalveolar lavage and transbronchial lung biopsy in cryptogenic organizing pneumonia. Eur Respir J 1996;9(12):2513–6.

[18] Miao L, Wang Y, Li Y, Ding J, Chen L, Dai J, et al. Lesion with morphologic feature of organizing pneumonia (OP) in CT-guided lung biopsy samples for diagnosis of bronchiolitis obliterans organizing pneumonia (BOOP): a retrospective study of 134 cases in a single center. J Thorac Dis 2014;6(9):1251–60.

[19] Lee JW, Lee KS, Lee HY, Chung MP, Yi CA, Kim TS, et al. Cryptogenic organizing pneumonia: serial high-resolution CT findings in 22 patients. AJR Am J Roentgenol 2010;195(4):916–22.

[20] Müller NL, Staples CA, Miller RR. Bronchiolitis obliterans organizing pneumonia: CT features in 14 patients. AJR Am J Roentgenol 1990;154(5):983–7.

[21] Bouchardy LM, Kuhlman JE, Ball WC, Hruban RH, Askin FB, Siegelman SS. CT findings in bronchiolitis obliterans organizing pneumonia (BOOP) with radiographic, clinical, and histologic correlation. J Comput Assist Tomogr 1993;17(3):352–7.

[22] Gruden JF, Webb WR, Warnock M. Centrilobular opacities in the lung on high-resolution CT: diagnostic considerations and pathologic correlation. AJR Am J Roentgenol 1994;162(3):569–74.

[23] Akira M, Yamamoto S, Sakatani M. Bronchiolitis obliterans organizing pneumonia manifesting as multiple large nodules or masses. AJR Am J Roentgenol 1998;170(2):291–5.

[24] Johkoh T, Muller NL, Cartier Y, Kavanagh PV, Hartman TE, Akira M, et al. Idiopathic interstitial pneumonias: diagnostic accuracy of thin-section CT in 129 patients. Radiology 1999;211(2):555–60.

[25] Lee KS, Kullnig P, Hartman TE, Müller NL. Cryptogenic organizing pneumonia: CT findings in 43 patients. AJR Am J Roentgenol 1994;162(3):543–6.

[26] Kim SJ, Lee KS, Ryu YH, Yoon YC, Choe KO, Kim TS, et al. Reversed halo sign on high-resolution CT of cryptogenic organizing pneumonia: diagnostic implications. AJR Am J Roentgenol 2003;180(5):1251–4.

[27] Hansell DM, Bankier AA, MacMahon H, McLoud TC, Muller NL, Remy J. Fleischner Society: glossary of terms for thoracic imaging. Radiology 2008;246(3):697–722.

[28] Godoy MC, Viswanathan C, Marchiori E, Truong MT, Benveniste MF, Rossi S, et al. The reversed halo sign: update and differential diagnosis. Br J Radiol 2012;85(1017):1226–35.

[29] Jara-Palomares L, Gomez-Izquierdo L, Gonzalez-Vergara D, Rodriguez-Becerra E, Marquez-Martin E, Barrot-Cortés E, et al. Utility of high-resolution computed tomography and BAL in cryptogenic organizing pneumonia. Respir Med 2010;104(11):1706–11.

[30] Cottin V, Cordier J-F. Cryptogenic organizing pneumonia. Semin Respir Crit Care Med 2012;33(5):462–75.

[31] Chung MP, Nam BD, Lee KS, Han J, Park JS, Hwang JH, et al. Serial chest CT in cryptogenic organizing pneumonia: Evolutional changes and prognostic determinants. Respirology 2018;23(3):325–30.

[32] Meyer KC, Raghu G, Baughman RP, Brown KK, Costabel U, du Bois RM, et al. An official American Thoracic Society clinical practice guideline: the clinical utility of bronchoalveolar lavage cellular analysis in interstitial lung disease. Am J Respir Crit Care Med 2012;185(9):1004–14.

[33] Patel RA, Sellami D, Gotway MB, Golden JA, Webb WR. Hypersensitivity pneumonitis: patterns on high-resolution CT. J Comput Assist Tomogr 2000;24(6):965–70.

[34] Silva CI, Churg A, Müller NL. Hypersensitivity pneumonitis: spectrum of high-resolution CT and pathologic findings. AJR Am J Roentgenol 2007;188(2):334–44.

[35] Arakawa H, Kurihara Y, Niimi H, Nakajima Y, Johkoh T, Nakamura H. Bronchiolitis obliterans with organizing pneumonia versus chronic eosinophilic pneumonia: high-resolution CT findings in 81 patients. AJR Am J Roentgenol 2001;176(4):1053–8.

[36] Mehrian P, Doroudinia A, Rashti A, Aloosh O, Dorudinia A. High-resolution computed tomography findings in chronic eosinophilic vs. cryptogenic organising pneumonia. Int J Tuberc Lung Dis 2017;21(11):1181–6.

[37] Fauci AS, Haynes BF, Katz P, Wolff SM. Wegener's granulomatosis: prospective clinical and therapeutic experience with 85 patients for 21 years. Ann Intern Med 1983;98(1):76–85.

[38] Holle JU, Gross WL, Latza U, Nölle B, Ambrosch P, Heller M, et al. Improved outcome in 445 patients with Wegener's granulomatosis in a German vasculitis center over four decades. Arthritis Rheum 2011;63(1):257–66.

[39] Omar M, Goli S, Ramnarine I, Sakhamuri S. Organizing pneumonia: contemplate beyond cryptogenic. Am J Med 2018;131(3):e81–5.

[40] Beasley MB, Franks TJ, Galvin JR, Gochuico B, Travis WD. Acute fibrinous and organizing pneumonia: a histological pattern of lung injury and possible variant of diffuse alveolar damage. Arch Pathol Lab Med 2002;126(9):1064–70.

[41] Tzouvelekis A, Koutsopoulos A, Oikonomou A, Froudarakis M, Zarogoulidis P, Steiropoulos P, et al. Acute fibrinous and organizing pneumonia: a case report and review of the literature. J Med Case Rep 2009;3:74.

[42] Poletti V, Casoni GL. Cryptogenic organising pneumonia or acute fibrinous and organising pneumonia? Eur Respir J 2005;25(6):1128.

[43] Maldonad F, Daniels C, Hoffman E, Yi ES, Ryu JH. Focal organizing pneumonia on surgical lung biopsy causes, clinicoradiologic features, and outcomes. Chest 2007;132(5):1579–83.

[44] Nizami IY, Kissner DG, Visscher DW, Dubaybo BA. Idiopathic bronchiolitis obliterans with organizing pneumonia. An acute and life-threatening syndrome. Chest 1995;108(1):271–7.

[45] Lazor R, Vandevenne A, Pelletier A, Leclerc P, Court-Fortune I, Cordier JF. Cryptogenic organizing pneumonia: characteristics of relapses in a series of 48 patients. The Groupe d'Etudes et de Recherche sur les Maladies "Orphelines" Pulmonaires (GERM"O"P). Am J Respir Crit Care Med 2000;162(2 Pt 1):571–7.

[46] Nishino M, Mathai SK, Schoenfeld D, Digumarthy SR, Kradin RL. Clinicopathologic features associated with relapse in cryptogenic organizing pneumonia. Hum Pathol 2014;45(2):342–51.

[47] Pardo J, Panizo A, Sola I, Queipo F, Martinez-Penuela A, Carias R. Prognostic value of clinical, morphologic, and immunohistochemical factors in patients with bronchiolitis obliterans-organizing pneumonia. Hum Pathol 2013;44(5):718–24.

[48] Kim M, Cha SI, Seo H, Shin KM, Lim JK, Kim H, et al. Predictors of relapse in patients with organizing pneumonia. Tuberc Respir Dis (Seoul) 2015;78(3):190–5.

[49] Ding Q-L, Lv D, Wang B-J, Ma H, Zhang Q-L, Yu Y-M, et al. Macrolide therapy in cryptogenic organizing pneumonia: a case report and review. Exp Ther Med 2015;9(3):829–34.

[50] Ichikawa Y, Ninomiya H, Katusuki M, Hotta M, Tanaka M, Oizumi K. Long-dose/long-term erythromycin for treatment of bronchiolitis obliterans organizing pneumonia (BOOP). Kurume Med J 1993;40(2):65–7.

[51] Pathak V, Kuhn JM, Durham C, Funkhouser WK, Henke DC. Macrolide use leads to clinical and radiological improvement in patients with cryptogenic organizing pneumonia. Ann Am Thorac Soc 2014;11(1):87–91.

[52] Shitenberg D, Fruchter O, Fridel L, Kramer MR. Successful rituximab therapy in steroid-resistant, cryptogenic organizing pneumonia: a case series. Respiration 2015;90(2):155–9.

[53] Barešić M, Bakula M, Anić B. Polymyositis with pulmonary and renal involvement refractory to combined immunosuppressive therapy treated with rituximab. Clin Exp Rheumatol 2016;34(4):720.

Chapter 2.4

Desquamative interstitial pneumonia

Desquamative interstitial pneumonia (DIP) is classified as a smoking-related idiopathic interstitial pneumonia (IIP) and a respiratory bronchiolitis associated with interstitial lung disease [1]. The term DIP is based on initial histological description of desquamation of cells, predicted to be epithelial cells, in the alveolar lumen [2]. Subsequently the specific pathological sign of DIP was defined as accumulation of macrophages, sometimes accompanied with giant cells, in the terminal and respiratory bronchioles, alveoli, and alveolar ducts, resulting from an inflammatory response to inhalation of tobacco smoke or other pollutants [3], and the term DIP was retained albeit not reflecting the pathological changes in the respiratory compartment. DIP is a rarer form of IIP than idiopathic pulmonary fibrosis, nonspecific interstitial pneumonia (NSIP), and cryptogenic organizing pneumonia. Although accurate information is not available, DIP is considered to account for less than 3% of all IIP cases [4].

The onset of DIP is most often noted in patients over 40 years of age who have a long history of tobacco smoking (up to 90%) [5]. Professional exposure to mineral or organic dusts may also predispose to the development of DIP [6]; dusts from cobalt, beryllium, asbestos, coal, aflatoxin-containing materials during textile production, wood, and potato chips, among others, are described as possible inhalation triggers of DIP [7–13]. This wide range of air pollutants capable of inducing DIP indicates its resemblance to hypersensitivity pneumonitis (HP), a distinct histological form of inflammatory response with potentially similar pathogenic mechanisms.

DIP was also described in patients with connective tissue diseases (CTDs), marijuana smokers, and those treated with sirolimus and certain other drugs [14–16]. Thus DIP, like most IIP types, can be a respiratory manifestation of systemic inflammatory reaction. Of note the diagnosis of DIP is twice more common in males than females, although several studies failed to find sex differences [17].

Morphology

DIP is characterized by severe damage to the alveoli, including the alveolar epithelium and the interstitium of the alveolar septa (Figs. 2.4.1–2.4.4) [2]. Accumulation of a large number of hyperpigmented alveolar macrophages in the alveolar lumen and respiratory bronchioles is typical for early-stage DIP; occasional eosinophils, plasma and giant cells in the cellular mass, and moderate thickening of the interalveolar septa due to inflammatory infiltration can also be observed [2,18]. At later stages, interstitial fibrosis may occur, and lymphoid follicles associated with respiratory bronchioles or septa are frequently defined [19]. The appearance of myofibroblastic foci and honeycombing is not typical for DIP, and their presence may indicate the presence of usual interstitial pneumonia.

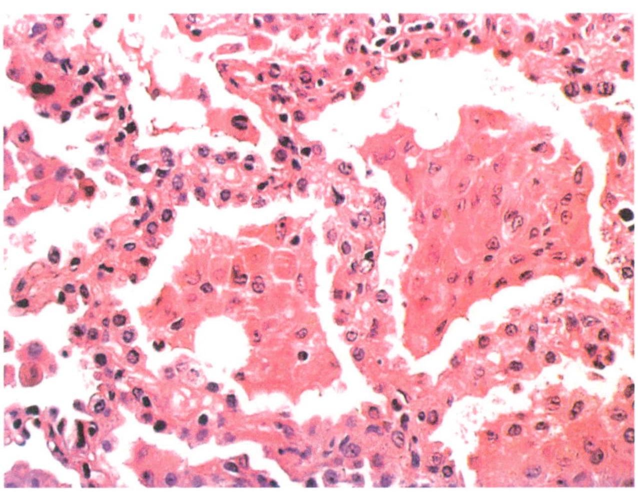

FIG. 2.4.1 Desquamative interstitial pneumonia. Alveolar lumens are filled with large macrophages with cytoplasmic *brown*-pigmented granules. Hyperplasia of alveolar cells. Hematoxylin and eosin staining, 400×.

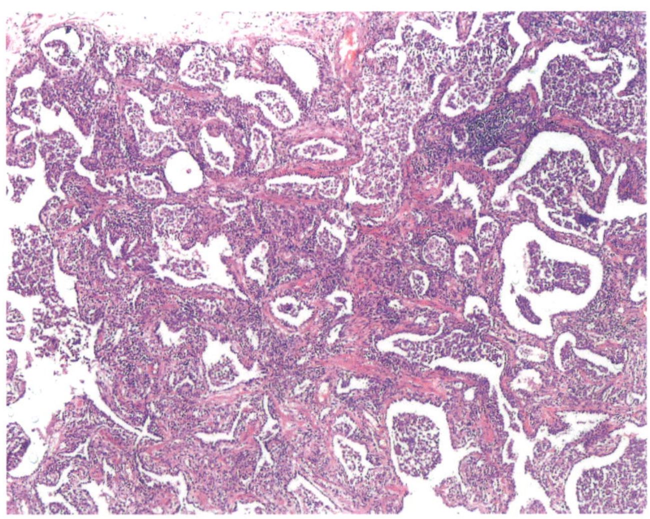

FIG. 2.4.2 Desquamative interstitial pneumonia. Formation of interstitial fibrosis and alveolar cell hyperplasia. Hematoxylin and eosin staining, 100×.

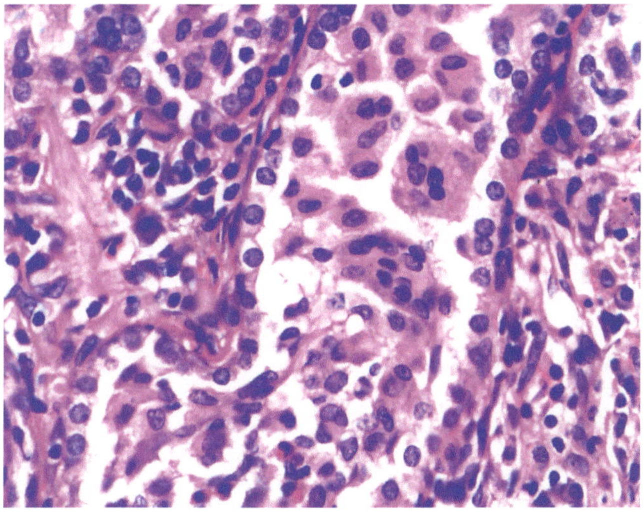

FIG. 2.4.3 Desquamative interstitial pneumonia. Alveolar lumens containing large macrophages with an admixture of single lymphoid elements and plasma cells. Alveolar cell hyperplasia. Hematoxylin and eosin staining, 600×.

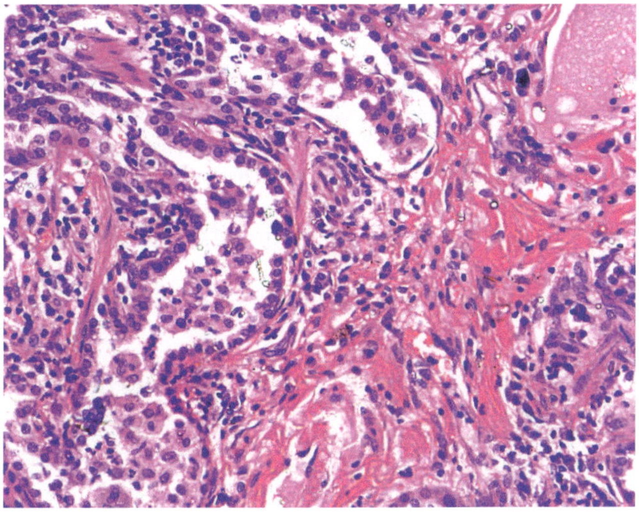

FIG. 2.4.4 Desquamative interstitial pneumonia with interstitial fibrosis and lymphohistiocytic infiltration of the interstitium. Hematoxylin and eosin staining, 400×.

DIP should be differentiated histologically with other interstitial pulmonary lesions notable for intra-alveolar accumulation of macrophages, namely, respiratory bronchiolitis–associated interstitial lung disease (RB-ILD), chronic eosinophilic pneumonia, and Langerhans cell histiocytosis. The histological manifestations of DIP resemble very closely to those of RB-ILD; however, the accumulation of macrophages in distal airway lumen, which exhibits a bronchiolocentric distribution in RB-ILD, is distributed diffusely and uniformly in DIP [8]. Additionally, lymphoid hyperplasia and signs of interstitial inflammation and fibrosis are not typical for RB-ILD [19].

The prevalence of eosinophils over macrophages is a characteristic of chronic eosinophilic pneumonia, which, unlike DIP, does not include brown pigmentation [19]. In Langerhans cell histiocytosis, peribronchial and periseptal granulomas containing histiocytes and eosinophils are observed in all cases [5].

Some phenotypes of pulmonary adenocarcinoma, as well as metastatic tumors (e.g., melanoma), can histologically manifest with intra-alveolar floating accumulation of tumor cells that mimic macrophages, which can lead to the erroneous diagnosis of DIP. In such cases, additional immunohistochemical analyses may be required to clarify the diagnosis [20,21].

Clinical presentation

The onset of DIP is usually slow. The first sign is dyspnea during physical exertion, which gradually progresses. The second most common symptom of DIP is dry or slowly productive cough [22]. Other nonspecific symptoms such as chest pain, hemoptysis, weight loss, and subfebrile fever can also be observed, albeit in a smaller percentage of patients with DIP [23]. Clubbing sign is observed in a substantial number of patients with DIP (25%–42%) [17,22]. By auscultation, end-inspiratory crackles are heard in the posterior-basal areas of the lungs of more than half of the patients [24]. However, up to 15% of patients with DIP have no obvious clinical manifestations of disease [22]. Although DIP progresses slowly in most cases, patients with rapid manifestation and a dramatic increase in symptoms that do not respond to steroid therapy and are fatal have also been described [8,25].

Diagnosis

Laboratory data of DIP are not specific. A study by Kawabata et al. [26] reported an average accelerated erythrocyte sedimentation rate of more than 50 mm/h and increased levels of lactate dehydrogenase (52%), total IgG (74%), and KL-6 (93%) in the serum of patients with DIP.

Regular spirometry provides minimal aid in establishing the diagnosis of DIP. Approximately 30% of patients show moderate restrictive changes [17]. Evaluation of the diffusion capacity of the lungs, which aids in assessing the status of the alveolar-capillary membrane, is of greater value. A decrease in this parameter is observed in most patients and generally reflects the prevalence of the pathological processes detected by radiological methods [17].

Cytological analysis of the bronchoalveolar lavage (BAL) fluid is not a highly specific diagnostic method. Usually, moderate eosinophilia (approximately 20%), neutrophilia (15%–30%), and an increase in the number of alveolar macrophages with brown inclusions are detected [27]. There is no clear consensus on the method of obtaining of histological material from patients with suspicious DIP. Surgical biopsy, albeit preferable over forceps transbronchial biopsy [28], should be performed only in patients with a progressive course of DIP; its value is significantly lower in mild and moderate forms of DIP with a favorable prognosis [29].

Transbronchial lung cryobiopsy (TBLC) facilitates the collection of tissue sufficient for the histological diagnosis of DIP. In a study by Dias et al., TBLC was effective in the diagnosis of all five patients with DIP [30]. However, although an apparent alternative to surgical biopsy, TBLC should be performed only at specialized centers with experience in such interventional procedures due to the high risk of complications.

High-resolution computed tomography

Due to its low incidence, high-resolution computed tomography (HRCT) findings of DIP in original studies are limited to approximately 40 patients per study [4]. The most significant HRCT sign observed in all patients with DIP is ground-glass opacity (GGO) (Figs. 2.4.5 and 2.4.6). The areas of GGO can be significantly diffuse or patchy and exhibit a predominant

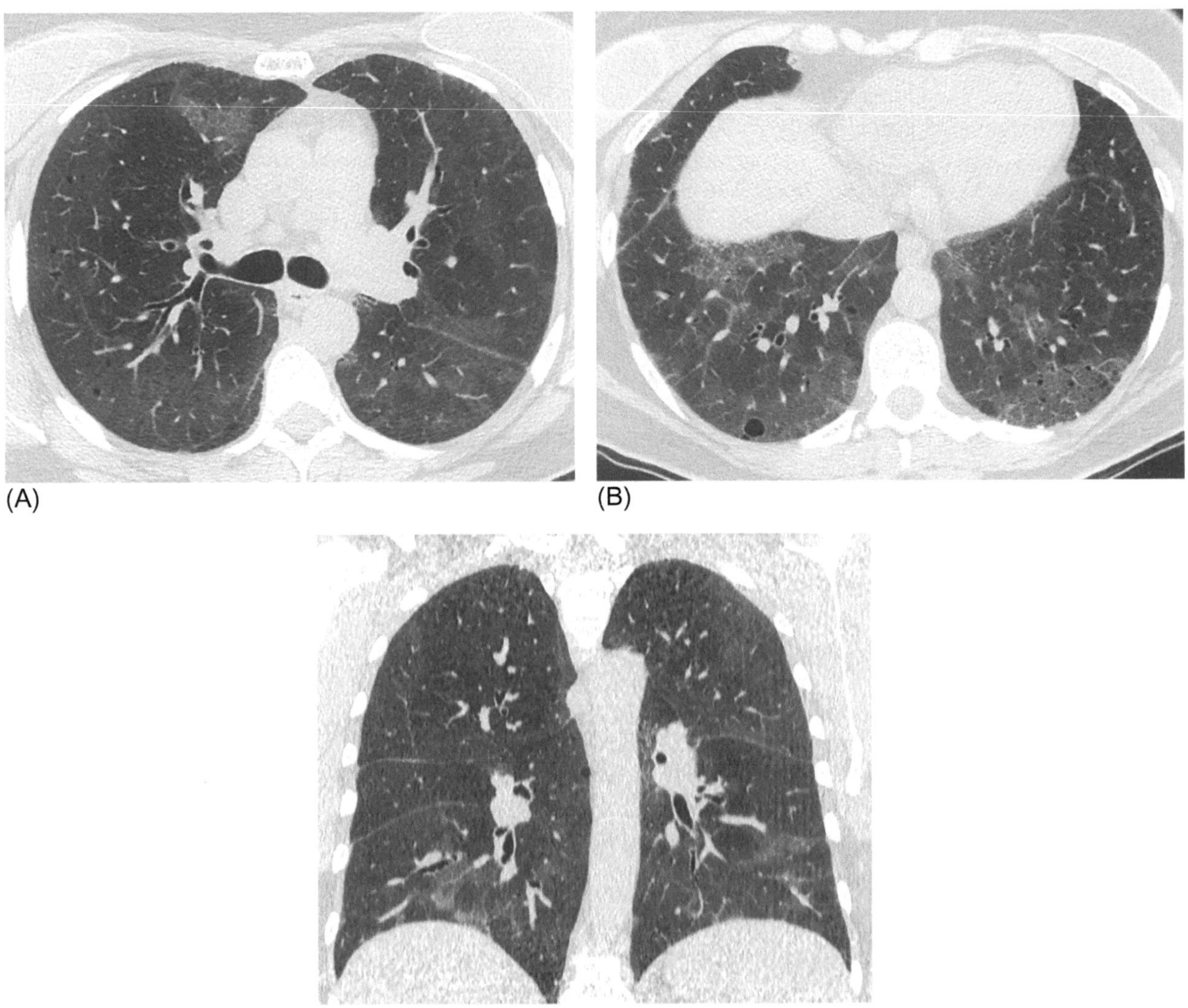

(A) (B) (C)

FIG. 2.4.5 Desquamative interstitial pneumonia. Bilateral patchy areas of ground-glass opacity, thickening of the bronchial walls (A and B). In posterior-basal areas inside the ground-glass opacity, small cysts are visualized on the right, and emphysema foci are observed on the left (B). Abnormalities are distributed mainly in the lower areas of the lungs (C).

bilateral localization in the basal and/or subpleural regions [31]. There is often a mosaic pattern of GGO distribution, with a clear border against the lobular areas of intact parenchyma, leading to an appearance resembling air traps (Fig. 2.4.6) [32].

Inside the areas of GGO, centrilobular emphysema, resembling bronchiolectasis can be visualized (Fig. 2.4.5), which was found in four of the eight patients included in a study by Akira et al. [33].

Irregular septal and extraseptal linear attenuation is also described in many patients with DIP [33]. Centrilobular nodules are observed frequently by HRCT in up to 45% of the patients [34]. In the Japanese population, small, thin-walled cysts were observed within the areas of GGO in 38%–75% of cases [26,33]; however, these HRCT findings were observed in less than 10% of cases in studies from Europe and the United States [17,31]. Cysts are usually thin-walled, their size does not exceed 2cm in diameter (Fig. 2.4.5B); additionally, cysts may disappear with complete restoration of the lung structure during treatment [33,35].

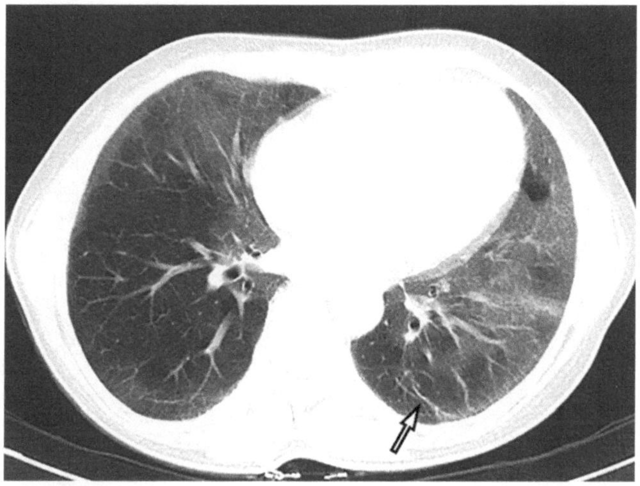

FIG. 2.4.6 Desquamative interstitial pneumonia. Bilateral diffuse areas of ground-glass opacity. Transseptal linear attenuation *(arrow)* and unaffected lobules resembling air traps are noticeable in the left lung lingula and the cardiodiaphragmatic angle of the right lung.

The subpleural areas of honeycombing in the basal regions are a possible but not frequent radiological finding that is observed primarily during the long-term course of DIP in patients who do not receive adequate therapy or in those who continue to be exposed to tobacco or did not fully respond to corticosteroids. Honeycombing in these patients does not exceed 10% of the entire volume of the lower lobes [31,36]. Among other radiological signs accompanying DIP, centriacinar or irregular emphysema and thickening of the bronchial walls reflect changes associated with smoking and are not directly associated with DIP [36].

Differential diagnosis

DIP should be considered in heavy smokers and those exposed to occupational or household pollutants that complain of slow progressive dyspnea and exhibit areas of GGO with no signs of fibrosis by chest HRCT. The differential diagnosis should include HP, NSIP, RB-ILD, and *Pneumocystis jirovecii* pneumonia (PP) (Table 2.4.1). NSIP is an interstitial lung disease that is significantly more common than DIP, and the cellular subtype of NSIP can resemble DIP clinically and radiologically. However, nonsmoking females are more likely to be diagnosed with NSIP, whereas DIP affects male smokers primarily. NSIP is often a secondary syndrome in patients with CTDs and drug-induced pulmonary lesions. Additionally the BAL fluid in NSIP demonstrates a moderate increase in lymphocyte and neutrophil fractions; conversely, brown-pigmented macrophages and eosinophils, and neutrophils to a lesser extent, are found in the BAL fluid of patients with DIP. Conversely, areas of GGO, which are found by HRCT in both NSIP and DIP, are more pronounced in the lower and peripheral areas. Reticular abnormalities of varying severity, thickening of the interlobular and intralobular septa in particular, are common HRCT signs of NSIP (Fig. 2.4.7) [37].

The HRCT signs specific for NSIP are subpleural sparing and the straight-edge sign, which are not registered in DIP. However, it is often impossible to differentiate between NSIP and DIP without a lung biopsy [38].

Subacute forms of HP are very difficult to differentiate from DIP if based solely on clinical and radiological patterns. Nodular or diffuse areas of GGO and lobular air traps with no obvious signs of fibrosis are also characteristic of subacute HP (Fig. 2.4.8) [39]. Several experts believe that differentiating HP and DIP only by HRCT characteristics is not possible [40]. Patients with HP are very frequently nonsmokers; in many cases the history includes regular contact with organic triggers of HP. Additionally the distribution of lesions in HP is usually more uniform, with occupation of the upper lobes, whereas areas of GGO are found not only in subpleural areas but also deep in the lung parenchyma [41]. An important criterion for the HP diagnosis is lymphocytosis above 50% in BAL fluid, which can be a decisive argument for HP in the absence of histological diagnostic confirmation.

Pneumonia caused by an infectious pathogen affecting mainly the interstitial tissue, especially PP, may have a HRCT pattern similar to that of DIP. Patchy or diffuse areas of GGO are also among the HRCT characteristics of PP; however, they are located primarily in the upper and middle fields of the lungs in contrast to DIP, in which they typically exhibit a basal distribution (Fig. 2.4.9).

TABLE 2.4.1 Differential diagnosis of desquamative interstitial pneumonia

	DIP	NSIP	HP	PP
Anamnesis	Heavy smoking, regular contact with inorganic pollutants	Primarily nonsmoking females, connective tissue disease, therapeutic drug intake	Contact with exogenous, often organic pollutants	HIV infection, intake of cytostatic drugs or steroids
Fever	Absent or subfebrile	Absent	Absent or subfebrile	>38°C
Blood tests	ESR, 40–60 mm/h	No abnormalities	Accelerated ESR	Lymphopenia ($CD4^+$ T-cell count <200 per μL), accelerated ESR
BAL	Eosinophilia, 15%–30%; brown-pigmented macrophages	Lymphocytosis >20%, moderate neutrophilia	Lymphocytosis >50%	+ PCR for *P. jirovecii*
GGO distribution	Basal and subpleural areas, often patchy	Basal and subpleural areas	Evenly or in the lower parts	Primarily upper and middle fields
Related changes on CT	Centrilobular nodules, emphysema	Subpleural sparing, reticular changes, straight-edge sign	Air traps, subpleural sparing, centrilobular nodules	Thickening of interlobular septa, intralobular nodules, air traps; consolidation areas with cavitation are possible

BAL, bronchoalveolar lavage; *CT*, computed tomography; *DIP*, desquamative interstitial pneumonia; *ESR*, erythrocyte sedimentation rate; *GGO*, ground-glass opacity; *HIV*, human immunodeficiency virus; *HP*, hypersensitivity pneumonitis; *NSIP*, nonspecific interstitial pneumonia; *PCR*, polymerase chain reaction; *PP*, *Pneumocystis jirovecii* pneumonia.

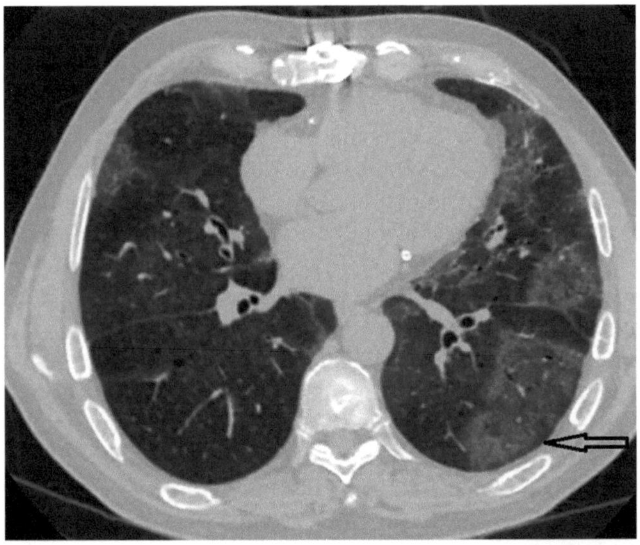

FIG. 2.4.7 Nonspecific interstitial pneumonia. Bilateral patchy areas of ground-glass opacity associated with reticular abnormalities. On the left, subpleural sparing is clearly visualized as a specific sign of nonspecific interstitial pneumonia *(arrow)*.

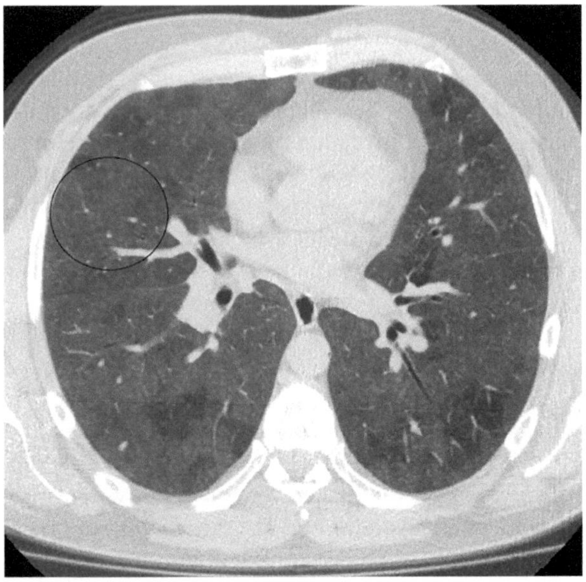

FIG. 2.4.8 Subacute hypersensitivity pneumonitis in a bird fancier. Diffuse areas of ground-glass opacity associated with air traps. Within areas of ground-glass opacity, multiple small ill-defined centrilobular nodules *(circle)* can be distinguished.

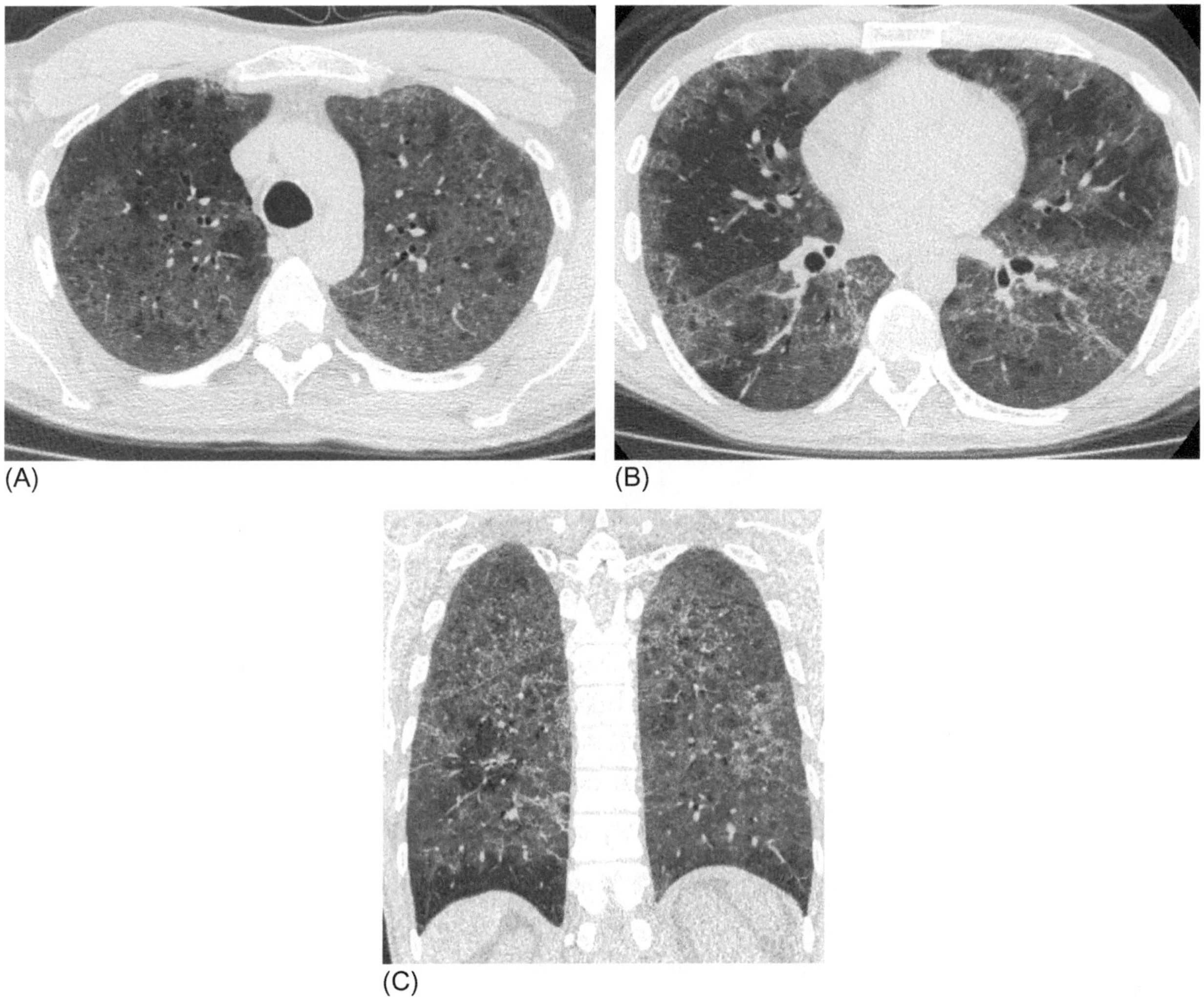

FIG. 2.4.9 *Pneumocystis jirovecii* pneumonia in a patient with acquired immunodeficiency syndrome. Bilateral diffuse (A) and patchy (B) areas of ground-glass opacity associated with air traps and reticular abnormalities. Basal segments are affected less significantly (C).

In PP, thickened interlobular septa are often found within the areas of GGO, which can create the crazy-paving pattern. Foci of consolidation can also be present, and thin-walled cysts in areas of interstitial inflammation can be detected in patients with long-term disease [42]. During differential diagnosis, it should be considered that PP develops in the presence of pronounced immunodeficiency and/or lymphopenia, often in combination with other opportunistic infections such as herpes zoster infection and oropharyngeal candidiasis. As a rule, these patients have pronounced fever accompanying dyspnea. The definitive diagnosis of PP relies on the visualization of pneumocystis in the sputum or the BAL fluid or on a positive polymerase chain reaction for *P. jirovecii*.

RB-ILD is very closely related to DIP; the two diagnoses are often combined together into one nosological group, since they share the common risk factor of smoking and exhibit similar clinical and radiographic manifestations. However, DIP usually presents with more pronounced symptoms and inflammatory changes in the blood. In addition, several HRCT findings differ between DIP and RB-ILD, including less pronounced areas of GGO that are more likely to be distributed in the upper lobes (Fig. 2.4.10) in RB-ILD, which tend to exhibit a basal distribution in DIP [43].

Treatment and prognosis

Since there is a direct relationship of smoking and environmental factors with the development of DIP, smoking cessation and termination of contact with pollutants is essential for treatment.

All studies conducted in patients with DIP confirm that systemic corticosteroids are highly efficient in 80%–90% of patients. Most studies use medium doses (0.5–0.75 mg/kg) of prednisone per day for 2–4 weeks, followed by a slow, stepwise decrease, sometimes with preceding pulse therapy [17,26]. However, these dosages have not been strictly established by

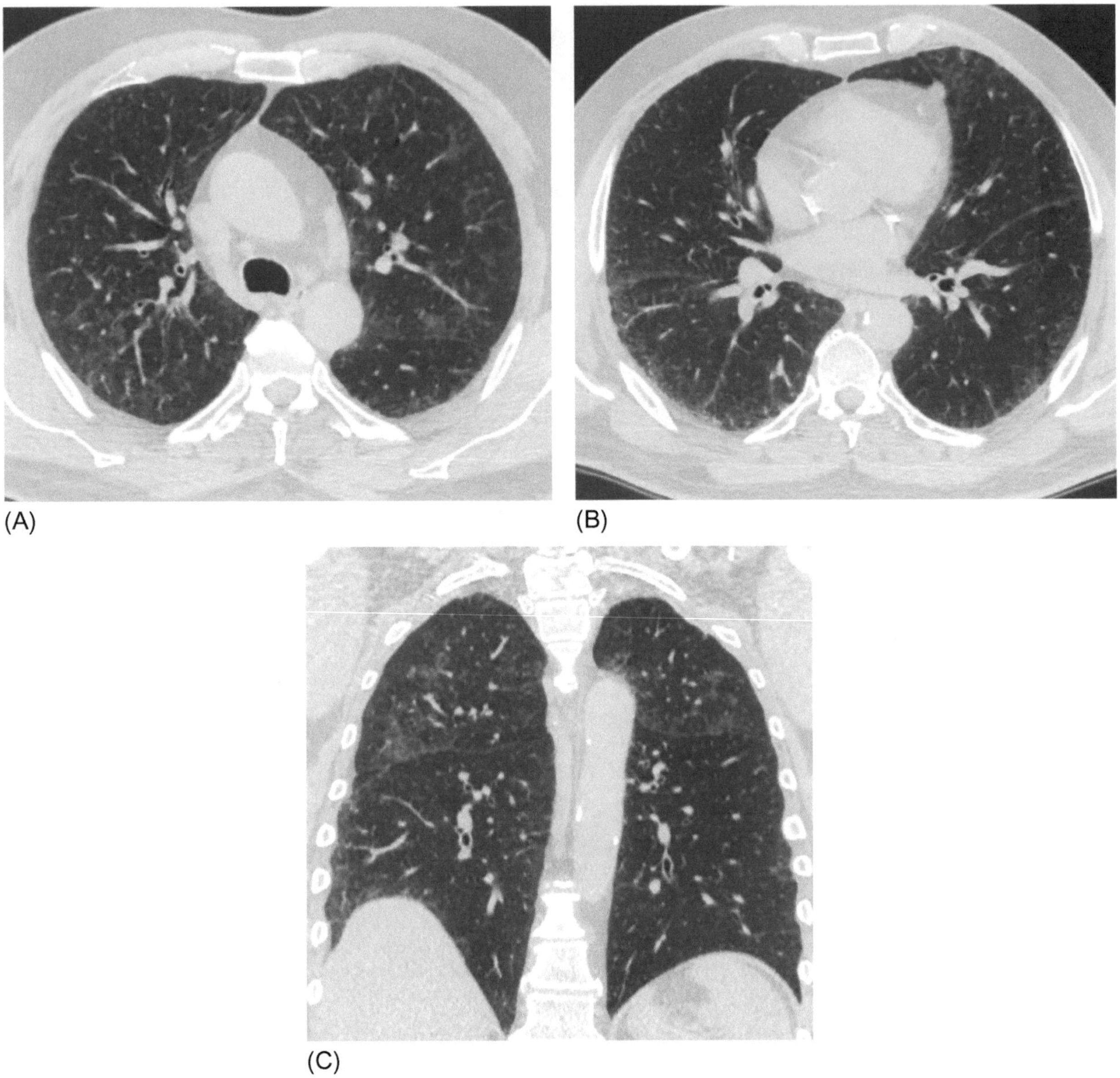

FIG. 2.4.10 Respiratory bronchiolitis associated with interstitial lung disease. Bilateral foci of ground-glass opacity, moderate reticular changes, thickened bronchial walls. Multiple small nodules and single Y-shaped structures (A and B) are visualized subpleurally. The abnormalities are visualized maximally in the upper lobes (C).

randomized clinical trials; therefore variations in one direction or another are acceptable, depending on the severity of the clinical and radiological manifestations.

The prognosis of DIP is favorable, with more than 90% 5-year rates, which corresponds approximately to the proportion of patients responding to steroid therapy [5]. Cases of spontaneous remission of DIP with smoking cessation or termination of contact with environmental factors have also been described [44].

Despite the generally favorable course of DIP and the possibility of complete discontinuation of steroids, a study by Kawabata et al. that followed up patients for 5 years reported that mild honeycombing developed in approximately 36% of the patients [26].

References

[1] Travis WD, Costabel U, Hansell DM, King Jr TE, Lynch DA, Nicholson AG, et al. An official American Thoracic Society/European Respiratory Society statement: update of the international multidisciplinary classification of the idiopathic interstitial pneumonias. Am J Respir Crit Care Med 2013;188(6):733–48.

[2] Liebow AA, Steer A, Billingsley JG. Desquamative interstitial pneumonia. Am J Med 1965;39:369–404.
[3] Godbert B, Wissler M-P, Vignaud J-M. Desquamative interstitial pneumonia: an analytic review with an emphasis on aetiology. Eur Respir Rev 2013;22(128):117–23.
[4] Carrington CB, Gaensler EA, Coutu RE, Fitzgerald MX, Gupta RG. Usual and desquamative interstitial pneumonia. Chest 1976;69(2 Suppl):261–3.
[5] Vassalo R, Ryu JH. Smoking-related interstitial lung diseases. Clin Chest Med 2012;33(1):165–78.
[6] Abraham JL, Hertzberg MA. Inorganic particles associated with desquamative interstitial pneumonia. Chest 1981;80(1 Suppl):67–70.
[7] Nemery B. Metal toxicity and the respiratory tract. Eur Respir J 1990;3(2):202–19.
[8] Craig PJ, Wells AU, Doffman S, Rassl D, Colby TV, Hansell DM, et al. Desquamative interstitial pneumonia, respiratory bronchiolitis and their relationship to smoking. Histopathology 2004;45(3):275–82.
[9] Freed JA, Miller A, Gordon RE, Fischbein A, Kleinerman J, Langer AM. Desquamative interstitial pneumonia associated with chrysotile asbestos fibres. Br J Ind Med 1991;48(5):332–7.
[10] Jelic TM, Estalilla OC, Sawyer-Kaplan PR, Plata MJ, Powers JT, Emmett M, et al. Coal mine dust desquamative chronic interstitial pneumonia: a precursor of dust-related diffuse fibrosis and of emphysema. Int J Occup Environ Med 2017;8(3):153–65.
[11] Lougheed MD, Roos JO, Waddell WR, Munt PW. Desquamative interstitial pneumonitis and diffuse alveolar damage in textile workers. Potential role of mycotoxins. Chest 1995;108(5):1196–200.
[12] Moon J, du Bois RM, Colby TV, Hansell DM, Nicholson AG. Clinical significance of respiratory bronchiolitis on open lung biopsy and its relationship to smoking related interstitial lung disease. Thorax 1999;54(11):1009–14.
[13] Kroll RR, Flood DA, Srigley J. Desquamative interstitial pneumonitis in a healthy non-smoker: a rare diagnosis. Can Respir J 2014;21(2):86–8.
[14] Nicholson AG, Colby TV, Wells AU. Histopathological approach to patterns of interstitial pneumonia in patient with connective tissue disorders. Sarcoidosis Vasc Diffuse Lung Dis 2002;19(1):10–7.
[15] Gill A. Bong lung: regular smokers of cannabis show relatively distinctive histologic changes that predispose to pneumothorax. Am J Surg Pathol 2005;29(7):980–2.
[16] Flores-Franco RA, Luevano-Flores E, Gaston-Ramirez C. Sirolimus associated desquamative interstitial pneumonia. Respiration 2007;74(2):237–8.
[17] Ryu JH, Myers JL, Capizzi SA, Douglas WW, Vassallo R, Decker PA. Desquamative interstitial pneumonia and respiratory bronchiolitis-associated interstitial lung disease. Chest 2005;127(1):178–84.
[18] Nicholson AG, Rice AJ. Interstitial lung diseases. In: Hasleton P, Flieder DB, editors. Spencer's pathology of the lung. 6th ed. New York: Cambridge Medicine; 2016. p. 366–408.
[19] Tazelaar HD, Wright JL, Churg A. Desquamative interstitial pneumonia. Histopathology 2011;58(4):509–16.
[20] Raparia K, Ketterer J, Dalurzo ML, Chang YH, Colby TV, Leslie KO. Lung tumors masquerading as desquamative interstitial pneumonia (DIP): report of 7 cases and review of the literature. Am J Surg Pathol 2014;38(7):921–4.
[21] Miyaoka M, Hatanaka K, Iwazaki M, Nakamura N. Pulmonary adenocarcinoma mimicking desquamative interstitial pneumonia: report of 2 cases with genetic analysis. Int J Surg Pathol 2018;26(7):655–9.
[22] Yousem SA, Colby TV, Gaensler EA. Respiratory bronchiolitis associated interstitial lung disease and its relationship to desquamative interstitial pneumonia. Mayo Clin Proc 1989;64(11):1373–80.
[23] Diken ÖE, Şengül A, Beyan AC, Ayten Ö, Mutlu LC, Okutan O. Desquamative interstitial pneumonia: risk factors, laboratory and bronchoalveolar lavage findings, radiological and histopathological examination, clinical features, treatment and prognosis. Exp Ther Med 2019;17(1):587–95.
[24] Mason RJ, Murray JF, Broaddus VC, Nadel JA. Textbook of respiratory medicine. 4th ed. USA: Elsevier Inc; 20051597–9.
[25] Gould TH, Buist MD, Meredith D, Thomas PD. Fulminant desquamative interstitial pneumonitis. Anaesth Intensive Care 1998;26(6):677–9.
[26] Kawabata Y, Takemura T, Hebisawa A, Sugita Y, Ogura T, Nagai S, et al. Desquamative interstitial pneumonia may progress to lung fibrosis as characterized radiologically. Respirology 2012;17(8):1214–21.
[27] Kawabata Y, Tamkemura T, Hebisawa A, Ogura T, Yamaguchi T, Kuriyama T, et al. Eosinophilia in bronchoalveolar lavage fluid and architectural destruction are features of desquamative interstitial pneumonia. Histopathology 2008;52(2):194–202.
[28] Kumar A, Cherian SV, Vassallo R, Yi ES, Ryu JH. Current concepts in pathogenesis, diagnosis, and management of smoking-related interstitial lung diseases. Chest 2018;154(2):394–408.
[29] Kligerman S, Franks TJ, Galvin JR. Clinical-radiologic-pathologic correlation of smoking-related diffuse parenchymal lung disease. Radiol Clin N Am 2016;54(6):1047–63.
[30] Dias C, Mota P, Neves I, Guimarães S, Souto Moura C, Morais A. Transbronchial cryobiopsy in the diagnosis of desquamative interstitial pneumonia. Rev Port Pneumol (2006) 2016;22(5):288–90.
[31] Hartman TE, Primack SL, Swensen SJ, Hansell D, McGuinness G, Müller NL. Desquamative interstitial pneumonia: thin-section CT findings in 22 patients. Radiology 1993;187(3):787–90.
[32] Wittram C, Mark EJ, McLoud TC. CT-histologic correlation of the ATS/ERS 2002 classification of idiopathic interstitial pneumonias. Radiographics 2003;23(5):1057–71.
[33] Akira M, Yamamoto S, Hara H, Sakatani M, Ueda E. Serial computed tomographic evaluation in desquamative interstitial pneumonia. Thorax 1997;52(4):333–7.
[34] Johkoh T, Muller NL, Cartier Y, Hartman TE, Akira M, Ichikado K, et al. Idiopathic interstitial pneumonias: diagnostic accuracy of thin-section CT in 129 patients. Radiology 1999;211(2):555–60.
[35] Koyama M, Johkoh T, Honda O, Tsubamoto M, Kozuka T, Tomiyama N, et al. Chronic cystic lung disease: diagnostic accuracy of high-resolution CT in 92 patients. AJR Am J Roentgenol 2003;180:827–35.
[36] Pandit-Bhalla M, Diethelm L, Ovella T, Sloop GD, Valentine VG. Idiopathic interstitial pneumonias: an update. J Thorac Imaging 2003;18(1):1–13.
[37] Johkoh T, Muller NL, Colby TV, Ichikado K, Taniguchi H, Kondoh Y, et al. Nonspecific interstitial pneumonia: correlation between thin-section CT findings and pathologic subgroups in 55 patients. Radiology 2002;225(1):199–204.

[38] Lynch DA, Travis WD, Müller NL, Galvin JR, Hansell DM, Grenier PA, et al. Idiopathic interstitial pneumonias: CT features. Radiology 2005;236(1):10–21.
[39] Lacasse Y, Girard M, Cormier Y. Recent advances in hypersensitivity pneumonitis. Chest 2012;142(1):208–17.
[40] Lynch DA, Newell JD, Logan PM, King Jr TE, Müller NL. Can CT distinguish hypersensitivity pneumonitis from idiopathic pulmonary fibrosis? AJR Am J Roentgenol 1995;165(4):807–11.
[41] Lacasse Y, Selman M, Costabel U, Dalphin JC, Ando M, Morell F, et alHP Study Group. Clinical diagnosis of hypersensitivity pneumonitis. Am J Respir Crit Care Med 2003;168(8):952–8.
[42] Kanne JP, Yandow DR, Meyer CA. Pneumocystis jiroveci pneumonia: high-resolution CT findings in patients with and without HIV infection. AJR Am J Roentgenol 2012;198(6):W555–61.
[43] Webb WR, Higgins CB. Thoracic imaging: pulmonary and cardiovascular radiology. 2nd ed. Philadelphia: Lippincott Williams & Wilkins; 2011443–5.
[44] Matsuo K, Tada S, Kataoka M, Okahara M, Hiramatsu J, Horiba M, et al. Spontaneous remission of desquamative interstitial pneumonia. Intern Med 1997;36(10):728–31.

Chapter 2.5

Respiratory bronchiolitis–associated interstitial lung disease

Respiratory bronchiolitis–associated interstitial lung disease (RB-ILD) is a form of idiopathic interstitial pneumonia characterized by the nodular accumulation of brown macrophages within and around respiratory bronchioles with simultaneous peribronchiolar inflammation and fibrosis [1,2]. RB-ILD must be differentiated from isolated respiratory bronchiolitis (RB), one of the histological markers of active smokers' lungs that are asymptomatic and are in almost all individuals with a long smoking history [3]. The respiratory bronchioles of RB contain pigmented macrophages but without signs of fibrosis in the acini. RB is not considered an independent disease, and it undergoes regression after smoking cessation [3]. RB-ILD and desquamative interstitial pneumonia (DIP) both belong to the group of idiopathic interstitial pneumonia associated with smoking [4]. Individuals who have never smoked (or who have not been exposed to other forms of smoke) do not present RB-ILD [5]. Although the occurrence of the disease is usually associated with a high smoking index (>30 packs/year), the smoking intensity is not directly associated with the severity of the pathological process in RB-ILD [3]. The spread of vaping has led to the appearance of RB-ILD induced by the use of electronic cigarettes [6], indicating the possible toxic effect of nicotine and not just of combustion products in the disease genesis.

Some researchers believe that RB-ILD is a milder version of DIP, but this form is considered independent in the modern classification of IIP because it has differences not only in the course and prognosis but also in its radiological and morphological patterns [1].

No accurate data on the epidemiology of RB-ILD exist, due to the significant amount of suspected asymptomatic cases. However, RB-ILD is found in 6% of interstitial lung diseases (ILD) according to surgical biopsies [2]. In the ILD register of a German clinic, the RB-ILD frequency is also approximately 6% [7]. Most patients with RB-ILD are men aged 30–50 years [8].

Morphology

Histologically, RB-ILD is characterized by a combination of chronic bronchiolitis (Figs. 2.5.1 and 2.5.2) with moderate peribronchiolar lymphohistiocytic interstitial inflammation filling bronchiolar and alveoli lumens with dusty brown macrophages [1]. In addition to cells, the respiratory bronchioles can be filled with mucus, and mild peribronchiolar fibrosis can also be present (Fig. 2.5.3). The bronchioles may contain patches of mucosal hyperplasia, squamous metaplasia, and dysplasia (Fig. 2.5.4).

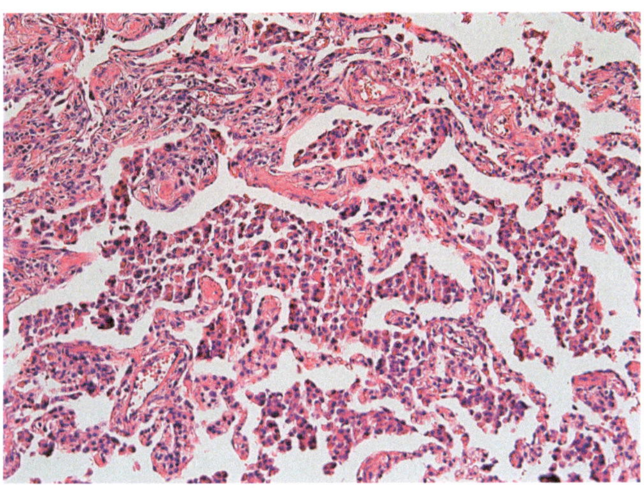

FIG. 2.5.1 Respiratory bronchiolitis. Accumulation of polynuclear macrophages with *brown pigment* in the lumen of alveoli. Hematoxylin and eosin staining, 100×.

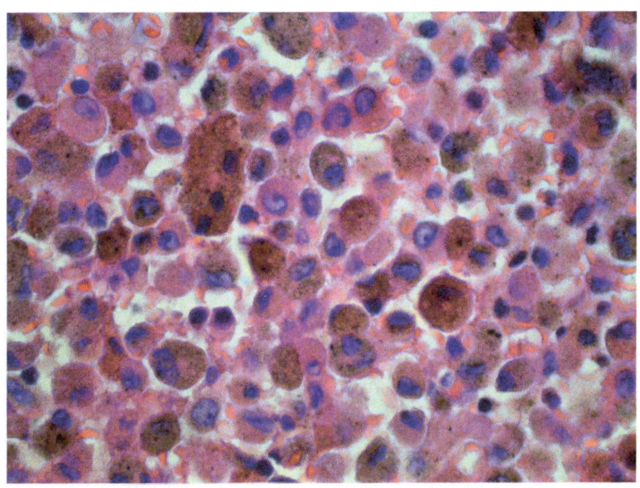

FIG. 2.5.2 Respiratory bronchiolitis. Single- and polynuclear macrophages with *brown pigment* in the lumen of the alveoli. Hematoxylin and eosin staining, 600×.

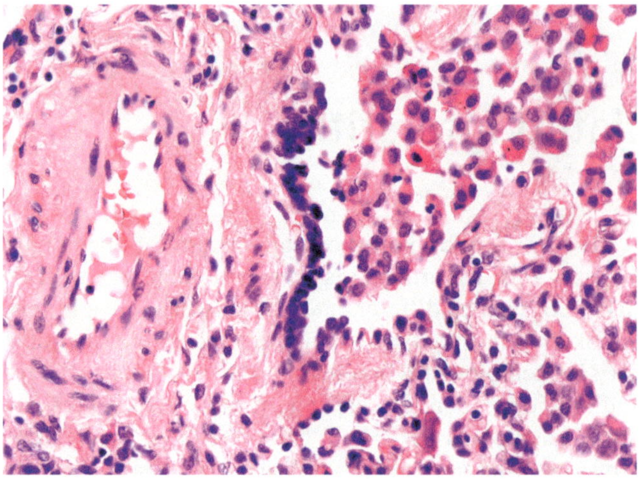

FIG. 2.5.3 Respiratory bronchiolitis associated with interstitial lung disease. Macrophages accumulated in central bronchioles. Peribronchiolar interstitial fibrosis. Sclerosis and hyalinosis of the walls of small branches of the pulmonary artery. Hematoxylin and eosin staining, 400×.

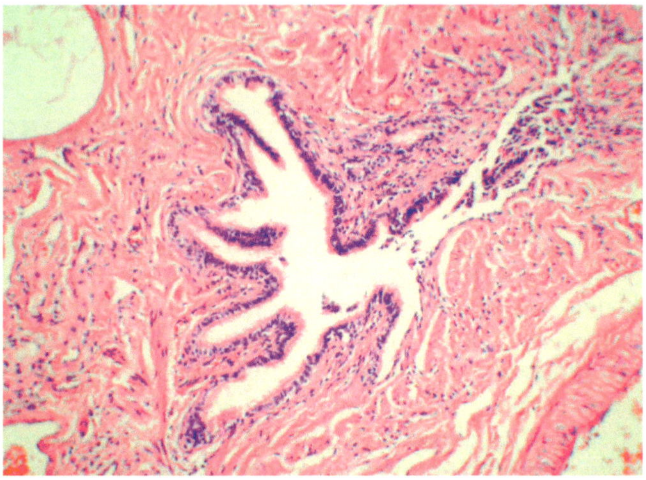

FIG. 2.5.4 Respiratory bronchiolitis associated with interstitial lung disease. Severe sclerosis of the walls of the respiratory bronchiole and nodule of squamous metaplasia of the bronchiolar epithelium. Hematoxylin and eosin staining, 200×.

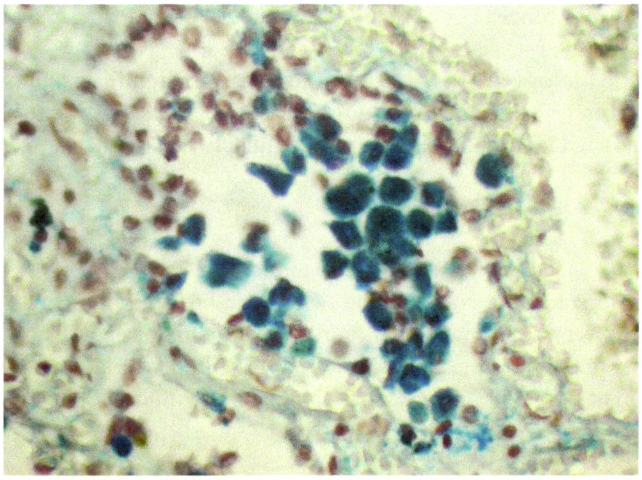

FIG. 2.5.5 Respiratory bronchiolitis associated with interstitial lung disease. Single- and polynuclear macrophages with smoker pigment in the lumen of the bronchiole. Perls reaction, 600×.

RB-ILD is very similar to DIP, but its macrophage reaction is more pronounced in the center of bronchioles, and the uniform lesions in alveolar spaces that are typical of DIP are not present [9]. The brown cytoplasmic pigment that fills macrophages is a product of phagocytosis of tobacco smoke particles and is similar to hemosiderin; therefore RB-ILD needs to be differentiated from alveolar hemorrhages characterized by clusters of sideroblasts and siderophages in the areas of hemorrhage, the presence of vasculitis, and other vascular lesions [10]. Perls' Prussian blue stains both the pigment of smokers and hemosiderin, so it cannot help differentiate them (Fig. 2.5.5) [11]. However, the smoker's pigment is usually present as fine pale-stained granules, while hemosiderin is darker and present in coarse granules [12].

Clinical presentation

The clinical presentation of RB-ILD is nonspecific with a slow or subacute onset, although rare cases of acute onset have been described [13]. Common manifestations include cough with sputum, moderately progressive dyspnea during physical exertion, and wheezing [8]. Less common symptoms include hemoptysis, weight loss, chest pain, or mild fever [2]. A significant proportion of patients has no clinical symptoms or does not attach importance to them, and the disease is accidentally detected during CT examinations. Auscultation of the lungs in approximately half of the patients reveals fine end-inspiratory crackles [14]. Concomitant chronic bronchitis can lead to dry expiratory rales. The appearance of the clubbing symptom in RB-ILD patients is not typical, although it may be present in patients with severe forms of the disease [15].

Diagnosis

Functional tests usually reflect the degree of involvement of the small airways in the process, and the changes range from normal to moderately pronounced obstructive and/or restrictive. In a study by Portnoy et al., 47% of patients showed moderate obstructive patterns, 31% had restrictive patterns, 9% had mixed disorders, and 13% had normal pulmonary function [8]. The measurement of the diffusion capacity of the lungs is an important study assessing the degree of respiratory disorders and monitoring the course of the disease; a diminished diffusion capacity is seen in the majority of patients with RB-ILD [14]. Patients with RB-ILD often have concomitant chronic obstructive pulmonary disease (COPD) and emphysema; therefore the interpretation of functional parameters to assess the severity of the process, including DLCO, must consider the disorders unrelated to the interstitial involvement.

Bronchoalveolar lavage is not critical for RB-ILD diagnosis, given the disease's characteristic increase in the total number of macrophages, especially pigmented ones, which is typical for all smoker patients [16]. However, the absence of such a finding after a correctly performed BAL makes the diagnosis of RB-ILD unlikely [1].

Transbronchial lung biopsy (TBLB) usually does not result in samples of tissue of sufficient size for diagnosis of RB-ILD, but may be useful to rule out other interstitial lung diseases (hypersensitivity pneumonitis and sarcoidosis) [17].

The benefits of *transbronchial cryobiopsy* have been demonstrated only in single patients, but this method may be an alternative to the surgical biopsy that is considered the most effective procedure for obtaining histological material [18]. Surgical biopsy should not be performed in patients with suspected RB-ILD, in the absence of a threat of rapid progression of the disease, or of severe pulmonary dysfunction, when the patient is motivated to stop smoking [1].

Radiological pattern

Radiography of the chest usually does not help in establishing the correct diagnosis. Most often, thickening of the walls of the central and peripheral bronchi is detected as a sign associated with smoking, and a slight increased attenuation of the lung parenchyma with diffuse distribution may be present. In 20% of patients, radiographs lack abnormal findings [19].

The high-resolution CT (HRCT) signs characteristic for RB-ILD include poorly differentiated centrilobular nodules and limited patches of ground-glass opacity (GGO) (Figs. 2.5.6 and 2.5.7). The intensity of GGO correlates with the degree of accumulation of macrophages in the alveolar spaces and alveolar ducts [20,21]. The distribution of GGO areas is usually bilateral and can be localized in both the upper and lower lung fields [22].

Centrilobular nodules, reflecting the macrophage response in the respiratory bronchioles, have a predominant pattern of localization in the upper lobes, with greater severity in the peripheral zones [19]. Moderately pronounced reticular

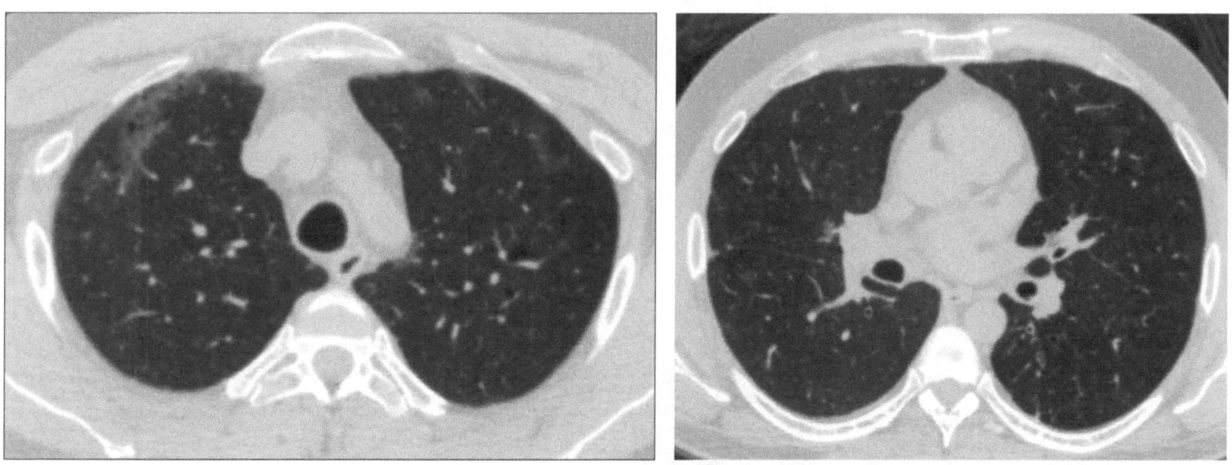

(A) (B)

FIG. 2.5.6 Respiratory bronchiolitis associated with interstitial lung disease. Bilateral small patchy areas of GGO and poorly differentiated nodules. Thickened bronchial walls (A and B). Abnormalities are most pronounced in the upper lobes. Foci of interstitial emphysema (A) are also visible.

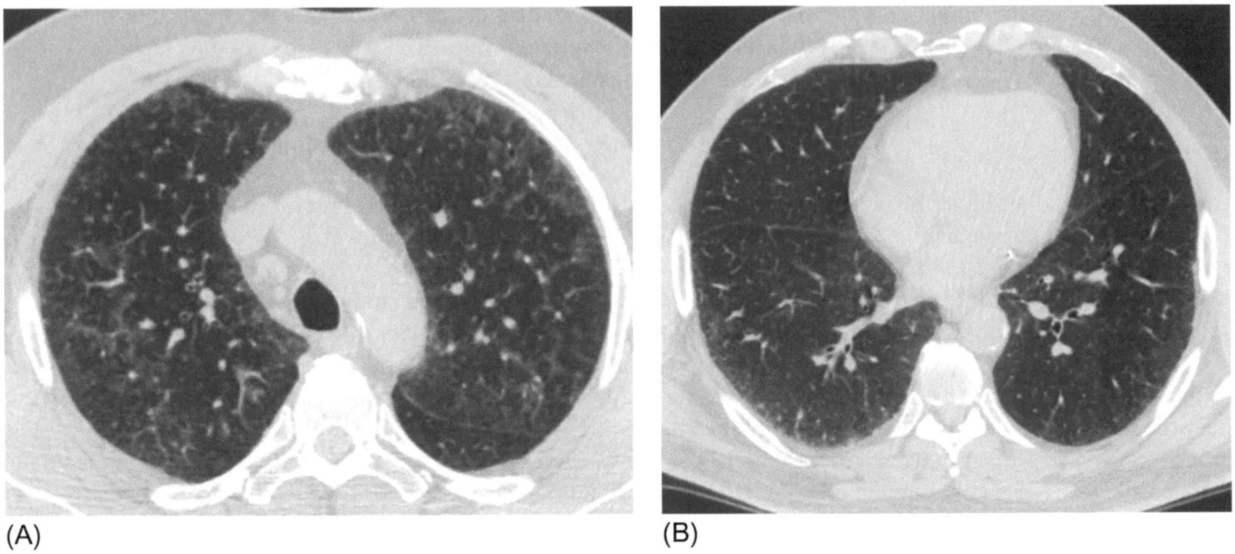

FIG. 2.5.7 Respiratory bronchiolitis associated with interstitial lung disease. Bilateral GGO foci, multiple centriacinar, and ill-defined nodules. Moderate reticular abnormalities (A and B). Predominant upper lobe distribution of pathological changes (A).

abnormalities are possible, but they do not dominate the common HRCT pattern [23]. The tree-in-bud sign, characteristic of infectious bronchiolitis, is rare in patients with RB-ILD [21]. The appearance of consolidation and honeycombing and traction bronchiectasis is atypical in patients with RB-ILD and helps differentiate them from patients with fibrosing types of interstitial pneumonia [23].

Additional frequent findings in patients with RB-ILD include thickening of the bronchial walls (in up to 90% of patients), centriacinar emphysema, and air traps that reflect an independent underlying disease, COPD or chronic bronchitis [24].

After smoking cessation the HRCT pattern improves, and abnormalities such as GGO and nodules decrease or disappear within a few months, but the signs of hyperinflation may increase [24].

Differential diagnosis

If RB-ILD is suspected, the differential diagnosis should include such diseases as DIP, hypersensitivity pneumonitis (HP), onset forms of nonspecific interstitial pneumonia (NSIP), and *Pneumocystis jirovecii* pneumonia (Chapter 2.4, Table 2.4.1).

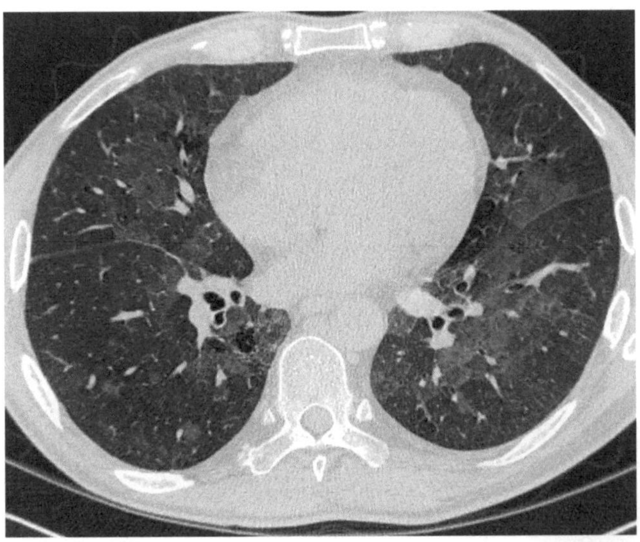

FIG. 2.5.8 Desquamative interstitial pneumonia. Patchy, merging areas of ground-glass opacity and mild reticular changes in the lower segments of both lungs.

RB-ILD is difficult to differentiate from DIP because both diseases have similar clinical and radiological characteristics and due to the possibility of the parallel presence in one patient of the histological signs of both diseases in different parts of the lungs [25]. In approximately half of the cases of DIP on HRCT, no centrilobular nodules are found, but in patients with RB-ILD, this sign is almost essential [26]. The prevalence and intensity of GGO in patients with DIP is higher than those in patients with RB-ILD, in cases with a severe course; clinically milder variants of DIP, as well as of RB-ILD, may have a very close HRCT pattern (Fig. 2.5.8). The appearance of honeycombing practically rules out the diagnosis of RB-ILD but may occur in a small number of DIP patients [27]. Finally the upper lobe distribution of abnormalities is often present in patients with RB-ILD and can help differentiate it from DIP, whereas in patients with DIP, most pathological changes occur in the lower zones [28]. The detection of histological RB signs (a frequent finding in TBLB) is not a differential sign of these diseases because it reflects only a specific morphological sign in smokers or ex-smokers.

HP may have a HRCT pattern similar to that of RB-ILD in the form of centrilobular nodules and patchy GGO areas (Fig. 2.5.9). However, HP usually develops in nonsmokers and results in a high level of BAL lymphocytosis. In addition, a history of professional or household inhalation contact with potentially dangerous allergens is usually traced.

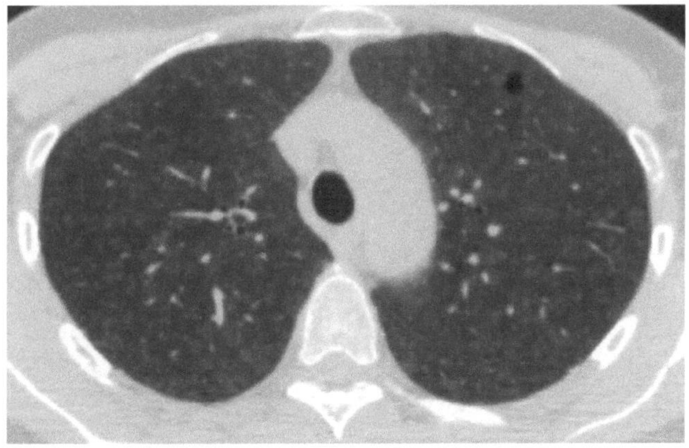

FIG. 2.5.9 Hypersensitivity pneumonitis in a bird fancier. Multiple small poorly defined centriacinar nodules.

The cell subtype of NSIP, especially during the onset of the disease or after treatment with steroids, may resemble that of RB-ILD in the radiological picture (Fig. 2.5.10). Never smokers (70%) and mainly women (2/3 of cases) predominate among NSIP patients [29]. The typical centrilobular nodules of RB-ILD are not usually present in NSIP, and GGOs are mostly present in the lower fields in NSIP. In addition, a specific, but not permanent, sign of NSIP is subpleural sparing; and as a rule, pronounced reticular changes occur. RB-ILD can be ruled out in cases with honeycombing traction bronchiectasis. NSIP often develops in patients with connective tissue diseases or as a manifestation of a drug-induced pulmonary disease.

Langerhans cell histiocytosis (LCH) and RB-ILD, being smoking-related diseases, may resemble each other on high-resolution CT at early granulomatous stages (Fig. 2.5.11). However, the nodules in LCH are usually denser, well-defined, often associated with central cavitation and cysts, and lack spotty GGO like those in RB-ILD. TBLB usually produces specific granulomas containing CD1a+ Langerhans cells [30].

Pneumocystis pneumonia (PP), especially during the initial stages, may have a similar radiological pattern to that in RB-ILD with GGO foci and mild intralobular nodules (Fig. 2.5.12). The disease develops in immunocompromised patients and is manifested primarily by fever. Dyspnea is associated with large lesion in the lung parenchyma.

Blood tests in patients with PP usually have highly increased ESR and moderately increased CRP levels with normal or reduced leukocyte count, often with absolute and relative lymphopenia. PCR analysis of the BAL fluid for *P. jirovecii* can confirm the diagnosis of PP. Other opportunistic pulmonary infections can also sometimes mimic RB-ILD, but the lung involvement is often asymmetrical (Fig. 2.5.13). If an infectious lesion is suspected, microbiological and PCR studies of the BAL fluid should be performed. Atypical forms of pulmonary sarcoidosis manifested by areas of GGO and centrilobular nodules can resemble interstitial pneumonia on the HRCT pattern (Fig. 2.5.14). In such cases, TBLB confirms the presence of a granulomatous lesion.

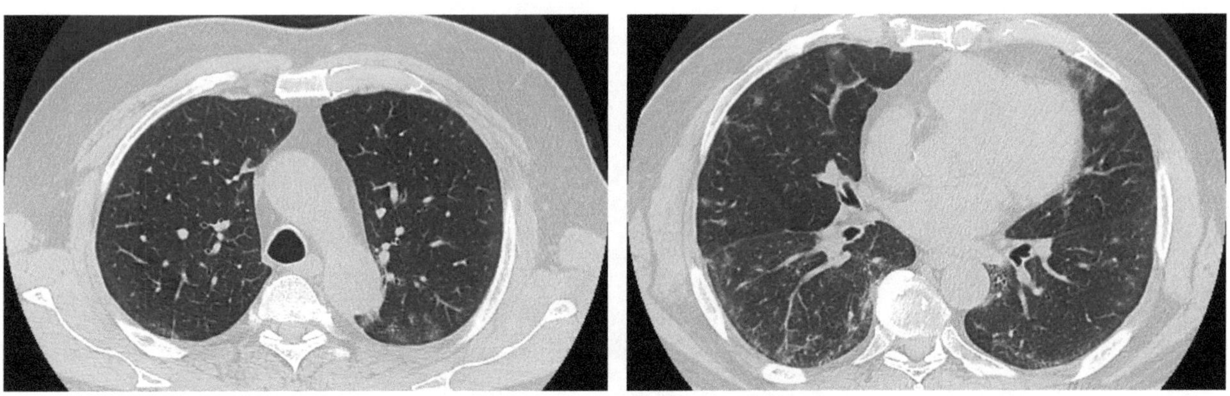

(A) (B)

FIG. 2.5.10 Nonspecific interstitial pneumonia, cellular subtype. Bilateral patchy areas of ground-glass opacity, primarily with basal and subpleural distributions (A and B). Moderately expressed reticular abnormalities mostly sparing the subpleural space *(arrows)*.

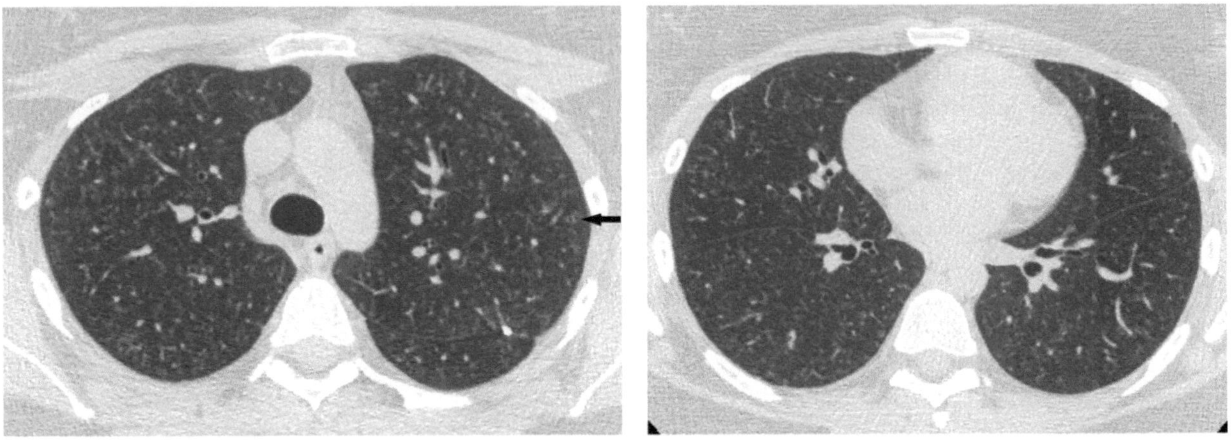

(A) (B)

FIG. 2.5.11 Early stage of Langerhans cell histiocytosis in a 20-year-old woman with a high smoking index. Multiple poorly differentiated and randomly distributed nodules in the upper lobes of the lungs, one of them with cavitation *(arrow)* (A). The lower lobes contain fewer nodules, the elements of tree-in-bud can be seen, and the cardiodiaphragmatic angles are lesion-free (B).

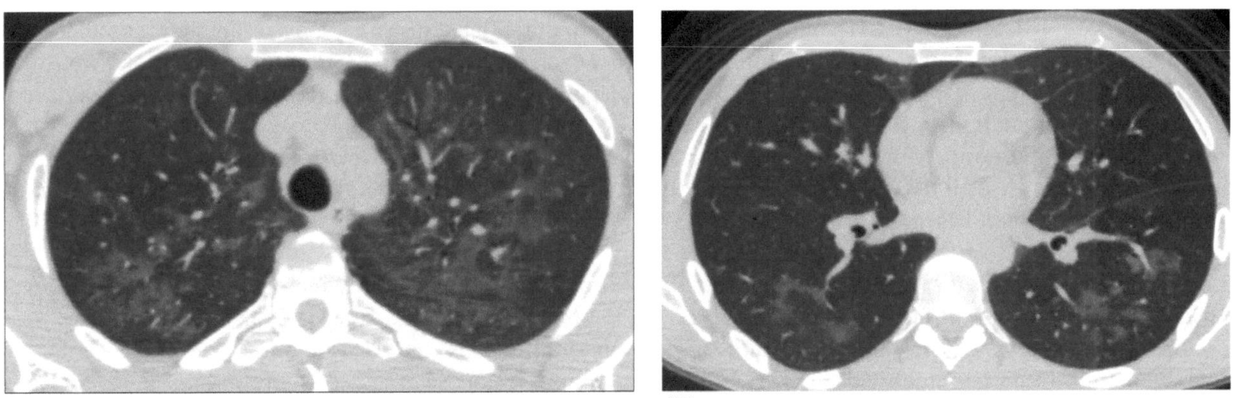

(A) (B)

FIG. 2.5.12 *Pneumocystis jirovecii* pneumonia in an HIV-infected patient. Bilateral patchy areas of GGO with primary involvement of the upper lobes (A and B).

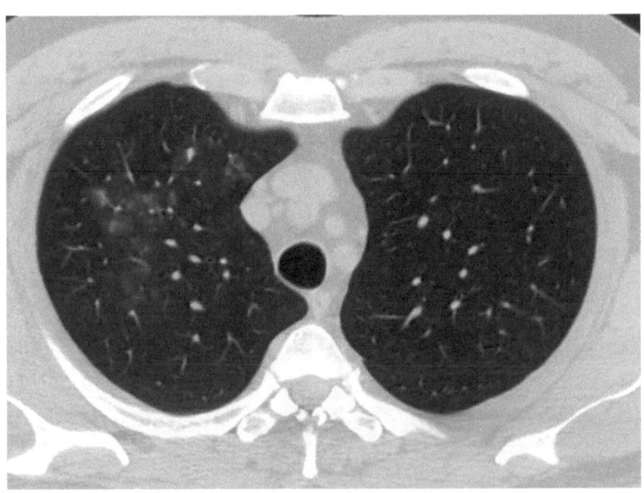

FIG. 2.5.13 Pulmonary mycosis in a man who pilots international airlines caused by *Penicillium marneffei*. Patchy foci of GGO in the upper lobe of the right lung.

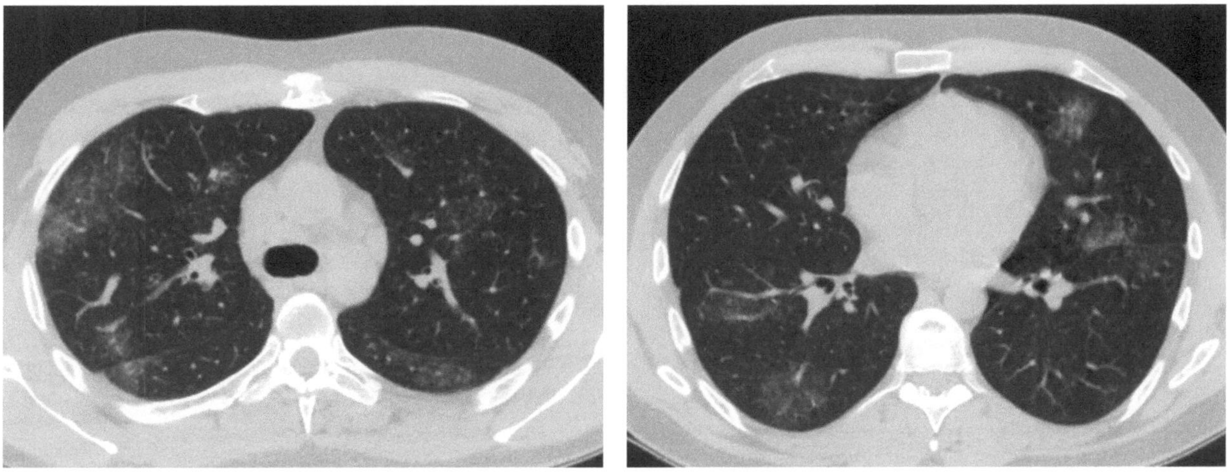

(A) (B)

FIG. 2.5.14 Pulmonary sarcoidosis. Bilateral patchy areas of GGO and multiple small ill-defined nodules with locally thickened interlobular septa (A and B). Enlarged intrathoracic lymphatic nodes are also visible (A).

Treatment and prognosis

Since smoking is a contributing factor for the disease, quitting is a prerequisite for patient management. The reversal of the disease and minimization of both clinical and radiological signs was thought to occur after smoking cessation, even in the absence of treatment measures [14,31]. But, a study performed by Portnoy et al. evaluating outcomes in patients with proved RB-ILD showed that 50% of patients experienced physical and functional deterioration during the follow-up, regardless of their smoking habit termination or treatment with corticosteroids and cytostatics [8]. Over 7 years, three patients died, two of them from non-small-cell lung cancer and one from progression of the interstitial disease; thus the overall survival rate for the period was 75% [8].

Systemic steroids are considered a possible drug therapy for RB-ILD, but it is unclear whether they can affect the course of the disease. In the work of Ryu JH et al., fact-based improvement after the start of treatment with medium doses of prednisone (35–40 mg/day) was observed in 64% of patients, but after 9 months the improvement remained only in 25% of patients, and the rest returned to their initial functional condition level [14]. Few reports on the use of immunosuppressive drugs in steroid-resistant forms exist [8], primarily with azathioprine, but without significant improvements. The absence of large-scale studies on this issue makes the treatment of RB-ILD a daunting task without guidelines for the selection of the starting doses, the duration of the course, or the criteria for stepwise reductions of drug loads.

References

[1] American Thoracic Society/European Respiratory Society. American Thoracic Society/European Respiratory Society international multidisciplinary consensus classification of the idiopathic interstitial pneumonias. Am J Respir Crit Care Med 2002;165(2):277–304.

[2] Moon J, du Bois RM, Colby TV, Hansell DM, Nicholson AG. Clinical significance of respiratory bronchiolitis on open lung biopsy and its relationship to smoking related interstitial lung disease. Thorax 1999;54(11):1009–14.

[3] Fraig M, Shreesha U, Savici D, Katzenstein AL. Respiratory bronchiolitis: a clinicopathologic study in current smokers, ex-smokers, and never-smokers. Am J Surg Pathol 2002;26(5):647–53.

[4] Travis WD, Costabel U, Hansell DM, King Jr TE, Lynch DA, Nicholson AG, et al. An official American Thoracic Society/European Respiratory Society statement: update of the international multidisciplinary classification of the idiopathic interstitial pneumonias. Am J Respir Crit Care Med 2013;188(6):733–48.

[5] Vassalo R, Ryu JH. Smoking-related interstitial lung diseases. Clin Chest Med 2012;33(1):165–78.

[6] Flower M, Nandakumar L, Singh M, Wyld D, Windsor M, Fielding D. Respiratory bronchiolitis-associated interstitial lung disease secondary to electronic nicotine delivery system use confirmed with open lung biopsy. Respirol Case Rep 2017;5(3):e00230.

[7] Theegarten D, Müller HM, Bonella F, Wohlschlaeger J, Costabel U. Diagnostic approach to interstitial pneumonias in a single centre: report on 88 cases. Diagn Pathol 2012;7:160.

[8] Portnoy J, Veraldi KL, Schwarz MI, Cool CD, Curran-Everett D, Cherniack RM, et al. Respiratory bronchiolitis-interstitial lung disease: long-term outcome. Chest 2007;131(3):664–71.

[9] Visscher DW, Myers JL. Histologic spectrum of idiopathic interstitial pneumonias. Proc Am Thorac Soc 2006;3(4):322–9.

[10] Nicholson AG, Rice AJ. Interstitial lung diseases. In: Ph H, Flieder DB, editors. Spencer's pathology of the lung. 6th ed. Cambridge Medicine; 2016. p. 366–408.

[11] Myers JL, Veal Jr CF, Shin MS, Katzenstein AL. Respiratory bronchiolitis causing interstitial lung disease: a clinicopathologic study of six cases. Am Rev Respir Dis 1987;135(4):880–4.

[12] Konopka KE, Myers JL. A review of smoking-related interstitial fibrosis, respiratory bronchiolitis, and desquamative interstitial pneumonia: overlapping histology and confusing terminology. Arch Pathol Lab Med 2018;142(10):1177–81.

[13] Mavridou D, Laws D. Respiratory bronchiolitis associated interstitial lung disease (RB-ILD): a case of an acute presentation. Thorax 2004;59(10):910–1.

[14] Ryu JH, Myers JL, Capizzi SA, Douglas WW, Vassallo R, Decker PA. Desquamative interstitial pneumonia and respiratory bronchiolitis-associated interstitial lung disease. Chest 2005;127(1):178–84.

[15] Sadikot RT, Johnson J, Loyd JE, Christman JW. Respiratory bronchiolitis associated with severe dyspnea, exertional hypoxemia, and clubbing. Chest 2000;117(1):282–5.

[16] Nakanishi M, Demura Y, Mizuno S, Ameshima S, Chiba Y, Miyamori I, et al. Changes in HRCT findings in patients with respiratory bronchiolitis-associated interstitial lung disease after smoking cessation. Eur Respir J 2007;29(3):453–61.

[17] Sieminska A, Kuziemski K. Respiratory bronchiolitis-interstitial lung disease. Orphanet J Rare Dis 2014;9:106.

[18] Bango-Álvarez A, Ariza-Prota M, Torres-Rivas H, Fernández-Fernández L, Prieto A, Sánchez I, et al. Transbronchial cryobiopsy in interstitial lung disease: experience in 106 cases – how to do it. ERJ Open Res 2017;3(1). pii:00148-2016.

[19] Heyneman LE, Ward S, Lynch DA, Remy-Jardin M, Johkoh T, Müller NL. Respiratory bronchiolitis, respiratory bronchiolitis-associated interstitial lung disease, and desquamative interstitial pneumonia: different entities or part of the spectrum of the same disease process? AJR Am J Roentgenol 1999;173(6):1617–22.

[20] Remy-Jardin M, Remy J, Boulenguez C, Sobaszek A, Edme JL, Furon D. Morphologic aspects of cigarette smoking on airways and pulmonary parenchyma in healthy adult volunteers: CT evaluation and correlation with pulmonary function tests. Radiology 1993;186(1):107–15.

[21] Palmucci S, Roccasalva F, Puglisi S, Torrisi SE, Vindigni V, Mauro LA, et al. Clinical and radiological features of idiopathic interstitial pneumonias (IIPs): a pictorial review. Insights Imaging 2014;5(3):347–64.

[22] Hartman TE, Tazelaar HD, Swensen SJ, Müller NL. Cigarette smoking: CT and pathologic findings of associated pulmonary diseases. Radiographics 1997;17(2):377–90.

[23] Dixon S, Benamore R. The idiopathic interstitial pneumonias: understanding key radiological features. Clin Radiol 2010;65(10):823–31.

[24] Park JS, Brown KK, Tuder RM, Hale VA, King Jr TE, Lynch DA. Respiratory bronchiolitis-associated interstitial lung disease: radiologic features with clinical and pathologic correlation. J Comput Assist Tomogr 2002;26(1):13–20.

[25] Caminati A, Cavazza A, Sverzellati N, Harari S. An integrated approach in the diagnosis of smoking-related interstitial lung diseases. Eur Respir Rev 2012;21(125):207–17.

[26] Johkoh T, Muller NL, Cartier Y, Kavanagh PV, Hartman TE, Akira M, et al. Idiopathic interstitial pneumonias: diagnostic accuracy of thin-section CT in 129 patients. Radiology 1999;211(2):555–60.

[27] Pandit-Bhalla M, Diethelm L, Ovella T, Sloop GD, Valentine VG. Idiopathic interstitial pneumonias: an update. J Thorac Imaging 2003;18(1):1–13.

[28] Webb WR, Müller NL, Naidich DP. High-resolution CT of the lung. 5th ed. Philadelphia, PA: Lippincot Williams and Wilkins; 2015242–5.

[29] Travis WD, Hunninghake G, King Jr TE, Lynch DA, Colby TV, Galvin JR, et al. Idiopathic nonspecific interstitial pneumonia: report of an American Thoracic Society project. Am J Respir Crit Care Med 2008;177(12):1338–47.

[30] Tazi A, Bonay M, Grandsaigne M, Battesti JP, Hance AJ, Soler P. Surface phenotype of Langerhans cells and lymphocytes in granulomatous lesions from patients with pulmonary histiocytosis X. Am Rev Respir Dis 1993;147(6 Pt 1):1531–6.

[31] Wells AU, Nicholson AG, Hansell DM, du Bois RM. Respiratory bronchiolitis-associated interstitial lung disease. Semin Respir Crit Care Med 2003; 24(5):585-94.

Chapter 2.6

Acute interstitial pneumonia

Acute interstitial pneumonia (AIP) is the most aggressive of all the types of idiopathic interstitial pneumonias; it has morphological signs of diffuse alveolar damage (DAD) [1]. AIP was first described in 1935 by L. Hamman and A. Rich as an acute diffuse interstitial pulmonary fibrosis resulting in lethal outcomes in all patients; the disease was called the Hamman-Rich syndrome [2]. Since DAD can be caused by a number of diseases and exogenous factors (Table 2.6.1), the diagnosis of AIP can only be considered after ruling out secondary origin. The triggering event for the severe immunoinflammatory reaction cascade in AIP seems to be a viral infection [3], in particular, a rhinovirus infection [4], but a possible role for *Pneumocystis jirovecii* is discussed due to its detection in a significant number of immunocompetent patients with AIP [5].

Morphology

DAD is the histopathologic substrate of AIP, and it results in injury to the alveolar–capillary membrane [6]. However, unlike DAD with known specific damaging events (Table 2.6.1), AIP's damaging causes are unknown. At least two phases are distinguishable in the development of DAD with AIP: an acute phase with hyaline membranes (clusters of fibrin, proteins,

TABLE 2.6.1 Causes of diffuse alveolar damage

General cause	Example
Infections	Influenza *Pneumocystis jirovecii* Fungi Sepsis Severe pneumonia
Connective tissue diseases and systemic vasculitis	Systemic lupus erythematosus Anti-Jo-1 syndrome Mixed connective tissue disease Rheumatoid arthritis Granulomatosis with polyangiitis
Idiopathic interstitial pneumonia	Acute exacerbation of idiopathic pulmonary fibrosis Acute exacerbation of nonspecific interstitial pneumonia Acute exacerbation of idiopathic pleuroparenchymal fibroelastosis Acute interstitial pneumonia
Drugs	Amiodarone Bleomycin Busulfan Gefitinib Imatinib Cocaine Melphalan Methotrexate Mitomycin Monoclonal antibodies Gold salts Cyclophosphamide Cytosine arabinoside
Other causes	Traumatic shock Acute pancreatitis Aspiration Hematopoietic stem cell transplantation
Unknown	Acute eosinophilic pneumonia

and necrotic epithelial cells [7]) lining extended alveolar ducts (a key feature of DAD) (Figs. 2.6.1–2.6.5), necrosis of the alveolar epithelium, edema, hemorrhage in the interstitial tissue and alveolar lumen and neutrophil clusters in capillaries and a subsequent phase with organization of fibrinous exudates, thickening of the alveolar septa caused by fibroblast proliferation and collagen accumulation, nodules of organizing pneumonia with marked diffuse lymphoplasmacytic and macrophage infiltration, proliferation of type II alveolocytes with reparation of the alveolar lining, and signs of destruction and macrophage infiltration in the remnants of hyaline membranes and fibrinous exudates (Figs. 2.6.6–2.6.8). In addition, thrombosis of small arteries [8] is often found, with the adjacent pleura thickened and infiltrated with lymphocytes, macrophages, and single neutrophils. Further development can be toward the resolution of pathological changes or transition to a third fibrous phase with the development of honeycombing [2,7].

Beasley et al. described a special variant of interstitial pneumonia called "acute fibrinous organizing pneumonia" with the development of severe respiratory failure and clinically similar to AIP but with a peculiar histological pattern presenting fibrin balls in alveolar lumens, along with organizing pneumonia and hyperplasia of type II alveolocytes [9]. However, in contrast to DAD, this variant does not present hyaline membranes [8].

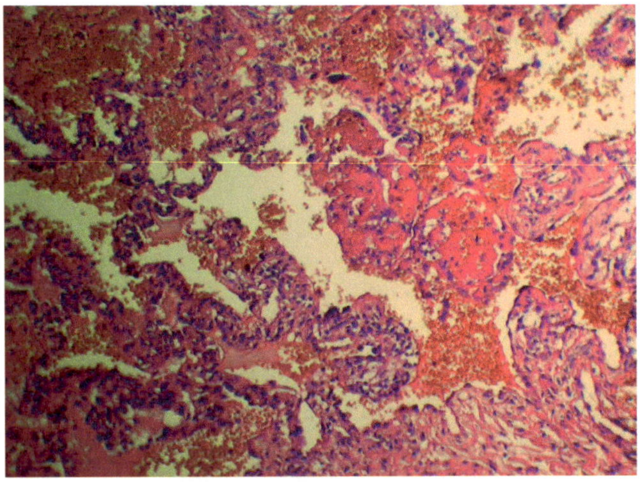

FIG. 2.6.1 Early-phase AIP. Hyaline membranes, interstitial lymphohistiocytic infiltrate with leukocyte admixture, and dystelectasis. Hematoxylin and eosin staining, 100×.

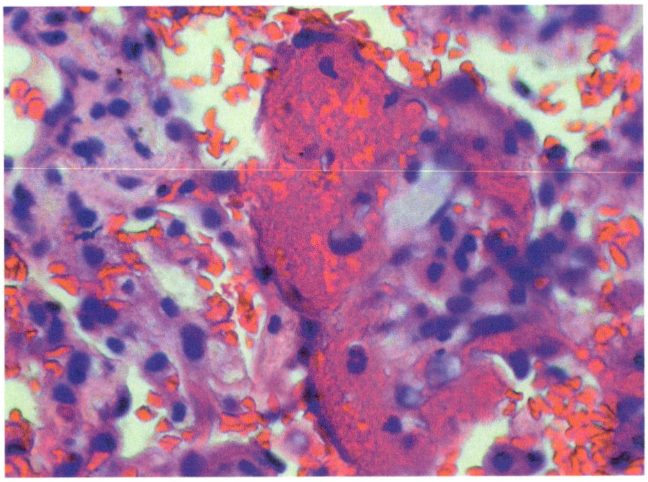

FIG. 2.6.2 Early-phase AIP. Hyaline membranes, interstitial lymphohistiocytic infiltrate with leukocyte admixture, and dystelectasis. Resorption of hyaline membranes. Hematoxylin and eosin staining, 600×.

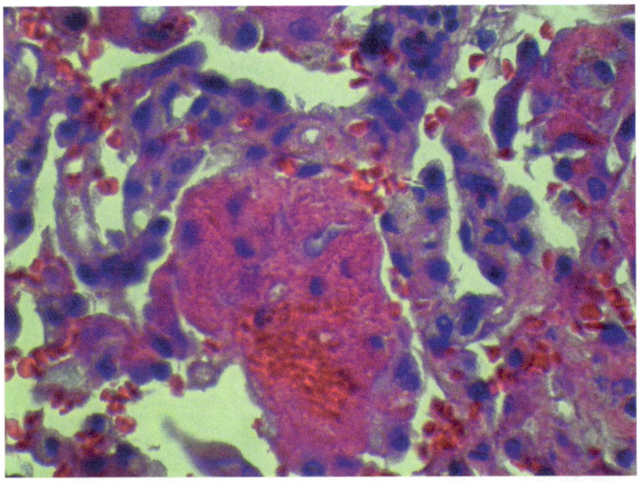

FIG. 2.6.3 Early-phase AIP. Hyaline membranes and lymphohistiocytic elements in the lumens and walls of the alveoli. Hematoxylin and eosin staining, 600×.

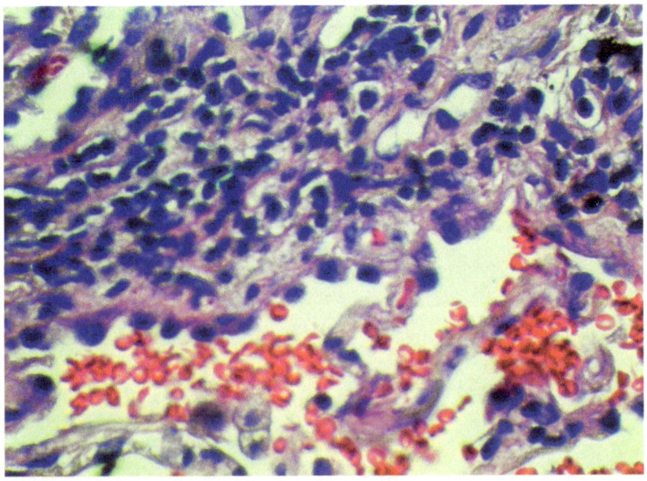

FIG. 2.6.4 Early-phase AIP. Lymphohistiocytic infiltration in the walls of the alveoli. Hematoxylin and eosin staining, 400×.

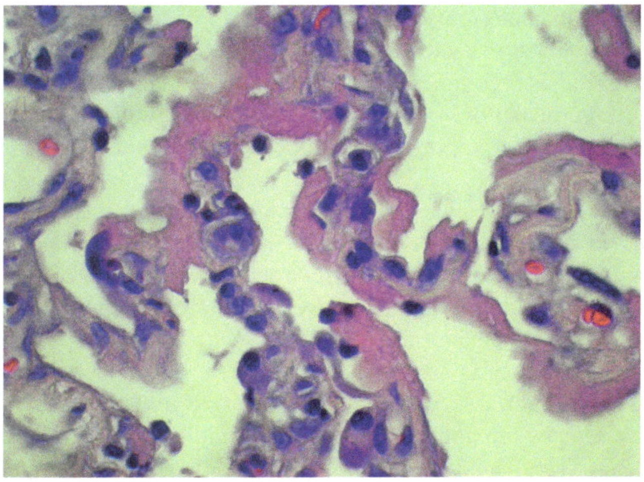

FIG. 2.6.5 Early-phase AIP. Hyaline membranes in the alveoli. Hematoxylin and eosin staining, 400×.

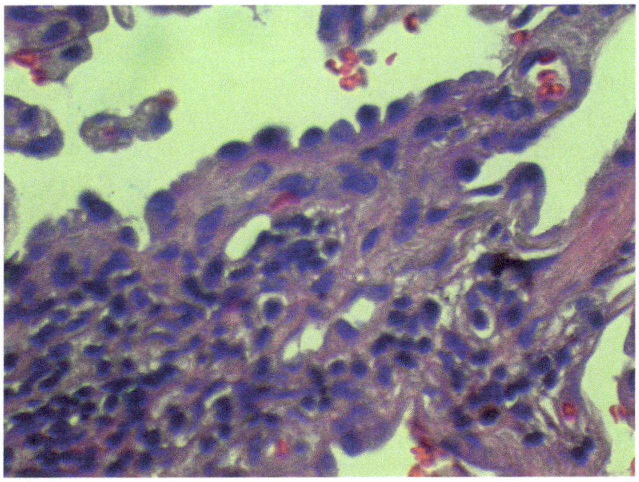

FIG. 2.6.6 Interstitial fibrosis and hyperplasia of two type of pneumocytes. Hematoxylin and eosin staining, 400×.

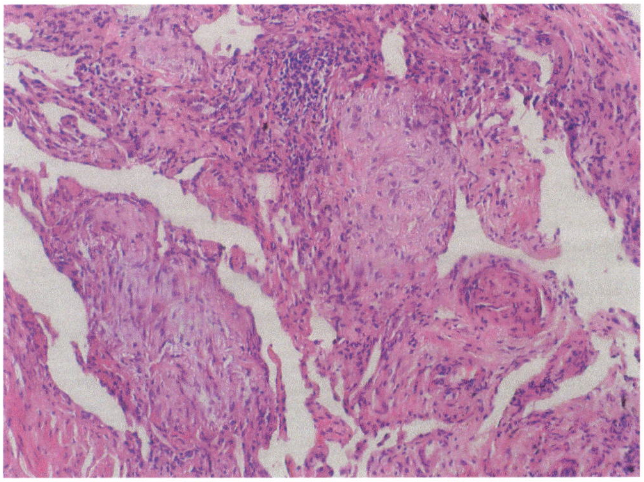

FIG. 2.6.7 Late-phase AIP. Nodules of organization and interstitial fibrosis. Hematoxylin and eosin staining, 200×.

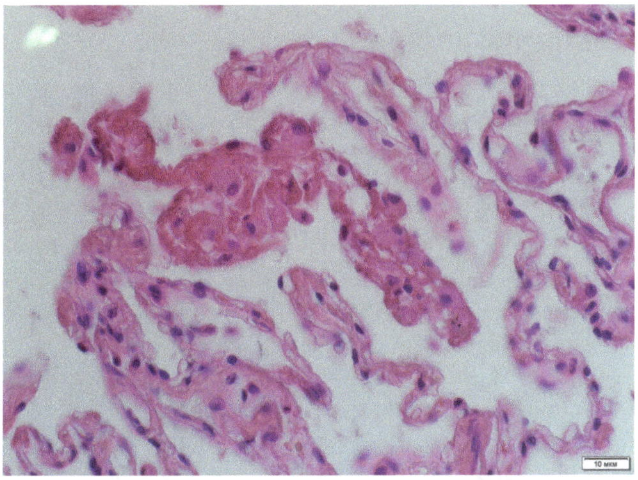

FIG. 2.6.8 Late-phase AIP. Resorption of hyaline membranes and fibrinous exudate by alveolar macrophages. Hematoxylin and eosin staining, 600×.

Clinical presentation

The disease can develop at any age (on average of 50 years) regardless of previous pathology, smoking history, or other exogenous factors [1].

At its onset the disease presents with clinical symptoms of a viral respiratory infection of the upper airways, arthralgia, myalgia, and dry cough [10]. At the time of hospitalization, fever above 38.3°C is noted in 75% of patients [4]. A progressive dyspnea develops quickly. From the onset of the prodromal stage to the development of severe respiratory failure, days to weeks can pass (usually 7–14 days) [11]. Vourlekis et al. described less active forms with a duration of up to 120 days from the first symptoms to invasive ventilation [4]. Respiratory hypoxemic failure often fails to respond to bronchodilators or corticosteroids. During advanced stages the disease acquires the clinical features of acute respiratory distress syndrome (ARDS) and requires invasive lung ventilation [12]. The auscultatory pattern is not specific and includes diffuse crackles (71%) or wheezing (30%) [4].

Diagnosis

Laboratory tests generally reflect the activity of the inflammatory process, namely, leukocytosis and accelerated erythrocyte sedimentation rate (ESR) with increased C-reactive protein (CRP) levels [4,10]. Given the importance of differentiating

AIP from DAD associated with other diseases and exogenous factors, the determination of the panel of immunologic markers of connective tissue diseases (CTD) and vasculitis, including rheumatoid factor, antinuclear antibodies, cyclic citrullinated peptide, and synthetase antibodies [13] upon hospital admission, is recommended. In our opinion, this list should be supplemented with antineutrophil cytoplasmic antibodies and serum procalcitonin. In addition, blood and sputum cultures, polymerase chain reaction (PCR) of sputum or tracheobronchial aspirate for opportunistic pathogens, *Mycoplasma pneumoniae*, and *Legionella pneumophila* should be studied.

Elevated neutrophil counts (up to 70%–80%) are often found in the bronchoalveolar lavage (BAL) fluid accompanied by erythrocytes and/or hemosiderin [14]. BAL is necessary to rule out respiratory infections and diseases with a specific cytological pattern (acute eosinophilic pneumonia, drug-induced pneumonitis, hypersensitivity pneumonitis, and alveolar hemorrhage).

The feasibility of transbronchial lung biopsy in patients with AIP should be considered given the risk of pneumothorax and the likelihood of developing life-threatening respiratory failure [15]. In general the decision is made individually. Due to the large area of lung lesions, a transbronchial lung biopsy from unchanged parenchyma is rare [16]. The choice of the correct diagnostic test determines the prognosis in patients. The lowest mortality from AIP was observed in the studies by Suh (20%) and Quefatieh (13%) in which aggressive diagnostic tactics were applied including a minithoracotomy in the first case and open lung biopsy in the second case with an urgent histological study, despite severe respiratory failure and invasive lung ventilation [17,18].

A functional study of the lungs is often not feasible due to the severity of the condition, but whenever possible, severe restrictive disorders have been detected with decreases in the diffusion capacity of the lungs [14].

Computed tomography

AIP is characterized by a pattern of diffuse parenchymal lesions with ground-glass opacity (GGO) and consolidation areas, mostly in the posterior and basal fields, although cases of more pronounced consolidation in the upper lobes have been also described. Within GGOs, thickening of interlobular septa and irregular linear opacities are often found (Fig. 2.6.9) [19]. At the same time, lobules free from damage creating a mosaic pattern can also be seen. The costodiaphragmatic angles remain intact often. GGOs are present in 100% of reports, while consolidations affect 66%–92% of patients [20,21].

In patients with AIP, pulmonary fibrosis begins to develop rapidly and is manifested radiographically by reticular changes, the appearance of subpleural honeycombing, and architectural distortion of the lung parenchyma with traction bronchiectases (Fig. 2.6.10). The architectural distortion results from the displacement or deformity of the interlobar fissures, interlobular septa, bronchi, or blood vessels [22].

Such changes appear from the second week after the onset of the symptoms and determine the prognosis. In a study by Ichikado et al. [22] comparing computed tomography signs between survivors and patients with AIP who died, significantly fewer patients with architectural distortion and traction bronchiectases were present in the group of survivors. The total parenchymal damage was also an outcome predictor.

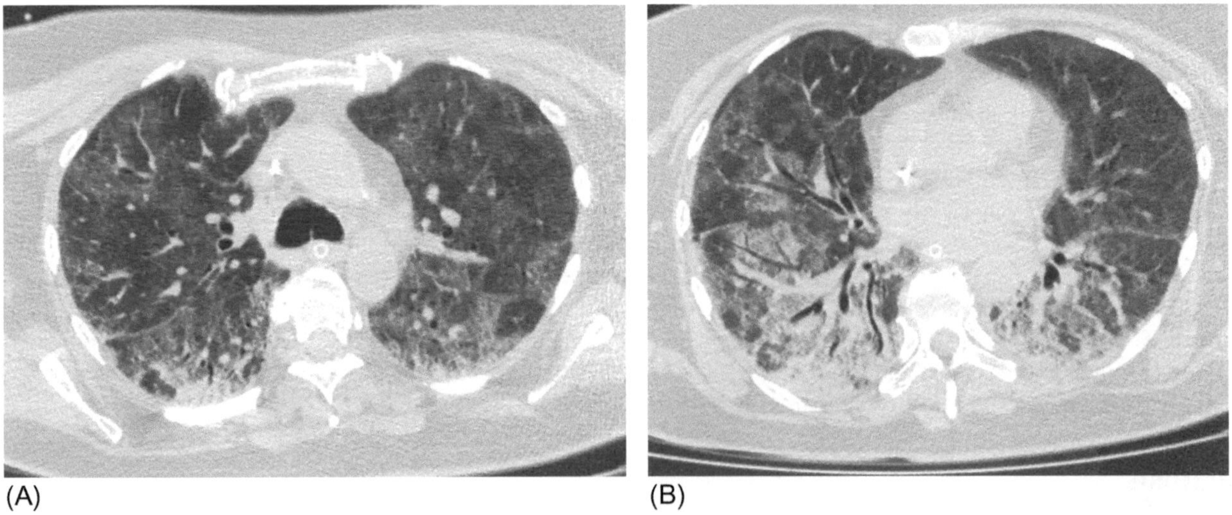

FIG. 2.6.9 AIP in a 51-year-old woman. Day 3 of invasive pulmonary ventilation. Extensive bilateral areas of GGO and subpleural consolidation (A and B), more pronounced in the lower lobes (B). The separated lobules (A) and the costodiaphragmatic angles (B) appear to be unaffected.

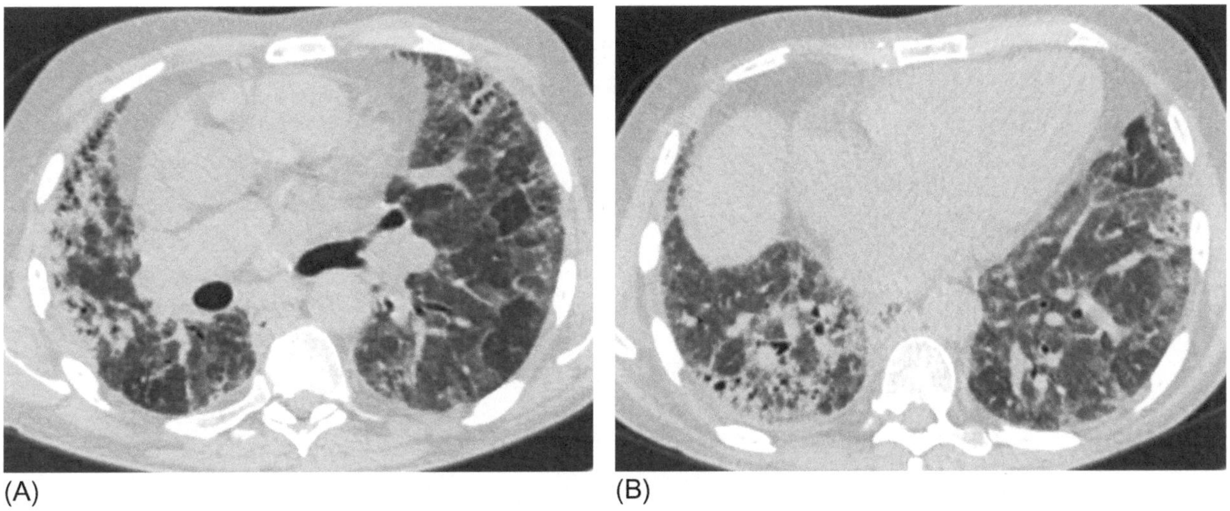

FIG. 2.6.10 AIP in a 68-year-old patient on day 19 from disease onset. Bilateral diffuse areas of GGO with a mosaic distribution. Reticular abnormalities. In the subpleural areas of consolidation, small subpleural cysts and bronchiolectases are seen as a manifestation of the initial honeycombing stage (A and B).

Approximately half of the patients may have pleural effusions, and hydropericardium is also detected in some patients [16]. In general the high-resolution computed tomography (HRCT) characteristics of AIP do not differ from the DAD signs caused by known causes (infections, CTD, and drugs), although traction bronchiectasis and honeycombing during the fibrous stage of the disease are more typical for AIP [23].

Differential diagnosis

AIP must be differentiated from diseases that present radiological and morphological patterns of DAD (Table 2.6.1) and from cryptogenic organizing pneumonia, acute fibrinous organizing pneumonia, and diffuse alveolar hemorrhages that can resemble AIP clinically (Table 2.6.2).

The most common cause of acute bilateral consolidation in the lungs accompanied by severe respiratory failure is a pulmonary infection. Acute onset, the presence of catarrhal symptoms, and bilateral pulmonary infiltrates on a radiograph are signs that can occur both in patients with bacterial pneumonia (BP) and AIP. In AIP cases, as in BP, blood tests reveal an acute inflammatory reaction with increased leukocyte levels, ESR, and CRP [14]. Severe BP is characterized by an increase in the level of blood urea nitrogen >7 mmol/L (one of the criteria for severity), which is not noted in AIP cases [24].

An important test to rule out bacterial infections is serum procalcitonin. Although the scientific literature does not describe the role of these tests in establishing the diagnosis of AIP, in our practice, the procalcitonin level was within normal limits in four patients with AIP. All available methods should be applied to verify a possible pathogen, including blood and bronchial aspirate cultures. HRCTs in patients with severe BP reveal mostly consolidation areas, which are often asymmetrical; GGO can occur, but it does not dominate, and most often, it represents an intermediate phase in the formation of new parenchymal consolidations. However, the HRCT patterns of BP and AIP can be very similar (Fig. 2.6.11). According to Tomiyama et al. [25], the most beneficial HRCT sign for differentiating BP from AIP is the presence of centrilobular nodules in BP that are not typical for AIP.

Reticular abnormalities are not characteristic of BP unless they develop in patients with underlying interstitial lung disease. Finally, correct antibacterial therapy leads rather quickly to symptom reduction in patients with BP.

ARDS caused by an infectious process (influenza virus and pneumocystis) can mimic AIP. During the influenza epidemic caused by the H1N1 strain, fulminant forms with the ARDS development were seen, and their clinical, radiological, and morphological signs were indistinguishable from those of AIP (Fig. 2.6.12) [26]. During the influenza epidemic season, the emergence of ARDS in pregnant women or people with obesity presenting high fever since the first days of the disease should make the doctor suspect influenza.

Severe *Pneumocystis jirovecii* pneumonia variants with subtotal pulmonary lesions can also cause diffuse alveolar damage and should be considered in the differential series of AIP (Fig. 2.6.13) [27].

Contemporary PCR diagnostic methods with pathogen identification in a few hours help in establishing a diagnosis. BAL is the most suitable material for PCR, especially since patients often have no sputum. However, the presence of

TABLE 2.6.2 Differential series of acute interstitial pneumonia

	AIP	BP	COP	Influenza pneumonia	AE ILD
Anamnesis	Onset of catarrhal phenomena in the upper respiratory airway	Preceding hypothermia, aspiration	Common with autoimmune diseases	During the influenza season, pregnancy, obesity	Dyspnea during preceding months (years)
Blood test	Leukocytosis, ESR 40–80 mm/h	Leukocytosis, shift of the formula toward immature forms. Accelerated ESR	Accelerated ESR	Leukopenia, lymphopenia	Leukocytosis
Urea nitrogen level (mmol/L)	Normal	>7.7	Normal	Normal	Normal
Procalcitonin (ng/mL)	<0.5	>1.0	<0.5	<1.0	<0.5
Bronchoalveolar lavage	Neutrophils >40% Erythrocytes	Neutrophils	Foamy macrophages, eosinophilia 2%–25%, lymphocytosis >25%, CD4/CD8 <0.9	Erythrocytes + PCR for influenza virus	Neutrophils
Ground-glass opacity	+++	+	+++	++	+++
Consolidation	+++	+++	+++	+++	++
Thickening of the interlobular septa	+++	−	++	+	+++
Centrilobular nodules	−	++	++	+	−
Architectural distortion, traction bronchiectasis	++	−	−	−	−
Response to antibiotics	−	+++	−	−	−
Response to systemic steroids	+	−	+++	−	++

AIP, acute interstitial pneumonia; *AE ILD*, acute exacerbation of chronic interstitial lung disease; *BP*, bacterial pneumonia; *COP*, cryptogenic organizing pneumonia; *ESR*, erythrocyte sedimentation rate.

pneumocystis organisms in the BAL fluid does not always signal the presence of *Pneumocystis* pneumonia. In immunocompetent patients, *P. jirovecii* may act as a trigger of the primary alveolar-capillary injury with subsequent development of DAD by AIP types [5]. The negative microbiological result of the BAL fluid analysis cannot completely rule out respiratory infections since the sensitivity of culture method for many infections does not exceed 70% [13].

Aggressive forms of cryptogenic organizing pneumonia (COP) or secondary organizing pneumonia can be very similar to AIP in their clinical and radiological characteristics (Fig. 2.6.14), although they have a morphological pattern different from that of DAD. COP is an inflammatory reaction of the terminal airways characterized by the organization of exudate and fibroblastic proliferation in the form of granulation tissue ingrowths into respiratory bronchioles, alveolar ducts, and alveoli [1]. The common course of COP is subacute, but dramatic cases of the disease with acute respiratory failure and manifestations of ARDS have been described [28,29]. Fulminant forms of COP may not fully correspond histologically to the classical histopathologic pattern of COP since the hyperplastic reaction of the bronchioles fails to develop in due time.

For COP, as well as for AIP, the lack of signs of respiratory infection, the absence of reaction to antibiotics, the development of bilateral consolidation areas, and GGOs are typical. One of the methods useful for differentiating between these two interstitial diseases is the analysis of BAL fluid. AIP is characterized by a pronounced neutrophil reaction (>40%), whereas in COP, a lymphocytosis greater than 25% is common with decreases in the CD4+/CD8+ ratio (lower than 0.9) and moderate eosinophilia (up to 25%). In one of the clinical descriptions of BAL in the case of an acute form of COP, only a critical decrease in the lymphocytic ratio to 0.14 was described [30]. Such imbalance is not typical for AIP. When considering radiological features, one of the most frequent signs for COP, detected during HRCT and in BP, is the presence of intralobular nodules, reflecting lymphoid infiltration of respiratory bronchioles. A reversed halo CT sign is not common at AIP, but it is also characteristic of COP (Fig. 2.6.14) [31]. COP tends to respond quickly to corticosteroid therapy, although steroid-resistant forms have also been reported. As mentioned, rare variants of interstitial lung disease with an acute course and severe respiratory failure, but without histological signs of DAD, exist and occupy an intermediate niche between COP and AIP; they respond quickly to corticosteroid therapy and are called acute fibrinous organizing pneumonia (Fig. 2.6.15) [9]. Similar clinical and radiological patterns with both COP and AIP characteristics mean the diseases can only be differentiated on the basis of their morphology.

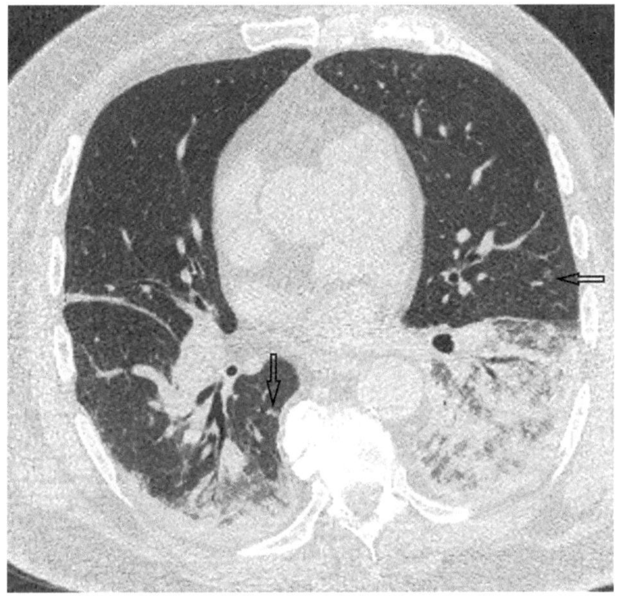

FIG. 2.6.11 Bilateral community-acquired pneumonia caused by *S. pneumoniae*. Bilateral subpleural heterogeneous areas of consolidation with adjacent areas of GGO (subpleural and peribronchovascular distribution). Separate nodules *(arrows)* are visible outside of affected areas. Bilateral minor pleural effusion.

Idiopathic acute eosinophilic pneumonia (AEP) is a rare disease of unknown origin, diagnosed based on prominent infiltration of alveoli and bronchioles with eosinophils causing diffuse alveolar damage, with a rapid development of acute hypoxemic respiratory failure, often requiring invasive lung ventilation [32]. Unlike the case in other eosinophilic pulmonary lesions (eosinophilic granulomatosis with polyangiitis and chronic eosinophilic pneumonia), eosinophilia in the blood rarely occurs with AEP, which hinders early diagnosis. The HRCT pattern resembles that of AIP, with diffuse nodular infiltration of GGO with consolidation areas and thickening of interlobular septa (Fig. 2.6.16), but a mostly upper lobe distribution of the abnormalities is often present (rare in AIP cases). Minor (including interlobar) pleural effusions are often

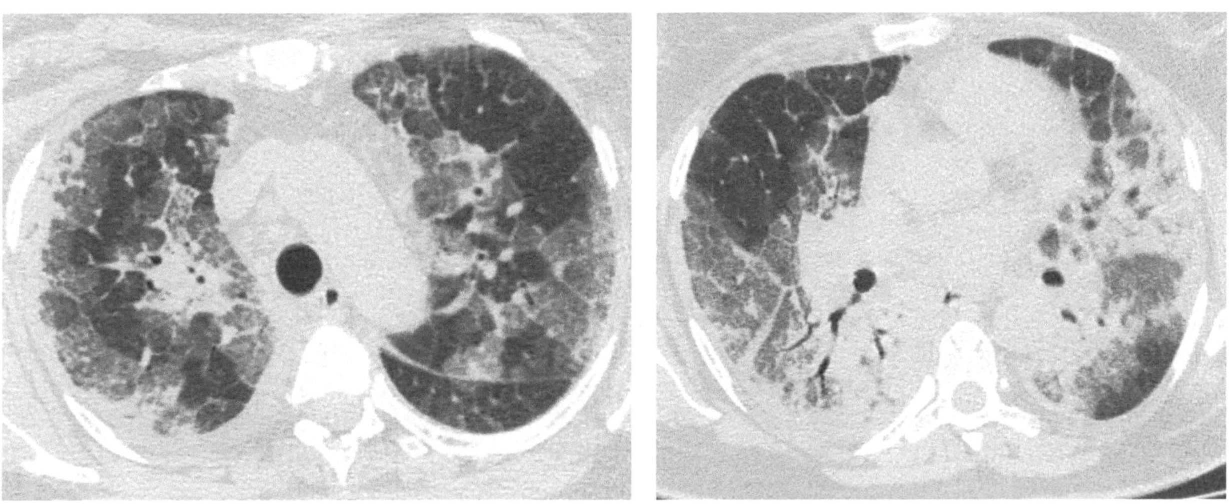

(A) (B)

FIG. 2.6.12 ARDS caused by the H1N1 influenza virus in a 38-year-old woman. Day 9 from the onset of respiratory symptoms. Bilateral subpleural and peribronchovascular areas of consolidation and GGOs (A and B). The GGO (with thickened interlobular septa) dominates over consolidation in the upper lobes (A). In the lower lobes, consolidation is more prominent than GGOs (B). Individual segments do not appear to be involved in the lesion.

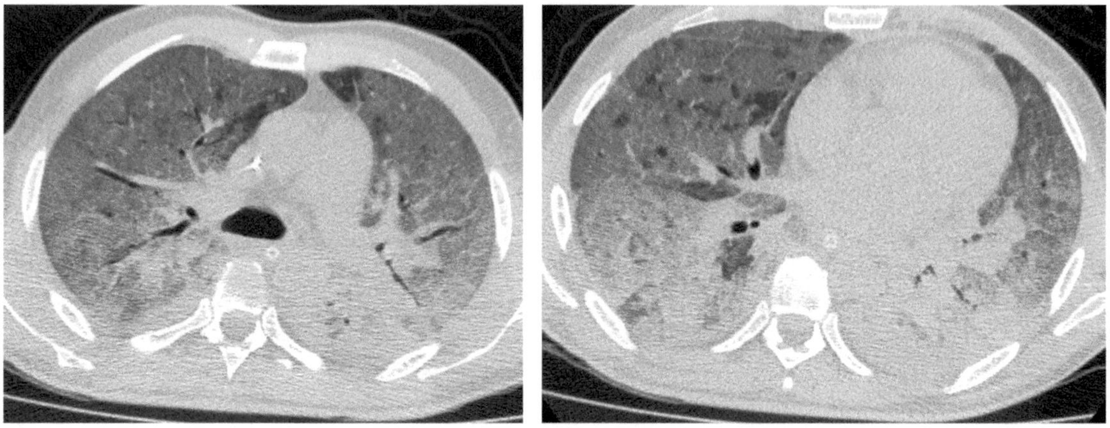
(A) (B)

FIG. 2.6.13 Total pneumonia caused by *P. jirovecii* in a patient with AIDS, who had not received antiretroviral therapy. Diffuse areas of GGO, subpleural consolidation pronounced in the lower lobes. Single lobular air traps (A and B) are visible inside affected areas.

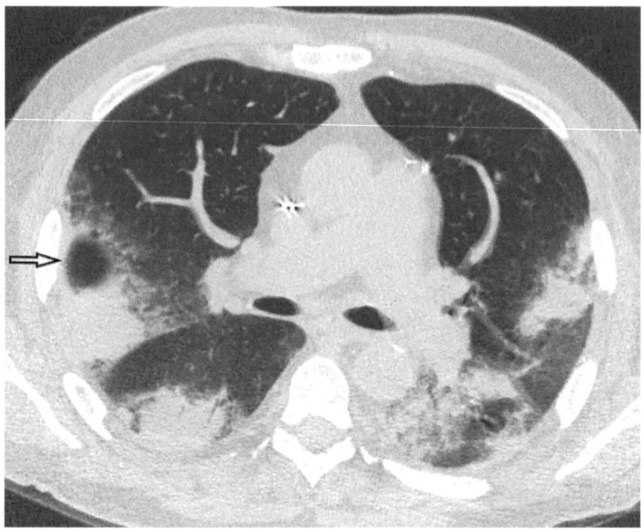

FIG. 2.6.14 COP of acute course. Massive bilateral consolidation areas surrounded by GGO with subpleural and peribronchovascular distribution. Atoll sign *(arrow)*.

seen (up to 70%) [17,25]. However, AEP diagnoses based only on the HRCT pattern are difficult to suspect. In a study by Tomiyama et al. [25], the correct radiological diagnosis of AEP was made only in 30% of cases. The primary minimally invasive diagnostic tool for AEP is BAL fluid analysis. An increase in eosinophil counts greater than 30% in the BAL fluid in combination with bilateral opacities in the lungs on HRCT with acute respiratory failure in the absence of an infectious trigger provides the basis for the AEP diagnosis [33].

AIP is very difficult to differentiate from an acute exacerbation of underlying interstitial lung diseases like idiopathic pulmonary fibrosis or nonspecific interstitial pneumonia if the diagnosis has not been established previously. Since all signs of acute exacerbations of interstitial lung diseases (AE ILD) (clinical, radiological, and biochemical) may overlap with AIP, the patient's history must be evaluated. Patients with AE ILD usually have respiratory symptoms long before the development of acute respiratory failure. At AE ILD, honeycombing and reticular abnormalities outside the areas of GGO and consolidation and the dominance of areas of GGO over consolidation are the frequent findings on HRCT (Fig. 2.6.17). The acute exacerbation of chronic interstitial lung disease generally responds to therapy with high doses of steroids and rarely requires invasive lung ventilation [34].

Diffuse alveolar hemorrhages can have an HRCT pattern similar to that of DAD (Fig. 2.6.18). However, the hemorrhages develop faster than AIP clinically, usually within a few hours; in addition to causing acute respiratory failure, usually (but not always), they are accompanied by hemoptysis, a drop in the level of hemoglobin, erythrocytes, and hematocrit and show a predominance of erythrocytes in the cellular spectrum of the BAL fluid.

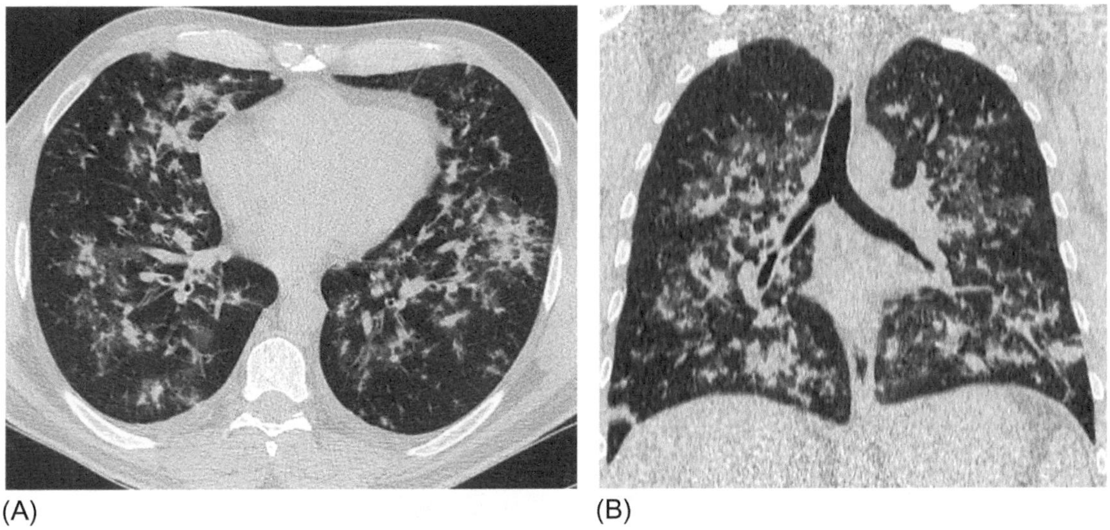

FIG. 2.6.15 Acute fibrinous and organizing pneumonia. Bilateral patchy areas of GGO and consolidation with random distribution (A). The perihilar pattern of distribution of the abnormalities can be seen on the coronal section (B).

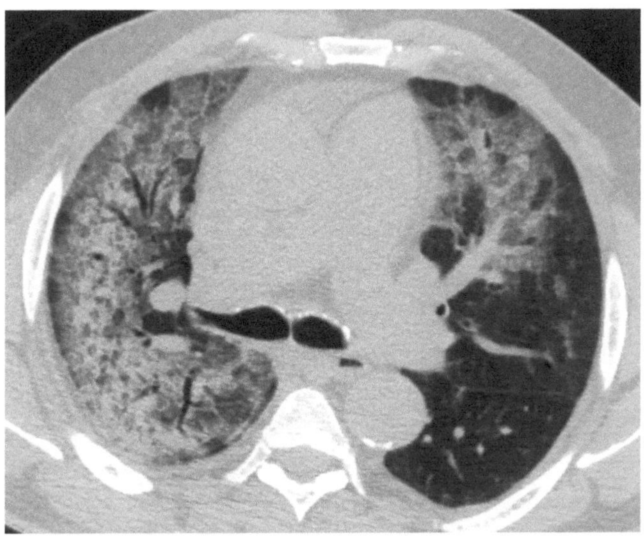

FIG. 2.6.16 AEP in a 69-year-old patient, diagnosed by the presence of eosinophilia in BAL fluid (45%) and transbronchial lung biopsy. Converging areas of GGO and consolidation (diffuse on the right and patchy on the left sides); thickened interlobular septa are visible inside the affected areas.

Thus the differential diagnosis of AIP is complicated; it requires ruling out many acute diseases accompanied by severe respiratory failure. The morphological confirmation of DAD may support AIP only after negative results of tests for pulmonary infections, connective tissue diseases and vasculitis, chronic interstitial lung diseases, eosinophilia, and hemorrhage of BAL fluid, as well as a lack of a history of taking potentially hazardous medicines. As a rule, a multidisciplinary approach is required for correct AIP diagnoses.

Treatment and prognosis

In general, AIP treatment approaches use high doses of steroids (2–4 mg/kg of prednisone with or without previous pulse therapy) [4,15]. The total duration of a course of corticosteroids in survivors lasts up to 4 months, and complete remission can be achieved with residual fibrotic changes in the lungs [17].

Immunosuppressants (cyclophosphamide) have been used in some patients [35]. Suzuka et al. used high doses of tacrolimus maintaining the trough concentration within a range of 15–20 ng/mL for periods as long as 3 months in patients with acute/subacute interstitial pneumonia associated with dermatomyositis, and they reported the death of only one patient out

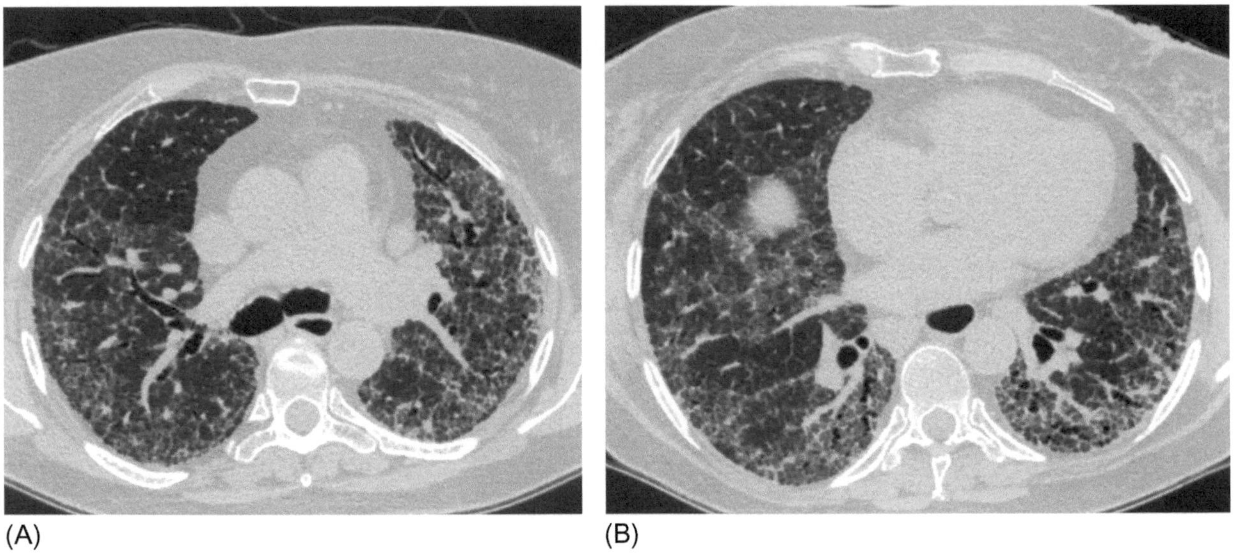

FIG. 2.6.17 Acute exacerbation of idiopathic pulmonary fibrosis. Bilateral areas of GGO associated with thickening of intralobular and interlobular septa, bronchiectases, and subpleural honeycombing. In segments without GGO, reticular abnormalities are also visible (A and B).

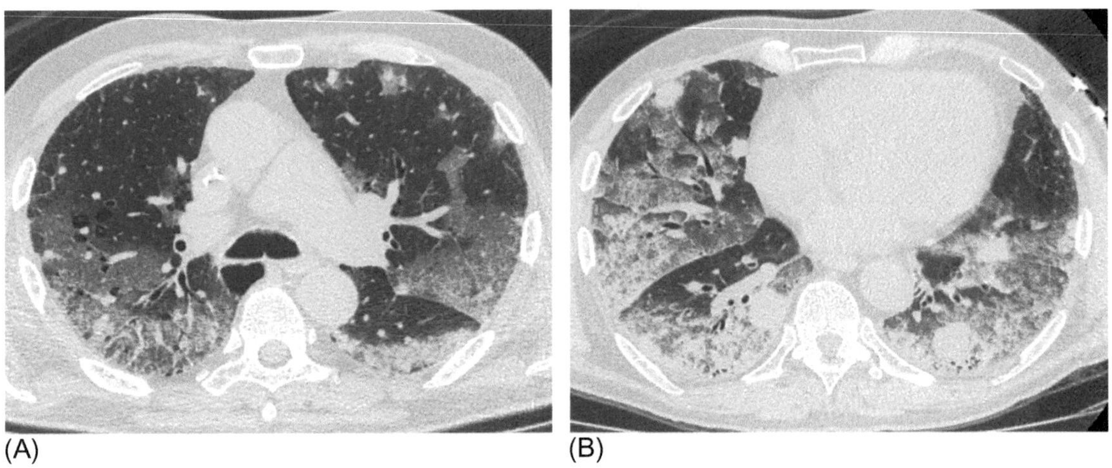

FIG. 2.6.18 Diffuse alveolar hemorrhages in a patient taking anticoagulants. Day 3 from the onset of respiratory symptoms. Bilateral areas of GGO and subpleural consolidation, associated with mild thickening of intralobular and interlobular septa (A and B), separated GGO foci in the upper lobes (A).

of 11 patients enrolled in the study [36]. Given the low incidence rate, no randomized studies evaluating the efficacy and safety of drug therapy in AIP exist. Extracorporeal membrane oxygenation (ECMO) can be a life-saving support in patients with potentially reversible acute respiratory failure. Gonzalves-Venade et al. reported the survival after ECMO of two of the three patients with AIP and refractory respiratory failure despite the use of high doses of corticosteroids, cyclophosphamide, lung protective ventilation, and prone position [35].

AIP is the most life-threatening idiopathic interstitial pneumonia with 60-day mortalities usually exceeding 70% [4,10,16]. Achievement of a survival rate of 80%–87% is associated with rapid diagnosis using surgical biopsy methods and treatment with high doses of steroids (which are probably more effective in the early phases of the disease than later) and with protective ventilation modes with a respiratory volume of 6–8 mL/kg and high PEEP [17,18].

References

[1] American Thoracic Society/European Respiratory Society. International multidisciplinary consensus classification of the idiopathic interstitial pneumonias. Am J Respir Crit Care Med 2002;165(2):277–304.

[2] Hamman L, Rich AR. Fulminating diffuse interstitial fibrosis of the lungs. Trans Am Clin Climat Assoc 1935;51154–63.

[3] Kaarteenaho R, Kinnula VL. Diffuse alveolar damage: a common phenomenon in progressive interstitial lung disorders. Pulm Med 2011;2011:531302.

[4] Vourlekis JS, Brown KK, Cool CD, Young DA, Cherniack RM, King Jr TE, et al. Acute interstitial pneumonitis. Case series and review of the literature. Medicine (Baltimore) 2000;79(6):369–78.

[5] Martínez-Rísquez MT, Friaza V, de la Horra C, Martín-Juan J, Calderón EJ, Medrano FJ. Pneumocystis jirovecii infection in patients with acute interstitial pneumonia. Rev Clin Esp 2018;218(8):417–20.

[6] Matsubara O, Tamura A, Ohdama S, Mark EJ. Alveolar basement membrane breaks down in diffuse alveolar damage: an immunohistochemical study. Pathol Int 1995;45(7):473–82.

[7] Katzenstein AL, Bloor CM, Leibow AA. Diffuse alveolar damage—the role of oxygen, shock, and related factors. A review. Am J Pathol 1976;85(1):209–28.

[8] Katzenstein AL, Myers JL, Mazur MT. Acute interstitial pneumonia. A clinicopathologic, ultrastructural, and cell kinetic study. Am J Surg Pathol 1986;10(4):256–67.

[9] Beasley MB, Franks TJ, Galvin JR, Gochuico B, Travis WD. Acute fibrinous and organizing pneumonia: a histological pattern of lung injury and possible variant of diffuse alveolar damage. Arch Pathol Lab Med 2002;126(9):1064–70.

[10] Olson J, Colby TV, Elliott CG. Hamman-Rich syndrome revisited. Mayo Clin Proc 1990;65(12):1538–48.

[11] Mason RJ, Murray JF, Broaddus VC, Nadel JA. Textbook of respiratory medicine. 5th ed. vol. 2. USA: Elsevier Inc; 2015. p. 1148.

[12] Mukhopadhyay S, Parambil JG. Acute interstitial pneumonia (AIP): relationship to Hamman-Rich syndrome, diffuse alveolar damage (DAD), and acute respiratory distress syndrome (ARDS). Semin Respir Crit Care Med 2012;33(5):476–85.

[13] Disayabutr S, Calfee CS, Collard HR, Wolters PJ. Interstitial lung diseases in the hospitalized patient. BMC Med 2015;13:245.

[14] Bonaccorsi A, Cancellieri A, Chilosi M, Trisolini R, Boaron M, Crimi N, et al. Acute interstitial pneumonia: report of a series. Eur Respir J 2003;21(1):187–91.

[15] Bradley B, Branley HM, Egan JJ, Greaves MS, Hansell DM, Harrison NK, et al. Interstitial lung disease guideline: the British Thoracic Society in collaboration with the Thoracic Society of Australia and New Zealand and the Irish Thoracic Society. Thorax 2008;63(Suppl 5):v51–8.

[16] Avnon LS, Pikovsky O, Sion-Vardy N, Almog Y. Acute interstitial pneumonia–Hamman-Rich syndrome: clinical characteristics and diagnostic and therapeutic considerations. Anesth Analg 2009;108(1):232–7.

[17] Suh GY, Kang EH, Chung MP, Lee KS, Han J, Kitaichi M, et al. Early intervention can improve clinical outcome of acute interstitial pneumonia. Chest 2006;129(3):753–61.

[18] Quefatieh A, Stone CH, DiGiovine B, Toews GB, Hyzy RC. Low hospital mortality in patients with acute interstitial pneumonia. Chest 2003;124(2):554–9.

[19] Palmucci S, Roccasalva F, Puglisi S, Torrisi SE, Vindigni V, Mauro LA, et al. Clinical and radiological features of idiopathic interstitial pneumonias (IIPs): a pictorial review. Insights Imaging 2014;5(3):347–64.

[20] Primack SL, Hartman TE, Ikezoe J, Akira M, Sakatani M, Müller NL. Acute interstitial pneumonia: radiographic and CT findings in nine patients. Radiology 1993;188(3):817–20.

[21] Johkoh T, Müller NL, Taniguchi H, Kondoh Y, Akira M, Ichikado K, et al. Acute interstitial pneumonia: thin-section CT findings in 36 patients. Radiology 1999;211(3):859–63.

[22] Ichikado K, Suga M, Müller NL, Taniguchi H, Kondoh Y, Akira M, et al. Acute interstitial pneumonia: comparison of high resolution computed tomography findings between survivors and non survivors. Am J Respir Crit Care Med 2002;165(11):1551–6.

[23] Silva CI, Müller NL. Idiopathic interstitial pneumonias. J Thorac Imaging 2009;24(4):260–73.

[24] Lim WS, van der Eerden MM, Laing R, Boersma WG, Karalus N, Town GI, et al. Defining community acquired pneumonia severity on presentation to hospital: an international derivation and validation study. Thorax 2003;58(5):377–82.

[25] Tomiyama N, Muller NL, Johkoh T, Honda O, Mihara N, Kozuka T, et al. Acute parenchymal lung disease in immunocompetent patients: diagnostic accuracy of high-resolution CT. AJR Am J Roentgenol 2000;174(6):1745–50.

[26] Fujita J, Tohyama M, Haranaga S, Cash HL, Higa F, Tateyama M. Hamman-Rich syndrome revisited: how to avoid misdiagnosis. Influenza Other Respir Viruses 2013;7(1):4–5.

[27] Askin FB, Katzenstein AL. Pneumocystis infection masquerading as diffuse alveolar damage: a potential source of diagnostic error. Chest 1981;79(4):420–2.

[28] Iannuzzi MC, Farhi DC, Bostrom PD, Petty TL, Fisher JH. Fulminant respiratory failure and death in a patient with idiopathic bronchiolitis obliterans. Arch Intern Med 1985;145(4):733–4.

[29] Perez de Llano LA, Soilan JL, Garcia Pais MJ, Mata I, Moreda M, Laserna B. Idiopathic bronchiolitis obliterans with organizing pneumonia presenting with adult respiratory distress syndrome. Respir Med 1998;92(6):884–6.

[30] Koinuma D, Miki M, Ebina M, Tahara M, Hagiwara K, Kondo T, et al. Successful treatment of a case with rapidly progressive bronchiolitis obliterans organizing pneumonia (BOOP) using cyclosporine A and corticosteroid. Intern Med 2002;41(1):26–9.

[31] Lee JW, Lee KS, Lee HY, Chung MP, Yi CA, Kim TS, et al. Cryptogenic organizing pneumonia: serial high-resolution CT findings in 22 patients. AJR Am J Roentgenol 2010;195(4):916–22.

[32] Tazelaar HD, Linz LJ, Colby TV, Myers JL, Limper AH. Acute eosinophilic pneumonia: histopathologic findings in nine patients. Am J Respir Crit Care Med 1997;155(1):296–302.

[33] De Giacomi F, Vassallo R, Yi ES, Ryu JH. Acute eosinophilic pneumonia. Causes, diagnosis, and management. Am J Respir Crit Care Med 2018;197(6):728–36.

[34] Kolb M, Bondue B, Pesci A, Miyazaki Y, Song JW, Bhatt NY, et al. Acute exacerbations of progressive-fibrosing interstitial lung diseases. Eur Respir Rev 2018;27(150). pii: 180071.

[35] Gonzalves-Venade G, Lacerda-Prнncipe N, Roncon-Albuquerque Jr R, Paiva JA. Extracorporeal membrane oxygenation for refractory severe respiratory failure in acute interstitial pneumonia. Artif Organs 2018;42(5):569–74.

[36] Suzuka T, Kotani T, Takeuchi T, Fujiki Y, Hata K, Yoshida S, et al. Efficacy and safety of oral high-trough level tacrolimus in acute/subacute interstitial pneumonia with dermatomyositis. Int J Rheum Dis 2019;22(2):303–13.

Chapter 2.7

Lymphoid interstitial pneumonia

Lymphoid (syn. lymphocytic) interstitial pneumonia (LIP) is a rare interstitial inflammatory disease characterized by polymorphic infiltration of the interstitial lung tissue by lymphocytes, histiocytes, and plasma cells. It was first described by Liebow and Carrington in 1966 [1].

Similar to other types of interstitial pneumonia, LIP is often a secondary pulmonary lesion in the Sjögren syndrome (25% of LIP cases), other connective tissue diseases (CTDs), and autoimmune thyroiditis. Moreover, LIP may present as a pulmonary reaction in allogeneic bone marrow transplantation and other autoimmune diseases; cytomegalovirus, herpetic, HIV, and hepatitis B infections; and in common variable immunodeficiency [2–5]. LIP is also a morphological substrate of interstitial pneumonia with autoimmune features [6].

Nevertheless, LIP can present as an independent nosological form based on its clinical, radiological, and histopathologic signs. However, in general, idiopathic LIP is much less common than secondary LIP and affects less than 20% of all LIP patients [7]. LIP, in addition to idiopathic pleuroparenchymal fibroelastosis, is one of the rarest variants of idiopathic interstitial pneumonia (IIP) [8].

LIP is considered a benign pulmonary lymphoid disorder, a classification that also includes follicular bronchiolitis, nodular lymphoid hyperplasia (pulmonary pseudolymphoma), inflammatory pseudotumor (plasma cell granuloma), immunoglobulin G4-related disease, Castleman disease, and posttransplantation lymphoproliferative disease [9].

In some cases, LIP may represent a transitional form of a malignant lymphoma [10]. However, in general, LIP is not considered a prelymphoma because less than 5% of LIP cases undergo lymphomatous transformation [11]. The association with many autoimmune disorders suggests an autoimmune pathogenesis of the disease, and the presence of familial cases of the disease indicates a genetic predisposition to the development of LIP [12]. Middle-aged women (40–50 years) are most commonly affected by LIP and develop the condition nearly three times more often than men [7]. Furthermore, LIP generally (up to 14% of cases) develops in children who were HIV-infected in utero [13].

In a study by Cha, 75% of patients with LIP never smoked, and the rest were former smokers [7].

Morphology

Morphology of LIP is characterized by the presence of polyclonal lymphoid infiltrates with an admixture of plasma cells and histiocytes in the pulmonary interstitium along the alveolar septa (Figs. 2.7.1 and 2.7.2) and in the peribronchiolar and perivascular tissues. These findings are often combined with hyperplasia of the bronchus-associated lymphoid tissue and lymphoid follicles [14]. The formation of germinal centers and the proliferation of both B-(CD19 and CD20) and CD8+ T-lymphocytes are noted in hyperplastic lymphoid tissue [15,16].

Typical signs also include the development of lesions on the branches of the medium-sized pulmonary artery in the form of arteritis with inflammatory infiltration of the walls, intimal fibrosis, fragmentation of elastic tissue, fibrosis, and calcification

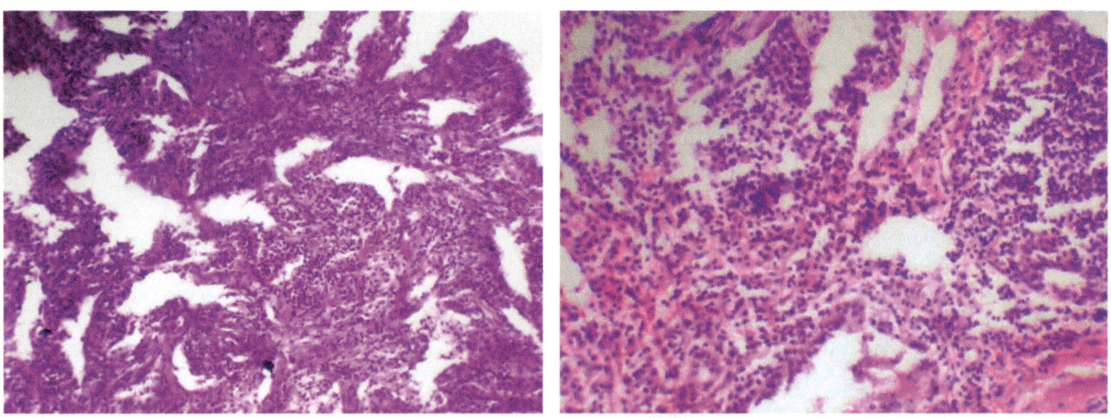

(A) (B)

FIG. 2.7.1 Lymphoid interstitial pneumonia. Diffuse interstitial infiltration with lymphocytes, plasma cells, and histiocytes along the alveolar septa (A and B). Hematoxylin and eosin staining, 200×.

of the media. Possible findings in LIP include foci of organizing pneumonia and poorly formed nonnecrotizing granulomas. Over time, honeycombing areas can appear [17].

In addition to interstitial inflammatory and fibrotic changes, the formation of amyloid deposits is possible when combined with the Sjögren syndrome [18].

Differential histopathologic diagnosis of LIP should be performed with B-cell lymphoma based on immunohistochemical findings [19]. LIP presents with polyclonal lymphoid infiltrates, whereas monoclonal B-cell proliferates are detected in lymphoma.

LIP can histologically resemble a cellular subtype of nonspecific interstitial pneumonia (NSIP), which also displays lymphohistiocytic infiltration of the interalveolar septa. However, in LIP, interstitial inflammation is more diffuse, whereas NSIP has unaffected lung parenchyma foci [3]. Multidisciplinary discussion often results in a change in the histopathologic diagnosis of LIP to the NSIP cellular subtype [8].

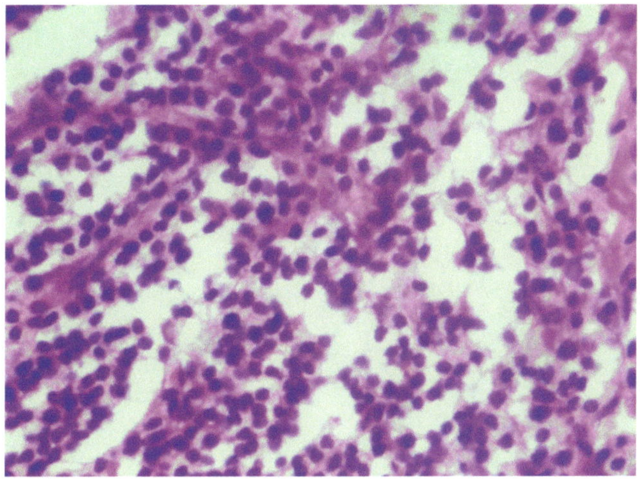

FIG. 2.7.2 Lymphoid interstitial pneumonia. Lymphohistiocytic infiltrate of the interstitium. Hematoxylin and eosin staining, 600×.

Clinical presentation

The disease has a slow onset. On average, there are more than 2 years from the appearance of the first symptoms to the established diagnosis [7]. Similar to other IIPs the most significant respiratory symptoms of LIP are dyspnea in physical exertion and cough. The most common general constitutional symptom is fatigue (88% of patients) [7]. Fever, weight loss, thoracalgia, and arthralgia syndrome can be observed [7,20]. The disease usually progresses gradually with a slow (over several months) increase in symptoms. Crackles can be auscultated in the lower fields. Clubbed fingers can sometimes be observed [3]. If LIP is a secondary disorder, the clinical manifestations of the underlying disease are usually identified.

Diagnosis

Mild anemia is often revealed in blood tests. Up to 75% of LIP cases are accompanied by polyclonal hypergammaglobulinemia or a monoclonal increase in IgG or IgM levels [3,21]. To rule out secondary LIP, laboratory study is required to evaluate a range of antinuclear antibodies, anticyclic citrullinated peptide antibodies, anti-SSA/Ro, anti-SSB/La antibodies, EBV titers, HIV, human T-cell lymphotropic virus, quantitative immunoglobulins, rheumatoid factor, serum and urine protein electrophoresis, thyroid function tests, and IgG4 [12].

LIP patients are characterized by a decrease in forced vital capacity (FVC) and forced expiratory volume in one second (FEV1), an increase in the FEV1/FVC ratio, and a decrease in the total lung capacity and diffusing lung capacity for carbon monoxide (DLCO), which corresponds directly to the volume of the pulmonary parenchyma lesion [7].

Moderate lymphocytosis (~30%) is usually revealed in an analysis of bronchoalveolar lavage (BAL) fluid with a decrease in the CD4+/CD8+ ratio but with normal CD3+ and CD21+ levels [7].

The histopathologic diagnosis of lymphoproliferative lung diseases (which often includes an immunohistochemical analysis) usually requires a sufficiently large amount of biopsy material. Therefore the diagnostic method of choice for lung biopsy is video-assisted thoracoscopy or video-assisted mediastinoscopy (transbronchial forceps biopsy has a lower diagnostic value) [9].

High-resolution computed tomography

The primary radiological findings of LIP detected by high-resolution computed tomography (HRCT) are as follows (Figs. 2.7.3–2.7.5) [22–24]:

- Bilateral diffuse areas of ground-glass opacity (GGO)
- Centrilobular nodules indicating the presence of lymphocytic bronchiolitis
- Formation of thin-walled cysts, mainly in the basal regions
- Well-defined nodules with perilymphatic distribution
- Thickening of the peribronchovascular interstitium
- Reticular abnormalities
- Intrathoracic lymphadenopathy

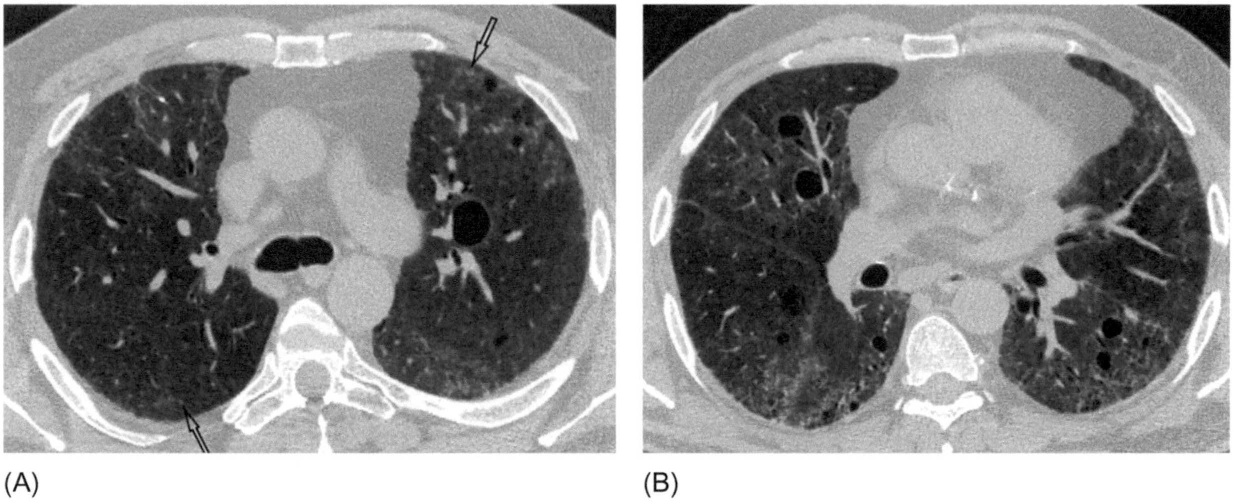

FIG. 2.7.3 Idiopathic lymphoid interstitial pneumonia. Bilateral diffuse areas of ground-glass opacity, with thin-walled cysts up to 25 mm in diameter, located both subpleurally and intraparenchymally (A and B). Numerous small centriacinar nodules are visualized in the upper lobes (A) *(arrows)*.

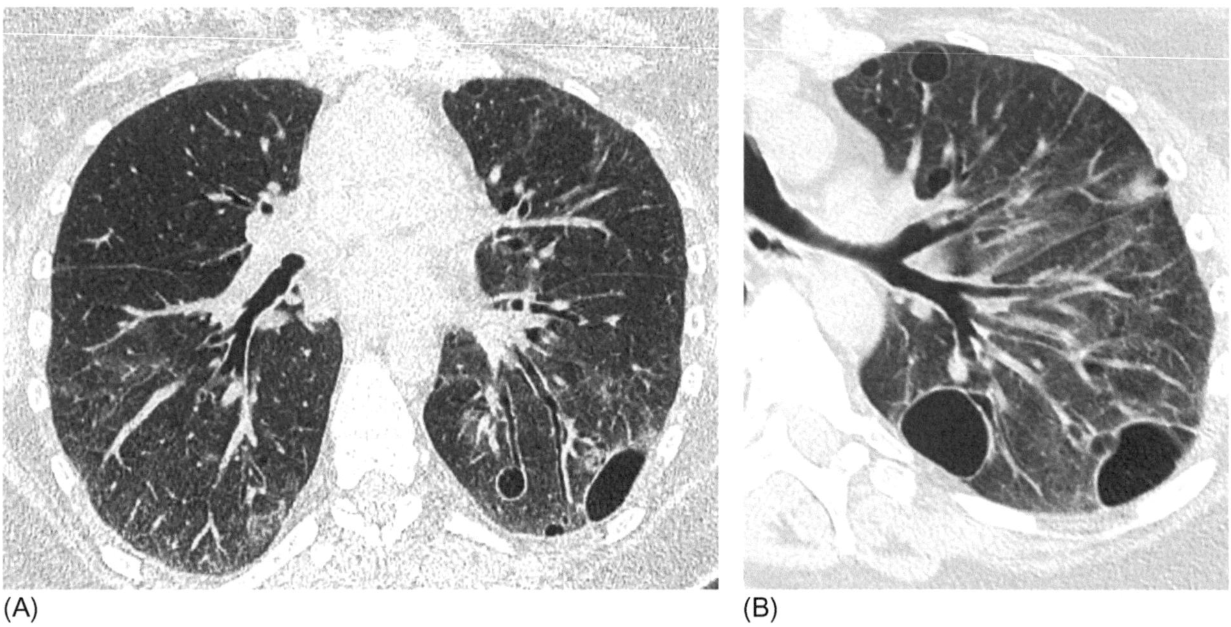

FIG. 2.7.4 Lymphoid interstitial pneumonia in a patient with the Sjögren syndrome. Bilateral patchy areas of ground-glass opacity, moderate reticular changes, and multiple cysts up to 40 mm in diameter (A and B). In the lower segments, subpleural poorly differentiated nodules (A) are visualized.

One of the most frequently cited studies included 22 LIP patients and revealed the presence of GGO and intralobular nodules in all patients. Subpleural nodules and thickening of the bronchovascular bundles were noted in 19 cases (86%), thickening of the interlobular septa was observed in 82% of patients, and cysts of different sizes were detected in 68% of cases [22]. The presence of cysts is a specific sign used to confidently differentiate LIP from other variants of IIP. Cysts vary in size from 1 to 30 mm in diameter, and they are often located within the lung parenchyma (as opposed to predominantly subpleural cysts in idiopathic pulmonary fibrosis (IPF)) [25].

The pathogenesis of cyst formation in LIP is associated with local obstruction of the terminal bronchioles due to bronchiolar and peribronchiolar lymphoid infiltration [24]. Infiltration of the walls of the respiratory bronchioles also explains the appearance of intralobular nodules (a sign of bronchiolitis) on the HRCT.

Mediastinal lymphadenopathy is a common finding in LIP (~70%); however, this sign accompanies other forms of IIP (IPF, NSIP, and respiratory bronchiolitis associated with interstitial lung disease) just as frequently [26].

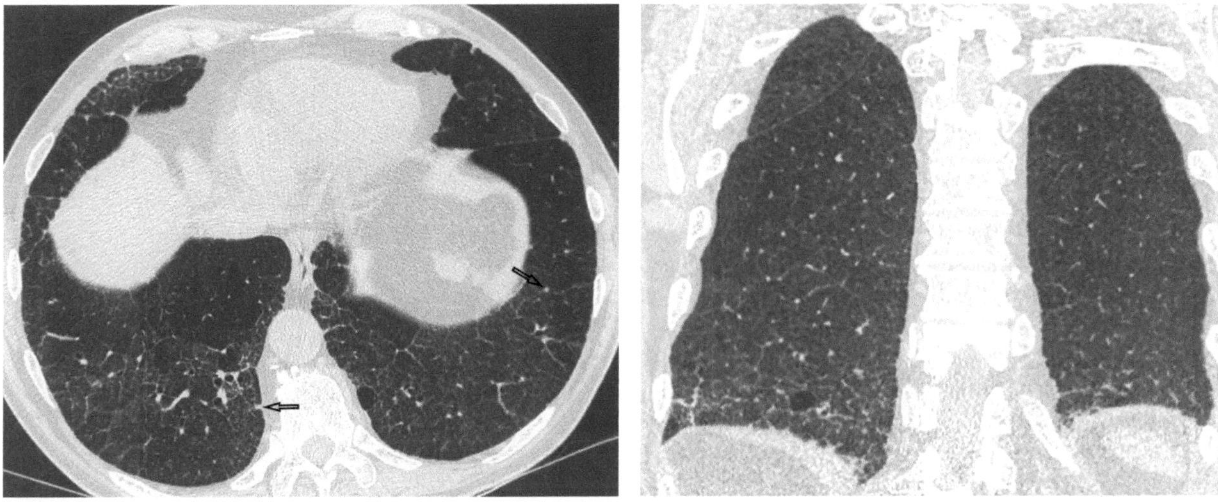

(A) (B)

FIG. 2.7.5 Idiopathic lymphoid interstitial pneumonia. The reticular pattern predominates. In addition, small poorly differentiated centrilobular nodules and well-defined nodules with a perilymphatic distribution *(arrows)* are observed, as is a single small cyst (A). Abnormalities are concentrated in the basal regions of the lungs (B).

During the long course of LIP, limited areas or foci of consolidation can be detected. Their occurrence may be due to the development of secondary amyloidosis, particularly if calcification is visualized inside. Small areas of honeycombing were also detected during the long-term follow-up of LIP patients in approximately 30% of cases (Fig. 2.7.6), although in general this feature is not typical for LIP [27].

Under the influence of therapy, the HRCT pattern usually changes; thus, in patients who respond to treatment, GGO areas are reduced or disappear, the number of intralobular nodules decreases, and the cysts remain in a stable condition. With disease progression, new cysts and areas of honeycombing are formed, and areas of consolidation appear [27].

However, it must be noted that there were less than 100 total patients in the studies that assessed how the radiological pattern of idiopathic LIP is formed; therefore it is possible that knowledge on this pathology will be further refined.

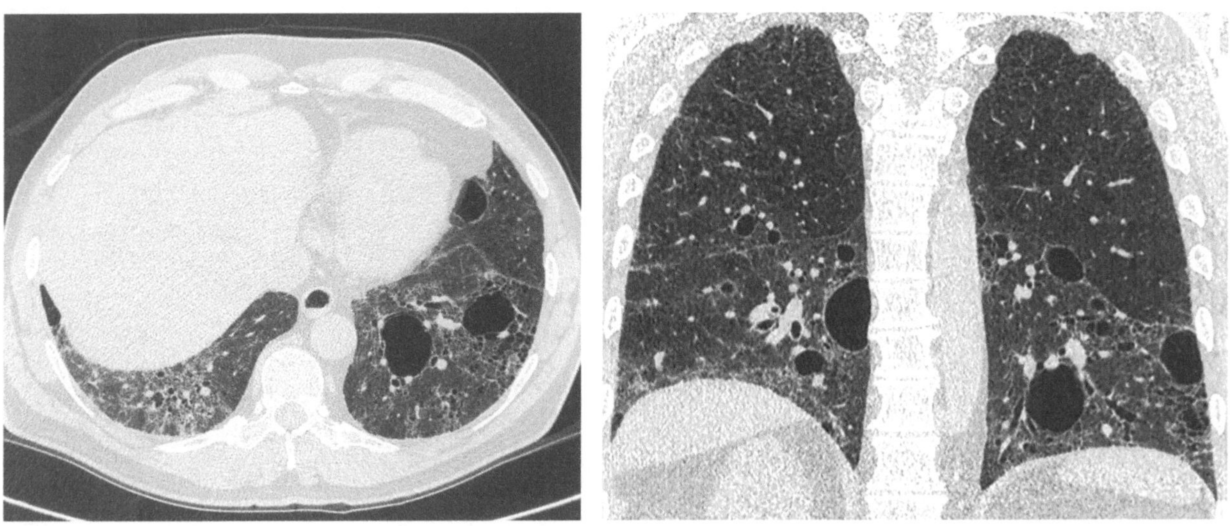

(A) (B)

FIG. 2.7.6 Idiopathic lymphoid interstitial pneumonia. On the right the area of diffuse GGO is visualized; on the left, there are reticular abnormalities, large thin-walled cysts, and areas of subpleural and intraparenchymal honeycombing (A). The coronal section shows abnormalities, most pronounced in the lower areas, and multiple cysts of different sizes (B).

Differential diagnosis

LIP must be differentiated from other IIPs, diseases characterized by the formation of cysts in the lungs, and nonneoplastic lymphoproliferative pulmonary diseases (Table 2.7.1). LIP should always be suspected in the presence of cystic changes in the lungs in combination with GGO. However, the absence of cysts (in 20%–30% of patients) makes the radiological diagnosis of LIP impossible without histological confirmation. Subacute hypersensitivity pneumonitis (HP), the cellular type of NSIP, and desquamate interstitial pneumonia (DIP) may have the same clinical and radiological features (GGO, intralobular nodules, and thickening of interlobular septa) as LIP (Fig. 2.7.7). Moreover, thin-walled cysts are also common in HP and DIP (Fig. 2.7.8). However, unlike LIP, the cysts in these other diseases are usually single and small in size (Fig. 2.7.9). In addition, NSIP, like LIP, often develops in combination with autoimmune diseases, and their morphological signs significantly overlap.

In addition to small-cyst honeycombing, the formation of larger cysts may be observed in IPF patients, which may sometimes require differentiation from LIP. The presence of honeycombing cannot completely rule out the diagnosis of

TABLE 2.7.1 Differential characteristics of DIP and similar diseases

	Histopathologic signs	HRCT signs	Additional characteristics
Lymphoid interstitial pneumonia	Diffuse infiltration of the interalveolar septa with lymphocytes, plasma cells, and histiocytes Hyperplasia of lymphoid follicles with germinal centers Lymphocytic infiltration of the walls of small vessels	Single or moderate number of thin-walled cysts 1–30mm in diameter, mainly in the basal regions Ground-glass opacity thickening of interlobular septa, consolidation areas, honeycombing in severe cases	Mostly nonsmoking women 40–50 years old Frequent association with autoimmune diseases or immunodeficiencies
Follicular bronchiolitis	Hyperplasia of lymphoid follicles with germinal centers, peribronchiolar lymphocytic infiltration	Multiple poorly differentiated centrilobular nodules, spotted areas of ground-glass opacity, air traps, small cysts, mild reticular abnormalities	Frequent association with autoimmune diseases or immunodeficiencies
NSIP (cellular subtype)	Uniform moderate interstitial lymphohistiocytic infiltration with alternating of affected and sparing fields Hyperplasia of type II pneumocystis in areas of inflammation	Diffuse or patchy GGO in the posterior-basal segments, subpleural sparing, moderate reticular signs Cysts are not characteristic	Mostly nonsmoking women over 40 years of age Frequent association with CTDs and drug-induced interstitial lung disease
Subacute hypersensitivity pneumonitis	Bronchiolocentric lymphocytic infiltration, nonnecrotizing poorly shaped granulomas in the peribronchiolar interstitium	Diffuse or spotty areas of GGO with subpleural sparing, air traps, centrilobular, poorly differentiated nodules Single thin-walled cysts up to 15mm in diameter	Regular contact with allergens (birds, fungi, wood, metal dust, and others) Lymphocytosis of BAL >50%
Desquamative interstitial pneumonia	Accumulation of hyperpigmented macrophages in the lumens of alveoli and bronchioles Moderate lymphohistiocytic infiltration of the interalveolar septa	Diffuse or spotty areas of GGO in the basal and subpleural segments Moderate reticular abnormalities Single thin-walled cysts up to 1cm in diameter	Active smoking or contact with pollutants
Pneumocystis pneumonia	Foamy intra-alveolar exudate, lymphoplasmacytic interstitial infiltration Pneumocystis are positive for Gomori methenamine silver, cresyl echt violet, toluidine blue O, or calcofluor white stain	Spotted or diffuse areas of ground-glass opacity, nodules, and air traps primarily in the upper and middle fields Cysts of the same localization are thin- and thick-walled, closely adjacent to the pleura	HIV infection, immunodeficiency, fever, weight loss, oropharyngeal candidiasis, herpes zoster The presence of *P. jirovecii* in BAL fluid and sputum

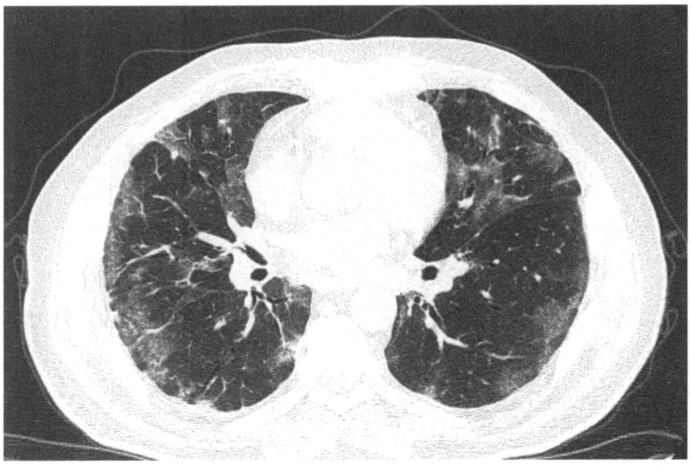

FIG. 2.7.7 Nonspecific interstitial pneumonia, cellular subtype. Bilateral spotted areas of GGO in the lower segments, associated with moderate reticular changes, and subpleural sparing.

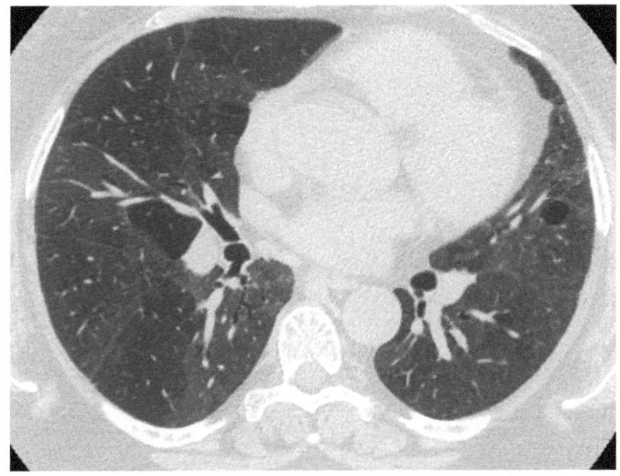

FIG. 2.7.8 Hypersensitivity pneumonitis in a bird fancier. Diffuse areas of GGO associated with air traps. On the left, a single thin-walled cyst is observed.

LIP. However, a pronounced reticular pattern, the absence of signs of bronchiolitis, and minimal expression of GGO usually help establish a correct radiological diagnosis of IPF (Fig. 2.7.10). In addition, large cysts in IPF are usually located subpleurally and in the areas of honeycombing, whereas in LIP, cysts are more often found inside the lung parenchyma within GGO areas [25].

Several infectious diseases (*Pneumocystis jirovecii* pneumonia (PP), tuberculosis, mycobacterioses, and fungal infections) that destroy lung tissue and form cysts may require a differential diagnosis with LIP. PP almost always occurs in patients with a severe immunodeficiency, often AIDS. PP is characterized by a pronounced febrile fever, the emergence of diffuse or nodular areas of GGO, and consolidation on HRCT. Cysts are usually formed in severe courses in areas of greatest attenuation. PP is characterized by primary lesions of the upper lobes, with further progression to the lower fields. Cysts are usually thick-walled in infectious diseases, which is uncharacteristic for LIP. Thin-walled cavities can also form, with prevailing upper lobe localization [28]. Since HIV can be an inducer of secondary LIP, it can sometimes be differentiated from PP only by the presence of *P. jirovecii* in sputum or BAL fluid or by morphological characteristics (Fig. 2.7.11). Tuberculosis, mycobacterioses, and PP usually affect the upper lobes most significantly, and there are generally many bronchogenic disseminated nodules around the thick-walled cavities.

For cystic lung diseases such as lymphangioleiomyomatosis (LAM) or Birt-Hogg-Dubé syndrome, the main differential HRCT sign that enables ruling out the diagnosis of LIP is the absence of signs of damage to the parenchyma outside of the cysts (Fig. 2.7.12). Langerhans cell histiocytosis can reveal oneself with polymorphic cysts in the lungs that usually do not exceed 10 mm in size and is generally accompanied by the presence of multiple nodules with a predominant upper lobe distribution; GGO can also be noted at times (Fig. 2.7.13). Pulmonary amyloidosis and the light chain deposition disease are characterized by the formation of cysts in the lungs; however, as a rule, there are multiple areas of consolidation, indicating intra-alveolar lipid, protein, and amyloid deposits.

Cystic pulmonary metastases usually appear as separated, thin-walled cavities in the unaltered lung tissue. It is extremely difficult to establish the exact nature of such changes without obtaining cytological or histological confirmation.

Diffuse cystic degeneration, sometimes observed in the terminal stage of LIP, can be indistinguishable from other cystic diseases or pulmonary emphysema [24].

The differential diagnosis of cystic lung lesions is presented in greater detail in Chapter 9.4.

Nodular lymphoid hyperplasia as a lymphoproliferative disease has a different HRCT pattern that involves the presence of consolidated foci 2–3 cm in diameter and nodules with a perilymphatic distribution [29]. Moderate reticular changes and GGO may also occur (Fig. 2.7.14).

Follicular bronchiolitis (FB) displays a similar HRCT and histological pattern to that of LIP; moreover, it is associated with the same autoimmune and immunodeficiency conditions as LIP. However, FB is characterized by the appearance of multiple centrilobular peribronchiolar nodules of 1–3 mm in the lower segments and the frequent emergence of air traps (Fig. 2.7.15) [30]. GGO and mild reticular abnormalities may also be found, which makes FB visually indistinguishable from LIP.

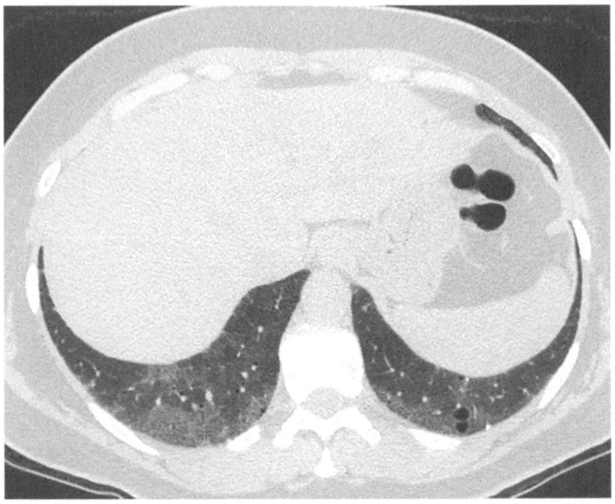

FIG. 2.7.9 Desquamative interstitial pneumonia in a 46-year-old woman. Bilateral subpleural areas of GGO and single small nodules. On the left, two small thin-walled cysts are observed.

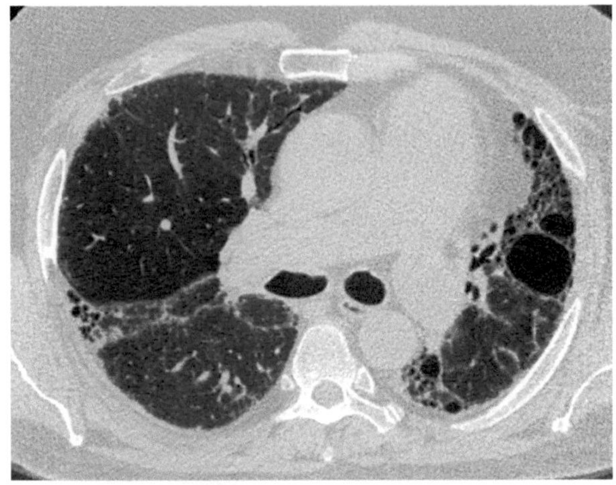

FIG. 2.7.10 Idiopathic pulmonary fibrosis. Bilateral subpleural areas of honeycombing and reticular abnormalities. Transseptal linear attenuations are noted on the left. Two thin-walled subpleural cysts are visualized in the lingular lobes in the honeycombing region.

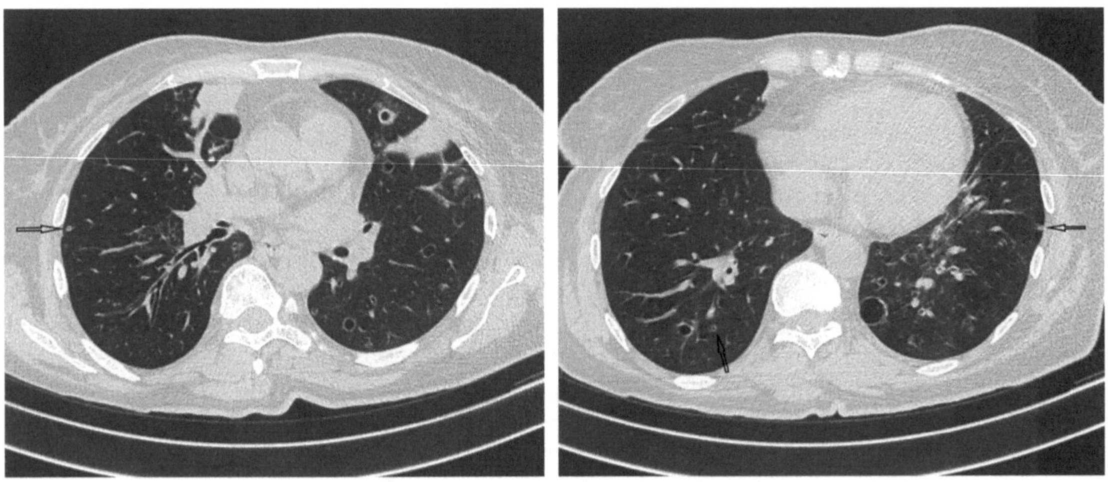

(A) (B)

FIG. 2.7.11 Atypical presentation of *Pneumocystis jirovecii* pneumonia in an AIDS patient. Bilateral subpleural areas of consolidation in the right middle and left upper lobes (A). Single small well-defined nodules *(arrows)*, random thin- and thick-walled cavities, and a minimally prominent sign of GGO (A and B).

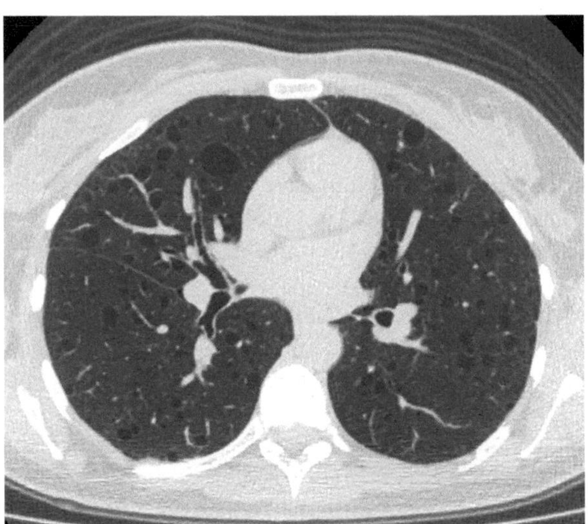

FIG. 2.7.12 Lymphangioleiomyomatosis. Bilateral multiple randomly distributed thin-walled cysts, some of which reach a diameter of 25 mm. There are no other signs of the lung tissue lesion.

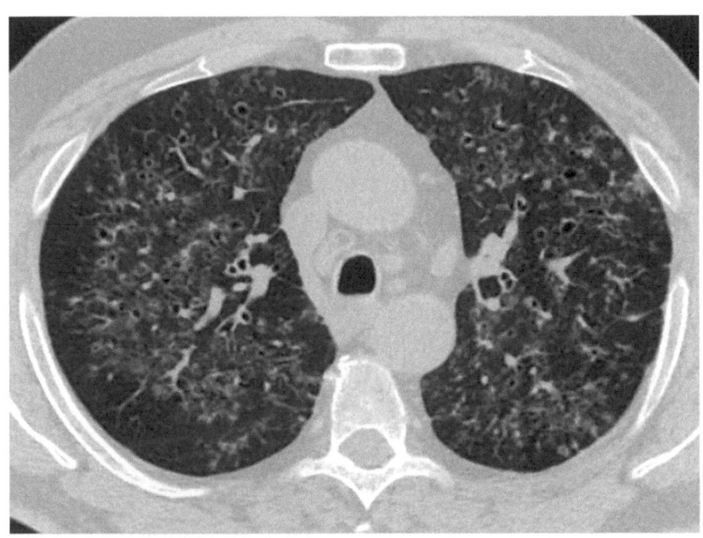

FIG. 2.7.13 Langerhans cell histiocytosis. Multiple bilateral thin- and thick-walled cavities, polymorphic nodules, and patchy GGO areas with parachilar distribution are visible.

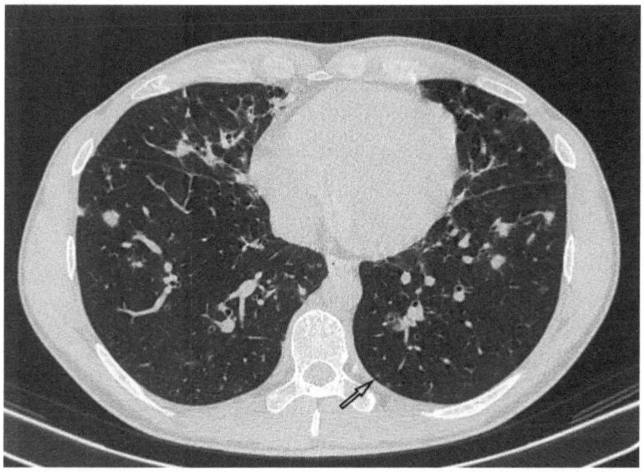

FIG. 2.7.14 Nodular lymphoid hyperplasia in a patient with primary variable immunodeficiency. Bilateral dense nodules up to 6 mm in diameter, moderate reticular abnormalities in the middle lobe and lingular lobes, and small poorly differentiated centriacinar nodules *(arrow)*.

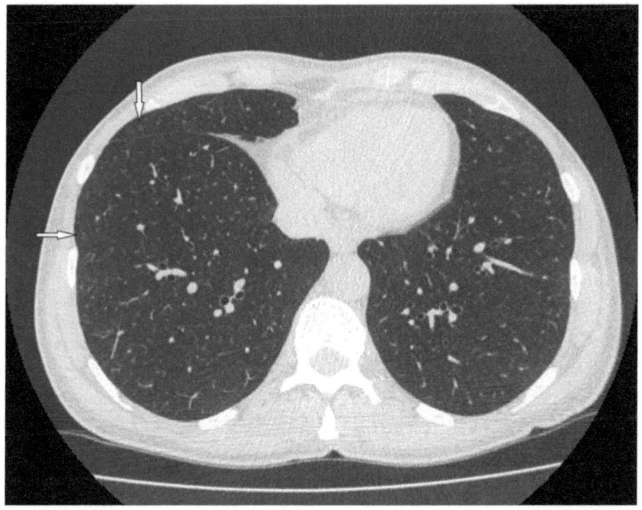

FIG. 2.7.15 Follicular bronchiolitis in a patient with the Sjögren syndrome. Small ill-defined nodules *(arrows)*, slight subpleural reticular changes.

Treatment and prognosis

Treatment of secondary LIP should be focused primarily on the treatment and control of the underlying disease [9]. Because of the low prevalence of idiopathic LIP, there are no evidence-based studies evaluating the efficacy of certain drugs, different doses, and courses of treatment. Depending on the severity of the disease course, the starting dose of prednisone can vary from 0.75 to 1 mg/kg of body weight for 3 months, with a possible subsequent dose reduction by 0.25 mg/kg/day every 6–12 weeks [11]. With ineffectiveness or the lack of effectiveness, single successful attempts of treating LIP with cyclophosphamide, azathioprine, methotrexate, cyclosporine, mycophenolate mofetil, and rituximab have been described [31–34].

A total of 40%–65% of LIP patients who receive systemic steroids have a good or moderate treatment outcome; however, in some patients, the disease progresses. The 5-year survival rate is approximately 70% [19]. The average survival rate at 14-year follow-up, according to Cha et al., reached 11.5 years from the time of establishing the diagnosis [7].

References

[1] Liebow A, Carrington CB. The interstitial pneumonias. In: Simon M, Potchen EJ, LeMay M, editors. Frontiers of pulmonary radiology. 1st ed. New York: Grune & Stratton; 1969. p. 102–41.

[2] Khardori R, Eagleton L, Soler N, McConnachie P. Lymphocytic interstitial pneumonitis in autoimmune thyroid disease. Am J Med 1991;90(5):649–52.

[3] American Thoracic Society/European Respiratory Society. American Thoracic Society/European Respiratory Society international multidisciplinary consensus classification of the idiopathic interstitial pneumonias. Am J Respir Crit Care Med 2002;165(2):277–304.

[4] Schlemmer F, Chevret S, Lorillon G, De Bazelaire C, Peffault de Latour R, Meignin V, et al. Late-onset noninfectious interstitial lung disease after allogeneic hematopoietic stem cell transplantation. Respir Med 2014;108(10):1525–33.

[5] Zdziarski P, Gamian A, Dworacki G. A case report of lymphoid interstitial pneumonia in common variable immunodeficiency: oligoclonal expansion of effector lymphocytes with preferential cytomegalovirus-specific immune response and lymphoproliferative disease promotion. Medicine (Baltimore) 2017;96(23):e7031.

[6] Fischer A, Antoniou KM, Brown KK, Cadranel J, Corte TJ, du Bois RM, et al. An official European Respiratory Society/American Thoracic Society research statement: interstitial pneumonia with autoimmune features. Eur Respir J 2015;46(4):976–87.

[7] Cha S-I, Fessler MB, Cool CD, Schwarz MI, Brown KK. Lymphoid interstitial pneumonia: clinical features, associations and prognosis. Eur Respir J 2006;28(2):364–9.

[8] Travis WD, Costabel U, Hansell DM, King TE, Lynch DA, Nicholson AG, et al. An official American Thoracic Society/European Respiratory Society statement: update of the international multidisciplinary classification of the idiopathic interstitial pneumonias. Am J Respir Crit Care Med 2013;188(6):733–48.

[9] Arcadu A, Moua T, Yi ES, Ryu JH. Lymphoid interstitial pneumonia and other benign lymphoid disorders. Semin Respir Crit Care Med 2016;37(3):406–20.

[10] Kradin RL, Young RH, Kradin LA, Mark EJ. Immunoblastic lymphoma arising in chronic lymphoid hyperplasia of the pulmonary interstitium. Cancer 1982;50(7):1339–43.

[11] Swigris JJ, Berry GJ, Raffin TA, Kuschner WG. Lymphoid interstitial pneumonia: a narrative review. Chest 2002;122(6):2150–64.
[12] Panchabhai TS, Farver C, Highland KB. Lymphocytic interstitial pneumonia. Clin Chest Med 2016;37(3):463–74.
[13] Nesheim SR, Lindsay M, Sawyer MK, Mancao M, Lee FK, Shaffer N, et al. A prospective population-based study of HIV perinatal transmission. AIDS 1994;8(9):1293–8.
[14] Travis WD, Galvin JR. Non-neoplastic pulmonary lymphoid lesions. Thorax 2001;56(12):964–71.
[15] Nicholson AG, Rice AJ. Interstitial lung diseases. In: Hasleton P, Flieder DB, editors. Spencer's pathology of the lung. 6th ed. New York: Cambridge Medicine; 2016. p. 366–408.
[16] Costabel U, du Bois RM, Egan JJ, editors. Diffuse parenchymal lung disease. Prog Respir Resvol. 36. Basel: Karger; 2007. 348 p.
[17] Guinee Jr DG. Update on nonneoplastic pulmonary lymphoproliferative disorders and related entities. Arch Pathol Lab Med 2010;134(5):691–701.
[18] Baqir M, Kluka EM, Aubry MC, Hartman TE, Yi ES, Bauer PR, et al. Amyloid-associated cystic lung disease in primary Sjögren's syndrome. Respir Med 2013;107(4):616–21.
[19] Nicholson AG, Wotherspoon AC, Diss TC, Hansell DM, Du Bois R, Sheppard MN, et al. Reactive pulmonary lymphoid disorders. Histopathology 1995;26(5):405–12.
[20] Mason RJ, Murray JF, Broaddus VC, Nadel JA. Textbook of respiratory medicine. 4th ed. vol. 2. USA: Elsevier Inc; 20051601–2.
[21] DeCoteau WE, Tourville D, Ambrus JL, Montes M, Adler R, Tomasi TBJ. Lymphoid interstitial pneumonia and autoerythrocyte sensitization syndrome. A case with deposition of immunoglobulins on the alveolar basement membrane. Arch Intern Med 1974;134(3):519–22.
[22] Johkoh T, Müller NL, Pickford HA, Hartman TE, Ichikado K, Akira M, et al. Lymphocytic interstitial pneumonia: thin-section ct findings in 22 patients. Radiology 1999;212(2):567–72.
[23] Ichikawa Y, Kinoshita M, Koga T, Oizumi K, Fujimoto K, Hayabuchi N. Lung cyst formation in lymphocytic interstitial pneumonia: CT features. J Comput Assist Tomogr 1994;18(5):745–8.
[24] Silva CIS, Flint JD, Levy RD, Müller NL. Diffuse lung cysts in lymphoid interstitial pneumonia: high-resolution CT and pathologic findings. J Thorac Imaging 2006;21(3):241–4.
[25] Seaman DM, Meyer CA, et al. Diffuse cystic lung disease at high-resolution CT. AJR 2011;196:1305–11.
[26] Souza CA, Muller NL, Lee KS, et al. Idiopathic interstitial pneumonias: prevalence of mediastinal lymph node enlargement in 206 patients. AJR Am J Roentgenol 2006;186:995–9.
[27] Johkoh T, Ichikado K, Akira M, Honda O, Tomiyama N, Mihara N, et al. Lymphocytic interstitial pneumonia: follow-up CT findings in 14 patients. J Thorac Imaging 2000;15(3):162–7.
[28] Kanne DR, Meyer CA. Pneumocystis jiroveci pneumonia: high-resolution CT findings in patients with and without HIV infection. AJR Am J Roentgenol 2012;198(6):W555–61.
[29] Abbondanzo SL, Rush W, Bijwaard KE, Koss MN. Nodular lymphoid hyperplasia of the lung: a clinicopathologic study of 14 cases. Am J Surg Pathol 2000;24(4):587–97.
[30] Howling SJ, Hansell DM, Wells AU, Nicholson AG, Flint JD, Müller NL. Follicular bronchiolitis: thin-section CT and histologic findings. Radiology 1999;212(3):637–42.
[31] Koss MN, Hochholzer L, Langloss JM, Wehunt WD, Lazarus AA. Lymphoid interstitial pneumonia: clinicopathological and immunopathological findings in 18 cases. Pathology 1987;19(2):178–85.
[32] Davies CW, Juniper MC, Gray W, Gleeson FV, Chapel HM, Davies RJ. Lymphoid interstitial pneumonitis associated with common variable hypogammaglobulinaemia treated with cyclosporin A. Thorax 2000;55(1):88–90.
[33] Devauchelle-Pensec V, Pennec Y, Morvan J, Pers JO, Daridon C, Jousse-Joulin S, et al. Improvement of Sjogren's syndrome after two infusions of rituximab (anti-CD20). Arthritis Rheum 2007;57(2):310–7.
[34] Yum HK, Kim ES, Ok KS, Lee HK, Choi SJ. Lymphocytic interstitial pneumonitis associated with Epstein-Barr virus in Systemic Lupus Erythematosus and Sjogren's syndrome. Complete remission with corticosteroid and cyclophosphamide. Korean J Intern Med 2002;17(3):198–203.

Chapter 2.8

Idiopathic pleuroparenchymal fibroelastosis

Idiopathic pleuroparenchymal fibroelastosis (IPPFE) is a rare form of idiopathic interstitial pneumonia included in this group in 2013 by the consensus of the American Thoracic Society and the European Respiratory Society [1]. Pleuroparenchymal fibroelastosis (PPFE) represents a specific pattern of pulmonary fibrosis localized mainly or exclusively in the upper lobes with the obligatory involvement of the pleura with subpleural and paraseptal intra-alveolar fibroelastosis [2]. The first publication on PPFE summarized 12 cases of upper lobe pulmonary fibrosis in 1992; and IPPFE is sometimes named *Amitani disease* after the name of the first Japanese author [3]. The term IPPFE, adopted by international experts, appeared only in 2004 [4]. Some cases of isolated progressive upper lobe pulmonary fibrosis had been described earlier; however, they were regarded as lung response to provoking factors, such as radiation, intake of anticancer drugs, lungs and bone marrow transplantations, and contact with asbestos and aluminum dust [5]. PPFE can also be associated with autoimmune diseases such as rheumatoid arthritis, ankylosing spondylitis, and ulcerative colitis [5]. In the absence of such associations, PPFE should be considered idiopathic. Thusen et al. reviewed all publications up to mid-2013 and showed that only 78 patients with proved PPFE aged from 13 to 85 years (mean age 49 years) had been described (without a gender preference and regardless of any smoking histories) [2]. The disease developed after lung transplantation in 47% of patients, after bone marrow transplantation in 6% of patients, and in 9% of patients with a history of pulmonary fibrosis, and about 10% of patients had taken immunosuppressants [2]. In a later study by Enomoto et al., 18% of 44 patients with IPPFE had a history of exposure to occupational dust, 16% had malignant neoplastic diseases, and 11% had a family history of interstitial lung disease [6].

IPPFE belongs to the group of rare interstitial pneumonias [1]; however, its prevalence seems higher than thought, and according to Nakatani et al. [7], 6% of patients with interstitial lung disease, who underwent surgical biopsy, were diagnosed as having IPPFE according to the results of a multidisciplinary discussion panel.

The pathogenesis of PPFE is unclear. Since a significant proportion of cases represent patients who have undergone lung or bone marrow transplantation, PPFE has been associated with a graft versus host disease [8,9]. In these cases the primary process is considered the occurrence of obliterating bronchiolitis with atelectasis of subpleural areas and a fibroproliferative reaction. Regarding idiopathic forms the possible triggers can be infections, allergens, or pneumotoxic drugs that cause alveolar damage or organizing pneumonia with simultaneous fibrinolytic and macrophage activity disorders and increased fibroblast activity [2].

Morphology

The current description of histopathologic abnormalities in PPFE, as in other fibrosing interstitial pneumonias, tends to include specific patterns. For a "definite pattern," a set of the following findings are necessary [2,10,11] (Figs. 2.8.1–2.8.4):

- Fibrosis of the visceral pleura in the upper lobes
- Marked homogeneous subpleural intra-alveolar fibrosis with septal elastosis (>80% of elastic fibers)

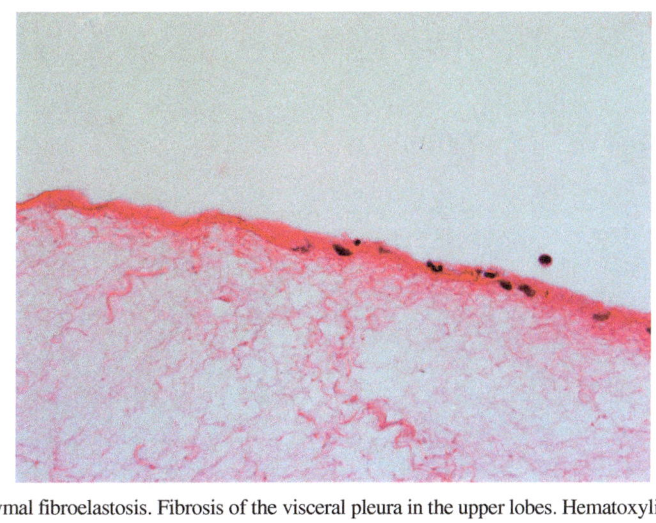

FIG. 2.8.1 Idiopathic pleuroparenchymal fibroelastosis. Fibrosis of the visceral pleura in the upper lobes. Hematoxylin and eosin (H&E) staining, 200×.

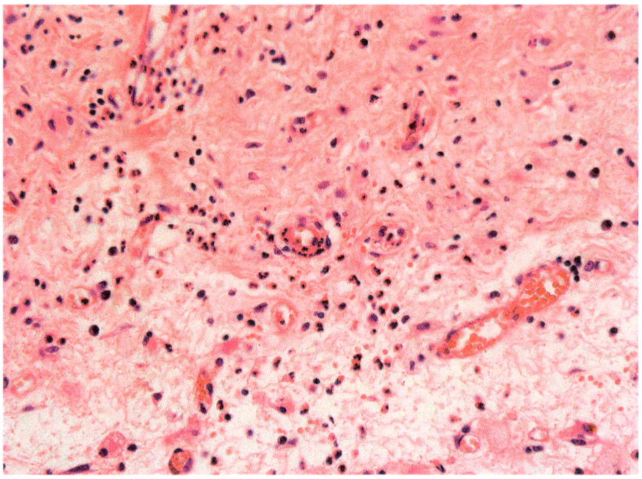

FIG. 2.8.2 Idiopathic pleuroparenchymal fibroelastosis. Fibrosis of the visceral pleura in the upper lobes. H&E stain, 200×.

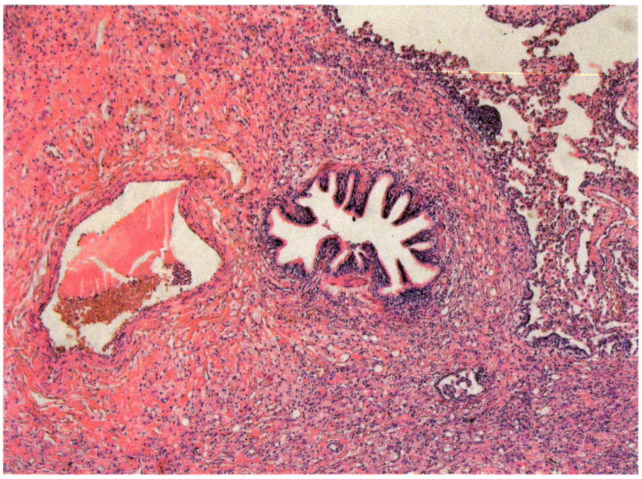

FIG. 2.8.3 Idiopathic pleuroparenchymal fibroelastosis. Subpleural intra-alveolar fibrosis. H&E stain, 100×.

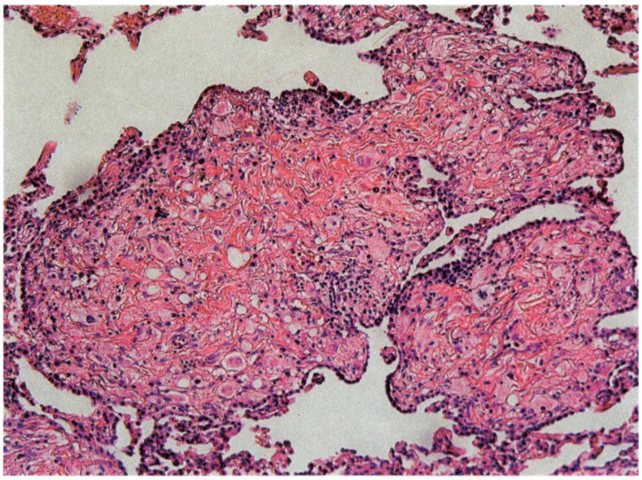

FIG. 2.8.4 Idiopathic pleuroparenchymal fibroelastosis. Subpleural intra-alveolar and peribronchial fibrosis and fibroblastic foci. H&E stain, 200×.

- Preserved pulmonary parenchyma at a distance from the visceral pleura
- Moderate irregular lymphoplasmacytic infiltration
- The presence of some fibroblastic foci

The "consistent pattern" of PPFE [2,10,11] includes the following:

- Intra-alveolar fibrosis unassociated with marked subpleural fibrosis
- Not predominantly subpleural
- Not on upper lobe biopsy

Along with the signs of PPFE, some researchers have described features of other interstitial lung diseases beyond the areas of fibroelastosis, such as usual interstitial pneumonia (up to 43%), nonnecrotizing granulomas (up to 35%), inflammation in areas of fibrosis (up to 36%), and/or vascular fibrointimal thickening in veins and arteries (up to 14%) [10,12]. The presence of granulomas in histological material is associated with a reduced risk of death in IPPFE patients [12].

Hirota et al. described four patients with established chronic fibrosing interstitial pneumonia, organizing pneumonia, and acute lung injury diagnosed based on surgical lung biopsies that were repeated after 3–12 years and revealed replacement of areas of inflammation and organization by subpleural fibroelastosis resulting in clinical diagnoses changes into PPFE [13].

Clinical presentation and course

As with other chronic idiopathic interstitial pneumonias (IIPs), most patients complain of unproductive cough and dyspnea on physical exertion, which increases with the progression of the disease. Chest pain is common and sometimes acute [14]. About 30% of patients have pneumothorax, and in some patients the pneumothorax becomes the first disease manifestation [5]. In 20% of patients, weight loss is noted [5]. Low body mass index (BMI) is generally typical for IPPFE patients; in the study by Enomoto et al., the BMI median was 17.2 kg/m^2 [6]. Since secondary PPFE develops along with serious pathological conditions, the symptoms of the underlying disease, rather than respiratory complaints, may dominate in the clinical presentation. In patients with invasive pulmonary aspergillosis or nontuberculous mycobacterioses, cases of PPFE may develop [6]. In these patients, symptoms that are generally atypical for the fibrotic process, such as cough with purulent sputum and hemoptysis, may be prominent [10]. Japanese authors have noted the asthenic body type and the flattened chest of patients (reduced ratio of the anteroposterior to the transverse diameter), and this anatomical feature is not found together with a "cobbler's chest" (pectus excavatum) [15,16]. In European and American studies, such constitutional symptoms have not been noted. The clubbing sign, characteristic of idiopathic pulmonary fibrosis, was described in a small number of patients with IPPFE (in up to 18% of them) [5,6]. Inspiratory crackles are auscultated in less than half of patients, and they appear only in advanced stages in the basal fields [4,5].

IPPFE usually has an unpredictable course; in asymptomatic patients, signs of progression may not be noted for many years; however, after the appearance of the first clinical symptoms, the disease often progresses rapidly [5]. A gradual course of the disease has also been described in which periods of prolonged slow decline in pulmonary function alternate with short periods of rapid functional decline [17]. There may be episodes of acute exacerbations triggered by surgical lung biopsy or pulmonary infection, which are also associated with concomitant histological patterns of usual interstitial pneumonia (UIP) [18,19].

Acute exacerbations occur in 11%–18% of patients and are accompanied by increased dyspnea, cough, and the appearance of areas of ground-glass opacity (GGO) and consolidation on high-resolution computed tomography (HRCT) images that span the lower areas of the lungs [6,19].

When monitoring IPPFE patients the most frequent complications include respiratory infections, pneumothorax, and pneumomediastinum [6].

Diagnosis

Laboratory abnormalities are uncommon in IPPFE patients; however, increased levels of antinuclear antibodies and rheumatoid factor can be detected [10,20].

With stable IPPF, no significant abnormalities are found in the bronchoalveolar lavage (BAL) fluid; however, this analysis (including culture and polymerase chain reaction tests) may be useful to rule out hypersensitivity pneumonitis, infectious diseases (aspergillosis and mycobacteriosis), and pulmonary sarcoidosis. Although cases of histological IPPFE diagnoses on transbronchial biopsy exist [21], surgical lung biopsy is considered the method of choice for obtaining histological material. Transbronchial cryobiopsy appears to be less informative, since samples do not include the pleura [12].

Functional tests in most patients reveal a restrictive pattern and a decrease in the *carbon monoxide diffusing capacity* (DLCO), which is sometimes quite significant [6,22].

HRCT characteristics

The HRCT signs of PPFE include the following (Figs. 2.8.5 and 2.8.6) [5,10,23,24]:

- Pleural thickening
- Subpleural nodular or extended consolidation areas, often with "corroded" internal contours
- Interlobular septa thickening
- Upper lobe volume losses
- Bullae and cysts in fibrosis areas
- Traction bronchiectasis and subpleural honeycombing

These abnormalities are always localized in the upper lobes, often increasing toward the lung apex. As the disease progresses, the signs of interstitial fibrosis appear also in the lower fields of the lungs, but do not reach the same severity as those in the upper lobes. Pleural plaques and calcification have also been described in the upper lobe visceral pleura [24].

With IPPFE, signs of GGO, pleural effusion, intralobular nodules, or intraparenchymal consolidation are scarce; however, these are typical for many others IIPs.

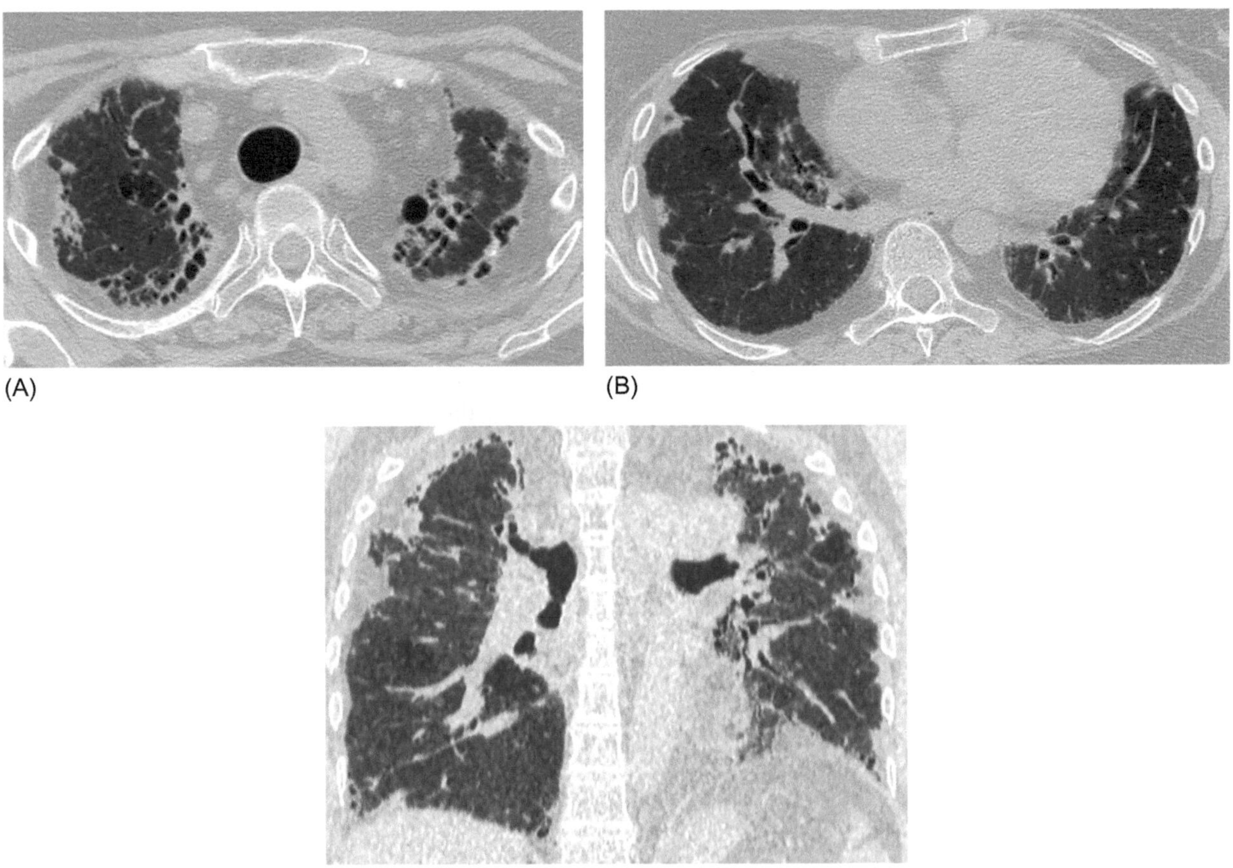

(A) (B) (C)

FIG. 2.8.5 Idiopathic pleuroparenchymal fibroelastosis. Thickening of the visceral pleura in the upper lobes, with subpleural fibrosis, honeycombing, and traction bronchiectasis (A). Moderate thickening of pleura with notched contours and irregular reticular abnormalities in the lower lobes (B). The coronal reconstruction reveals more damage to the upper lobes than lower lobes (C).

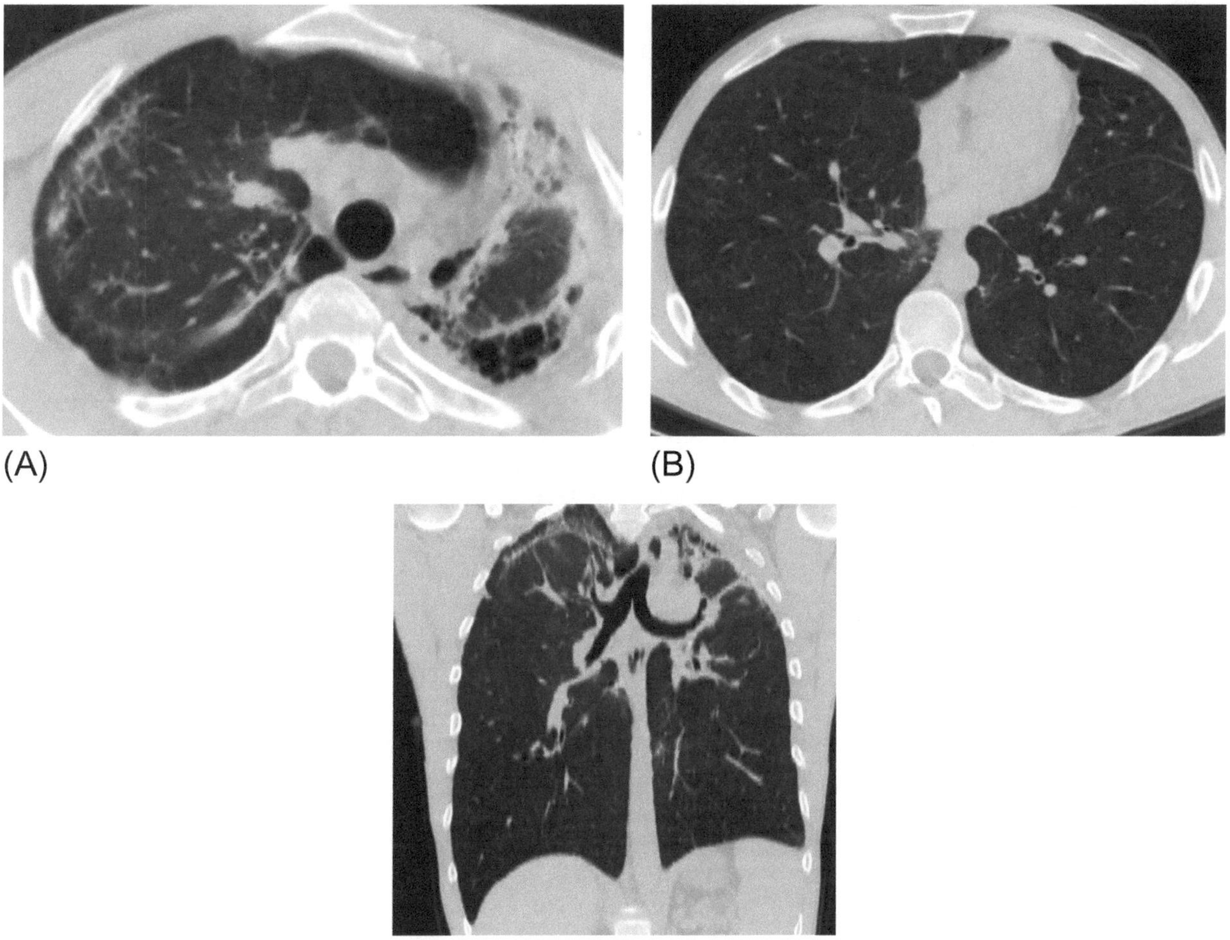

(A) (B) (C)

FIG. 2.8.6 Idiopathic pleuroparenchymal fibroelastosis. Thickening of the visceral pleura with decreased volume of the left upper lobe and honeycombing; pronounced reticular abnormalities, linear attenuation associated with the pleura with thin intersections on the right lobe (A). Moderate reticular abnormalities in the lower lobes (B). Upper lobe distribution of fibrous abnormalities on the coronal section (C).

Enomoto et al. [6] proposed the following criteria for IPPFE diagnosis without morphological confirmation:

(1) Radiological PPFE pattern on the HRCT ("definite PPFE"), based on the criteria of Reddy et al.: (a) bilateral subpleural consolidation with or without pleural thickening in the upper lobes and (b) absent or minimal involvement of the lower parts of the lungs.
(2) Radiologically confirmed progression of the disease (expansion of the areas of consolidation with or without pleural thickening and/or a decrease in the volume of the upper lobes).
(3) Other lung diseases capable of causing upper lobe fibrosis (IIP, connective tissue diseases, chronic hypersensitivity pneumonitis, pulmonary infections, pneumonoconiosis, and sarcoidosis) have been ruled out.

Differential diagnosis

IPPFE must be differentiated from idiopathic pulmonary fibrosis (IPF), fibrotic forms of hypersensitivity pneumonitis (HP), pulmonary sarcoidosis with upper lobe lesions, and infectious diseases of the lungs (such as tuberculosis, mycobacteriosis, and aspergillosis) (Table 2.8.1). IPPFE is most similar to IPF. In IPF, as in IPPFE, subpleural fibrosis and honeycombing are often detected, as well as reticular abnormalities, including thickening of interlobular and intralobular septa in the upper lobes. Most HRCT IPF features occur in the basal fields, and the disease progression results in the spread of the fibrosis from bottom to top. With IPPFE the situation is reversed, most fibrous manifestations occur in

TABLE 2.8.1 Differential group of IPPF

	IPPFE	IPF	HP (chronic course)	Sarcoidosis	Mycobacterioses
Special aspects of patients	Any gender and age	Most often men over 50 years, smoking, GERD history	Most often women who never smoked, young and middle age, long contact with antigen	Most often young and middle age. Inhalation of pollutants is often revealed in history	Patients with bronchiectasis, immunodeficiency
Clinical presentation	Low body mass index	Clubbing and velcro crackles	Velcro crackles	Usually no complaints or mild weakness	Purulent sputum, hemoptysis, fever, often bronchial obstruction
BAL	No abnormalities	Neutrophilia; eosinophilia is possible	Lymphocytosis >50%, foamy macrophages	Moderate lymphocytosis	+ PCR for mycobacteria
CT pattern	Areas of consolidation of subpleural localization. Thickening of interlobular septa, traction bronchiectasis, cysts, loss volume of the upper lobes	Honeycombing with a minimum of GGO, pronounced reticular abnormalities, traction bronchiectasis	Lobular air traps, intralobular nodules, GGO, reticular abnormalities, and honeycombing	Dense well-defined nodules foci of consolidation, masses. Linear attenuation. Pleural pseudoplaques. Moderate reticular abnormalities. Intrathoracic and hilar lymphadenopathy	Bronchiectasis. Polymorphic nodules and areas of consolidation inside the parenchyma. The destruction cavities are often seen. Areas of GGO. Intrathoracic lymphadenopathy
Distribution of abnormal signs	Mostly or exclusively upper lobes	Subpleural, primarily lower lobes	Uniform distribution, maybe more pronounced in the upper lobes	Perilymphatic and peribronchovascular. It can tend toward the upper and middle areas	Predominantly upper lobes, peribronchovascular areas

IPPFE, idiopathic pleuroparenchymal fibroelastosis; *IPF*, idiopathic pulmonary fibrosis; *HP*, hypersensitivity pneumonitis; *GERD*, gastroesophageal reflux disease; *PCR*, polymerase chain reaction; *BAL*, bronchoalveolar lavage; *CT*, computed tomography.

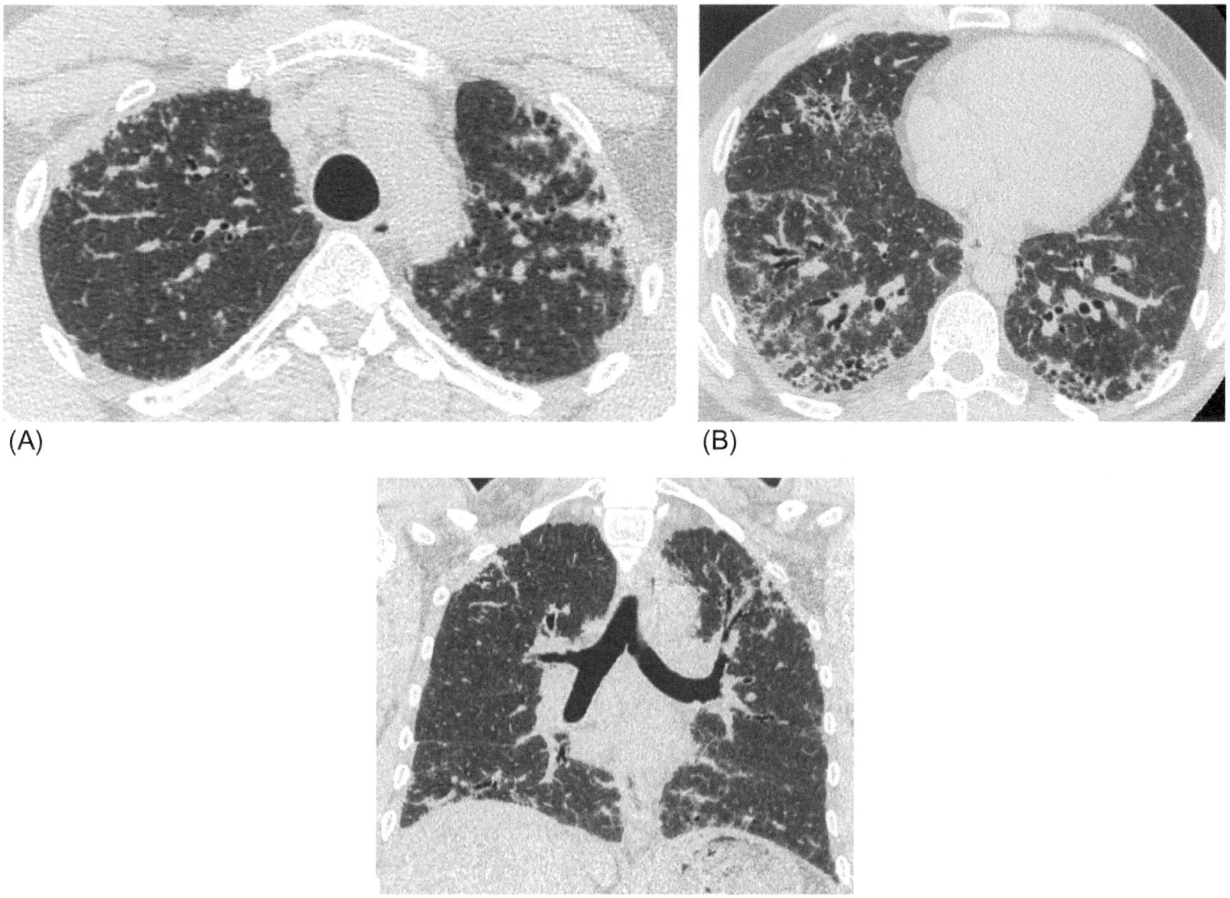

(A) (B) (C)

FIG. 2.8.7 Idiopathic pulmonary fibrosis proved by surgical biopsy. Moderate thickening of the visceral pleura with subpleural fibrosis and limited honeycombing in the upper lobes. Irregular reticular abnormalities (A). Subpleural honeycombing, thickening of the intralobular septa in the subpleural areas, moderate pleural thickening, and traction bronchiectasis in the lower fields (B). The coronal reconstruction shows a relatively uniform lesion in both the upper and lower lobes (C).

the upper lobes, and only severe cases show lower part involvement. However, with IPF, a more uniform distribution of abnormalities is possible in both the upper and lower fields (Fig. 2.8.7), and an atypical predominant upper lobe distribution is possible too. In these cases, in the absence of previous radiological images, the diagnoses of IPF and IPPFE are extremely difficult to differentiate based only on HRCT features. In a study, Rosenbaum et al. revealed that signs of lesion in the lower lobes are no exception in IPPFE cases [25]. In their series of five IPPFE patients, they detected histological signs of the disease in both the upper and lower fields, suggesting that the disease is actually more diffuse than previously thought [25]. Moreover, radiological patterns of IPPFE and IPF often overlap in the same patients, as evidenced in a study by Kato et al. in which 15 of 36 patients with IPPFE had HRCT patterns of definite UIP and seven had possible UIP pattern in the lower lobes of the lungs [26]. In contrast to IPF, patients with HRCT patterns of IPPFE plus UIP are characterized by more pronounced apical abnormalities with loss of upper lobe volume rather than by fibrotic abnormalities in the lower lobes [26].

For IPPFE, only small-cyst honeycombing is infrequent; large and medium cysts are formed often in the areas of the most pronounced fibrosis. Areas of subpleural consolidation, reflecting the ingrowth of fibrous tissue from the pleura to the lung parenchyma (often observed in patients with IPPFE), are not only less typical for IPF but also possible. In some cases, chronic HP may resemble IPPFE, if reticular abnormalities are more pronounced in the upper lobes (Fig. 2.8.8). Regarding HRCT images, lobular air traps are specific signs of HP, especially in the upper lobes, and subpleural sparing. Intralobular ill-defined nodules, characteristic of subacute HP, are rarely detected during the chronic course. In cases of HP, BAL lymphocytosis is common, if the patients have not been treated by steroids. In 10% of ankylosing spondylitis (AS) cases, interstitial lung disease with damage to the upper lobes may develop in the form of secondary PPFE (Fig. 2.8.9) [27].

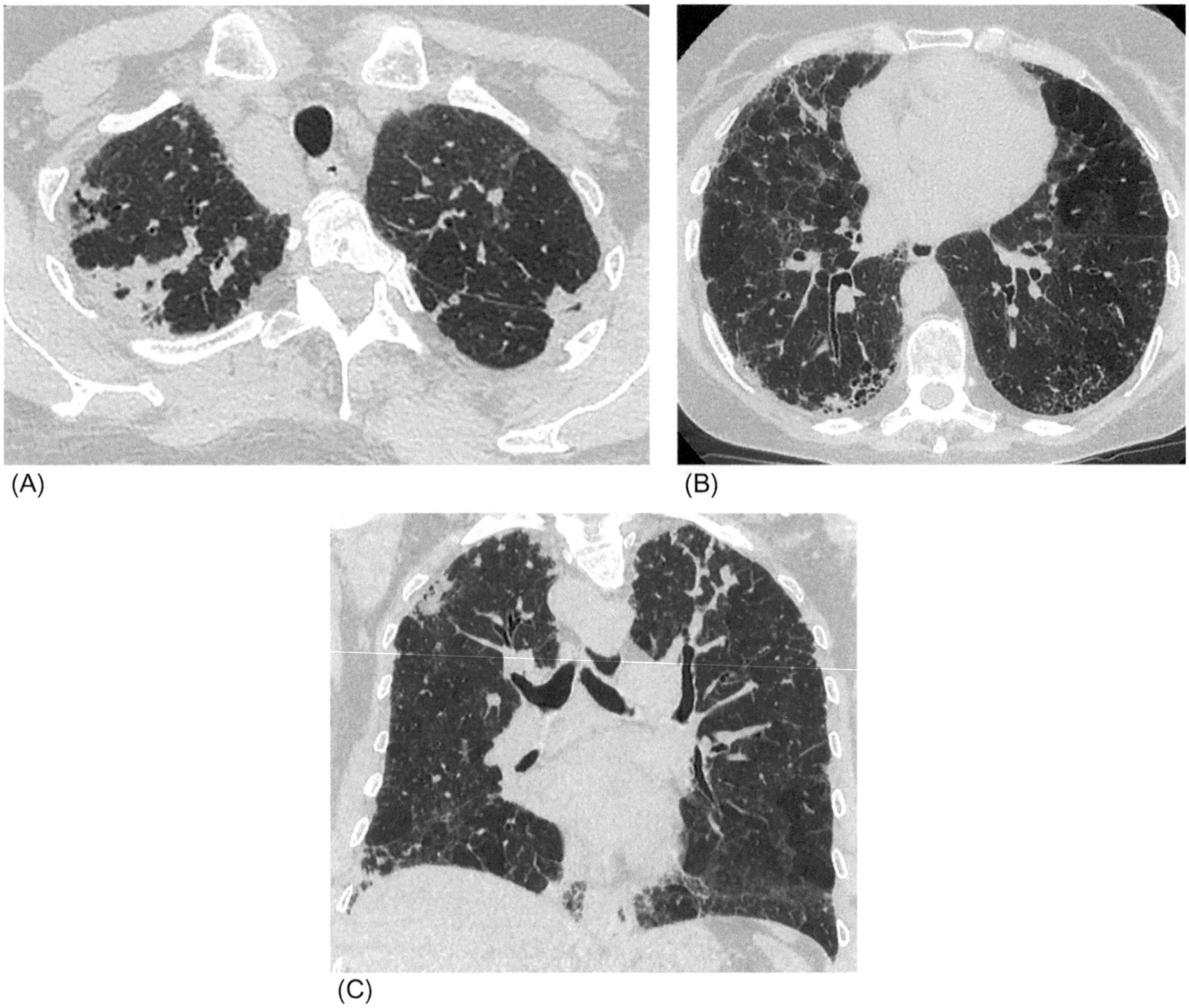

FIG. 2.8.8 Chronic hypersensitivity pneumonitis in a mushroom farm worker (a woman). Thickened pleura from which fibrous bands extend into the parenchyma of the upper lobes. Moderate reticular abnormalities (A). Patchy GGO with a mosaic pattern of distribution, multiple lobular air traps, subpleural honeycombing, and reticular abnormalities in the lower fields (B). On the coronal reconstruction, subpleural fibrosis predominates in the upper lobes with multiple small, poorly differentiated nodules (C).

The formation of cavities that are often colonized by *Aspergillus* fungi, the detection rate of which reaches 50%–60%, is a common sign of pulmonary AS manifestations [28]. Interstitial lung disease in AS has a subclinical course; dyspnea and a decrease in DLCO are generally uncharacteristic [29]. However, the symptoms of the underlying disease such as back pain (mainly in the lumbosacral region), stiffness and deformity of the spine, and sacroiliitis are common, according to X-ray or MRI studies. Sometimes apical fibrosis precedes the clinical manifestations of a systemic disease, and then it is almost impossible to distinguish it from PPFE, especially if genetic diagnostics (HLA-B27 allele) are not used [30]. Pulmonary sarcoidosis often has a preferred distribution in the upper and middle lung fields with formation of areas of consolidation and masses. However, these abnormalities usually have not only subpleural but also peribronchovascular distributions and are accompanied by multiple dense nodules and intrathoracic lymphadenopathy (Fig. 2.8.10). Pulmonary infections may have HRCT patterns similar to those of IPPFE (Fig. 2.8.11); however, they are characterized by the presence of symptoms of active inflammation, purulent sputum, hemoptysis, abnormalities in blood tests, and detection of microflora in sputum and BAL fluid. Nontuberculous mycobacterioses and aspergillosis often develop in patients with structural defects of the bronchi such as bronchiectasis or cysts, and tuberculosis develops in patients with immunodeficiency or in cases of regular contact with infected people.

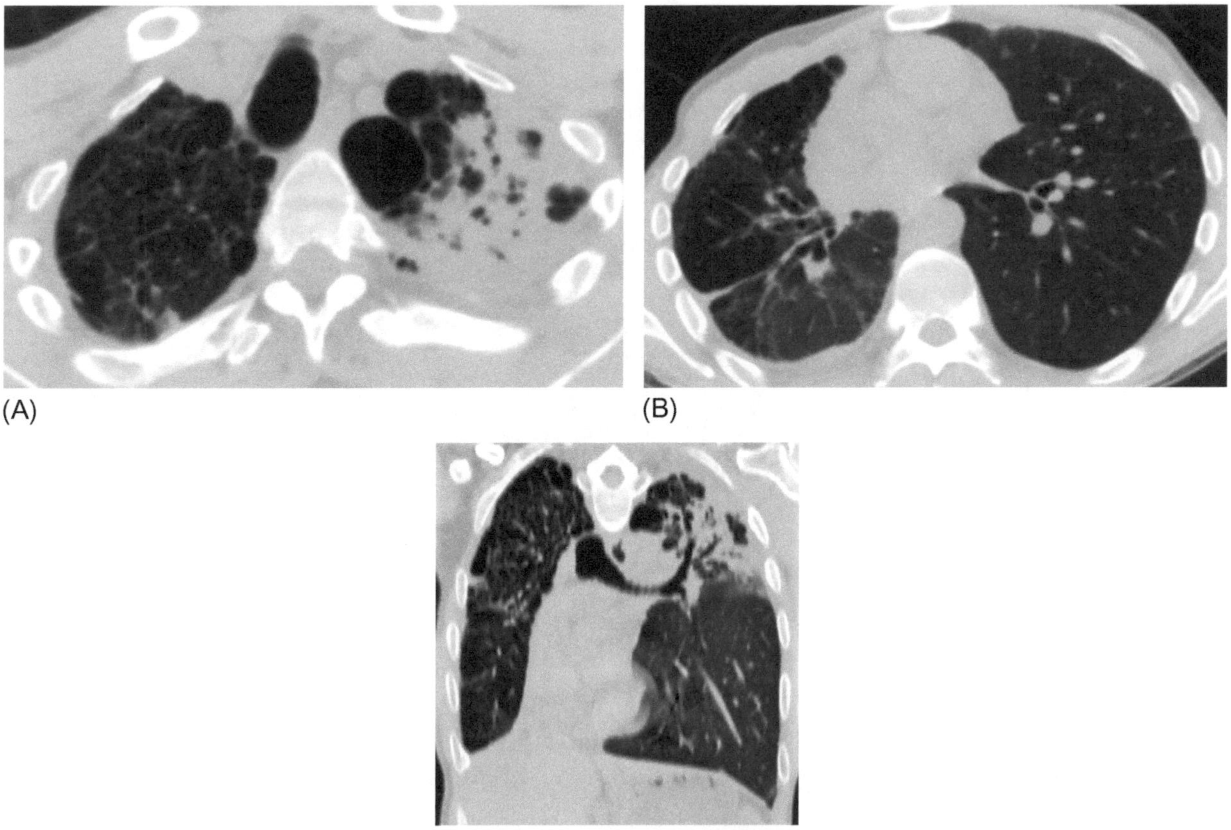

(A) (B) (C)

FIG. 2.8.9 Ankylosing spondylitis in a 31-year-old nonsmoking patient. Massive consolidation area merging with the visceral pleura and multiple cavities on the left field. Bullous-cystic abnormalities and severe reticular signs on the right (A). Irregular thickening of the visceral and interlobar pleura and areas of GGO on the lower right area and mild reticular abnormalities on the lower left area (B). The coronal section reveals predominantly upper lobe abnormalities (C).

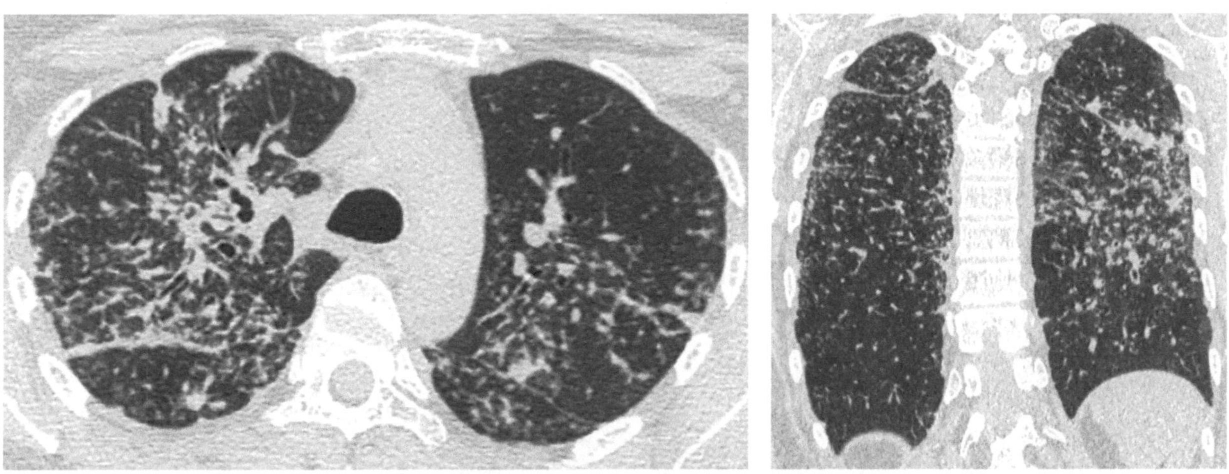

(A) (B)

FIG. 2.8.10 Sarcoidosis of the lungs and intrathoracic lymph nodes. Bilateral subpleural nodules and bands, sometimes merging with the pleura (A). The lesion is maximally represented in the upper and middle areas, along the interlobar pleura (B).

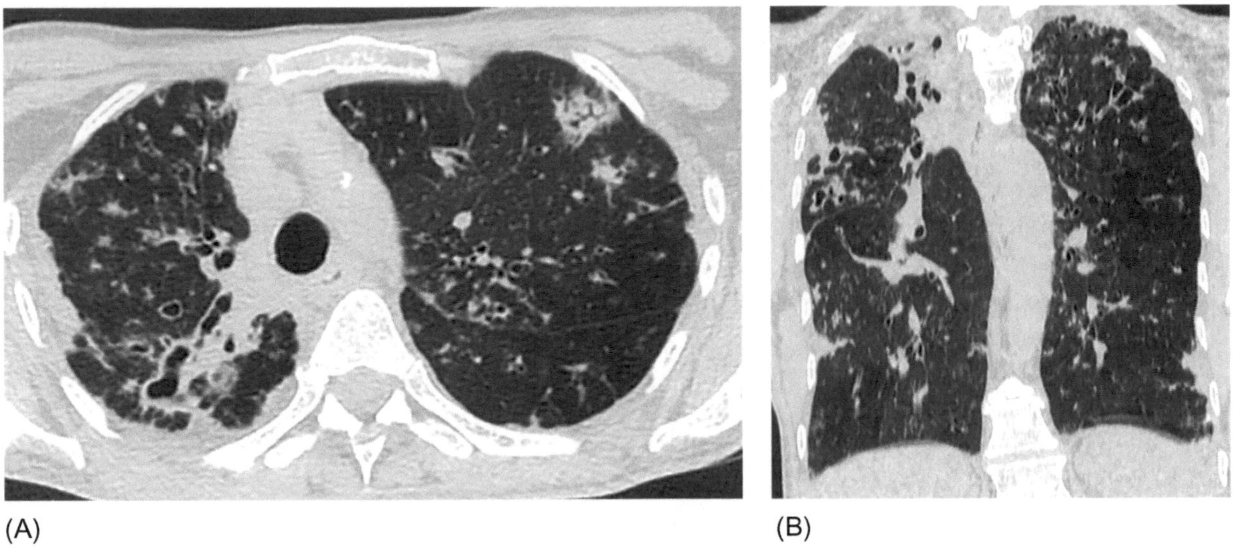

FIG. 2.8.11 Nontuberculous mycobacteriosis caused by *M. abscessus*. In the upper lobes the pleura are unevenly thick; subpleural areas of consolidation and multiple randomly distributed nodules are visible. Multiple bronchiectasis on the right field (A). Bilateral upper lobe bronchiectasis is visible on the coronal slice; the abnormalities are most pronounced in the upper lobes (B).

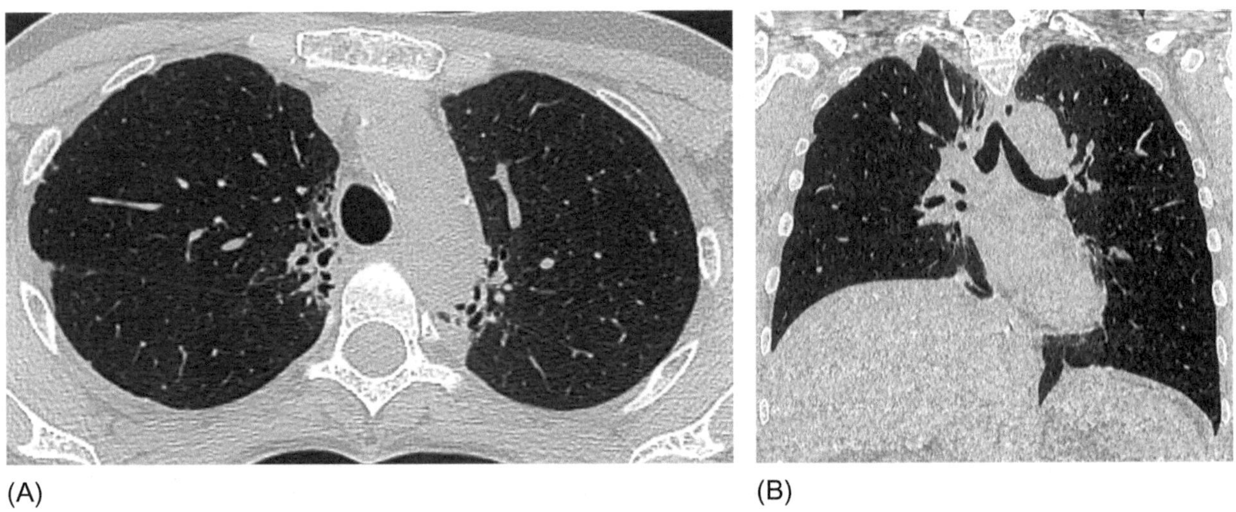

FIG. 2.8.12 Radiation fibrosis in a patient with mediastinal lymphoma. Thickened pleura, subpleural fibrosis, and honeycombing in the paramediastinal areas of the upper lobes (A). The rest fields of the lungs look intact (B).

Pleural thickening and subpleural fibrosis of the pulmonary parenchyma in some diseases (such as asbestosis and radiation pneumonitis) (Fig. 2.8.12) may be similar to IPPFE; however, as a rule, these abnormalities are stable and are rarely manifested clinically.

Treatment and prognosis

An IPPFE therapy has not been developed. Reddy et al., who described the greatest number of cases, adhered to the same principles as for IPF treatment before the era of antifibrotic drugs, namely, monotherapy with corticosteroids or their combination with immunosuppressive drugs or *N*-acetylcysteine. However, they used doses of steroids lower than those commonly used in IPF [10]. In later publications the efficacy of a steroid and immunosuppressive therapy in these patients has been questioned [5], and reports on the use of antifibrotic drugs in IPPFE are controversial. Sato et al. demonstrated the efficacy of pirfenidone for slowing the decline in lung function in a patient with IPPFE who also had an UIP pattern in the lower lobes of the lungs [31]. However, in a study by Enomoto et al. [6], during the 2-year follow-up of eight patients receiving pirfenidone, no clinical effects were noticed by the physicians. Watanabe et al. did not find a significant effect of this drug on patient survival [22]. Nasser et al. treated five patients with IPPFE using nintedanib (150 mg two times a day), including two patients who received pirfenidone before, and noted stabilization of the FVC in all the patients [32]. During the acute exacerbation, as with other types of fibrosing interstitial pneumonia, pulse therapy with methylprednisolone (1000 mg/day for 3 consecutive days) can be successfully applied, followed by oral prednisone (20 mg/day) and cyclosporine A (100 mg/day) [19].

Early studies noted a slow progression of the disease with a median survival of about 11 years [5]. Later studies have presented less optimistic results, with an average survival rate of 24 months from the time of diagnosis [26] and survivals of 88.5%, 45.5%, and 28.9% for the 1-year, 3-year, and 5-year rates, respectively [6]. Cases with severe course and rapid lung function decline have also been reported [33]. According to Kato et al., the factor determining the lowest survival for patients with IPPFE was the presence of UIP signs in the lower lobes (according to the HRCT data) [26], while Enomoto et al. did not notice a difference in the prognosis of patients who had a "pure" PPFE pattern compared with the prognosis of those who additionally had a radiological pattern of definite or possible UIP [6]. A low muscle mass and muscle attenuation of erector-spinae muscles, detected by CT, are also predictors of low survival rates in patients with IPPFE [34].

According to different studies, other factors determining a negative prognosis include the male gender; low FVC rates; and a high gender, age, pulmonary function (GAP) index [12,35].

At an advanced stage, IPPFE progresses significantly faster than IPF, the annual decrease in FEV1 amounts to an average of 0.28 and 0.11 L, respectively [35].

References

[1] Travis WD, Costabel U, Hansell DM, King Jr TE, Lynch DA, Nicholson AG, et al. An official American Thoracic Society/European Respiratory Society statement: update of the international multidisciplinary classification of the idiopathic interstitial pneumonias. Am J Respir Crit Care Med 2013;188(6):733–48.

[2] Thüsen J. Pleuroparenchymal fibroelastosis: its pathological characteristics. Curr Respir Med Rev 2013;9(4):238–47.

[3] Amitani R, Niimi A, Kuze F. Idiopathic pulmonary upper lobe fibrosis. Kokyu 1992;11:693–9.

[4] Frankel SK, Cool CD, Lynch DA, Brown KK. Idiopathic pleuroparenchymal fibroelastosis: description of a novel clinicopathologic entity. Chest 2004;126(6):2007–13.

[5] Watanabe K. Pleuroparenchymal fibroelastosis: its clinical characteristics. Curr Respir Med Rev 2013;9(4):229–37.

[6] Enomoto Y, Nakamura Y, Satake Y, Sumikawa H, Johkoh T, Colby TV, et al. Clinical diagnosis of idiopathic pleuroparenchymal fibroelastosis: a retrospective multicenter study. Respir Med 2017;133:1–5.

[7] Nakatani T, Arai T, Kitaichi M, Akira M, Tachibana K, Sugimoto C, et al. Pleuroparenchymal fibroelastosis from a consecutive database: a rare disease entity? Eur Respir J 2015;45(4):1183–6.

[8] Ofek E, Sato M, Saito T, Wagnetz U, Roberts HC, Chaparro C, et al. Restrictive allograft syndrome post lung transplantation is characterized by pleuroparenchymal fibroelastosis. Mod Pathol 2013;26(3):350–6.

[9] Fujikura Y, Kanoh S, Kouzaki Y, Hara Y, Matsubara O, Kawana A. Pleuroparenchymal fibroelastosis as a series of airway complications associated with chronic graft-versus-host disease following allogeneic bone marrow transplantation. Intern Med 2014;53(1):43–6.

[10] Reddy TL, Tominaga M, Hansell DM, von der Thusen J, Rassl D, Parfrey H, et al. Pleuroparenchymal fibroelastosis: a spectrum of histopathological and imaging phenotypes. Eur Respir J 2012;40(2):377–85.

[11] Portillo K, Guasch Arriaga I, Ruiz-Manzano J. Pleuroparenchymal fibroelastosis: is it also an idiopathic entity? Arch Bronconeumol 2015;51(10):509–14.

[12] Khiroya R, Macaluso C, Montero MA, Wells AU, Chua F, Kokosi M, et al. Pleuroparenchymal fibroelastosis: a review of histopathologic features and the relationship between histologic parameters and survival. Am J Surg Pathol 2017;41(12):1683–9.

[13] Hirota T, Yoshida Y, Kitasato Y, Yoshimi M, Koga T, Tsuruta N, et al. Histological evolution of pleuroparenchymal fibroelastosis. Histopathology 2015;66(4):545–54.

[14] Bonifazi M, Montero MA, Renzoni EA. Idiopathic pleuroparenchymal fibroelastosis. Curr Pulmonol Rep 2017;6(1):9–15.

[15] Iesato K, Ogasawara T, Masuda A, et al. Idiopathic pulmonary upper lobe fibrosis; clinical and pathological features. Rinsho Houshasen 2005;50:13–25.

[16] Kusagaya H, Nakamura Y, Kono M, Kaida Y, Kuroishi S, Enomoto N, et al. Idiopathic pleuroparenchymal fibroelastosis: consideration of a clinicopathological entity in a series of Japanese patients. BMC Pulm Med 2012;12:72.

[17] Yoshida Y, Nagata N, Tsuruta N, Kitasato Y, Wakamatsu K, Yoshimi M, et al. Heterogeneous clinical features in patients with pulmonary fibrosis showing histology of pleuroparenchymal fibroelastosis. Respir Investig 2016;54(3):162–9.

[18] Oda T, Ogura T, Kitamura H, Hagiwara E, Baba T, Enomoto Y, et al. Distinct characteristics of pleuroparenchymal fibroelastosis with usual interstitial pneumonia compared with idiopathic pulmonary fibrosis. Chest 2014;146(5):1248–55.

[19] Miyamoto A, Uruga H, Morokawa N, Moriguchi S, Takahashi Y, Ogawa K, et al. Various bronchiolar lesions accompanied by idiopathic pleuroparenchymal fibroelastosis with a usual interstitial pneumonia pattern demonstrating acute exacerbation: a case study. Intern Med 2018. https://doi.org/10.2169/internalmedicine.1649-18.

[20] Kokosi MA, Nicholson AG, Hansell DM, Wells AU. Rare idiopathic interstitial pneumonias: LIP and PPFE and rare histologic patterns of interstitial pneumonias: AFOP and BPIP. Respirology 2016;21(4):600–14.

[21] Kushima H, Hidaka K, Ishii H, Nakao A, On R, Kinoshita Y, et al. Two cases of pleuroparenchymal fibroelastosis diagnosed with transbronchial lung biopsy. Respir Med Case Rep 2016;19:71–3.

[22] Watanabe S, Waseda Y, Takato H, Matsunuma R, Johkoh T, Egashira R, et al. Pleuroparenchymal fibroelastosis: distinct pulmonary physiological features in nine patients. Respir Investig 2015;53(4):149–55.

[23] Becker CD, Gil J, Padilla M. Idiopathic pleuroparenchymal fibroelastosis: an unrecognized or misdiagnosed entity? Mod Pathol 2008;21(6):784–7.

[24] Piciucchi S, Tomassetti S, Casoni G, Sverzellati N, Carloni A, Dubini A, et al. High resolution CT and histological findings in idiopathic pleuroparenchymal fibroelastosis: features and differential diagnosis. Respir Res 2011;12:111.

[25] Rosenbaum JN, Butt YM, Johnson KA, Meyer K, Batra K, Kanne JP, et al. Pleuroparenchymal fibroelastosis: a pattern of chronic lung injury. Hum Pathol 2015;46(1):137–46.

[26] Kato M, Sasaki S, Kurokawa K, Nakamura T, Yamada T, Sasano H, et al. Usual interstitial pneumonia pattern in the lower lung lobes as a prognostic factor in idiopathic pleuroparenchymal fibroelastosis. Respiration 2018;6:1–10.

[27] Hasiloglu ZI, Havan N, Rezvani A, Sariyildiz MA, Erdemli HE, Karacan I. Lung parenchymal changes in patients with ankylosing spondylitis. World J Radiol 2012;4(5):215–9.

[28] Pamuk ON, Harmandar O, Tosun B, Yörük Y, Cakir N. A patient with ankylosing spondylitis who presented with chronic necrotising aspergillosis: report on one case and review of the literature. Clin Rheumatol 2005;24(4):415–9.

[29] Hunninghake GW, Fauci AS. Pulmonary involvement in the collagen vascular diseases. Am Rev Respir Dis 1979;119(3):471–503.

[30] Costantino F, Breban M, Garchon HJ. Genetics and functional genomics of spondyloarthritis. Front Immunol 2018;9:2933.

[31] Sato S, Hanibuchi M, Takahashi M, Fukuda Y, Morizumi S, Toyoda Y, et al. A patient with idiopathic pleuroparenchymal fibroelastosis showing a sustained pulmonary function due to treatment with pirfenidone. Intern Med 2016;55(5):497–501.

[32] Nasser M, Chebib N, Philit F, Senechal A, Traclet J, Zarza V, et al. Treatment with nintedanib in patients with pleuroparenchymal fibroelastosis. Eur Respir J 2017;50:PA4876.

[33] Watanabe K, Nagata N, Kitasato Y, Wakamatsu K, Nabeshima K, Harada T, et al. Rapid decrease in forced vital capacity in patients with idiopathic pulmonary upper lobe fibrosis. Respir Investig 2012;50(3):88–97.

[34] Suzuki Y, Yoshimura K, Enomoto Y, Yasui H, Hozumi H, Karayama M, et al. Distinct profile and prognostic impact of body composition changes in idiopathic pulmonary fibrosis and idiopathic pleuroparenchymal fibroelastosis. Sci Rep 2018;8(1):14074.

[35] Shioya M, Otsuka M, Yamada G, Umeda Y, Ikeda K, Nishikiori H, et al. Poorer prognosis of idiopathic pleuroparenchymal fibroelastosis compared with idiopathic pulmonary fibrosis in advanced stage. Can Respir J 2018;2018:6043053.

Chapter 3

Hypersensitivity pneumonitis

Alexander Averyanov[a,b], Evgeniya Kogan[c], Victor Lesnyak[d], Olesya Danilevskaya[e]

[a]Clinical Department, Pulmonology Research Institute under FMBA of Russia, Moscow, Russia, [b]Pulmonary Division, Federal Research Clinical Center under FMBA of Russia, Moscow, Russia, [c]Anatomic Pathology Department, Sechenov University, Moscow, Russia, [d]Radiology Department, Federal Research Clinical Center under FMBA of Russia, Moscow, Russia, [e]Endoscopy Department, Pulmonology Research Institute under FMBA of Russia, Moscow, Russia

Hypersensitivity pneumonitis (HP), also known as extrinsic allergic alveolitis, is a diffuse granulomatous inflammatory pulmonary disease, caused by the inhalation of antigen-containing organic particles or chemical compounds [1]. More than 200 different substances capable of causing HP have been described. These substances can be divided into three groups: microorganisms (bacteria and fungi), animal proteins, and chemical products [2]. HP develops more often in workers in industries that use organic substances or are associated with the production of air pollutants (Table 3.1). Household cases of HP are often determined by unusual conditions of the living environments and hobbies of patients (Table 3.2). Annually the number of potential etiologic factors is increased. For example, in recent years, cases caused by the use of mosquito fumigators have been described; similarly, cases stemming from working with MDF panels, argan-derived products, and electronic cigarettes have been reported [7–10].

The exact prevalence of HP is unknown. The disease most often develops via repeated contact with an allergen. Thus agricultural workers, food production workers, and amateur poultry are most susceptible to HP. In recent years, there has been a significant increase in cases of HP due to the use of air-conditioning and ventilation systems, poorly ventilated wet rooms (baths and indoor swimming pools), and even shower caps contaminated with hydrophilic bacteria, fungi, and nontuberculous mycobacteria [2]. Epidemiological studies conducted in Sweden and Finland revealed that the frequency of HP was 23 cases per 100,000 individuals and 44 cases per 100,000 agricultural workers. In the United Kingdom, this indicator was significantly lower: 1:100,000 in the general population [11]. The average annual increase in the incidence rate of HP caused by production factors is approximately one case per 1 million workers [12]. HP is one of the most frequent interstitial lung diseases; its frequency in this group is approximately 29% [13]. It is notable that smokers are much less likely to develop HP than nonsmokers with similar exposure to allergens. The protective role of smoking against the onset of HP is most likely associated with the stimulation of the macrophage response and a decrease in lymphocytic activity [14].

Pathogenesis

In acute HP, there is type 3 immune response (immune-complex mediated), while subacute and chronic HP are characterized by type 4 immune reaction (T lymphocytes mediated) in the distal parts of the respiratory tract in response to the inhalation of small allergen particles capable to reach the terminal bronchioles [3,15].

The role of genetic factors in the pathogenesis of HP is discussed. Patients with chronic HP showed a significantly higher rate of MUC5B promoter polymorphism than healthy individuals [16]. Similarly to idiopathic pulmonary fibrosis (IPF), the presence of age-adjusted short telomere lengths has been revealed in HP patients [17]. Patients with pigeon-breeder's disease showed an increased TNF-2(−)(308) allele frequency versus the control group [18]. However, the role of genetics in the development of this disease has not been well studied.

Morphology

The morphological characteristics of HP depend largely on the nature of the course.

Difficult to Diagnose Rare Diffuse Lung Disease. https://doi.org/10.1016/B978-0-12-815375-8.00003-0

TABLE 3.1 The most frequent occupational inducers of HP (extract from [3–6] with additions)

Profession	Source	Antigen
Agriculture	Grain	*Aspergillus fumigatus* *Sitophilus granarius*
	Hay/silage/composite	*Thermophilic actinomycetes* *Saccharopolyspora rectivirgula* *Thermoactinomyces vulgaris* *Lichtheimia corymbifera* *Eurotium amstelodami* *Wallemia sebi*
	Mushrooms	*Penicillium citrinum* *Lycoperdon* species
	Vegetables, fruits, flowers	*Aspergillus fumigatus*, *Aspergillus* spp. *Botrytis cinerea*, *Penicillium* sp., *Paecilomyces* sp.
	Tobacco	*Aspergillus fumigatus*
	Sugarcane	*Thermoactinomyces sacchari*
	Birds	Excreta, feather
	Pituitary snuff	Pork and ox proteins
Food production	Cheeses	*Penicillium roqueforti* *Penicillium* casei *Penicillium camemberti* *Penicillium chrysogenum* *Acarus siro*
	Dry sausages	*Penicillium camemberti*
	Wheat flour	*Sitophilus*
	Malt	*Aspergillus fumigatus* *Aspergillus clavatus*
	Soybean	*Aspergillus oryzae* Soybean hull antigens
	Grape mold	*Botrytis cinerea*
	Fish meal	Fish meal dust
	Coffee and tea	Coffee and tea dust
	Veterinary feed	Soybean hulls antigens
Dust manufacture	Wood dust	*Penicillium chrysogenum* *Cryptostroma corticale* *Paecilomyces* sp. *Rhizopus* sp. *Trichoderma koningii* *Alternaria* *Mucor* *Pullaria*
	Plaster	Thermophilic or Aspergillus
	Metal dust	Cobalt, beryllium
	Pearl, mollusks shell	Dust from shellfish, shells, pearls
Chemical production	Paints, plastics, rubber and other polymers, fiberglass, insecticides, fungicides	Isocyanates, anhydrides, alginic acid
	Detergents	Dimethyl phthalate Pyrethrum
Medicine, biology, pharmacy, veterinary medicine	Material of dentures	Methylmethacrylate
	Antibiotics	Penicillin
	Laboratory reagents	Pauly's reagent

TABLE 3.1 The most frequent occupational inducers of HP (extract from [3–6] with additions)—cont'd

Profession	Source	Antigen
Humid or aerosolized working environments and low-ventilation environments	Cold storage Large closed tanks, Basements, sewage Poorly air-conditioned premises	*Thermophilic actinomycetes* *Saccharopolyspora rectivirgula* *Thermoactinomyces vulgaris* *Thermoactinomyces candidus* *Micromycetes* Penicillium sp. *Cephalosporium* *Aspergillus fumigatus* *Mucor* sp. Candida albicans *Acremonium* sp. *Saccharomyces* sp. *Alternaria* sp. *Fusarium* *Aureobasidium pullulans* *Debaryomyces hansenii* *Paecilomyces nivea* *Paecilomyces variotii* *Bacteria* *Pseudomonas* sp. *Cytophaga* sp. *Staphylococcus* sp. *Bacillus* sp. *Stenotrophomonas* sp. *Acinetobacter* sp. *Flavobacterium multivorum* *Yersinia pseudotuberculosis* *Aerobacterium liquefasciens* *Klebsiella oxytoca* *Sphingobacterium spiritivorum* *Mycobacteria* *Mycobacterium immunogenum* *Mycobacterium chelonae*
Other	Wind musical instruments	*Mycobacterium chelonae, Fusarium* spp., *Candida* alb. *Ulocladium botrytis, Cladosporium* sp.
	Silk	Silkworm larvae cocoon fluff
	Pelts hair and fur	Animal proteins

Morphological changes in the lungs in a case of *acute HP* may be represented by a variety of processes: a combination of bronchiolitis and bronchocentric lymphocytic alveolitis; obliterating bronchiolitis; accumulation of edematous fluid and fibrinous exudate in the alveoli with an admixture of eosinophils, neutrophils, and erythrocytes; and leukocytoclastic capillaritis (Figs. 3.1–3.5) [19].

The detection of giant cells, clusters of histiocytes, and noncaseating granulomas is possible, although the latter finding is more characteristic of subacute forms of the disease [20].

The morphology of the *subacute variant* of HP is characterized by a combination of noncaseating granulomas located in the pulmonary interstitium and walls of bronchioles, lymphocytic alveolitis, and bronchiolitis (Figs. 3.6 and 3.7).

Granulomas with HP as a manifestation of type IV hypersensitivity are typical, although not absolutely a mandatory sign, and are found in approximately 66%–70% of cases [19]. Granulomas without clear boundaries consist of large histiocytes, epithelioid cells, and giant multinucleate Langhans cells. There is no fibrosis on the periphery of the granulomas. Occasionally, crystals of cholesterol, Schaumann bodies, and asteroid corpuscles are observed in giant multinucleate cells. In patients with "hot tub lung"—associated with the colonization of the lower respiratory tract by *Mycobacterium avium* complex—granulomas are larger, with clear contours, without necrosis in the center, and may resemble sarcoid granulomas [21]. In general, *Lymphocytic alveolitis* is localized around the affected bronchioles. It is represented by lymphoplasmacytic

TABLE 3.2 Possible household inducers of HP (extract from [3–6] with additions)

Living conditions	Source	Antigen
Contact with birds, animals	Birds, tamed rodents, bats	Avian, bat, handler's animal droppings, feathers, serum proteins
Wet, poorly ventilated premises	Saunas, indoor swimming pools, shower cabins, bathtubs, basements, attics, old houses, mold on the walls	*Thermophilic actinomycetes* *Mycobacterium* sp. *Aureobasidium* sp. *Epicoccum nigrum* *Phoma violacea* *Klebsiella oxytoca* *Aspergillus* sp. *Penicillium* sp. *Trichosporon* sp.
Hobbies associated with dust formation, working with earth	Wood or other organic dust, soil	Penicillium sp. Aspergillus sp. *Cryptostroma corticale* *Chrysonilia sitophila* *Paecilomyces* sp. *Mycobacterium* sp.
Other	Feather bedding or down comforters Smoking electronic cigarettes, hookahs	Avian feathers' antigens Chemical components of electronic cigarettes, bacterial colonization of hookah tanks and pipes

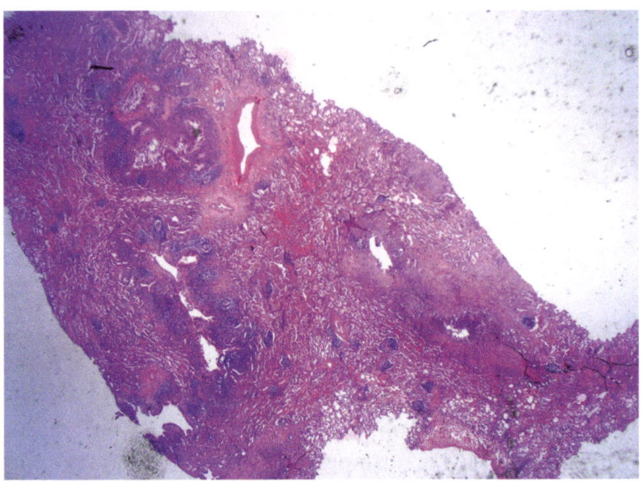

FIG. 3.1 Acute hypersensitive pneumonitis (HP). Foci of bronchiolocentric alveolitis with bronchiolitis in the center of the lesion and noncaseating granulomas and hyperplasia of the bronchus-associated lymphoid tissue. Hematoxylin and eosin staining, 1.25×.

infiltrates with an admixture of neutrophils and eosinophils in the alveolar septa (Fig. 3.3) and the formation of Masson's bodies in the organization of intra-alveolar exudates. *Bronchiolitis* is characterized by the presence of lymphoplasmacytic wall infiltration, foci of squamous metaplasia of the epithelium, ulceration foci, and hyperplasia of broncho-associated lymphoid tissue (i.e., lymphoid tissue associated with bronchioles mucosa). In addition, the organization of exudates in the lumens of bronchioles and bronchiolar passages is characterized by the development of granulation tissue. In some cases the histopathological picture may resemble cellular nonspecific interstitial pneumonia [22].

Chronic HP is characterized by pronounced interstitial centrilobular (airway-centered) fibrosis or branching pattern fibrosis, following the bronchovascular bundles with foci of the carnification and obliterating bronchiolitis up to the development of the "honeycomb lung" (Figs. 3.8–3.11). *Interstitial fibrosis* resembles that observed in common interstitial pneumonia and nonspecific interstitial pneumonia. However, unlike the listed diseases, it has a bronchiolocentric nature

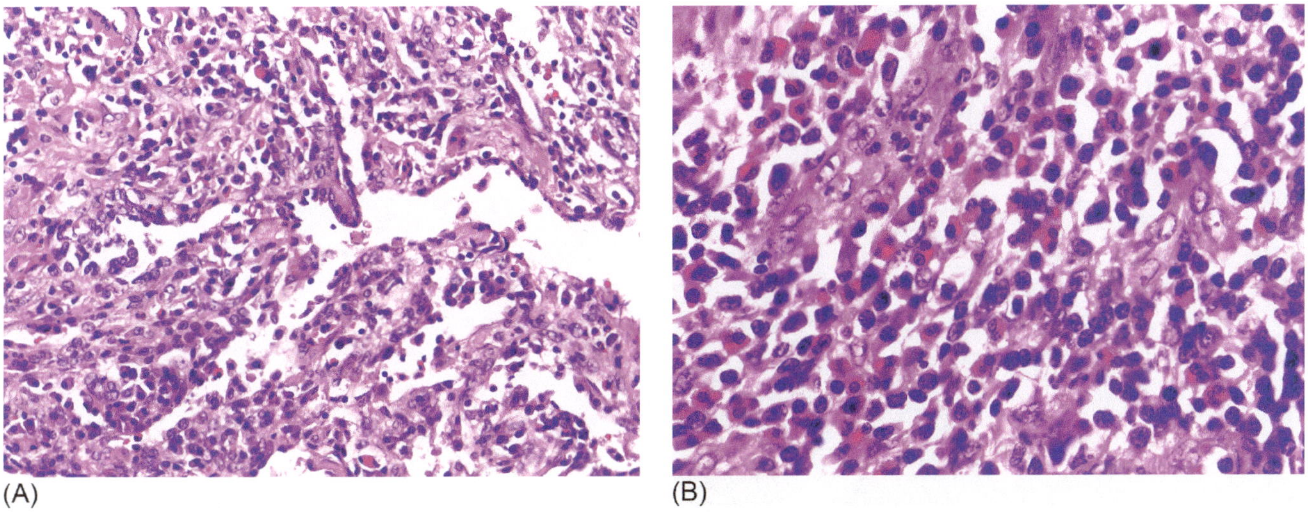

FIG. 3.2 (A) Acute HP. Bronchiolocentric lymphocytic alveolitis with lymphoid infiltration of the alveolar septa. Alveoli with exudates containing a large number of eosinophils and lymphohistiocytic elements. Alveolar septa with inflammatory lymphohistiocytic infiltrate and an admixture of eosinophils. Hematoxylin and eosin staining, 100×. (B) Acute HP. Interstitial component. Alveolar septa with an inflammatory lymphohistiocytic infiltrate and an admixture of eosinophils. Hematoxylin and eosin staining, 600×.

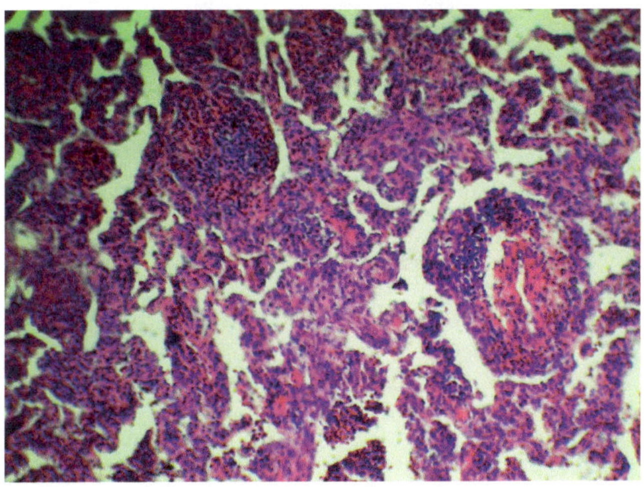

FIG. 3.3 Acute HP. Lymphocytic alveolitis with noncaseating granulomas in the walls of alveoli, bronchioles, and peribronchial tissue. Hematoxylin and eosin staining, 100×.

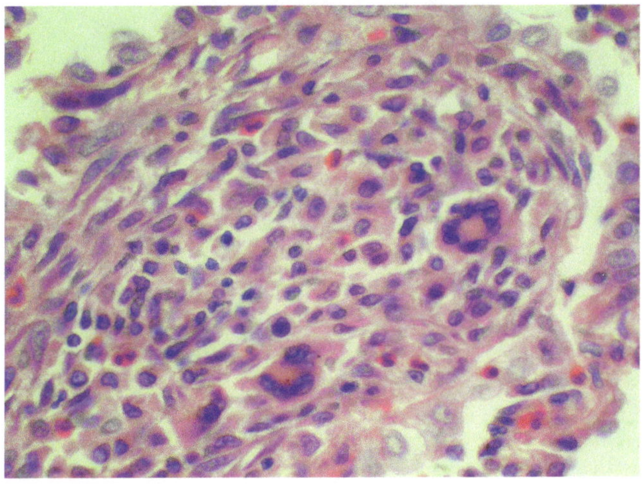

FIG. 3.4 Acute HP. Destructive bronchiolitis with noncaseating granulomas with giant multinucleate cells of Langhans cell type, large histiocytes, eosinophils, and lymphoid and fibroblastic elements. Hematoxylin and eosin staining, 400×.

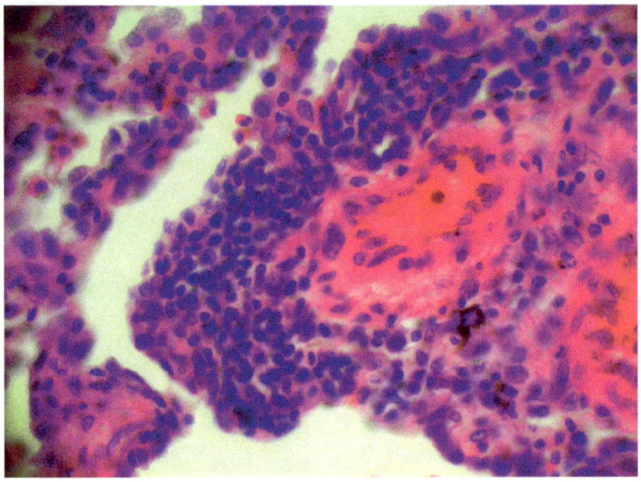

FIG. 3.5 Acute HP. Destructive bronchiolitis with a giant multinucleate cell of the Langhans type with cytoplasmic crystalloid inclusions. Hematoxylin and eosin staining, 600×.

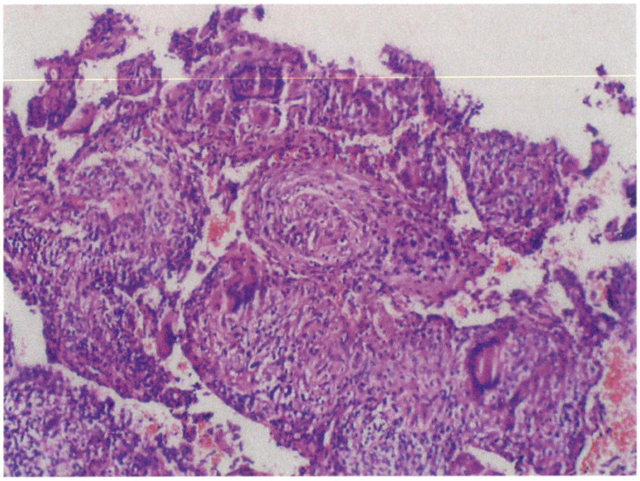

FIG. 3.6 Subacute HP. Bronchiolitis obliterans with Masson's bodies and noncaseating granulomas with epithelioid and giant multinucleate Langhans cells. Hematoxylin and eosin staining, 200×.

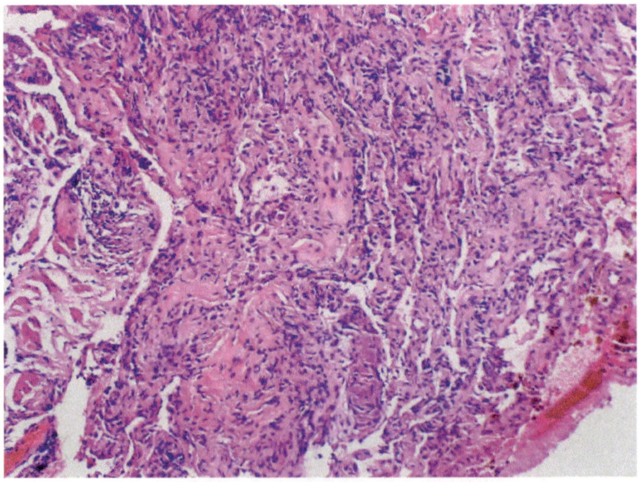

FIG. 3.7 Subacute HP. The focus of carnification is the alveoli, with organized exudate, and the alveolar septa, with an inflammatory lymphohistiocytic infiltrate and an admixture of eosinophils. Hematoxylin and eosin staining, 400×.

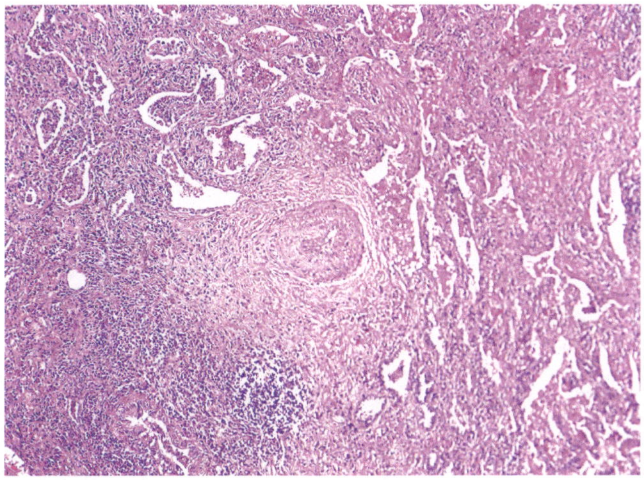

FIG. 3.8 Chronic HP. Bronchiole-centric sclerosis with granulomatous inflammation in the interstitium. Hematoxylin and eosin staining, 100×.

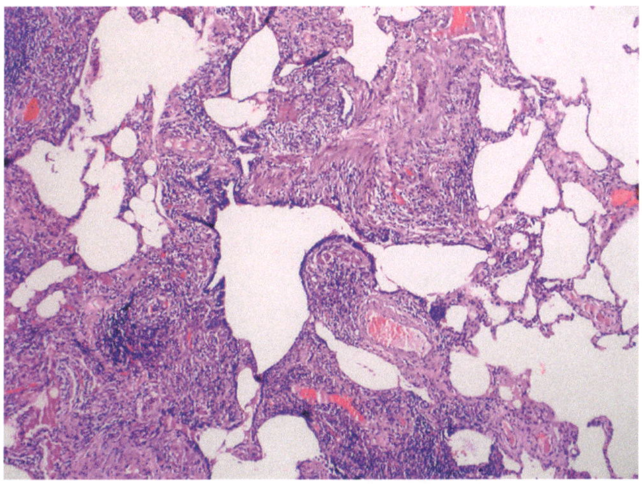

FIG. 3.9 Chronic HP. Bronchiole-centric sclerosis and carnification with granulation tissue in the bronchiole lumen and Masson's bodies in the lumens of the alveoli. Hematoxylin and eosin staining, 100×.

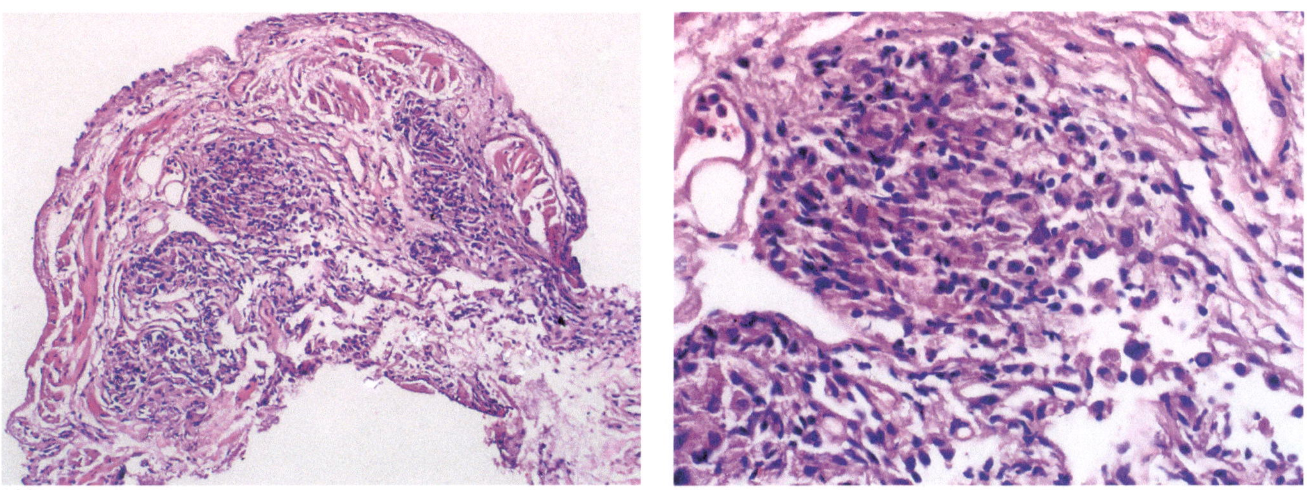

(A) (B)

FIG. 3.10 (A) Chronic HP. Granulomas in the bronchiole wall with wall sclerosis. Hematoxylin and eosin staining, 100×. (B) Chronic HP. Noncaseating granulomas in the bronchiolar wall, constructed from epithelioid, giant multinucleate cells such as Langhans cells, macrophages, and lymphocytes. Hematoxylin and eosin staining, 400×.

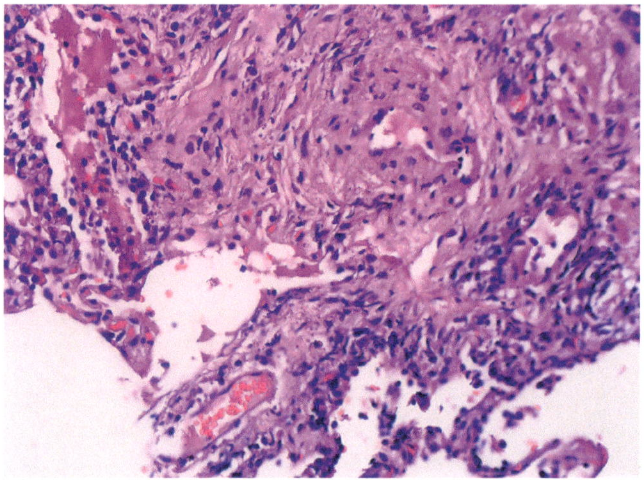

FIG. 3.11 Chronic HP. Foci of carnification with arteriopathy of a small branch of the pulmonary artery. Hematoxylin and eosin staining, 400×.

(i.e., it is represented by foci of fibrosis with the scleroid bronchiole in the center). Furthermore, areas with active manifestations in the form of poorly formed noncaseating granulomas, obliterating bronchiolitis, and lymphocytic alveolitis with the presence of giant cells are possible findings. Secondary pulmonary hypertension and a pulmonary heart with typical restructuring of the branches of the pulmonary artery and microcirculation vessels may develop in patients [19,23].

Morphological material for the diagnostics of HP can be obtained by transbronchial forceps biopsy (TBB), lung cryobiopsy (LCB) during bronchoscopy, or surgical biopsy. Cryobiopsy is a minimally invasive method enabling the extraction of a sufficient amount of material for research. However, it is thought that the diagnosis of HP cannot be ruled out based on the absence of its symptoms in TBB or LCB [24]. There is additional complexity in diagnosis due to the mosaic histological pattern, especially in chronic HP, when in different fields of the lung, both typical HP manifestations and signs of common and nonspecific interstitial pneumonia are simultaneously observed [25]. Thus a definitive diagnosis may only be reached through surgical biopsy performed in different lobes of the lung [3,24].

Differential diagnosis of HP according to lung biopsy data

In the study of biopsy material, differential diagnosis should be considered for other granulomatous diseases and idiopathic interstitial pneumonia (Table 3.3). For this purpose a combination of clinical, laboratory, radiological, and morphological studies should be utilized. This multidisciplinary approach has become a mainstay for the diagnosis of HP, especially the chronic form.

TABLE 3.3 Differential diagnosis of HP based on lung biopsy data [19,22,26]

	HP	Sarcoidosis	UIP	NSIP	COP
Etiology	Allergen	Unknown	Unknown	Unknown	Unknown
Noncaseating granuloma	+ CD4+ Lymphocytes	+ CD8+ Lymphocytes in the periphery of the granulomas; CD4+ in the center of the granulomas	−	−	−
Bronchiolitis obliterans	+	−	−	−	+
Bronchiolocentric lymphocytic alveolitis	+	−	−	−	−
Bronchiolocentric interstitial sclerosis	+	−	−	−	−
Carnification	+	−	−	−	+

UIP, usual interstitial pneumonia; *NSIP*, nonspecific interstitial pneumonia; *COP*, cryptogenic organizing pneumonia; *HP*, hypersensitivity pneumonitis.

Using lung biopsy materials, Takemura et al. showed that in HP (unlike in usual interstitial pneumonia (UIP), the presence of signs such as bronchiolitis, centrilobular (bronchiolocentric) fibrosis, foci of carnification, granulomatosis, giant multinucleate cells, and lymphocytic alveolitis is significantly more frequent [26]. In addition, the presence of fibroblastic foci is a distinctive characteristic of UIP versus HP [19]. For nonspecific interstitial pneumonia (NSIP) and cryptogenic organizing pneumonia, granulomatous inflammation with the presence of noncaseating granulomas and centrilobular (bronchiolocentric) fibrosis is not a typical finding. For sarcoidosis, "stamped," clearly delineated granulomas containing CD4+ and CD8+ lymphocytes, often surrounded by a zone of hyalinosis and sclerosis, are commonly observed.

Clinical presentation and course

Conventionally, three variants of the course of HP have been identified, namely, acute, subacute, and chronic [27]. Manifestations of the acute form of HP include dry cough, dyspnea, fever, and chills. These develop within a few hours after contact with a large dose of antigen and are often prone to spontaneous remission following the termination of the exposure to the allergen. Single episodes of acute HP may be similar to manifestations of respiratory viral infection and often remain unrecognized. In addition, especially in patients with an allergological anamnesis, symptoms of bronchial hyperresponsiveness can simultaneously occur, namely, wheezing and paroxysmal cough. In the absence of pathological changes observed on radiographs, a differential diagnosis of bronchial asthma may be considered [3].

The subacute form of HP develops more slowly, after repeated inhalations of moderate doses of the "malfactor" allergen. This group of HP patients is the most numerous. At the onset of the subacute course of the disease, subfebrile fever and general fatigue may occur, followed by a gradual increase in dyspnea and unproductive (low-productive) cough. On HRCT, there is a characteristic pattern of the form of poorly differentiated centriacinar foci or ground-glass opacity with air trapping and mosaic attenuation.

Chronic forms typically develop with prolonged exposure to low doses of the allergen, and the onset of the disease is usually difficult to establish. In more than half of the cases, it is not possible to identify the allergen responsible for the development of the disease [28]. Persistent cough and dyspnea are initially mild and slowly increase. Hence the disease is often detected at the stage of fibrotic changes in the lungs (the main criterion of chronic HP). In addition to respiratory symptoms, clubbed fingers and weight loss may occur in the long course of the disease. At the late stages of the disease, when honeycombing develops, HP can be very difficult to differentiate from other forms of fibrotic interstitial pneumonia, primarily IPF and NSIP.

Currently, there are no clear criteria for the forms of HP, and the subacute variants of the disease have significant overlap in terms of clinical and radiological characteristics with the acute and chronic forms. Thus, in 2017, the experts of American Thoracic and European Respiratory Society proposed a revised approach to the classification to HP with the separation of only two clinical, radiological, and morphological variants. These forms are the acute (duration of symptoms <6 months and potentially completely reversible) and chronic (duration >6 months with the risk of progression, despite treatment and termination of contact with allergens) [24]. In addition, it was proposed that the term "cryptogenic" HP be used for patients with histologically confirmed HP but without a determined antigen trigger.

Some experts believe that HP is a syndrome rather than a separate nosological form, reflecting an immune response to various inhalation factors, with a heterogeneous clinical picture, course, and prognosis [2].

Similarly to IPF the course of chronic HP can be complicated by periods of exacerbation. These may involve acute respiratory deterioration, with the appearance of new areas of ground-glass opacity and consolidation on the HRCT and neutrophil shift in the BAL fluid [29]. Such events usually occur in patients with a similar pattern to UIP. In cases with fibrotic changes, there are histological signs of diffuse alveolar injury or organized pneumonia [30].

Diagnosis

Household and occupational history is of paramount importance for the diagnosis of HP. Working with products of biological and chemical origin, under conditions of dustiness, inadequate ventilation, air-conditioning, and rooms with high humidity, increases the likelihood of disease development. As for the household and social risk factors, treating physicians should consider the presence of birds, household humidifiers, the correct functioning of ventilation systems, the frequency of hot baths, visits to saunas, indoor swimming pools, public showers, signs of mold on the walls or in the water supply systems, the presence of flower pots with soil at home, the origin of composition of blankets and pillows, and bird feathers and fluff, which can induce the development of the disease.

To facilitate the identification of potential exogenous triggers of the disease, it is convenient to use special questionnaires, such as the Interstitial and Diffuse Lung Disease Patient Questionnaire [31].

At the early stages of the disease, the disorders are usually undetectable. Although not present in all patients, the only possible symptom is expiratory rales in the lungs. As chronic respiratory failure progresses, more obvious symptoms may appear, such as cyanosis, clubbing, Velcro crackles, and "inspiratory squeaks," which usually represent the development of the honeycomb lung.

Functional tests can be normal in the acute forms of HP [32]. In the subacute and chronic course, restrictive or obstructive disorders are usually detected. In general a reduction of the diffusion capacity of the lungs and FVC reflects the volume and severity of lesions in the interstitial tissue and can be used to monitor the course of the disease and evaluate the efficacy of the therapy [3].

Bronchoalveolar lavage (BAL) is an important method for the diagnosis of HP. An increase in the lymphocyte fraction of >40%–50%, with a general increase in cytosis, is an important diagnostic marker of the disease [33]. Nevertheless, it is necessary to correlate the BAL findings with the clinical and radiological picture, since high BAL fluid lymphocytosis can also occur in individuals without signs of an interstitial process in the lungs [34]. In addition, BAL fluid lymphocytosis is not a prerequisite of HP and may be absent in a certain number of patients [35]. In smoking patients with significant fibrotic changes in the lungs receiving treatment with corticosteroids, lymphocytosis of the BAL fluid may be less pronounced.

The role of immunophenotyping of T lymphocytes and the ratio of CD4+/CD8+ in the diagnosis and differential diagnosis of HP is discussed. In general, CD8+ levels usually increase in this disease [33]. However, Caillaud et al. showed that a low CD4+/CD8+ ratio was detected in only 34% of patients. Thus this ratio cannot be used as a criterion for ruling out the diagnosis of HP [35]. In patients with acute HP, neutrophil BAL can be observed within the first week after the onset of the disease. However, subsequently, the level of the lymphocyte fraction increases to the typical level [3].

In the blood serum the presence of IgG antibodies specific to the antigen possibly responsible for triggering HP is evidence of sensitization. The presence of antibodies, in combination with other symptoms, confirms the diagnosis of this allergic disease. Nevertheless, it is not recommended to depend exclusively on the presence or absence of antibodies for the diagnosis of HP, especially the chronic course, since these tests are not characterized by high specificity or sensitivity [36]. In asymptomatic individuals constantly exposed to contact with birds, specific precipitating antibodies are detected in approximately half of the cases [37].

One of the directions for the diagnosis of HP is the ex vivo testing of the antigen-stimulated proliferation of lymphocytes. According to numerous studies the sensitivity and specificity of this method reaches 90%–100% and, unlike the definition of antibodies, is not observed in asymptomatic patients exposed to allergens [36,38].

In HP patients, inhalational challenge tests (i.e., specific inhalation challenge) with extracts of allergens serve as evidence for the development of a pathological immune response. The sensitivity and specificity of these tests for avian and fungal sensitization are 73% and 84%, respectively, whereas for other kinds of allergens, they are significantly lower [39]. In general, these tests are not standardized, bear the risk of the disease recurrence, and are not recommended for use in in routine practice [3].

High-resolution computed tomography

Data of HRCT for the different variants of the HP course vary considerably. In acute/subacute HP, two main patterns are identified, such as foci or significant areas of ground-glass opacity with mosaic attenuation (≤80% of cases) (Figs. 3.12 and 3.13), or poorly differentiated multiple centrilobular nodules (Fig. 3.14) [40]. Researchers identified zones of consolidation, although this sign is less typical [41]. Of note the acute forms of HP may not be detected using HRCT, especially if the study is conducted a few days after the termination of contact with the allergen.

HRCT findings usually tend to be localized in the upper and middle fields.

The ground-glass opacity reflects the morphological pattern of the lymphocytic alveolitis [42]. An important HRCT sign of subacute HP is the appearance of lobular zones of air trapping. This, in combination with areas of ground-glass opacity, creates a picture of mosaic attenuation. It is explained by the development of focal bronchiolitis with obstruction of the distal airways, air retention in the alveoli, and mosaic perfusion, with shunting of the blood bypassing areas of hyperinflation [43] (Fig. 3.13). Occasionally, mosaic attenuation may resemble a cut of cheese, the so-called headcheese sign.

In contrast to other radiological findings, the detection of lobular areas of hyperinflation (air trapping) is a specific sign of HP rarely found in other diffuse interstitial lung diseases. An exception is *Pneumocystis jirovecii* pneumonia (PP) in which similar changes can be detected. In general the number of such sites may be small, with improved detection through expiratory scanning [42].

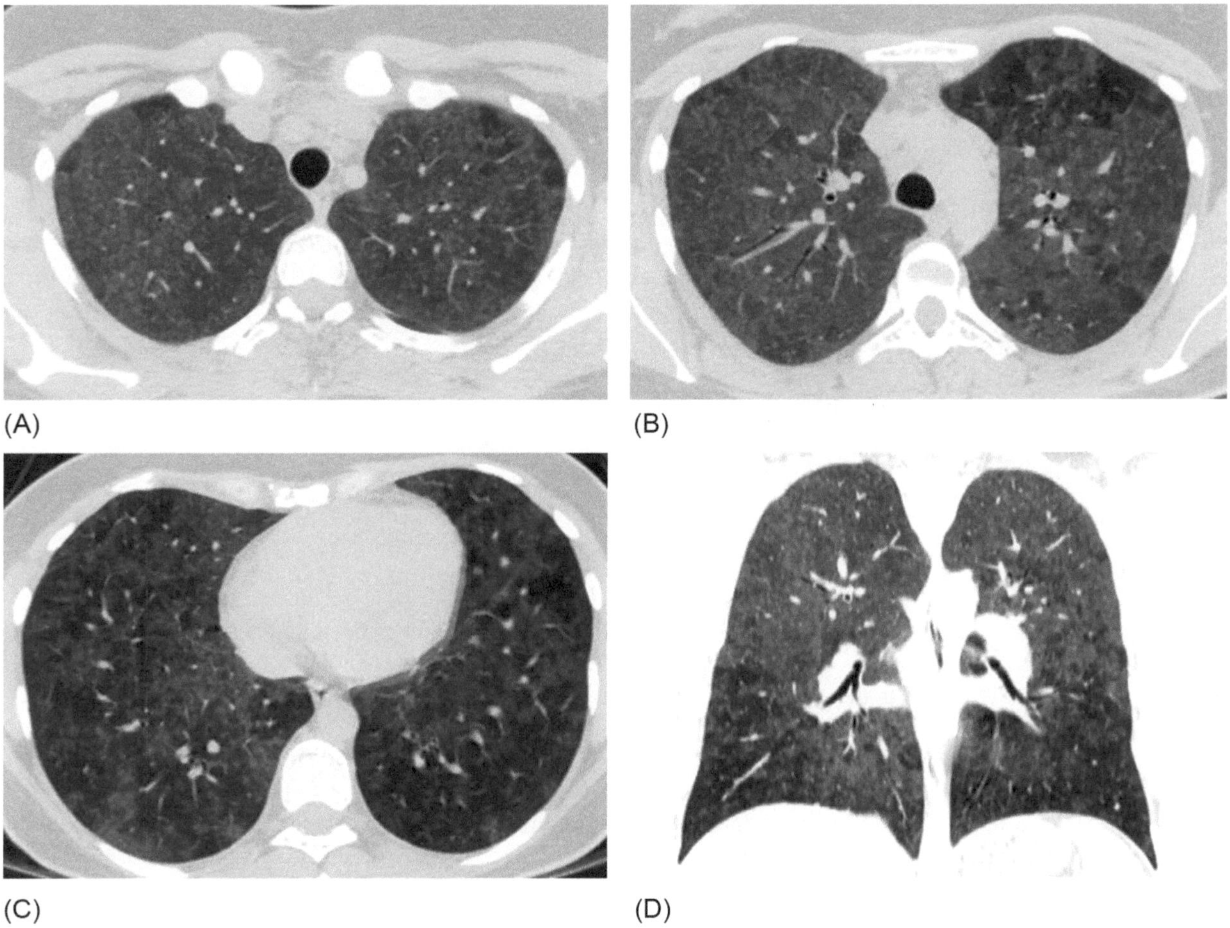

FIG. 3.12 Subacute HP, caused by household contact with parrots (A, B, and C). Diffuse zones of ground-glass opacity, inside of which numerous poorly defined nodules and air trapping are visualized. (D) Changes are mainly located in the upper and middle parts of the lungs. Costodiaphragmatic angles are minimally involved in the pathological process.

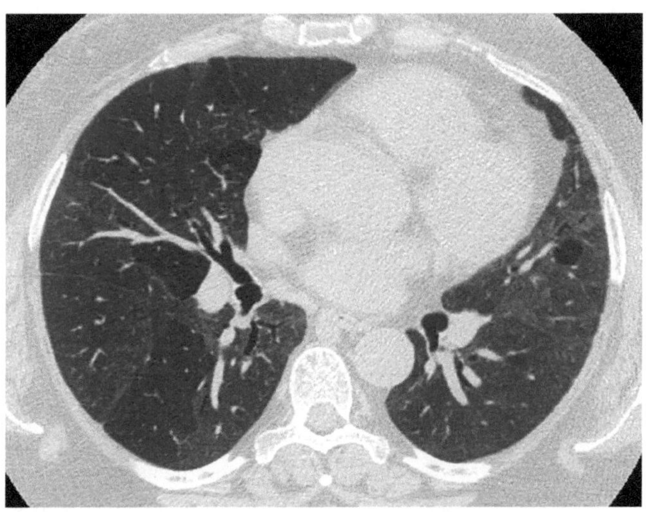

FIG. 3.13 Subacute HP, caused by *A. fumigatus* in a worker in a poorly ventilated cold store. Widespread areas of ground-glass opacity and areas of hyperinflation with hypoperfusion. Thin-walled cyst in the left lung.

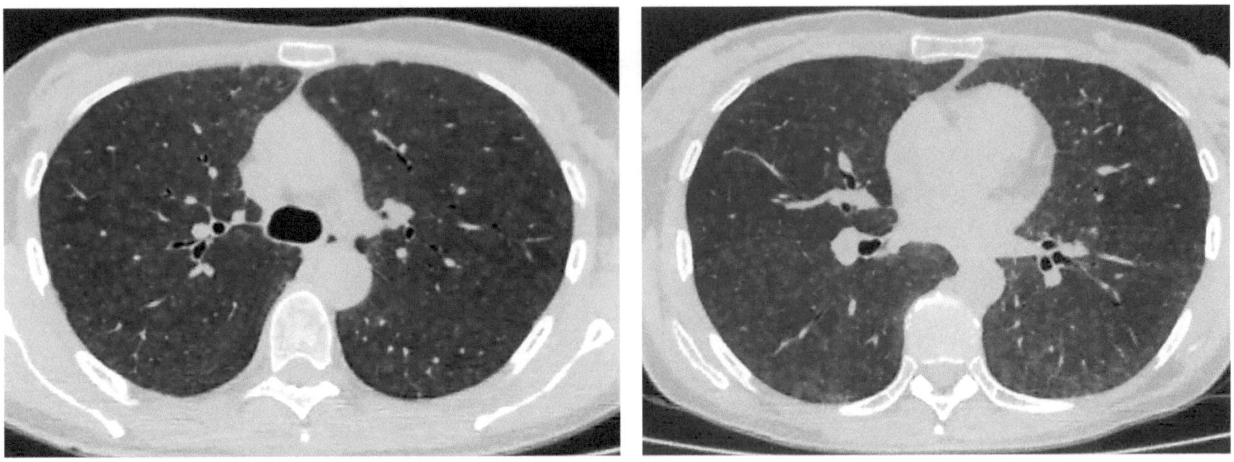

FIG. 3.14 Subacute HP in a bird-fancier. Multiple small, poorly defined centrilobular foci, diffusely distributed in the lung parenchyma, creating the "graininess" effect observed in the scan (A and B).

A frequent radiological sign of subacute HP is the formation of centrilobular ill-defined nodules, which may be combined with ground-glass opacity and represent a separate CT pattern. The size of the nodules varies from 1 to 5mm with blurred outlines. These are diffusely distributed in the pulmonary parenchyma, with a slight increase from the tops to the bases of the lungs (Fig. 3.14).

At the core of this radiographic phenomenon, there is the infiltration of lymphocytes in the walls of bronchioles and the peribronchiolar space [41]. Innumerable centrilobular nodules may be the only CT finding of HP.

In subacute HP the appearance of consolidation sites of small dimensions with ragged contours surrounded by areas of ground-glass opacity (a halo sign) is possible. The presence of consolidation indicates the development of bronchiolitis obliterans with organizing pneumonia as a variant of immunologic fibroproliferative reaction of the lung tissue [42].

Finally, in a small proportion of patients (~13%), thin-walled cysts (with size ranging from 3 to 25mm in diameter) are detected, which have a random distribution (Fig. 3.15). Their development is associated with local bronchiolar obstruction due to lymphocytic bronchiolitis, similar to that observed in idiopathic lymphocytic pneumonia [44].

The chronic course of HP is characterized by the same findings as those observed in the subacute forms. However, additional signs of interstitial fibrosis, such as thickening of the interlobular and intralobular septa, traction bronchiectasis, and honeycombing, are invariably present (Figs. 3.16–3.18). These changes are usually present in all lobes, with their intensity increasing from the top to the basal parts. In 20% of patients, fibrotic findings are distributed mainly in the upper lobes or evenly at all levels of the lungs, which rare occur in IPF [32].

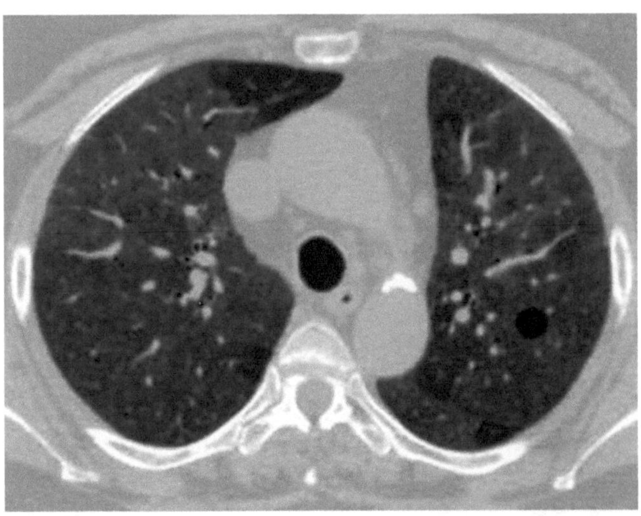

FIG. 3.15 Subacute HP in a poultry farm worker. Diffuse bilateral areas of ground-glass opacity, multiple small nodules are visible. Separate lobules seem lucent. A small, thin-walled cyst in the S2 of the left lung.

In the advanced stages of HP, total replacement of the normal parenchyma with fibrous tissue can be observed. This process renders the differentiation of HP from other fibrotic interstitial pneumonia very difficult radiologically and morphologically [45]. An interesting feature of the distribution of fibrosis zones in chronic HP is often the relative sparing of costophrenic angles, which is not typical in IPF [46].

The increase in intrathoracic lymph nodes is observed in 27%–53% of patients with subacute and chronic HP. Usually the number of such lymph nodes is small, and their size does not exceed 15mm [47]. In chronic HP, there may be signs of emphysema not associated with smoking. In such cases, centriacinar emphysema, less often paraseptal and bullous emphysema, predominates mainly in the upper lobes of the lungs [48].

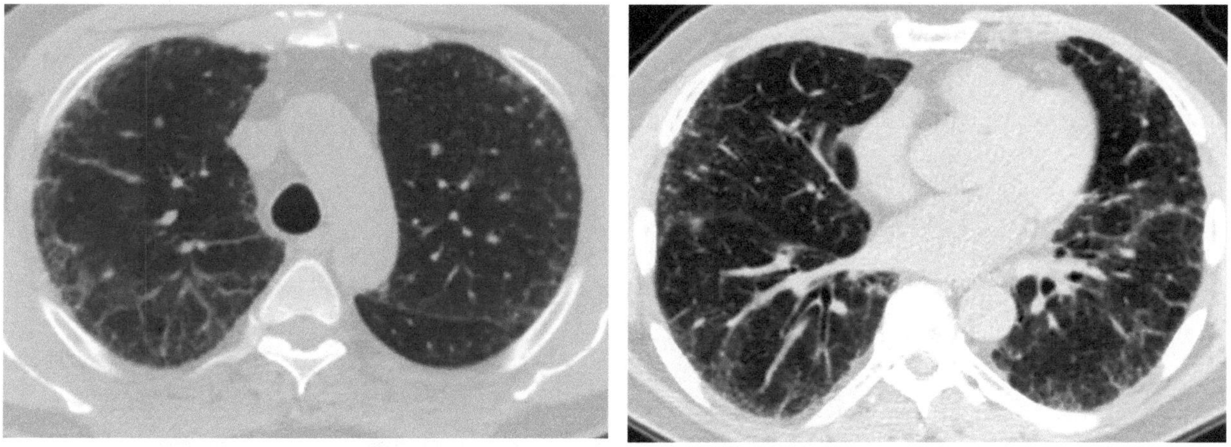

(A) (B)

FIG. 3.16 Chronic HP due to prolonged contact with birds. Bilateral subpleural areas of ground-glass opacity and thickened intralobular and interlobular septa. Thickening of the bronchial walls. The changes are distributed evenly in both the upper and lower fields (A and B).

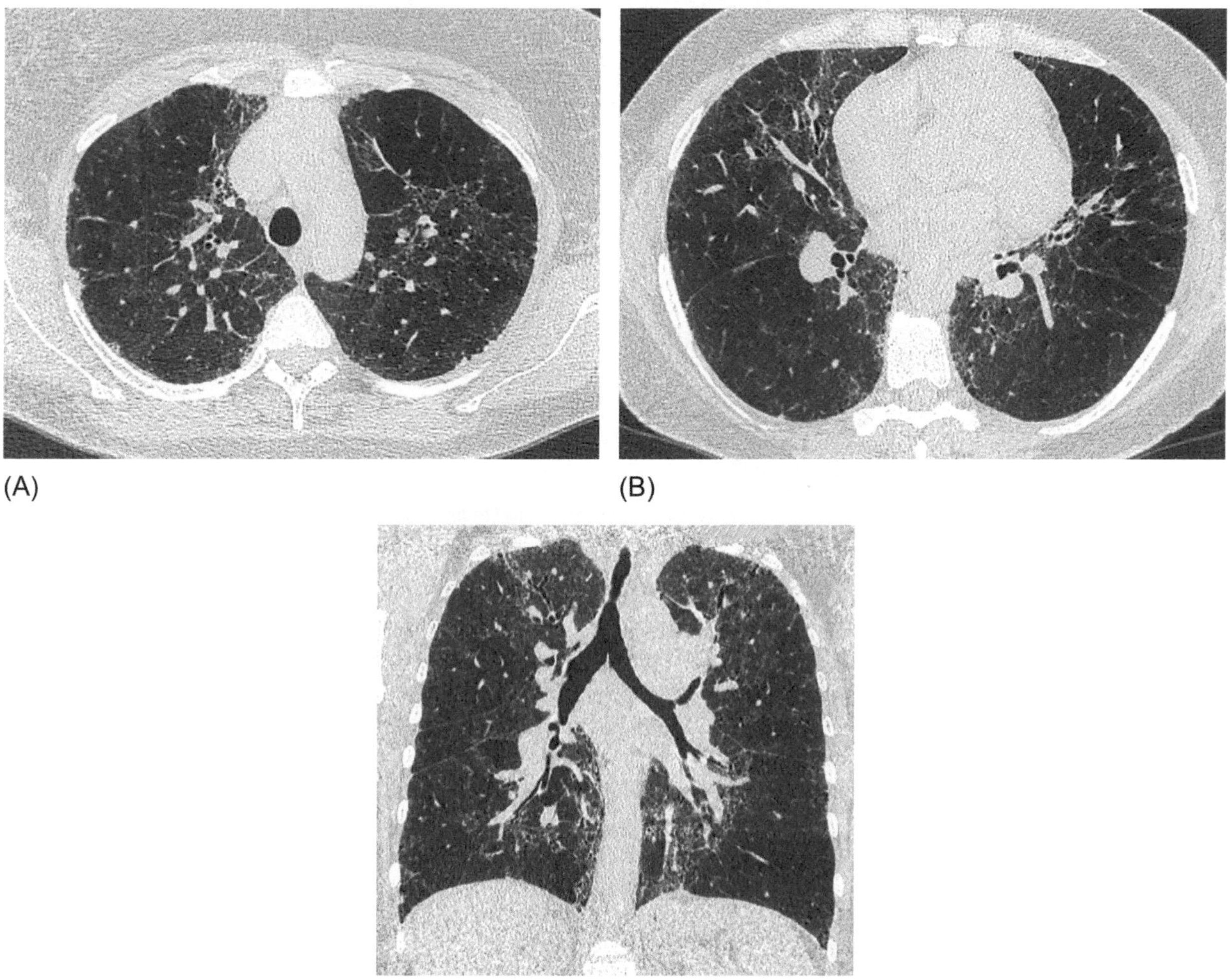

(A) (B)

(C)

FIG. 3.17 Chronic HP in a bird-fancier. Diffuse areas of ground-glass opacity with air trapping in the upper lobes (A). Uneven reticular changes (A and B). Honeycombing with predominant distribution in the lower lobes. Traction bronchiectasis in the both lungs (C).

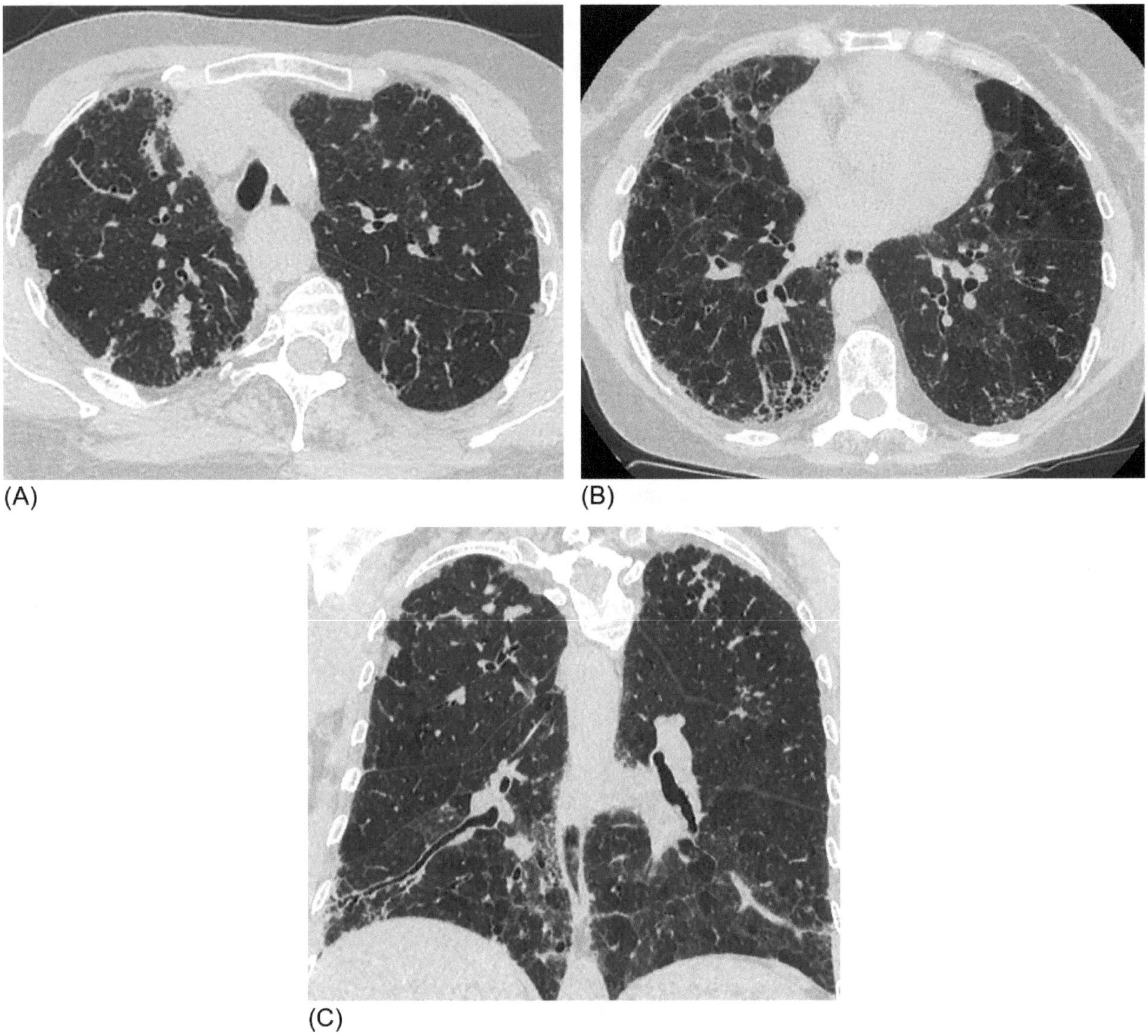

FIG. 3.18 Chronic HP in a wood manufacturing worker. Thickening of interlobular and intralobular septa, thickening of the bronchial walls, zones of subpleural fibrosis, and consolidation are visible (A). In the lower lobes, there is patchy ground-glass opacity with lobular hyperinflation and separate sections of the subpleural honeycombing (B). Changes are more pronounced in the upper and lower fields, whereas the middle zones are affected to a lesser extent (C).

The contemporary diagnostic concept of HP identifies typical CT patterns of the acute/subacute course (upper- and middle-lobe predominant ground-glass opacities, poorly defined centrilobular nodules, mosaic attenuation, air trapping, or, more rarely, consolidation) and chronic course (upper- and middle-lobe predominant fibrosis, peribronchovascular fibrosis, honeycombing, mosaic attenuation, air trapping, centrilobular nodules, and relative sparing of the bases). These patterns, combined with the history of exposure to the allergen and/or the presence of specific IgG and BAL fluid lymphocytosis, are robust indicators for the clinical diagnosis of HP [24].

Probe-based confocal laser endomicroscopy (pCLE)

In two patients with HP examined by pCLE, 35 bronchopulmonary regions were analyzed, 174 informative images were obtained, and 1202 measurements were performed. In the visual fields with a normal structure of the alveoli, in 54.3% of the regions examined, there were alveolar macrophages in an amount from single cells to covering a half of the field of vision, and their giant forms are often noted endomicroscopically in 37.1% (Fig. 3.19A). In general the interalveolar septa are moderately thickened in 31.4% of the regions under study. However, there are individual fields of vision in 5.7% of regions, with interalveolar septa thickened by more than sixfold (Fig. 3.19E). Using alveoloscopy a significant increase

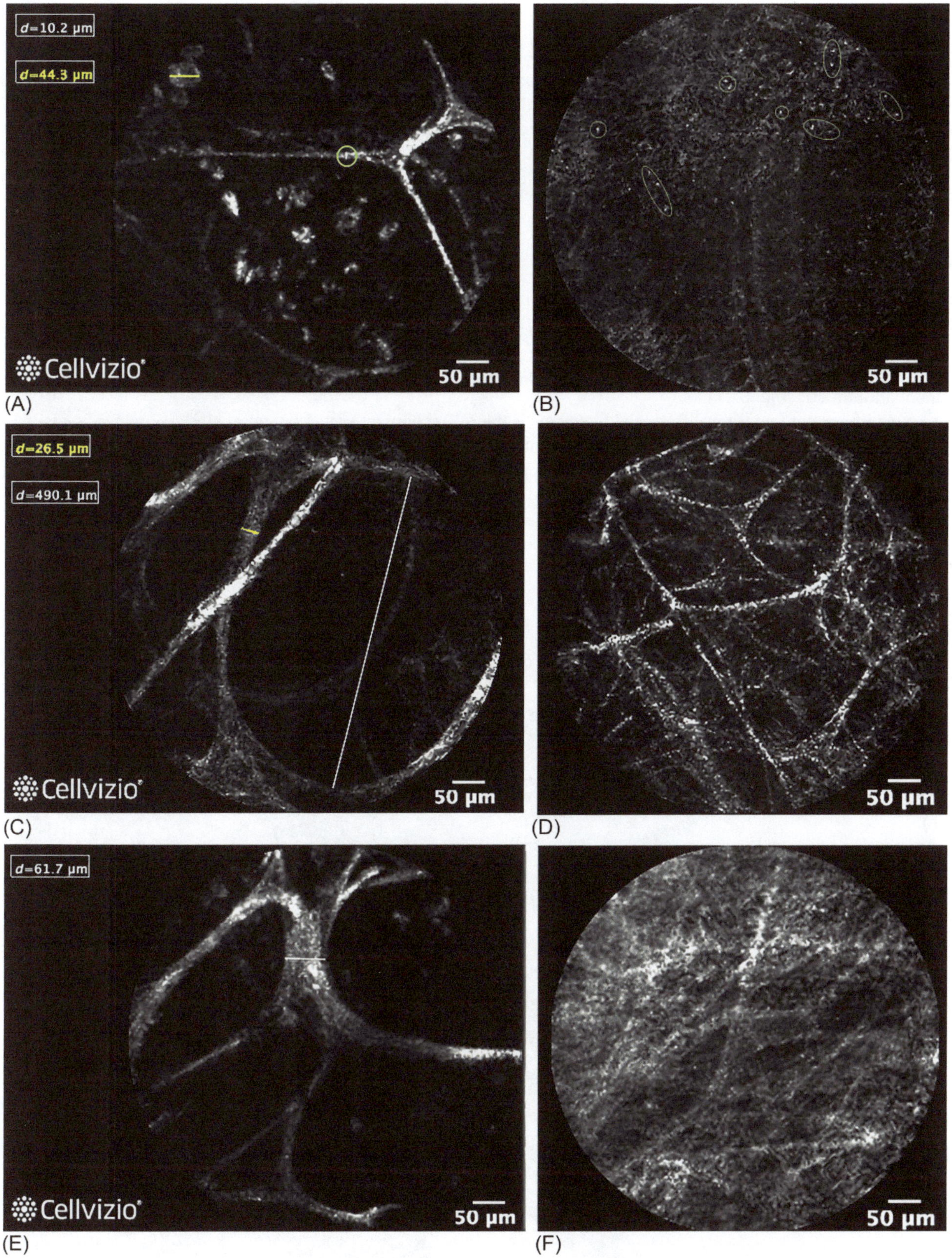

FIG. 3.19 PCLE of the distal airways in patients with subacute HP. (A) Normal pattern with the usual thickness of elastic fibers, moderate amount (1–2 points) of alveolar macrophages (some of them larger than 40 µm). (B) Terminal bronchiole with brightly fluorescent small (3–4 µm) rounded structures are distinct. (C) Alveoli have a regular rounded shape, while the walls are thickened by 2.5- to 3-fold, and the diameter is significantly increased. (D) Moderately pronounced phenomena of dystelectasis at normal thickness of interalveolar septa. (E) Thickening of interalveolar septa to a maximum of 61.7 µm with preserved alveolar structures. (F) The absence of differentiation of the acinus structures. The field of vision is completely filled with highly autofluorescent elements of connective tissue.

in the diameter of the alveolar structures with interalveolar septa thickened by two- to fourfold was observed in 5.7% of the regions (Fig. 3.19C). This observation may indicate the initial stage of "honeycombing lung" formation. In 33.3% of the examined zones, in the part of the fields of vision, the phenomenon of dystelectasis is noted (Fig. 3.19D). Out of 20% of the bronchopulmonary zones, where signs of replacement of normal alveoli with connective tissue elements are noted with the partly saved alveoli, in the fourth part of them, the differentiation of alveolar structures is completely absent (Fig. 3.19F). Small, round, bright fluorescent elements are revealed in 25.7% of bronchopulmonary zones in the distal parts of the terminal bronchioles (Fig. 3.19B).

Differential diagnosis

Acute forms of HP (AHP) must be differentiated with infectious lesions of the lungs (e.g., mycoplasma, PP, and viral diseases) in which similar CT findings, such as scattered areas of ground-glass opacity and centrilobular nodules caused by respiratory bronchiolitis, may be present (Figs. 3.20–3.22).

In PP, even with a diffuse lesion, clearly demarcated intact secondary lobules are possible to detect that resembles mosaic attenuation in HP (Fig. 3.21).

The most important factor for differential diagnosis is the recent contact with a potential allergen, which is easily traceable in AHP. Opportunistic infections develop in patients with severe immunodeficiency (HIV, treatment with immunosuppressants or steroids, and exposure to radiation). Infections causing interstitial inflammation in the lung parenchyma are usually accompanied by acute-phase reactions in blood tests, a significant increase in the level of C-reactive protein (often >100 mg/L). Moreover, in the case of PP or cytomegalovirus pneumonia, infections are often accompanied by lymphopenia. Detection of pneumocystis, cytomegalovirus, or mycoplasma pneumonia in the BAL fluid through PCR indicates

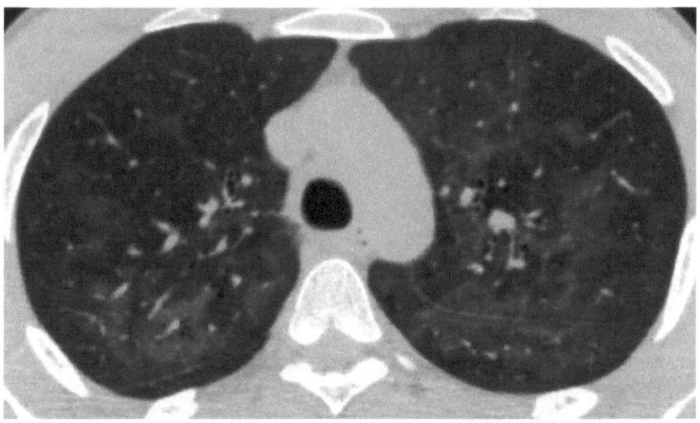

FIG. 3.20 *Pneumocystis jirovecii* pneumonia. Patchy areas of ground-glass opacity are fused together. Single poorly differentiated small nodules.

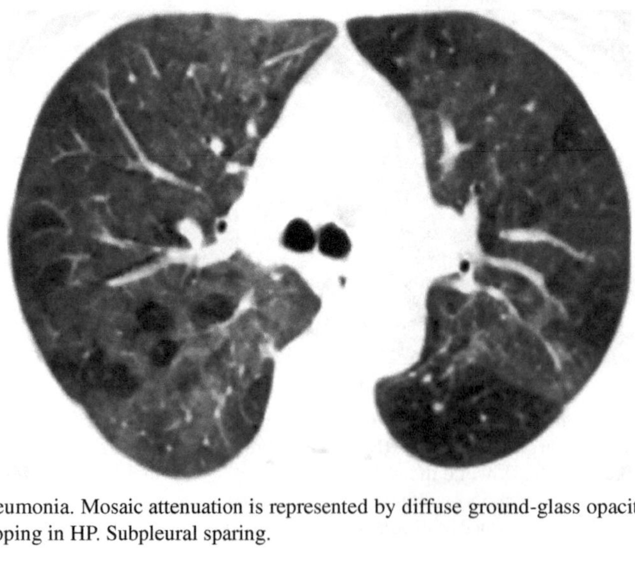

FIG. 3.21 *Pneumocystis jirovecii* pneumonia. Mosaic attenuation is represented by diffuse ground-glass opacity with lobular areas of the parenchyma that is not damaged, resembling air trapping in HP. Subpleural sparing.

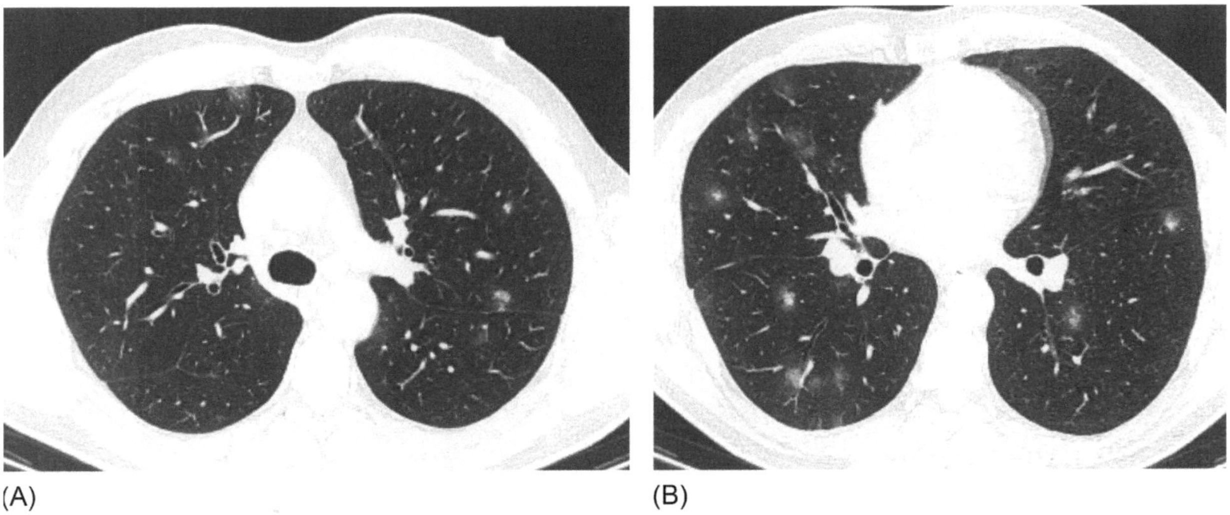

FIG. 3.22 Community-acquired pneumonia in a 27-year-old patient, caused by *M. pneumoniae*. Bilateral multiple small foci of ground-glass opacity and consolidation with a halo sign (A and B).

TABLE 3.4 Differential signs of HP

	HP (subacute and chronic course)	NSIP	IPF	DIP/RB-ILD
History	Contact with allergens Nonsmokers Middle-aged and young patients	Middle-aged and elderly patients, more often nonsmoking females History of arthralgia	Age >50 years, smoking history	The majority are heavy smokers
Visual signs	Clubbing in chronic course	Rarely clubbing	Clubbing	Nonspecific
BAL	Lymphocytosis >50%, "foamy macrophages"	Moderate lymphocytosis Moderate neutrophilia	Neutrophilia, possible eosinophilia Lymphocytosis is not typical	Eosinophilia 15%–30%, macrophages contain brown pigmentation
Radiographic markers	Lobular areas of decreased attenuation, mosaic attenuation, intralobular nodules, ground-glass opacity. Reticular changes and honeycomb lung in chronic forms. Uniform distribution of affected areas. Predominance of changes in the upper lobes is possible	Absence of strictly specific CT signs Ground-glass opacity, traction bronchiectasis Subpleural sparing	Subpleural reticular pattern, traction bronchiectasis, subpleural honeycombing with a minimum of ground-glass opacity and consolidation	Ground-glass opacity with basal and subpleural distribution Intralobular poorly defined nodules. In RB-ILD the changes are more pronounced in the upper parts

HP, hypersensitivity pneumonitis; *NSIP*, nonspecific interstitial pneumonia; *DIP*, desquamative interstitial pneumonia; *RB-ILD*, respiratory bronchiolitis associated with interstitial lung disease.

the presence of acute infection. The rapid reverse improvement of clinical and radiological symptoms (if contact with the allergen) is discontinued—typical of AHP—but not for infectious interstitial pneumonias.

The differential diagnosis of subacute and chronic HP is more difficult, especially if it is not possible to identify an obvious association with an allergen. Such forms should be distinguished from NSIP, desquamative interstitial pneumonia (DIP), respiratory bronchiolitis associated with ILD (RB-ILD), and IPF (Table 3.4). According to Silva et al., it is possible to differentiate chronic HP from NSIP and IPF using CT findings in only 50% of patients [49]. These investigators proposed that subpleural honeycombing, the reticular pattern, centrilobular nodules, ground-glass opacity, and costodiaphragmatic "sparing" may be present in HP and NSIP (Fig. 3.23) [49].

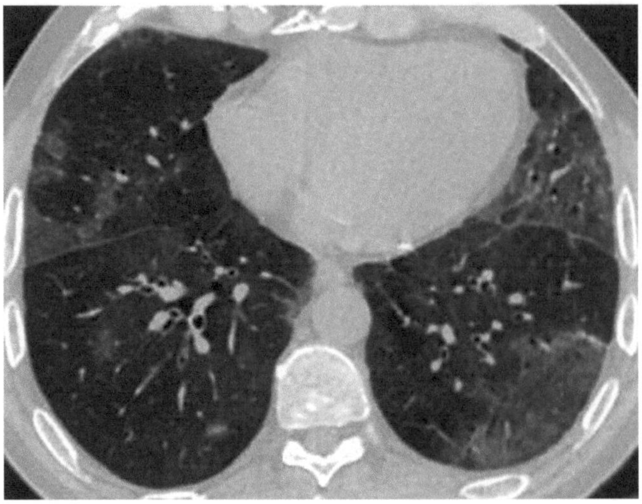

FIG. 3.23 Nonspecific interstitial pneumonia. Patchy areas of ground-glass opacity with a predominance of basal and subpleural distribution are visible. Subpleural sparing.

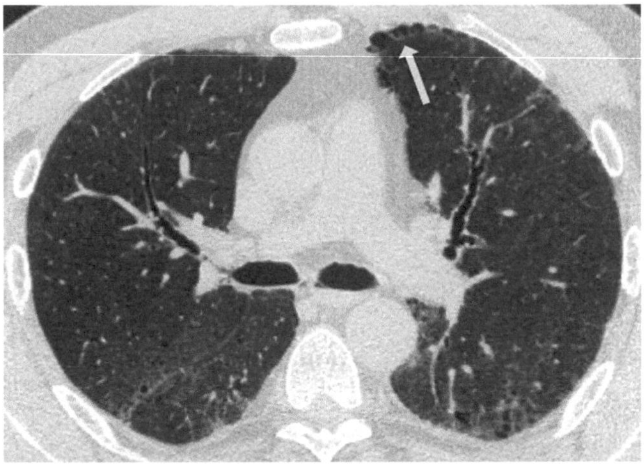

FIG. 3.24 Initial manifestations of idiopathic pulmonary fibrosis. Separate subpleural foci of ground-glass opacity, small thickening of interlobular septa, fibrous cords in the S6 on both sides, and traction bronchiectasis. Paraseptal emphysema in the cortical layer S3 of the left lung *(arrow)*.

The most specific CT findings of chronic HP—rarely observed in other types with similar pathology—are air trapping, mosaic attenuation, and a more even distribution of the lesion with capture of the upper lobes. In IPF, unlike HP, ground-glass opacity (except for cases of acute exacerbation), the appearance of intralobular nodules, and consolidation (Figs. 3.24–3.25) are not typical.

Reticular abnormalities are usually most pronounced in IPF and chronic HP, whereas they are characteristic of cellular subtype of NSIP to a lesser extent. Clinically and radiologically the subacute course of HP may be similar to DIP and RB-ILD. Nevertheless, DIP and RB-ILD develop almost exclusively in heavy smokers, whereas HP often develops in nonsmokers. In DIP and RB-ILD, brown macrophages and eosinophilia are usually found in the BAL fluid. In HP the presence of lymphocytosis is revealed. Ground-glass opacity with basal and subpleural distribution are the main CT findings of DIP (Fig. 3.26).

In DIP the emergence of consolidation, lobular hyperinflation, the halo sign, and honeycombing, often observed in HP, is not typical. Nevertheless, it has been suggested that it is not possible to reliably differentiate between DIP and HP merely through CT [50]. RB-ILD is another disease related to intensive smoking, which may require a differential diagnosis to HP. CT scans of such patients invariably reveal the presence of centrilobular nodules and foci of ground-glass opacity, localized primarily in the upper lobes (Fig. 3.27). Similarly to DIP a history of smoking, the characteristic BAL, and a lack of association with exogenous allergens enable physicians to establish the correct diagnosis without lung biopsy.

Finally, diffuse alveolar hemorrhages with centrilobular ill-defined nodules may resemble the granular pattern (Fig. 3.28) or mosaic attenuation (Fig. 3.29) of HP.

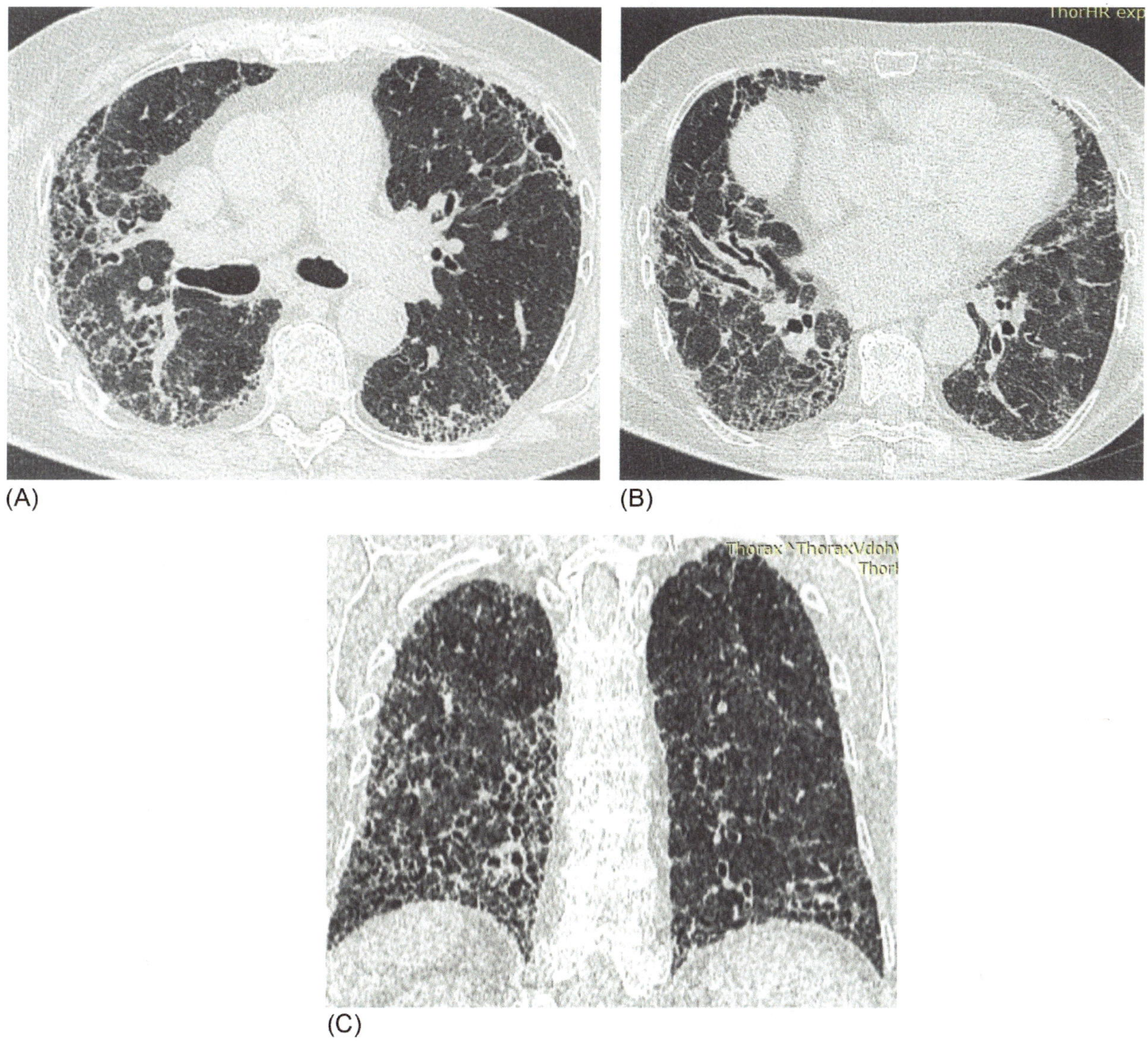

FIG. 3.25 Advanced idiopathic pulmonary fibrosis. Reticular abnormalities and honeycombing are maximal in the subpleural and basal zones and more pronounced the right lung. Traction bronchiectasis (A–C).

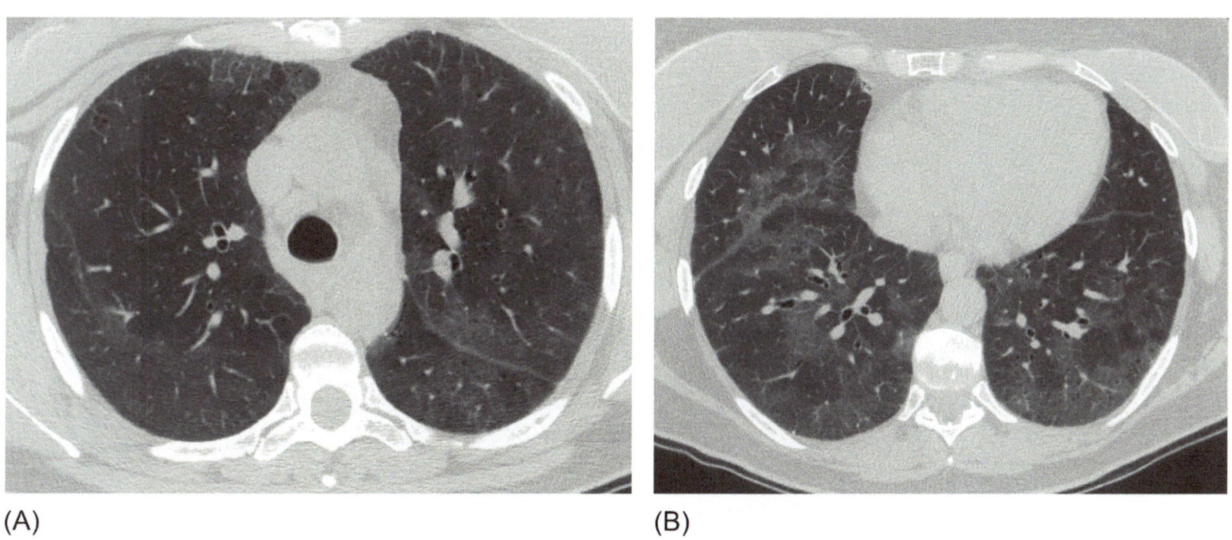

FIG. 3.26 Desquamative interstitial pneumonia. Bilateral patchy areas of ground-glass opacity with a predominance of lower-lobe distribution in which the moderately thickened interlobular septa and emphysema foci are visible. The alternation of ground-glass opacity with intact pulmonary parenchyma creates the effect of mosaic attenuation, especially in the lower lobes (A and B).

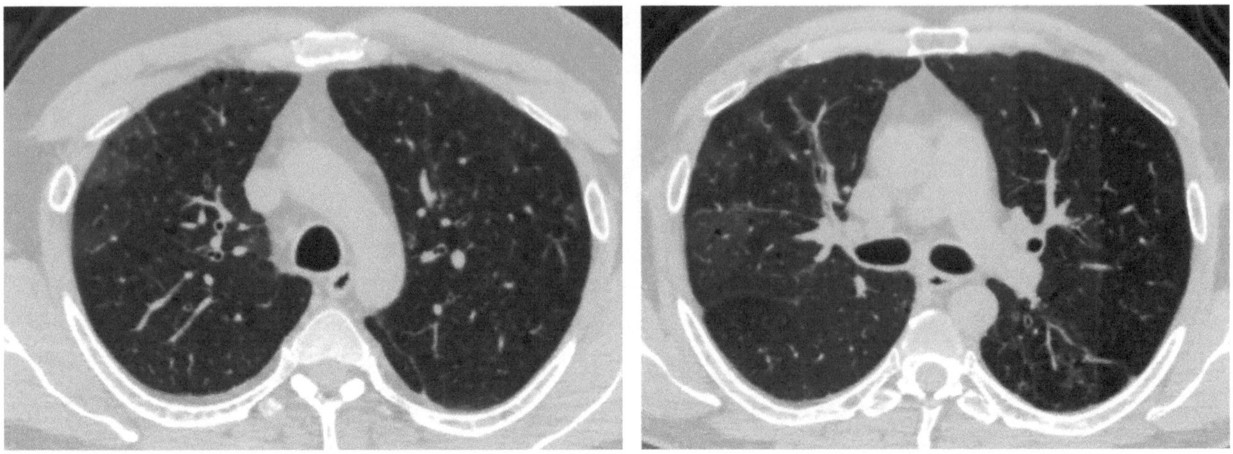

(A) (B)

FIG. 3.27 Respiratory bronchiolitis associated with interstitial lung disease. Multiple small poorly differentiated nodules. Foci of ground-glass opacity present mainly in the right lung. Thickening and deformity of bronchial walls (A and B).

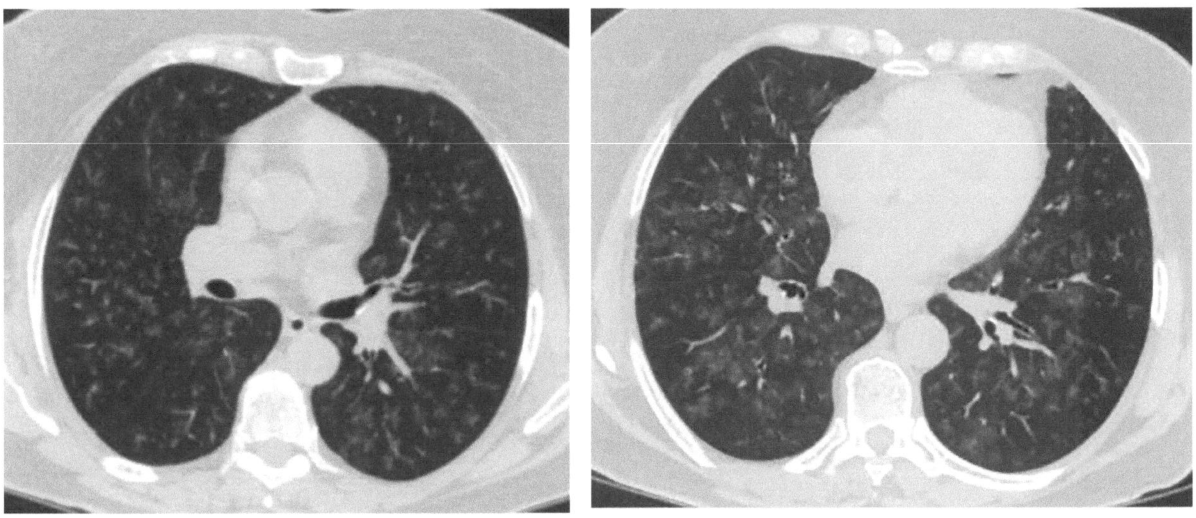

(A) (B)

FIG. 3.28 Microscopic polyangiitis associated with diffuse alveolar hemorrhages. Multiple bilateral, poorly defined polymorphic intralobular nodules. Subpleural sparing (A and B).

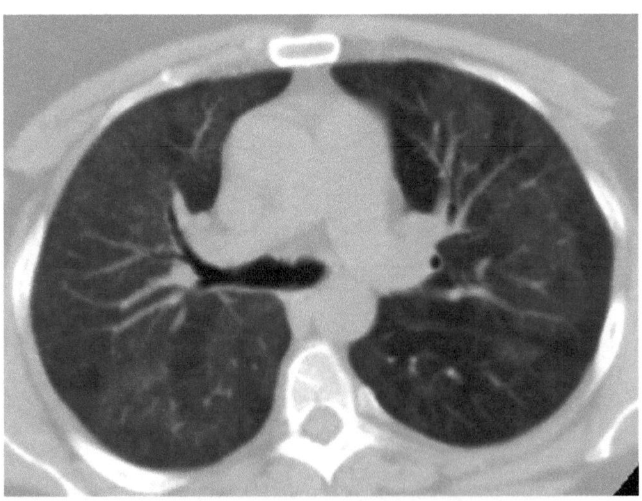

FIG. 3.29 Systemic lupus erythematosus complicated by diffuse alveolar hemorrhages. Diffuse zones of ground-glass opacity with mosaic attenuation and thickened interlobular septa. Subpleural sparing.

Treatment and prognosis

The main target in the treatment of HP is the elimination of the allergen. Removal of the antigenic source from housing accommodations does not immediately lead to the disappearance of allergens from the air. Craig et al. showed that bird antigens persisted in the atmosphere of housing accommodations in a significant concentration up to 6 months (in one case up to 18 months) [51]. An additional challenge is the persistent sensitization of lymphocytes and the production of antibodies in the respiratory tract, which is persistent for ≤5 years after the termination of direct contact with the allergen. This can lead to progression and relapse of the disease, even after the interruption of exposure to the allergen [52].

Treatment of acute HP is usually limited to the administration of a short course of systemic steroids. In many patients, remission can occur spontaneously without any treatment after the termination of contact with the allergen.

The subacute and chronic forms of HP require the administration of systemic steroids. The standard regimen recommended by the American Thoracic Society includes prednisone at a dose of 0.5 mg/kg/day for 6–8 weeks, followed by a gradual decrease to a dose of 10 mg/day. Treatment with steroids can be discontinued only in the absence of clinical symptoms and functional disorders [3]. In subacute HP the usual duration of steroid therapy is 3–6 months. In patients with chronic HP, the question of the possibility of withdrawal of systemic steroids is solved individually. It is recommended that physicians strive to reduce the dose to 20 mg per day during the first 3 months of treatment. Long-term treatment of a dose >20 mg requires the prevention of pneumocystis infection [53].

In cases of delayed treatment initiation for chronic HP, response to steroid therapy is observed in approximately 58% of patients [54]. Such patients should be treated with immunosuppressive agents, namely, azathioprine (1.5–2 mg/kg/day) or mycophenolate mofetil (1000–3000 mg/day), for 6–18 months. This treatment demonstrates a significant improvement in DLCO and a similar profile of side effects [55].

According to individual clinical reports, rituximab and leflunomide may be used as alternatives to azathioprine and mycophenolate mofetil [56].

The off-label administration of antifibrotic drugs (e.g., pirfenidone and nintedanib) prior to the results of the ongoing studies is acceptable in patients with a progressive increase in honeycombing and reticular changes and a UIP-like pattern [24].

Treatment of HP with steroid inhalers is generally not justified by controlled studies, although it has been described in a few clinical cases [57].

In the absence of repeated contact with the allergen, the course of acute HP is usually favorable, with patients achieving complete remission. In the subacute and chronic courses, the prognosis is usually determined by the extent of fibrotic changes. Poor prognosis for survival is linked to patients with BAL fluid neutrophilia, development of honeycombing, and bronchiectasis [29,58]. The presence of histological signs of UIP, fibrotic nonspecific IP, and peribronchiolar fibrosis renders the prognosis less favorable. Thus, in UIP pattern, the average period to lung transplantation or death is 3.6 years [59]. In general the course of chronic HP is similar to that observed for IPF. An insufficient response to steroids, possible periods of exacerbation with increased clinical and radiological signs of the disease, and a similar morphological pattern align this variant of HP with IPF.

In the chronic course of HP, as in IPF, lung cancer develops in approximately 10% of patients, most often occurring in the areas of honeycombing [60].

References

[1] Mohr LC. Hypersensitivity pneumonitis. Curr Opin Pulm Med 2004;10(5):401–11.

[2] Mason RJ, Murray JF, Broaddus VC, Nadel JA. Murray & Nadel's textbook of respiratory medicine. 6th ed. vol. 2. USA: Elsevier Inc.; 2015. p. 1153–64.

[3] Selman M, Pardo A, King Jr TE. Hypersensitivity pneumonitis: insights in diagnosis and pathobiology. Am J Respir Crit Care Med 2012;186(4):314–24.

[4] Dobashi K, Akiyama K, Usami A, Yokozeki H, Ikezawa Z, Tsurikisawa N, et al. Japanese guidelines for occupational allergic diseases 2017. Allergol Int 2017;66(2):265–80.

[5] Zacharisen MC, Fink JN. Hypersensitivity pneumonitis and related conditions in the work environment. Immunol Allergy Clin North Am 2011;31(4):769–86.

[6] Dalphin J-C, Gondouin A. Rare causes and the spectrum of hypersensitivity pneumonitis. In: Cottin V, Cordier J-F, Richeldi L, editors. Orphan lung diseases. A clinical guide to rare lung disease. London: Springer-Verlag; 2015. p. 457–72.

[7] Roshnee G, Cao GQ, Chen H. Hypersensitivity pneumonitis due to residential mosquito-coil smoke exposure. Chin Med J (Engl) 2011;124(12):1915–8.

[8] Toribio R, Cruz MJ, Morell F, Muñoz X. Hypersensitivity pneumonitis related to medium-density fiberboard. Arch Bronconeumol 2012;48(1):29–31.

[9] Paris C, Herin F, Reboux G, Penven E, Barrera C, Guidat C, et al. Working with argan cake: a new etiology for hypersensitivity pneumonitis. BMC Pulm Med 2015;15:18.

[10] Sommerfeld CG, Weiner DJ, Nowalk A, Larkin A. Hypersensitivity pneumonitis and acute respiratory distress syndrome from e-cigarette use. Pediatrics 2018;141(6) pii: e20163927.

[11] Solaymani-Dodaran M, West J, Smith C, Hubbard R. Extrinsic allergic alveolitis: incidence and mortality in the general population. QJM 2007;100(4):233–7.

[12] McDonald JC, Chen Y, Zekveld C, Cherry NM. Incidence by occupation and industry of acute work related respiratory diseases in the UK, 1992–2001. Occup Environ Med 2005;62(12):836–42.

[13] Lacasse Y, Selman M, Costabel U, Dalphin JC, Ando M, Morell F, et al, HP Study Group. Clinical diagnosis of hypersensitivity pneumonitis. Am J Respir Crit Care Med 2003;168(8):952–8.

[14] Blanchet MR, Israël-Assayag E, Cormier Y. Inhibitory effect of nicotine on experimental hypersensitivity pneumonitis in vivo and in vitro. Am J Respir Crit Care Med 2004;169(8):903–9.

[15] Barrera L, Mendoza F, Zuciga J, Estrada A, Zamora AC, Melendro EI, et al. Functional diversity of T-cell subpopulations in subacute and chronic hypersensitivity pneumonitis. Am J Respir Crit Care Med 2008;177(1):44–55.

[16] Ley B, Newton CA, Arnould I, Elicker BM, Henry TS, Vittinghoff E, et al. The MUC5B promoter polymorphism and telomere length in patients with chronic hypersensitivity pneumonitis: an observational cohort-control study. Lancet Respir Med 2017;5(8):639–47.

[17] Snetselaar R, van Moorsel CHM, Kazemier KM, van der Vis JJ, Zanen P, van Oosterhout MFM, et al. Telomere length in interstitial lung diseases. Chest 2015;148(4):1011–8.

[18] Camarena A, Juárez A, Mejía M, Estrada A, Carrillo G, Falfán R, et al. Major histocompatibility complex and tumor necrosis factor-alpha polymorphisms in pigeon breeder's disease. Am J Respir Crit Care Med 2001;163(7):1528–33.

[19] English JC. Hypersensitivity pneumonitis. Chapter 12, In: Ph H, Flieder DB, editors. Spencer's pathology of the lung. 6th ed., vol. 1. New York: Cambridge University Press; 2013. p. 439–74.

[20] Hariri LP, Mino-Kenudson M, Shea B, Digumarthy S, Onozato M, Yagi Y, et al. Distinct histopathology of acute onset or abrupt exacerbation of hypersensitivity pneumonitis. Hum Pathol 2012;43(5):660–8.

[21] Khoor A, Leslie KO, Tazelaar HD, Helmers RA, Colby TV. Diffuse pulmonary disease caused by nontuberculous mycobacteria in immunocompetent people (hot tub lung). Am J Clin Pathol 2001;115(5):755–62.

[22] Grunes D, Beasley MB. Hypersensitivity pneumonitis: a review and update of histologic findings. J Clin Pathol 2013;66(10):888–95.

[23] Reyes CN, Wenzel FG, Lawton BR, Emanuel DA. The pulmonary pathology of farmer's lung disease. Chest 1982;81(2):142–6.

[24] Vasakova M, Morell F, Walsh S, Leslie K, Raghu G. Hypersensitivity pneumonitis: perspectives in diagnosis and management. Am J Respir Crit Care Med 2017;196(6):680–9.

[25] Trahan S, Hanak V, Ryu JH, Myers JL. Role of surgical lung biopsy in separating chronic hypersensitivity pneumonia from usual interstitial pneumonia/idiopathic pulmonary fibrosis: analysis of 31 biopsies from 15 patients. Chest 2008;134(1):126–32.

[26] Takemura T, Akashi T, Kamiya H, Ikushima S, Ando T, Oritsu M. Pathological differentiation of chronic hypersensitivity pneumonitis from idiopathic pulmonary fibrosis/usual interstitial pneumonia. Histopathology 2012;61(6):1026–35.

[27] Richerson HB, Bernstein IL, Fink JN, Hunninghake GW, Novey HS, Reed CE. Guidelines for the clinical evaluation of hypersensitivity pneumonitis: report of the Subcommittee on Hypersensitivity Pneumonitis. J Allergy Clin Immunol 1989;84(5 Pt 2):839–44.

[28] Fernandez Perez ER, Swigris JJ, Forssen AV, Tourin O, Solomon JJ, Huie TJ, et al. Identifying an inciting antigen is associated with improved survival in patients with chronic hypersensitivity pneumonitis. Chest 2013;144(5):1644–51.

[29] Miyazaki Y, Tateishi T, Akashi T, Ohtani Y, Inase N, Yoshizawa Y. Clinical predictors and histologic appearance of acute exacerbations in chronic hypersensitivity pneumonitis. Chest 2008;134(6):1265–70.

[30] Olson AL, Huie TJ, Groshong SD, Cosgrove GP, Janssen WJ, Schwarz MI, et al. Acute exacerbations of fibrotic hypersensitivity pneumonitis: a case series. Chest 2008;134(4):844–50.

[31] http://www.chestnet.org/Search#q/q=Lung%20Disease%20Questionaire.

[32] Lynch DA, Rose CS, Way D, King Jr TE. Hypersensitivity pneumonitis: sensitivity of high-resolution CT in a population-based study. Am J Roentgenol 1992;159(3):469–72.

[33] Meyer KC, Raghu G, Baughman RP, Brown KK, Costabel U, du Bois RM, et al. An official American Thoracic Society clinical practice guideline: the clinical utility of bronchoalveolar lavage cellular analysis in interstitial lung disease. Am J Respir Crit Care Med 2012;185(9):1004–14.

[34] Lacasse Y, Girard M, Cormier Y. Recent advances in hypersensitivity pneumonitis. Chest 2012;142(1):208–17.

[35] Caillaud DM, Vergnon JM, Madroszyk A, Melloni BM, Murris M, Dalphin JC, French Group of Environmental Immunoallergic Bronchopulmonary Diseases. Bronchoalveolar lavage in hypersensitivity pneumonitis: a series of patients. Inflamm Allergy Drug Targets 2012;11(1):15–9.

[36] Inase N, Unoura K, Miyazaki Y, Yasui M, Yoshizawa Y. Measurement of bird specific antibody in bird-related hypersensitivity pneumonitis. Nihon Kokyuki Gakkai Zasshi 2011;49(10):717–22.

[37] Cormier Y, Létourneau L, Racine G. Significance of precipitins and asymptomatic lymphocytic alveolitis: a 20-yr follow-up. Eur Respir J 2004;23(4):523–5.

[38] Bouic PJ, Nel NC, Beer PM, Joubert JR. In vitro reactivities of blood lymphocytes from symptomatic and asymptomatic pigeon breeders to antigen and mitogens. Int Arch Allergy Appl Immunol 1989;89(2–3):222–8.

[39] Muñoz X, Sánchez-Ortiz M, Torres F, Villar A, Morell F, Cruz MJ. Diagnostic yield of specific inhalation challenge in hypersensitivity pneumonitis. Eur Respir J 2014;44(6):1658–65.

[40] Tateishi T, Ohtani Y, Takemura T, Akashi T, Miyazaki Y, Inase N, et al. Serial high-resolution computed tomography findings of acute and chronic hypersensitivity pneumonitis induced by avian antigen. J Comput Assist Tomogr 2011;35(2):272–9.
[41] Hansell DM, Wells AU, Padley SP, Müller NL. Hypersensitivity pneumonitis: correlation of individual CT patterns with functional abnormalities. Radiology 1996;199(1):123–8.
[42] Silva CI, Churg A, Müller NL. Hypersensitivity pneumonitis: spectrum of high-resolution CT and pathologic findings. Am J Roentgenol 2007;188(2):334–44.
[43] Patel RA, Sellami D, Gotway MB, Golden JA, Webb WR. Hypersensitivity pneumonitis: patterns on high-resolution CT. J Comput Assist Tomogr 2000;24(6):965–70.
[44] Franquet T, Hansell DM, Senbanjo T, Remy-Jardin M, Müller NL. Lung cysts in subacute hypersensitivity pneumonitis. J Comput Assist Tomogr 2003;27(4):475–8.
[45] Wright JL, Tazelaar HD, Churg A. Fibrosis with emphysema. Histopathology 2011;58(4):517–24.
[46] Hypersensitivity pneumonitis and eosinophilic lung diseases. In: Webb WR, Muller NL, Naidich DP, editors. High-resolution CT of the lung. 5th ed. Philadelphia: Lippincott Williams and Wilkins; 2015. p. 379.
[47] Cormier Y, Brown M, Worthy S, Racine G, Müller NL. High-resolution computed tomographic characteristics in acute farmer's lung and in its follow up. Eur Respir J 2000;16(1):56–60.
[48] Baqir M, White D, Ryu JH. Emphysematous changes in hypersensitivity pneumonitis: a retrospective analysis of 12 patients. Respir Med Case Rep 2018;24:25–9.
[49] Silva CI, Muller NL, Lynch DA, Curran-Everett D, Brown KK, Lee KS, et al. Chronic hypersensitivity pneumonitis: differentiation from idiopathic pulmonary fibrosis and nonspecific interstitial pneumonia by using thin-section CT. Radiology 2008;246(1):288–97.
[50] Lynch DA, Newell JD, Logan PM, King Jr TE, Müller NL. Can CT distinguish hypersensitivity pneumonitis from idiopathic pulmonary fibrosis? AJR Am J Roentgenol 1995;165(4):807–11.
[51] Craig TJ, Hershey J, Engler RJ, Davis W, Carpenter GB, Salata K. Bird antigen persistence in the home environment after removal of the bird. Ann Allergy 1992;69(6):510–2.
[52] Yoshizawa Y, Miyake S, Sumi Y, Hisauchi K, Sato T, Kurup VP. A follow-up study of pulmonary function tests, bronchoalveolar lavage cells, and humoral and cellular immunity in bird fancier's lung. J Allergy Clin Immunol 1995;96(1):122–9.
[53] Salisbury ML, Myers JL, Belloli EA, Kazerooni EA, Martinez FJ, Flaherty KR. Diagnosis and treatment of fibrotic hypersensitivity pneumonia: where we stand and where we need to go. Am J Respir Crit Care Med 2017;196(6):690–9.
[54] Yoshizawa Y, Ohtani Y, Hayakawa H, Sato A, Suga M, Ando M. Chronic hypersensitivity pneumonitis in Japan: a nationwide epidemiologic survey. J Allergy Clin Immunol 1999;103(2 Pt 1):315–20.
[55] Morisset J, Johannson KA, Vittinghoff E, Aravena C, Elicker BM, Jones KD, et al. Use of mycophenolate mofetil or azathioprine for the management of chronic hypersensitivity pneumonitis. Chest 2017;151(3):619–25.
[56] Lota HK, Keir GJ, Hansell DM, Nicholson AG, Maher TM, Wells AU, et al. Novel use of rituximab in hypersensitivity pneumonitis refractory to conventional treatment. Thorax 2013;68(8):780–1.
[57] Agache IO, Rogozea L. Management of hypersensivity pneumonitis. Clin Transl Allergy 2013;3(1):5.
[58] Walsh SL, Sverzellati N, Devaraj A, Wells AU, Hansell DM. Chronic hypersensitivity pneumonitis: high resolution computed tomography patterns and pulmonary function indices as prognostic determinants. Eur Radiol 2012;22(8):1672–9.
[59] Wang P, Jones KD, Urisman A, Elicker BM, Urbania T, Johannson KA. Pathologic findings and prognosis in a large prospective cohort of chronic hypersensitivity pneumonitis. Chest 2017;152(3):502–9.
[60] Kuramochi J, Inase N, Miyazaki Y, Kawachi H, Takemura T, Yoshizawa Y. Lung cancer in chronic hypersensitivity pneumonitis. Respiration 2011;82(3):263–7.

Chapter 4

Airspace-predominant diseases

Alexander Averyanov[a,b], Evgeniya Kogan[c], Victor Lesnyak[d], Olesya Danilevskaya[e]

[a]*Clinical Department, Pulmonology Research Institute under FMBA of Russia, Moscow, Russia,* [b]*Pulmonary Division, Federal Research Clinical Center under FMBA of Russia, Moscow, Russia,* [c]*Anatomic Pathology Department, Sechenov University, Moscow, Russia,* [d]*Radiology Department, Federal Research Clinical Center under FMBA of Russia, Moscow, Russia,* [e]*Endoscopy Department, Pulmonology Research Institute under FMBA of Russia, Moscow, Russia*

Chapter 4.1

Pulmonary alveolar proteinosis

Pulmonary alveolar proteinosis (PAP) is a rare lung disease characterized by the accumulation of a protein-lipid complex in the alveoli due to impaired utilization of surfactant by alveolar macrophages [1]. The condition was first described by Rosen et al. in 1958 [2]. Some experts consider PAP to be a syndrome that comprises various heterogeneous disorders manifested by the accumulation of surfactant components in the alveolar space for various reasons [3]. According to epidemiological studies, PAP occurs with a frequency of two to four cases per 1 million adults. A higher prevalence of up to 6.2 cases per million has been described in some populations [4]. An analysis of 410 patients with PAP showed that the patients were predominantly of middle age (mean age 39 years for men and 35 years for women), that the disease was 2.65 times more common in male smokers than in women smokers, and that among nonsmokers there were fewer men than women (in the ratio 0.69:1) [5].

There are three forms of PAP: congenital, secondary (pseudoproteinosis), and autoimmune (previously referred to as an idiopathic form). The congenital form is caused by mutations of the genes that encode the structure of α and β subunits of granulocyte-macrophage colony-stimulating factor (GM-CSF) and is more common in children [1]. The secondary form develops due to alveolar clearance being damaged by macrophages resulting from an underlying disease (such as blood disorders, severe immunodeficiency, or tumors) or the inhalation of inorganic dust, toxic vapors, NO_2, and other pollutants (see Table 4.1.1) [6].

In 90% of cases, PAP is considered to be an autoimmune disease, because patients have elevated titers of antibodies to GM-CSF [7]. An autoimmune attack downregulates the clearance activity of alveolar GM-CSF-dependent macrophages and disrupts surfactant homeostasis [1,6]. In addition, a so-called metabolic dysfunction of the pulmonary surfactant that results from genetically determined erroneous surfactant synthesis has recently been identified [3]. Thus PAP may develop not only as a result of clearance disruption but also due to abnormal surfactant production.

Morphology

Alveolar proteinosis is diagnosed by characteristic macroscopic and microscopic features. At macroscopic examination the lungs are heavy and dense, and sometimes foci of consolidation are observed. Cut surfaces drain viscous yellowish fluid. Microscopic examination shows areas of acellular intra-alveolar Schiff-positive lipoprotein deposits that are separated by broad areas of intact lung tissue (Figs. 4.1.1–4.1.4). Lipoprotein masses have a fine-granular structure and are stained with eosin. This amorphous material fills alveolar sacs, alveolar ducts, and respiratory bronchioles with minimal interstitial inflammation [8]. However, if the process is long term, interstitial fibrosis develops and giant cell reaction with clusters of multinucleated foreign-body giant cells (Fig. 4.1.5). The presence of surfactant and surfactant apoprotein can be confirmed by immunohistochemical study [9]. Electron microscopy shows concentrically multilaminated bodies containing surfactant, which are produced by type 2 pneumocytes [10]. The condition may be complicated by a secondary infection with an inflammatory response in the lung tissue.

Difficult to Diagnose Rare Diffuse Lung Disease. https://doi.org/10.1016/B978-0-12-815375-8.00004-2

TABLE 4.1.1 Causes of secondary alveolar proteinosis [1,3,6, supplemented]

Type of disorder	Causes
Hematologic disorders	Fanconi anemia Myelodysplastic syndrome Acute myeloid leukemia Chronic lymphocytic leukemia Chronic myelogenous leukemia Lymphoma
Hereditary disorders	Congenital lysinuria (aminoaciduria and lysinuric protein intolerance) Niemann–Pick disease, type C2
Immune deficiency	Thymic alymphoplasia IgA deficiency Immunosuppression after organ transplantation Severe combined immunodeficiency
Autoimmune disorders	Psoriasis Amyloidosis IgG monoclonal gammopathy Connective tissue diseases Transplanted lung Ulcerative colitis
Neoplastic disorders	Lung cancer Mesothelioma Melanoma Glioblastoma metastases in the lungs
Inhalation of pollutants	Coal dust Cotton Cement dust Titanium dust Aluminum dust Indium oxide Cellulose NO_2
Infections	HIV Nocardia *Cytomegalovirus* *Pneumocystis jirovecii* *Histoplasma capsulatum* *Cryptococcus neoformans* *Mycobacterium tuberculosis* *Mycobacterium avium* complex
Other lung disorders	Hypersensitivity pneumonitis Alveolar microlithiasis

Clinical presentation

The clinical presentation of PAP is nonspecific. Most patients complain of a slowly progressing shortness of breath and an unproductive cough. The presence of sputum and fever is usually associated with the development of lower respiratory tract infections. Between 10% and 30% of patients are asymptomatic at the time of diagnosis. For the remainder, it typically takes between 7 and 11 months between the appearance of first symptoms and the establishment of a PAP diagnosis [7]. Although this disorder most often develops gradually, cases with an acute manifestation of symptoms and the rapid development of dyspnea have been described [11]. Physical examination and lung auscultation generally do not reveal any special features, except in severe cases, where there can be evidence of respiratory failure and basilar crackles.

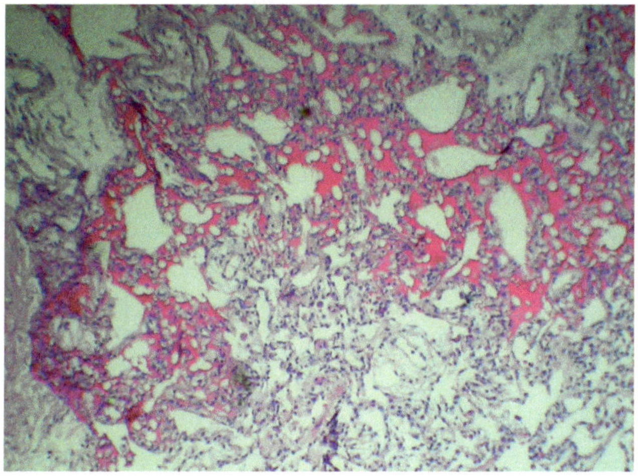

FIG. 4.1.1 PAP. Accumulation of homogeneous granular masses in alveolar space and the presence of interstitial inflammatory infiltration. H&E stain, 100×.

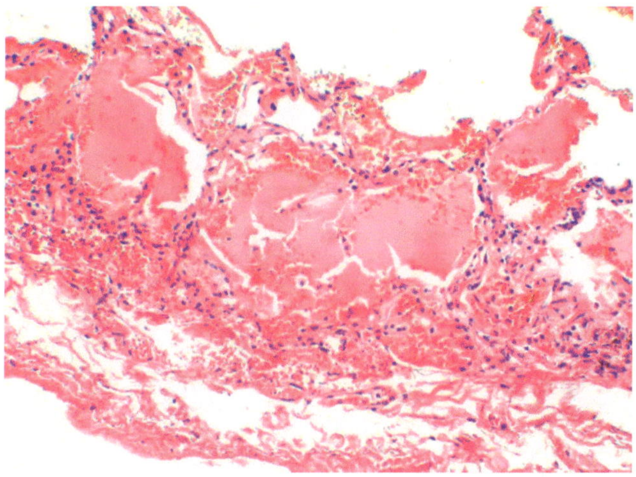

FIG. 4.1.2 PAP. Accumulation of homogeneous granular masses in alveolar sacs under the pleura and interstitial infiltration by granulocytes and eosinophils. H&E stain, 200×.

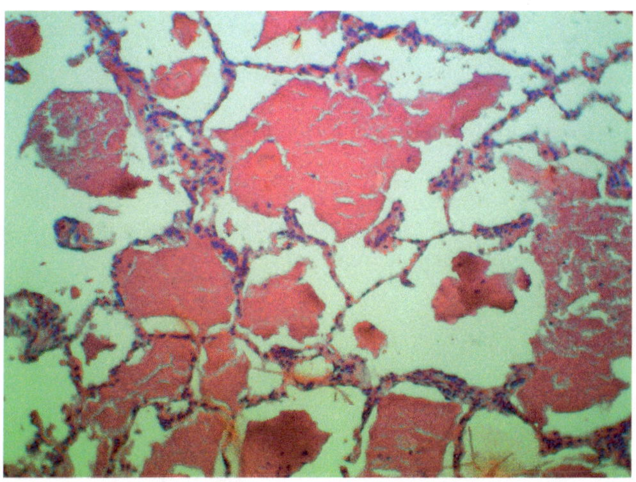

FIG. 4.1.3 PAP. Accumulation of homogeneous lipoprotein masses in alveolar space. H&E stain, 200×.

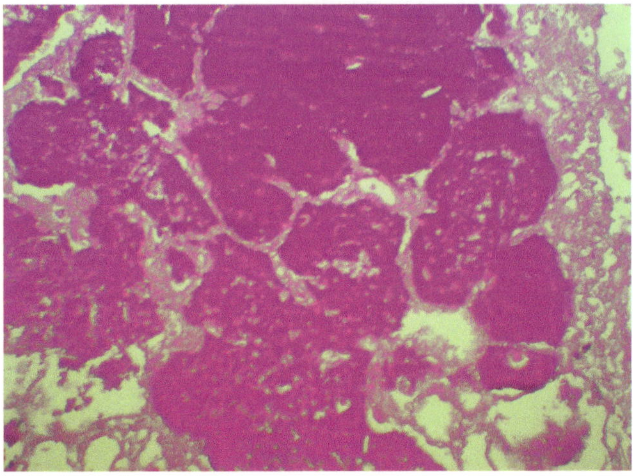

FIG. 4.1.4 PAP. Granular Schiff-positive masses filling alveoli. PAS reaction, 200×.

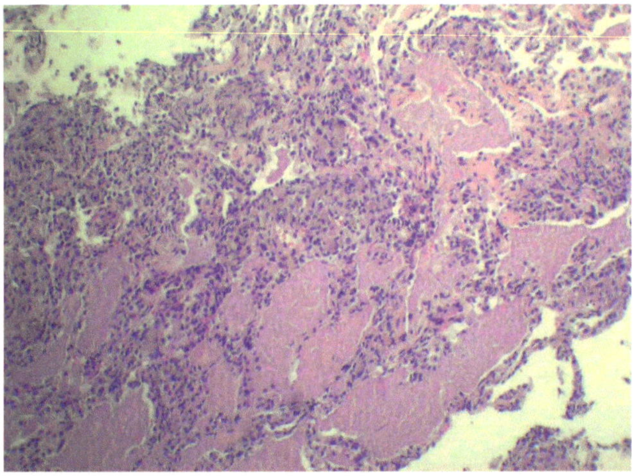

FIG. 4.1.5 PAP. Patchy alveolar filling by homogeneous mass accompanied with inflammatory infiltration and mild interstitial fibrosis. H&E stain, 200×.

Diagnosis

The chest radiograph can detect changes in the lungs even at a preclinical stage of PAP. In general, bilateral multifocal patchy ill-defined airspace opacities are observed; these are often interpreted as a manifestation of pulmonary infection (Fig. 4.1.6).

Characteristic features for PAP observed on high-resolution computer tomography (HRCT) are the presence of ground-glass opacity (GGO) and thickening of the interlobular septa [3] (Fig. 4.1.7).

The GGO has the following characteristics:

- It is always bilateral.
- Areas of GGO are sharply demarcated from areas of intact parenchyma and have an irregular shape, which gives them the appearance of a "geographic map."
- There is an absence of any preferences of distribution.
- The process may be patchy or diffuse.

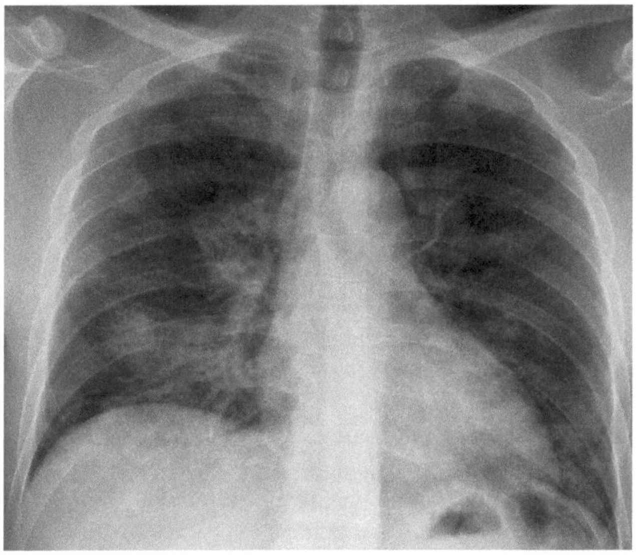

FIG. 4.1.6 Chest radiograph of a patient with autoimmune pulmonary alveolar proteinosis. Numerous asymmetrical patchy areas of airspace opacities can be observed in both lungs.

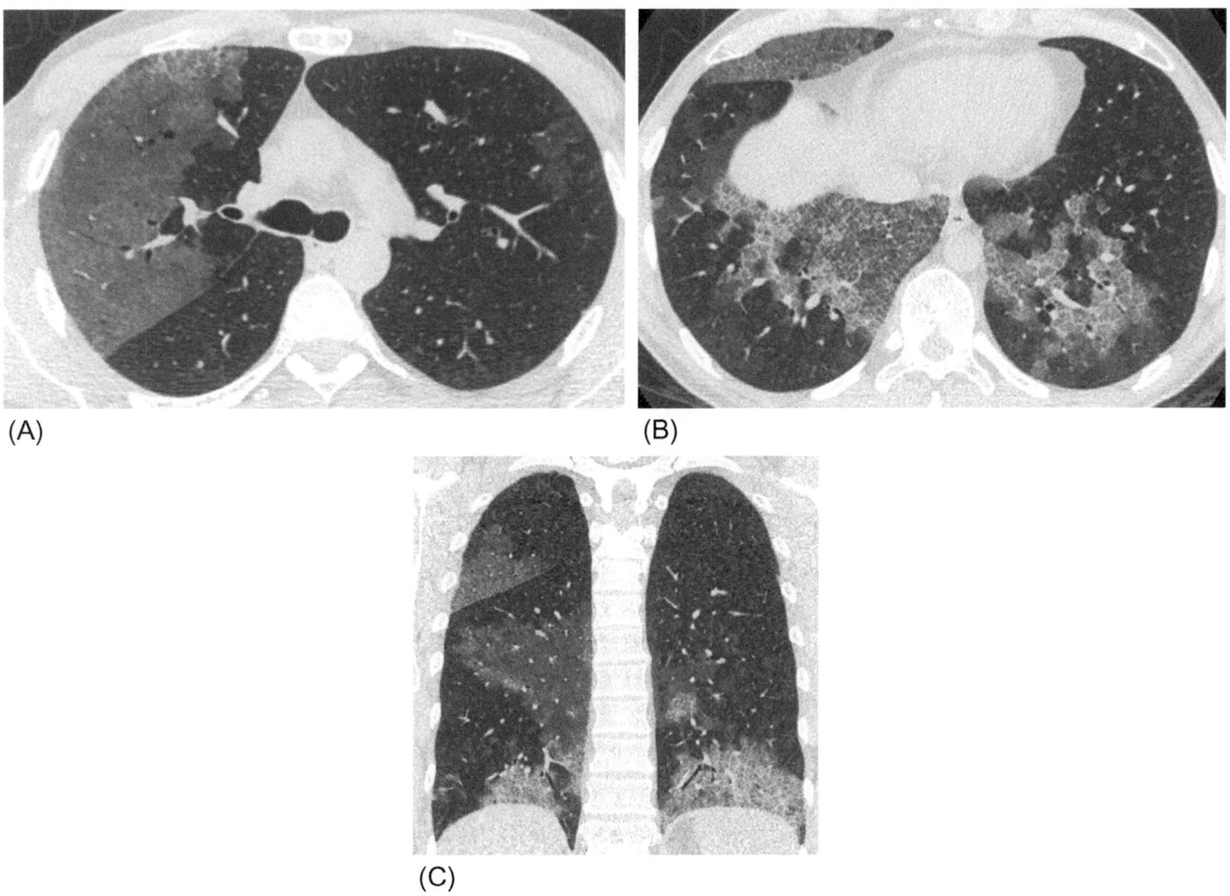

FIG. 4.1.7 Autoimmune PAP. Bilateral areas of GGO demarcated from normal lung tissue. Interlobular septal thickening is visible mainly in the basal lung segments (A, B, and C).

Haruyuki Ishii et al. compared GGO signs in 42 patients with autoimmune or secondary PAP. They found limited "geographic map" areas with a subpleural location in 71% of patients with autoimmune forms of PAP but only in 24% of patients with secondary PAP. A diffuse distribution of GGO was more typical for patients with secondary PAP; this was seen in 61% of these patients, compared with 19% of patients with autoimmune PAP [12]. A case of unilateral PAP with a limited subpleural area of GGO, originally interpreted as bronchioloalveolar adenocarcinoma, has also been described [13].

Thickening of interlobular septa has the following characteristics on HRCT:

- It is only observed in the GGO areas.
- It has the visual appearance of crazy paving sign (Fig. 4.1.8).

Thickening of the interlobular septa is due to the accumulation of lipoprotein complexes, septal edema, and macrophage infiltration. It has been shown that crazy paving is characteristic for autoimmune PAP, where it is present in 71% of cases [12]. However, only 14% of patients with secondary PAP exhibited this signs.

Although the crazy paving sign was first described in patients with PAP and was at the time considered a specific feature of this disorder, it was later established that this sign occurs in a number of other pathological conditions, which have to be differentiated from PAP. Among these are interstitial pneumonias, minimally invasive adenocarcinoma/lepidic invasive adenocarcinoma (formerly bronchioloalveolar adenocarcinoma), hypersensitivity pneumonitis, lipoid pneumonia, drug-induced lung disease, and a number of other disorders (Table 4.1.2).

Because such a wide range of conditions can be accompanied by radiological sign of crazy paving and often by a similar clinical presentation to PAP, HRCT cannot be considered as the sole and decisive method for a PAP diagnosis. However, it may reveal additional features that are not characteristic for PAP. For example, the concurrent presence of honeycombing, lymphadenopathy, nodules, or pleural effusion practically allows PAP to be excluded as a diagnosis if the patient does not have a concomitant lung disease.

As surfactant detritus accumulates in the alveoli, thick interlobular septa merge with filled alveolar spaces and form foci of consolidation (Fig. 4.1.9).

The analysis of bronchoalveolar lavage (BAL) fluid should be considered an important diagnostic tool. Patients with PAP have opalescent, turbid, white or yellowish fluid that settles into a dense sediment (Fig. 4.1.10). The protein concentration is 10–100 times higher than normal. The solid phase is Schick-positive. Antibodies to GM-CSF are present in the BAL fluid and blood serum of almost all patients with autoimmune PAP. A characteristic of PAP is the presence of foamy macrophages, which contain eosinophilic granules and extracellular hyaline bodies [1].

The regular comprehensive blood test panel does not usually show any significant abnormalities, although a patient in our practice with an aggressive course of PAP exhibited a low-grade fever and an increased level of C-reactive protein

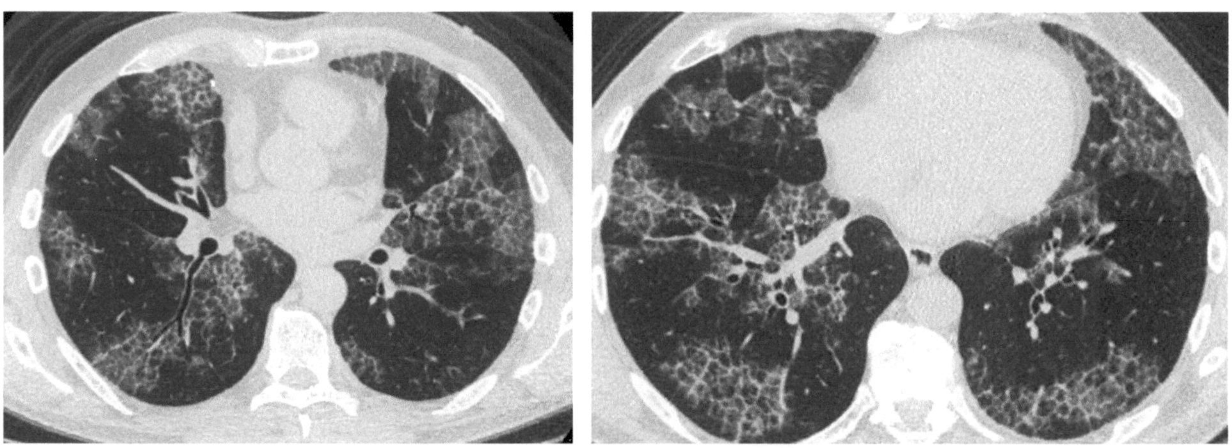

(A) (B)

FIG. 4.1.8 Autoimmune PAP. Thickening of the interlobular septa creates a pattern similar to crazy paving within the patchy areas of GGO. Abnormalities give the impression of a "geographic map" (A and B).

TABLE 4.1.2 Conditions with a possible sign of crazy paving [14,15, supplemented]

Acute conditions	Subacute/chronic conditions
Pulmonary edema	Pulmonary alveolar proteinosis
Acute respiratory distress syndrome	Idiopathic pulmonary fibrosis
Pulmonary infection (bacterial, viral, mycoplasma, and *Pneumocystis jirovecii*)	Nonspecific interstitial pneumonia
Alveolar hemorrhage	Lung lesions in connective tissue diseases
Acute interstitial pneumonia	Bronchiolitis obliterans with organizing pneumonia
Radiation pneumonitis	Subacute and chronic hypersensitivity pneumonitis
Acute eosinophilic pneumonia	Eosinophilic granulomatosis with poliangiitis
	Minimally invasive adenocarcinoma/lepidic invasive adenocarcinoma (formerly bronchioloalveolar adenocarcinoma)
	Lymphangitis carcinomatosa
	Chronic eosinophilic pneumonia
	Lipoid pneumonia
	Sarcoidosis
	Alveolar microlithiasis
	Aspiration of barium
	Drug-induced pneumonitis (amiodarone, methotrexate, and bleomycin)
	Diffuse pulmonary lymphangiomatosis

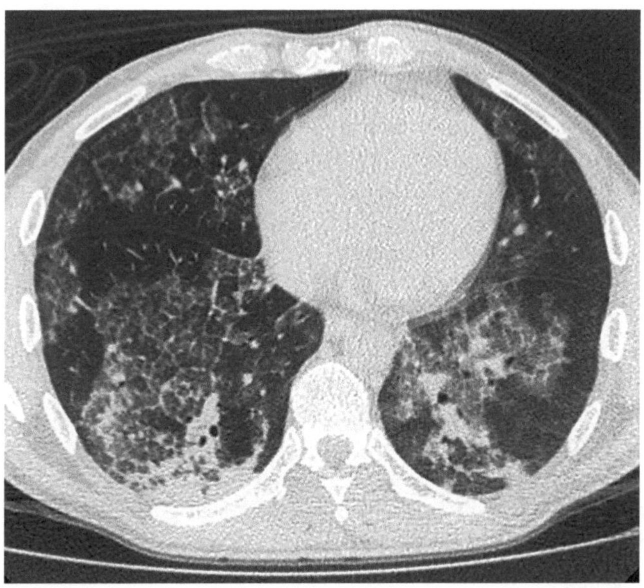

FIG. 4.1.9 PAP. Subpleural consolidation and the areas of GGO with thickened interlobular septa in the basal segments. There is a sharp border between most of affected and intact areas.

FIG. 4.1.10 Bronchoalveolar lavage fluid of a PAP patient.

(CRP) in the absence of infection. The metabolic panel often shows an elevated level of lactate dehydrogenase (LDH), proportional to the severity of the condition [5]. However, such an increase is not specific to PAP and is frequently observed in other lung disorders, such as *Pneumocystis jirovecii* pneumonia and acute respiratory distress syndrome. If there is a reason to suspect PAP, an immunosorbent assay should be performed to determine the presence of serum antibodies to GM-CSF. A diagnostic threshold for autoimmune PAP is an antibody concentration >19 mg/mL [16].

Probe-based confocal laser endomicroscopy of distal airways

PAP is one of a limited group of lung disorders for which probe-based confocal laser endomicroscopy of the distal airways has a real value both at initial diagnosis and when monitoring its course. We examined six patients with autoimmune PAP, 121 bronchopulmonary regions were analyzed, 1017 informative images were obtained, and 4853 measurements were performed. In all cases, we detected intra-alveolar floating fluorescent complexes that are not found in other conditions (Figs. 4.1.11 and 4.1.12A–E). These complexes were detected in 74.4% of the examined regions. The interalveolar septa were moderately thickened in 19% of the regions under study (Fig 4.1.12B). Some yellow fibers of interalveolar septum were disrupted in 4.1% of the regions examined (Fig 4.1.12F). In 2.5% of the regions, alveolus was not seen because of their replacement with the brightly fluorescent elements of connective tissue (Fig 4.1.12G).

The intra-alveolar inclusions decreased significantly after whole lung BAL treatment. Alveoscopy detected fluorescent masses even in areas of the parenchyma where there were no signs of pathology on HRCT, although to a lesser extent than in the areas that were radiographically visualized (Fig. 4.1.13). Based on these findings, we concluded that PAP is not a patchy process, but that it diffusely affects the lung parenchyma with varying severity of alveolar accumulation of amorphous complexes and that it may proceed subclinically and being radiologically indistinguishable. Another conclusion is that confocal laser endomicroscopy of distal airways is more sensitive than HRCT, at least for the diagnosis and evaluation of PAP, especially in the early stages [17].

In the past the gold standard of PAP diagnostics has been lung tissue biopsy, obtained either transbronchially or with video-assisted thoracoscopic surgery. However, the diagnosis PAP can now be established with a high degree of reliability

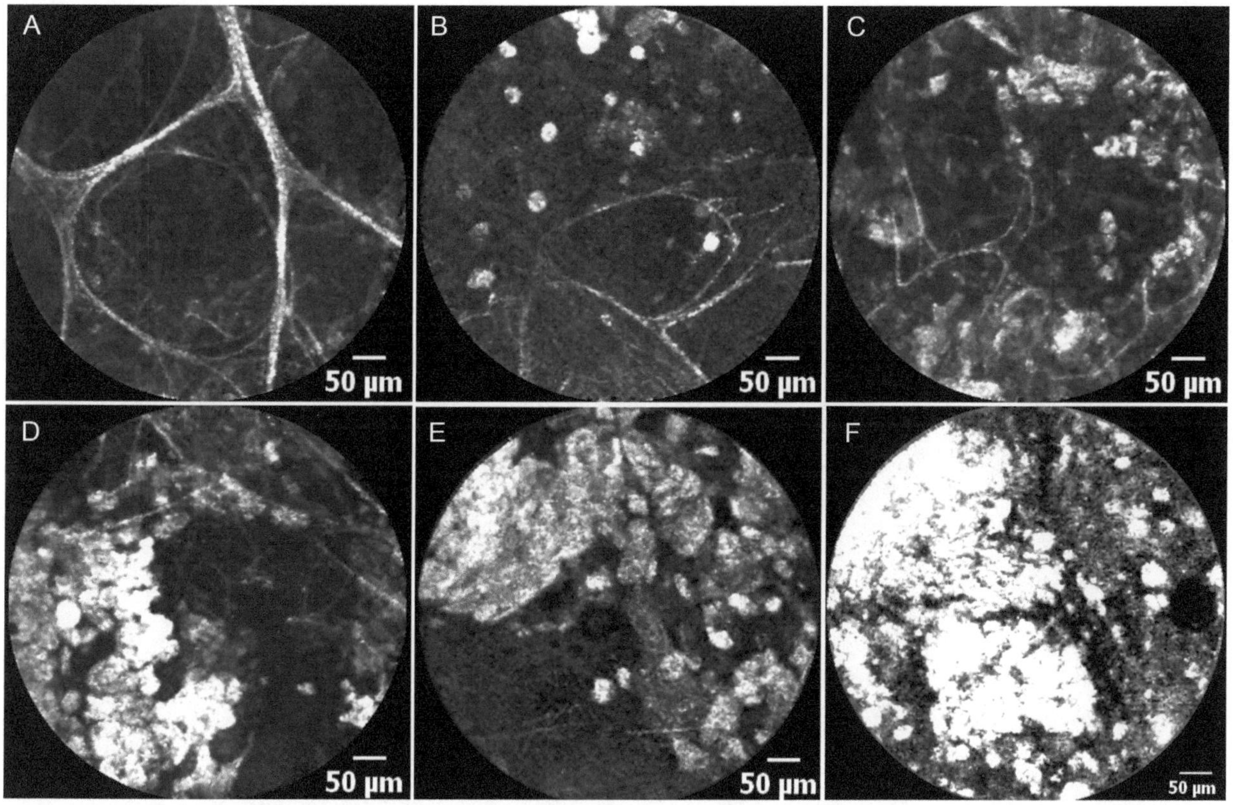

FIG. 4.1.11 PAP. Probe-based confocal laser endomicroscopy images (pCLE). (A) Normal alveoli. Different degrees of accumulation in the alveoli of floating fluorescent complexes, from single bodies (B) to complete filling (F). (C–E) Images demonstrate partial filling of the alveoli with intraluminal complexes.

without lung biopsy given the presence of three criteria: a characteristic HRCT pattern, protein material in BAL fluid, and a significant antibody titer to GM-CSF [1].

Differential diagnosis

It has been mentioned that it is necessary to differentiate PAP from a number of conditions with a subacute or chronic course and similar radiographic patterns, especially those that present the sign of "crazy paving." The most common disorders that disguise PAP are lepidic invasive adenocarcinoma (formerly bronchioloalveolar adenocarcinoma) (Fig. 4.1.14), hypersensitivity pneumonitis (Figs. 4.1.15 and 4.1.16), *Pneumocystis jirovecii* pneumonia (Fig. 4.1.17), and lipoid pneumonia (Figs. 4.1.18 and 4.1.19).

Their distinctive features are presented in Table 4.1.3.

The disorder with clinical and radiological characteristics most similar to PAP is lipoid pneumonia, which may be either exogenous or endogenous. Exogenous forms are more frequent and are associated with the aspiration of various fatty substances used for irrigation of the pharynx, administration into the nasal passages, or tracheostomy care. The inhalation of oil-based suspensions may also be a cause of lipoid pneumonia (LP). The highest prevalence of this condition is seen during early childhood and in the elderly, when there are problems with swallowing and the risk of aspiration [18]. Because there is an accumulation of lipids in the alveoli in LP, as in PAP, the clinical and radiological presentation may be very similar (Fig. 4.1.18). However, it is more common in LP for the process to be unilateral, or predominantly unilateral (more frequently in the right lower lobe), and consolidation areas may appear even in cases with smaller lesions, mostly in the basal and spinal segments (Fig. 4.1.19). In PAP, such problems develop at a later stage, when a large volume of lung parenchyma is involved in the process.

Sometimes it is necessary to differentiate PAP from diffuse alveolar hemorrhage (DAH), in which spots of GGO with thickened interalveolar septa also may be observed (Fig. 4.1.20). However, it is characteristic for DAH to take a more acute

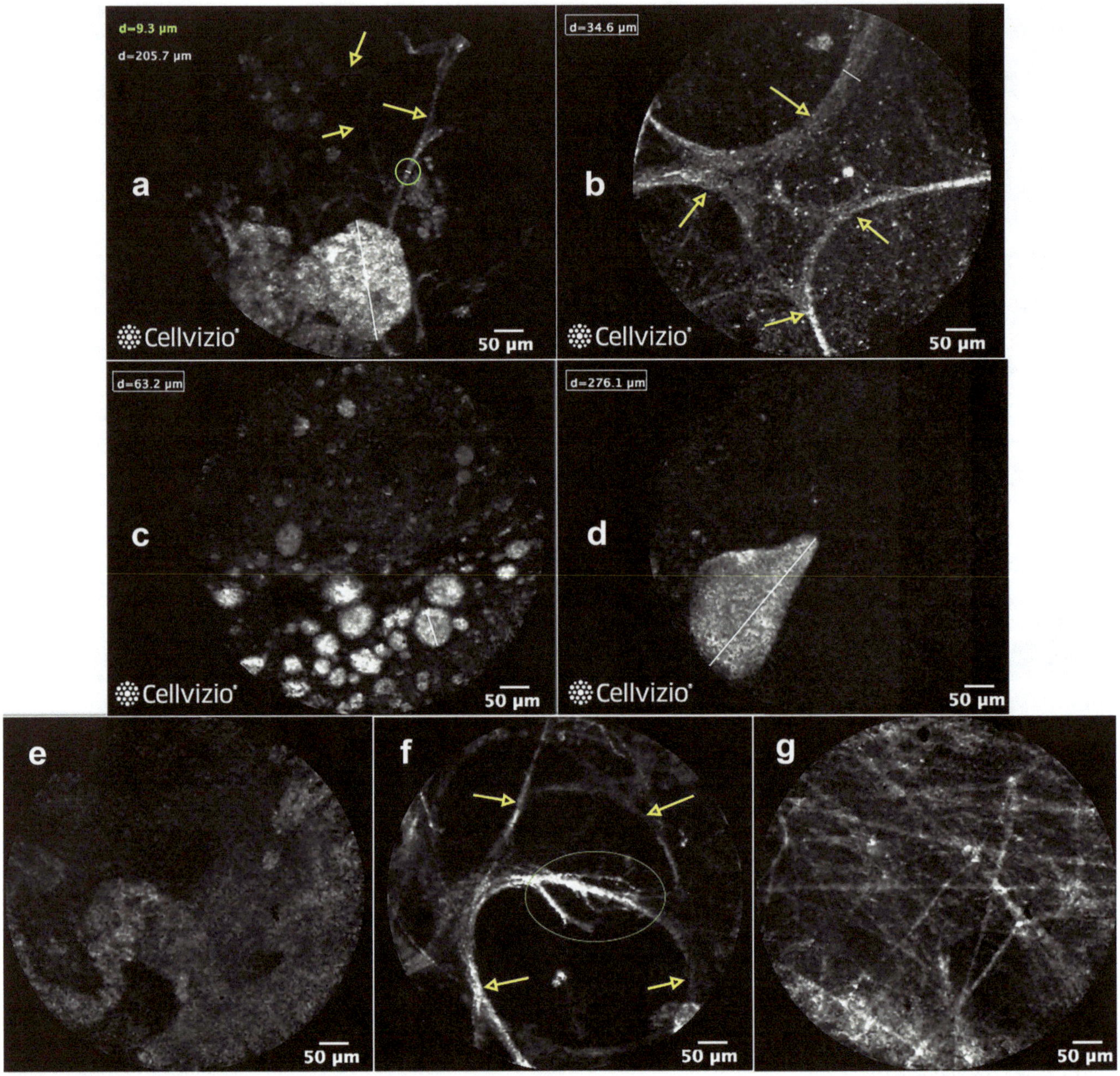

FIG. 4.1.12 PAP. PCLE of distal airways. (A) Floating intra-alveolar complexes (more than 200 μm in diameter) on the background of normal alveoli. (B) Interalveolar septa are thickened by two to three times; all visible alveoli are filled with moderately fluorescent liquid. (C and D) Floating fluorescent complexes of different shape and size in alveoli. (E) Heterogeneous masses with the moderate level of fluorescence, alveoli are not visualized. (F) Normal alveoli with single alveolar macrophages, some yellow fibers of interalveolar septum are disrupted (ellipse). (G) Alveolus is not differentiated; brightly fluorescent elements of connective tissue are seen. *Yellow arrows* mark the interalveolar septa.

course, with hemoptysis, and, in most cases, anemia (see Chapter 4.3). It is difficult to differentiate pulmonary involvement in systemic diseases (e.g., connective tissue disorders) from secondary alveolar proteinosis, which may develop in patients with autoimmune disease (Fig. 4.1.21). In such cases a final diagnosis is possible only after morphological study of biopsy material.

It is essential to keep in mind that PAP does not necessarily present the specific CT sign of "crazy paving," especially in the early stages. In such cases, it is difficult to establish the diagnosis without a lung biopsy. It is likely that the development of confocal laser endomicroscopy of distal airways will provide a tool that helps toward making a diagnosis of PAP without invasive procedures.

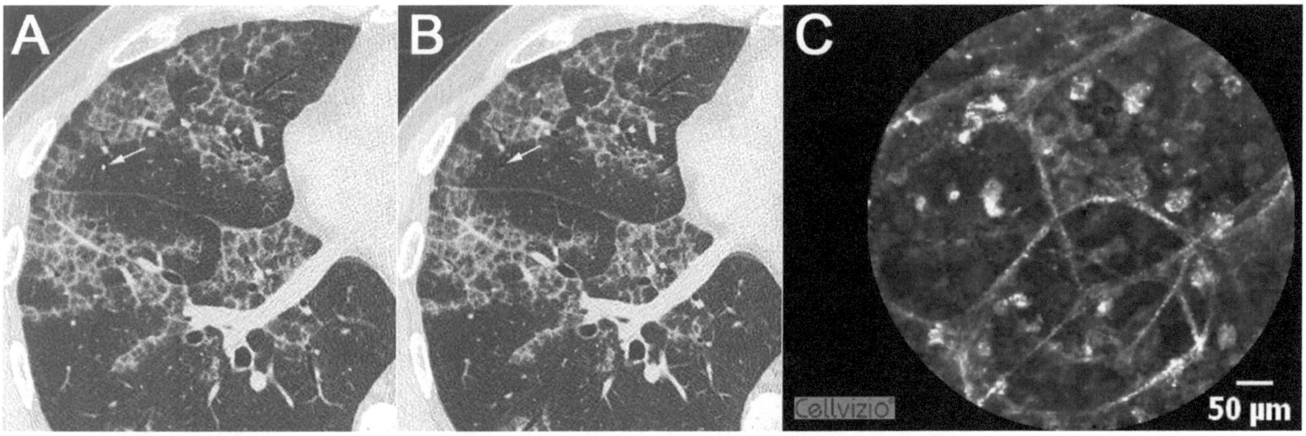

FIG. 4.1.13 PAP. The correspondence between HRCT and pCLE images. (A) The last scan of the RS4, where a miniprobe is still visible *(arrow)*. (B) The next 1.0-mm step without the probe *(arrow)* demonstrates the focus of the unaffected lung parenchyma by HRCT. (C) The pCLE image displays fluorescent floating complexes at the point of the probe, whereas the HRCT image does not show pathology.

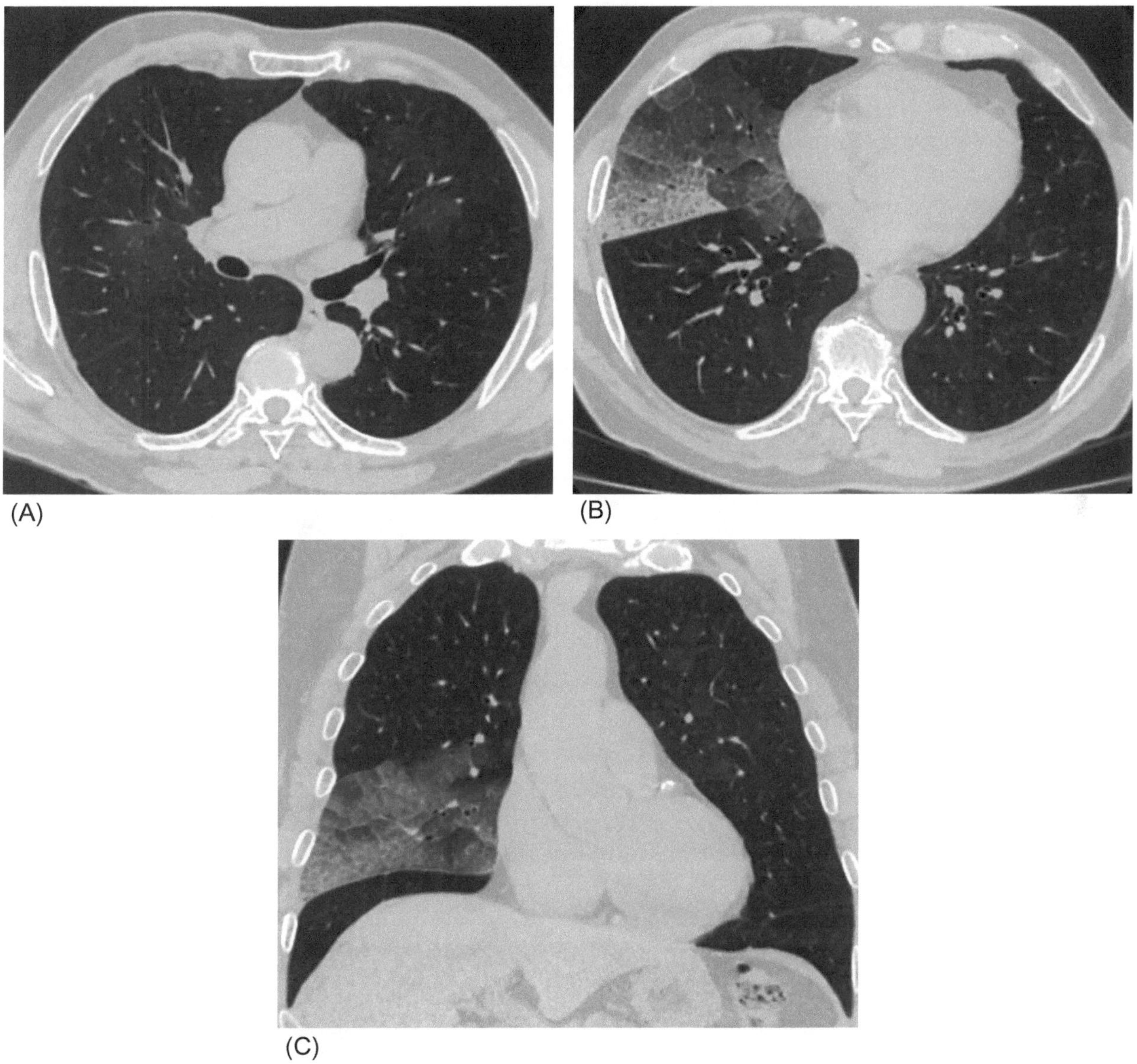

FIG. 4.1.14 Lepidic invasive adenocarcinoma. Bilateral mild areas of GGO (A). The crazy paving sign in the fourth and fifth segments of the right lung (B and C).

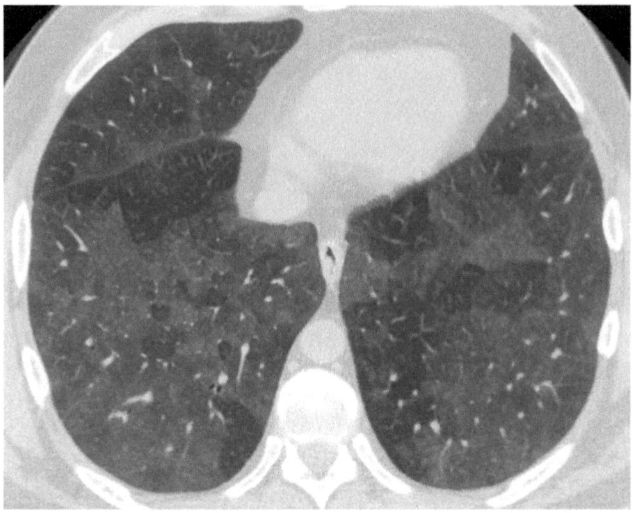

FIG. 4.1.15 Subacute hypersensitivity pneumonitis. Large areas of GGO, lobular areas of hyperinflation with sharp borders, similar to the "geographic map" in PAP. Thickening of intralobular septa; in some areas, there is also mild thickening of interlobular septa.

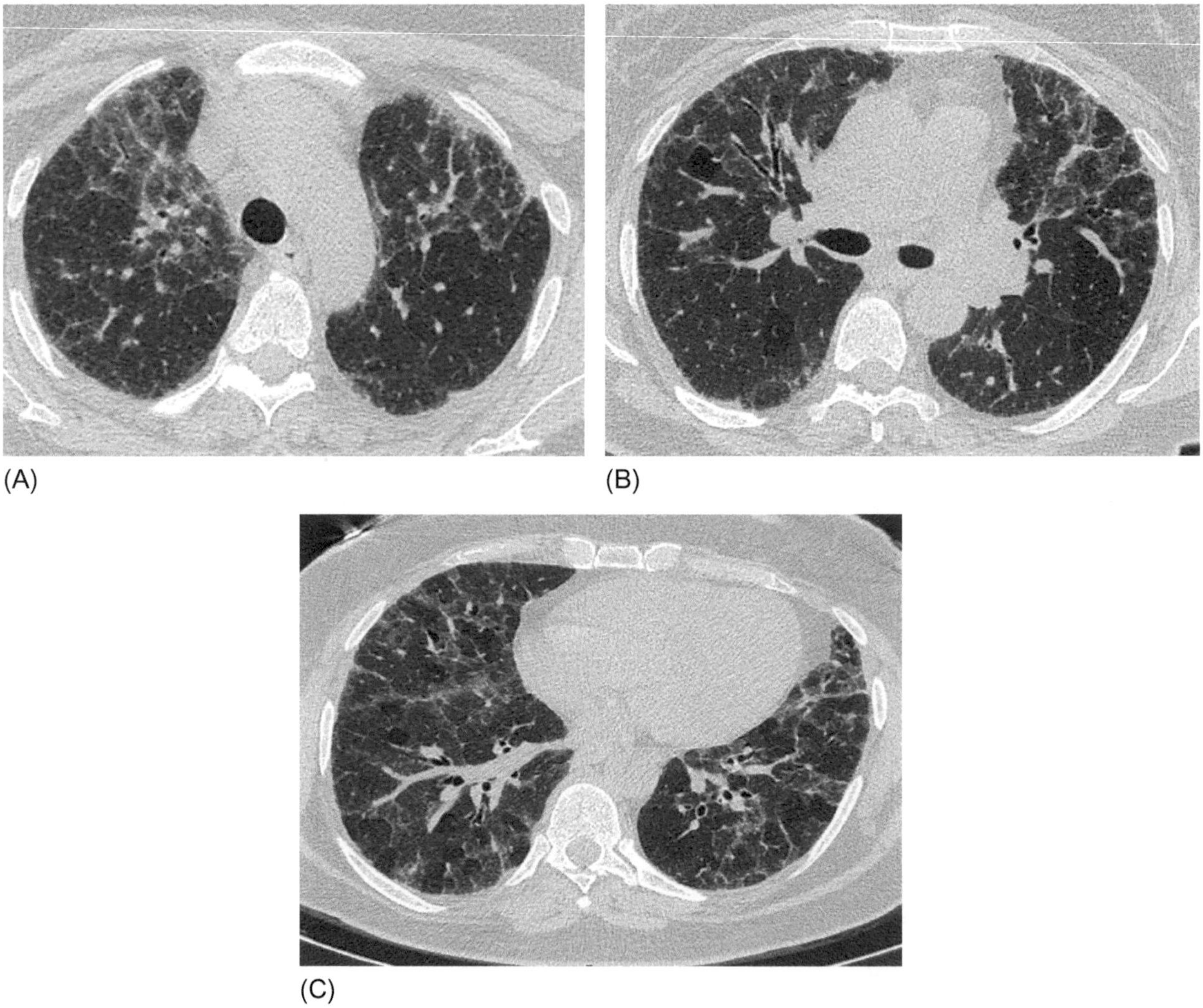

FIG. 4.1.16 Chronic hypersensitivity pneumonitis. Thickened intra- and interlobular septa are visible in the background of GGO areas; multiple air trappings, as a manifestation of lymphocytic bronchiolitis (A and B).

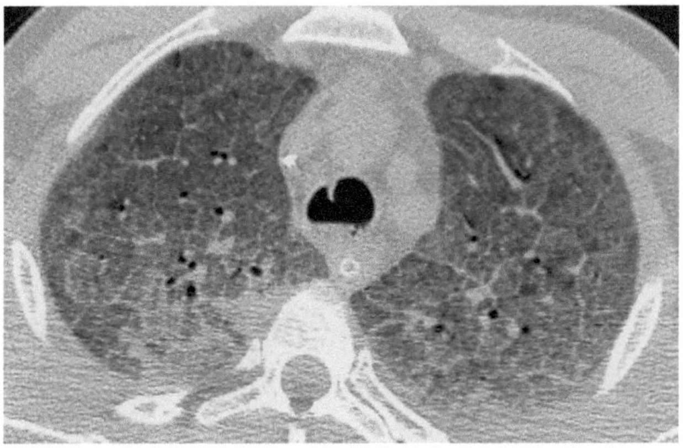

FIG. 4.1.17 *Pneumocystis jirovecii* pneumonia. Diffuse intensive GGO with thickened interlobular septa and consolidation areas in S2 of the right lung.

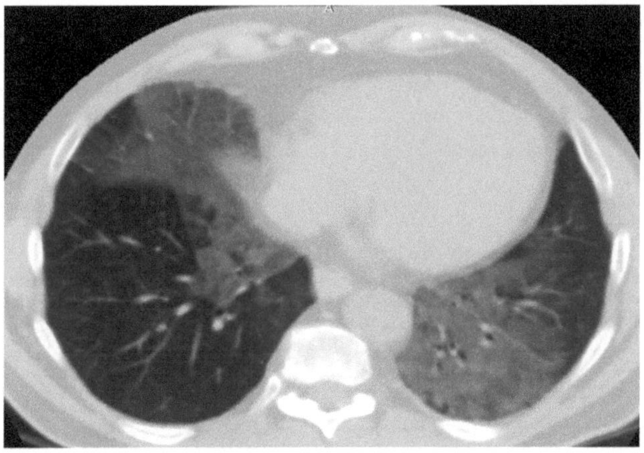

FIG. 4.1.18 Exogenous lipoid pneumonia in a 76-year-old male patient who used mineral oil for tracheostomy care. Large areas of GGO in the middle right lobe and the left lower lobe, with moderate thickening of the interlobular septa.

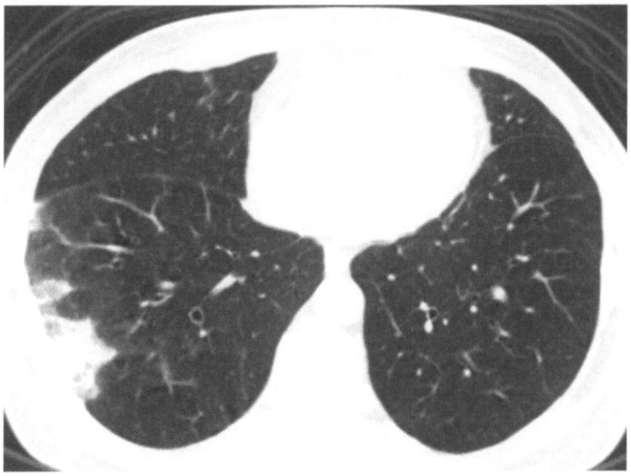

FIG. 4.1.19 Exogenous lipoid pneumonia. Unilateral subpleural areas of consolidation are surrounded by GGO in a 56-year-old female patient who used oil emulsion for oropharynx irrigation.

TABLE 4.1.3 Pulmonary alveolar proteinosis differential diagnosis

	PAP	LIA	HP	PP	LP
Patient type	Middle-aged patients, predominantly male smokers	Smokers over 40 years	Predominantly nonsmoking women who have contact with organic pollutants (birds, fungi, etc.)	Young and middle-aged patients with HIV or patients of any age who receive immunosuppressants	Elderly patients who use oil-based suspensions for inhalation or insufflation and to have impaired swallowing
Symptoms and course of development	Slowly increasing dyspnea	Cough, which may be with abundant sputum, hemoptysis, increasing shortness of breath, weight loss	Shortness of breath, dry cough, slight fever (not always)	Fever, weight loss, progressive shortness of breath	Chronic cough, slowly progressive shortness of breath
Blood test results	Normal, possible increased LDH, increased GM-CSF antibody titer	Anemia, increased ESR, increased CRP	Slightly increased ESR, increased CRP and LDH	Moderately increased ESR, increased CRP and LDH	Normal or slightly increased ESR
BAL fluid	Milky, opalescent fluid with a high protein content	Neutrophils and cancer cells	Lymphocytosis >40%–50%	Positive PCR for *P. jirovecii*	Milky fluid, the presence of lipid–laden macrophages, slight eosinophilia
CT results	Bilateral, sharply demarcated GGOs with thickening of interlobular septa. Foci of consolidation are present in severe cases	Bilateral areas of consolidation, surrounded by GGO and crazy paving. Air bronchogram, pseudocysts, centrilobular nodules, lymphadenopathy	Lobular areas of air trapping, centrilobular ill-defined nodules, patchy or diffuse GGO	Bilateral areas of GGO, predominantly in the upper and middle zones. Thickening of the interlobular septa. Possible areas of consolidation and cystic changes	Areas of consolidation, not infrequently unilateral, often in the basal segments, surrounded by GGO areas and crazy paving

PAP, pulmonary alveolar proteinosis; *LIA*, lepidic invasive adenocarcinoma (formerly bronchioloalveolar adenocarcinoma); *HP*, hypersensitivity pneumonitis; *PP*, *Pneumocystis jirovechii* pneumonia; *LP*, lipoid pneumonia; *ESR*, erythrocyte sedimentation rate; *LDH*, lactate dehydrogenase.

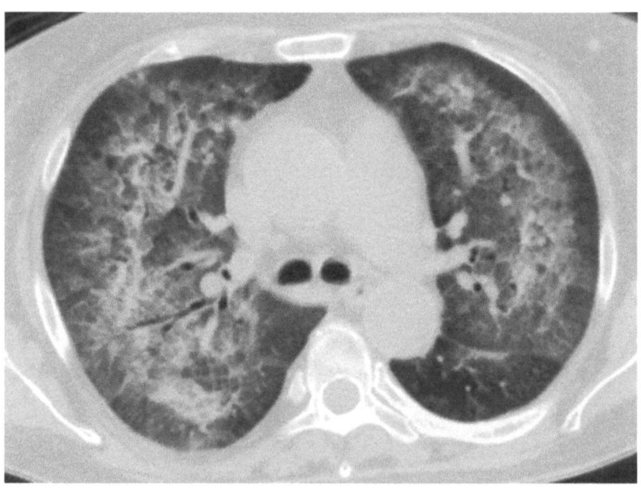

FIG. 4.1.20 Diffuse alveolar hemorrhage in ANCA-associated vasculitis. Patchy areas of GGO associated with thickening of the interlobular and intralobular septa. Subleural sparing is visible.

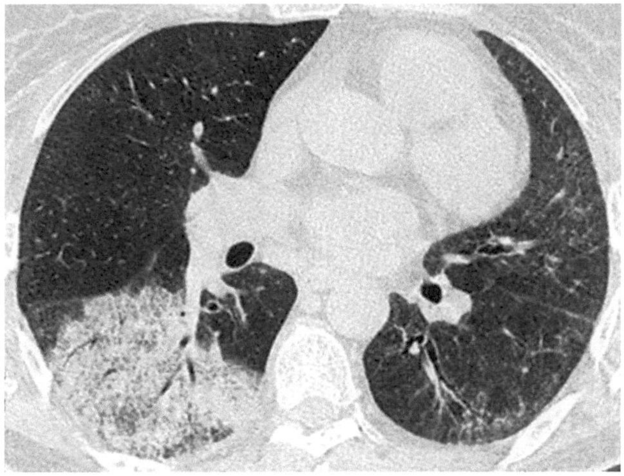

FIG. 4.1.21 Secondary PAP in a patient with dermatomyositis/polymyositis and interstitial lung disease (nonspecific interstitial pneumonia). Clearly demarcated GGO area with thickened intralobular septa is on the right; reticular abnormalities with traction bronchiectasis in the left low lobe.

Treatment and prognosis

An efficient, low-cost, albeit old, method for the treatment of patients with PAP is a whole lung lavage (WLL). The procedure is performed under general anesthesia in a supine or lateral position, with separate intubation for the right and left lungs. During the procedure, one lung is ventilated with O_2 at a concentration sufficient to maintain normal gas exchange parameters, and the other lung is washed with a sterile isotonic solution of sodium chloride, warmed to 37°C [19]. The total volume of liquid needed is between 5 and 40 L, depending on the size of the area that needs to be washed (a segment, lobe, or the whole lung); the average is 15.4 L per lung [19]. The procedure takes between 2 and 6 h. Manual percussion or high-frequency airway clearance is used frequently during the procedure [20]. The result of effective therapeutic WLL is a collection of turbid liquid in which a precipitate forms after standing. After treatment, patients have reduced dyspnea and improved respiratory function and blood gas levels; the chest radiograph and HRCT show positive dynamics of changes in the lungs. Clinical, functional, and radiological improvement are observed in most patients, with the most rapid and significant improvement observed in PaO_2 [5]. According to our data, HRCT within 5 days after the procedure do not show significant positive dynamics (Fig. 4.1.22), and so follow-up with HRCT should be conducted no sooner than one month after WLL. After the therapeutic WLL the protein-lipid complex begins to accumulate again in the lung at different rates. The therapeutic effect of properly performed WLL lasts, on average,

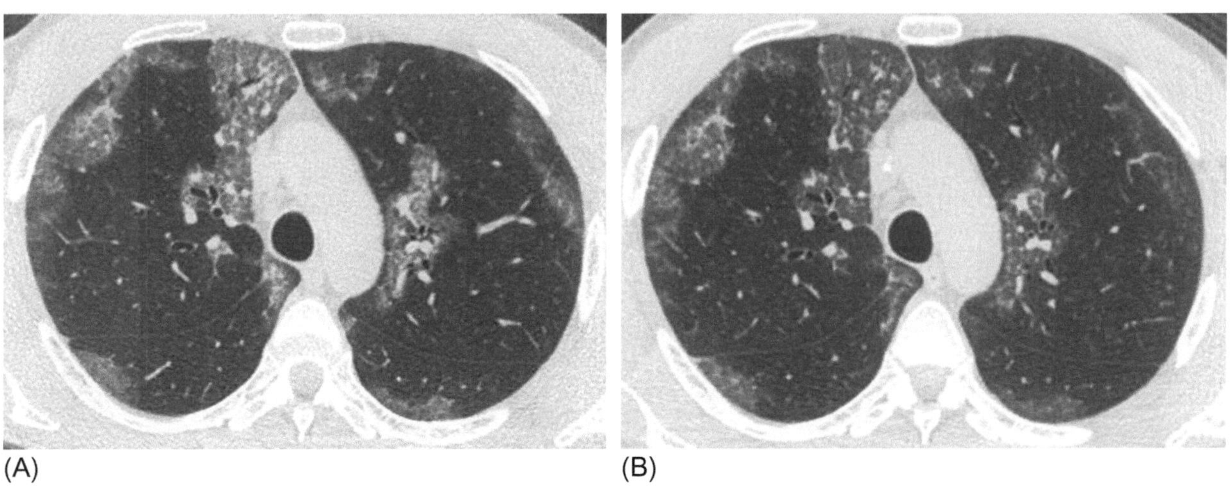

FIG. 4.1.22 HRCT monitoring of the effectiveness of therapeutic BAL before (A) and 5 days after (B) the procedure in a 37-year-old patient. A mild decrease lung attenuation in the ground-glass areas.

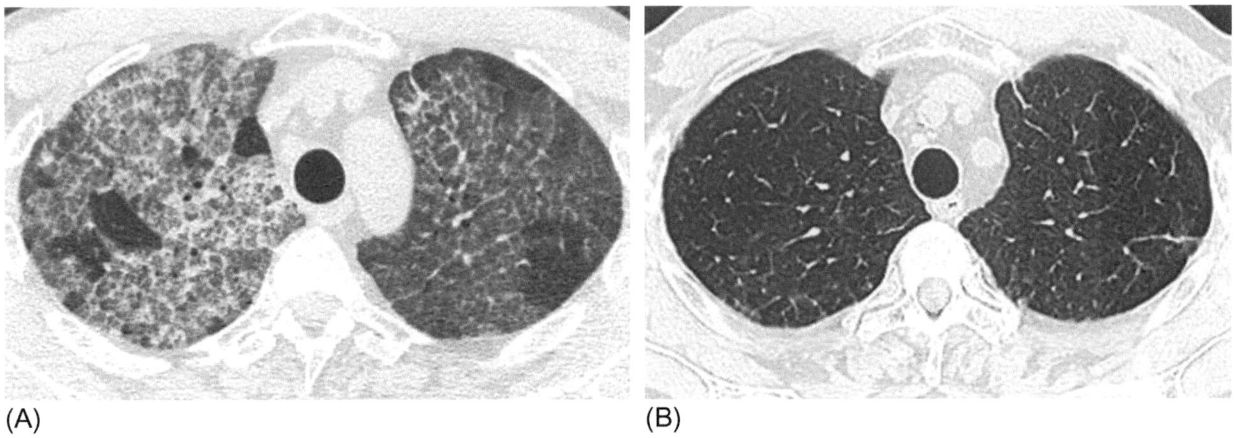

FIG. 4.1.23 Autoimmune PAP follow-up in a 47-year-old patient: (A) Before WLL. Diffuse area of GGO and associated reticular opacities. (B) Twenty-four months from WLL of only the right lung. Interlobular septal thickening with mild foci of GGO in both lungs.

for about 15 months [4]. We observed one patient whose HRCT signs of PAP were minimal in both lungs two years after a lavage of only right lung (Fig. 4.1.23). It is possible this was a case of the spontaneous remission of PAP, which has being described in some patients [5,21].

A relatively new direction in PAP treatment is the use of GM-CSF. As discussed earlier the presence of antibodies to GM-CSF inhibits its biological activity in patients with PAP. A meta-analysis of the efficacy of subcutaneous and inhalation forms of GM-CSF application has demonstrated a positive effect of this therapy in 43%–96% of cases, with a moderate effect observed in 56% of patients. Relapse after treatment was seen in 30% of patients. It should be noted that it takes on

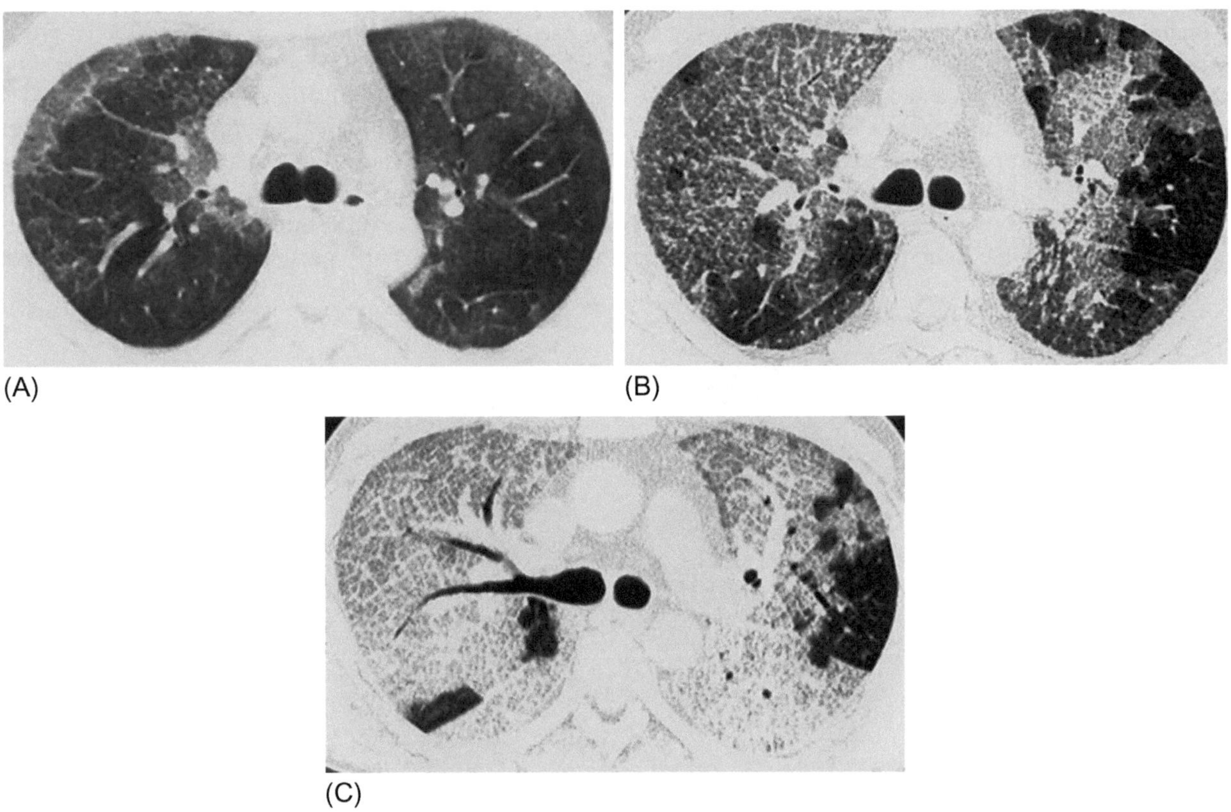

FIG. 4.1.24 Progressive course of PAP in a 37-year-old patient, who didn't take any treatment. First CT scan. Moderate subpleural GGO areas and reticular abnormalities (A). Follow-up at 3 years. Areas of extensive GGO and associated interlobular septal thickening (B). Follow-up at 7 years. Diffuse subtotal areas of increased attenuation—GGO with interlobular thickening turning into consolidation (C).

average 8 weeks from the beginning of treatment to obtain a clinical effect, and different doses and duration of treatment are needed to achieve results [22]. Follow-up of 30 months from the end of therapy with patients who received inhaled GM-CSF showed that the clinical and radiological effects persisted in 66% of cases [23].

The first pilot study of the efficacy of rituximab in 10 patients with PAP demonstrated a significant improvement in gas exchange and HRCT results, with minimal changes in functional parameters (diffusing lung capacity and 6-min walk distance); the study also showed a significant improvement in alveolar clearance in the patients treated with rituximab [24,25]. Veno-venous extracorporeal membrane oxygenation can be used in severe cases with progressive respiratory failure, including for gas exchange support during WLL [26]. Lung transplantation is performed if all conservative measures have failed. However, the recurrence of the disease is often developed in the lung allograft [8]. A new promising direction in the treatment of PAP has recently been demonstrated using a humanized immunodeficiency mouse model of hereditary disease. Happle et al. showed that human-induced pluripotent stem cell-derived macrophages, transplanted intrapulmonarily and engrafted into the lung parenchyma, become similar to the primary host macrophages and could uptake surfactant, reducing lipoid accumulation in the respiratory tract [27].

The prognosis of PAP is generally favorable, especially if the diagnosis was established before the onset of severe respiratory failure. However, the condition slowly progresses in the absence of treatment and/or the continuation of smoking (Fig. 4.1.24). In early studies the 5-year survival rate was estimated at 75% [5]. In the study by Inoue [4], no deaths were recorded among autoimmune PAP patients during a 5-year follow-up period. It is generally believed that patients with autoimmune PAP have phagocytosis deficiency and therefore are more prone to respiratory infections, which is the second leading cause of death in PAP patients after respiratory failure [3].

References

[1] Borie R, Danel C, Debray M-P, Taille C, Dombret MC, Aubier M, et al. Pulmonary alveolar proteinosis. Eur Respir Rev 2011;20(120):98–107.

[2] Rosen SH, Castelman B, Liebow AA. Pulmonary alveolar proteinosis. N Engl J Med 1958;258(23):1123–42.

[3] Trapnell BC, Luisetti M. Pulmonary alveolar proteinosis syndrome. In: Broaddus VC, Mason RJ, Ernst JD, et al., editors. Murray and Nadel's textbook of pulmonary medicine. 6th ed. Philadelphia: Elsevier Saunders; 2016. p. 1260–74.

[4] Inoue Y, Trapnell BC, Tazawa R, Arai T, Takada T, Hizawa N, et al. Characteristics of a large cohort of autoimmune pulmonary alveolar proteinosis patients in Japan. Am J Respir Crit Care Med 2008;177(7):752–62.

[5] Seymour JF, Presneill JJ. Pulmonary alveolar proteinosis: progress in the first 44 years. Am J Respir Crit Care Med 2002;166(2):215–35.

[6] Khan A, Agarwal R. Pulmonary alveolar proteinosis. Respir Care 2011;56(7):1016–28.

[7] Campo I, Mariani F, Rodi G, Paracchini E, Tsana E, Piloni D, et al. Assessment and management of pulmonary alveolar proteinosis in a reference center. Orphanet J Rare Dis 2013;8:40.

[8] Kumar A, Abdelmalak B, Inoue Y, Culver DA. Pulmonary alveolar proteinosis in adults: pathophysiology and clinical approach. Lancet Respir Med 2018;6(7):554–65.

[9] Singh G, Katyal SL, Bedrossian CW, Rogers RM. Pulmonary alveolar proteinosis. Staining for surfactant apoprotein in alveolar proteinosis and in conditions simulating it. Chest 1983;83(1):82–6.

[10] Verbeken EK, Demedts M, Vanwing J, Deneffe G, Lauweryns JM. Pulmonary phospholipid accumulation distal to an obstructed bronchus: a morphologic study. Arch Pathol Lab Med 1989;113(8):886–90.

[11] Cohen ES, Elpern E. Silver MR pulmonary alveolar proteinosis causing severe hypoxemic respiratory failure treated with sequential whole-lung lavage utilizing venovenous extracorporeal membrane oxygenation: a case report and review. Chest 2001;120(3):1024–6.

[12] Ishii H, Trapnell B, Tazawa R, Inoue Y, Akira M, Kogure Y, et al. Comparative study of high-resolution CT findings between autoimmune and secondary pulmonary alveolar proteinosis. Chest 2009;136(5):1348–55.

[13] Sunadome H, Nohara J, Noguchi T, Matsui C, Kono T, Terada Y. A case of pulmonary alveolar proteinosis that showed solitary ground-glass opacity in the subpleural area. Nihon Kokyuki Gakkai Zasshi 2010;48(7):516–9.

[14] De Wever W, Meersschaert J, Coolen J, Verbeken E, Verschakelen JA. The crazy-paving pattern: a radiological-pathological correlation. Insights Imaging 2011;2(2):117–32.

[15] Rossi SE, Erasmus JJ, Volpacchio M, Franquet T, Castiglioni T, McAdams HP. "Crazy-paving" pattern at thin-section CT of the lungs: radiologic-pathologic overview. Radiographics 2003;23(6):1509–19.

[16] Uchida K, Nakata K, Suzuki T, Luisetti M, Watanabe M, Koch DE, et al. Granulocyte/macrophage colony-stimulating factor autoantibodies and myeloid cell immune functions in healthy individuals. Blood 2009;113(11):2547–56.

[17] Danilevskaya O, Averyanov A, Lesnyak V, Chernyaev A, Sorokina A. Confocal laser endomicroscopy for diagnosis and monitoring of pulmonary alveolar proteinosis. J Bronchology Interv Pulmonol 2015;22(1):33–40.

[18] Marchiori E, Zanetti G, Mano CM, Hochhegger B. Exogenous lipoid pneumonia. Clinical and radiological manifestations. Respir Med 2011;105(5):659–66.

[19] Michaud G, Reddy C, Ernst A. Whole-lung lavage for pulmonary alveolar proteinosis. Chest 2009;136(6):1678–81.

[20] Campo I, Luisetti M, Griese M, Trapnell BC, Bonella F, Grutters J, et al. Whole lung lavage therapy for pulmonary alveolar proteinosis: a global survey of current practices and procedures. Orphanet J Rare Dis 2016;11(1):115.

[21] Bonella F, Bauer PC, Griese M, Ohshimo S, Guzman J, Costabel U. Pulmonary alveolar proteinosis: new insights from a single-center cohort of 70 patients. Respir Med 2011;105(12):1908–16.

[22] Khan A, Agarwal R, Aggarwal AN. Effectiveness of granulocyte-macrophage colony-stimulating factor therapy in autoimmune pulmonary alveolar proteinosis: a meta-analysis of observational studies. Chest 2012;141(5):1273–83.

[23] Tazawa R, Inoue Y, Arai T, Takada T, Kasahara Y, Hojo M, et al. Duration of benefit in patients with autoimmune pulmonary alveolar proteinosis after inhaled granulocyte-macrophage colony-stimulating factor therapy. Chest 2014;145(4):729–37.

[24] Kavuru MS, Malur A, Marshall I, Barna BP, Meziane M, Huizar I, et al. An open-label trial of rituximab therapy in pulmonary alveolar proteinosis. Eur Respir J 2011;38(6):1361–7.

[25] Malur A, Kavuru MS, Marshall I, Barna BP, Huizar I, Karnekar R. Rituximab therapy in pulmonary alveolar proteinosis improves alveolar macrophage lipid homeostasis. Respir Res 2012;13:46.

[26] Sivitanidis E, Tosson R, Wiebalck A, Laczkovics A. Combination of extracorporeal membrane oxygenation (ECMO) and pulmonary lavage in a patient with pulmonary alveolar proteinosis. Eur J Cardiothorac Surg 1999;15(3):370–2.

[27] Happle C, Lachmann N, Ackermann M, Mirenska A, Göhring G, Thomay K, et al. Pulmonary transplantation of human induced pluripotent stem cell-derived macrophages ameliorates pulmonary alveolar proteinosis. Am J Respir Crit Care Med 2018;198(3):350–60.

Chapter 4.2

Alveolar microlithiasis

Alveolar microlithiasis (AML) is a rare autosomal-recessive disease characterized by widespread deposition of spherical calcium and phosphorus deposits in the intra-alveolar space [1]. By 2015, there were 1022 reported cases, predominantly in Asia (56.3%) and Europe (27.8%). AML is more common in males and clinically manifests in the second and third decades of life in most cases, although AML in newborns and elderly patients up to 84 years in age were also described [2].

AML results from a mutation in *SLC34A2*, which regulates the phosphorus transport in the alveoli [3]. Aberrant gene function leads to impaired capture of phosphorus, which is formed during the degradation of the surfactant phospholipids. The subsequent accumulation of excessive phosphorus in combination with calcium leads to the formation of intra-alveolar salt deposits called calciferites, calcospherites, or microliths [4]. The incidence of mutant *SLC34A2* alleles in the general population is very low, less than 0.008 [4].

Despite the evident genetic cause, most of the known AML cases are considered sporadic (62.7%), with only 37.2% of all AM cases reported as familial [2].

Morphology

At macroscopic examination, both lungs are invariably affected, which exhibit dense consistency due to reduced alveolar space and are often with a granular parietal pleural surface. On palpation the tissue contains inclusions resembling sand grains. The total weight of the lungs is increased and can reach 5 kg in some cases [5].

Microscopic examination of the lung tissue reveals concentrically laminated calcifications with small-sized radial striations ranging from a few microns to several millimeters. These acid-Schiff-positive microcalcifications are localized in the alveolar lumen and the pulmonary interstitium, as well as in the walls of blood vessels, bronchi, and visceral pleural layers (Figs. 4.2.1–4.2.4) [6]. Calcium deposition can lead to the development of interstitial fibrosis and progression of pulmonary and heart failure. In some cases, foci of ossification may occur.

The differential diagnosis of AML includes other lung diseases accompanied by calcification and ossification, mainly dystrophic and metastatic lung calcification, as well as diffuse ossification of the lungs.

Dystrophic calcification results from healing of foci of necrosis, degenerations, and sclerosis, which occurs most commonly in tuberculosis (Ghon lesions and Aschoff-Puhl foci, Figs. 4.2.5 and 4.2.6) and histoplasmosis, as well as in viral and parasitic pneumonia. Separate foci of dystrophic calcification are accompanied by the formation of bone and bone marrow tissue, which differ from AML by their larger size and amorphous structure, as well as the presence of signs related to the main pulmonary pathology.

Metastatic calcification in the lungs (MCL) can develop in patients with hypercalcemia as a consequence of chronic diseases of the thyroid and parathyroid glands, intestines, and kidneys; sarcoidosis; and liver transplantation; this type is termed benign metastatic calcification. Conversely, hypercalcemia due to multiple bone metastases in patients with

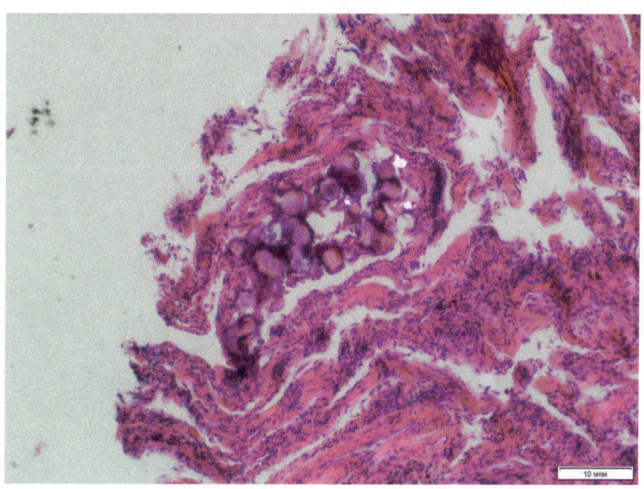

FIG. 4.2.1 Pulmonary alveolar microlithiasis. Laminated calcifications in the lumen of the alveoli. Hematoxylin and eosin (H&E) staining, 100×.

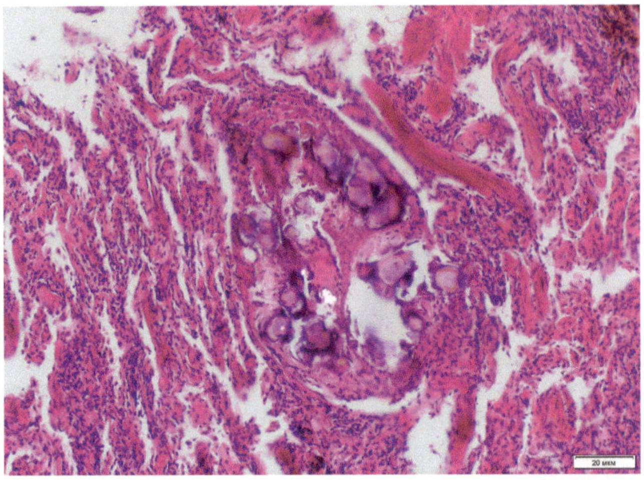

FIG. 4.2.2 Pulmonary alveolar microlithiasis. Laminated calcifications in the alveolar lumen. Interstitial sclerosis in the adjacent lung tissue. H&E stain, 100×.

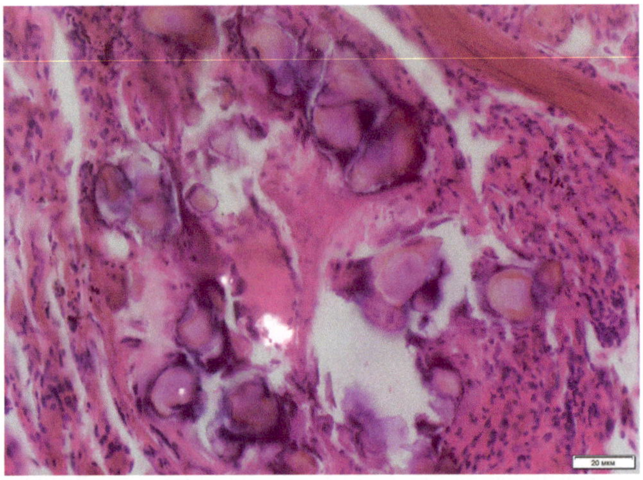

FIG. 4.2.3 Pulmonary alveolar microlithiasis. Laminated calcifications in the alveolar lumen. Interstitial sclerosis in the adjacent lung tissue. H&E stain, 600×.

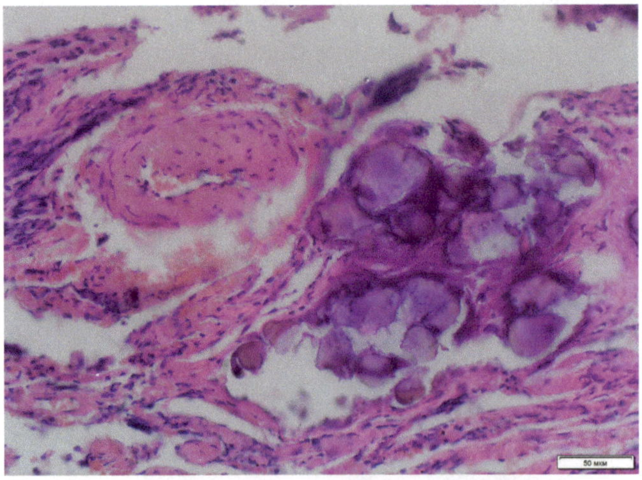

FIG. 4.2.4 Pulmonary alveolar microlithiasis. Laminated calcifications in the pulmonary interstitium of the perivascular tissue. H&E stain, 600×.

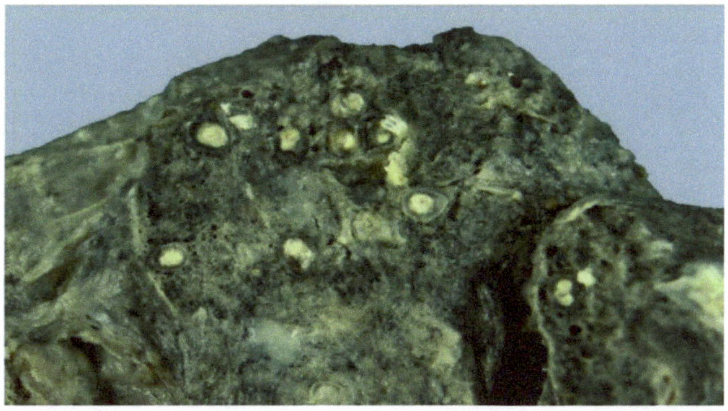

FIG. 4.2.5 Dystrophic calcification in tuberculosis. Gross specimen. Petrification in the right lung apex is white with a stonelike density (Aschoff-Puhl foci).

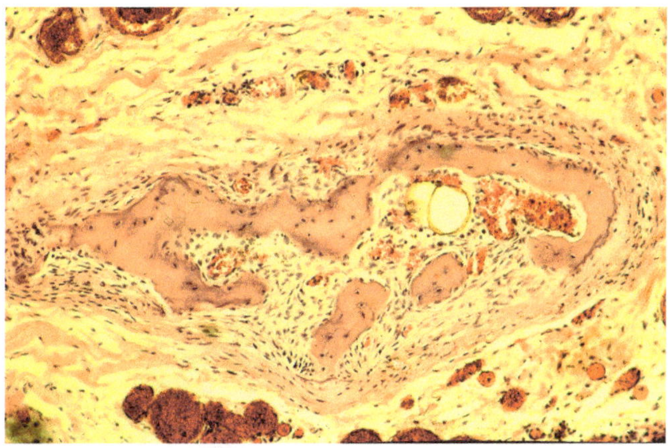

FIG. 4.2.6 Dystrophic calcification in tuberculosis (Ghon lesion). Fragments of the fibrous capsule around the petrification with trabeculae of bone and bone marrow tissue, 200×.

multiple myeloma or lung cancer or breast cancer metastasis leads to the development of malignant metastatic calcification. Microscopic examination of amorphous deposits of small-sized calcium salts is found in the pulmonary interstitium and the walls of the alveoli and bronchi (Fig. 4.2.7). MCL differ from AML by their larger size and amorphous structure, the presence of signs of the underlying disease that are accompanied with calcification in other organs (i.e., the kidney, stomach, and myocardium), and hypercalcemia.

Diffuse ossification of the lungs (DOL) is a rare disease of unknown origin, which is characterized by unilateral or bilateral lesions in the lower lobes of the lungs, accompanied with the formation of bone-tissue foci with bone marrow in the pulmonary interstitium [7,8]. There are two forms of DOL: dendriform parenchymal osteodystrophy (DPOD) (Figs. 4.2.8 and 4.2.9) and nodular parenchymal pulmonary osteodystrophy (NPPO) [7,9]. NPPO is characterized by the formation of foci in the lung tissue; histological examination reveals round, ossified lesions inside the affected alveoli [9]. DPOD is characterized by the presence of dendriform or coral-like formations growing in a serpentine form along the alveolar septa with invasion into the alveolar space [9].

Clinical presentation

The clinical manifestations of AML depend on the extent of the lung parenchymal involvement. More than half of the AML cases reported were asymptomatic and diagnosed during routine evaluation [10]. The most common symptoms are dyspnea, dry cough, chest pain, hemoptysis, and fatigue [6]. In the advanced disease stage with respiratory failure, finger clubbing may be present [2]. During auscultation, crackles and rhonchi in the lower fields are often heard.

Extrapulmonary manifestations of AML were rarely reported, which are usually observed as extrapulmonary microliths in the seminal vesicles and epididymides, accompanied by azoospermia. Calcification in the kidneys, prostate, pericardium, gall bladder, and other organs described in AML patients is already larger accumulations of calcium and cannot be

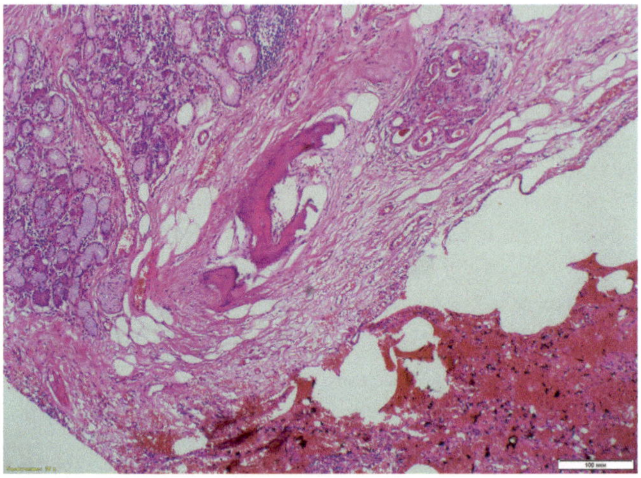

FIG. 4.2.7 Metastatic calcification in hyperplasia of the parathyroid glands. Petrification in the bronchus wall with tissue ossification. H&E staining, 200×.

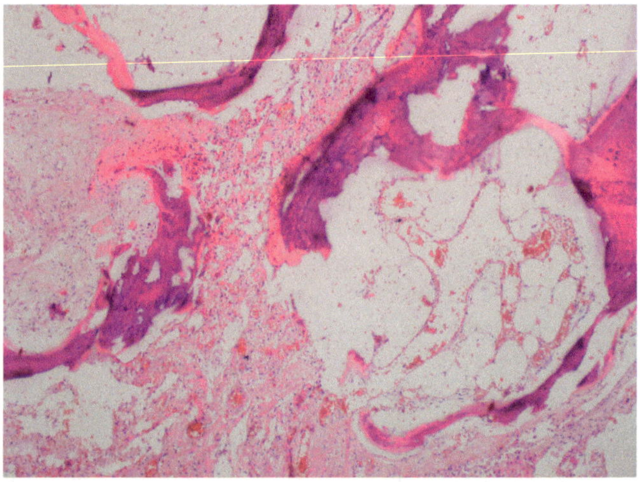

FIG. 4.2.8 Diffuse parenchymal lung osteodystrophy, dendritic form. Multiple petrification lesions in the pulmonary interstitium with bone marrow tissue. H&E staining, 100×.

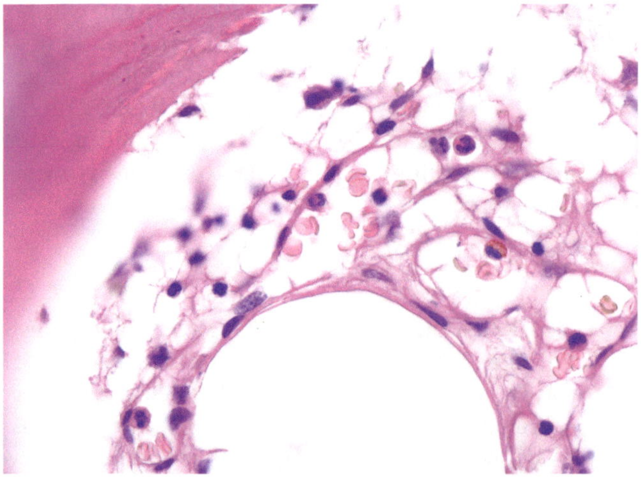

FIG. 4.2.9 Diffuse parenchymal lung osteodystrophy, dendritic form. Foci of bone marrow tissue and trabeculae of bone in the pulmonary interstitium. H&E staining, 600×.

unambiguously interpreted solely as a manifestation of AML [6]. The disease course of AML is usually benign; clinical and radiological symptoms may remain stable for many months and even years. According to Klikovits et al. the average disease duration from the detection of radiological changes to the onset of clinical symptoms was 25.2 years [11]. The maximum follow-up duration of an AML patient was 58 years [2]. However, rapidly progressive forms leading to the development of respiratory failure within one year after diagnosis were also reported [12]. In one study the course of AML was determined to be more aggressive in smokers than in those who never smoked [5]. The average life expectancy of AML patients ranges between 10 and 15 years from the time of definitive diagnosis [13].

Diagnosis

Diagnosis is based on the analysis of chest X-ray and HRCT images and biopsy material. Bronchoscopy with transbronchial biopsy and bronchoalveolar lavage (BAL) are minimally invasive methods for AML diagnosis. Microliths are often found in the BAL fluid sediment [14], and the volume of histological material is usually sufficient to detect typical acid-Schiff-positive calciospherocitis.

Radiography

The radiological diagnosis of AML is very difficult during the early disease stage, that is, precalcification. All patients are asymptomatic, and the number and the density of microliths are still insufficient for their definitive identification as calcifications; on radiographs and high-resolution computed tomography (HRCT) images, multiple focal reductions in transparency with ground-glass opacity are identified. These presentations, described exclusively in pediatric cases, can probably be revealed at the time of disease onset in adults [15]. At later stages the phosphorus and calcium salts accumulate in the alveolar space, a presentation described by many authors as sandstorm lung on chest radiographs, which includes bilateral, diffuse scattered areas of micronodular calcifications, with increasing intensity toward the lower fields of the lungs (Fig. 4.2.10) [16]. Due to the high density, the contours of the lungs fuse with the shadow of the heart, diaphragm, and chest wall, and the pleural sinuses become impossible to visualize [6]. A distinctive feature of AML is the dissociation between the radiological and clinical data. Despite the massive involvement of the pulmonary parenchyma by radiography, patients may experience minimal clinical symptoms.

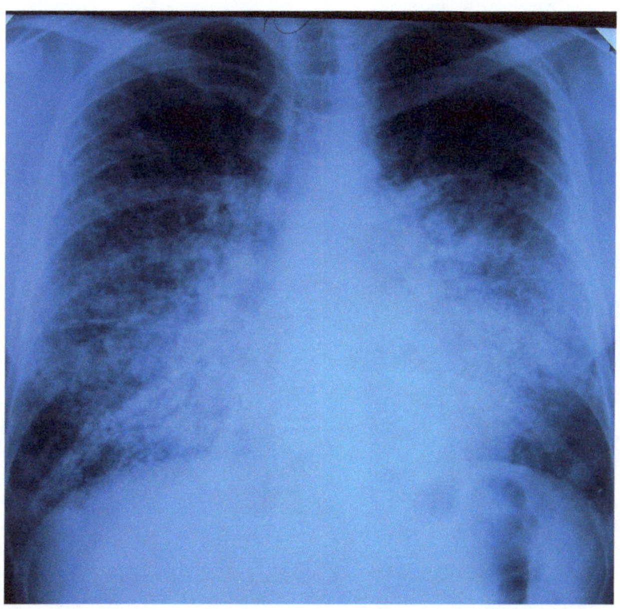

FIG. 4.2.10 Chest radiograph in a patient with alveolar microlithiasis (AML). Multiple bilateral dense foci, merging together, primarily in the middle and basal zones, to create the impression of a snow storm. *(Case courtesy of Dr. M.V. Samsonova and Prof. A.L. Chernyaev, Pulmonology Scientific and Research Institute, Moscow, Russia.)*

Computed tomography

HRCT is the most important primary diagnostic tool for AML. The changes detected by HRCT are so specific that many authors deny the need for morphological verification of AML [17]. We propose that this provision relates only to the later disease stages. At earlier stages (Fig. 4.2.11), HRCT may not be the definitive diagnostic method. In three studies including a total of 33 AML patients [17–19], the following features were identified: innumerous parenchymal calcified nodules about 1 mm in diameter (70%–100%), ground-glass opacity (90%–100%), innumerous small subpleural nodules (50%–100%); subpleural cysts (50%–100%) as a reflection of concomitant interstitial fibrosis, pleural thickening, and subpleural linear calcification (69%–90%); consolidation areas, primarily in the lower posterior fields (46%–100%); and mosaic pattern (crazy paving, 40%–69%). Following close contact of the separately located nodules, a large-scale calcification of more than the typical size of 1–2 mm can be observed by HRCT. In the study by Sumikawa et al. [18], all 10 patients had thickening of the interlobular septa, which was not described as a separate finding in other studies, which however identified calcifications along the interlobular septa in 46%–50% of the cases. The perilobular or centrilobular arrangement of the calcifications is typical, whereas the perilymphatic pattern is less common (Fig. 4.2.11), with a predominantly subpleural and peribronchovascular distribution [20]. Some studies noted the characteristic symptom of "black pleura", which is the radiolucent line between the calcined lung parenchyma and the adjacent ribs, which is readily visible on X-ray; in HRCT images, it resembles a chain of subpleural cysts (Fig. 4.2.12) [17]. At the late stages of the disease, due to

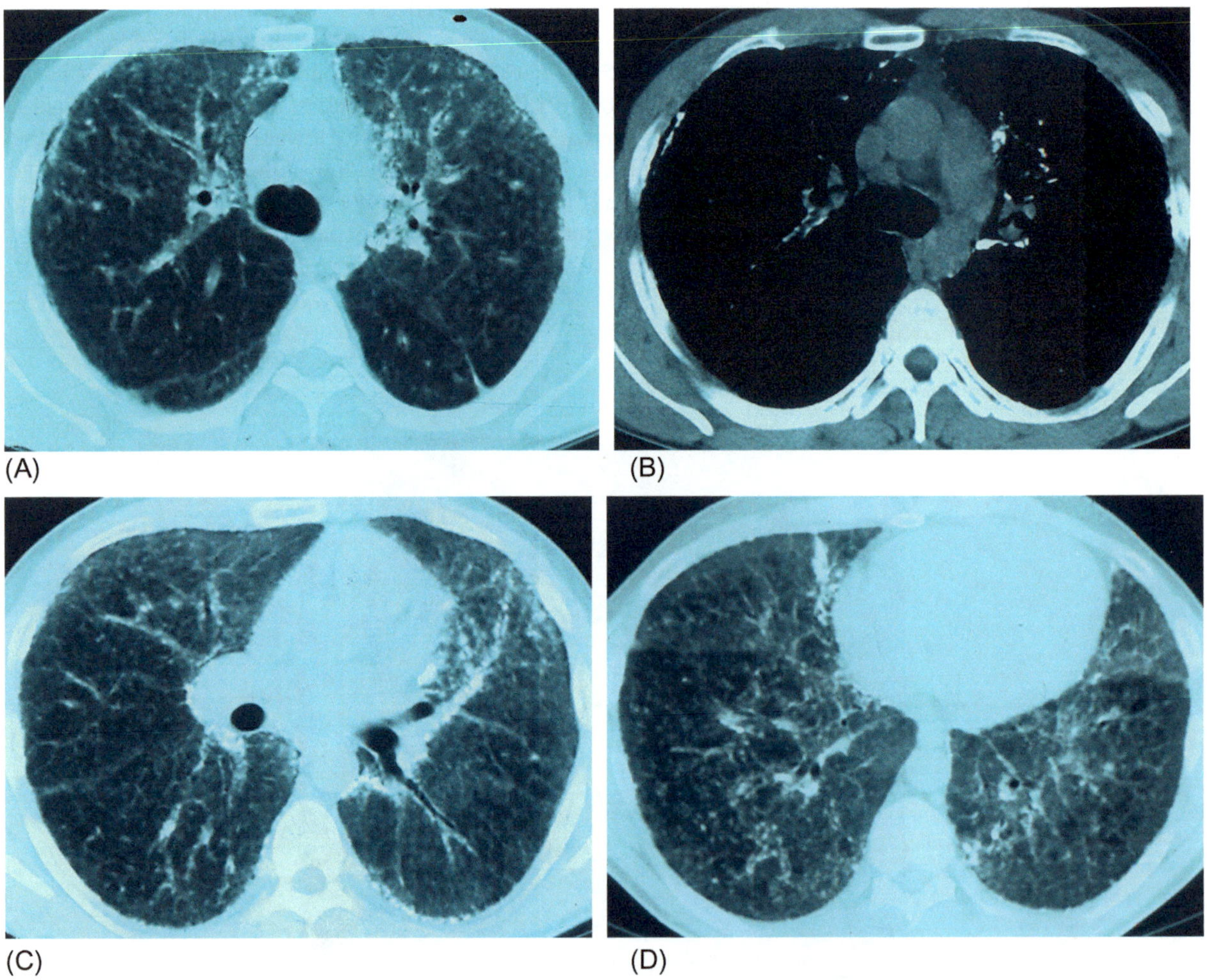

FIG. 4.2.11 AML. Innumerous bilateral centrilobular nodules, 1–2 mm in diameter, of different density ranging from ground-glass opacity to bone-tissue-like attenuation. In comparable sections in the pulmonary (A) and mediastinal lymph nodes (B), maximum calcification is observed in the peribronchovascular and subpleural regions. (A, C, D) Distinct perilymphatic distribution pattern. In some areas, larger calcifications are noted, which are arranged in chains along the visceral pleura and large bronchi. Mosaic thickening of interlobular septa. (A) The beginning of the formation of small subpleural cysts. *(Case courtesy of Dr. M.V. Samsonova and Prof. A.L. Chernyaev, Pulmonology Scientific and Research Institute, Moscow, Russia.)*

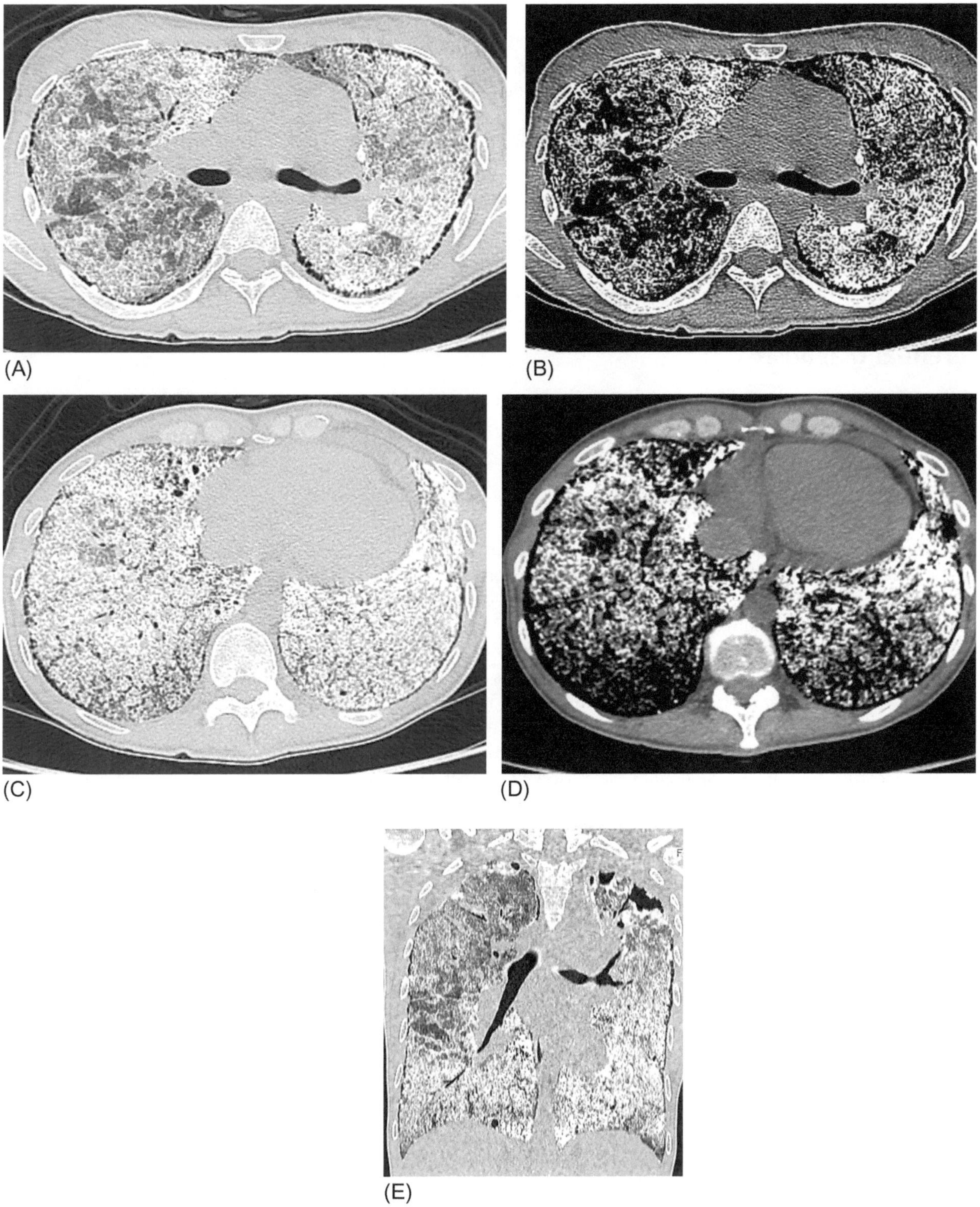

FIG. 4.2.12 AML, advanced stage. (A–D) Diffuse filling of the lung parenchyma with small calcifications, sometimes merging into conglomerates. Multiple subpleural small cysts merging together. (E) Subtotal replacement of the lung parenchyma with calcined foci, in a presentation termed white lungs. Bullous changes in the upper lobes of the lungs. Traction bronchiectasis is also visible. The "black pleura" sign is presented. *(Case courtesy of Prof. S.N. Avdeev, Sechenov First Moscow State Medical University, Moscow, Russia.)*

total calcification and fibrosis of the surrounding tissue, the so-called sign of white lungs appears; paraseptal and bullous emphysema in the upper lobes are also often detected, and pneumothorax is possible (Fig. 4.2.12) [21].

Additional imaging methods are magnetic resonance imaging (MRI) and chest ultrasonography. MRI in AML patients reveals increased signal intensity in the lower fields of the lungs on T1-weighted images [22]. This effect is due to the accumulation of microliths and differs from the increased T-signal, which is observed in pulmonary fibrosis. Ultrasonography reveals the irregularity and thickening of the pleura, as well as innumerous small subpleural echoic foci without significant acoustic shadowing [23,24].

Functional diagnostic, which is not essential for the initial diagnosis, is nonetheless important for assessing the severity of respiratory failure and monitoring the disease course. Usually, either a normal or a restrictive functional pattern is detected. The diffusion capacity of the lungs decreases with the progressing of the disease. Among other diagnostic tools, an increase in serum levels of surfactant proteins A and D can be noted in AML patients; however, these changes can be observed in other alveolar lesions and are more likely to be utilized for monitoring the disease progression [25].

Differential diagnosis

The differential diagnosis is performed primarily to rule out diseases that can manifest with innumerous small calcified nodules by chest radiography or CT, including, most frequently, metastatic calcification in the lungs (MCL), followed by certain pulmonary infections, recurrent diffuse alveolar hemorrhage, occupational diseases, and diffuse pulmonary ossification (Table 4.2.1).

MCL is a rare metabolic disorder usually caused by diseases accompanied by hypercalcemia and manifests with the deposition of calcium salts in the interstitium [26]. MCL is most commonly caused by chronic renal failure; however, other benign and malignant diseases with chronically increased serum calcium and phosphorus levels can lead to their deposition in the lungs. These conditions include primary and secondary hyperparathyroidism; overdose of calcium and vitamin D; sarcoidosis; milk-alkali syndrome; osteoporosis; osteitis deformans and complications of renal, liver, and heart transplantation; cardiac surgery; multiple myeloma; parathyroid carcinoma; leukemia; lymphoma; breast carcinoma; synovial carcinoma; choriocarcinoma; malignant melanoma; and hypopharyngeal squamous carcinoma [26]. Similar to AML, MCL may be asymptomatic or present with clinical manifestations that are not different from those of AML.

However, the HRCT characteristics of MCL differ significantly from those of AML. First the distribution pattern predominantly involves the upper lobes of the lungs. Second the multiple calcified nodules are significantly larger in size compared with those in AML and range from 3 to 10 mm. Third the nodules often look fluffy, mimicking airspace nodules, rather than appearing completely calcified as that observed in AML. Finally, in MCL, simultaneously calcified nodules are often found in the vessels of the soft tissues of the chest and other organs including the heart, bronchial walls, small pulmonary arteries, superior vena cava, and dura of the dorsal spine [27,28].

An additional differential test is the determination of calcium and phosphorus levels in serum, which are within the normal range in AML. Pulmonary hemosiderosis (idiopathic or secondary) may manifest with the accumulation of hemosiderin and calcium deposition, which creates a pattern of innumerous, high-level foci that is sometimes called endogenous pneumoconiosis [27]. As a rule, thickened interlobular septa are also present. However, a history of recurrent diffuse alveolar hemorrhage in pulmonary hemosiderosis aids in establishing the correct diagnosis.

Postinfectious innumerous small calcifications in the lung parenchyma can be formed as a result of primary focal necrotic injury, such that may occur in disseminated tuberculosis, histoplasmosis, and varicella pneumonia [20]. In the first two instances, calcification occurs not only in the lungs but also in the hilar or mediastinal lymph nodes. These calcifications are usually larger in size (2–5 mm) [27], their total number is significantly less than that in AML, and the typical basal distribution pattern typical for AML is not observed (Fig. 4.2.13). If patients do not have a history of any of these diseases, circulating antibodies to pathogens (histoplasmosis and varicella) or the T-SPOT.TB test for suspicious history of tuberculosis should be determined.

One of the rare radiological patterns of silicosis and coal worker pneumoconiosis is diffuse innumerous calcined parenchymal nodules, which tend to distribute in the middle and upper fields of the lungs (Fig. 4.2.14) [27].

Patients exposed to the dust particles of iron oxide (siderosis), tin oxide (stannosis), and barium (baritosis) can manifest with a similar pattern. The size of the nodules is less than 5 mm, and their number is generally much lower than that observed in AML [28,29]. In typical cases, lung metastases of osteogenic sarcoma or chondrosarcoma manifest as calcified lesions and nodes, with partial or total calcification (Fig. 4.2.15) [29].

Another similar to AML condition is diffuse ossification of the lungs (DOL). DOL is a very rare disease in which ectopic formation of bone-tissue fragments with bone marrow occurs in alveolar spaces [30]. Over the 160 years since its first

TABLE 4.2.1 Differential range of alveolar microlithiasis

	AML	MCL	Hemosiderosis	Postinfectious	Pneumonoconiosis	DOL
Anamnesis	Family history in 30%	Renal failure, multiple myeloma, hyperparathyroidism, etc.	Recurrent alveolar hemorrhages	Miliary tuberculosis, severe histoplasmosis, varicella pneumonia, immunodeficiency	Work with coal dust and its derivatives, dust particles of iron oxide, tin or barium oxide	Mostly male over 50 years of age, gastroesophageal reflux disease, congestive left ventricular failure, fibrous lung diseases
Focal size (mm)	1–2	3–10	2–5	2–5	<5	1–5
Distribution pattern	Basal preference	Upper zonal preference	Perihilar	Random or upper lobe distribution	Preferably in upper and middle lobes	Basal preference
Calcification of hilar lymph nodes	Not typical	Frequent	Not typical	Frequent	Frequent	Not typical
Serum calcium level	Normal	Increased	Normal	Normal	Normal	Normal

AML, alveolar microlithiasis; *MCL*, metastatic calcification in the lungs; *DOL*, diffuse ossification in the lungs.

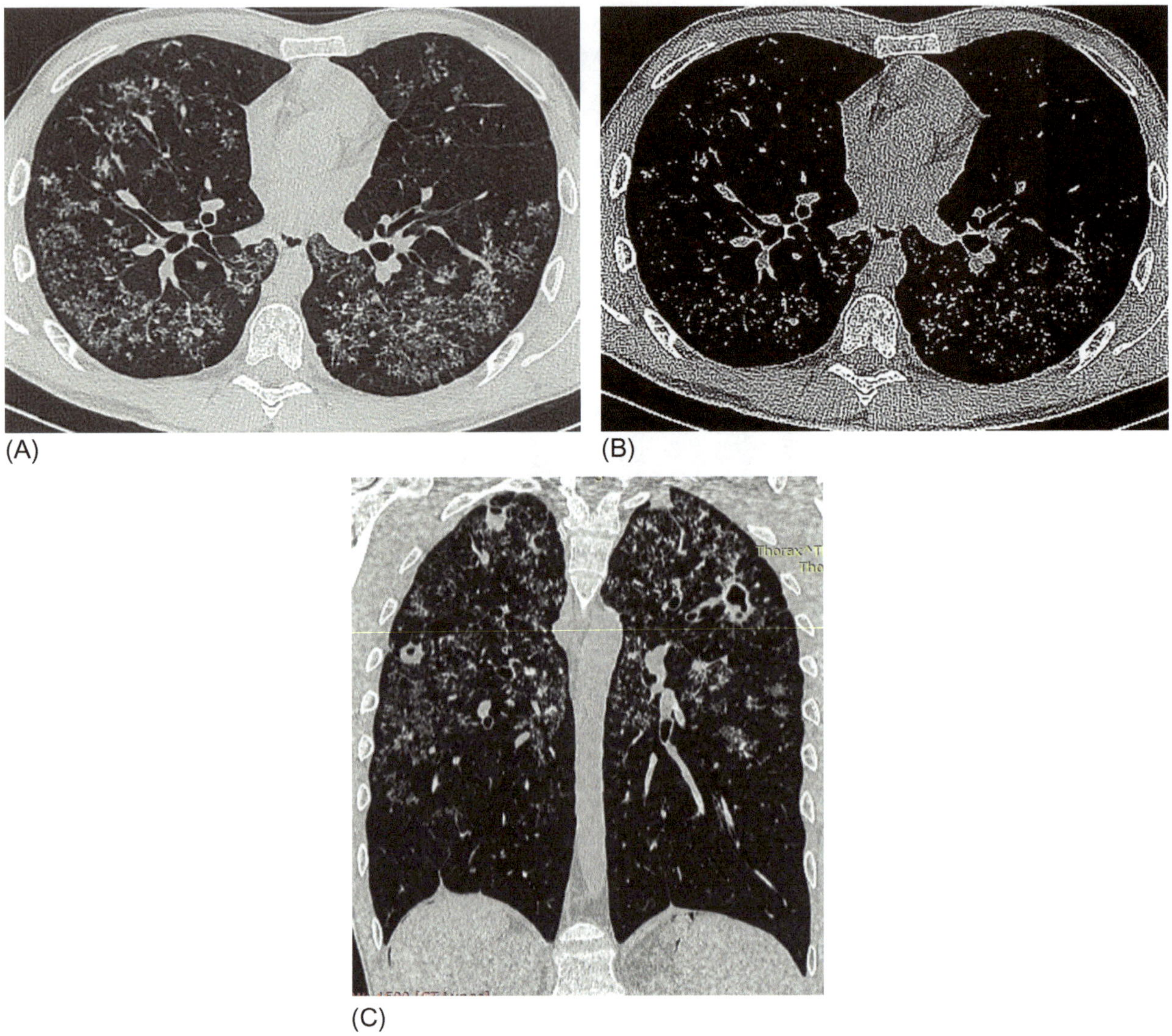

FIG. 4.2.13 Disseminated pulmonary tuberculosis. Multiple bilateral small nodules, sometimes merging together (A), some of which exhibit bone-tissue density (B). In coronal reconstruction, thin- and thick-walled cavitary formations are determined in the upper fields (C).

description by Luschka in 1856 [31], about 100 individual cases were reported, with most as autopsy observations. The pathogenesis of DOL is unclear; we suggest that it might result from epithelial-mesenchymal transition with the subsequent transformation of mesenchymal cells into bone tissue. This probable pathogenic mechanism is supported by the frequent detection of DOL in fibrous intestinal lung diseases, in which epithelial-mesenchymal transition plays a key role in disease pathogenesis [32]. DOL was found in biopsy specimens of 39% of patients with idiopathic pulmonary fibrosis, 12% of those with nonspecific interstitial pneumonia, and 19% of those with chronic hypersensitivity pneumonitis [33].

Two morphological and radiological patterns of DOL are described: dendriform, which is characterized by rare bone islets along the terminal vascular and bronchial bundles, and nodular, which depicts round bone particles deposited in the alveoli [30]. The nodular shape is associated exclusively with congestive heart failure and manifested by dense nodules that tend to confluence and the occasional presence of trabeculae [34]. HRCT findings of dendriform DOL, which are not associated with interstitial lung diseases, include reticular changes (thickening of the intralobular and interlobular septa, often with a chain of small nodules inside the interlobular septa) without traction bronchiectasis, and multiple nodules, a few millimeters in size, exhibiting the density of bone, with predominant posterior basal and subpleural distribution [35]. However, the number of the nodules is significantly lower than that in AML, and there are always zones free from pathological changes, even with progressive respiratory failure [36].

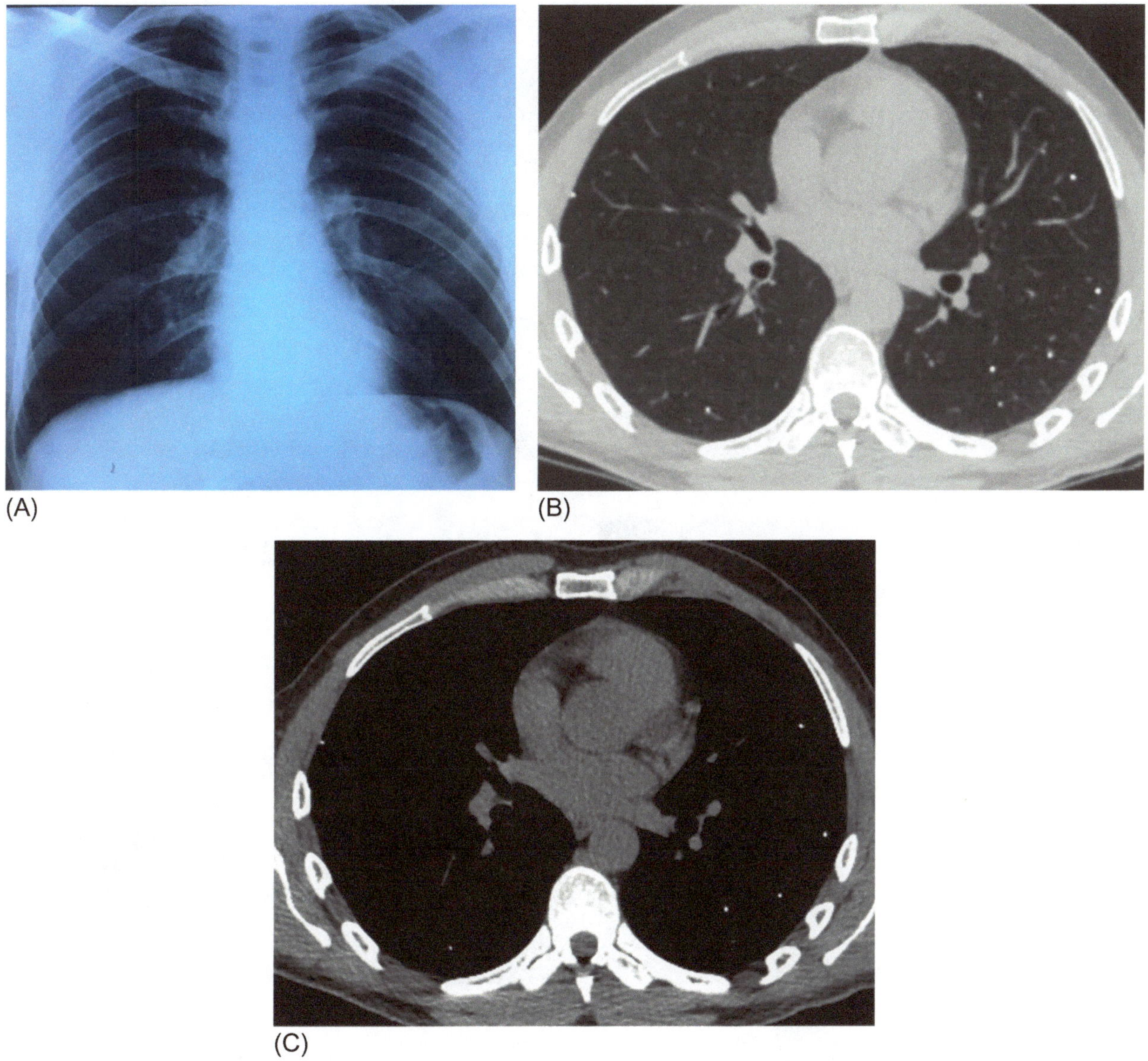

FIG. 4.2.14 Microcalcification in the lungs of a patient working with dispersed silica gel on chest X-ray (A) and HRCT (B and C). Separately located, small (1–2 mm), dense nodules with maximum distribution in the middle zones of the lungs.

At the early stages of AML when small parenchymal nodules without significant calcification prevail by HRCT, the differential diagnosis should include pulmonary tuberculosis, sarcoidosis, and Langerhans cell histiocytosis. In such cases the definitive diagnosis relies on morphological evaluation.

Treatment

To date, there are no treatments that can alter the disease progression [2]. Attempts at therapeutic BAL or steroid therapy were generally ineffective, although several studies reported positive results with various treatment approaches, including steroid inhalers [37,38].

Accumulating data suggest that AML patients might benefit from disodium etidronate, which can reduce the deposition of calcium and phosphorus salts in the alveoli and improve the functional and radiological parameters in patients [39].

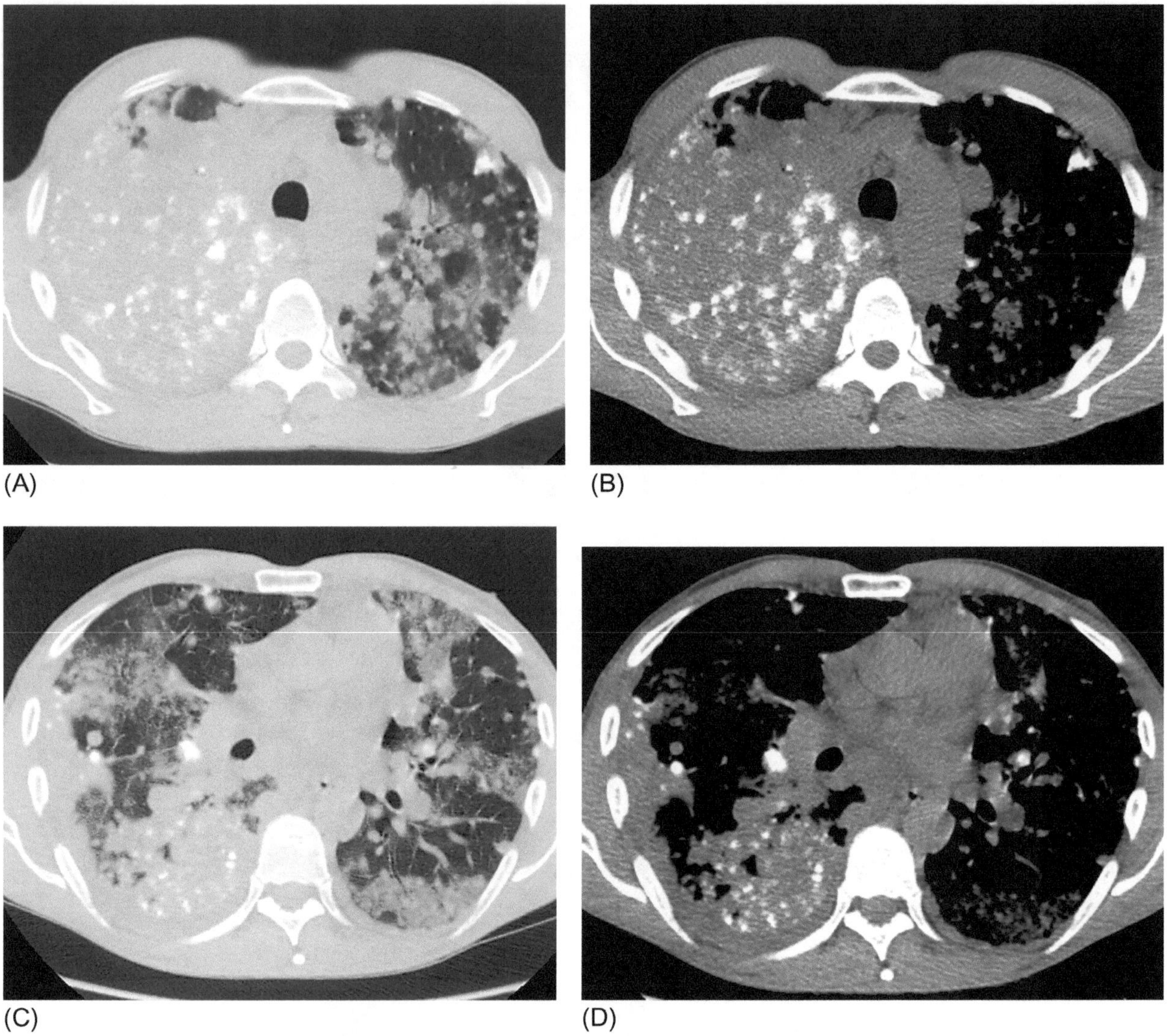

FIG. 4.2.15 Osteogenic sarcoma with metastases to the lungs. Bilateral foci of consolidation and mass, with multiple calcifications of different sizes, up to a maximum of 1 cm in diameter, located within the lungs (A–D).

The only definitive treatment modality in AML patients at the terminal stage of the disease is lung transplantation. Many experts suggest that severe respiratory failure, need for oxygen therapy, and decompensated right ventricular failure are clear indications for transplantation in AML patients [11]. The accumulation experience in AML patients who underwent transplantation indicates a good postoperative survival with a maximum follow-up period of 15 years [2]. Despite the genetic origin of the disease, none of the published reports states the return of the disease to the transplanted lungs [2,11].

References

[1] Barnard NJ, Crocker PR, Blainey AD, Davies RJ, Ell SR, Levison DA. Pulmonary alveolar microlithiasis. A new analytical approach. Histopathology 1987;11(6):639–45.

[2] Castellana G, Castellana G, Gentile M, Castellana R, Resta O. Pulmonary alveolar microlithiasis: review of the 1022 cases reported worldwide. Eur Respir Rev 2015;24(138):607–20.

[3] Poelma DL, Ju MR, Bakker SC, Zimmermann LJ, Lachmann BF, van Iwaarden JF. A common pathway for the uptake of surfactant lipids by alveolar cells. Am J Respir Cell Mol Biol 2004;30(5):751–8.

[4] Huqun IS, Miyazawa H, Ishii K, Uchiyama B, Ishida T, et al. Mutations in the SLC34A2 gene are associated with pulmonary alveolar microlithiasis. Am J Respir Crit Care Med 2007;175(3):263–8.
[5] Lauta VM. Pulmonary alveolar microlithiasis: an overview of clinical and pathological features together with possible therapies. Respir Med 2003;97(10):1081–5.
[6] Francisco FA, Silva JL, Hochhegger B, Zanetti G, Marchiori E. Pulmonary alveolar microlithiasis. State-of-the-art review. Respir Med 2013;107(1):1–9.
[7] Nicholson AG, Rice AJ. Interstitial lung diseases. In: Ph H, Flieder DB, editors. Spencer's pathology of the lung. 6th ed. New York: Cambridge University Press; 2013. p. 392–4.
[8] Jaderborg JM, Dunton RF. Rare clinical diagnosis of dendriform pulmonary ossification. Ann Thorac Surg 2001;71(6):2009–11.
[9] Lara JF, Catroppo JF, Kim D, da Costa D. Dendriform pulmonary ossification, a Form of diffuse pulmonary ossification. Report of a 26-year autopsy experience. Arch Pathol Lab Med 2005;129(3):348–53.
[10] Mariotta S, Ricci A, Papale M, De Clementi F, Sposato B, Guidi L, et al. Pulmonary alveolar microlithiasis: report on 576 cases published in the literature. Sarcoidosis Vasc Diffuse Lung Dis 2004;21(3):173–81.
[11] Klikovits T, Slama A, Hoetzenecker K, Waseda R, Lambers C, Muracoezy G, et al. A rare indication for lung transplantation – pulmonary alveolar microlithiasis: institutional experience of five consecutive cases. Clin Transpl 2016;30(4):429–34.
[12] Stamatopoulos A, Patrini D, Mitsos S, Khiroya R, Borg E, Hayward M, et al. An unusual late onset of pulmonary alveolar microlithiasis: a case report and literature review. Respir Med Case Rep 2017;22:24–7.
[13] Corut A, Senyigit A, Ugur SA, Altin S, Ozcelik U, Calisir H, et al. Mutations in SLC34A2 cause pulmonary alveolar microlithiasis and are possibly associated with testicular microlithiasis. Am J Hum Genet 2006;79(4):650–6.
[14] Palombini BC, da Silva Porto N, Wallau CU, Camargo JJ. Bronchopulmonary lavage in alveolar microlithiasis. Chest 1981;80(2):242–3.
[15] Schmidt H, Lörcher U, Kitz R, Zielen S, Ahrens P, König R. Pulmonary alveolar microlithiasis in children. Pediatr Radiol 1996;26(1):33–6.
[16] Brown K, Mund DF, Aberle DR, Batra P, Young DA. Intrathoracic calcifications: radiographic features and differential diagnoses. Radiographics 1994;14(6):1247–61.
[17] Francisco FA, Rodrigues RS, Barreto MM, Escuissato DL, Araujo Neto CA, Silva JL, et al. Can chest high-resolution computed tomography findings diagnose pulmonary alveolar microlithiasis? Radiol Bras 2015;48(4):205–10.
[18] Sumikawa H, Johkoh T, Tomiyama N, Hamada S, Koyama M, Tsubamoto M, et al. Pulmonary alveolar microlithiasis: CT and pathologic findings in 10 patients. Monaldi Arch Chest Dis 2005;63(1):59–64.
[19] Marchiori E, Gonçalves CM, Escuissato DL, Teixeira KI, Rodrigues R, Barreto MM, et al. Pulmonary alveolar microlithiasis: high-resolution computed tomography findings in 10 patients. J Bras Pneumol 2007;33(5):552–7.
[20] Chan ED, Morales DV, Welsh CH, McDermott MT, Swarz MI. Calcium deposition with or without bone formation in the lung. Am J Respir Crit Care Med 2002;165(12):1654–69.
[21] Samano MN, Waisberg DR, Canzian M, Campos SV, Pêgo-Fernandes PM, Jatene FB. Lung transplantation for pulmonary alveolar microlithiasis: a case report. Clinics (Sao Paulo) 2010;65(2):233–6.
[22] Hoshino H, Koba H, Inomata S, Kurokawa K, Morita Y, Yoshida K, et al. Pulmonary alveolar microlithiasis: high-resolution CT and MR findings. J Comput Assist Tomogr 1998;22(2):245–8.
[23] Resorlu M, Toprak CA, Aylanc N, Karatag O. Ultrasonography and computed tomography findings in pulmonary alveolar microlithiasis. Rofo 2018;190(11):1063–4.
[24] Rea G, Sperandeo M, Sorrentino N, Stanziola AA, D'Amato M, Bocchino M. Chest ultrasound findings in pulmonary alveolar microlithiasis. J Med Ultrason (2001) 2015;42(4):591–4.
[25] Takahashi H, Chiba H, Shiratori M, Tachibana T, Abe S. Elevated serum surfactant protein A and D in pulmonary alveolar microlithiasis. Respirology 2006;11(3):330–3.
[26] Belém LC, Zanetti G, Souza AS, Hochhegger B, Guimarães MD, Nobre LF, et al. Metastatic pulmonary calcification: state-of-the-art review focused on imaging findings. Respir Med 2014;108(5):668–76.
[27] Marchiori E, Souza Jr AS, Franquet T, Müller NL. Diffuse high-attenuation pulmonary abnormalities: a pattern-oriented diagnostic approach on high-resolution CT. AJR Am J Roentgenol 2005;184(1):273–82.
[28] Hartman TE, Müller NL, Primack SL, Johkoh T, Takeuchi N, Ikezoe J, et al. Metastatic pulmonary calcification in patients with hypercalcemia: findings on chest radiographs and CT scans. AJR Am J Roentgenol 1994;162(4):799–802.
[29] Ciccarese F, Bazzocchi A, Ciminari R, Righi A, Rocca M, Rimondi E, et al. The many faces of pulmonary metastases of osteosarcoma: retrospective study on 283 lesions submitted to surgery. Eur J Radiol 2015;84(12):2679–85.
[30] Konoglou M, Zarogoulidis P, Baliaka A, Boutsikou E, Dramba V, Tsakiridis K, et al. Lung ossification: an orphan disease. J Thorac Dis 2013;5(1):101–4.
[31] Lushka H. Verastigte knochenbildung im parenchym der lung. Virchows Arch 1856;10:500–5.
[32] Jolly MK, Ward C, Eapen MS, Myers S, Hallgren O, Levine H, et al. Epithelial-mesenchymal transition, a spectrum of states: role in lung development, homeostasis, and disease. Dev Dyn 2018;247(3):346–58.
[33] Egashira R, Jacob J, Kokosi MA, Brun AL, Rice A, Nicholson AG, et al. Diffuse pulmonary ossification in fibrosing interstitial lung diseases: prevalence and associations. Radiology 2017;284(1):255–63.
[34] Woolley K, Stark P. Pulmonary parenchymal manifestations of mitral valve disease. Radiographics 1999;19(4):965–72.
[35] Gruden JF, Green DB, Legasto AC, Jensen EA, Panse PM. Dendriform pulmonary ossification in the absence of usual interstitial pneumonia: CT features and possible association with recurrent acid aspiration. AJR Am J Roentgenol 2017;209(6):1209–15.

[36] Matsuo H, Handa T, Tsuchiya M, Kubo T, Yoshizawa A, Nakayama Y, et al. Progressive restrictive ventilatory impairment in idiopathic diffuse pulmonary ossification. Intern Med 2018;57(11):1631–6.

[37] Caputi M, Guarino C, Cautiero V, Castellano G, Perna A. Diagnostic role of BAL in pulmonary alveolar microlithiasis. Arch Monaldi Mal Torace 1990;45(5):353–64.

[38] Ganesan N, Ambroise MM, Ramdas A, Kisku KH, Singh K, Varghese RG. Pulmonary alveolar microlithiasis: an interesting case report with systematic review of Indian literature. Front Med 2015;9(2):229–38.

[39] Ozcelik U, Yalcin E, Ariyurek M, Ersoz DD, Cinel G, Gulhan B, et al. Long-term results of disodium etidronate treatment in pulmonary alveolar microlithiasis. Pediatr Pulmonol 2010;45(5):514–7.

Chapter 4.3

Diffuse alveolar hemorrhage

Diffuse alveolar hemorrhage (DAH) is a rare but potentially life-threatening condition that is characterized by diffuse blood leakage from the pulmonary microvessels into the alveolar spaces because of the disruption of the vascular wall [1]. Alveolar hemorrhage syndrome is a synonym of DAH.

There are several diverse causes of DAH (Table 4.3.1), with over 100 diseases and conditions being described, that lead to the development of alveolar hemorrhages. Most DAH cases are presented as clinical cases; however, several studies have also been published; the most extensive study was conducted by de Prost et al. who analyzed 112 DAH patients. They showed that DAH was caused by autoimmune diseases (35%), such as microscopic polyangiitis; granulomatosis

TABLE 4.3.1 Possible causes of DAH (reproduced with permission of the © ERS 2019, with supplements[1])

Group	Frequent variants	Rare variants
Systemic vasculitis	Granulomatosis with polyangiitis (Wegener's granulomatosis); microscopic polyangiitis	Henoch-Schonlein purpura, eosinophilic granulomatosis with polyangiitis, Behcet's disease, mixed cryoglobulinemia associated with the hepatitis C virus, pauci-immune pulmonary capillaritis (with or without antineutrophil antibodies), polyarteritis nodosa associated with hepatitis B, Takayasu disease
Connective tissue diseases	Systemic lupus erythematosus	Rheumatoid arthritis, systemic sclerosis, inflammatory myopathies, mixed connective tissue disease
Other autoimmune diseases	Anti-glomerular basement membrane disease (Goodpasture syndrome)	Pauci-immune glomerulonephritis, immune complex glomerulonephritis, hemolytic uremic syndrome, IgA-associated nephropathy, coeliac disease, inflammatory bowel diseases, cows' milk intolerance, IgG4-related lung disease
Infections	Leptospirosis	Invasive aspergillosis, systemic candidiasis, strongyloidiasis, staphylococcal infection, legionellosis, *Mycoplasma pneumoniae*, cytomegalovirus, influenza virus, herpes virus, Hantavirus, HIV, malaria, *Strongyloides stercoralis*, *Stachybotrys chartarum*
Medications	Propylthiouracil, non-steroidal anti-inflammatory drugs	Alemtuzumab, abciximab, transretinoic acid, aminoglutethimide, amiodarone, amphotericin B, azathioprine, carbamazepine, carbimazole, cyclosporin, clomiphene, cytarabine, dextran, dihydralazine, d-penicillamine, dimethylsulfoxide, everolimus fludarabine, gemcitabine, glibenclamide, methotrexate, mitomycin, moxalactam, nitrofurantoin, nitric oxide, phenytoin, quinidine, rituximab, sirolimus, sunitinib, tirofiban
Toxins	Cocaine	Trimellitic anhydride, pyromellitic dianhydride isocyanates, hydrocarbon derivatives
Intravascular metastases		Angiosarcoma, Kaposi sarcoma, choriocarcinoma, epithelioid hemangioendothelioma, multiple myeloma, renal cell carcinoma
Transplantation	Bone marrow transplantation	Solid organ transplantation
Hemostasis disorders	Drug-induced (anticoagulants, thrombolytics, and antithrombocytic)	Disseminated intravascular blood coagulation, thrombocytopenia, antiphospholipid syndrome, thrombocytopenic purpura, hemophilia
Diseases of the pulmonary vessels		Idiopathic and thromboembolic pulmonary hypertension, pulmonary veno-occlusive disease, pulmonary capillary hemangiomatosis
Congestive heart failure	Ischemic heart disease with left ventricular failure	Cardiac failures, cardiomyopathy, myocarditis, myxoma of the left heart
Others		Idiopathic pulmonary hemosiderosis, acute respiratory distress syndrome, pulmonary amyloidosis, lymphangioleiomyomatosis, sarcoidosis, idiopathic pulmonary fibrosis, barotrauma, fat embolism of the pulmonary artery, myeloid leukemia, obstructive sleep apnea syndrome

with polyangiitis (GPA); anti-glomerular basement membrane disease (Goodpasture syndrome); congestive heart failure (29%); mixed conditions (26%), including infections, barotrauma, thrombocytopenia; and drug and toxic damage. Finally, 14 patients (12.5%) with no obvious causes of alveolar hemorrhage were regarded idiopathic and virtually analogous to idiopathic pulmonary hemosiderosis [2]. In a less representative and earlier study conducted by Travis et al., 11 patients with DAH had GPA, and four had Goodpasture syndrome, connective tissue diseases, and idiopathic pulmonary hemosiderosis [3]. Furthermore, Jennings reported pauci-immune capillaritis in 8 of the 29 DAH patients [4].

DAH is a rare clinical condition; therefore, to facilitate the differentiation of its causes and management tactics, some authors suggest the isolation of two variants of alveolar hemorrhages, namely, autoimmune and nonimmune, depending on the pathogenesis. Autoimmune diseases include those in the first three groups (Table 4.3.1); all the others are nonimmune diseases [5].

Morphology

Common histopathologic signs of DAH include initial intra-alveolar presence of erythrocytes and fibrin; the subsequent appearance of hemosiderin-loaded macrophages that appear 48–72 h after the hemorrhage (Fig. 4.3.1) represents an obligatory finding. The causes of DAH are very diverse; therefore a microscopic examination enables the detection of additional histological signs that most commonly include pulmonary capillaritis, diffuse alveolar damage, and bland pulmonary hemorrhage. Pulmonary capillaritis is characterized by neurophilic interstitial infiltration, edema, fibrinoid necrosis of the alveolar and capillary walls, and leukocyte fragmentation [6]. In recurrent DAH, interstitial fibrosis is formed in the zones with the most severe hemorrhages (Fig. 4.3.2).

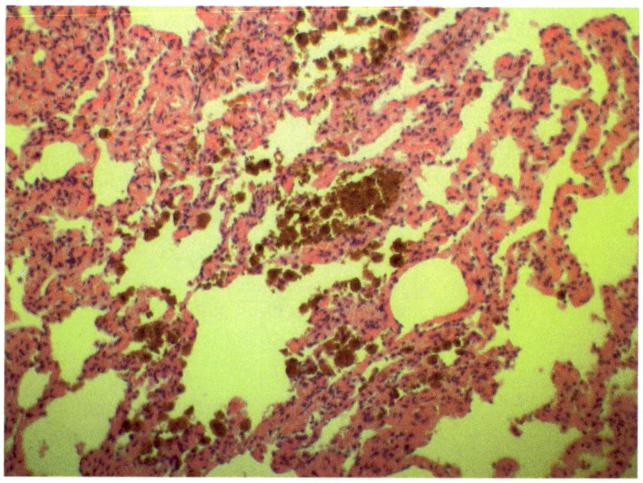

FIG. 4.3.1 Idiopathic pulmonary hemosiderosis. Macrophages loaded with granules of brown pigment (hemosiderin) in the alveoli lumen. Hematoxylin and eosin (H&E) staining, 200×.

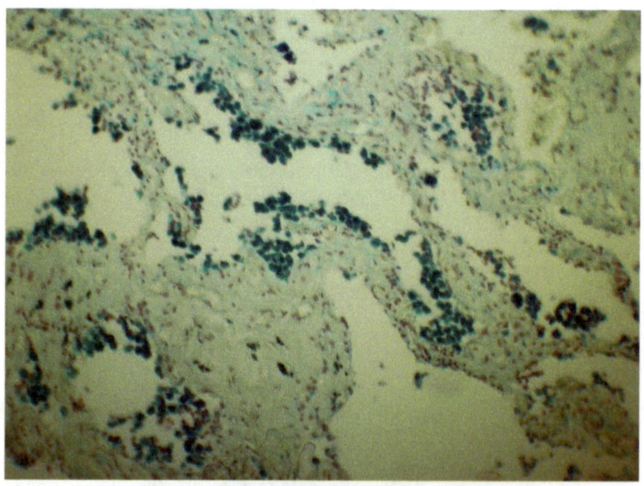

FIG. 4.3.2 Goodpasture syndrome. Hemosiderin-loaded macrophages (*brown* pigment granules) in the alveoli lumen and interstitial fibrosis. Perls' reaction, 200×.

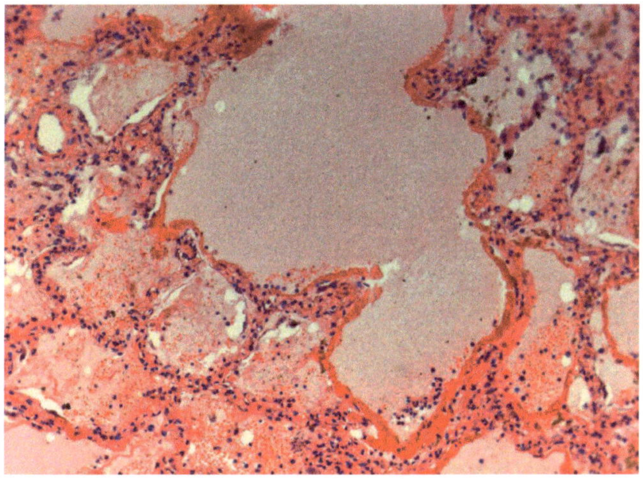

FIG. 4.3.3 H1N1 influenza (day 5 of the disease). Diffuse alveolar damage with DAH. Intra-alveolar edema, hyaline membranes, and erythrocytes in the alveoli lumen. H&E stain, 200×.

In case of diffuse alveolar damage (DAD) during the acute period, in addition to signs of capillaritis and general manifestations of DAH, there are intra-alveolar hyaline membranes, fibrinocellular alveolar edema, denudation of the basement membrane, and microthrombosis (Fig. 4.3.3). Subsequently the formation of alveolar exudate and proliferation of the alveolar epithelium are observed [7].

Bland pulmonary hemorrhage is characterized by the presence of classical signs of DAH (erythrocytes, fibrin, and hemosiderin-laden macrophages in alveolar spaces) without any sign of capillaritis and DAD [6].

Clinical presentation

One of the common clinical symptoms of DAH is hemoptysis; it may be profuse and intense or minor and slow progressing. For autoimmune diseases, a more gradual development of symptoms is characteristic; however, nonimmune conditions that lead to DAHs proceed more aggressively [5]. Up to one-third of DAH cases occur without hemoptysis [8]. Other respiratory clinical manifestations that occur with varying severity are shortness of breath, chest pain, and a nonproductive cough. Respiratory failure can be severe enough to warrant invasive mechanical ventilation (in up to 17% of patients). The severity of the condition in 77% of the cases requires admission to the intensive care unit [2]. The other common symptoms accompanying DAH include fever (usually subfebrile), fatigue, and other signs of blood loss. Some patients have arthralgias and myalgias [9]. Blood tests often reveal anemia and decreased hematocrit value that reflect the volume of blood released into the alveolar space.

DAH is generally a secondary process; therefore symptoms of the underlying disease also occur. However, the degree of their severity is often low and is masked by the dominant signs of alveolar hemorrhages. In addition, DAH may be the first clinical manifestation of systemic diseases that requires an in-depth diagnostic search [10].

A comparison of the autoimmune and nonimmune cases of DAH showed that patients with autoimmune mechanisms had slower symptom progression (i.e., the time from the onset of first symptoms to hospitalization was 19 days for the autoimmune cases and 4 days for the nonimmune cases) and a higher prevalence of anemia, proteinuria, and hematuria. These patients also had low hemoglobin indexes (the highest recorded patient value was 89 g/L), and 75% patients had microhematuria. By contrast, only 2 of the 34 patients with a nonimmune condition had microhematuria and a hemoglobin level <89 g/L [5].

Diagnosis

DAH diagnosis was made based on a clinical presentation with evidence of alveolar hemorrhages confirmed with characteristic radiographic features and bronchoalveolar lavage (BAL) findings. As per R. Lazor, DAH should be considered for the differential diagnosis when at least two of the three main manifestations of DAH, including hemoptysis, anemia, and bilateral characteristic interstitial changes in the lungs, are present [1].

Bronchoscopy and *BAL* are recommended to be performed for all patients with suspected DAH within the first 48h of symptom onset. The BAL technique involves the administration of 4–5 physiological saline portions of 30–60mL each to obtain 100–300mL of lavage fluid when the bronchoscope is inserted into the "wedge position" [11]. The diagnosis of DAH is confirmed if the hemorrhagic character of the fluid is retained or increased in all the portions. If BAL is conducted during the subacute stage, the presence of a large number of siderophages is expected [8]. The second important task of bronchoscopy with suspected DAH is to rule out the infectious process and analyze the bronchial contents and the BAL fluid for bacterial and viral pathogens, fungi, and *Pneumocystis jirovecii* pneumonia. Most experts agree that transbronchial lung biopsy is not necessary for confirming DAH; however, it can be performed if the primary disease is undiagnosed [12]. Surgical lung biopsy may not always be necessary for diagnosing the primary disease because first, the histological picture is often nonspecific and, second, several processes that manifest in DAH have characteristic clinical and laboratory markers that are sufficient for diagnosis [13].

Spirometry is not significant for establishing the diagnosis. In the chronic course of DAH, a restrictive pattern may appear owing to the development of pulmonary fibrosis. Evaluation of a diffusion test wherein DLCO increase is commonly observed is more informative. This is attributable to the accumulation of erythrocytes in the alveoli and the increased absorption of CO by hemoglobin [14]. For the same reason (absorption by erythrocytes) an exhaled nitric oxide level is decreased. In contrast to the study of diffusion capacity, the test for exhaled NO can be performed in the intensive care units with portable devices that increase its diagnostic value [8].

High-resolution computed tomography

The HRCT picture in DAH is nonspecific and includes diffuse or patchy areas of reduced attenuation that can vary in intensity from light ground-glass opacity to consolidation, depending on the degree to which the alveoli is filled with blood (Figs. 4.3.4 and 4.3.5) [15].

In some patients, the radiographic pattern of DAH is limited only by centrilobular nodules of decreased attenuation (Fig. 4.3.6); however, it can be combined with consolidation and ground-glass opacity (Fig. 4.3.5).

The prediction of the distribution of changes is difficult; however, most authors emphasize the presence of perihilar distribution, the subpleural sparing, and the unaffected apices of the lungs and costophrenic angles [16] (Figs. 4.3.5, 4.3.7, and 4.3.8).

However, these aspects cannot be prioritized while establishing a diagnosis because often diffuse lesion of the lungs with the capture of these zones is also observed. Pleural effusion is not typical of DAH and can occur in patients with heart failure or kidney failure (Fig. 4.3.8).

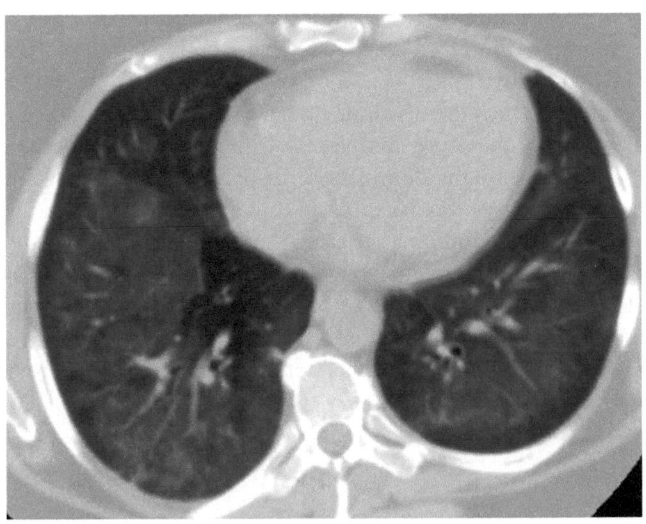

FIG. 4.3.4 DAH in a patient with systemic lupus erythematosus. Diffuse bilateral ground-glass opacity with thickening of the interlobular septa and subpleural sparing.

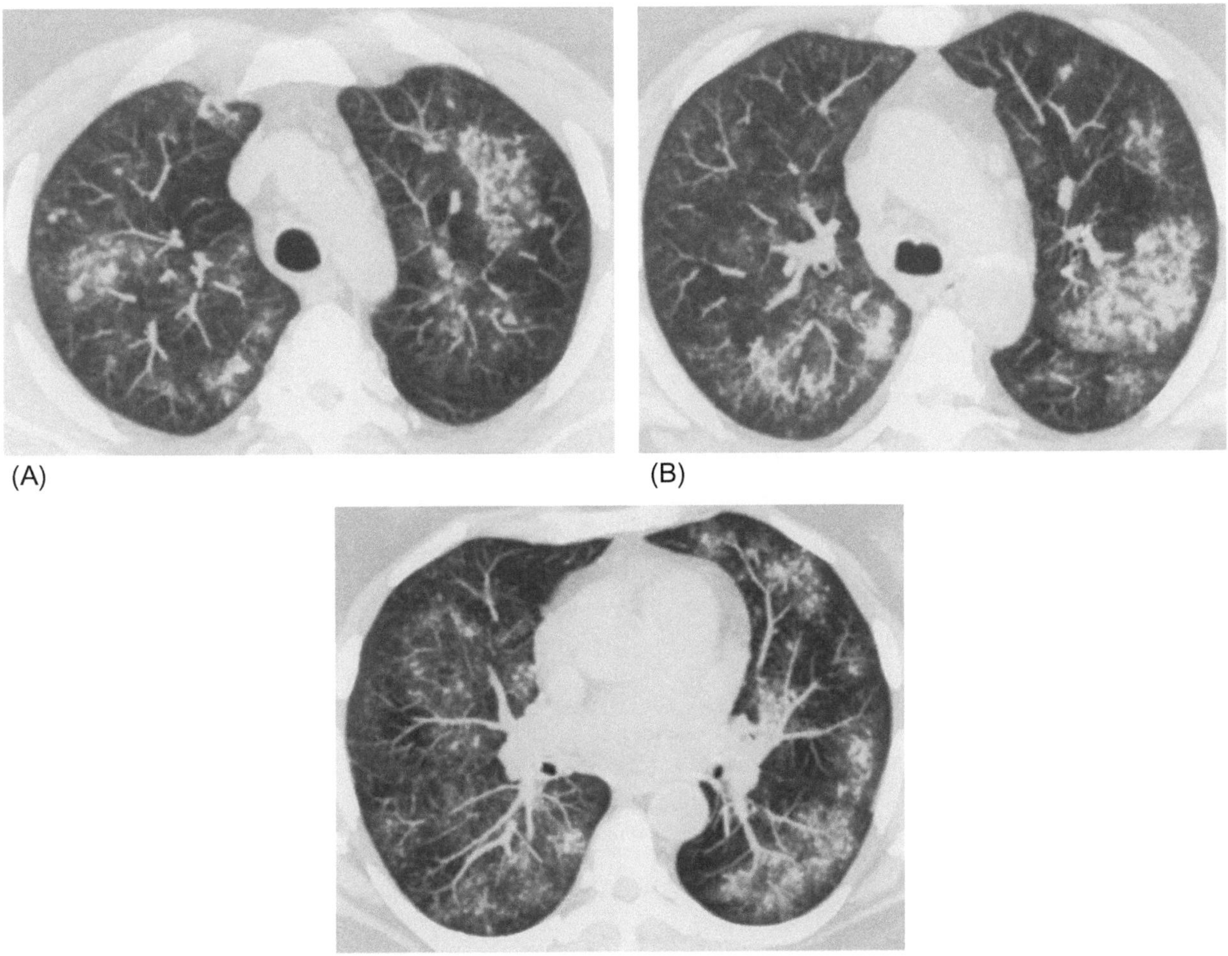

(A) (B) (C)

FIG. 4.3.5 Microscopic polyangiitis with DAH. (A, B, and C) Patchy areas of reduced attenuation and multiple polymorphic nodules. The changes only mildly affect the subpleural divisions.

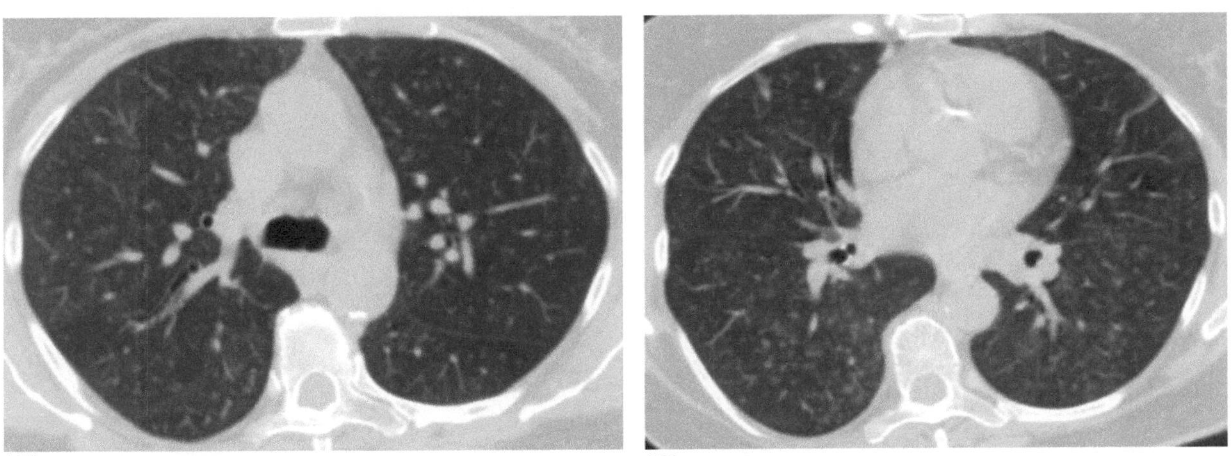

(A) (B)

FIG. 4.3.6 Initial manifestations of diffuse alveolar hemorrhage in a patient with myeloid leukemia and blast crisis. Multiple ill-defined intralobular nodules with diffuse distribution.

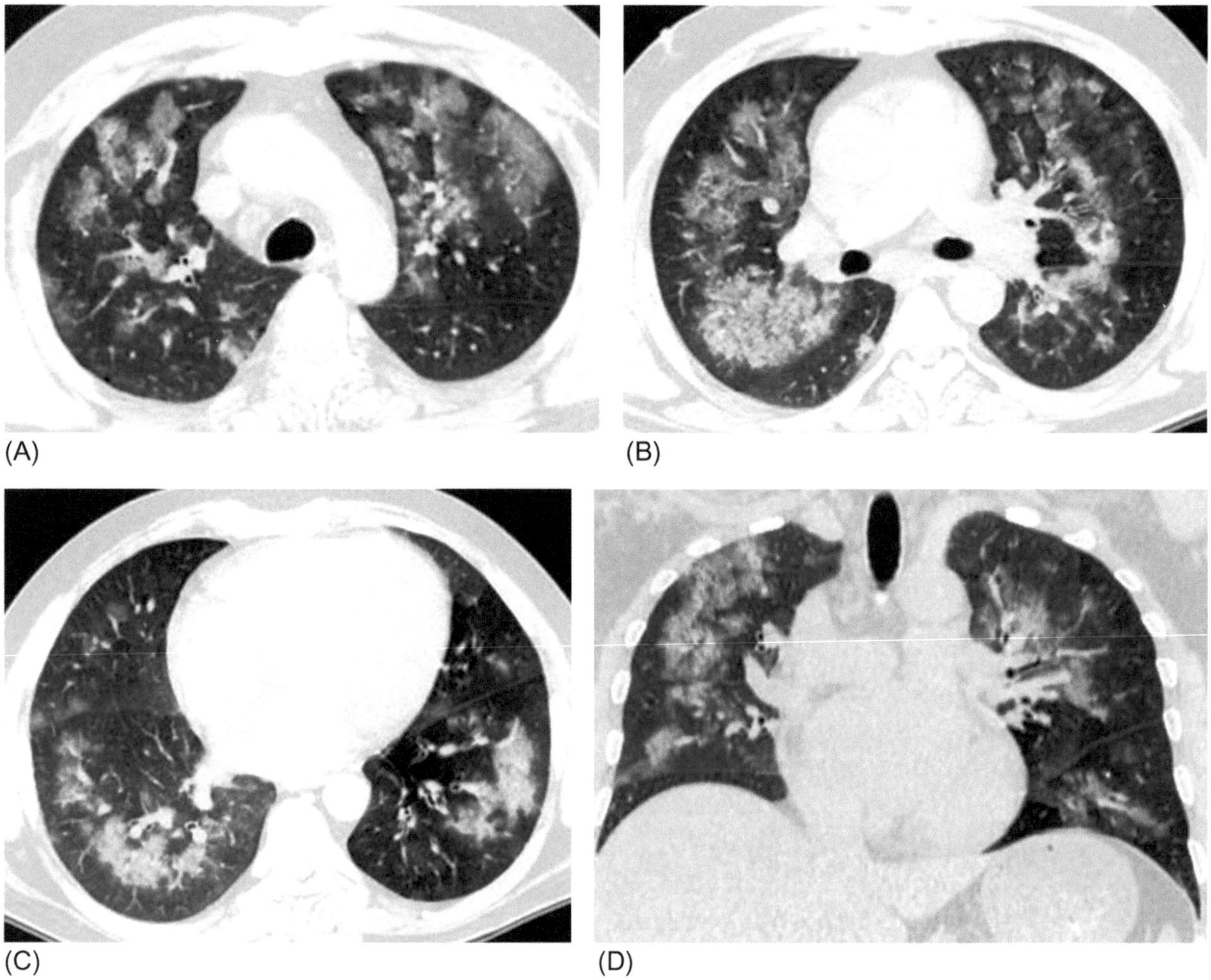

FIG. 4.3.7 DAH in a 30-year-old patient. Bilateral symmetrical patchy areas of ground-glass opacity with thickening of the interlobular septa and perihilar distribution (A–D).

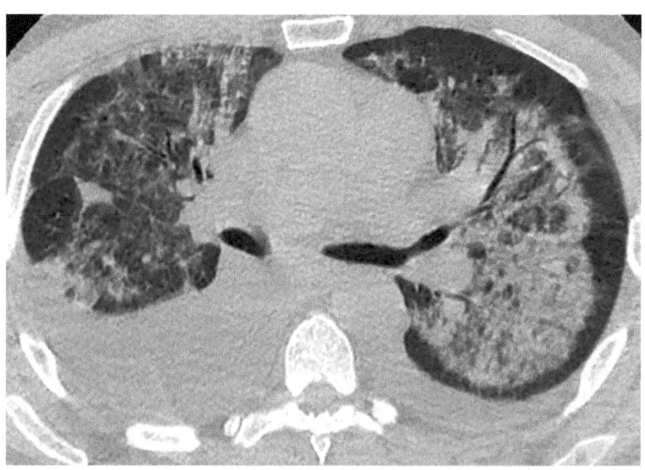

FIG. 4.3.8 Diffuse alveolar hemorrhage in a patient with severe left ventricular failure. Bilateral diffuse increased lung attenuation from the degree of ground-glass opacity on the right to consolidation on the left. Perihilar distribution. Bilateral hydrothorax. Changes are indistinguishable from those of pulmonary edema.

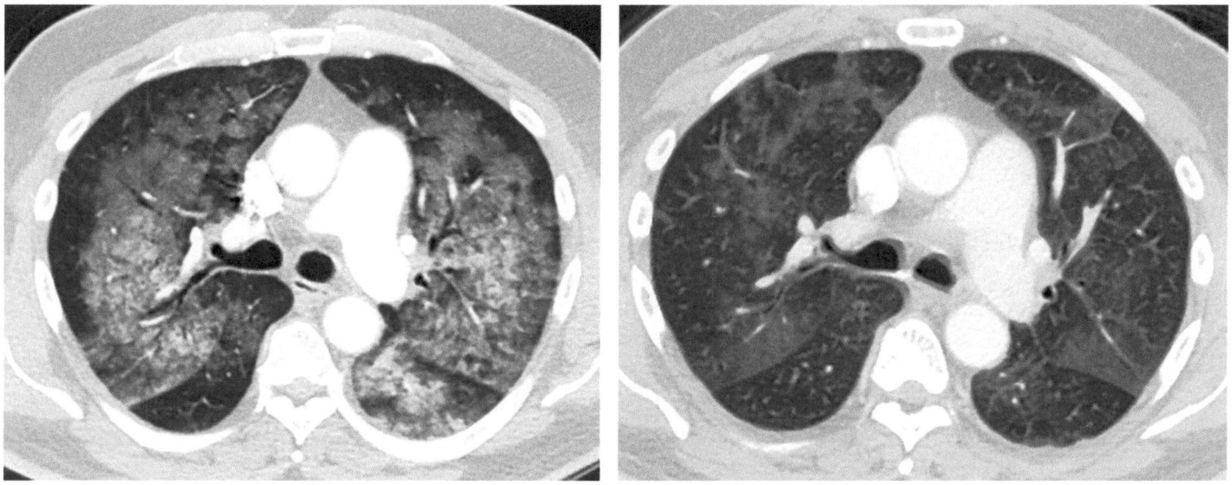

(A) (B)

FIG. 4.3.9 DAH in a patient taking anticoagulants. (A) Day 2 of the disease. Bilateral symmetrical areas of increased lung attenuation. Subpleural sparing. (B) Day 10 of the disease, separate flaps of ground-glass opacity with slight thickening of the intralobular septa persist.

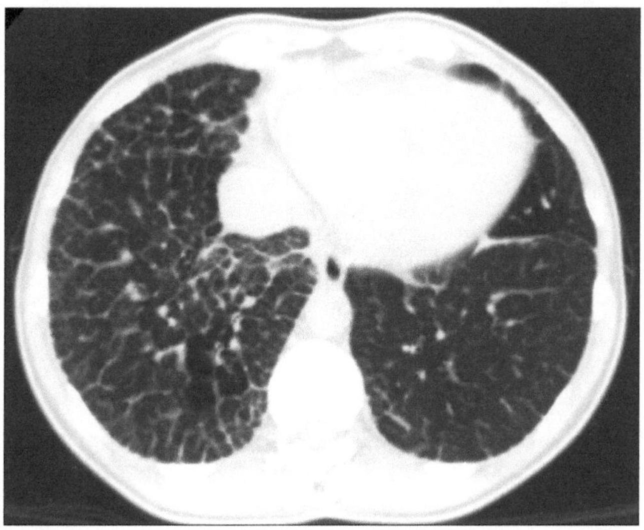

FIG. 4.3.10 Patient with idiopathic pulmonary hypertension and recurrent alveolar hemorrhages. Pronounced thickening of the interlobular septa, mainly in the right lower lobe, as a manifestation of interlobular fibrosis.

Few days after the DAH episode, there is accumulation of siderophages in the lung parenchyma and interstitium; this changes the radiological presentation in terms of the thickening of the interlobular septa and resorption of the consolidated areas, creating a picture of crazy paving. These changes persist till 10–14 days after the hemorrhage (Fig. 4.3.9). Repeated hemorrhages usually cause the deposition of hemosiderin in the interstitial space with the development of reticular changes and even sections of the honeycombing that may resemble idiopathic pulmonary fibrosis or nonspecific interstitial pneumonia [16] (Fig. 4.3.10).

Differential diagnosis

The first question that should be addressed while making a differential diagnosis is whether DAH is present. The acute or subacute bilateral, generalized process in the lungs with HRCT signs of interstitial and alveolar lesions should always be considered in terms of the possibility of DAH. The absence of hemoptysis makes this diagnosis less likely; however, it does not rule it out. Only the absence of red blood cells and/or siderophages in the BAL fluid can allow the exclusion of DAH diagnosis. However, if BAL is performed with 1–2 portions, rather than consistently with 4–5 portions from one segment, erythrocytosis of the lavage fluid may not necessarily be associated with DAH; it may be a consequence of blood aspiration with local pulmonary hemorrhage. Another important aspect in the assessment of BAL composition is staining macrophages by Perls to identify hemosiderin. Rabe et al. showed that in 11% of DAH cases with severe respiratory failure and the microscopic absence of red blood cells in the BAL fluid, hemosiderin staining yielded positive results in a significant number of alveolar macrophages [17].

Primarily, DAH should be differentiated from other acute lung diseases that have a similar clinical and radiological pattern, such as infections, especially pneumocystic pneumonia, hypersensitivity pneumonitis (HP), acute respiratory distress syndrome (ARDS), acute interstitial pneumonia, cardiogenic pulmonary edema, and diffuse pulmonary lymphangiomatosis.

For example, intralobular foci of opacification and unaffected subpleural zones, characteristic findings in DAH are also typically seen in HP (Fig. 4.3.11). The distribution pattern of cardiogenic pulmonary edema can also resemble DAHs, tending toward a symmetrical perihilar distribution (Fig. 4.3.12). However, as a rule, with cardiogenic pulmonary edema, there is previous history of heart failure; commonly there is pleural effusion and cardiomegaly, and the edema zones are subject to the law of gravitation and rapid dynamics during therapy. In *Pneumocystis jirovecii* pneumonia, diffuse zones of ground-glass opacity with subpleural sparing may be present (Fig. 4.3.13); however, the patient usually has a history of HIV or immunosuppressant therapy, fever, and the presence of *Pneumocystis jirovecii* in the BAL fluid. Diffuse pulmonary lymphangiomatosis (DPL) is an orphan disease that is based on the hereditary pathology of the lymphatic vessels of the lungs; it can be acute with rapidly progressive respiratory failure caused by pulmonary lymphedema [18] and have a CT

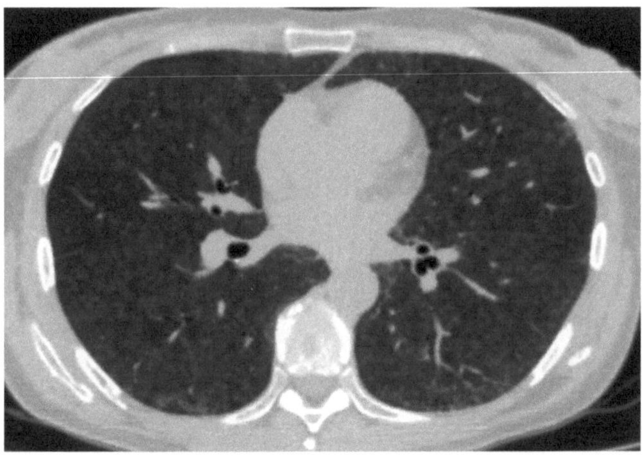

FIG. 4.3.11 Development of subacute hypersensitivity pneumonitis because of contact with birds. Multiple poorly defined centrilobular nodules that sometime merge with each other.

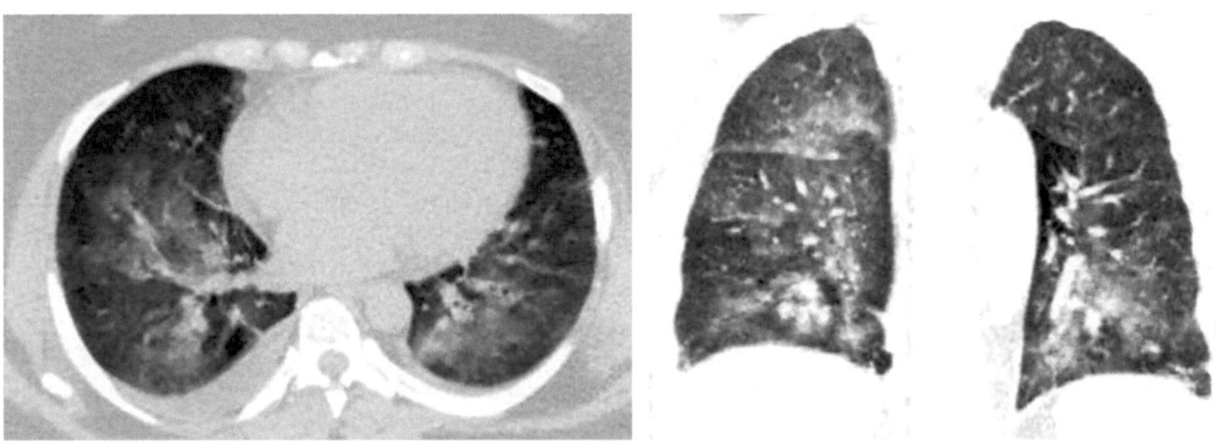

(A) (B)

FIG. 4.3.12 Pulmonary edema in a 46-year-old patient with acute myocarditis. (A and B) Bilateral symmetrical zones of increased lung attenuation, thickening of the interlobular and interacinar septa, thickening of the vascular-bronchial bundles, and subpleural sparing. Bilateral minor hydrothorax. (B) The changes are more pronounced in the lower lobes owing to gravitational forces.

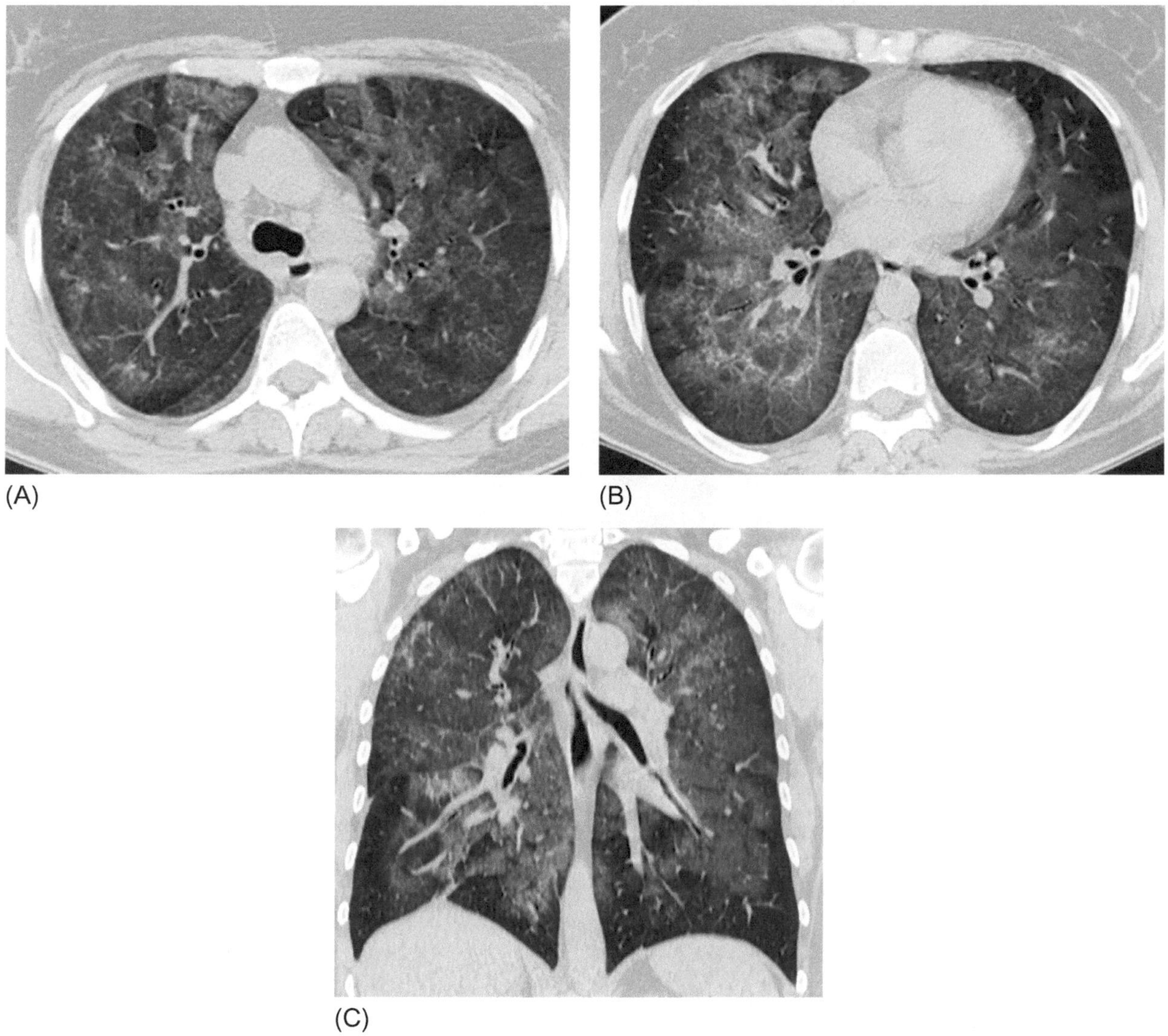

FIG. 4.3.13 *Pneumocystis jirovecii* pneumonia, as the first manifestation of AIDS in a HIV patient. (A–C) Bilateral zones of diffuse ground-glass opacity with thickened interlobular and intralobular septa, ill-defined centriacinar nodules, and subpleural sparing. (A) In the anterior regions, air traps that are not characteristic of DAH are visible. (C) Basal segments are predominantly free from damage.

scan that resembles that of a patient with DAH. Furthermore, chylothorax, usually considered an typical sign of DPL [19], may be absent (Fig. 4.3.14). However, in this case, there was a vivid clinical symptom of profuse lymphoptisis, with signs of plastic bronchitis that enabled immediate identification of this pathology [20].

In any such case, serial BAL enabled the exclusion or confirmation of DAH. Additional criteria for the diagnosis of DAH are normal or increased DLCO and decreased level of exhaled NO. It can be more difficult to differentiate DAH as an isolated morphological substrate from diseases that are accompanied by DAD. For example, acute interstitial pneumonia; lung damage in influenza, especially those caused by the H1N1 virus; and ARDS are often accompanied by alveolar hemorrhages, along with other signs of DAD (Fig. 4.3.15). In the study of BAL in influenza A/H1N1 and severe pneumonia patients, lavage fluid usually contains erythrocytes; however, in general, neutrophilia and lymphocytosis are predominantly present [11,21]. Nevertheless the dominance of DAH is possible, especially with influenza infection. In this case the BAL fluid acquires a pronounced hemorrhagic character [22].

The differential diagnosis of the cause of DAH can be more challenging. A detailed anamnesis can often provide information to the clinician about the likely underlying diagnosis. In most cases of systemic vasculitis and connective tissue diseases, the preceding extrapulmonary symptoms are characteristic, such as skin manifestations and lesions in the upper

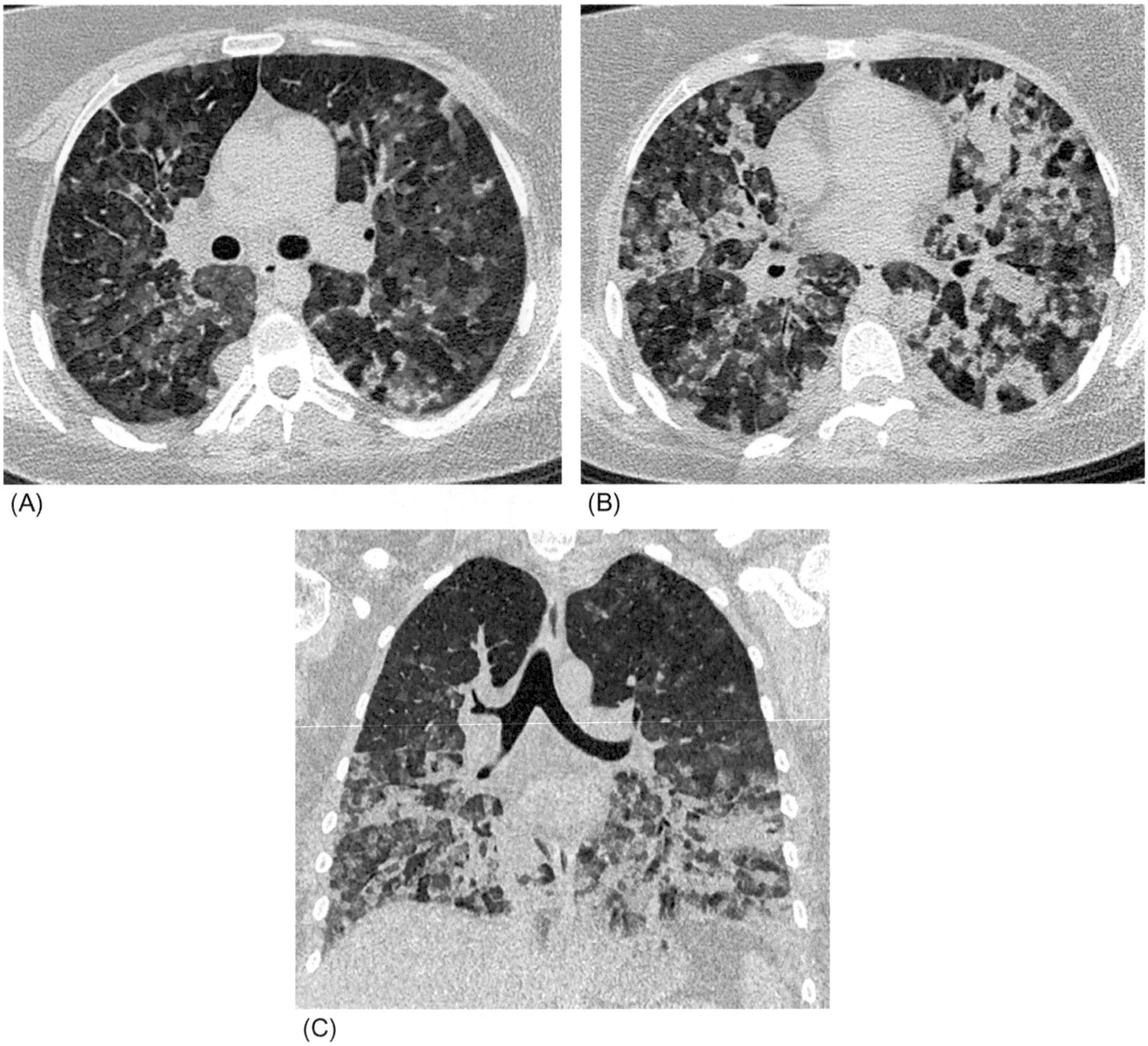

FIG. 4.3.14 Diffuse pulmonary lymphangiomatosis with pulmonary lymphedema in a 24-year-old patient. (A–C) Bilateral multiple patchy areas of ground-glass opacity (more expressed at the upper level) and consolidation (mainly in the lower segments). Thickened interlobular and intralobular septa. Maximum changes are expressed in the central and lower segments; the subpleural and upper zones are affected to a lesser degree.

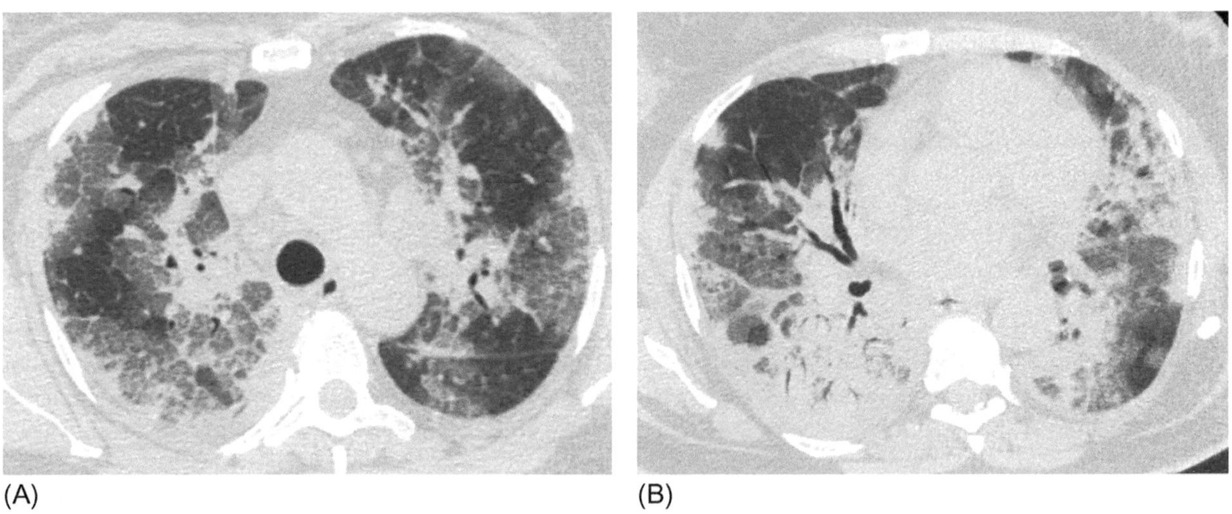

FIG. 4.3.15 H1N1 influenza with diffuse alveolar damage and DAH. Bilateral ground-glass opacity with thickened interlobular septa. (A) In the lower lobes, there is ground-glass opacity and consolidation with symptoms of bronchogram. (B) Changes present with predominance in the posterior and subpleural areas.

TABLE 4.3.2 List of laboratory tests for diagnosing the cause of DAH

Disease	Test
Systemic vasculitis	Antiproteinase 3 antibodies, antimyeloperoxidase antibody, blood eosinophils, eosinophilic cationic protein, cryoglobulins
Connective tissue diseases	ACCP, rheumatoid factor, antinuclear antibodies, lupus anticoagulant, anti-double-stranded DNA antibodies
Other autoimmune diseases	Circulating immune complexes, anti-GBM antibodies, antibodies to cardiolipin, beta-2 glycoprotein 1, antibodies to gliadin and tissue transglutaminase, level of complement (C3 and C4), immunoglobulins, IgE common and specific to cow milk, IgG4
Infectious diseases	Antibodies to leptospire, markers of hepatitis B and C, HIV PCR to the influenza virus, anti-H1N1, influenza A and B antibodies
Hemostasis disorders	Platelets, coagulogram

ACCP, antibodies to cyclic citrullinated peptide; *GMB*, glomerular basement membrane.

respiratory tract, kidneys, nervous system, joints, and muscles. With coagulopathies, bleeding from other sources (e.g., uterine bleeding and nasal bleeding) or thrombosis and miscarriage with an antiphospholipid syndrome are traced based on patient recollection. The history of bone marrow transplantation and intake of any medications must necessarily be mentioned when collecting history from patients with suspected DAH. If at the stage of history taking and examination it is not possible to reach a target search for the cause of alveolar hemorrhages, it is necessary to prescribe several laboratory studies consistently or simultaneously (in critical condition) that can help identify the "guilty" diseases, including those of subclinical forms. In addition to the general tests that assess the functional state of the organs, hemostasis, and inflammatory status, they include antinuclear and anti-DNA antibodies, anti-glomerular basal membrane antibodies, antibodies to the neutrophils cytoplasm (p-ANCA and c-ANCA), circulating immune complexes, anti-cyclic citrullinated peptide antibodies (ACCP), lupus anticoagulant, antibodies to cardiolipin, beta-2 glycoprotein 1, antibodies to gliadin and tissue transglutaminase, the level of complement and cryoglobulin, immunoelectrophoresis, total and specific IgE for cow milk, and IgG4 (Table 4.3.2). The frequent association of leptospirosis with DAH requires the study of the antibodies for this pathogen. The BAL fluid not only should be subjected to cytological analysis with Perls' staining but also must be examined for infectious agents, including opportunistic infections. In DAH patients with kidney damage who do not have a definite diagnosis, a kidney biopsy is recommended because it provides more diagnostic information than a lung biopsy. If all the possible causes of DAH have been ruled out, a diagnosis of *idiopathic pulmonary hemosiderosis* (IPHS) can be made. This is a rare pathology with an incidence of 0.2–1.2 cases per one million population every year [1]. Most IPHS patients are young children, and IPHS accounts for about 8% of all the interstitial lung diseases in children [23]. Adults are affected by IPHS less frequently, and most patients are men aged <30 years. Approximately 50% of the patients have increased IgA levels, and IPHS occurs along with coeliac disease [24]. Clinically, IPHS is manifested by repeated episodes of DAH with respiratory failure, anemia, and hemoptysis.

Treatment and prognosis

The treatment of DAH depends on the underlying disease. For DAH associated with autoimmune diseases, steroid pulse therapy, immunosuppressants, and plasmapheresis are usually administered. Thrombocytopenia and coagulopathy may require replacement therapy with platelets and fresh frozen plasma. Patients with acute life-threatening DAH usually require mechanical lung ventilation. Successful use of recombinant activated factor VII (RAF VII) in DAH has been reported. Mandal et al. were able to arrest massive pulmonary hemorrhage in a patient with microscopic polyangiitis and DAH with intravenous bolus administration of two doses of this drug at 90 μg/kg at 2-h intervals [25]. The administration of RAF VII to the lower respiratory tract of a patient with systemic lupus erythematosus also led to successful arrest of continued alveolar hemorrhages [26]. Successful treatment results have been reported for the intrabronchial administration of RAF VII in six children with DAH of different origins. The development of alveolar hemorrhages was halted in all the patients, with no significant adverse effects [27].

Although conventional hemostatics are used in most DAH cases, limited evidence exists in support of their efficiency.

The prognosis in DAH patients depends on the underlying disease. Ioachimescu believes that patients with systemic lupus erythematosus and anti-glomerular basement membrane disease (about 80%) have the highest 5-year survival rate, while those with IPHS (5%–15%) have the lowest 5-year survival [6].

References

[1] Lazor R. Alveolar haemorrhage syndromes. In: Cordier JF, editor. Orphan lung diseases. ERS monograph. 2011. p. 16–32.

[2] de Prost N, Parrot A, Cuquemelle E, Picard C, Antoine M, Fleury-Feith J, et al. Diffuse alveolar hemorrhage in immunocompetent patients: etiologies and prognosis revisited. Respir Med 2012;106(7):1021–32.

[3] Travis WD, Colby TV, Lombard C, Carpenter HA. A clinicopathologic study of 34 cases of diffuse pulmonary hemorrhage with lung biopsy confirmation. Am J Surg Pathol 1990;14(12):1112–25.

[4] Jennings CA, King Jr TE, Tuder R, Cherniak RM, Schwarz MI. Diffuse alveolar hemorrhage with underlying isolated, pauciimmune pulmonary capillaritis. Am J Respir Crit Care Med 1997;155(3):1101–9.

[5] de Prost N, Parrot A, Cuquemelle E, Picard C, Cadranel J. Immune diffuse alveolar hemorrhage: a retrospective assessment of a diagnostic scale. Lung 2013;191(5):559–63.

[6] Colby TV, Fukuoka J, Ewaskow SP, Helmers R, Leslie KO. Pathologic approach to pulmonary hemorrhage. Ann Diagn Pathol 2001;5(5):309–19.

[7] Beasley MB. The pathologist's approach to acute lung injury. Arch Pathol Lab Med 2010;134(5):719–27.

[8] Ioachimescu OC, Stoller JK. Diffuse alveolar hemorrhage: diagnosing it and finding the cause. Cleve Clin J Med 2008;75(4). 258, 260, 264-5 passim.

[9] Serisier DJ, Wong RC, Armstrong JG. Alveolar hemorrhage in antiglomerular basement membrane disease without detectable antibodies by conventional assays. Thorax 2006;61(7):636–9.

[10] Lara AR, Schwarz MI. Diffuse alveolar hemorrhage. Chest 2010;137(5):1164–71.

[11] Meyer KC, Raghu G, Baughman RP, Brown KK, Costabel U, du Bois RM, et al. An official American Thoracic Society clinical practice guideline: the clinical utility of bronchoalveolar lavage cellular analysis in interstitial lung disease. Am J Respir Crit Care Med 2012;185(9):1004–14.

[12] Collard H, King T, Schwarz M. Diffuse alveolar hemorrhage and rare infiltrative disorders of the lung. In: Mason R, Broaddus VC, Martin T, King T, Schraufnagel D, Murray J, et al., editors. Murray and Nadel's textbook of respiratory medicine. 5th ed. Philadelphia: Elsevier Health Sciences; 2010. p. 1449–68.

[13] Schreiber J, Knolle J, Kachel R, Schück R. Differential diagnosis of diffuse pulmonary haemorrhage. Pneumologie 2006;60(6):347–54.

[14] Ewan PW, Jones HA, Rhodes CG, Hughes JM. Detection of intrapulmonary hemorrhage with carbon monoxide uptake. Application in Goodpasture's syndrome. N Engl J Med 1976;295(25):1391–6.

[15] Castañer E, Alguersuari A, Andreu M, Gallardo X, Spinu C, Mata JM. Imaging findings in pulmonary vasculitis. Semin Ultrasound CT MR 2012;33(6):567–79.

[16] Webb R, Higgins C. Thoracic imaging: pulmonary and cardiovascular radiology W&K. 2nd ed. Philadelphia, PA: Lippincott Williams & Wilkins; 2011.

[17] Rabe C, Appenrodt B, Hoff C, Ewig S, Klehr HU, Sauerbruch T, et al. Severe respiratory failure due to diffuse alveolar hemorrhage: clinical characteristics and outcome of intensive care. J Crit Care 2010;25(2):230–5.

[18] Du MH, Ye RJ, Sun KK, Li JF, Shen DH, Wang J, et al. Diffuse pulmonary lymphangiomatosis: a case report with literature review. Chin Med J (Engl) 2011;124(5):797–800.

[19] Kadakia KC, Patel SM, Yi ES, Limper AH. Diffuse pulmonary lymphangiomatosis. Can Respir J 2013;20(1):52–4.

[20] Nair LG, Kurtz CP. Lymphangiomatosis presenting with bronchial cast formation. Thorax 1996;51(7):765–6.

[21] Faverio P, Aliberti S, Ezekiel C, Messinesi G, Brenna A, Pesci A. Influenza A/H1N1 severe pneumonia: novel morphocytological findings in bronchoalveolar lavage. Interdiscip Perspect Infect Dis 2014;2014:470825.

[22] Yokoyama T, Tsushima K, Ushiki A, Kobayashi N, Urushihata K, Koizumi T, et al. Acute lung injury with alveolar hemorrhage due to a novel swine-origin influenza A (H1N1) virus. Inter Med 2010;49(5):427–30.

[23] Boccon-Gibod L, Couvreur J. Results of lung biopsy in interstitial pneumopathies in children. A report on 100 cases. Ann Med Interne (Paris) 1979;130(11):501–6.

[24] Wright PH, Menzies IS, Pounder RE, Keeling PW. Adult idiopathic pulmonary haemosiderosis and coeliac disease. Q J Med 1981;50(197):95–102.

[25] Mandal SK, Sagar G, Sahoo M, Jasuja S. Recombinant activated factor VII for diffuse alveolar hemorrhage in microscopic polyangiitis. Indian J Nephrol 2012;22(2):130–2.

[26] Alabed IB. Treatment of diffuse alveolar hemorrhage in systemic lupus erythematosus patient with local pulmonary administration of factor VIIa (rFVIIa): a case report. Medicine (Baltimore) 2014;93(14):e72.

[27] Park JA, Kim BJ. Intrapulmonary recombinant factor VIIa for diffuse alveolar hemorrhage in children. Pediatrics 2015;135(1):e216–20.

Chapter 5

Amyloidosis

Alexander Averyanov[a,b], Evgeniya Kogan[c], Victor Lesnyak[d], Igor E. Stepanyan[e]

[a]Clinical Department, Pulmonology Research Institute under FMBA of Russia, Moscow, Russia, [b]Pulmonary Division, Federal Research Clinical Center under FMBA of Russia, Moscow, Russia, [c]Anatomic Pathology Department, Sechenov University, Moscow, Russia, [d]Radiology Department, Federal Research Clinical Center under FMBA of Russia, Moscow, Russia, [e]Central TB Research Institute, Moscow, Russia

Amyloidosis represents a group of heterogeneous diseases in which amyloid, consisting of improperly packed fibrillar proteins, is deposited in various organs [1]. It is a rare disease in Europe, with about 0.4 new cases per 100,000 population detected annually [2]. The prevalence of the most common form, systemic immunoglobulin light-chain (AL) amyloidosis, is 3–5 cases per 1 million population [3]. Cases of pulmonary amyloidosis appear to be quite rare, accounting for only 1% of diffuse lung diseases, although the true frequency of amyloidosis in the lung remains unknown [4].

Amyloidosis is classified according to the distribution (local or systemic) and type of amyloid precursor proteins involved in the pathological process. Amyloid deposits in various types of amyloidosis are formed by 36 different fibrillar proteins, with 15 types found in systemic amyloidosis, 19 in local, and 2 in amyloidosis associated with deposition of light- or heavy (AL or AH)-chain immunoglobulins. According to the National Amyloidosis Centre of the United Kingdom, five forms of the disease were the most frequent among 5100 patients with systemic amyloidosis [5]:

1. Primary or AL amyloidosis accounts in 68% of cases.
2. Secondary or amyloid protein A (AA) disease in 12%.
3. Amyloidosis transthyretin (ATTR), which may be congenital (mutant, in 6.6%) or wild type (acquired, in 3.2%).
4. Dialysis-associated or β-2 microglobulin amyloidosis in 1.8%.
5. Fibrinogen A α-chain amyloidosis in 1.7%.

Lung lesions are most common in AL, AA, and ATTR types of systemic amyloidosis [6]. A local form of amyloidosis caused by the deposition of a surfactant lipoprotein in the lungs has also been described [4].

Amyloid deposits result from monoclonal hyperproduction of proteins prone to abnormal packaging and aggregation and from defective or incomplete proteolytic degradation of extracellular proteins. The majority of amyloid deposits are apparently formed from heterogeneous components of blood plasma, including both fibrillar proteins and nonfibrillar substances such as glycosaminoglycans and serum amyloid P component (SAP) [7]. AA amyloid is formed from the acute-phase protein serum amyloid A protein (SAA) that is synthesized in the liver under the influence of pro-inflammatory cytokines and chemokines (TNF-α, IL-6, and IL-1) [8]. Thus any chronic inflammatory disease with increased SAA levels can theoretically lead to secondary AA amyloidosis. Localized amyloidosis is associated with in situ production of amyloidogenic light chains by clonal B cells [5]. Amyloid deposits exert pressure on the tissues, damaging them and disrupting the normal function of the organs.

Morphology

In the lungs, amyloid appears as dense, amorphous, eosinophilic masses that also contain lymphocytes and plasma cells. These deposits occur primarily along the reticular fibers in the walls of blood vessels, alveolar septa, walls of the bronchi and bronchioles, and alveolar lumens (Fig. 5.1) [9].

The gold standard for morphological identification of amyloid is the detection of a characteristic color with Congo red stain (Fig. 5.2) and red-green birefringence under cross polarized light (Fig. 5.3). Light-chain deposition disease (LCDD) may histologically resemble nodular pulmonary amyloidosis [10]. However, the deposits in LCDD are Congo red-negative

Difficult to Diagnose Rare Diffuse Lung Disease. https://doi.org/10.1016/B978-0-12-815375-8.00005-4

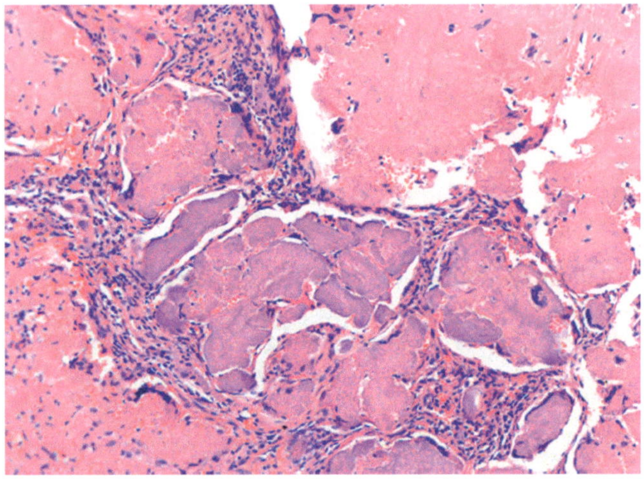

FIG. 5.1 Amyloidosis of the lungs seen in an open lung biopsy specimen with accumulation of amorphous masses in the alveolar lumens. Hematoxylin and eosin stain, 100×. *(Case courtesy of Prof. A.L. Chernyaev, Pulmonology Scientific and Research Institute, Moscow, Russia.)*

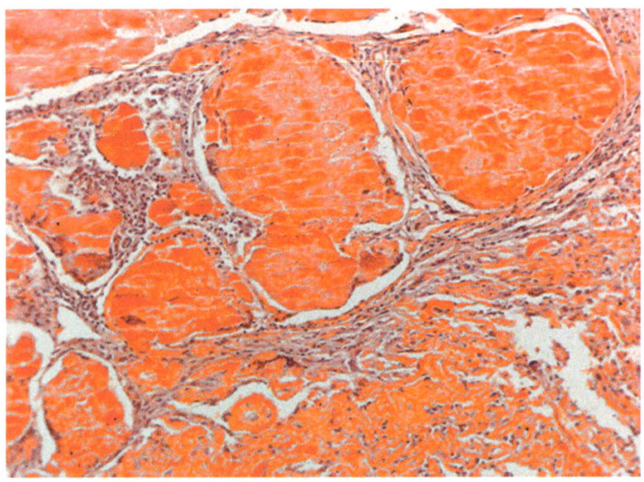

FIG. 5.2 Amyloidosis of the lungs in an open lung biopsy specimen showing deposition of amyloid in the interstitium and the alveolar lumens. Congo red stain, 100×. *(Case courtesy of Prof. A. L. Chernyaev, Pulmonology Scientific and Research Institute, Moscow, Russia.)*

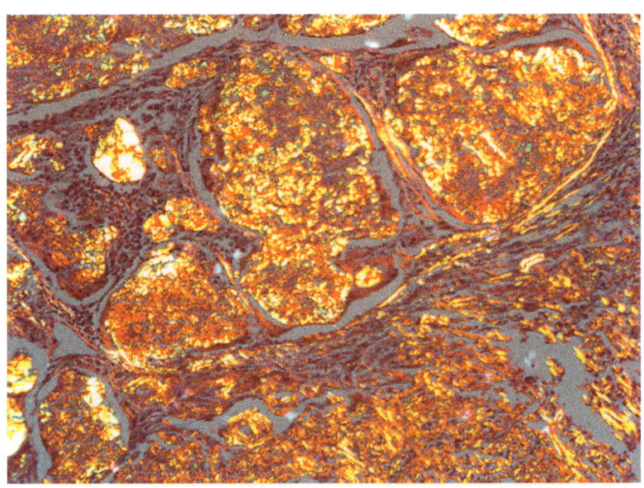

FIG. 5.3 Amyloidosis of the lungs. Birefringence of amyloid masses is seen. Polarization microscopy, 100×. *(Case courtesy of Prof. A. L. Chernyaev, Pulmonology Scientific and Research Institute, Moscow, Russia.)*

and appear on electron microscopy as granular rather than fibrillar material [9]. When stained with hematoxylin and eosin, amyloid fibrils may look like collagen, such that diffuse alveolar-septal amyloidosis is sometimes mistaken for fibrosing interstitial pneumonia [1].

Clinical presentation

Systemic amyloidosis usually manifests clinically in individuals over 50 years of age, while local forms may occur earlier [11]. Systemic amyloidosis may present with generalized symptoms (fatigue and weight loss) or, more commonly, with evidence of organ damage (nephrotic syndrome, restrictive cardiomyopathy, hepatomegaly with elevated liver enzymes, macroglossia, onychodystrophy, periorbital purpura, hemorrhagic diathesis, or neuropathy) [5,12].

Lung involvement occurs in approximately 50% of patients with amyloidosis, mainly in systemic forms of the disease [13]. The symptoms of pulmonary amyloidosis are nonspecific and depend on the volume and form of the amyloid lesions. The nodular pattern of amyloidosis usually presents at the age of 60–70 years and occurs more often in men than in women. Solitary nodes in the lungs are usually asymptomatic and are detected incidentally during radiological examination of the chest [1]. The primary symptom caused by significant deposits of amyloid in the lung parenchyma is dyspnea. In localized and tracheobronchial forms of amyloidosis, dyspnea may be associated with obstruction of large airways with amyloid masses and is usually accompanied by stridor or inspiratory rales on auscultation [1,13].

The tracheobronchial variant of amyloidosis is the rarest form, with reported cases presenting from age 48 to 57 years and with an equal frequency in both sexes. The symptoms depend on the degree of airway damage. Proximally located lesions mostly lead to airway obstruction manifested by dyspnea, cough, hoarseness, and airflow limitation on pulmonary function testing. Hemoptysis often occurs. In cases of distal lesions, symptoms are less pronounced, and spirometry findings may be normal [14]. Local obstruction of a large bronchus can lead to atelectasis and recurrent pneumonia in that lung segment [13,15]. Since dyspnea in systemic amyloidosis is often a manifestation of cardiac involvement, the myocardium should always be assessed in patients with pulmonary amyloidosis.

Diagnosis

High-resolution computed tomography

As a rule, pulmonary amyloidosis is suspected because of abnormalities seen on chest high-resolution computed tomography (HRCT). The most common HRCT findings in amyloidosis of the lungs include the following [16,17]:

- Multiple foci 2–15 mm in size (78%).
- Uneven thickening of interlobular septa and linear intralobular induration (28%).
- Mediastinal lymphadenopathy (28%–75%).
- Zones of consolidation (50%) and masses greater than 3 cm (38%).
- Ground-glass opacities (28%).
- Pleural thickening (31%).
- Calcified nodules or masses (44%).
- Cysts greater than 10 mm in diameter (38%).

More rarely seen, signs of pulmonary amyloidosis are pleural effusion (9%) and traction bronchiectasis [17]. Some authors describe honeycombing [16], but in our experience, only one patient had a single row of merging multiple cysts without a multilayered stratum, which is usually the case with classical honeycombing (Fig. 5.4).

There are four possible patterns of thoracic amyloidosis, depending on the characteristics and location of the protein deposits [18]:

1. The diffuse alveolar and septal pattern (Figs. 5.4 and 5.5) is characterized by nodular thickening along the interlobular and intralobular septa, subpleural deposition of amyloid in the form of consolidation zones, and ground-glass opacities. As a rule, the small (2–4 mm in diameter) nodules mimic sarcoidosis and miliary tuberculosis. The diffuse interstitial pattern is less common than other types and appears only in systemic amyloidosis [19].
2. In the tracheobronchial pattern, amyloid accumulates mainly in the walls of the trachea and bronchi, causing focal thickening and calcification. Such changes are typical of local forms of amyloidosis and are usually not found in the diffuse variants. Bronchial stenosis may be accompanied by atelectasis and recurrent pneumonia [20].
3. The nodular pattern occurs in approximately 45% of cases of amyloidosis in the lung and is characterized by single or multiple clearly defined, round foci up to 15 mm in diameter in the parenchyma, although some can be as large

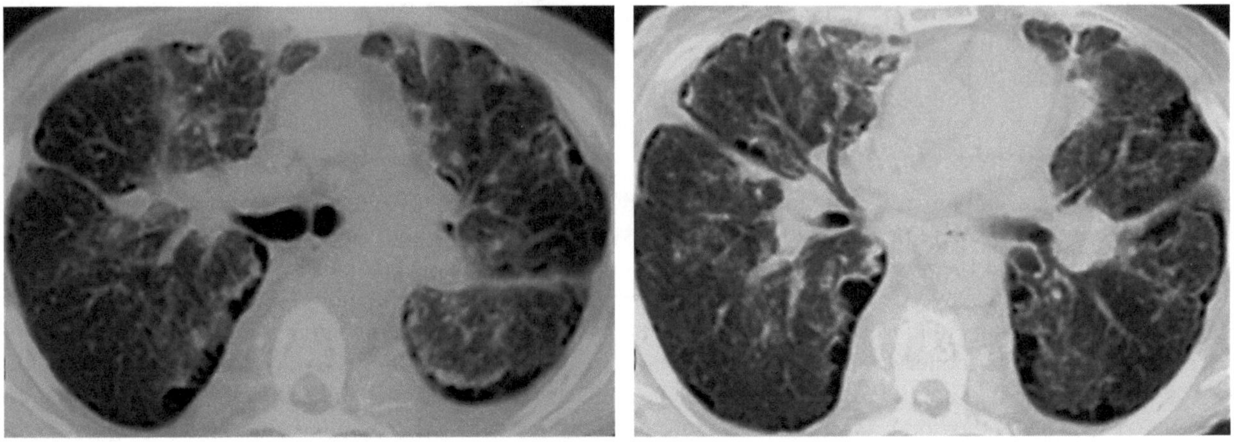

(A) (B)

FIG. 5.4 (A and B) Amyloidosis of the lungs in a diffuse alveolar-interstitial pattern. Features include diffuse; irregular thickening of the interlobar and visceral pleura with marginal calcification; thickening of the interlobular septa; multiple subpleural, single-row, merging cysts; nodules of various shapes located separately; and bronchopulmonary lymphadenopathy.

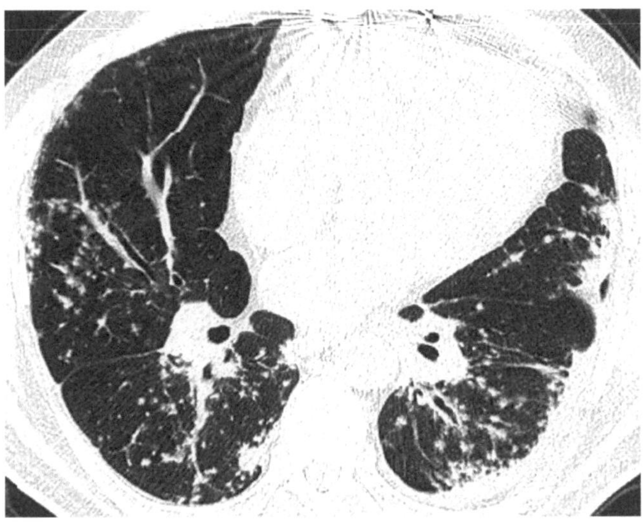

FIG. 5.5 Amyloidosis of the lungs in a patient with rheumatoid arthritis showing a diffuse interstitial pattern. Multiple small nodules are present in a perilymphatic distribution; the foci merge with each other in some places. There is subpleural consolidation. The radiological presentation is very similar to that of sarcoidosis.

as 15 cm. Amyloid nodules are predominantly subpleural, and they usually slowly increase in size [21]. Solitary amyloid nodes are found only in localized forms and may cavitate [17].

4. Lymphadenopathy of the intrathoracic and hilar lymph nodes is observed in approximately 75% of patients with AL amyloidosis and is usually associated with nodules or alveolar and septal lesions. However, isolated involvement of mediastinal lymph nodes without signs of parenchymal disease, especially at disease onset, has been reported, though rarely [13]. A characteristic finding in amyloidosis is calcification in amyloid nodules and lymph nodes, which occurs in about 20%–50% of cases of pulmonary amyloidosis [19,22].

Approximately 20% of patients have a mixed pattern of distribution of pathological changes (Figs. 5.6 and 5.7) [17]. The tracheobronchial pattern rarely occurs in diffuse disease, and the nodular pattern is primarily found in local forms. By contrast, diffuse interstitial lesions and lymphadenopathy are most often found in patients with systemic amyloidosis [16].

Thin-walled cysts in the lungs can be caused by amyloidosis. However, they appear predominantly in patients whose amyloidosis is associated with chronic inflammatory diseases, particularly Sjögren syndrome. The cause of a cyst may not be amyloidosis itself but rather lymphocytic interstitial pneumonia- or mucosa-associated lymphoma [23,24].

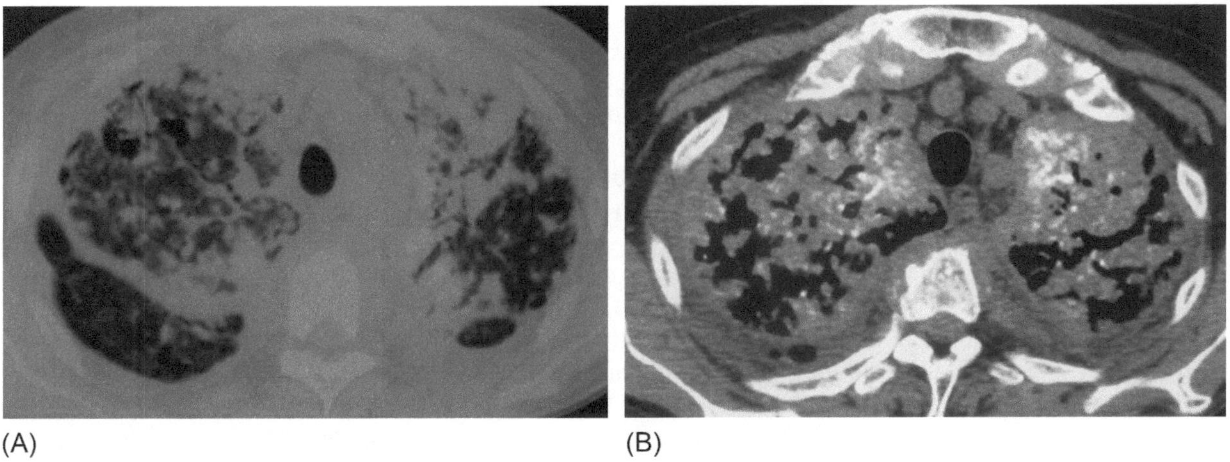

(A) (B)

FIG. 5.6 Amyloidosis of the lungs, mixed pattern. (A) Confluent consolidation zones are located mainly subpleurally and in peribronchovascular areas. (B) Numerous calcifications are seen within amyloid deposits. There is intrathoracic lymphadenopathy.

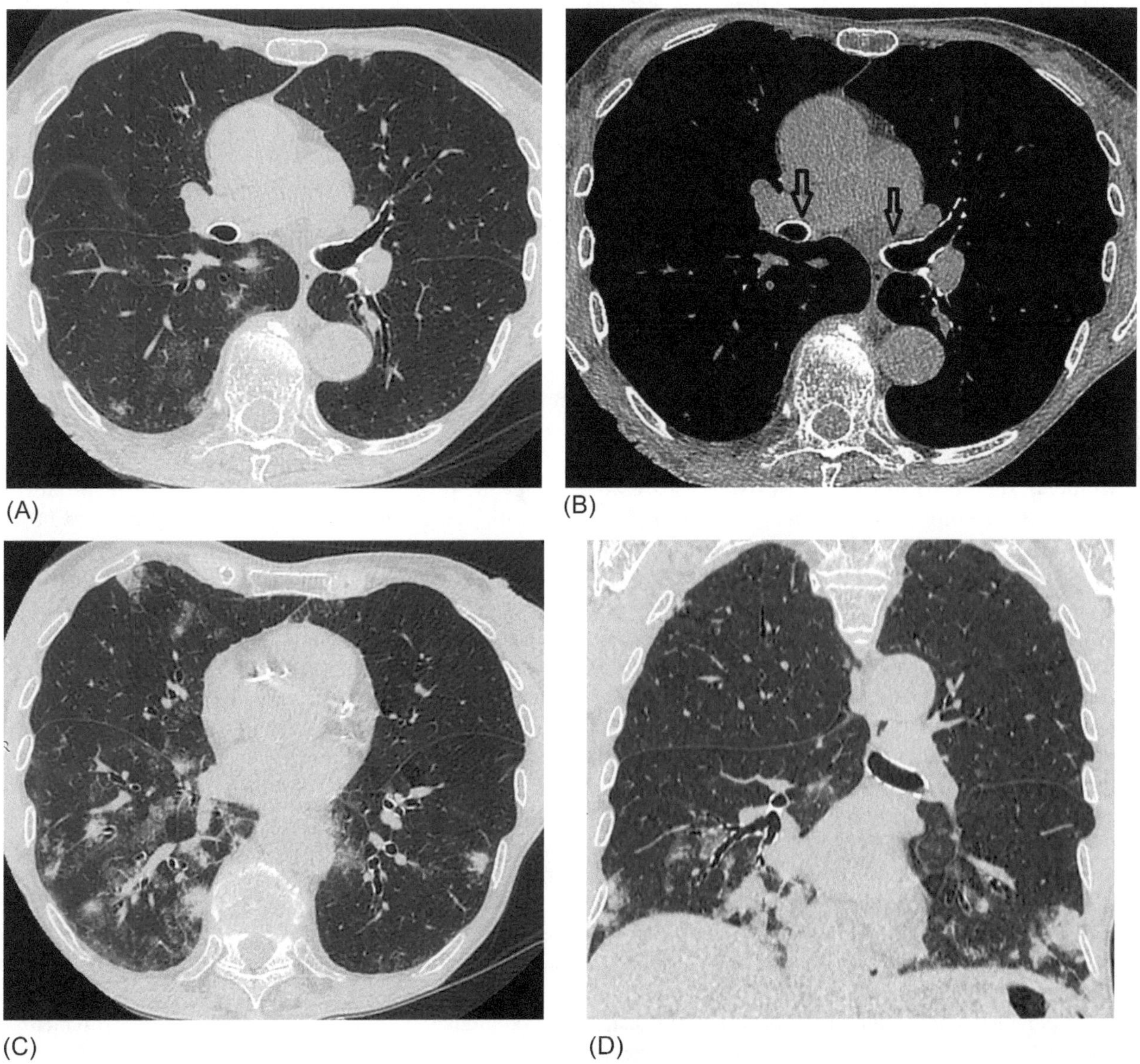

(A) (B) (C) (D)

FIG. 5.7 Amyloidosis AA, mixed pattern. Features include (A) small foci of ground-glass opacities and consolidation in the right lung, (B) calcification of bronchial walls *(arrows)*, (C) multiple foci of consolidation with a halo sign and patchy areas of ground-glass opacity, and (D) nodular areas of consolidation in the lower zones of the lungs and calcification of the walls of large bronchi. *(Case courtesy of Prof. A.L. Chernyaev and Dr. M.V. Samsonova, Pulmonology Scientific and Research Institute, Moscow, Russia.)*

The diagnosis of pulmonary amyloidosis can be strongly suspected based on a characteristic patterns on thoracic HRCT, but it requires histological and immunohistochemical verification. Due to an increased risk of bleeding during biopsy of parenchymal organs, tissue specimens can be obtained from the minor salivary glands, gastric mucosa, rectum, or abdominal fat [25,26], eliminating the need for transbronchial or open lung biopsy. However, for localized tracheobronchial amyloidosis, direct forceps biopsy is fully justified [14]. On visual examination of the bronchial mucosa, waxy or yellowish erythematous nodules are noted that bleed easily on contact and biopsy [27].

In most cases, the diagnosis of amyloidosis is confirmed by histological examination of biopsy specimens stained with Congo red and examined with a Puchtler polarization microscope or with thioflavine fluorescence microscopy. Identification of the amyloid-forming proteins is fundamental for choosing appropriate treatment. Immunohistochemical analysis of amyloid deposits is the simplest and the most often available method to achieve this aim [28]. However, while it is highly effective for AA and most forms of ATTR, the results are often uncertain for AL amyloidosis, which most commonly affects the lungs [5]. A new gold standard for typing amyloid fibrillae is mass spectrometric proteomic analysis [5,29]. After identifying the type of amyloid protein (most often AL or AA), a search should be made for other diseases possibly causing the amyloidosis. For secondary AA amyloidosis, the most frequent underlying conditions are rheumatoid arthritis, bronchiectasis and other chronic infectious diseases, periodic fever syndromes, and Castleman disease [30]. AL amyloidosis may be caused by other diseases with abnormal light-chain production, such as multiple myeloma or monoclonal gammopathy of undetermined significance [5,27]. Therefore, immunoelectrophoresis of urine and serum proteins and determination of serum lambda and kappa light chains are important tests for determining monoclonal protein production [31].

Another study useful in the diagnosis of amyloidosis and for the assessment of its extent and the organs involved is scintigraphy with radiolabeled ^{123}I antibodies to serum amyloid P (SAP), a component of both AA and AL [32]. A more advanced method is dual-modality ^{123}I-labeled SAP single-photon emission CT that enables clearer localization of isotope-captured amyloid foci. This study is also useful in monitoring response to therapy [5].

According to the recommendations of the Nomenclature Committee of the International Society of Amyloidosis, the diagnosis should include the name of the amyloid fibrillar protein (A), the precursor protein, and the specific clinical form of the disease [4]. If hereditary ATTR is suspected, genetic testing for TTR mutations is necessary, the majority of which are heterozygous. The Val30Met mutation is present in more than 90% of cases, while Aps38Tyr, Ile107Val, Val71Ala, and Val121Il are much less common [33].

Differential diagnosis

In patients with disorders such as multiple myeloma, uncontrolled chronic inflammatory diseases (especially Sjögren syndrome, rheumatoid arthritis, autoimmune intestinal diseases, or bronchiectasis), or Castleman disease, the development of pulmonary nodular or interstitial changes can be a manifestation of amyloidosis. The tracheobronchial pattern of amyloidosis should be differentiated from diseases that affect the proximal portions of the lower respiratory tract and are associated with stenosis, namely, relapsing polychondritis, granulomatosis with polyangiitis, and bronchial sarcoidosis [34].

Tracheobronchopathia osteochondroplastica (TBOH) is a rare disease characterized by the formation of cartilage and bone tissue in the submucosal layer of the respiratory tract (Fig. 5.8). Not only does it resemble the CT and endobronchial picture of amyloidosis, but also it can itself be associated with amyloidosis [35]. In contrast to amyloidosis of the trachea and large bronchi, in relapsing polychondritis and TBOH, the posterior membranous part of the trachea is not affected [36].

Nodular forms of amyloidosis should be differentiated from bronchogenic carcinomas, neuroendocrine lung tumors, lymphogenous metastases to the lungs, and tumorlike sarcoidosis variants. The final judgment on the nature of lung nodules is often possible based only on histological and immunohistochemical studies. Amyloid also accumulates radioactive fluorodeoxyglucose, although less intensively than tumors. Therefore the results of positron emission spectroscopy-CT should be interpreted with due diligence in favor of one or the other of those two diagnoses [37].

The differential diagnosis of diffuse alveolar and septal forms of amyloidosis with calcification is limited to the range of disseminated diseases having a perilymphatic distribution pattern, namely, sarcoidosis (Fig. 5.9), tuberculosis, pneumoconiosis, and lymphogenous metastases to the lungs [38]. Diffuse interstitial forms of pulmonary amyloidosis are the most difficult to distinguish from other interstitial lung diseases, as this type of amyloidosis is quite rare and also because the radiographic pattern closely resembles that of pulmonary sarcoidosis, which is the diagnosis in most such cases. Pneumoconioses (Fig. 5.10) also have a pattern similar to that of diffuse interstitial amyloidosis, so a careful occupational history is mandatory. The perilymphatic pattern of the distribution of small nodules in amyloidosis together with mediastinal lymphadenopathy may resemble lymphogenous metastases in the lungs.

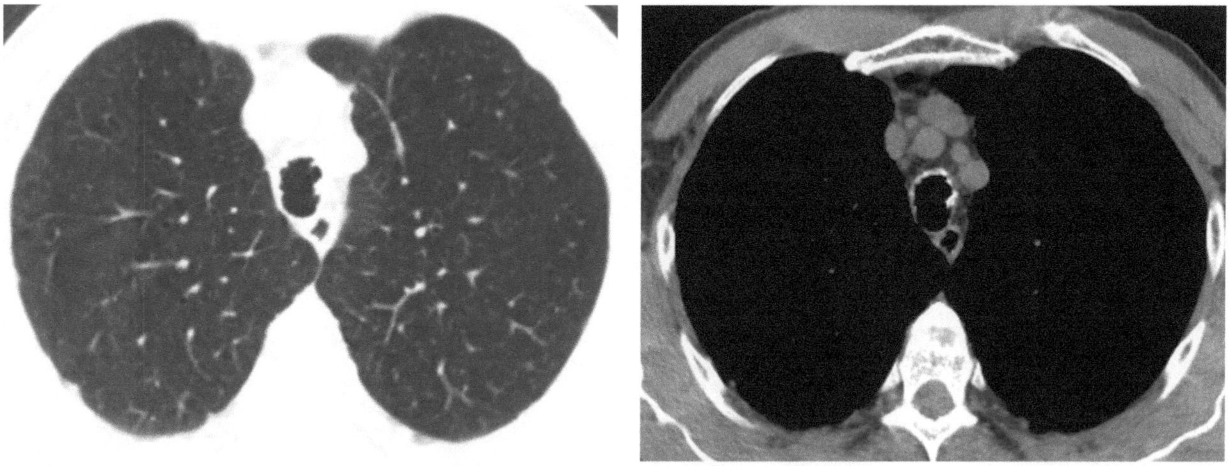

(A) (B)

FIG. 5.8 Tracheobronchopathia osteochondroplastica. There is circular, uneven induration of the submucosal layer of the trachea (A) with calcification (B) and corrosion of the internal contours protruding into the lumen. The posterior tracheal wall remains intact.

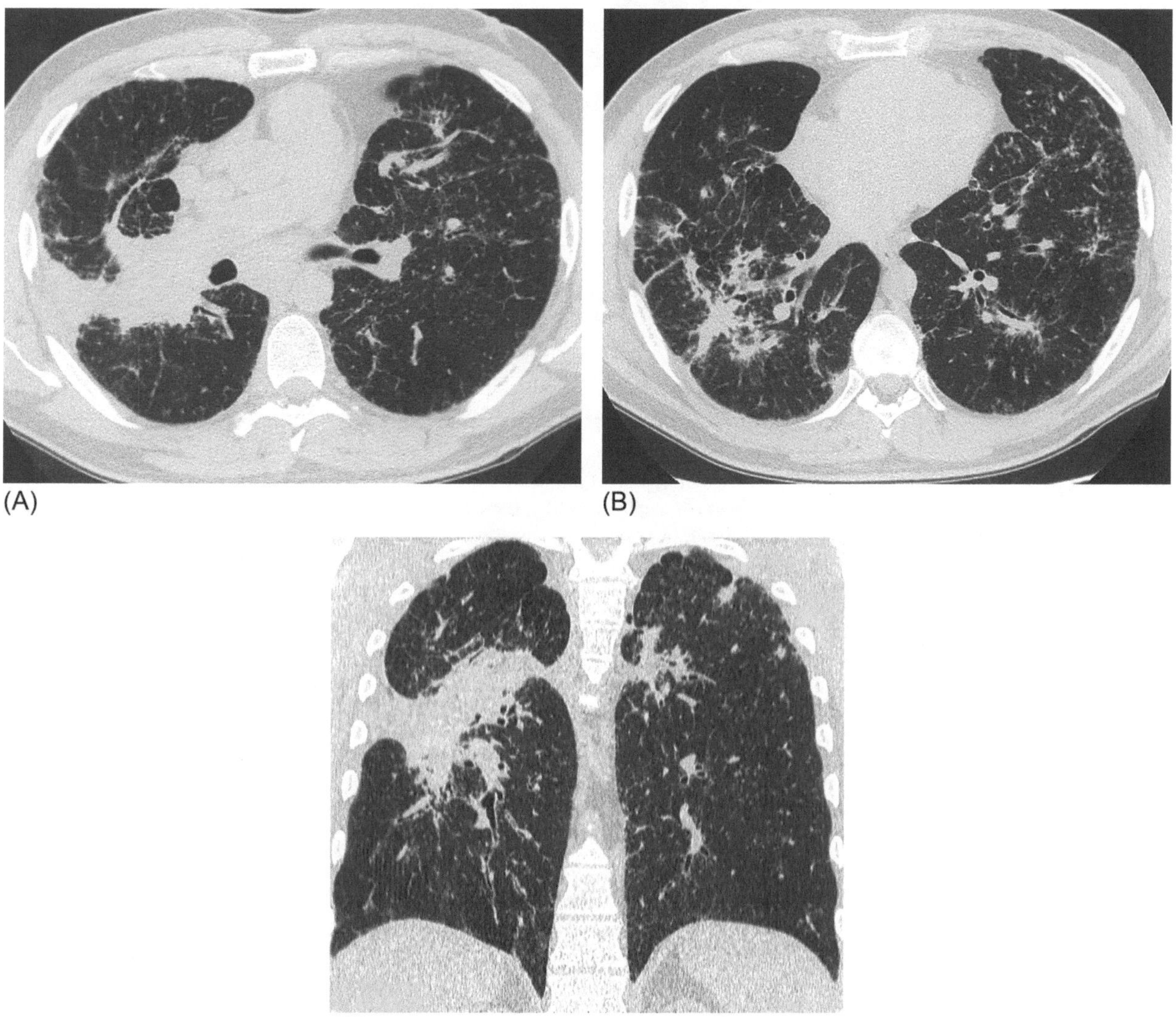

(A) (B) (C)

FIG. 5.9 Pulmonary sarcoidosis. (A) On the right, there is a mass extending from the hilum to the pleura. There are multiple small nodules bilaterally. (B) Features include radiating peribronchovascular foci of consolidation, multiple 1–3-mm nodules, and separate thickened interlobular septa. (C) In the coronal reconstruction, a perilymphatic pattern of abnormalities is clearly visible.

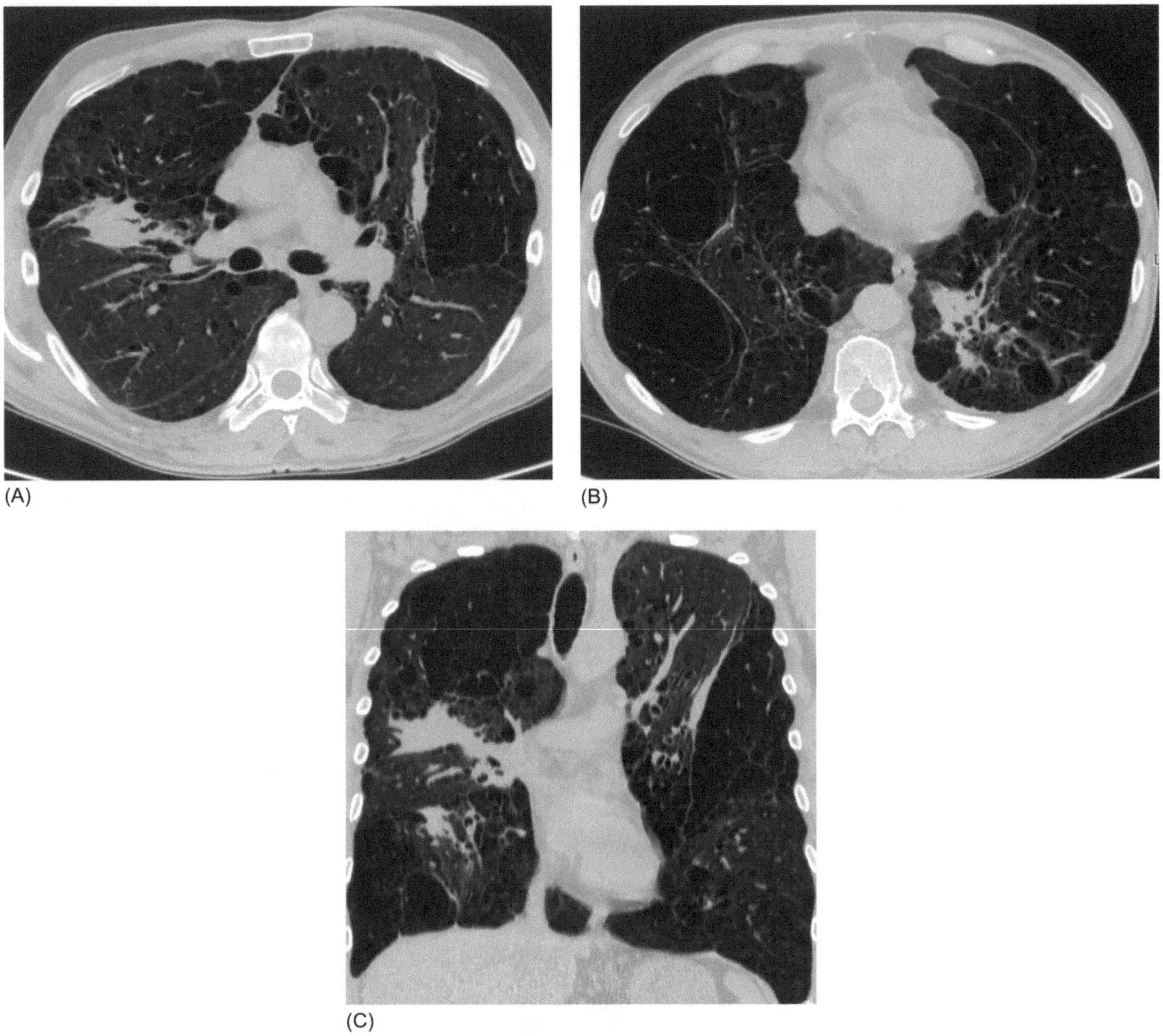

FIG. 5.10 Hard metal pneumoconiosis caused by prolonged inhalation of iron oxide dust. (A) Zones of consolidation are present along the interlobar pleura bilaterally, and there is severe centriacinar and paraseptal emphysema. (B) There are several bullae on the right; multiple areas of centriacinar emphysema; and dense, thickened interlobular septa. On the left, there is a focus of consolidation around the bronchovascular bundle, along with uneven thickening of the interlobular septa and linear areas of fibrosis. (C) In the coronal reconstruction, the perilymphatic distribution of parenchymal consolidation is clearly visible.

In patients with chronic renal failure who are hemodialysis-dependent, metastatic calcification in the lungs (MCL) can develop, characterized by the deposition of calcium and phosphorus salts in the small vessels and interstitium. Such changes are usually observed in the peripheral zones of the upper lobe parenchyma. The nodules have a predominantly centrilobular location, and not all have calcium densities. Sometimes, there is no visible calcification at all, and ground-glass opacities predominate [39]. Since pulmonary amyloidosis also often develops in patients with kidney lesions, differentiating it from MCL is necessary. In addition to the difference in the focal distribution pattern (acinar in MCL and perilymphatic in amyloidosis), radionuclide scintigraphy with technetium diphosphonate reveals isotope accumulation in zones of pulmonary calcification in MCL, even in the absence of obvious deposition of calcium salts on CT.

When there is the simultaneous presence of thin-walled cysts and nodules in the lungs, possible diagnoses include amyloidosis, LCDD, lymphocytic interstitial pneumonia, and Birt-Hogg-Dubé syndrome (see Chapter 9). LCDD is most similar to amyloidosis pathogenetically, as it involves conglomerates of immunoglobulin light chains deposited in the tissues and is often associated with multiple myeloma. However, the protein accumulations in LCDD are granular rather than fibrillar and do not have the characteristic birefringence with Congo red staining under cross polarized light. The cysts in LCDD are round and are usually larger the lower in the lungs they appear. Nodules are dense and homogeneous and have clear contours that are often irregular in shape. They are caused by lymphocytic infiltration and can be as large as 20mm in diameter. In general, however, the cysts are larger than focal infiltrates or nodules [40].

Treatment

Local forms of the disease may be monitored without treatment if the patient is asymptomatic. The tracheobronchial form of amyloidosis may require endobronchial cryodestruction or laser excision of masses that occlude the airway [41]. Newer treatments for this form of amyloidosis include external beam radiation therapy and endobronchial brachytherapy. The latter has demonstrated efficacy in early studies with minimal risk of complications or significant recurrence [42,43].

Treatment of diffuse forms of amyloidosis depends on the severity of the disease and the type of amyloid precursor protein. When treatment is effective, the amyloid deposits may decrease, leading to restoration of lung function [31].

Treatment for AL amyloidosis is the most well developed. The aim of chemotherapy is the inhibition of abnormal clonal secretion of protein by plasma cells, similar to the goal in treating multiple myeloma [44]. The combination of high-dose dexamethasone and immunosuppressant melphalan is currently regarded as the optimal chemotherapy regimen in patients who are not candidates for autologous stem cell transplantation (SCT) [45]. SCT is the treatment of choice for patients with good physical status, a cardiac ejection fraction of more than 50%, lung diffusion capacity greater than 50%, and no renal failure. Chemotherapy should be reserved for more severe cases [46]. In AL amyloidosis, other drugs used to treat multiple myeloma are also effective, such as bortezomib and lenalidomide, particularly for recurrent or refractory disease [47]. Several newer antiplasmacytic agents are also effective for refractory or recurrent AL amyloidosis, including pomalidomide, ixazomib, and daratumumab [48].

Treatment of secondary AA amyloidosis is aimed primarily at stabilizing the inflammatory manifestations of the underlying disease, suppressing SAA production to decrease the deposition of amyloid, and improving the function of affected organs [49]. There are as yet no drugs effective for AA amyloidosis. In a phase III trial, eprodisate (1,3-propanedisulfonate), which binds SAA sulfate, did not achieve the study targets for reducing mortality and disease progression, although it had positive effects in reducing amyloidogenesis in animals [50].

A promising approach is the use of anti-SAP antibodies, which have demonstrated improved clearance of amyloid deposits [51]. Therapy of hereditary amyloidosis is not sufficiently developed. Diflunisal and tafamidis are TTR stabilizers that can inhibit the development of neurological manifestations of the disease [52]. The new TTR gene silencing drugs inotersen and patisiran have completed phase III clinical trials, achieving the target end points for efficacy, and are now in the registration phase [52]. With rapidly progressing and refractory forms of ATTR amyloidosis in young patients with a Val30Met mutation, liver transplantation is recommended [53].

Prognosis depends on the degree of involvement of the lung parenchyma. The median survival rate of untreated patients with diffuse AL amyloidosis is about 13 months from the time of diagnosis [11]. Combined chemotherapy with melphalan and corticosteroids prolongs survival to 17 months [54]. The survival rate in tracheobronchial amyloidosis after endobronchial correction, particularly with cryodestruction and stenting, is about 9 years [14]. Lung damage in systemic amyloidosis is usually not critical for estimating prognosis, since survival primarily depends on damage to the heart and kidneys.

References

[1] Khoor A, Colby TV. Amyloidosis of the lung. Arch Pathol Lab Med 2017;141(2):247–54.

[2] Pinney JH, Smith CJ, Taube JB, Lachmann HJ, Venner CP, Gibbs SD, et al. Systemic amyloidosis in England: an epidemiological study. Br J Haematol 2013;161(4):525–32.

[3] Kyle RA, Linos A, Beard CM, Linke RP, Gertz MA, O'Fallon WM, et al. Incidence and natural history of primary systemic amyloidosis in Olmsted County, Minnesota, 1950 through 1989. Blood 1992;79(7):1817–22.

[4] Sipe JD, Benson MD, Buxbaum JN, Ikeda SI, Merlini G, Saraiva MJ, et al. Amyloid fibril proteins and amyloidosis: chemical identification and clinical classification. International Society of Amyloidosis 2016 Nomenclature Guidelines. Amyloid 2016;23(4):209–13.

[5] Wechalekar AD, Gillmore JD, Hawkins PN. Systemic amyloidosis. Lancet 2016;387(10038):2641–54.

[6] Scala R, Maccari U, Madioni C, Venezia D, La Magra LC. Amyloidosis involving the respiratory system: 5-year's experience of a multi-disciplinary group's activity. Ann Thorac Med 2015;10(3):212–6.

[7] Pepys MB. Amyloidosis. Annu Rev Med 2006;57:223–41.

[8] Yamada T. Serum amyloid A (SAA): a concise review of biology, assay methods and clinical usefulness. Clin Chem Lab Med 1999;37(4):381–8.

[9] Milani P, Basset M, Russo F, Foli A, Palladini G, Merlini G. The lung in amyloidosis. Eur Respir Rev 2017;26(145). pii: 170046.

[10] Khoor A, Myers JL, Tazelaar HD, Kurtin PJ. Amyloid-like pulmonary nodules, including localized light-chain deposition: clinicopathologic analysis of three cases. Am J Clin Pathol 2004;121(2):200–4.

[11] Kyle RA, Gertz MA. Primary systemic amyloidosis: clinical and laboratory features in 474 cases. Semin Hematol 1995;32(1):45–59.

[12] Berk JL, O'Regan A, Skinner M. Pulmonary and tracheobronchial amyloidosis. Semin Respir Crit Care Med 2002;23(2):155–65.

[13] Utz JP, Swensen SJ, Gertz MA. Pulmonary amyloidosis: the Mayo Clinic experience from 1980 to 1993. Ann Intern Med 1996;124(4):407–13.

[14] O'Regan A, Fenlon HM, Beamis Jr. JF, Steele MP, Skinner M, Berk JL. Tracheobronchial amyloidosis: the Boston University experience from 1984 to 1999. Medicine (Baltimore) 2000;79(2):69–79.

[15] Cordier JF, Loire R, Brune J. Amyloidosis of the lower respiratory tract. Clinical and pathologic features in a series of 21 patients. Chest 1986;90(6):827–31.

[16] Webb WR, Muller NL, Naidich DP. Miscellaneous infiltrative lung diseases. In: Webb WR, Muller NL, Naidich DP, editors. High-resolution CT of the lung. 5th ed. Philadelphia: Lippincott Williams and Wilkins; 2015. p. 419–23.

[17] Baumgart JV, Stuhlmann-Laeisz C, Hegenbart U, Nattenmüller J, Schönland S, Krüger S, et al. Local vs. systemic pulmonary amyloidosis-impact on diagnostics and clinical management. Virchows Arch 2018;473(5):627–37.

[18] Loizos S, Shiakalli Chrysa T, Christos GS. Amyloidosis: review and imaging findings. Semin Ultrasound CT MR 2014;35(3):225–39.

[19] Pickford HA, Swensen SJ, Utz JP. Thoracic cross-sectional imaging of amyloidosis. Am J Roentgenol 1997;168(2):351–5.

[20] Kirchner J, Jacobi V, Kardos P, Kollath J. CT findings inextensive tracheobronchial amyloidosis. Eur Radiol 1998;8(3):352–4.

[21] Gillmore JD, Hawkins PN. Amyloidosis and the respiratory tract. Thorax 1999;54(5):444–51.

[22] Van Geluwe F, Dymarkowski S, Crevits I, De Wever W, Bogaert J. Amyloidosis of the heart and respiratory system. Eur Radiol 2006;16(10):2358–65.

[23] Jeong YJ, Lee KS, Chung MP, Han J, Chung MJ, Kim KI, et al. Amyloidosis and lymphoproliferative disease in Sjögren syndrome: thin-section computed tomography findings and histopathologic comparisons. J Comput Assist Tomogr 2004;28(6):776–81.

[24] Baqir M, Kluka EM, Aubry MC, Hartman TE, Yi ES, Bauer PR, et al. Amyloid-associated cystic lung disease in primary Sjögren's syndrome. Respir Med 2013;107(4):616–21.

[25] Foli A, Palladini G, Caporali R, Verga L, Morbini P, Obici L, et al. The role of minor salivary gland biopsy in the diagnosis of systemic amyloidosis: results of a prospective study in 62 patients. Amyloid 2011;18(Suppl. 1):80–2.

[26] Fernández de Larrea C, Verga L, Morbini P, Klersy C, Lavatelli F, Foli A, et al. A practical approach to the diagnosis of systemic amyloidoses. Blood 2015;125(14):2239–44.

[27] Cottin V, Cordier JF, Richeldi L, editors. Orphan lung diseases: a clinical guide to rare lung disease. London: Springer-Verlag; 2015. p. 85.

[28] Schonland SO, Hegenbart U, Bochtler T, Mangatter A, Hansberg M, Ho AD, et al. Immunohistochemistry in the classification of systemic forms of amyloidosis: a systematic investigation of 117 patients. Blood 2012;119(2):488–93.

[29] Vrana JA, Gamez JD, Madden BJ, Theis JD, Bergen 3rd HR, Dogan A. Classification of amyloidosis by laser microdissection and mass spectrometry-based proteomic analysis in clinical biopsy specimens. Blood 2009;114(24):4957–9.

[30] Lachmann HJ, Goodman HJB, Gilbertson JA, Gallimore JR, Sabin CA, Gillmore JD, et al. Natural history and outcome in systemic AA amyloidosis. N Engl J Med 2007;356(23):2361–71.

[31] Pepys MB. Pathogenesis, diagnosis and treatment of systemic amyloidosis. Philos Trans R Soc Lond B Biol Sci 2001;356(1406):203–10.

[32] Hawkins PN, Lavender JP, Pepys MB. Evaluation of systemic amyloidosis by scintigraphy with 123 I-labeled serum amyloid P component. N Engl J Med 1990;323(8):508–13.

[33] Lavigne-Moreira C, Marques VD, Magno Gonçalves MV, de Oliveira MF, Tomaselli PJ, Nunez J, et al. The genetic heterogeneity of hereditary transthyretin amyloidosis in a sample of the Brazilian population. J Peripher Nerv Syst 2018;23(2):134–7.

[34] Kwong JS, Muller NL, Miller RR. Diseases of the trachea and main-stem bronchi: correlation of CT with pathologic findings. Radiographics 1992;12(4):645–57.

[35] Sakula A. Tracheobronchopathia osteochondroplastica: its relationship to primary tracheobronchial amyloidosis. Thorax 1968;23(1):105–10.

[36] Renapurkar RD, Kanne JP. Metabolic and storage lung diseases: spectrum of imaging appearances. Insights Imaging 2013;4(6):773–85.

[37] Seo JH, Lee SW, Ahn BC, Lee J. Pulmonary amyloidosis mimicking multiple metastatic lesions on F-18 FDG PET/CT. Lung Cancer 2010;67(3):376–9.

[38] Savader SJ, Nokes SR, Chappel G. Computed tomography of multiple nodular pulmonary amyloidosis. Comput Radiol 1987;11(3):111–5.

[39] Webb R, Higgins C. Thoracic imaging: pulmonary and cardiovascular radiology. W&K 2nd ed. Philadelphia, PA: Lippincott Williams & Wilkins; 2011.

[40] Sheard S, Nicholson AG, Edmunds L, Wotherspoon AC, Hansell DM. Pulmonary light-chain deposition disease: CT and pathology findings in nine patients. Clin Radiol 2015;70(5):515–22.

[41] Breuer R, Simpson GT, Rubinow A, Skinner M, Cohen AS. Tracheobronchial amyloidosis: treatment by carbon dioxide laser photoresection. Thorax 1985;40(11):870–1.

[42] Neben-Wittich MA, Foote RL, Kalra S. External beam radiation therapy for tracheobronchial amyloidosis. Chest 2007;132(1):262–7.

[43] Moore A, Kramer MR, Silvern D, Shtraichman O, Allen AM. Endobronchial brachytherapy—a novel approach for the management of airway amyloidosis. Brachytherapy 2018;17(6):966–72.

[44] Ryšavá R. AL amyloidosis: advances in diagnostics and treatment. Nephrol Dial Transplant 2018. Epub Oct 8.

[45] Palladini G, Perfetti V, Obici L, Caccialanza R, Semino A, Adami F, et al. Association of melphalan and high-dose dexamethazone is effective and well tolerated in patients with AL (primary) amyloidosis who are ineligible for stem cell transplantation. Blood 2004;103(8):2936–8.

[46] Mahmood S, Palladini G, Sanchorawala V, Wechalekar A. Update on treatment of light chain amyloidosis. Haematologica 2014;99(2):209–21.

[47] Comenzo RL. Managing systemic light-chain amyloidosis. J Natl Compr Canc Netw 2007;5(2):179–87.

[48] Milani P, Merlini G, Palladini G. Novel therapies in light chain amyloidosis kidney. Int Rep 2017;3(3):530–41.

[49] Papa R, Lachmann HJ. Secondary, AA, amyloidosis. Rheum Dis Clin North Am 2018;44(4):585–603.

[50] Rumjon A, Coats T, Javaid MM. Review of eprodisate for the treatment of renal disease in AA amyloidosis. Int J Nephrol Renov Dis 2012;5:37–43.

[51] Richards DB, Cookson LM, Barton SV, Liefaard L, Lane T, Hutt DF, et al. Repeat doses of antibody to serum amyloid P component clear amyloid deposits in patients with systemic amyloidosis. Sci Transl Med 2018;10(422). pii:eaan3128.

[52] Finsterer J, Iglseder S, Wanschitz J, Topakian R, Löscher WN, Grisold W. Hereditary transthyretin-related amyloidosis. Acta Neurol Scand 2019;139(2):92–105.

[53] Stangou AJ, Hawkins PN, Heaton ND, Rela M, Monaghan M, Nihoyannopoulos P, et al. Progressive cardiac amyloidosis following liver transplantation for familial amyloid polyneuropathy: implications for amyloid fibrillogenesis. Transplantation 1998;66(2):229–33.

[54] Kyle RA, Gertz MA, Greipp PR, Witzig TE, Lust JA, Lacy MQ, et al. A trial of three regimens for primary amyloidosis: colchicine alone, melphalan and prednisone, and melphalan, prednisone, and colchicine. N Engl J Med 1997;336(17):1202–7.

Chapter 6

Granulomatosis with polyangiitis

Alexander Averyanov[a,b], Evgeniya Kogan[c], Victor Lesnyak[d], Igor E. Stepanyan[e]

[a]*Clinical Department, Pulmonology Research Institute under FMBA of Russia, Moscow, Russia,* [b]*Pulmonary Division, Federal Research Clinical Center under FMBA of Russia, Moscow, Russia,* [c]*Anatomic Pathology Department, Sechenov University, Moscow, Russia,* [d]*Radiology Department, Federal Research Clinical Center under FMBA of Russia, Moscow, Russia,* [e]*Central TB Research Institute, Moscow, Russia*

Granulomatosis with polyangiitis (GPA), an autoimmune necrotizing granulomatous vasculitis, is a rare disease from the group of systemic vasculitides that is associated with the production of antineutrophil cytoplasmic antibodies (ANCAs), which also includes eosinophilic granulomatosis with polyangiitis (EGPA) and microscopic polyangiitis (MPA) [1–3]. The disease was first described as an independent nosological form in 1936 by the German pathologist F. Wegener, who was eponymous to the disease. In 2011 the American College of Rheumatologists, the European League Against Rheumatism, and the American Society of Nephrology decided to abandon the use of eponyms for systemic vasculitis and Wegener granulomatosis in particular, in favor of the term "granulomatosis with polyangiitis" to reflect the pathological features of the disease [4].

The nature of GPA is unknown, but it is believed that bacterial (especially *Staphylococcus aureus*) and viral infections of the respiratory tract, together with environmental factors (mercury and lead), in genetically predisposed individuals may trigger the disease [2,5]. In 60%–70% of patients with GPA, the nasopharyngeal mucosa was found to be chronically colonized with *S. aureus*, which is associated with more frequent relapses of GPA [6].

The incidence of GPA varies from 3 per 1 million population to 3.3 per 100,000 population; the prevalence reaches 42 per 100,000 population. The age of patients varies from 30 to 70 years old. Children and young people under the age of 25 years of age are rarely ill. Men and women are equally affected. The disease most often develops in Caucasian people [7–10].

ANCAs have the largest pathogenetic role in the development of GPA. The mechanisms of their production are still unknown. ANCAs cause the activation of neutrophils and monocytes, which leads to necrosis of the vessel wall. Additionally, neutrophils are a target of autoimmune reactions and effector cells that cause damage to the vascular endothelium. The inferior T- and B-cell regulation contributes to the autoimmune response of ANCA. [2]. Vasculitis with GPA occurs more regularly (in 75%–80% of cases) due to the production of ANCAs against proteinase 3 and occurs less often (in 10%–25% of cases) from the production of antibodies to myeloperoxidase of neutrophils and monocytes [11]. A pool of inflammatory cytokines released by these cells stimulates a cascade of immune reactions with predominantly small vascular lesions. Polyangiitis in GPA is characterized by a more systemic type of lesion and a greater predisposition to granuloma formation than other forms of ANCA-associated vasculitis [12,13]. Phenotypic differences of GPA are also determined depending on the type of ANCA antibodies; for example, antibodies to myeloperoxidase (p-ANCA)-positive vasculitis are more common in older patients, especially in female patients. Conversely, c-ANCA positive vasculitis is more common at a younger age, having a predominance of male patients [14].

The role of heredity in the development of GPA has not been proved. However, there is evidence of a link between the deficiency of alpha-1 antitrypsin and histocompatibility antigens HLA-DPB1, HLA-DR4, and HLA-DR13 and having a predisposition toward the development of this disease [15–17].

Morphology

GPA is a combination of necrotizing granulomatosis and systemic necrotizing vasculitis with the possibility of affecting blood vessels of different sizes, including capillaries, arterioles, arteries, and the aorta and veins [18].

The local form of GPA almost always affects the upper respiratory tract (rhinitis, sinusitis, and perforation of the nasal septum). Damage to the eyes (uveitis, iridocyclitis, episcleritis, scleritis, periorbital granuloma, and exophthalmos), ears

Difficult to Diagnose Rare Diffuse Lung Disease. https://doi.org/10.1016/B978-0-12-815375-8.00006-6

(otitis and eustachitis), skin (papules, purpura, and ulcers), and oral mucosa (ulcerative stomatitis with the systemic form of GPA) is also possible. In the systemic form of GPA, the lungs, in addition to the listed organs, are involved in 90% of cases, while the kidneys are involved in 80% of cases [19].

The heart (pericarditis, coronaritis, myocardial infarction, and mitral and aortic valve damage) and the nervous system (asymmetrical polyneuropathy) can also be involved. Isolated variants of GPA that occur only with lung damage are rare.

Macroscopic manifestations of GPA have their own special aspects in different organs. Multiple nodes; cavities; foci of dark red hemorrhages; and foci of pneumonia with signs of carnification, fibrinous pleurisy, and pleural adhesions are found in the lungs. Bronchopulmonary and peribronchial lymph nodes are usually not involved in the pathological process.

Microscopy is often critical in establishing the diagnosis of GPA. Histological examination reveals a variety of changes that occur at different frequencies. Based on the accumulation of related experience, morphological changes are divided into "major and minor diagnostic criteria of GPA" [18]. The triad of characteristic signs is related to "major diagnostic criteria," namely, necrosis of the pulmonary parenchyma, which can involve all pulmonary structures and vessels; necrotic polymorphocellular granulomas; and vasculitis (Fig. 6.1). It was revealed that in the material of open lung biopsies, necrosis in combination with vasculitis in GPA is detected in 89% of the biopsy samples; granulomatosis and necrosis are seen in 90% of the biopsy samples; while necrosis, vasculitis, and granulomatosis are found in 91% of biopsy samples [20].

The "minor diagnostic criteria" include the following findings: microabscesses; acute bronchiolitis, bronchiolitis obliterans with organizing pneumonia, focal interstitial fibrosis, lipoid pneumonia, xanthogranulomatosis, infiltration with an admixture of eosinophils, alveolar hemorrhages, bronchocentric granulomatosis, and sarcoid granulomas [18].

Necrotic granulomas in the lungs have an irregular shape and contain a variety of cells, such as neutrophilic leukocytes, lymphocytes, plasma cells, macrophages, giant multinucleated histiocytes, and eosinophils. Histiocytes can form characteristic palisades around foci of necrosis (Fig. 6.2).

Vasculitis with GPA is destructive in nature and develops in vessels of various types, namely, in arteries, veins, and capillaries (Fig. 6.3). In the vessels, fibrinoid necrosis, polymorphocellular infiltration of their walls, and thrombosis can be detected.

Areas of necrosis are found both in granulomas and in lung tissue and sometimes resemble a geographic map (Fig. 6.4). Foci of necrosis probably arise as a result of heterolysis in the activation of neutrophilic leukocytes and resemble microabscesses. In addition to heterolysis in the formation of necrosis foci, ischemia may be of importance, with the development of foci of infarcts of the lung. This can be connected with the blocking of the blood supply through vascular shunts from branches of the bronchial artery due to widespread vasculitis in the system of the pulmonary arteries and also of the bronchial arteries.

According to the clinical and morphological characteristics, specific forms of GPA are distinguished, such as fulminant, bronchocentric, eosinophilic, and hemorrhagic.

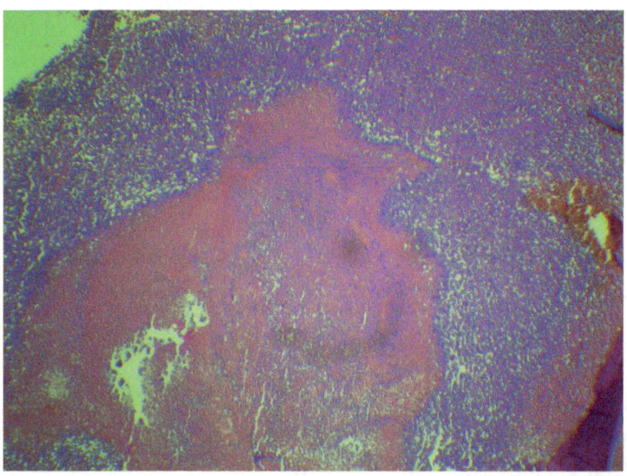

FIG. 6.1 GPA. Zone of necrosis, permeated with nuclear dust with granulomatous tissue around the periphery. Hematoxylin and eosin (H&E) staining, 1.25×.

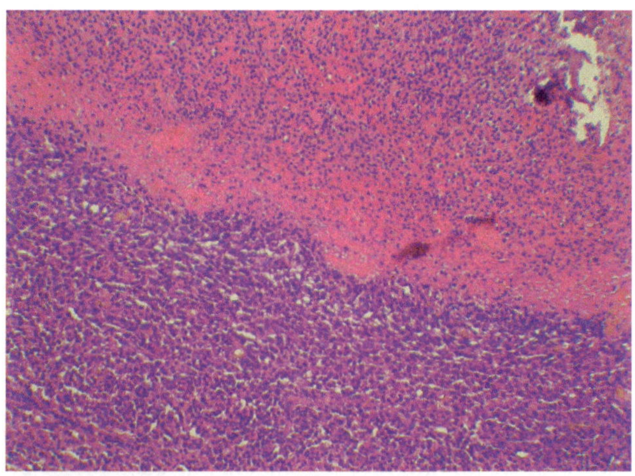

FIG. 6.2 GPA. The boundary of the zone of necrosis and granulomatous tissue with large histiocytes. H&E staining, 200×.

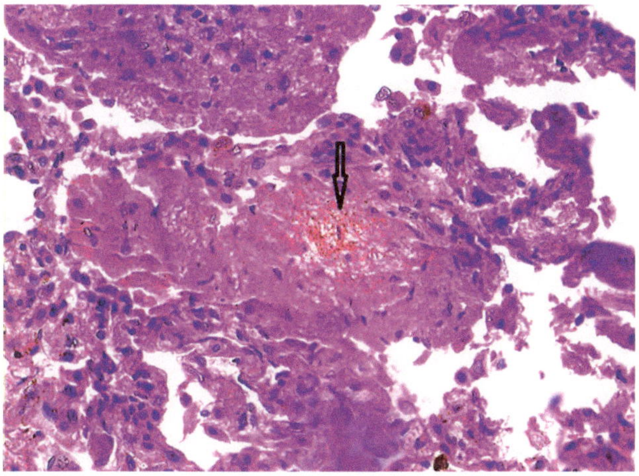

FIG. 6.3 GPA. Destructive vasculitis *(arrow)* in the lung tissue affected area. H&E staining, 400×.

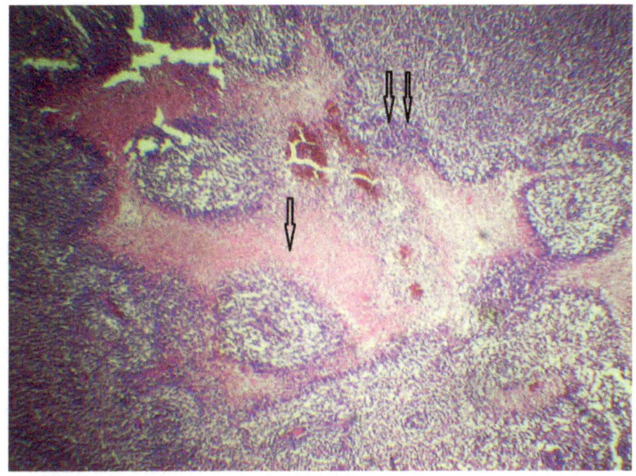

FIG. 6.4 GPA. Geographic necrosis *(arrow)* with a granulomatous tissue around the periphery *(double arrow)*. H&E staining, 1.25×.

The differential diagnosis for GPA primarily includes systemic vasculitis from the ANCA-positive group, namely, microscopic polyangiitis, eosinophilic granulomatosis with polyangiitis, and giant-cell arteritis, as well as noninfectious and infectious pulmonary granulomatosis, sarcoidosis, and Henoch-Schonlein purpura. Diagnostics should be based on a combination of clinical, laboratory, and morphological data. In addition, it is necessary to consider the presence of ANCA in combination with the "major diagnostic morphological criteria for GPA" and exclusion of clinical and laboratory data of the aforementioned diseases.

For histological confirmation of the diagnosis, a biopsy of the nasal mucosa, skin, or kidney is usually performed. The diagnostic value of the material obtained by transbronchial biopsy does not exceed 7%, whereas by surgical biopsy, this figure approaches 90% [21]. However, the indications for surgical lung biopsy in GPA are confined to cases of ANCA-negative vasculitis with minimal involvement of other organs.

Clinical presentation

The systemic nature of GPA determines the variety of symptoms in many organs. Nevertheless, there is a certain pattern of localization of changes.

Upper respiratory tract lesions very often dominate at the onset of the disease and are observed in 90% of GPA patients. Patients may complain of persistent blood-tinged discharge from the nose, nasal congestion, the formation of crusts in the nasal passages, and ulcers of the mucous membranes. The erosion or destruction of the bones of the nasal sinuses, perforation of the nasal septum, and saddle nose deformity are also possible [22].

The involvement of auditory system (otitis media, mastoiditis, and hearing loss) is observed in 30%–50% of cases. In 60%–80% of patients, a nasopharyngeal lesion is noted [22].

Eye lesions are observed in 20%–50% of patients with GPA. A pseudotumor of the orbit develops in 25% of cases due to the formation of a periorbital granuloma. Proptosis as a result of retroorbital granulomatous inflammation can disrupt the blood supply to the optic nerve. Blindness due to uveitis, vasculitis, or compression of the optic nerve may develop in 2%–9% of patients. Conjunctivitis and scleritis are rarely noted [22].

Granulomatous lesions of the mucous membranes of the trachea and the main bronchi are characterized by the formation of necrotizing plaques in GPA, which leads to the formation of hypopharyngeal stenoses in 10%–30% of cases. The length of tracheal stenosis below the glottis usually does not exceed 3–5 cm, but the distal part of the trachea and main bronchi also may be involved in the process [4,22,23]. Clinically, a trachea or main bronchus lesion is manifested by strident breathing and inspiratory dyspnea with heavy coughing and hemoptysis with scarlet blood.

A lung lesion in GPA is diagnosed in 90% of patients. Generally, it is combined with the involvement of other organs, but in 9% of patients, it may be the only manifestation of the disease [19]. In the lung parenchyma, infiltrates with cavities are usual. A frequent symptom is hemoptysis. Dyspnea appears with a significant decrease in the respiratory surface or lesions of the pulmonary artery. However, pulmonary symptoms may be absent, even with a large lesion (up to 34% of cases) [24]. Auscultatory findings are often absent. Pleural lesions are manifested by pain in the chest and exudation into the pleural cavity (usually unilateral).

The third most frequent target organ is the kidney, where glomerulonephritis develops with microhematuria, proteinuria, and kidney failure [25].

Other organs are affected less frequently in GPA and have nonspecific manifestations, including skin rash, arthritis, episcleritis, scleritis, uveitis, keratitis, thrombosis of veins and retinal arteries, and various disorders of the central nervous system. Common symptoms of vasculitis include persistent fever, weight loss, myalgia, arthralgia, and anorexia [25].

The course of GPA, as a rule, is progressive, even with adequate therapy. Members of the European Vasculitis Study Group [26] have proposed identifying five stages of the disease, depending on the degree of involvement of vital organs and response to treatment. They are limited, early generalized, active generalized, severe, and refractory. Modern approaches to GPA therapy are based on this classification.

Diagnosis

Diagnostic criteria for GPA were developed and proposed by the American College of Rheumatology in 1990 and still remain relevant currently, despite significant progress in the field of medicine [27]. They are as follows:

1. Inflammation of the nose and mouth (purulent, bloody discharge from the nose, crusting, and ulcers in the nasal cavity).
2. Changes in the lungs on a radiological examination (nodules, infiltrates, and cavities in the lungs).
3. Changes in the urine (hematuria of more than five erythrocytes in the field of vision and red blood cell casts in the urine sediment).
4. Biopsy data (granulomatous inflammation in the artery wall or in the perivascular and extravascular space).

The presence of two or more criteria allows diagnosis with a sensitivity of 88% and a specificity of 92% [27].

Later, for the diagnostics of ANCA-associated vasculitis, simplified, so-called surrogate criteria were proposed [28], which for GPA included signs of upper and lower respiratory tract lesions; a radiograph showing infiltrates, nodes, or cavitary lesions in the lungs, which last for at least 1 month; stenosis of the bronchi; bloody discharge from the nose and crusting for 1 month or the formation of ulcers in the nose; chronic sinusitis, otitis media, or mastoiditis for 3 months; retroorbital pseudotumor; hypopharyngeal laryngostenosis; saddle nose deformity; or destructive changes in the nose and paranasal sinuses.

For the diagnosis of GPA, one marker is sufficient, provided that other causes of the disease are excluded [28].

The most important diagnostic test for GPA is the detection of ANCAs by the indirect immunofluorescence and enzyme-linked immunosorbent assay methods. An increased level of ANCA to proteinase 3 (c-ANCA) is revealed in 90% of patients with ANCA-associated GPA, and an increase in p-ANCA is revealed in only 5% of patients [29]. In cases of restricted forms, ANCA in the blood can be increased slightly or be absent. A normal level of c-ANCA is insufficient to rule out a diagnosis of GPA in the presence of other clinical and histological evidences [11]. In accordance with the latest international consensus on the clinical suspicion of ANCA-associated vasculitis, the first negative ANCA test (especially in the case of a screening test) must be rechecked in a reference laboratory [30].

Opinions on the significance of ANCA as a marker of disease activity are contradictory. A number of studies have indicated a direct correlation between the ANCA level in the blood and the disease activity and the possibility of its use for monitoring the disease [11,31]. Conversely, in the study by Finkielman et al., an increase in the level of ANCA was associated with relapse of the GPA in only 40% of patients [32].

It should be noted that positive ANCA titers can be detected not only in ANCA-associated vasculitis but also in autoimmune bowel diseases, some tumors in the hematopoietic system, a number of infectious diseases, and drug-induced vasculopathy [33] (see section "Differential diagnosis").

Among other laboratory changes, depending on the severity of the course of GPA, development of complications, and signs of kidney damage in the blood, there can be signs of anemia, increased erythrocyte sedimentation rate (ESR), and C-reactive protein (CRP), hypoproteinemia, decreased glomerular filtration, increased serum urea, and increased creatinine.

Analysis of bronchoalveolar lavage (BAL) fluid does not play a decisive role in the diagnostics of GPA. In most cases, an increase in the level of neutrophils to 61% and lymphocytosis (on average 40%) is observed during the relapse of the process, with an average ratio CD4+/CD8+ of 4.1 during the period of clinical and laboratory activity remission [34]. Nevertheless, BAL can play an important role in ruling out diseases with a similar clinical and radiological presentation, such as pulmonary infections (including tuberculosis, mycobacterioses, and fungal lesions) and tumor infiltrates.

Computed tomography

The characteristic changes seen with GPA, namely, destruction of the bones of the nasal sinuses and perforation of the nasal septum, can be detected on a radiograph or computed tomography of the nasal sinuses in 80% of cases (Fig. 6.5) [35].

At the initial stages of GPA, up to 45% of patients have changes on radiographs of the chest, primarily being in the appearance of single or multiple rounded infiltrative formations, sometimes with signs of inner decay.

The most common CT signs of GPA are [36] nodules and nodal zones of consolidation in the lung parenchyma with irregular distribution, peribronchovascular location of these masses (Fig. 6.6), cavities in infiltration zones with thick or thin walls (Fig. 6.7), the phenomenon of ground-glass opacity around the consolidation (a halo sign), centrilobular nodules, and thickening of the bronchial walls.

Except for the presence of cavities in infiltration zones, CT features are nonspecific; however, as with any other pathology, there are additional signs that can help in ascertaining the correct diagnosis.

Nodules and masses in the parenchyma are found in 90% of patients with lung lesions (Fig. 6.6) [37]. Their average size is usually from 2 to 4 cm, with a range from a few millimeters to 10 cm [38]. In 85% of cases, nodules and nodes in the lungs are multiple; in 67% of cases, they are bilateral; and in 89% of patients, they are subpleural [37]. The contours of these nodes are often even; the shape is round or oval but may be irregular (Fig. 6.6). Generally, several small nodules are detected; the presentation of many small centrilobular nodules resembling a CT scan of lung tuberculosis (Fig. 6.7), hypersensitivity pneumonitis, or mycoplasma pneumonia may occur [19]. Cavities are present in approximately half of patients with nodal infiltrates with a diameter of more than 2 cm (Figs. 6.7 and 6.8) [39]. The appearance of decay in the infiltrates is due to the development of necrotizing vasculitis involving the small and medium arteries [40]. The cavities can have different sizes, wall thicknesses, and internal contours; they can undergo changes when the process progresses or during immunosuppressive therapy (Fig. 6.9). In addition, they can be infected with bacterial or fungal flora, demonstrating features of lung abscesses (Fig. 6.10) [19]. The halo sign in patients with GPA, usually reflects the appearance of hemorrhagic filling of lung tissue [41]. The emergence of ground-glass opacity foci not related to the infiltrates is generally also associated with alveolar hemorrhages due to local lesions in small vessels (Figs. 6.7 and 6.11) [42]. Abundant alveolar hemorrhages can lead to the filling of alveolar spaces and the formation of zones of consolidation resembling pneumonic ones. Approximately, 10% of patients develop diffuse alveolar hemorrhages, which can be displayed on the HRCT by a subtotal distribution of ground-glass opacity, while the subpleural zones usually remain intact (Fig. 6.12) [39]. In general, the ground-glass opacity is found in 30%–50% of patients with GPA and lung damage [43,44].

The involvement of lower airways (most often the trachea) is observed in 10%–55% of patients with GPA [45]. This usually occurs in patients with multisystemic lesions, but cases of an isolated process in the trachea have also been described [46]. The affected place can be widespread, but more often, it is located in an area of 2–4 cm with a typical distribution in the subpharyngeal segment. The thickening of airways is usually circular with a smooth or tuberous inner surface.

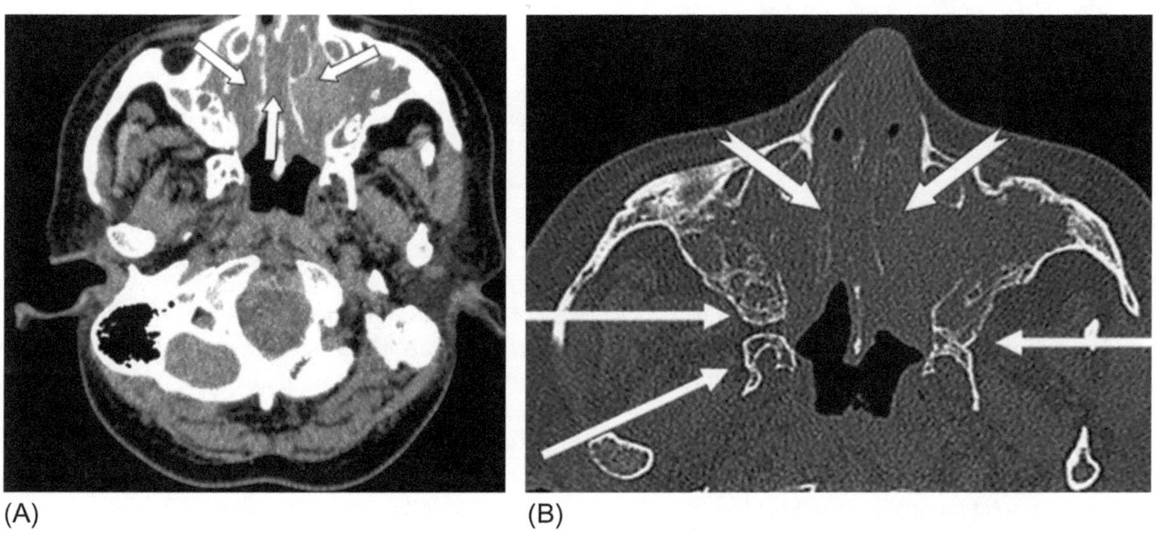

FIG. 6.5 GPA. (A) Necrotic masses filling the nasal cavity and lumens of the maxillary sinuses *(arrows)*. (B) Destruction of the bony elements of the nasal septum, the medial walls of the maxillary sinuses *(short arrows)*, deformity and bony growths of the posteroexternal walls, and pterygoid processes of the main bone *(long arrows)*.

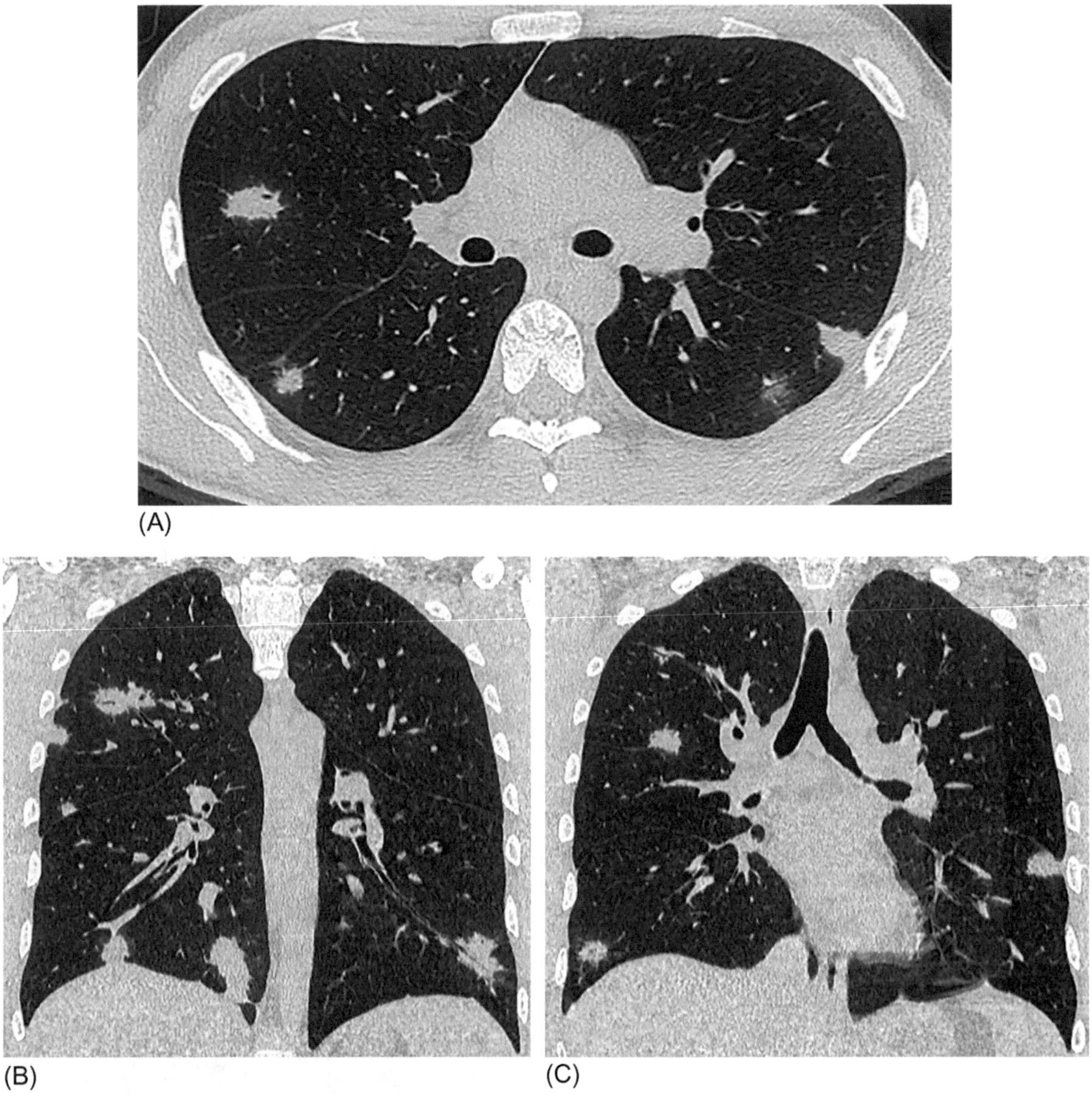

FIG. 6.6 GPA. Multiple nodules and masses with random distribution in the lungs. Some of them have a rounded shape, while others have an irregular shape with clear or blurred boundaries, mainly with subpleural or peribronchovascular location. Most nodes have a spicular sign (A, B, and C).

An important focus is the involvement of the posterior membranous trachea in the process, which is a specific sign of GPA, unlike recurrent polychondritis and tracheobronchopathia osteochondroplastica [47]. On the airway mucosa, there is often characteristic ulceration with bleeding. Up to 70% of patients with pulmonary manifestations of GPA have thickening of the walls of the segmental and subsegmental bronchi [37]. The consequent large bronchi stenosis may be the appearance of lobar and segmental atelectasis in the corresponding zones. Bronchiectasis is a possible, but rarer, finding with GPA (10%–20%) [38,48]. A higher incidence of bronchiectasis is associated with an increase in the level of p-ANCA and a more frequent lesion of the peripheral nerves [49].

Pleural effusion (usually unilateral) occurs in 10%–20% of patients, and it can be not only a consequence of granulomatous inflammation in the pleura (about 6%) but also a manifestation of renal failure [20].

Intrathoracic lymphadenopathy is not typical for GPA; its emergence indicates a secondary infection or an alternative diagnosis (lymphoma and sarcoidosis).

Under the influence of treatment, the CT manifestations of GPA change; in approximately 50% of cases, nodules disappear; in another 40% of cases, they decrease in size, and the diameter and wall thickness of the cavities are reduced (Fig. 6.9) [38,48].

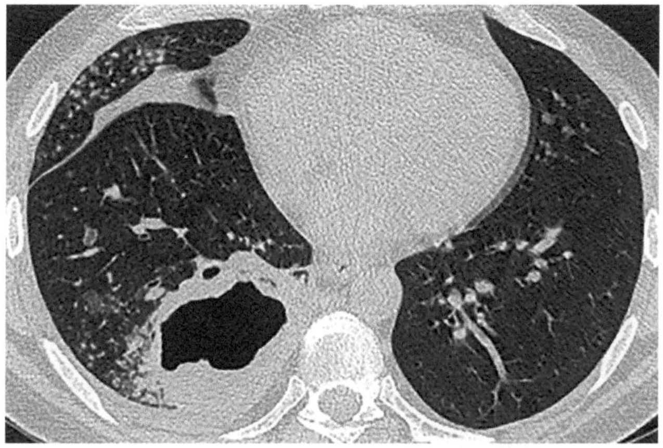

FIG. 6.7 GPA. A large cavity with thick walls in the lower lobe of the right lung, surrounded by shadow of ground-glass opacity (probably alveolar hemorrhages) and merging foci. In the middle lobe on the right, there is a subpleural consolidation layer and multiple small nodules of 1–2 mm in diameter.

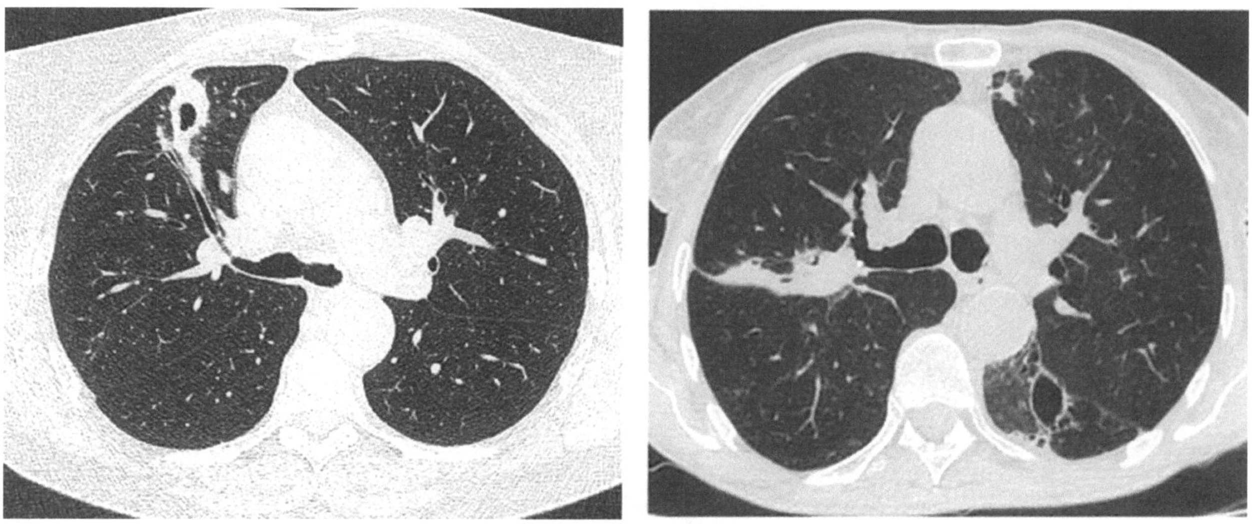

(A) (B)

FIG. 6.8 GPA. (A) A thick-walled cavity in the middle lobe of the right lung. Around it, there is a shadow of ground-glass opacity and small nodules. (B) In the lower lobe of the left lung, there is a thin-walled cavity surrounded by multiple small cavities and spicular nodules in the upper lobe on the left; on the right, there is a heterogeneous zone of consolidation along the interlobar pleura.

Secondary radiological changes in the lungs can be associated with the development of infectious complications under immunosuppression. Infection of the cavities with Aspergillus spp. or bacterial pathogens forming abscesses is noted most commonly (Fig. 6.10) [39].

Differential diagnosis

The range of diseases to be differentiated from lung damage with GPA is quite extensive and includes diseases that have a similar clinical presentation and radiographic pattern (nodular infiltrates with cavities), such as tumors (Figs. 6.13 and 6.14), septic metastases (Fig. 6.15), tuberculosis (Fig. 6.16), and pulmonary mycobacteriosis (Fig. 6.17); it also includes aspergillosis (Fig. 6.18), echinococcosis (Fig. 6.19), organizing pneumonia (Fig. 6.20), and ANCA-associated vasculitis.

The main differential diagnostic criteria of diseases with a similar radiographic pattern are presented in Table 6.1.

Given the infrequency of isolated lung lesions in GPA, the greatest difficulty in differential diagnosis is caused by cases of other diseases when there are symptoms in the upper respiratory tract resembling those of the GPA clinical presentation.

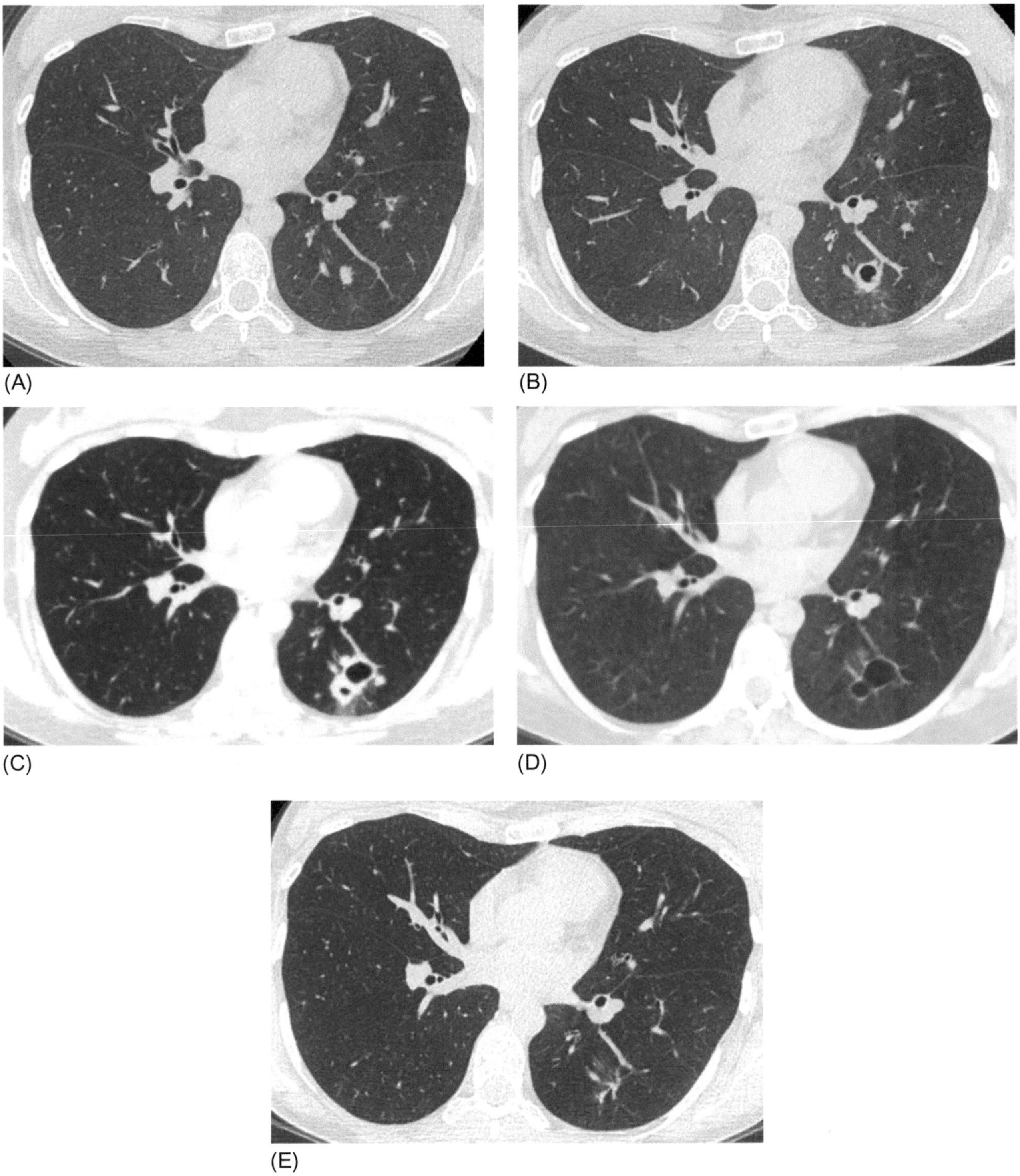

FIG. 6.9 GPA. Dynamics of a nodular lesion in the lower lobe of the left lung. (A) Oval focus of consolidation of 12×6mm and two foci of 3–4mm. The diagnosis is not established yet. (B) Three months after the observation, the cavity in the node thickness appeared. (C) Six months after the observation, the zone of consolidation expanded, and a new focus appeared. Combined therapy with corticosteroids and cyclophosphamide was started. (D) Four months after the onset of treatment, a residual cyst was retained in place of the thick-walled cavity. Remission of GPA was achieved. (E) Twelve months after the onset of treatment, there is a site of radiant fibrosis in place of the cyst. Remission continues.

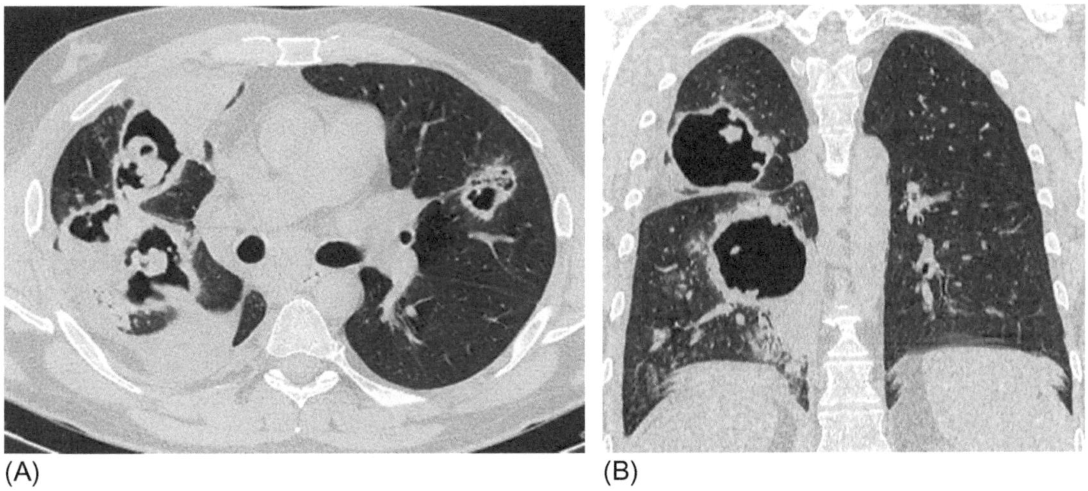

FIG. 6.10 Combination of GPA with pulmonary aspergillosis. (A) Extensive zones of consolidation in the right lung with the formation of giant cavities, within which the tuberous dense masses are visible, which is mycelium of *Aspergillus niger* (the air-crescent sign). (B) Around the cavities, there are multiple small nodules that can both manifest fungal invasion and vasculitis.

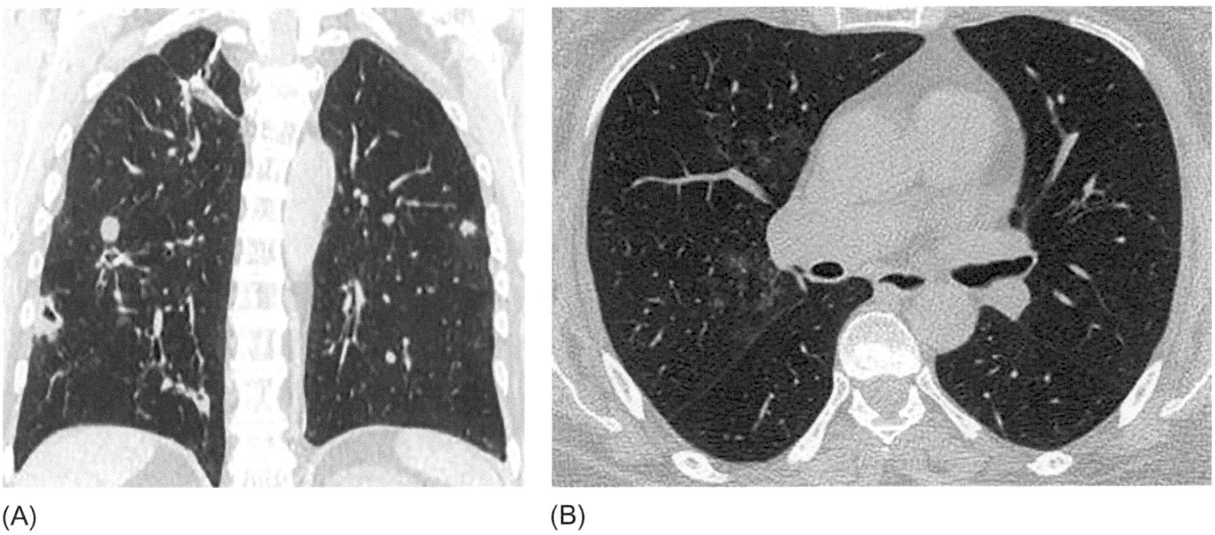

FIG. 6.11 GPA. (A) Coronal reconstruction shows multiple nodules and thick-walled cavities with an irregular margins in both lungs. (B) On the axial section, foci of ground-glass opacity are visible, which probably represent alveolar hemorrhages.

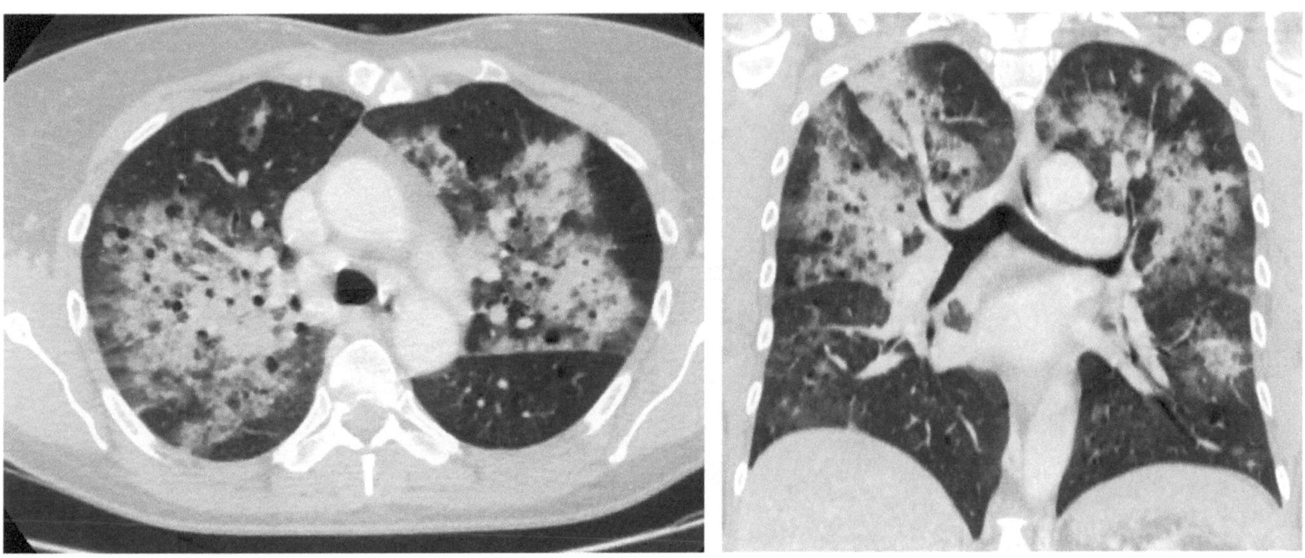

FIG. 6.12 ANCA-associated vasculitis. Diffuse alveolar hemorrhages. Bilateral, diffuse areas of ground glass opacity and consolidation with parahilar distribution. The lesion is maximally expressed in the upper lobes and central zones. (A and B) Coronal reconstruction shows two rounded nodules in the upper lobe of the left lung.

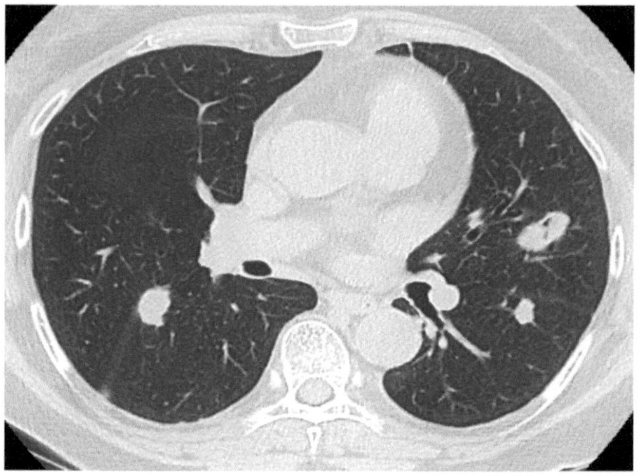

FIG. 6.13 Multiple metastases of ovarian cancer in the lungs. Dense nodes with sharp contours, one of them with the cavity. Metastases are usually located below the carina level.

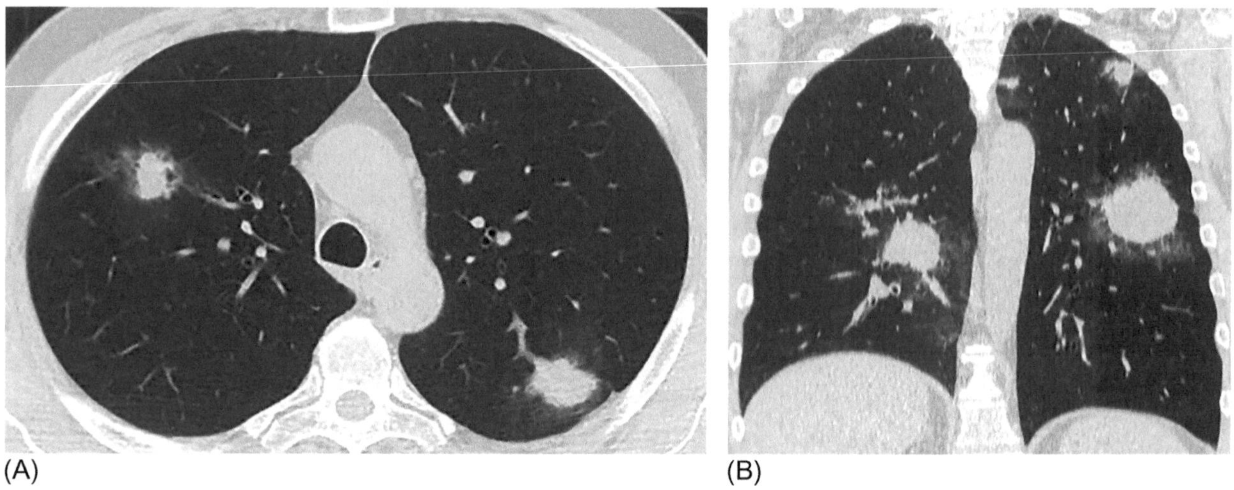

FIG. 6.14 T-cell lymphoma. Several nodes in both lungs surrounded by ground-glass opacity. Unlike GPA, large masses do not have internal cavities (A and B).

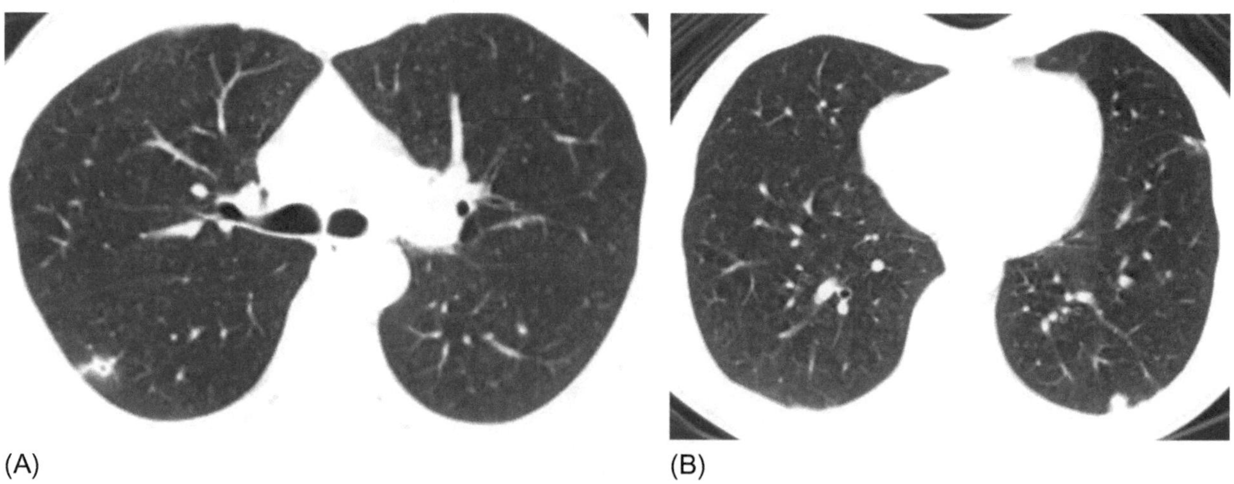

FIG. 6.15 Septic metastases (one of them with cavity) in a patient with a staphylococcal infection of soft tissues (A and B).

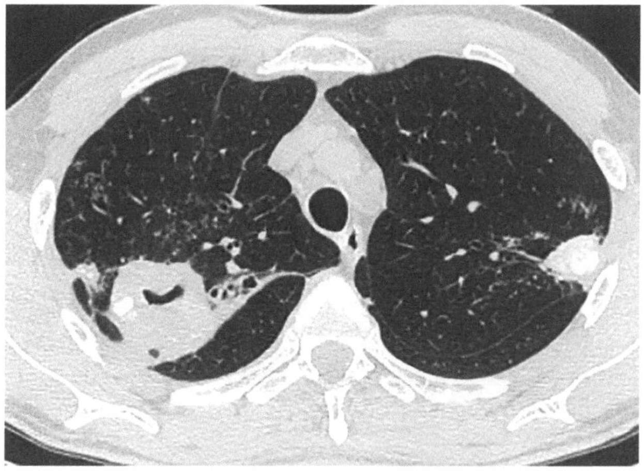

FIG. 6.16 Infiltrative tuberculosis. Large masses in the upper lobes of the lungs with calcification inside; on the right, with the cavity, multiple polymorphic nodules are visualized in the adjacent pulmonary tissue.

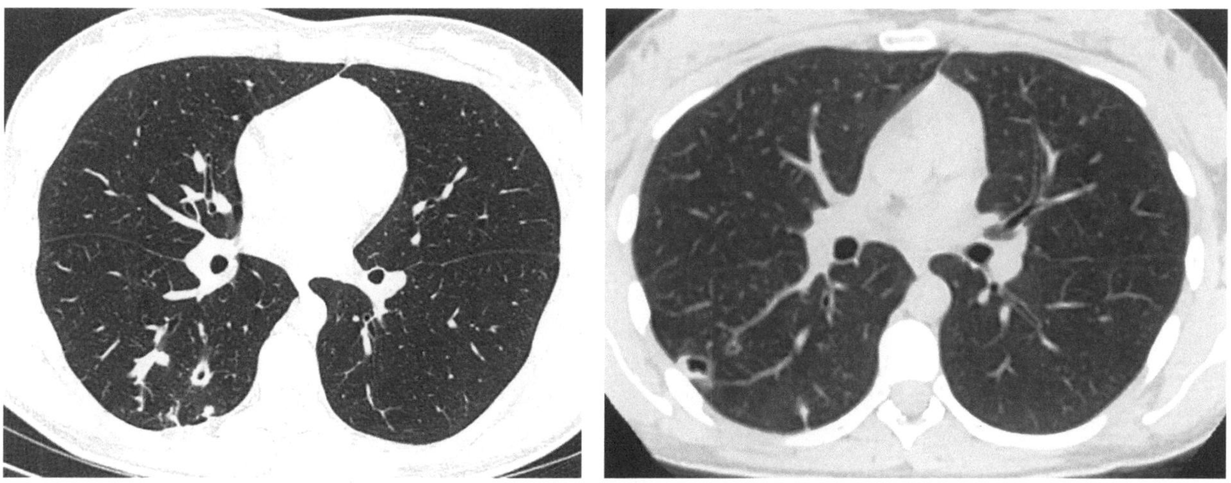

(A) (B)

FIG. 6.17 Atypical presentation of *Mycobacterium chelonae* infection in immunocompetent patient. Thick-wall cavities and single nodules in the right low lobe (A and B).

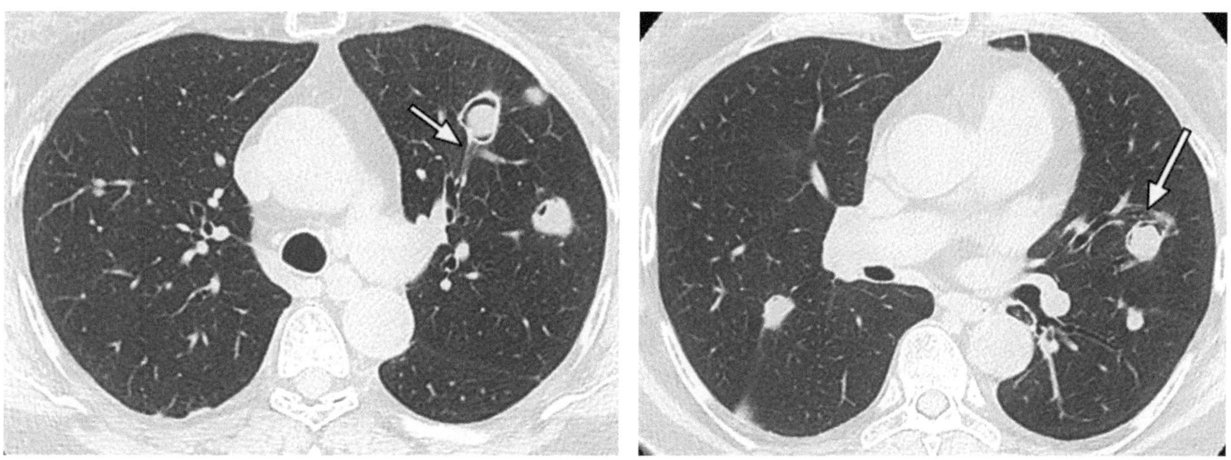

(A) (B)

FIG. 6.18 Multiple pulmonary aspergillomas. Polymorphic nodes and cavities, inside which the round opacity (mycetoma) are seen, loosely adjacent to the inner wall (the air-crescent sign). In some cases, there is a junction between the cavity and the bronchus *(arrows)* (A and B).

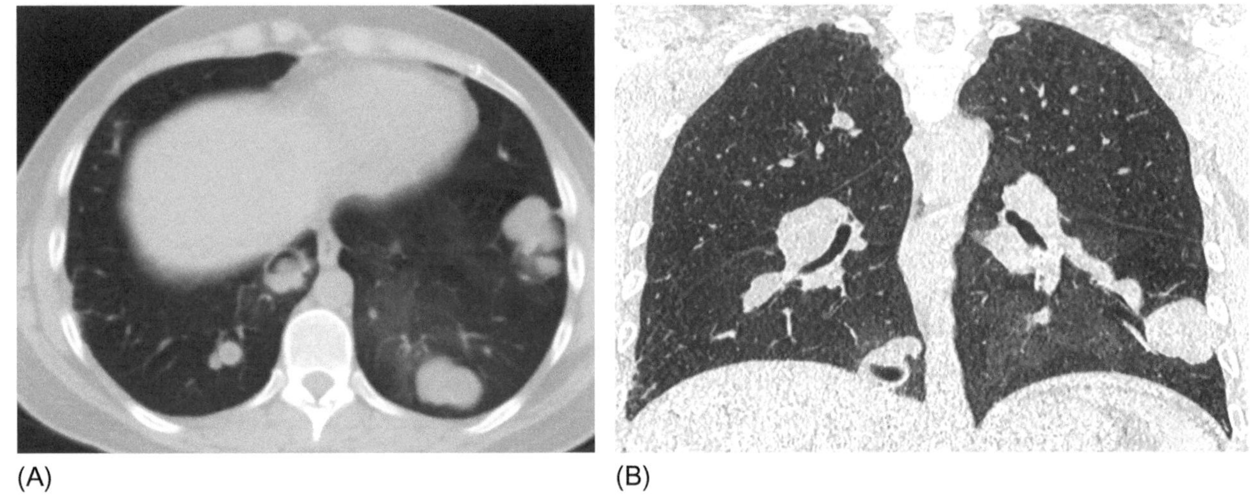

FIG. 6.19 Pulmonary echinococcosis. (A) Multiple nodes, with sharp contours and different sizes in diameter from 6 mm to 48 × 27 mm, filled inside with liquid. (B) In S7 on the right gas-liquid levels are visualized.

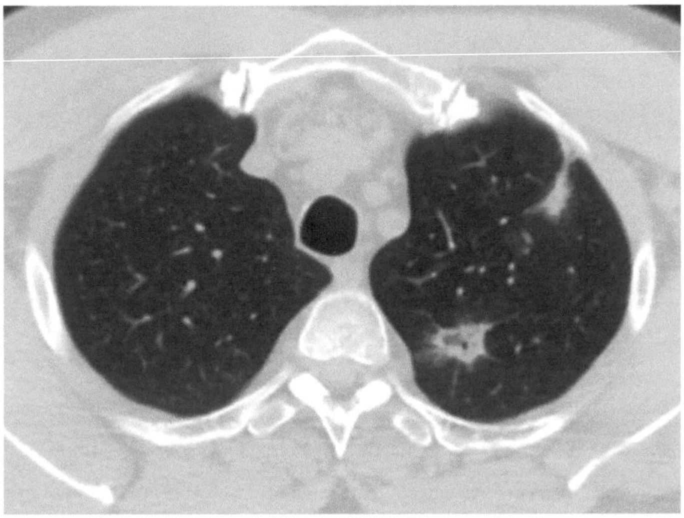

FIG. 6.20 Cryptogenic organizing pneumonia confirmed by surgical biopsy. Two irregular spiculated nodules, one of which contains air bronchogram are visible in the left upper lobe.

Most often, they are concomitant chronic diseases of the nasopharynx (chronic sinusitis, etc.), but sometimes the symptoms can be like GPA (formation of crusts in the nose and nosebleeds). In doubtful cases, it is necessary to perform a biopsy of the nasopharyngeal mucosa.

In countries with a high prevalence of tuberculosis, the differential diagnosis of pulmonary tuberculosis and GPA remains an urgent problem, since pulmonary manifestations of GPA are often initially diagnosed as tuberculosis [50]. Major problems in the differential diagnosis between pulmonary tuberculosis and the refractory form of GPA may occur when a new infiltrate with cavitation appears in the lungs of patients with GPA receiving immunosuppressant therapy. Generally, a thorough search for a causative agent of tuberculosis is required, the effectiveness of which is significantly increased in such patients by the use of endobronchial methods of obtaining the material [34]. Another special aspect of pulmonary tuberculosis that can hinder differential diagnosis of GPA cases is the possible appearance in the blood of ANCA in persons having tuberculosis [51].

Other diseases that include an increase in the ANCA titer and that may result in infiltrates in the lungs include Crohn's disease, ulcerative colitis, rheumatoid arthritis, non-Hodgkin lymphoma, myelodysplasia, and infective endocarditis [33,52–54]. It is especially difficult to differentiate ANCA-associated vasculitis from subacute bacterial endocarditis (SBE), in which ANCA is detected in 18%–24% of cases (anti-PR3-ANCA, rarely anti-MPO-ANCA, or a combination of these that

TABLE 6.1 Differential diagnosis of GPA

	GPA	Tumors of the lungs and metastases to the lungs	Septic metastases	Tuberculosis	Organizing pneumonia
Age	Older than 30 years	Any	Any	Any	Any
Medical history	Rhinitis, sinusitis, nasal bloody discharge	Malignant neoplasms	Trauma, previous episode of infection, surgical intervention	Contacts with tuberculosis patients, immunodeficiency	Onset usually after a respiratory infection
Clinical presentation	Cough, hemoptysis, shortness of breath, chest pain	Weight loss, nonproductive cough	Fever, cough with purulent sputum	Fever, weight loss, sweating, coughing, hemoptysis	Fever, shortness of breath, nonproductive cough
Procalcitonin (ng/mL)	<0.5	<0.5	>1	<1	<0.5
Elevated c-ANCA	+++	–	+	+	–
Bronchoalveolar lavage fluid	Lymphocytosis in the period of low activity CD4/CD8 > 3, neutrophilia in the high activity phase	Tumor cells	Neutrophilia	Moderate lymphocytosis, neutrophilia CD4/CD8 < 2 + PCR to *M. tuberculosis*	Foamy macrophages, eosinophilia 2%–25%, lymphocytosis >25%
CT signs	Nodules and masses of various sizes up to 10 cm in diameter, often with cavities in large nodes. The halo sign. Pleural effusion is possible. Thickening of the walls of the trachea and large bronchi	Multiple bilateral nodules of different sizes with usual localization below carina (for metastases). Cavities are generally not characteristic, signs of invasive growth are possible (halo and spiculation) Lymphadenopathy	Several foci of consolidation up to 10 mm in diameter with or without cavities Lymphadenopathy	Infiltrates with cavities, mainly of upper-lobe localization. Signs of bronchogenic seeding; there are small nodules around the infiltrate Centrilobular nodules Lymphadenopathy	Bilateral subpleural consolidation areas without cavities, changing the size and configuration, bordering with the areas of ground-glass opacity. An atoll sign is very common, imitating cavities

–, Not typical; +, possible sign; ++, frequent sign; +++, typical sign.

is more frequent) [54,55]. The multiorgan lesions, including those in the kidneys and lungs, where septic metastases can be discovered that are indistinguishable in their characteristics from nodes with GPA, laboratory signs of systemic inflammation (increased CRP and ESR), and fever that is resistant to standard antibiotic therapy, are all signs that are similar for both diseases, so that a positive titer to ANCA can make the diagnosis of GPA more likely. However, for a patient with infective endocarditis, if the wrong diagnosis is made and the wrong treatment, such as a immunosuppressive therapy, is prescribed, this can become critical. Level of procalcitonin in a prolonged course of SBE can also be within the limits of reference values. If the diagnosis is not clear, repeated blood cultures and mandatory transesophageal echocardiography are necessary if transthoracic access does not reveal signs of endocarditis.

It is not difficult to differentiate GPA from EGPA; since the latter is characterized by high eosinophilia in the blood and BAL, it is not characterized by nodal consolidation with cavities in the lung parenchyma; there is severe bronchial asthma in almost all cases; and symptoms from the upper respiratory tract are allergic in nature (rhinosinusitis polyposa and allergic rhinitis).

Microscopic polyangiitis (MPA) is characterized by an aggressive course of glomerulonephritis, frequent epistaxis, and frequent development of alveolar hemorrhages. Unlike GPA, MPA is not a granulomatosis, and it does not form a nodal

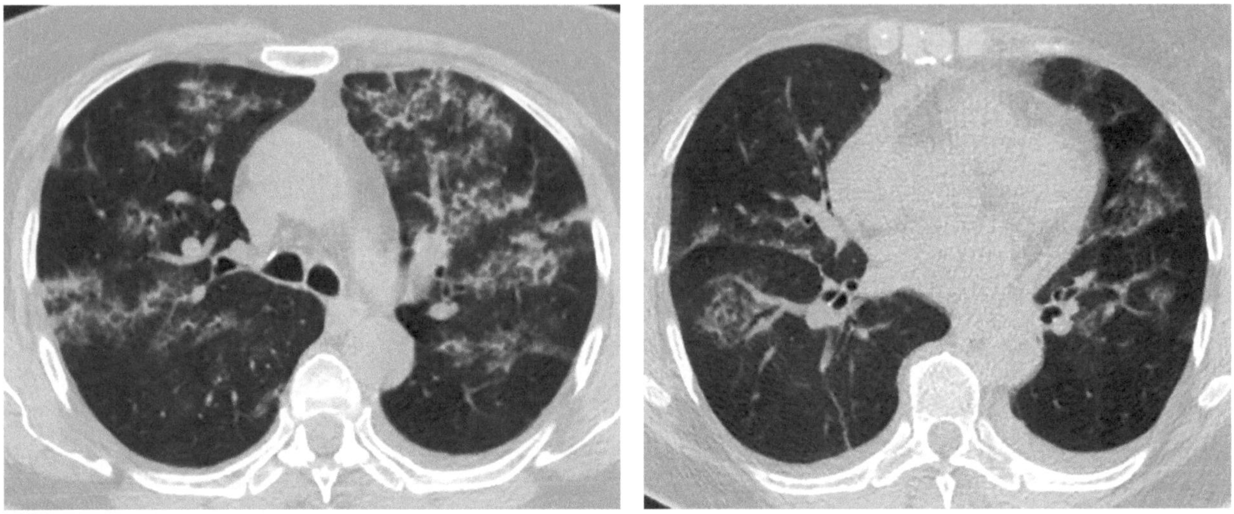

(A) (B)

FIG. 6.21 Microscopic polyangiitis with diffuse pulmonary hemorrhages. Bilateral symmetrical zones of ground-glass opacity associated with thickened interlobular and intralobular septa. Unlike GPA, there are no foci of consolidation with cavities (A and B).

consolidation in the lungs with cavities (Fig. 6.21). In addition, in 50%–75% of cases of MPA, the level of p-ANCA is increased, while an increase in the ratio of c-ANCA to proteinase 3 is more common for GPA [56].

Some forms of small vessels vasculitis cannot always be clearly differentiated; in such cases the use of the term ANCA-associated vasculitis is utilized.

The use of certain medications and narcotic agents (propylthiouracil, hydralazine, minocycline, and levamisole-adulterated cocaine) can induce the development of ANCA-associated vasculopathies mimicking vasculitis and causing alveolar hemorrhages [57,58]. Therefore careful history taking, including questions on prior drug therapy and drug addiction, can help in differentiating true vasculitis from vasculopathy.

Treatment and prognosis

The main principle of GPA treatment and that of other ANCA-associated vasculitides is a two-phase approach, consisting of the first stage, which is remission induction therapy, and after its achievement a shift to long-term maintenance therapy [59,60]. To select a remission induction scheme, the severity and activity of the disease are first assessed using different scales, the Birmingham Vasculitis Activity Score [61] and the European Vasculitis Study Group disease staging [26] (Table 6.2). Monotherapy with corticosteroids or immunosuppressants is acceptable for localized forms with low activity of microscopic polyangiitis and EGPA, while for GPA, even at an early stage, it is recommended to use combination therapy of

TABLE 6.2 Schemes of remission induction, depending on GPA severity [62–69]

Limited	Early generalized	Active generalized	Severe	Refractory
Localization of the process only in the upper respiratory airway. No systemic symptoms or impaired function of organs	The appearance of general symptoms; there are no signs of severe organ dysfunction	General symptoms with severe functional impairment of target organs	Severe life-threatening kidney damage with creatinine >500 mmol/L	Progression of the disease, the lack of response to therapy
CS+CYC or CS+MT or CS+AZA	CS+CYC or CS+MT or CS+AZA	CS+CYC or CS+RTX	CS+PE+CYC or CS+PE+RTX	No standard therapy Alternative therapies are possible, such as immunoglobulin, stem cells

CS, systemic corticosteroids; *AZA*, azathioprine; *MT*, methotrexate; *CYC*, cyclophosphamide; *RTX*, rituximab; *PE*, plasma exchange.

glucocorticosteroids with immunosuppressants. The same approach is used for early generalized forms in which remission can be effectively achieved with an oral combination of systemic steroids with one of the cytostatic agent (cyclophosphamide, methotrexate, and azathioprine) [62,63]. For remission induction treatment of the active generalized forms, a course of intravenous pulse therapy with corticosteroids in combination with cyclophosphamide or rituximab is usually utilized. The standard dose for prednosone in such cases is 1-mg/kg body weight per day or from 1 to 3 bolus injections of methylprednisolone of 7.5–15 mg/kg body weight per day [64]. After 2–4 weeks of treatment, the dose is reduced by 10% every 1–2 weeks, reaching an average dose of 0.5 mg/kg/day by month 3, followed by a decrease and a possible withdrawal by month 6. However, there is no consensus on the duration of the remission induction period in patients with ANCA-associated vasculitis [65]. Cyclophosphamide, which is the most commonly used immunosuppressant for GPA treatment, is administered either intravenously with pulse doses (15 mg/kg every 14 days for 1 month and then every 3 weeks) or orally (2 mg/kg/day with a maximum of 200 mg/day) considering the age and the glomerular filtration rate [66]. Rituximab is an alternative to traditional immunosuppressants, especially in patients with severe generalized forms of GPA who have already received a cumulative dose of cyclophosphamide. In combination with systemic corticosteroids, it is noninferior to cyclophosphamide with a similar incidence of side effects [67,68]. According to the European Medicines Agency and FDA recommendations, the remission induction dose of rituximab should be 375-mg/m^2 body surface area infused once a week for 4 weeks.

In cases of severe life-threatening complications such as renal failure or diffuse alveolar hemorrhages, combined drug therapy is supplemented by plasma exchange [69]. For refractory forms of GPA, there are no generally accepted methods of treatment. It is generally agreed that such patients should be treated at specialty centers. Usually, the guidelines being studied are used for such patients, such as new immunobiologic drugs (Ofatumumab, Abatacept, Belimumab, and Avacopan), mesenchymal stem cells, and immunoglobulins [70–72].

In 70%–90% of cases, the first phase of GPA treatment is able to achieve remission of the disease within a period of 6 months [62,73]. The second phase of treatment is maintenance therapy that is designed to prolong remission and prevent relapses. For patients in which remission induction was performed with cyclophosphamide, maintenance therapy with azathioprine (2 mg/kg/day) per os, or methotrexate (0.3 mg/kg/week), is preferred, possibly in combination with low doses of corticosteroids [74]. If they are not tolerated, or if serious side effects develop, maintenance therapy with mycophenolate mofetil is possible, although its effectiveness against relapse is lower than that of the previously listed agents [75]. Maintenance therapy should be conducted for at least 18–24 months [62].

If remission induction was achieved using rituximab, the maintenance therapy can be continued with its use at doses of either 1000 mg every 4–6 months or 500 mg every 6 months (in combination or without low doses of GCS), or it can be switched to a maintenance regimen as after cyclophosphamide therapy [76,77].

Due to the high risk of development of pneumocystosis infection in persons receiving immunosuppressive therapy with cyclophosphamide or rituximab, cotrimoxazole at a dose of 160 mg, trimethoprim, and 800 mg of sulfamethoxazole daily, 3 days/week, is recommended for the prevention of infectious complications, during and several months after withdrawal of immunosuppressive drugs [64].

In patients receiving adequate therapy, a 5-year survival rate reaches 72.4%–79.9% [38,47]. Patients with refractory forms of GPA have an unfavorable prognosis.

References

[1] Specks U, Wheatley CL, McDonald TJ, Rohrbach MS, DeRemee RA. Anticytoplasmic autoantibodies in the diagnosis and follow-up of Wegener's granulomatosis. Mayo Clin Proc 1989;64(1):28–36.

[2] Jarrot PA, Kaplanski G. Pathogenesis of ANCA-associated vasculitis: an update. Autoimmun Rev 2016;15(7):704–13.

[3] Gómez-Puerta JA, Hernández-Rodríguez J, López-Soto A, Bosch X. Antineutrophil cytoplasmic antibody-associated vasculitides and respiratory disease. Chest 2009;136(4):1101–11.

[4] Falk RJ, Gross WL, Guillevin L, Hoffman G, Jayne DR, Jennette JC, et al. Granulomatosis with polyangiitis (Wegener's): an alternative name for Wegener's granulomatosis. Ann Rheum Dis 2011;70(4):704.

[5] Cartin-Ceba R, Peikert T, Specks U. Pathogenesis of ANCA-associated vasculitis. Curr Rheumatol Rep 2012;14(6):481–93.

[6] Popa ER, Stegeman CA, Abdulahad WH, van der Meer B, Arends J, Manson WM, et al. Staphylococcal toxic-shock-syndrome-toxin-1 as a risk factor for disease relapse in Wegener's granulomatosis. Rheumatology (Oxford) 2007;46(6):1029–33.

[7] Watts RA, Al-Taiar A, Scott DG, Macgregor AJ. Prevalence and incidence of Wegener's granulomatosis in the UK general practice research database. Arthritis Rheum 2009;61(10):1412–6.

[8] Mahr AD, Neogi T, Merkel PA. Epidemiology of Wegener's granulomatosis: lessons from descriptive studies and analyses of genetic and environmental risk determinants. Clin Exp Rheumatol 2006;24(2 Suppl 41):S82–91.

[9] Berti A, Cornec D, Crowson CS, Specks U, Matteson EL. The epidemiology of antineutrophil cytoplasmic autoantibody-associated vasculitis in Olmsted County, Minnesota: a twenty-year US population-based study. Arthritis Rheumatol 2017;69(12):2338–50.

[10] Iudici M, Quartier P, Terrier B, Mouthon L, Guillevin L, Puéchal X. Childhood-onset granulomatosis with polyangiitis and microscopic polyangiitis: systematic review and meta-analysis. Orphanet J Rare Dis 2016;11(1):141.

[11] Savige J, Gillis D, Benson E, Davies D, Esnault V, Falk RJ, et al. International consensus statement on testing and reporting of antineutrophil cytoplasmic antibodies (ANCA). Am J Clin Pathol 1999;111(4):507–13.

[12] Kallenberg CG. Pathogenesis of PR3-ANCA associated vasculitis. J Autoimmun 2008;30(1–2):29–36.

[13] Miloslavsky EM, Lu N, Unizony S, Choi HK, Merkel PA, Seo P, et al. Myeloperoxidase-antineutrophil cytoplasmic antibody (ANCA)-positive and ANCA-negative patients with granulomatosis with polyangiitis (Wegener's): distinct patient subsets. Arthritis Rheumatol 2016;68(12):2945–52.

[14] Chen M, Yu F, Zhang Y, Zhao MH. Antineutrophil cytoplasmic autoantibodies-associated vasculitis in older patients. Medicine (Baltimore) 2008;87(4):203–9.

[15] Mahr AD, Edberg JC, Stone JH, Hoffman GS, St Clair EW, Specks U, et al. Alpha-antitrypsin deficiency-related alleles Z and S and risk of Wegener's granulomatosis. Arthritis Rheum 2010;62(12):3760–7.

[16] Gao Y, Schmitz S, Yuang G, Hogan SL, Bunch D, Hu Y, et al. DRB1*15 allele is a risk factor for PR3-ANCA disease in African Americans. J Am Soc Nephrol 2011;22(6):1161–7.

[17] Stassen PM, Cohen-Tervaert JW, Lems SP, Hepkema BG, Kallenberg CG, Stegeman CA. HLA-DR4, DR13 (6) and the ancestral haplotype A1B8DR3 are associated with ANCA-associated vasculitis and Wegener's granulomatosis. Rheumatology (Oxford) 2009;48(6):622–5.

[18] Hasleton P, Flieder DB, editors. Spencer's pathology of the lung. 6th ed. New York: Cambridge University Press; 2013.

[19] Ananthakrishnan L, Sharma N, Kanne JP. Wegener's granulomatosis in the chest: high-resolution CT findings. AJR Am J Roentgenol 2009;192(3):676–82.

[20] Travis WD, Hoffman GS, Leavitt RY, Pass HI, Fauci AS. Surgical pathology of the lung in Wegener's granulomatosis. Review of 87 lung biopsies from 67 patients. AJR Am J Surg Pathol 1991;15(4):315–33.

[21] Hoffman GS, Kerr GS, Leavitt RY, Hallahan CW, Lebovics RS, Travis WD, et al. Wegener's granulomatosis: an analysis of 158 patients. Ann Intern Med 1992;116(6):488–98.

[22] Lynch 3rd JP, Tazelaar H. Wegener granulomatosis (granulomatosis with polyangiitis): evolving concepts in treatment. Semin Respir Crit Care Med 2011;32(3):274–97.

[23] Lynch III JP, Fishbein MC, White ES. Pulmonary vasculitis. In: Costabel U, du Bois RM, Egan JJ, editors. Diffuse parenchymal lung disease. Basel: Karger; 2007. p. 196–211.

[24] Zycinska K, Wardyn KA, Zycinski Z, Zielonka TM. Association between clinical activity and high-resolution tomography findings in pulmonary Wegener's granulomatosis. J Physiol Pharmacol 2008;59(Suppl 6):833–8.

[25] Lutalo P, D'Cruz D. Diagnosis and classification of granulomatosis with polyangiitis (aka Wegener's granulomatosis). J Autoimmun 2014;48–49:94–8.

[26] Jayne D. Update on the European Vasculitis Study Group trials. Curr Opin Rheumatol 2001;13(1):48–55.

[27] Leavitt RY, Fauci AS, Bloch DA, Michel BA, Hunder GG, Arend WP, et al. The American College of Rheumatology 1990 criteria for the classification of Wegener's granulomatosis. Arthritis Rheum 1990;33(8):1101–7.

[28] Watts R, Lane S, Hanslik T, Hauser T, Hellmich B, Koldingsnes W, et al. Development and validation of a consensus methodology for the classification of the ANCA-associated vasculitides and polyarteritis nodosa for epidemiological studies. Ann Rheum Dis 2007;66(2):222–7.

[29] Basu N, Watts R, Bajema I, Baslund B, Bley T, Boers M, et al. EULAR points to consider in the development of classification and diagnostic criteria in systemic vasculitis. Ann Rheum Dis 2010;69(10):1744–50.

[30] Bossuyt X, Cohen Tervaert JW, Arimura Y, Blockmans D, Flores-Suárez LF, Guillevin L, et al. Position paper: revised 2017 international consensus on testing of ANCAs in granulomatosis with polyangiitis and microscopic polyangiitis. Nat Rev Rheumatol 2017;13(11):683–92.

[31] Hagen EC, Daha MR, Hermans J, Andrassy K, Csernok E, Gaskin G, et al. The diagnostic value of standardized assays for anti-neutrophil cytoplasmic antibodies (ANCA) in idiopathic systemic vasculitis: results of an international collaborative study. Kidney Int 1998;53(3):743–53.

[32] Finkielman JD, Lee AS, Hummel AM, Viss MA, Jacob GL, Homburger HA, et al. ANCA are detectable in nearly all patients with active severe Wegener's granulomatosis. Am J Med 2007;120(7):643.e9–643.e14.

[33] Weiner M, Segelmark M. The clinical presentation and therapy of diseases related to anti-neutrophil cytoplasmic antibodies (ANCA). Autoimmun Rev 2016;15(10):978–82.

[34] Schnabel A, Reuter M, Gloeckner K, Müller-Quernheim J, Gross WL. Bronchoalveolar lavage cell profiles in Wegener's granulomatosis. Respir Med 1999;93(7):498–506.

[35] Lohrmann C, Uhl M, Warnatz K, Kotter E, Ghanem N, Langer M. Sinonasal computed tomography in patients with Wegener's granulomatosis. J Comput Assist Tomogr 2006;30(1):122–5.

[36] Webb WR, Müller NL, Naidich DP. High-resolution CT of the lung. 5th ed. Philadelphia: Lippincott Williams and Wilkins; 2015, p. 646–8.

[37] Lee KS, Kim TS, Fujimoto K, Moriya H, Watanabe H, Tateishi U, et al. Thoracic manifestation of Wegener's granulomatosis: CT findings in 30 patients. Eur Radiol 2003;13(1):43–51.

[38] Lohrmann C, Uhl M, Kotter E, Burger D, Ghanem N, Langer M. Pulmonary manifestations of Wegener granulomatosis: CT findings in 57 patients and a review of the literature. Eur J Radiol 2005;53(3):471–7.

[39] Cordier JF, Valeyre D, Guillevin L, Loire R, Brechot JM. Pulmonary Wegener's granulomatosis: a clinical and imaging study of 77 cases. Chest 1990;97(4):906–12.

[40] Hansell DM. Small-vessel diseases of the lung: CT-pathologic correlates. Radiology 2002;225(3):639–53.

[41] Kim Y, Lee KS, Jung KJ, Han J, Kim JS, Suh JS. Halo sign on high-resolution CT: findings in spectrum of pulmonary diseases with pathologic correlation. J Comput Assist Tomogr 1999;23(4):622–6.

[42] Stokes TC, McCann BG, Rees RT, Sims EH, Harrison BD. Acute fulminating intrapulmonary haemorrhage in Wegener's granulomatosis. Thorax 1982;37(4):315–6.
[43] Farrelly CA. Wegener's granulomatosis: a radiological review of the pulmonary manifestations at initial presentation and during relapse. Clin Radiol 1982;33(5):545–51.
[44] Sheehan RE, Flint JD, Müller NL. Computed tomography features of the thoracic manifestations of Wegener granulomatosis. J Thorac Imaging 2003;18(1):34–41.
[45] Castañer E, Alguersuari A, Gallardo X, Andreu M, Pallardó Y, Mata JM, et al. When to suspect pulmonary vasculitis: radiologic and clinical clues. Radiographics 2010;30(1):33–53.
[46] Daum TE, Specks U, Colby TV, Edell ES, Brutinel MW, Prakash UB, et al. Tracheobronchial involvement in Wegener's granulomatosis. Am J Respir Crit Care Med 1995;151(2 Pt 1):522–6.
[47] Martinez F, Chung JH, Digumarthy SR, Kanne JP, Abbott GF, Shepard JA, et al. Common and uncommon manifestations of Wegener granulomatosis at chest CT: radiologic-pathologic correlation. Radiographics 2012;32(1):51–69.
[48] Komócsi A, Reuter M, Heller M, Muraközi H, Gross WL, Schnabel A. Active disease and residual damage in treated Wegener's granulomatosis: an observational study using pulmonary high-resolution computed tomography. Eur Radiol 2003;13(1):36–42.
[49] Néel A, Espitia-Thibault A, Arrigoni PP, Volteau C, Rimbert M, Masseau A, et al. Bronchiectasis is highly prevalent in anti-MPO ANCA-associated vasculitis and is associated with a distinct disease presentation. Semin Arthritis Rheum 2018;48(1):70–6.
[50] Cansu DÜ, Özbülbül NI, Akyol G, Arık D, Korkmaz C. Do pulmonary findings of granulomatosis with polyangiitis respond to anti-tuberculosis treatment? Rheumatol Int 2018;38(6):1131–8.
[51] Ghosh K, Pradhan V, Ghosh K. Background noise of infection for using ANCA as a diagnostic tool for vasculitis in tropical and developing countries. Parasitol Res 2008;102(5):1093–5.
[52] Bossuyt X. Serologic markers in inflammatory bowel disease. Clin Chem 2006;52(2):171–81.
[53] Philipponnet C, Garrouste C, Le Guenno G, Cartery C, Guillevin L, Boffa JJ, et al. Antineutrophilic cytoplasmic antibody-associated vasculitis and malignant hemopathies, a retrospective study of 16 cases. Joint Bone Spine 2017;84(1):51–7.
[54] Mahr A, Batteux F, Tubiana S, Goulvestre C, Wolff M, Papo T, et al. Brief report: prevalence of antineutrophil cytoplasmic antibodies in infective endocarditis. Arthritis Rheumatol 2014;66(6):1672–7.
[55] Langlois V, Lesourd A, Girszyn N, Ménard JF, Levesque H, Caron F, et al. Antineutrophil cytoplasmic antibodies associated with infective endocarditis. Medicine (Baltimore) 2016;95(3):e2564.
[56] Cornec D, Cornec-Le Gall E, Fervenza FC, Specks U. ANCA-associated vasculitis—clinical utility of using ANCA specificity to classify patients. Nat Rev Rheumatol 2016;12(10):570–9.
[57] Khan TA, Cuchacovich R, Espinoza LR, Lata S, Patel NJ, Garcia-Valladares I, et al. Vasculopathy, hematological, and immune abnormalities associated with levamisole-contaminated cocaine use. Semin Arthritis Rheum 2011;41(3):445–54.
[58] Molloy ES, Langford CA. Vasculitis mimics. Curr Opin Rheumatol 2008;20(1):29–34.
[59] Ntatsaki E, Carruthers D, Chakravarty K, D'Cruz D, Harper L, Jayne D, et al. BSR and BHPR guideline for the management of adults with ANCA-associated vasculitis. Rheumatology (Oxford) 2014;53(12):2306–9.
[60] Ozaki S. ANCA-associated vasculitis: diagnostic and therapeutic strategy. Allergol Int 2007;56(2):87–96.
[61] Mukhtyar C, Lee R, Brown D, Carruthers D, Dasgupta B, Dubey S, et al. Modification and validation of the Birmingham vasculitis activity score (version 3). Ann Rheum Dis 2009;68(12):1827–32.
[62] Jayne D, Rasmussen N, Andrassy K, Bacon P, Tervaert J, Dadoniene J, et al, European Vasculitis Study Group. A randomized trial of maintenance therapy for vasculitis associated with antineutrophil cytoplasmic autoantibodies. N Engl J Med 2003;349(1):36–44.
[63] de Groot K, Rasmussen N, Bacon PA, Tervaert JW, Feighery C, Gregorini G, et al. Randomized trial of cyclophosphamide versus methotrexate for induction of remission in early systemic antineutrophil cytoplasmic antibody-associated vasculitis. Arthritis Rheum 2005;52(8):2461–9.
[64] Mukhtyar C, Guillevin L, Cid MC, Dasgupta B, de Groot K, Gross W, et al, European Vasculitis Study Group. EULAR recommendations for the management of primary small and medium vessel vasculitis. Ann Rheum Dis 2009;68(3):310–7.
[65] Walsh M, Merkel PA, Mahr A, Jayne D. Effects of duration of glucocorticoid therapy on relapse rate in antineutrophil cytoplasmic antibody-associated vasculitis: a meta-analysis. Arthritis Care Res 2010;62(8):1166–73.
[66] Pagnoux C. Updates in ANCA-associated vasculitis. Eur J Rheumatol 2016;3(3):122–33.
[67] Stone JH, Merkel PA, Spiera R, Seo P, Langford CA, Hoffman GS, et al. Rituximab versus cyclophosphamide for ANCA-associated vasculitis. N Engl J Med 2010;363(3):221–32.
[68] Jones RB, Tervaert JW, Hauser T, Luqmani R, Morgan MD, Peh CA, et al. Rituximab versus cyclophosphamide in ANCA-associated renal vasculitis. N Engl J Med 2010;363(3):211–20.
[69] de Joode AA, Sanders JS, Smid MW, Stegeman CA. Plasmapheresis rescue therapy in progressive systemic ANCA-associated vasculitis: single-center results of stepwise escalation of immunosuppression. J Clin Apher 2014;29(5):266–72.
[70] McAdoo SP, Bedi R, Tarzi R, Griffith M, Pusey CD, Cairns TD. Ofatumumab for B cell depletion therapy in ANCA-associated vasculitis: a single-centre case series. Rheumatology (Oxford) 2016;55(8):1437–42.
[71] Langford CA, Monach PA, Specks U, Seo PC, Cuthbertson D, McAlear CA, et al. An open-label trial of abatacept (CTLA4-IG) in nonsevere relapsing granulomatosis with polyangiitis (Wegener's). Ann Rheum Dis 2014;73(7):1376–9.
[72] Clinicaltrials n.d. A study to evaluate the safety and efficacy of CCX168 in subjects with ANCA-associated vasculitis. Available from: https://clinicaltrials.gov/ct2/show/NCT01363388?term=ccx168&rank=3.

[73] de Groot K, Harper L, Jayne DR, Flores Suarez LF, Gregorini G, Gross WL, et al, European Vasculitis Study Group. Pulse versus daily oral cyclophosphamide for induction of remission in antineutrophil cytoplasmic antibody—associated vasculitis: a randomized trial. Ann Intern Med 2009;150(10):670–80.

[74] Pagnoux C, Mahr A, Hamidou MA, Boffa JJ, Ruivard M, Ducroix JP, et al, French Vasculitis Study Group. Azathioprine or methotrexate maintenance for ANCA-associated vasculitis. N Engl J Med 2008;359(26):2790–803.

[75] Hiemstra TF, Walsh M, Mahr A, Savage CO, de Groot K, Harper L, et al, European Vasculitis Study Group. Mycophenolate mofetil vs azathioprine for remission maintenance in antineutrophil cytoplasmic antibody-associated vasculitis: a randomized controlled trial. JAMA 2010;304(21):2381–8.

[76] Singer O, McCune WJ. Update on maintenance therapy for granulomatosis with polyangiitis and microscopic polyangiitis. Curr Opin Rheumatol 2017;29(3):248–53.

[77] Brown K, Frankel S, Cool C. Pulmonary vasculitis. In: Murray & Nadel's textbook on respiratory medicine. 6th ed. Elsevier; 2015. p. 1066–80.

Chapter 7

Eosinophilic lung disease

Alexander Averyanov[a,b], Evgeniya Kogan[c], Victor Lesnyak[d], Igor E. Stepanyan[e], Olesya Danilevskaya[f]

[a]Clinical Department, Pulmonology Research Institute Under FMBA of Russia, Moscow, Russia, [b]Pulmonary Division, Federal Research Clinical Center Under FMBA of Russia, Moscow, Russia, [c]Anatomic Pathology Department, Sechenov University, Moscow, Russia, [d]Radiology Department, Federal Research Clinical Center Under FMBA of Russia, Moscow, Russia, [e]Central TB Research Institute, Moscow, Russia, [f]Endoscopy Department, Pulmonology Research Institute Under FMBA of Russia, Moscow, Russia

Chapter 7.1

Eosinophilic granulomatosis with polyangiitis (Churg-Strauss syndrome)

Eosinophilic granulomatosis with polyangiitis (EGPA) is a systemic necrotizing ANCA-associated vasculitis that affects the small- and medium-sized vessels and is manifested by severe asthma and eosinophilia [1].

The first description of EGPA was made in 1951 by J. Churg and L. Strauss as an allergic granulomatosis with periarteritis and angiitis [2]. Until 2012 the eponym "Churg-Strauss syndrome" [3] was the generally accepted term to designate EGPA.

EGPA is a rare disease, with a prevalence of only 10–13 cases per 1 million people [4–6], and the annual incidence rate does not exceed 6.8 cases per million [7]. The morbidity rate among men and women is equal, and the mean age of EGPA patients is 38–49 years [8, 9].

The etiology of EGPA is unknown. About a third of patients have a history of allergies, mainly to indoor allergens. It is theorized that disease triggers can be mold and yeast fungi, inhalations of cocaine, bird allergens, vaccination against influenza and hepatitis B, and certain medications [10, 11]. The relationship between the intake of leukotriene receptor inhibitors and development of EGPA has been previously discussed [12]. However, this statement remains in dispute [1, 12], and thus some authors recommend avoiding the prescription of leukotriene receptor inhibitors to EGPA patients [13].

The presence of eosinophilia and asthma in the majority of EGPA patients supports the theory of allergic genesis of systemic vasculitis. An increased level of antibodies to the cytoplasm of neutrophils in a significant proportion of patients strongly suggests an autoimmune component in disease pathogenesis.

Morphology

Microscopic manifestations of lung lesions in EGPA consist of a combination of necrotizing eosinophilic cell vasculitis, the presence of inflammatory infiltrates with an admixture of eosinophils in lung tissue, extravascular granulomas, and typical structural changes, which are characteristic of asthma [2, 14]. However, the full morphological symptom complex is observed in about 20% of cases. Eosinophilic vasculitis develops in the arteries and veins and affects the medium- and small-sized vessels down to the capillaries. In addition, fibrinoid necrosis and inflammatory infiltration with an admixture of eosinophils have been observed in the vessel walls (Figs. 7.1.1–7.1.2), which can lead to thrombosis and pulmonary tissue infarction. Pulmonary capillaritis may be accompanied by alveolar hemorrhages (Fig. 7.1.3). Interstitial fibrosis can develop in place of necrosis and hemorrhage.

Difficult to Diagnose Rare Diffuse Lung Disease. https://doi.org/10.1016/B978-0-12-815375-8.00007-8

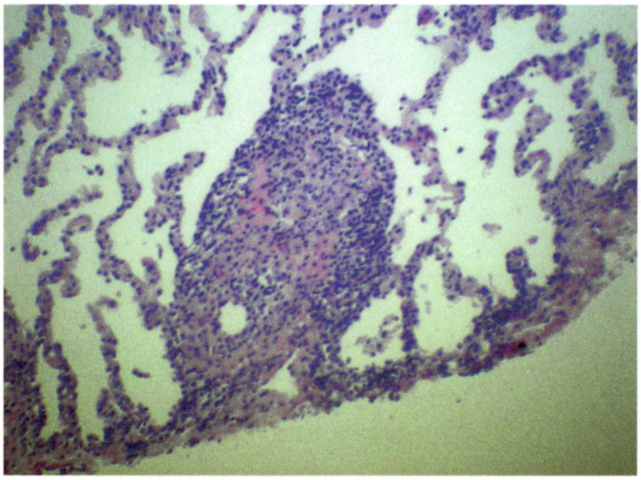

FIG. 7.1.1 EGPA. Destructive and productive vasculitis of the pulmonary small artery with the presence of eosinophils in the inflammatory infiltrate. Hematoxylin and eosin staining, 200×.

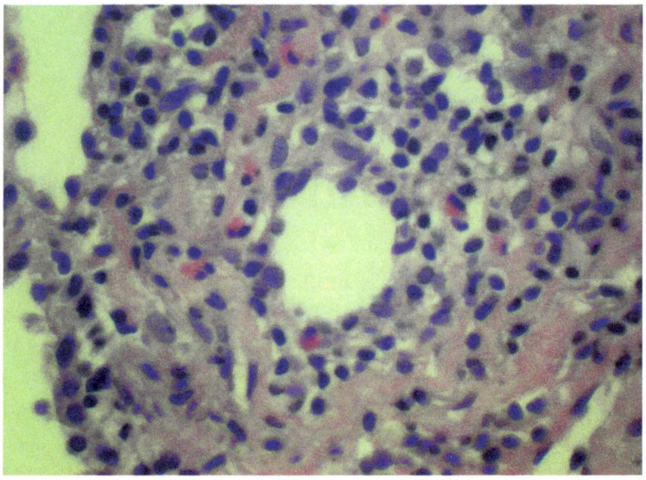

FIG. 7.1.2 EGPA. Destructive and productive vasculitis of the pulmonary small artery with the presence of eosinophils in the composition of the inflammatory infiltrate. Hematoxylin and eosin staining, 400×.

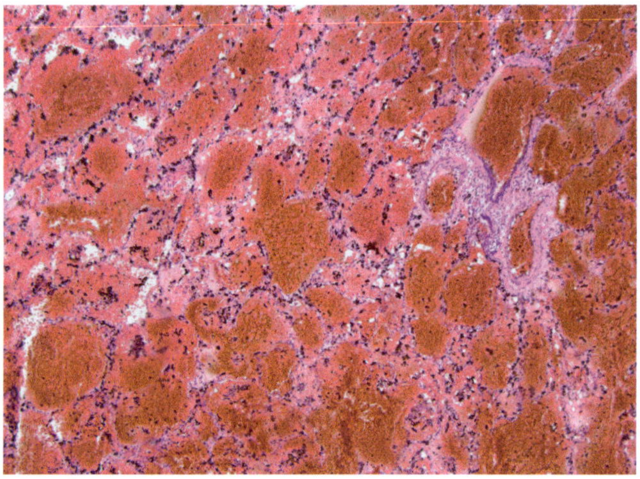

FIG. 7.1.3 EGPA. Pulmonary hemorrhages, the alveolar lumens are filled with red blood cells due to capillaritis. Hematoxylin and eosin staining, 200×.

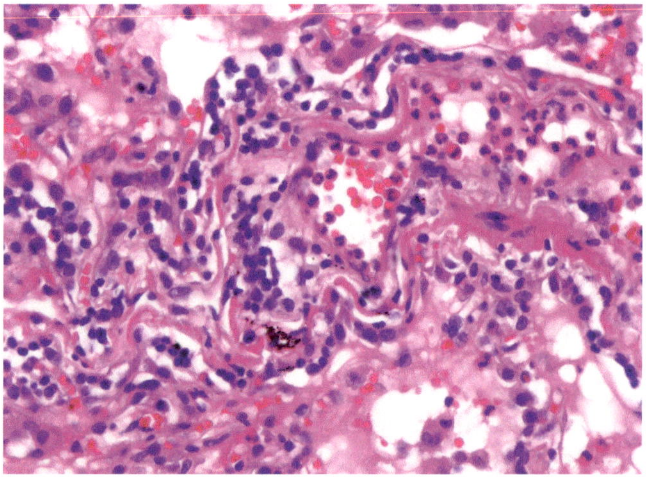

FIG. 7.1.4 EGPA. Eosinophilic inflammatory infiltrates in the walls and lumens of the alveoli. Hematoxylin and eosin staining, 250×.

Eosinophilic inflammatory infiltrates are also found in the alveoli and bronchial walls (Fig. 7.1.4). The bronchi contain viscous mucus, a thickened hyalinized basement membrane, hypertrophy of the bronchial glands, and smooth muscle fibers (Fig. 7.1.5).

Granulomas consist of histiocytes, multinucleated cells, and eosinophils and also contain a zone of necrosis in their center (Fig. 7.1.6). Granulomas are revealed rarely in patients treated with systemic corticosteroids [15].

In the early stages of the disease, eosinophilic infiltrates appear in the lungs with no signs of vasculitis. Then, eosinophilic pneumonia, granulomatosis, and destructive thrombovasculitis develop. With an aggressive course of treatment, the process involves the skin, lungs, central and peripheral nervous systems, kidneys, and heart.

Differential diagnosis in histological examination should be performed with chronic eosinophilic pneumonia, granulomatosis with polyangiitis (GPA), and pulmonary eosinophilia due to fungal and parasitic diseases and as medicinal vasculitis. Eosinophilic pneumonia differs from EGPA in the absence of necrotic vasculitis in biopsy samples. However, productive vasculitis can also be detected in eosinophilic pneumonia. Unlike GPA the granulomatous tissue comprises a large number of eosinophils, and also, there are no "geographic" foci of necrosis. In the case of fungal and parasitic diseases and drug-induced injury of the lungs (especially the intake of estrogen and carbamazepine), diagnosis should be

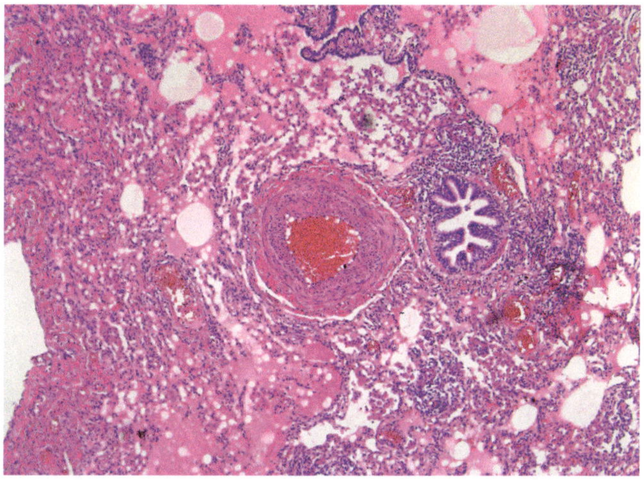

FIG. 7.1.5 EGPA. The wall of the small bronchus with thickened hyalinized basement membrane, hyperplasia of mucus-producing cells, hypertrophy of the bronchial glands, smooth muscle fibers, and inflammatory infiltrate in the wall. Hematoxylin and eosin staining, 200×.

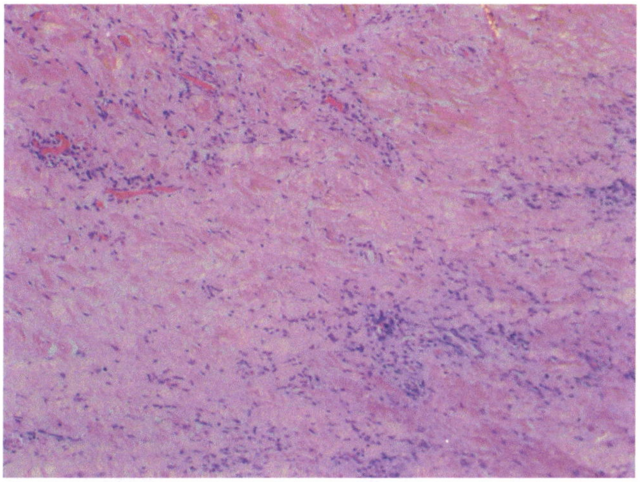

FIG. 7.1.6 Focus of hyalinosis with inflammatory infiltration. Hematoxylin and eosin staining, 100×.

made based on a combination of clinical, anamnestic, and laboratory data, since these conditions may be accompanied by granulomatous vasculitis and eosinophilic cell infiltrates.

Clinical presentation

Generally, three consecutive phases are distinguished in the course of EGPA [16]:

1. *Phase 1*: The prodromal phase may last for many years and manifest itself as atopic asthma and/or allergic rhinitis. Asthma is usually severe and requires the administration of high doses of inhaled corticosteroids. However, asthma is not always controlled without systemic steroids. Chronic allergic rhinitis or rhinosinusitis occurs in 3/4 of patients and may be accompanied by nasal polyposis and significant nasal obstruction. Eosinophils are usually present in the nasal secretion.
2. *Phase 2*: The phase of blood eosinophilia and tissue eosinophilic infiltrates. Some changes in the lungs occur in almost all EGPA patients [17]. Clinically, they are difficult to differentiate as asthma symptoms dominate.
3. *Phase 3*: Full-scaled vasculitis when common symptoms appear, such as fever, weight loss, arthralgia, and myalgia. There are also nonrespiratory symptoms of the disease, namely, paresthesia, erythematous rash on the skin, and signs of glomerulonephritis [18]. Cardiac disease can range from asymptomatic to full-scale eosinophilic myocarditis with severe heart failure [19]. It usually takes 3–6 years from the first symptoms of asthma and rhinitis to the onset of systemic vasculitis. A fulminant course is also possible with the rapid development of severe damage to multiple organs [20]. Some researchers have identified two phenotypes of the disease, determined by the presence or absence of ANCA (Table 7.1.1) [21].

TABLE 7.1.1 Phenotypes of EGPA

	Vasculitis phenotype	Eosinophilic infiltrative phenotype
Incidence	40%	60%
p-ANCA	+	–
Differences in the clinical presentation	Glomerulonephritis Neuropathy Purpura Histological signs of vasculitis	Eosinophilic myocarditis Eosinophilic infiltrates in the lungs Fever

Diagnostic tests

Blood eosinophilia is an obligate sign of EGPA. The diagnostically significant level is 10% of the total number of leukocytes; however, as a rule, before starting treatment, the proportion of eosinophils in the blood is 30%–40% [22]. Generally, there are direct correlations between the severity of clinical symptoms and the number of eosinophils in the blood. In all patients with lung lesions, pronounced eosinophilia of the BAL fluid is detected and often exceeds 30% [23]. Total IgE is increased in most patients; however, the level does not correlate with disease activity [20]. The appearance in the blood of increased antibody titers to the cytoplasm of neutrophils, mainly to myeloperoxidase (p-ANCA), is a significant marker of the disease. This finding is observed in approximately 40% of patients with a full-scale presentation of the disease. It is interesting to note that the level of p-ANCA has no direct correlation with the severity of the EGPA course; however, it is a strong predictor of relapses [24, 25].

Computed tomography

Changes in HRCT scans of the thoracic organs in EGPA patients are noted in various studies in 76%–100% of patients [17, 18].

The CT semiotics of changes in the lungs in the case of EGPA is nonspecific and, thus, cannot be used as a basis of the diagnosis. The main findings are summarized as follows (Fig. 7.1.7 and 7.1.8):

- Ground-glass opacity areas
- Foci of consolidation
- Small nodules

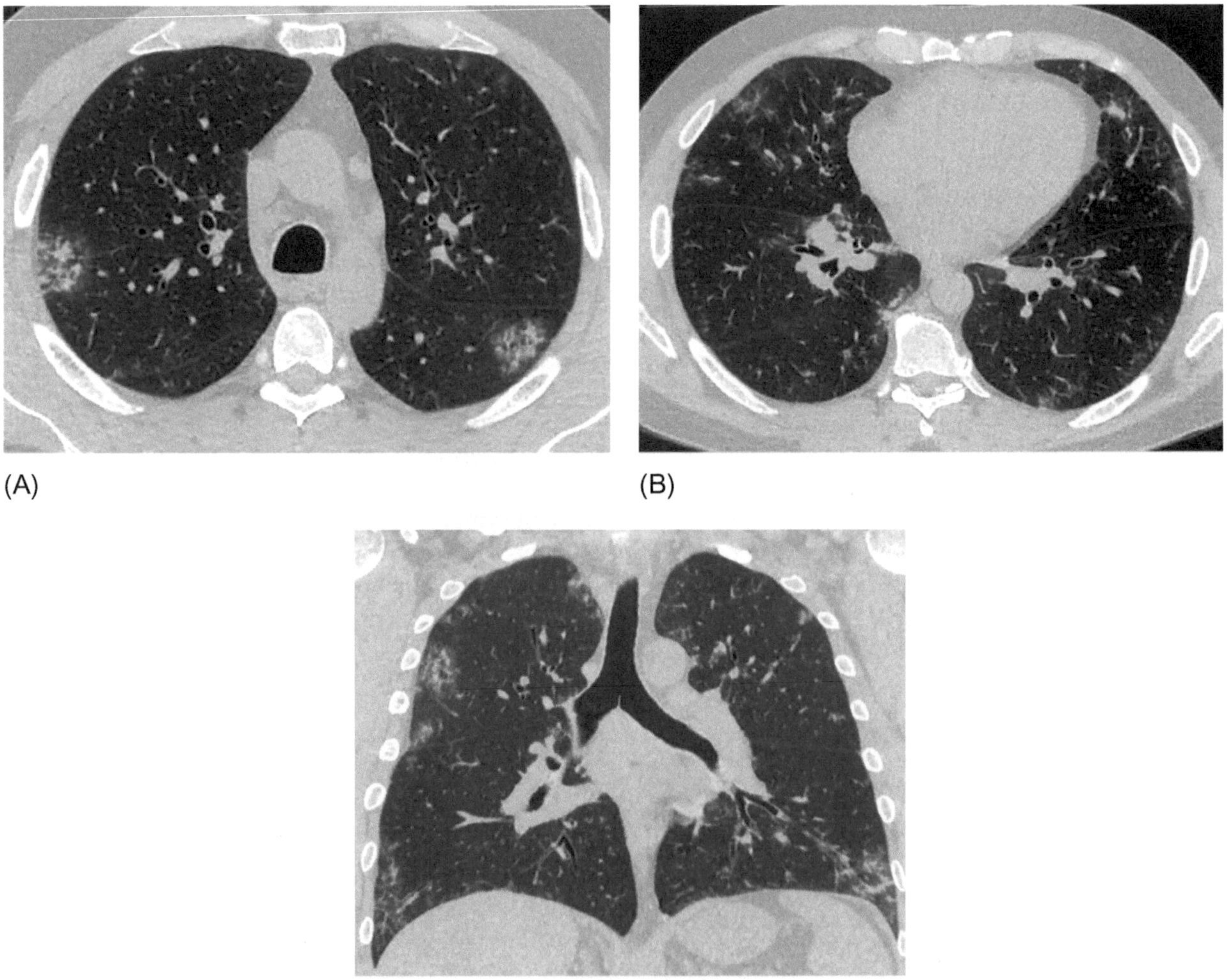

(A) (B) (C)

FIG. 7.1.7 EGPA. Bilateral round infiltrates of uneven attenuation and nodules in subpleural zones. Thickened bronchial walls and separated interlobular septa (A). Foci of ground-glass opacity and consolidation with subpleural distribution. The tree-in-bud sign (B). Abnormalities are symmetrically distributed in both upper and lower lobes (C).

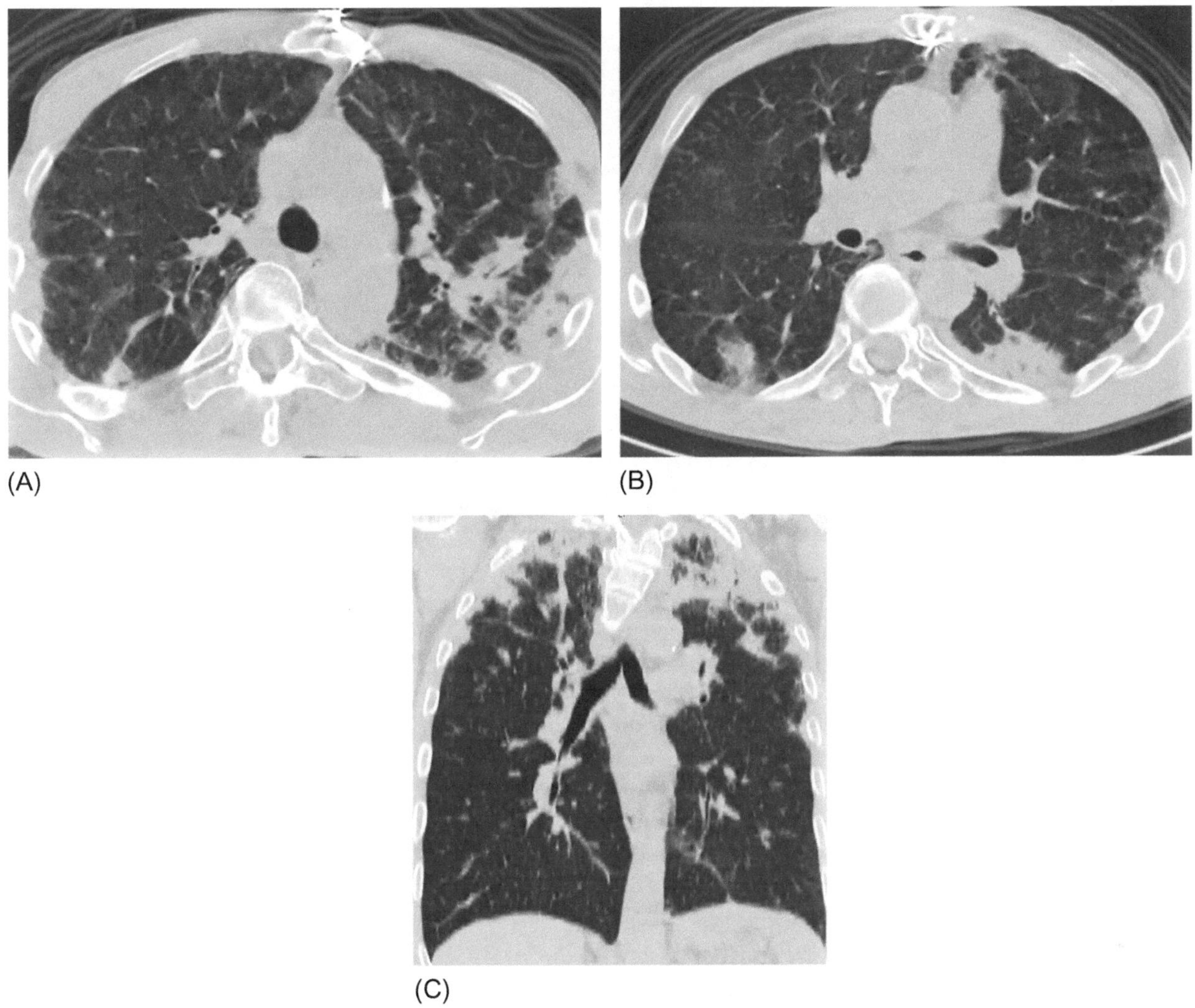

FIG. 7.1.8 EGPA. Bilateral consolidation zones located mostly subpleural. Part of the interlobular septa (A and B) is thickened. Pathological lesions are mostly distributed in the upper lobes (C).

- Thickening of interlobular septa
- Thickening of bronchial walls

Szczeklik et al. analyzed the data from six studies assessing CT semiotics in the case of EGPA and concluded that there is a rather large variation in the incidence of various radiological signs (Table 7.1.2) [17].

The most common CT finding of EGPA is the ground-glass opacity sign. The distribution of GGO and consolidation is usually subpleural (60%) or patchy (40%) [26, 27]. These changes are quickly reversed with proper treatment. Nodules with EGPA do not exceed 10mm in diameter, and sometimes, zones of hyperlucency can be seen within them, such as areas of cavitation due to necrotizing vasculitis. However, in general, the formation of cavities in zones of infiltration are not typical for EGPA, unlike, for example, granulomatosis with polyangiitis (GPA) [18].

Pleural effusion occurs in about 20% of EGPA patients and may have an eosinophilic characteristic but may also be a manifestation of cardiac or renal failure indicating a lesion to the corresponding organs [28].

Although rare, another possible finding from HRCT in EGPA patients may be a tree-in-bud sign and the reversed halo sign [26].

Pericardial effusion is a rare but life-threatening condition in EGPA patients. Sometimes, pericardial effusion and heart failure become the dominant symptoms in clinical presentation of the disease [29].

The CT semiotics usually reflect certain histological characteristics of the lesion zone in the lungs. According to Kim et al. [26] and Silva et al. [30], the signs of ground-glass opacity and consolidation correspond to the histological pattern of eosinophilic pneumonia, necrotizing granulomatosis, or necrotizing vasculitis. Nodules of up to 10mm corresponded to eosinophilic lymphocytic bronchiolitis, while larger foci corresponded to areas of necrotizing granulomatosis with alveolar hemorrhages

TABLE 7.1.2 The incidence of CT signs of EGPA in different studies (17 with additions)

	Patients		Signs on high-resolution computed tomography (%)						
	N	Mean age (years)	Ground-glass opacity	Consolidation	Nodules	TIS	TBW	Pleural effusion	LAP
Worthy (1998)	17	47	59	–	23.5	ND	35.3	11.8	ND
Choi (2000)	9	35	100	55.5	89	22.2	77.8	22.2	44.4
Silva (2005)	7	49	71.4	57.1	85.7	57.1	57.1	57.1	28.6
Kim (2007)	25	45	40	32	48	32	40	ND	ND
Furuiye (2010)	16	57	100	75	56	ND	37.5	18.7	50
Szczeklik (2010)	15	40.5	66.6	60	26.6	66.6	66.6	17.6	0.7
Li (2016)	43	53	79.1	ND	ND	ND	51.2	ND	ND

LAP, lymphadenopathy; *ND*, no data; *TIS*, thickening of interlobular septa; *TBW*, thickening of the bronchial walls.

and surrounding eosinophilic infiltration. Thickening of the bronchial walls corresponds to their eosinophilic and lymphocytic infiltration, and the thickening of interlobular septa corresponds to edema and infiltration with eosinophils [26, 30].

Bronchoscopy and bronchoalveolar lavage (BAL) are mandatory diagnostic tools in the case of eosinophilic lung lesions. Normally the content of eosinophils in BAL fluid should not exceed 1% [31]. In the study of Schnabel et al. [32], in EGPA patients, in the active phase of the disease, the median eosinophil level was 34% (interquartile range 8%–71%), while in the partial remission phase, it was only 0.5%. This is generally lower than observed in idiopathic chronic eosinophilic pneumonia for which the conservative cutoff for inclusion in clinical trials is 40% of eosinophils [33]. Transbronchial lung biopsy in EGPA is rarely performed due to insufficient informational value of the material obtained.

Probe-based confocal laser endomicroscopy

Our experience in alveoloscopy in one patient with EGPA with lung lesion did not establish any signs specific to only this disease. The endomicroscopic pattern of the distal respiratory tract was characterized by thickening of the interalveolar septa up to 14–20 μm in 8 of the 15 areas examined (Fig. 7.1.9A). Distelectatic changes (Fig. 7.1.9C and E) were found in 4 of the 15 areas. Cellular elements in the lumen of the alveoli, which fluoresce at a wavelength of 488 μm, were rarely registered, while moderate quantity was seen in 2 of 15 areas. There was a notable visualization of a large number of vessels with a pronounced fluorescence intensity of their walls (94 ± 19.5 relative units); at least 2–3 of these vessels were found within one region of interest in 80% of cases (Figs. 7.1.9C and 1.40). Such a phenomenon could reflect pulmonary vasculitis. In 7 of the 15 areas, small rounded, brightly fluorescent elements 3–7.5 μm in diameter were found in the lumen of the respiratory bronchioles (Fig. 7.1.9D); these are most likely protein elements and markers of inflammation that reflect the activity of eosinophilic bronchiolitis. In 5 of the 15 areas, a viscous moderately fluorescent secretion (Fig. 7.1.9B) was present in the lumen of the alveoli in differing amounts, which represent either an eosinophilic exudate or bronchial secretion; this is likely due to hyperproduction that could partially enter the alveolar passages.

Diagnostic criteria

Since 1990 most researchers use the criteria of the American College of Rheumatology [14] to diagnose EGPA, which include six characteristics:

(1) Asthma
(2) Blood eosinophilia >10%
(3) Mono- or polyneuropathy
(4) Migrating infiltrates in the lungs
(5) Rhinosinusopathy
(6) Extravascular eosinophilic infiltrates in the study of biopsy samples

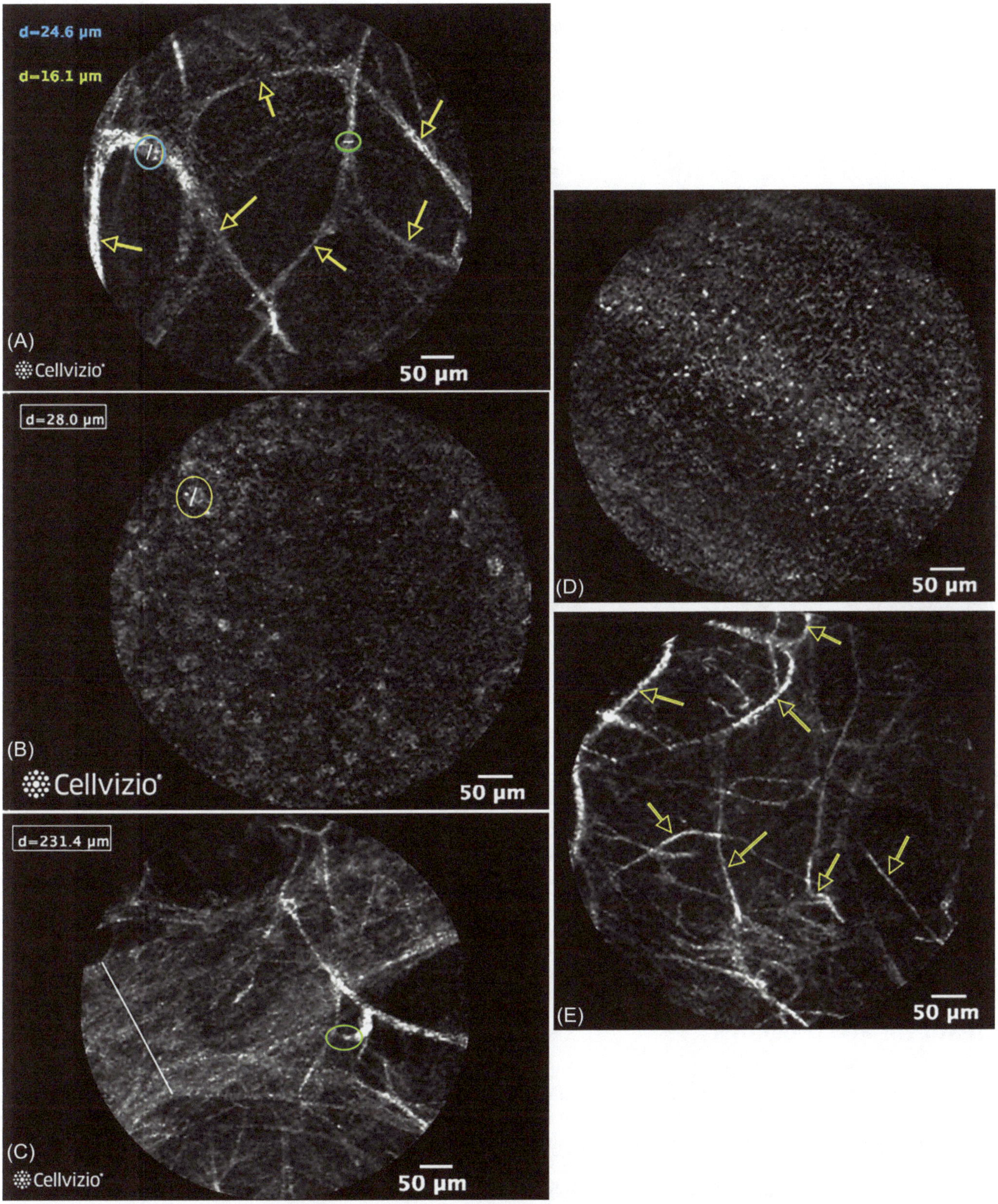

FIG. 7.1.9 Endomicroscopic distal airway pattern in an EGPA patient. Alveoli are of regular round shape. The interalveolar septa are thickened in different degrees (A). Viscous secretion in the lumen of the alveoli complicates the visualization of the alveolar structures. Alveolar macrophages are determined against the background of the secretion (B). About half of the field of vision (FOV) is occupied by a vessel with a diameter of 231.4 μm. In the right half of the FOV, the free edge of the fragmented interalveolar septum (ellipse) (C) is clearly visible. Brightly fluorescent elements are observed in the lumen of the terminal bronchioles (D). The shape of the alveolar cavities is impaired due to their partial collapse (E). The elastic fibers of interalveolar septa are indicated with yellow arrows. Some measurement areas are marked with ellipses.

TABLE 7.1.3 The proposed nomenclature for EGPA, hypereosinophilic asthma, and hypereosinophilic asthma with systemic manifestation

Hypereosinophilic asthma	EGPA	Hypereosinophilic asthma with systemic manifestations
Asthma + blood hypereosinophilia $>1.5\times10^9$/L or >10% leukocytes	*Definite vasculitis features* – Biopsy-proved necrotizing vasculitis of any organ – Biopsy-proved necrotizing glomerulonephritis or crescentic glomerulonephritis – Alveolar hemorrhage – Palpable purpura – Myocardial infarction due to proved coronaritis *Definite surrogates of vasculitis* – Hematuria associated with red casts or >10% dysmorphic erythrocytes or hematuria and 2+ proteinuria on urinalysis – Leukocytoclastic capillaritis and/or eosinophilic infiltration of the arterial wall at biopsy *Mononeuritis or mononeuritis multiplex* *ANCA with at least one extrathoracic non-ENT manifestation of disease*	Hypereosinophilic asthma Any systemic manifestation other than polyangiitis or surrogate of vasculitis or mononeuritis and the absence of ANCA

Non-ENT, other than ear, nose, and throat disease.
(Extract from Cottin V, Bel E, Bottero P, Dalhoff K, Humbert M, Lazor R, et al. Revisiting the systemic vasculitis in eosinophilic granulomatosis with polyangiitis (Churg-Strauss): a study of 157 patients by the Groupe d'Etudes et de Recherche sur les Maladies Orphelines Pulmonaires and the European Respiratory Society Taskforce on eosinophilic granulomatosis with polyangiitis (Churg-Strauss). Autoimmun Rev 2017;16(1):1–9.)

The presence of at least four of these criteria makes the diagnosis of EGPA highly probable with a sensitivity of 85% and specificity of 99% [14].

Later, simpler criteria were proposed [22]:

(1) Asthma
(2) Blood eosinophilia
(3) Eosinophilic and granulomatous inflammation in the airways
(4) Necrotizing vasculitis of small and medium vessels

For a long time, it was believed that the diagnosis of EGPA does not always require histological verification and thus relied only on the presence of characteristic clinical and laboratory signs [13]. However, in 2017, the experts of the French group of orphan lung diseases and the European Respiratory Society published results of an assessment of systemic manifestations in 157 patients with an established diagnosis of EGPA, according to the criteria earlier. The study determined that only 59% of patients had certain or strong surrogate signs of polyangiitis, mononeuritis, or positive ANCA. For the remaining patients, it was proposed to recategorize the term of EGPA in favor of a new nosological form, namely, hypereosinophilic asthma with (any) systemic (nonvasculitic) manifestations (Table 7.1.3) [34]. However, this topic is still under discussion.

Treatment and prognosis

Like other ANCA-associated vasculitis, the treatment plan for EGPA patients includes the induction of remission and further maintenance. As a basis for choosing induction therapy, it has been suggested to consider other factors of an unfavorable prognosis on a Five-Factor Score (FFS), including age over 65 years; signs of cardiac lesions; renal failure (creatinine more than 1.70 mg/dL); involvement of gastrointestinal tract; and the absence of lesion of the ear, nose, or throat [35]. The presence of at least one of these factors ($FFS \geq 1$) requires the prescription of combined induction therapy with glucocorticosteroids and cytotoxic agents (cyclophosphamide first) at a dose of 0.6–0.7 g/m^2 intravenously on the day 1, 15, and 30 and then every 3 weeks [35]. In addition, the manifestations that are not included in the said scale, such as diffuse alveolar hemorrhages, eye lesion, and severe peripheral neuropathy, can also be an indication for administering cytostatics [36]. In the absence of all these factors, the first-line drugs for the treatment of EGPA are glucocorticosteroids, and the recommended induction dose of prednisone is 1 mg/kg/day for 2–4 weeks [1, 3]. In the case of a more severe course but without unfavorable prognosis factors, intravenous methylprednisolone pulse therapy at a dose of 7.5–15 mg/kg/day for 1–3 days is recommended [36].

Remission is achievable in more than 90% of EGPA patients. However, relapses occur in 40%–50% of patients [15, 34].

Systemic glucocorticosteroids are used as maintenance therapy in EGPA patients without a life-threatening condition. In cases of recurrent disease and if it is not possible to reduce the dose of prednisone lower than 7.5 mg/day after 3 or 4 months of treatment, a combination of glucocorticosteroids with azathioprine, methotrexate, mycophenolate mofetil, or leflunomide is used [1, 3]. The duration of maintenance therapy should be at least 18–24 months [36].

Rituximab represents chimerical monoclonal antibodies against CD-20 B lymphocytes and also reduces the production of interleukin-5 by T lymphocytes. It is prescribed to ANCA-positive patients refractory to other types of therapy of EGPA or with renal diseases, both for induction and for remission maintenance. The efficiency of the medication exceeds 80% [1, 37, 38]. To prevent severe bronchospasm, which can be caused by rituximab, before the drug intake, the pulse administration of glucocorticosteroids is performed [39]. Mepolizumab is a human anti-IL-5 monoclonal antibody, and the first pilot study in 10 patients with relapsing EGPA is currently under way. This study administered high doses of systemic steroids in eight cases to achieve sustained remission and reduction of doses of prednisone down to 7.5 mg/day [40]. In a later study of 136 EGPA patients, mepolizumab allowed for protocol-defined remission in 53% of patients and reduction of the dose of prednisone to 4 mg/day or lower in 44% of patients [41].

In general, plasma exchange has not proved to be effective for EGPA patients but can be considered as a method of treatment in individual ANCA-positive patients with antiglomerular basement membrane disease or rapidly progressive glomerulonephritis [1].

Prescription of intravenous immunoglobulin may be considered in addition to corticosteroids and immunosuppressants in EGPA patients who are refractory to other types of therapy, or during pregnancy, or with drug-induced hypogammaglobulinemia [1].

Omalizumab may have a sparing effect with systemic corticosteroids with refractory or recurrent EGPA; however, reducing the dose of corticosteroids increases the risks of severe flares of the disease [42].

The prognosis of EGPA with adequate treatment is generally favorable as the 5-year survival rate has been observed to be 95% [43], and the 10-year survival rate reaches 79% [15]. Patients with the above five factors (5FFS) have the worst prognosis [35].

A prerequisite for successful management of EGPA patients is a multidisciplinary approach that is provided in specialized clinics [44].

References

[1] Groh M, Pagnoux C, Baldini C, Bel E, Bottero P, Cottin V, et al. Eosinophilic granulomatosis with polyangiitis (Churg-Strauss) (EGPA) Consensus Task Force recommendations for evaluation and management. Eur J Intern Med 2015;26(7):545–53.

[2] Churg J, Strauss L. Allergic granulomatosis, allergic angiitis and periarteritis nodosa. Am J Pathol 1951;27(2):277–301.

[3] Jennette JC, Falk RJ, Bacon PA, Basu N, Cid MC, Ferrario F, et al. 2012 Revised international chapel hill consensus conference nomenclature of vasculitides. Arthritis Rheum 2013;65(1):1–11.

[4] Dunogué B, Pagnoux C, Guillevin L. Churg-Strauss syndrome: clinical symptoms, complementary investigations, prognosis and outcome, and treatment. Semin Respir Crit Care Med 2011;32(3):298–309.

[5] Mahr A, Guillevin L, Poissonnet M, Aymé S. Prevalences of polyarteritis nodosa, microscopic polyangiitis, Wegener's granulomatosis, and Churg–Strauss syndrome in a French urban multiethnic population in 2000: a capture–recapture estimate. Arthritis Rheum 2004;51(1):92–9.

[6] Mohammad AJ, Jacobsson LT, Mahr AD, Sturfelt G, Segelmark M. Prevalence of Wegener's granulomatosis, microscopic polyangiitis, polyarteritis nodosa and Churg–Strauss syndrome within a defined population in southern Sweden. Rheumatology (Oxford) 2007;46(8):1329–37.

[7] Herlyn K, Hellmich B, Gross WL, Reinhold-Keller E. Stable incidence of systemic vasculitides in Schleswig-Holstein, Germany. Dtsch Arztebl Int 2008;105(19):355–61.

[8] Keogh KA, Specks U. Churg-Strauss syndrome: clinical presentation, antineutrophil cytoplasmic antibodies, and leukotriene receptor antagonists. Am J Med 2003;115(4):284–90.

[9] Guillevin L, Cohen P, Gayraud M, Lhote F, Jarrousse B, Casassus P. Churg-Strauss syndrome. Clinical study and long-term follow-up of 96 patients. Medicine (Baltimore) 1999;78(1):26–37.

[10] Bottero P, Bonini M, Vecchio F, Grittini A, Patruno GM, Colombo B, et al. The common allergens in the Churg-Strauss syndrome. Allergy 2007;62(11):1288–94.

[11] Vanoli M, Gambini D, Scorza R. A case of Churg-Strauss vasculitis after hepatitis B vaccination. Ann Rheum Dis 1998;57(4):256–7.

[12] Harrold LR, Patterson MK, Andrade SE, Dube T, Go AS, Buist AS, et al. Asthma drug use and the development of Churg–Strauss syndrome (CSS). Pharmacoepidemiol Drug Saf 2007;16(6):620–6.

[13] Cottin V, Cordier JF. Eosinophilic lung diseases. Immunol Allergy Clin North Am 2012;32(4):557–86.

[14] Masi AT, Hunder GG, Lie JT, Michel BA, Bloch DA, Arend WP, et al. The American College of Rheumatology 1990 criteria for the classification of Churg-Strauss syndrome (allergic granulomatosis and angiitis). Arthritis Rheum 1990;33(8):1094–100.

[15] Comarmond C, Pagnoux C, Khellaf M, Cordier JF, Hamidou M, Viallard JF, et al. Eosinophilic granulomatosis with polyangiitis (Churg–Strauss): clinical characteristics and long-term follow-up of the 383 patients enrolled in the French Vasculitis Study Group cohort. Arthritis Rheum 2013;65(1):270–81.

[16] Lanham JG, Elkon KB, Pusey CD, Hughes GR. Systemic vasculitis with asthma and eosinophilia: a clinical approach to the Churg-Strauss syndrome. Medicine (Baltimore) 1984;63(2):65–81.

[17] Szczeklik W, Sokołowska B, Mastalerz L, Grzanka P, Górka J, Pacułt K, et al. Pulmonary findings in Churg-Strauss syndrome in chest X-rays and high resolution computed tomography at the time of initial diagnosis. Clin Rheumatol 2010;29(10):1127–34.

[18] Choi YH, Im JG, Han BK, Kim JH, Lee KY, Myoung NH. Thoracic manifestation of Churg-Strauss syndrome: radiologic and clinical findings. Chest 2000;117(1):117–24.

[19] Ginsberg F, Parrillo JE. Eosinophilic myocarditis. Heart Fail Clin 2005;1(3):419–29.

[20] Chumbley LC, Harrison Jr EG, RA DR. Allergic granulomatosis and angiitis (Churg-Strauss syndrome). Report and analysis of 30 cases. Mayo Clin Proc 1977;52(8):477–84.

[21] Kallenberg CG. Churg-Strauss syndrome: just one disease entity? Arthritis Rheum 2005;52(9):2589–93.

[22] Jennette JC, Falk RJ, Andrassy K, Bacon PA, Churg J, Gross WL, et al. Nomenclature of systemic vasculitides. Proposal of an international consensus conference. Arthritis Rheum 1994;37(2):187–92.

[23] Wallaert B, Gosset P, Prin L, Bart F, Marquette CH, Tonnel AB. Bronchoalveolar lavage in allergic granulomatosis and angiitis. Eur Respir J 1993;6(3):413–7.

[24] Sablé-Fourtassou R, Cohen P, Mahr A, Pagnoux C, Mouthon L, Jayne D, et al. Antineutrophil cytoplasmic antibodies and the Churg-Strauss syndrome. Ann Intern Med 2005;143(9):632–8.

[25] Sinico RA, Di Toma L, Maggiore U, Bottero P, Radice A, Tosoni C, et al. Prevalence and clinical significance of antineutrophil cytoplasmic antibodies in Churg-Strauss syndrome. Arthritis Rheum 2005;52(9):2926–35.

[26] Kim YK, Lee KS, Chung MP, Han J, Chong S, Chung MJ, et al. Pulmonary involvement in Churg-Strauss syndrome: an analysis of CT, clinical, and pathologic findings. Eur Radiol 2007;17(12):3157–65.

[27] Worthy SA, Müller NL, Hansell DM, Flower CD. Churg-Strauss syndrome: the spectrum of pulmonary CT findings in 17 patients. AJR Am J Roentgenol 1998;170(2):297–300.

[28] Tanizawa K, Kaji Y, Tanaka E, Inoue T, Sakuramoto M, Minakuchi M, et al. Massive eosinophilic pleural effusion preceding vasculitic symptoms in Churg-Strauss syndrome. Intern Med 2010;49(9):841–5.

[29] Matsuo S, Sato Y, Matsumoto T, Naiki N, Horie M. Churg-Strauss syndrome presenting with massive pericardial effusion. Heart Vessels 2007;22(2):128–30.

[30] Silva CI, Muller NL, Fujimoto K, Johkoh T, Ajzen SA, Churg A. Churg-Strauss syndrome: high resolution CT and pathologic findings. J Thorac Imaging 2005;20(2):74–80.

[31] Meyer KC, Raghu G, Baughman RP, Brown KK, Costabel U, du Bois RM, et al. An official American Thoracic Society clinical practice guideline: the clinical utility of bronchoalveolar lavage cellular analysis in interstitial lung disease. Am J Respir Crit Care Med 2012;185(9):1004–14.

[32] Schnabel A, Csernok E, Braun J, Gross WL. Inflammatory cells and cellular activation in the lower respiratory tract in Churg-Strauss syndrome. Thorax 1999;54(9):771–8.

[33] Marchand E, Reynaud-Gaubert M, Lauque D, Durieu J, Tonnel AB, Cordier JF. Idiopathic chronic eosinophilic pneumonia. A clinical and follow-up study of 62 cases. Medicine (Baltimore) 1998;77(5):299–312.

[34] Cottin V, Bel E, Bottero P, Dalhoff K, Humbert M, Lazor R, et al. Revisiting the systemic vasculitis in eosinophilic granulomatosis with polyangiitis (Churg-Strauss): a study of 157 patients by the Groupe d'Etudes et de Recherche sur les Maladies Orphelines Pulmonaires and the European Respiratory Society Taskforce on eosinophilic granulomatosis with polyangiitis (Churg-Strauss). Autoimmun Rev 2017;16(1):1–9.

[35] Guillevin L, Pagnoux C, Seror R, Mahr A, Mouthon L, Le Toumelin P, French Vasculitis Study Group (FVSG). The Five-Factor Score revisited: assessment of prognoses of systemic necrotizing vasculitides based on the French Vasculitis Study Group (FVSG) cohort. Medicine (Baltimore) 2011;90(1):19–27.

[36] Pagnoux C, Groh M. Optimal therapy and prospects for new medicines in eosinophilic granulomatosis with polyangiitis (Churg-Strauss syndrome). Expert Rev Clin Immunol 2016;12(10):1059–67.

[37] Mohammad AJ, Hot A, Arndt F, Moosig F, Guerry MJ, Amudala N, et al. Rituximab for the treatment of eosinophilic granulomatosis with polyangiitis (Churg–Strauss). Ann Rheum Dis 2016;75(2):396–401.

[38] Guerry MJ, Brogan P, Bruce IN, D'Cruz DP, Harper L, Luqmani R, et al. Recommendations for the use of rituximab in anti-neutrophil cytoplasm antibody-associated vasculitis. Rheumatology (Oxford) 2012;51(4):634–43.

[39] Bouldouyre MA, Cohen P, Guillevin L. Severe bronchospasm associated with rituximab for refractory Churg–Strauss syndrome. Ann Rheum Dis 2009;68(4):606.

[40] Moosig F, Gross WL, Herrmann K, Bremer JP, Hellmich B. Targeting interleukin-5 in refractory and relapsing Churg–Strauss syndrome. Ann Intern Med 2011;155(5):341–3.

[41] Wechsler ME, Akuthota P, Jayne D, Khoury P, Klion A, Langford CA, et al. Mepolizumab or placebo for eosinophilic granulomatosis with polyangiitis. N Engl J Med 2017;376(20):1921–32.

[42] Jachiet M, Samson M, Cottin V, Kahn JE, Le Guenno G, Bonniaud P, et al. Anti-IgE monoclonal antibody (omalizumab) in refractory and relapsing eosinophilic granulomatosis with polyangiitis (Churg-Strauss): data on seventeen patients. Arthritis Rheumatol 2016;68(9):2274–82.

[43] Ribi C, Cohen P, Pagnoux C, Mahr A, Arène JP, Lauque D, et al. Treatment of Churg-Strauss syndrome without poor-prognosis factors: a multicenter, prospective, randomized, open-label study of seventy-two patients. Arthritis Rheum 2008;58(2):586–94.

[44] Moosig F, Bremer JP, Hellmich B, Holle JU, Holl-Ulrich K, Laudien M, et al. A vasculitis centre based management strategy leads to improved outcome in eosinophilic granulomatosis and polyangiitis (Churg–Strauss, EGPA): monocentric experiences in 150 patients. Ann Rheum Dis 2013;72(6):1011–7.

Chapter 7.2

Idiopathic chronic eosinophilic pneumonia

Idiopathic chronic eosinophilic pneumonia (ICEP) is a rare disease of unknown origin, characterized by general and respiratory symptoms and eosinophilia of the blood and/or alveolar eosinophilic accumulation with a chronic or subacute course [1]. Carrington et al. were first to describe ICEP in nine females with fever, dyspnea, weight loss, and eosinophilic pulmonary infiltrates [2].

ICEP is a rare disease. In Iceland, its prevalence is 0.23 cases per 100,000 people [3]. ICEP accounts for <2.5% of all interstitial lung diseases [4]. The disease usually develops around ages 40–50 years, and it occurs in females twice as frequently as in males [1]. In 60.5%–94.7% of cases, ICEP develops in nonsmokers and those patients who have never smoked [5].

Eosinophilic pneumonia (EP) can be induced by heterogeneous factors, such as the intake of medications (nonsteroidal antiinflammatory drugs and antibiotics) or helminths; moreover, it can be associated with mycoses, rheumatoid arthritis, T-cell lymphomas, and nontuberculous mycobacterial pulmonary disease. In these cases, EP is considered secondary, unlike ICEP that is treated as an independent disease of unknown etiology, after ruling out other causes of eosinophilia [6, 7].

Morphology

ICEP is characterized by the development of infiltrates located primarily in the peripheral subpleural regions of the lungs. They capture alveolar septa and lumens and consist of eosinophilic granulocytes, lymphocytes, macrophages, and plasma cells with possible signs of organized EP and eosinophilic microabscesses (Figs. 7.2.1–7.2.4) [2, 8].

Prolonged disease-related changes are often accompanied by mild interstitial fibrosis [8]. Aside from eosinophilia, there are frequently histological signs of bronchiolitis or nonnecrotizing vasculitis [9]. Because of its resemblance to other EPs, ICEP is diagnosed after ruling out such diseases as acute eosinophilic pneumonia; simple pulmonary eosinophilia (the Löffler syndrome); tropical eosinophilia; secondary eosinophilia with parasitic, fungal, and bacterial infections; allergic bronchopulmonary aspergillosis; drug-related diseases; and hypersensitivity pneumonitis.

Clinical presentation

Many patients present with a history of allergic diseases, namely, nasal polyposis, allergic rhinitis (up to 37%), atopic dermatitis (up to 15%), and drug allergies. In 33%–61% of cases, ICEP develops in patients with histories of asthma, although in 15% of cases, asthma onset coincides with the appearance of eosinophilic infiltrates in the lungs [5, 10].

Symptoms of the disease are not specific; their frequency varies significantly according to different studies; most cases include prolonged cough (65%–96%) with sputum discharge, dyspnea (21%–91%), fever (29%–77%), night sweats, general fatigue, asthenia (6%–88%), loss of appetite, and weight loss (2%–75%) [5]. The course is usually subacute or chronic, developing over a course of a few weeks, when symptoms first appear to a few months with peak disease activity. Some patients may have thoracalgia, arthralgia, skin rashes, and minor pericardial effusion, which make the clinical picture very similar to eosinophilic granulomatosis with polyangiitis (EGPA) [11]. During auscultation, wheezes and moist crackles are heard in about 1/3 of patients [1]. In contrast to acute eosinophilic pneumonia, ICEP, even at the height of clinical manifestations, is usually not accompanied by severe respiratory failure, which requires respiratory support.

Blood testing typically reveals inflammatory changes, namely, increased ESR and C-reactive protein. In 2/3 of the patients, there are increases in the level of total IgE [12]. Blood eosinophilia (if not treated with systemic glucocorticoids) is a very significant sign of ICEP. Usually the number of eosinophils varies from 5000 to 6000/mm^3; the eosinophil fraction amounts to 20%–30% and even higher. If a small number of patients with ICEP (5%–12%) do not exhibit blood eosinophilia [7], elaborating BAL eosinophilia will help [1, 9]. Active degranulation of eosinophils significantly increases urine concentrations of eosinophilic neurotoxins [13].

In 50%–70% of patients with a primary diagnosis of ICEP, impaired pulmonary ventilation and decreased diffusion lung capacity are revealed [5]. Different patients may exhibit obstructive, restrictive, or mixed patterns of lung

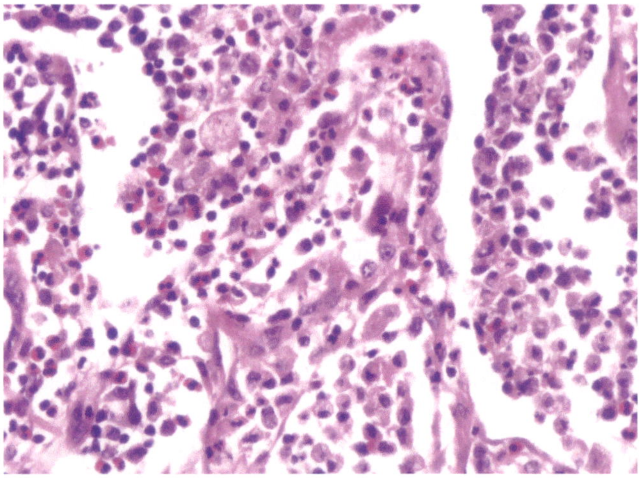

FIG. 7.2.1 ICEP. A large number of eosinophils in the inflammatory infiltrate. Hematoxylin and eosin staining, 400×.

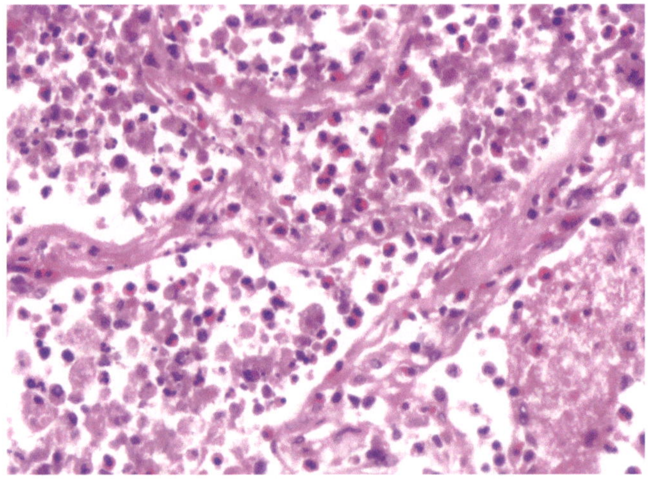

FIG. 7.2.2 ICEP. The focus of pneumonia with eosinophils within inflammatory infiltrate and the onset of fibrosis of alveolar septa. Hematoxylin and eosin staining, 400×.

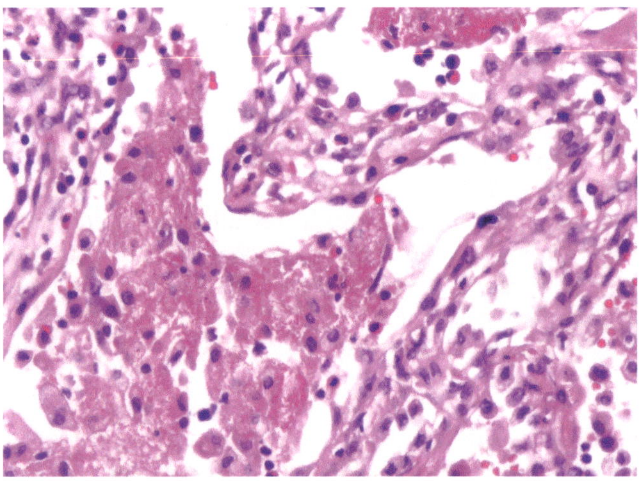

FIG. 7.2.3 ICEP. Focus of eosinophilic pneumonia with the organization of fibrinous exudate. Hematoxylin and eosin staining, 400×.

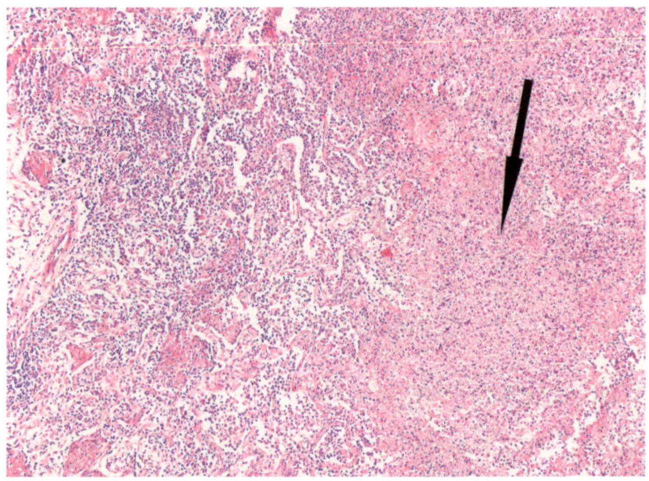

FIG. 7.2.4 ICEP. Focus of eosinophilic pneumonia with a focus of necrosis (arrow), fibrinous exudate in the lumens of alveoli, and foci of organization. Hematoxylin and eosin staining, 100×.

function damage. Despite the disappearance of pulmonary infiltrates, in 36% of patients, certain functional disorders, often obstructive, persist [5]. In patients with ICEP, high initial levels of eosinophils in BAL predict the persistence of obstructive symptoms [14].

Radiological imaging

On chest radiographs, areas of peripheral consolidation are usually found, generally in the upper lobes of the lungs [15]. The following primary findings can be detected on the chest high-resolution computer tomography [15, 16]:

- Widespread bilateral areas of homogeneous consolidation, located along the periphery of the lung parenchyma and subpleurally (where the distribution is usually symmetrical); however, consolidation is more often localized in the upper and middle zones (Fig. 7.2.5). The sign of air bronchogram is often observed. Sometimes the foci of consolidation may have a rounded shape and radiant contours and resemble tumor nodes (Fig. 7.2.6).
- Ground-glass opacity, sometimes associated with thickening of the interlobular septa (crazy paving sign) (Fig. 7.2.7).
- Small airspace nodules (see Fig. 7.2.7).
- Linear opacities (often when patients are being treated with steroids during foci of consolidation resolution) (Fig. 7.2.7).

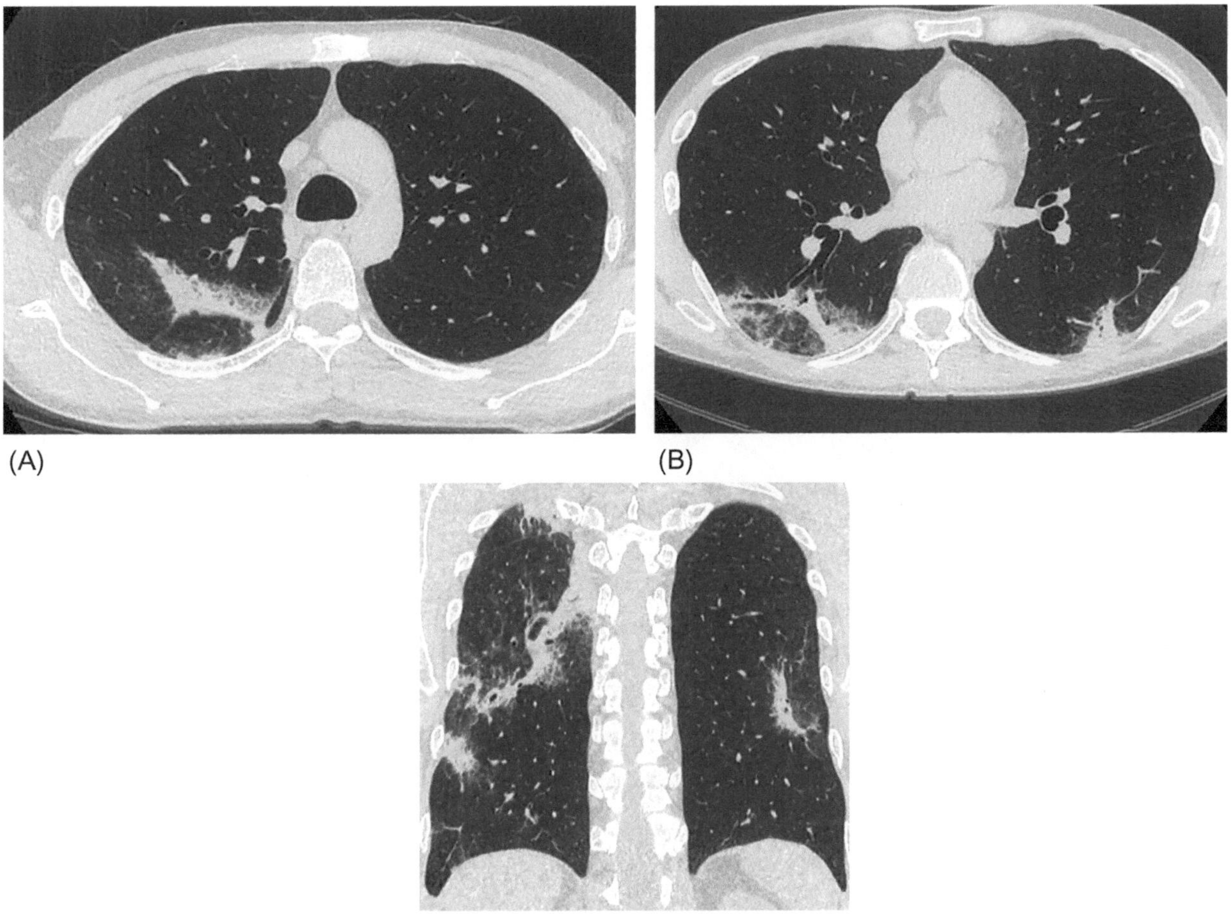

(A) (B) (C)

FIG. 7.2.5 ICEP. Subpleural zones of consolidation and ground-glass opacity, inside of which thickened interlobular septa are visible (A and B). Reversed halo sign is visible in the right lung (B). Changes are most pronounced in the upper and middle segments (C).

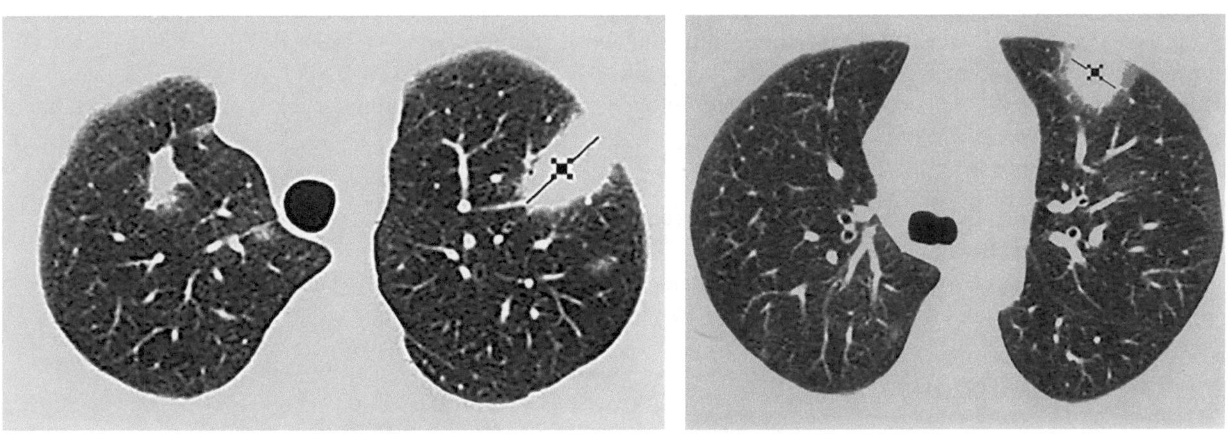

(A) (B)

FIG. 7.2.6 ICEP. Foci of consolidation of a homogeneous structure with irregular and spiculated contours, resemble a tumor lesion (A and B).

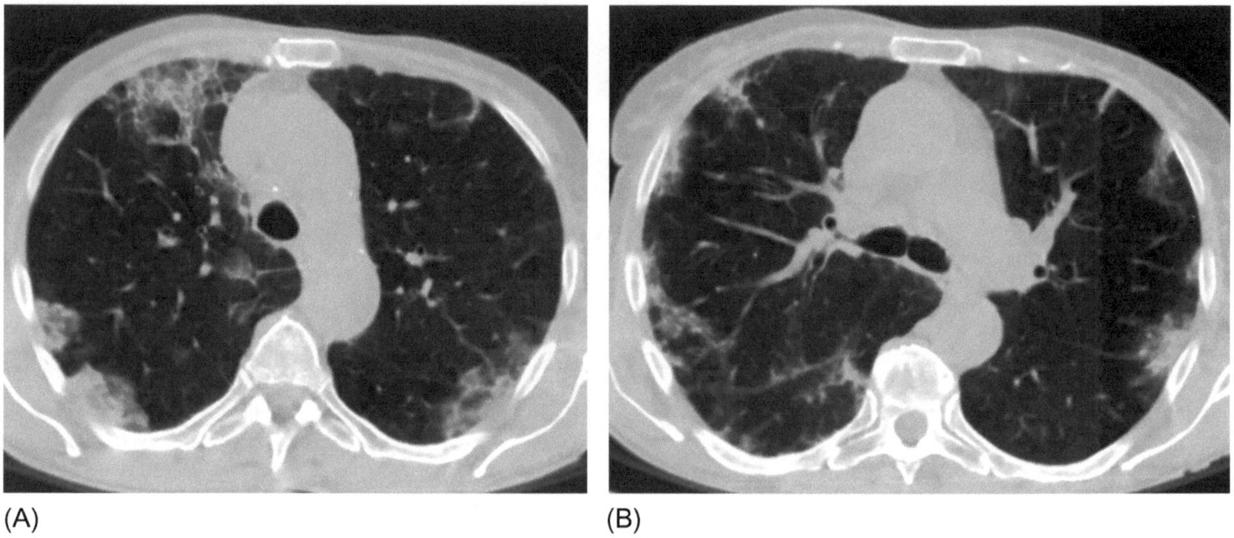

FIG. 7.2.7 ICEP. Subpleural spotted areas of consolidation and ground-glass opacities, separate nodules and linear seals are visible (A and B). Inside the ground-glass opacities, there are thickened interlobular septa.

Increases in intrathoracic lymph nodes and pleural effusion are less typical for ICEP [1, 15, 17]. An unusual (but possible) HRCT pattern of ICEP revealed dominance of the reversed halo sign, particularly when localized to the upper lobes of the lungs (Fig. 7.2.8) [18].

Typically, CT signs of ICEP rapidly diminish with systemic steroid treatment, but in the absence of adequate therapy, they can persist, changing their configuration [17].

Bronchoalveolar lavage (BAL)

BAL fluid eosinophilia is a key diagnostic sign of ICEP and can help the patient avoid lung biopsy. These levels can vary from 37% to 58% among other cells [5] with increase in total cytosis and possible moderate neutrophilia and lymphocytosis [1]. Eosinophil degranulation increases eosinophilic cationic proteins and eosinophil-derived neurotoxins in BAL fluid [19].

Increased expression of some cytokines and chemokines in ICEP can potentially assist differential diagnosis of this disease. Walker et al. found that IL-2R expressed by CD4 cells and significantly elevated concentrations of IL-2 and IL-5 in BAL fluid in five patients with ICEP compared with healthy volunteers and patients with other eosinophilic lung diseases [20].

Other possible biomarkers of ICEP in BAL fluid include IL-18, IL-25, and elevated interleukin-18 levels [21, 22]. Kobayashi et al. revealed increased uric acid, adenosine triphosphate, and IL-33 in BAL fluid from patients with ICEP. These increases correlated with the number of eosinophils and IL-5 concentrations [23]. Recently a significant increase in the level of human intelectin-1 (hITLN-1), Ca^{2+}-dependent lectin, was found in BAL fluid in patients with CEP and hypersensitive pneumonitis. In contrast, those with cryptogenic organized pneumonia or usual and nonspecific interstitial pneumonia remained within normal limits [24].

The criteria for diagnosing ICEP include [11]

(1) respiratory symptoms that persist for at least 2–4 weeks,
(2) eosinophilia in BAL fluid (>40%) and/or blood (>1000/mm^3),
(3) the presence of zones of alveolar consolidation with air bronchograms located mainly in the peripheral fields of the lungs,
(4) ruling out of other known variants of pulmonary eosinophilia.

Treatment and prognosis

ICEP usually responds well to treatment with systemic corticosteroids with complete regression of clinical, laboratory, and radiological manifestations. Pathological findings on CT can disappear within a week after the onset of appropriate therapy [1]. The starting dose of 0.5 mg/kg of prednisone for 2 weeks is recommended most commonly, then 0.25 mg/kg for another 2 weeks, followed by a gradual decrease to complete withdrawal after 6 months from the onset of treatment, absent relapse [11].

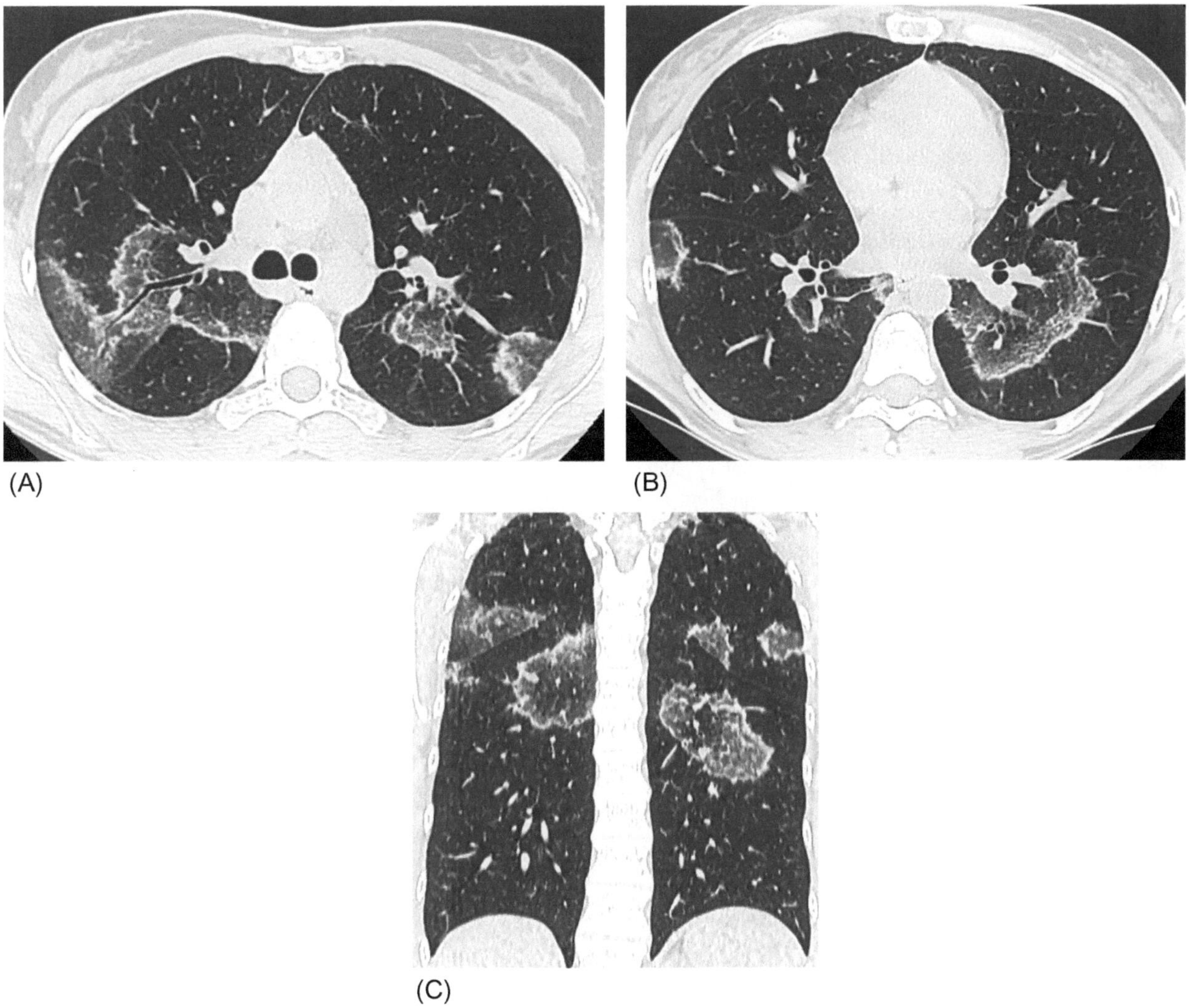

FIG. 7.2.8 ICEP. Uncommon pattern of distribution and shape of abnormalities. Peribronchovascular and subpleural extensive areas of ground-glass opacities surrounded by a thin layer of consolidation-reversed halo sign (A, B, and C). Only the upper and middle zones of the lungs are affected (C).

In cases of relapse, treatment is continued for up to 1 year. The initial dose of prednisone for relapse treatment is usually 20 mg/day [1]. As long as the patient continues to exhibit asthma symptoms, medium and high doses of inhaled corticosteroids are required. Inhaled corticosteroids were also successfully used to prevent relapse of ICEP after the withdrawal of systemic corticosteroids [25]. There are reports of isolated cases of successful treatment of ICEP with omalizumab, which is monoclonal IgE antibody, and mepolizumab, which is monoclonal interleukin-5 antibody. These medications are considered alternatives to corticosteroids in patients with contraindications for such treatments [26, 27].

ICEP is characterized by a long course and relapses that develop in about half of the patients after corticosteroids are withdrawn or their dose is reduced [28]. Fatal outcomes are extremely rare.

References

[1] Marchand E, Reynaud-Gaubert M, Lauque D, Durieu J, Tonnel AB, Cordier JF. Idiopathic chronic eosinophilic pneumonia: a clinical and follow-up study of 62 cases. The Groupe d'Etudes et de Recherche sur les Maladies "Orphelines" Pulmonaires (GERM"O"P). Medicine (Baltimore) 1998;77(5):299–312.

[2] Carrington C, Addington W, Goff A, Madoff IM, Marks A, Schwaber JR, et al. Chronic eosinophilic pneumonia. N Engl J Med 1969;280(15):787–98.

[3] Sveinsson OA, Isaksson HJ, Gudmundsson G. Chronic eosinophilic pneumonia in Iceland: clinical features, epidemiology and review. Laeknabladid 2007;93(2):111–6.

[4] Thomeer MJ, Costabe U, Rizzato G, Poletti V, Demedts M. Comparison of registries of interstitial lung diseases in three European countries. Eur Respir J Suppl 2001;32:114s–8s.

[5] Suzuki Y, Suda T. Long-term management and persistent impairment of pulmonary function in chronic eosinophilic pneumonia: a review of the previous literature. Allergol Int 2018;67(3):334–40.

[6] Fox B, Seed WA. Chronic eosinophilic pneumonia. Thorax 1980;35(8):570–80.

[7] Marchand E, Cordier JF. Idiopathic chronic eosinophilic pneumonia. Orphanet J Rare Dis 2006;1:11–4.

[8] Mochimaru H, Kawamoto M, Fukuda Y, Kudoh S. Clinicopathological differences between acute and chronic eosinophilic pneumonia. Respirology 2005;10(1):76–85.

[9] Jederlinic PJ, Sicilian L, Gaensler EA. Chronic eosinophilic pneumonia. A report of 19 cases and a review of the literature. Medicine (Baltimore) 1988;67(3):154–62.

[10] Marchand E, Etienne-Mastroianni B, Chanez P, Lauque D, Leclerc P, Cordier JF, Groupe d'Etudes et de Recherche sur les Maladies Orphelines Pulmonaires. Idiopathic chronic eosinophilic pneumonia and asthma: how do they influence each other. Eur Respir J 2003;22(1):8–13.

[11] Cottin V, Cordier JF. Eosinophilic lung diseases. Immunol Allergy Clin North Am 2012;32(4):557–86.

[12] Naughton M, Fahy J, FitzGerald MX. Chronic eosinophilic pneumonia: a long-term follow-up of 12 patients. Chest 1993;103(1):162–5.

[13] Cottin V, Deviller P, Tardy F, Cordier JF. Urinary eosinophil-derived neurotoxin/protein X: a simple method for assessing eosinophil degranulation in vivo. J Allergy Clin Immunol 1998;101(1 Pt 1):116–23.

[14] Durieu J, Wallaert B, Tonnel AB. Long-term follow-up of pulmonary function in chronic eosinophilic pneumonia. Groupe d'Etude en Pathologie Interstitielle de la Sociétè de Pathologie Thoracique du Nord. Eur Respir J 1997;10(2):286–91.

[15] Ebara H, Ikezoe J, Johkoh T, Kohno N, Takeuchi N, Kozuka T, et al. Chronic eosinophilic pneumonia: evolution of chest radiograms and CT features. J Comput Assist Tomogr 1994;18(5):737–44.

[16] Jeong YJ, Kim KI, Seo IJ, Lee CH, Lee KN, Kim KN, et al. Eosinophilic lung diseases: a clinical, radiologic, and pathologic overview. Radiographics 2007;27(3):617–37.

[17] Mayo JR, Müller NL, Road J, Sisler J, Lillington G. Chronic eosinophilic pneumonia: CT findings in six cases. AJR Am J Roentgenol 1989;153(4):727–30.

[18] Rea G, Dalpiaz G, Vatrella A, Damiani S, Marchiori EJ. The reversed halo sign: also think about chronic eosinophilic pneumonia. J Bras Pneumol 2017;43(4):322–3.

[19] Jorens PG, Van Overveld FJ, Van Meerbeeck JP, Van Alsenoy L, Gheuens E, Vermeire PA. Evidence for marked eosinophil degranulation in a case of eosinophilic pneumonia. Respir Med 1996;90(8):505–9.

[20] Walker C, Bauer W, Braun RK, Menz G, Braun P, Schwarz F, et al. Activated T cells and cytokines in bronchoalveolar lavages from patients with various lung diseases associated with eosinophilia. Am J Respir Crit Care Med 1994;150(4):1038–48.

[21] Katoh S, Matsumoto N, Matsumoto K, Fukushima K, Matsukura S. Elevated interleukin-18 levels in bronchoalveolar lavage fluid of patients with eosinophilic pneumonia. Allergy 2004;59(8):850–6.

[22] Katoh S, Ikeda M, Matsumoto N, Shimizu H, Abe M, Ohue Y, et al. Possible role of IL-25 in eosinophilic lung inflammation in patients with chronic eosinophilic pneumonia. Lung 2017;195(6):707–12.

[23] Kobayashi T, Nakagome K, Noguchi T, Kobayashi K, Ueda Y, Soma T, et al. Elevated uric acid and adenosine triphosphate concentrations in bronchoalveolar lavage fluid of eosinophilic pneumonia. Allergol Int 2017;66S:S27–34.

[24] Shinohara T, Tsuji S, Okano Y, Machida H, Hatakeyama N, Ogushi F. Elevated levels of intelectin-1, a pathogen-binding lectin, in the BAL fluid of patients with chronic eosinophilic pneumonia and hypersensitivity pneumonitis. Intern Med 2018;10.

[25] Chan C, DeLapp D, Nystrom P. Chronic eosinophilic pneumonia: adjunctive therapy with inhaled steroids. Respir Med Case Rep 2017;22:11–4.

[26] Kaya H, Gümüş S, Uçar E, Aydoğan M, Muşabak U, Tozkoparan E, et al. Omalizumab as a steroid-sparing agent in chronic eosinophilic pneumonia. Chest 2012;142(2):513–6.

[27] To M, Kono Y, Yamawaki S, Soeda S, Katsube O, Kishi H, et al. A case of chronic eosinophilic pneumonia successfully treated with mepolizumab. J Allergy Clin Immunol Pract 2018;6(5):1746–1748.e1.

[28] Ishiguro T, Takayanagi N, Uozumi R, Tada M, Kagiyama N, Takaku Y, et al. The long-term clinical course of chronic eosinophilic pneumonia. Intern Med 2016;55(17):2373–7.

Chapter 7.3

Differential diagnosis of eosinophilic pulmonary diseases

Eosinophilic pulmonary disease (EPD) is a heterogeneous group of disorders including the accumulation of eosinophils in the lung parenchyma [1]. All have common characteristics, such as eosinophilia of the blood and/or bronchoalveolar lavage (BAL), the presence of areas of increased pulmonary attenuation on high-resolution computed tomography (HRCT), and eosinophilic infiltration of the lung tissue, which can be detected during biopsy [1, 2]. Idiopathic and secondary EPDs are usually identified [3] (Table 7.3.1). In addition, some lung diseases may be accompanied by moderate tissue or blood eosinophilia (organizing pneumonia [OP], hypersensitivity pneumonitis, Langerhans cell histiocytosis, idiopathic interstitial pneumonia, and so forth) [4]. However, in such cases, eosinophils do not have a leading role in the pathogenesis and, except for OP, are not considered in the differential range of classic EPD.

In the first stage of the differential diagnosis of EPD, determining whether the disorder is secondary or idiopathic is necessary. For this purpose a thorough history is required (history of taking medications and food supplements, eating raw meat and fish products, contact with animals, staying in epidemically unfavorable areas, the presence of other diseases, and so forth). It is necessary to conduct serological diagnostics of widespread toxocariasis, which can cause severe eosinophilia and, at the same time, can be asymptomatic [5]. Strongyloidiasis, which is prone to progression under treatment with glucocorticosteroids even several decades after infection, can cause eosinophilia [6]. Therefore antibodies to *Strongyloides stercoralis* are recommended for determination in patients with eosinophilia of unknown origin. The study of feces for helminth infections also should be included in the list of diagnostic tests for eosinophilia of unknown origin [7]. Fungal infections, especially *Aspergillus* spp., can induce blood eosinophilia and diseases, such as allergic bronchopulmonary aspergillosis (ABPA) and bronchocentric granulomatosis (BG). Therefore detection of their possible presence (IgE and IgG to *Aspergillus* spp. combined with the definition of *Aspergillus* spp. in sputum and/or BAL fluids) is usually required in the diagnostic search for possible causes of EPD. High blood levels of vitamin B12, alkaline phosphatase, and tryptase are very typical for the myeloproliferative variant of the hypereosinophilic syndrome (HES). Therefore these common tests and antineutrophil cytoplasmic antibodies (ANCA) are usually included in the list of primary laboratory tests in the case of tissue and peripheral eosinophilia [1, 7] (Table 7.3.2).

When examining an EPD patient, attention should be paid to the presence of lesions of other organs, systemic manifestations that are typical for EGPA and HES. With parasitic invasions, damage to the liver, lymphadenopathy of the mediastinum, and abdominal lymph nodes may occur. The characteristics of EPD manifestations in HRCT also may be the key to a correct diagnosis (Table 7.3.3). In general, special aspects of clinical and radiological pattern of pulmonary eosinophilia generally enable to establish a correct diagnosis, avoiding lung biopsy in most cases [2]. Even in the absence of clinical data,

TABLE 7.3.1 Eosinophilic pulmonary diseases

Idiopathic eosinophilic pulmonary diseases	Secondary eosinophilic pulmonary diseases
Simple pulmonary eosinophilia Eosinophilic granulomatosis with polyangiitis Chronic eosinophilic pneumonia Acute eosinophilic pneumonia Hypereosinophilic syndrome Idiopathic hypereosinophilic obliterating bronchiolitis	*Infection-induced pulmonary eosinophilia* (A) Parasitic (ascariasis, strongyloidiasis, toxocariasis, paragonimiasis, opisthorchiasis, and so forth) (B) Fungal (allergic bronchopulmonary aspergillosis and bronchiolocentric granulomatosis) (C) Other infections *Drug-, toxic-, or radiation-induced pulmonary eosinophilia* – Antibiotics (nitrofurans, daptomycin, vancomycin, tetracyclines, sulfonamides, penicillins, linezolid, and so forth) – Chemotherapeutic (methotrexate, dapsone, sulfasalazine, pyrimethamine, allopurinol, carbamazepine, nonsteroidal antiinflammatory, and so forth) – Toxic (L-tryptophan, aniline derivatives, and so forth) – Radiation pneumonitis (after treatment of breast cancer or lymphoma) *Associated with malignant diseases* – Leukemia – Lymphoma – Lung cancer

(Extract from Cottin V, Cordier JF. Eosinophilic lung diseases. In: Murray & Nadel's textbook of respiratory medicine. 6th ed. Philadelphia: Elsevier Saunders; 2016. p. 1221–42; Allen J, Wert M. Eosinophilic pneumonias. J Allergy Clin Immunol Pract 2018;6(5):1455–61.)

TABLE 7.3.2 Differential signs of eosinophilic lung diseases

	EGPA	ICEP	ABPA	BG	SPE	HES	DIEP
Anemia, thrombocytopenia	+	–	–	–	–	+	–
Increased tryptase and vitamin B12 levels	–	–	–	–	–	++	–
Increased alkaline phosphatase, LDH, liver transaminases	–	–	–	–	–	+	+
Positive p-ANCA	++	–	–	–	–	–	–
High IgE	++	+	+++	+	+	+	+
Asthma	+++	++	+++	+	–	+	–
Lesion of the upper respiratory tract	+++	++	+	–	–	+	–
Other extrapulmonary lesions	++	–	–	–	–	+++	+

–, Not typical; +, possible symptom; ++, frequent symptom; +++, typical symptom.
EGPA, eosinophilic granulomatosis with polyangiitis; *ICEP*, idiopathic chronic eosinophilic pneumonia; *ABPA*, allergic bronchopulmonary aspergillosis; *BG*, bronchocentric granulomatosis; *SPE*, simple pulmonary eosinophilia; *HES*, hypereosinophilic syndrome; *DIEP*, drug-induced eosinophilic pneumonia.

TABLE 7.3.3 Differential HRCT and morphological signs of eosinophilic pulmonary disease

	EGPA	ICEP	ABPA	BG	SPE	HES	DIP
Ground-glass opacity	+++	++	+	+	+++	++	++
Consolidation	++	+++	+	+	+	+	++
Nodules	++	+	++	+	+	++ <10mm	+
Bronchiectasis	–	–	+++	++	–	–	–
Halo sign	+	+	–	–	++	++	
Atoll sign	–	+	–	–	–	–	+
Migration of lesions	+	++	–	–	++	–	+
Distribution	Subpleural, random	Upper lobes	Central	Unilateral, upper lobes	Peripheral, upper lobes	Random	Peripheral, more often upper and middle fields
Interstitial and alveolar eosinophil infiltration	++	+++	–	–	++	+	++
Vasculitis with necrosis	+++	–	–	–	–	–	+
Granulomas	++ Extravascular	–	+	+++ Peribronchial and peribronchiolar location	++	–	+
Organizing pneumonia	+	+++	–	–	–	+	++
Interstitial fibrosis	+	+	–	–	–	–	+
Eosinophil infiltration of the bronchial walls	+	+	+++	++	–	+	–

–, Not typical; +, possible sign; ++, frequent sign; +++, typical sign.
ABPA, allergic bronchopulmonary aspergillosis; *BG*, bronchocentric granulomatosis; *DIP*, drug-induced pneumonitis; *EGPA*, eosinophilic granulomatosis with polyangiitis; *HES*, hypereosinophilic syndrome; *ICEP*, idiopathic chronic eosinophilic pneumonia; *SPE*, simple pulmonary eosinophilia.

the specificity of radiographic manifestations in some pulmonary eosinophilia is quite high. In a blind assessment of chest HRCT in patients with seven eosinophilic lung diseases, Johkoh et al. [8] reported that the correct diagnosis was made in 61% of patients, with the highest number of adequate responses (78%) in cases of idiopathic chronic eosinophilic pneumonia (ICEP), ABPA (84%), and acute eosinophilic pneumonia (AEP, 81%), whereas the least (17%) were noted in the proportion of patients with a simple pulmonary eosinophilia (the Löffler syndrome).

Thus, for ICEP, rather common areas of consolidation, localized primarily in the upper lobes and subpleurally, are typical. Consolidation in untreated ICEP patients usually prevails over ground-glass opacity (Fig. 7.3.1), unlike eosinophilic granulomatosis with polyangiitis (EGPA), in which the opposite is noted (Fig. 7.3.2). Another CT difference between these diseases is considered to be the presence of nodules in EGPA, which are generally rarely seen in ICEP [9] (Fig. 7.3.2). Also, with ICEP, an "atoll sign" (the reversed halo sign), which is not typical for EGPA, can be detected (Fig. 7.3.3). However, in the presence of consolidation zones, it is not possible to distinguish ICEP from EGPA according to the HRCT data (Fig. 7.3.4). If at the same time the diagnostic titer has no p-ANCA and there are no clear signs of systemic involvement, it is possible to differentiate between EPGA and CEP only morphologically.

The greatest problem is the radiographic difference between ICEP (especially with minimal blood eosinophilia) and OP, the histological signs of which are found in 7%–10% of ICEP patients [10, 11]. The presence of such an overlap can lead to obtaining histological material by transbronchial biopsy from the OP sites and, accordingly, the direction of diagnostic

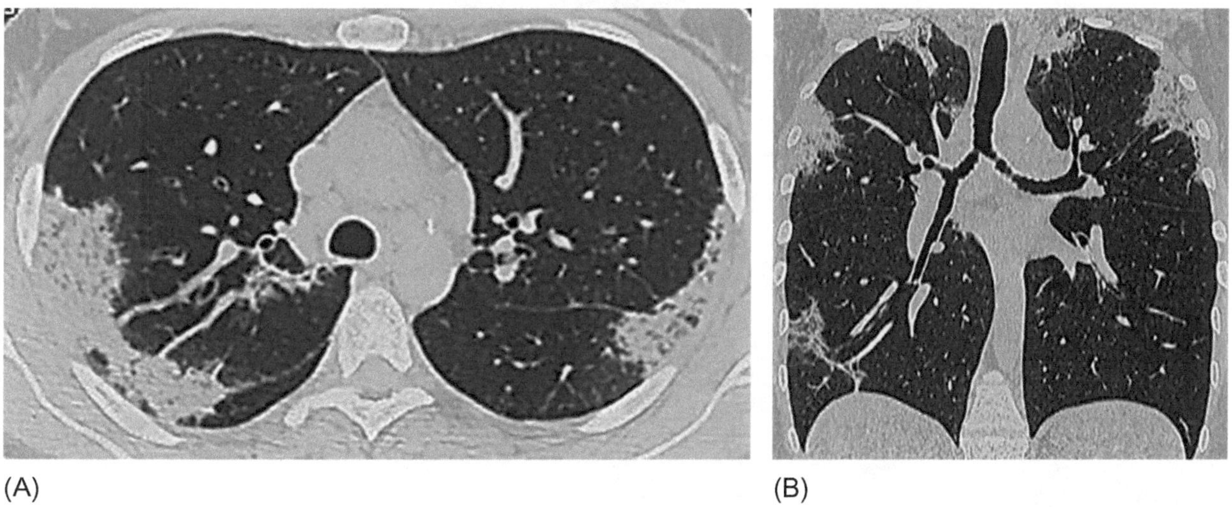

FIG. 7.3.1 ICEP. Bilateral symmetric subpleural consolidation zones with irregular, well-defined contours (A). Lesion areas are distributed mainly in the upper lobes (B).

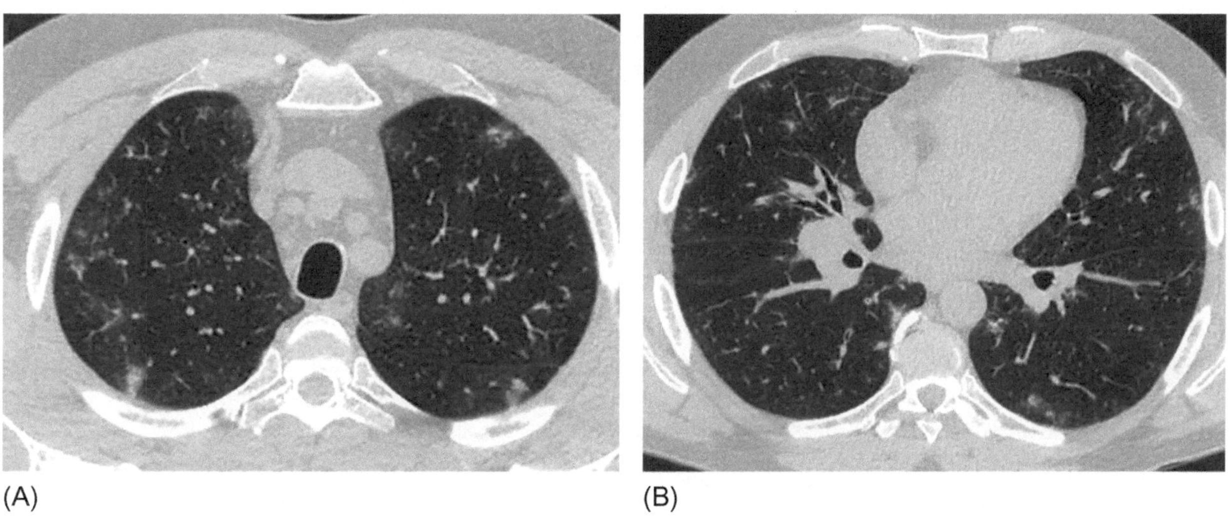

FIG. 7.3.2 EGPA. Multiple patches of ground-glass opacity and dense nodules, mainly with subpleural distribution in the upper (A) and lower (B) lung zones. A moderately pronounced symptom of tree-in-bud.

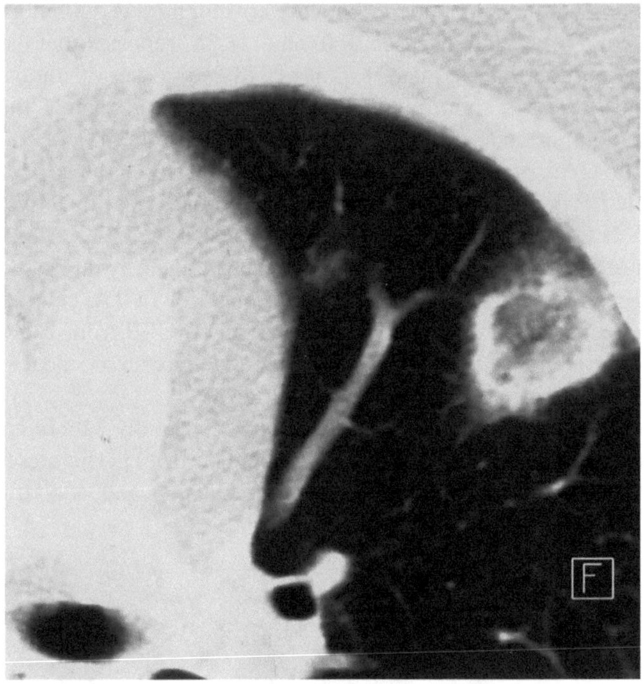

FIG. 7.3.3 ICEP. Ground-glass opacity focus surrounded by a crown of consolidation (reversed halo sign).

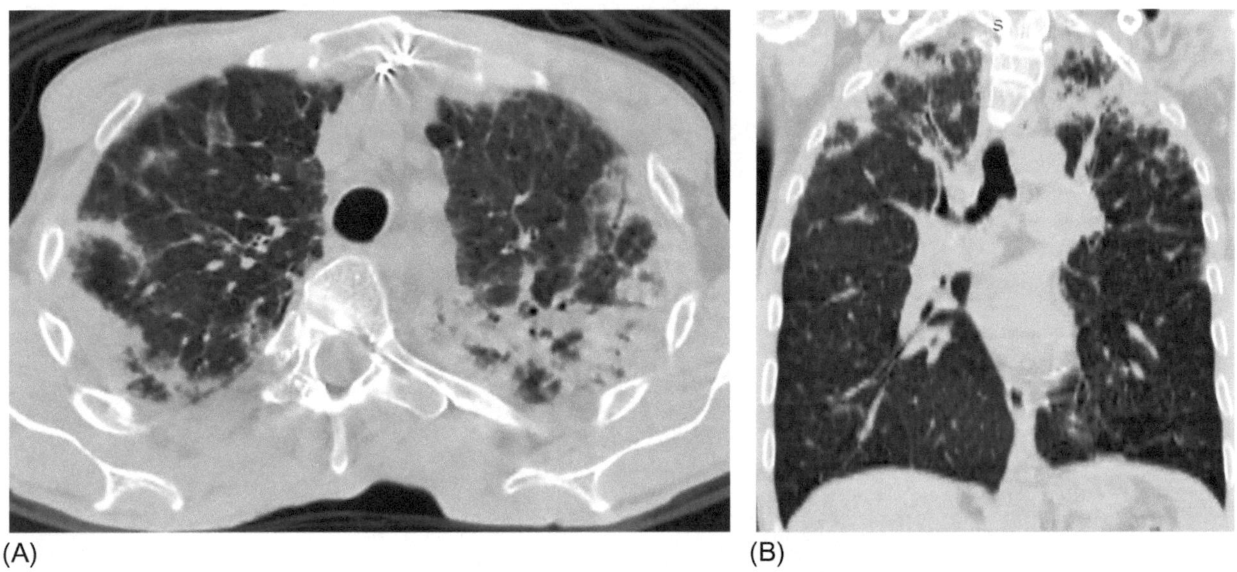

FIG. 7.3.4 EGPA. Bilateral subpleural consolidation areas of irregular shape. Thickened interlobular septa and thickened and indurated bronchial walls (A). The lesion is most pronounced in the upper lobes (B). The abnormalities are indistinguishable from the ICEP characteristics.

thought toward this pathology. Since both eosinophilic pneumonia and OP often act as secondary inflammatory reactions of the lung tissue to damage (drugs, infections, and so forth), anamnestic data usually do not help to establish the diagnosis. Arakawa et al. [12], comparing CT findings of ICEP and OP, found that, according to the main characteristics, which are the typical consolidation zones and ground-glass opacity, these conditions are almost indistinguishable; however, OP has more typical peribronchial distribution of abnormalities and bronchodilation, as well as the presence of centrilobular nodules (Fig. 7.3.5). In a later study, significant differences in addition to the listed signs were found in relation to the upper-lobe localization of changes (observed in 51% of ICEP patients and 4.5% of OP patients), thickening of the bronchial walls (25.5% and 6.4%, respectively), and ground-glass opacity dominance over consolidation (76.6% and 38.3%, respectively)

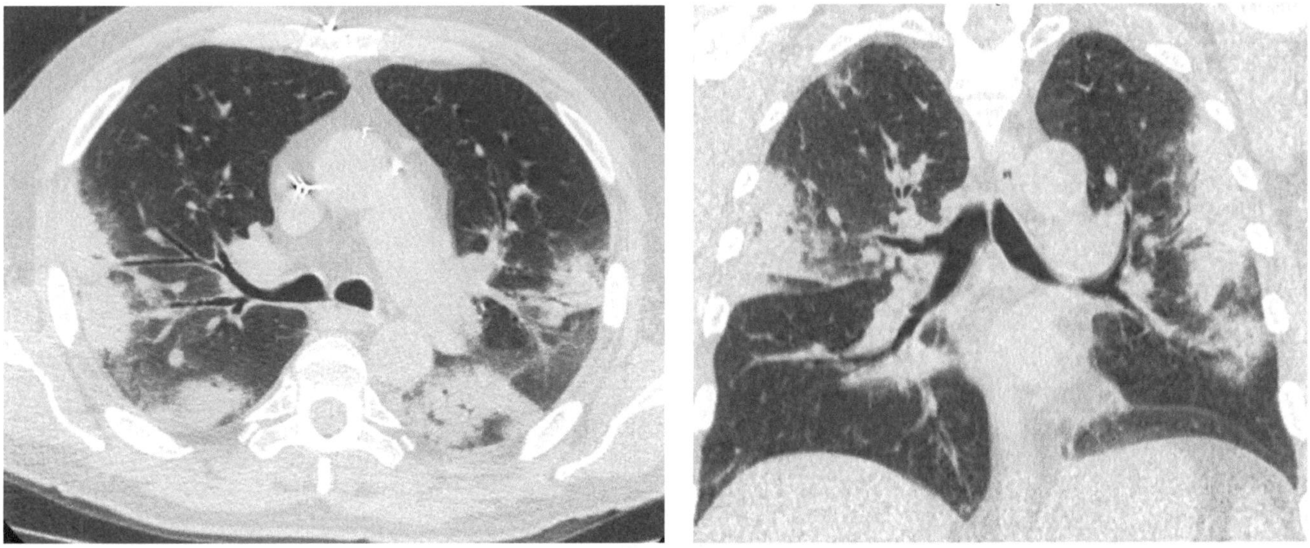

(A) (B)

FIG. 7.3.5 Cryptogenic organizing pneumonia. Bilateral consolidation areas surrounded by ground-glass opacity (halo sign) with subpleural and peribronchovascular distribution (A). The lesion areas are most represented in the middle lung zones. The volume of consolidation is substantially greater than the volume of ground-glass opacity (B).

[13]. Kim et al. [14] believed that the reversed halo sign may serve as a differential CT sign between bronchiolitis obliterans OP and ICEP (according to them, at the latter, it is not detected). According to Mehrian et al. [13], this sign is, indeed, significantly less likely to occur with ICEP emerging in approximately 15% of cases (Fig. 7.3.3).

The decisive method of differential diagnosis of ICEP (with minimal blood eosinophilia) and OP, as well as with other diseases (pulmonary infections and hypersensitive pneumonitis) that may have a similar clinical and radiological pattern and an increase in the eosinophil fraction in the blood (often as a reaction to treatment), is a cytological analysis of BAL fluid, where it is possible to detect a high content of eosinophils. In this case, not only the absolute level >40% but also the excess of the eosinophil fraction over lymphocytes and neutrophils is significant [15].

HES is a rare disorder based on monoclonal proliferation with eosinophil bone marrow cell lines (myeloid variant) or T-lymphocytic germ producing eosinophilopoietic chemokines, especially IL-5 (lymphocytic variant) [16]. The criteria for HES diagnosis are (1) peripheral blood eosinophilia ($>1.5\times10^9$ eosinophils/L) on two examinations at >1 month intervals and/or (2) tissue hypereosinophilia defined as >20% eosinophils in the bone marrow, a local marked increase in tissue eosinophils, and/or marked deposition of eosinophil-derived proteins even in the absence of eosinophils [17].

As with EGPA, there are signs of a multiorgan eosinophilic lesion in HES; these are the diseases that should be differentiated between each other first and foremost. The process most often involves the heart (up to 58%), skin (37%–69%), lungs (25%–67%), gastrointestinal tract (up to 38%), and central and peripheral nervous systems and the vascular bed with active thrombogenesis [18, 19].

It should be noted that asthma and damage to the mucous membranes of the ear, throat, and nose, in fact obligatory for EGPA, are revealed in HES not more often than in 27% of cases [20]. In addition, the overlap forms of HES and EGPA are described [7].

In a BAL fluid, in HES, eosinophilia is most pronounced among all EPDs and can reach 73%, whereas with EGPA, it rarely exceeds 40% [21, 22].

CT signs of a lung lesion in HES are nonspecific and include patchy areas of ground-glass opacity or (less commonly) consolidation, primarily with subpleural distribution, mediastinal lymphadenopathy (less often), small nodules, and pleural effusion [1, 20].

Analysis of the bone marrow with the identification of a monoclonal germ or genetic analysis for the most frequent mutations underlying the disease, namely, *FIP1L1-PDGFRA* gene rearrangement and JAK2 mutations [7], can be decisive in the diagnosis of HES.

ABPA develops in patients with central bronchiectasis in cystic fibrosis and with non-cystic fibrosis origin as a result of a hypersensitive reaction to their colonization with fungi *Aspergillus* spp. [23]. Criteria for ABPA diagnosis are shown in Table 7.3.4.

TABLE 7.3.4 Criteria for the allergic bronchopulmonary aspergillosis diagnosis

Predisposing conditions	Asthma Cystic fibrosis
Obligatory criteria (both should be present)	Type I *Aspergillus* skin test positive (immediate cutaneous hypersensitivity to *Aspergillus* antigen) or elevated IgE levels against *Aspergillus fumigatus* Elevated total IgE levels (>1000 IU/mL)[a]
Other criteria (at least two of three)	Presence of precipitating or IgG antibodies against *A. fumigatus* in serum Radiographic pulmonary opacities consistent with ABPA Total eosinophil count >500 cells/L in steroid naive patients (may be historical)

[a]*In the case of all other criteria, the level of IgE may be <1000 IU/mL.*
(Extract from Agarwal R, Chakrabarti A, Shah A, Gupta D, Meis JF, Guleria R, et al. Allergic bronchopulmonary aspergillosis: review of literature and proposal of new diagnostic and classification criteria. Clin Exp Allergy 2013;43(8):850–73.)

Among the HRCT signs of ABPA, the presence of central bronchiectasis, which can be either free or filled with mucus plugging, is essential. Around bronchiectasis, foci of consolidation and atelectasis, formed because of obstruction of the bronchi, are often revealed. Centrilobular nodules in a tree-in-bud sign are usually observed as a manifestation of obstructive bronchiolitis (Fig. 7.3.6) [1]. However, there is a so-called serological ABPA, which has clinical and laboratory signs of disease, but without bronchiectasis on CT [23]. With the development of pneumonia caused by bacterial flora, consolidation and ground-glass opacity areas may appear on HRCT with ABPA, which makes it very similar to EGPA and CEP.

BG is a rare eosinophilic disease based on necrotizing granulomatosis around the bronchioles and small bronchi. Approximately one-third of patients have symptoms of asthma, and half have eosinophilia of the blood [3]. Since BG is a tissue reaction to fungal invasion, including in ABPA patients, these diseases can be difficult to differentiate. On CT scan of the chest, unilateral (rarely bilateral) single or multiple foci of consolidation are usually revealed, and atelectases may occur. With a long course of BG, the peripheral bronchiectases are formed. Changes are localized mainly in the upper lung lobes and are often unilateral (Fig. 7.3.7) [1, 24]. BG in the presence of asthma and bronchiectasis cannot be differentiated from ABPA without morphological verification, since in both cases, hypersensitization to *Aspergillus* spp. is revealed.

AEP is often considered with other EPD. However, although this disease is based on diffuse alveolar damage by eosinophils, the absence of blood eosinophilia, an acute course with increasing respiratory failure, motivated us to describe this disease in the differential range of acute interstitial pneumonia (see Chapter 2.6).

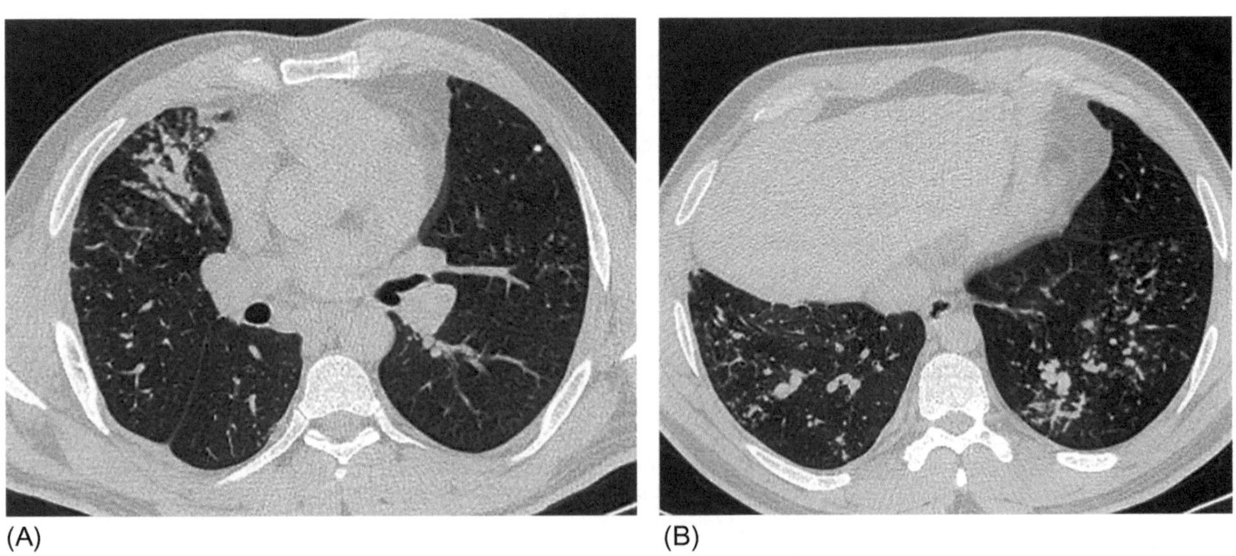

FIG. 7.3.6 Allergic bronchopulmonary aspergillosis. Central bronchiectasis in the right middle lobe, filled with mucous contents, a tree-in-bud sign in the left lung (A). Multiple mucus-filled small bronchiectasis in the lower lobes, surrounded by multiple nodules (B). *(Case courtesy of Prof. S.N. Avdeev, Sechenov First Moscow State Medical University, Moscow, Russia).*

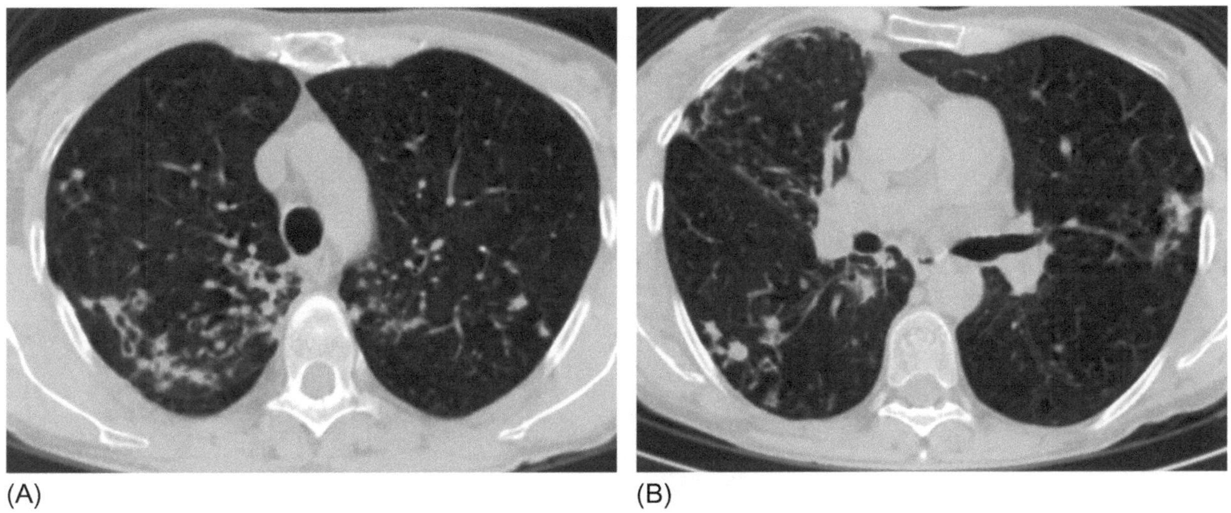

FIG. 7.3.7 Bronchocentric granulomatosis. Predominantly right-sided small bronchiectasis, some of which are filled with bronchial secretions; thickened bronchial walls, multiple small nodules, and thickened intralobular septa are visible (A and B).

Drug-induced eosinophilic pneumonia (DIEP), although it is EPD, is actually a very rare condition. Bartal et al. [25] counted in the system review only 228 cases described between 1990 and 2017. There are two main clinical and morphological patterns of DIEP, acute and CEP, with the corresponding characteristics of the clinical course and radiological manifestations. Most often the first one developed in cases of daptomycin intake, whereas the second developed in cases of mesalamine administration [25]. Unlike idiopathic AEP and CEP, with drug-induced variants, a lower degree of BAL eosinophilia is noted, which is often <20% [1] (Fig. 7.3.8).

To facilitate the differential diagnosis of EPD, two important questions must be answered: (1) does the patient have asthma, and (2) are there signs of an extrapulmonary lesion (Fig. 7.3.9). The presence of asthma enables to rule out simple pulmonary eosinophilia (SPE), parasitic invasions (PI), and DIEP for which this disease is not typical. The remaining eosinophilic lung diseases may be accompanied by symptoms of asthma and require analysis of additional differential signs. The presence of polyorgan (non-ENT) involvement enables to rule out ICEP, ABPA, and BG for which systemicity is not typical, whereas for EGPA and HES, this feature is characteristic. In some patients with ICEP and ABPA, allergic inflammation of the upper respiratory airways, but not of parenchymal organs, may occur. The absence of multisystemic

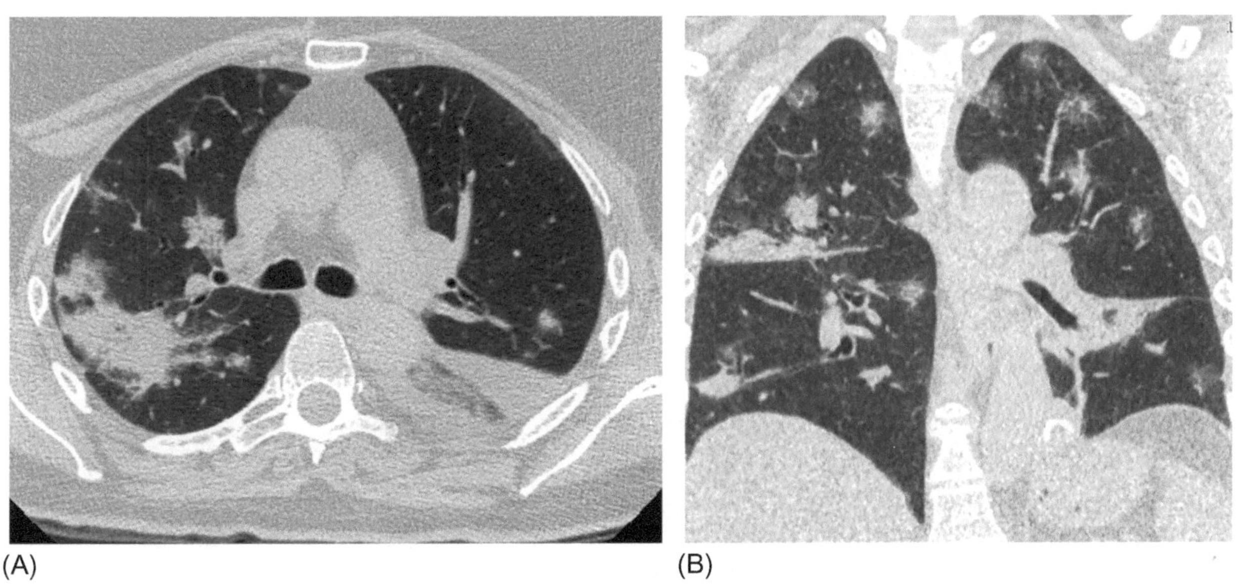

FIG. 7.3.8 Eosinophilic pneumonia caused by sulfasalazine. Bilateral patchy areas of consolidation and ground-glass opacity with subpleural and peribronchovascular distribution (A and B). The lesion is most pronounced in the upper lobes; there is the pattern of chronic eosinophilic pneumonia (B).

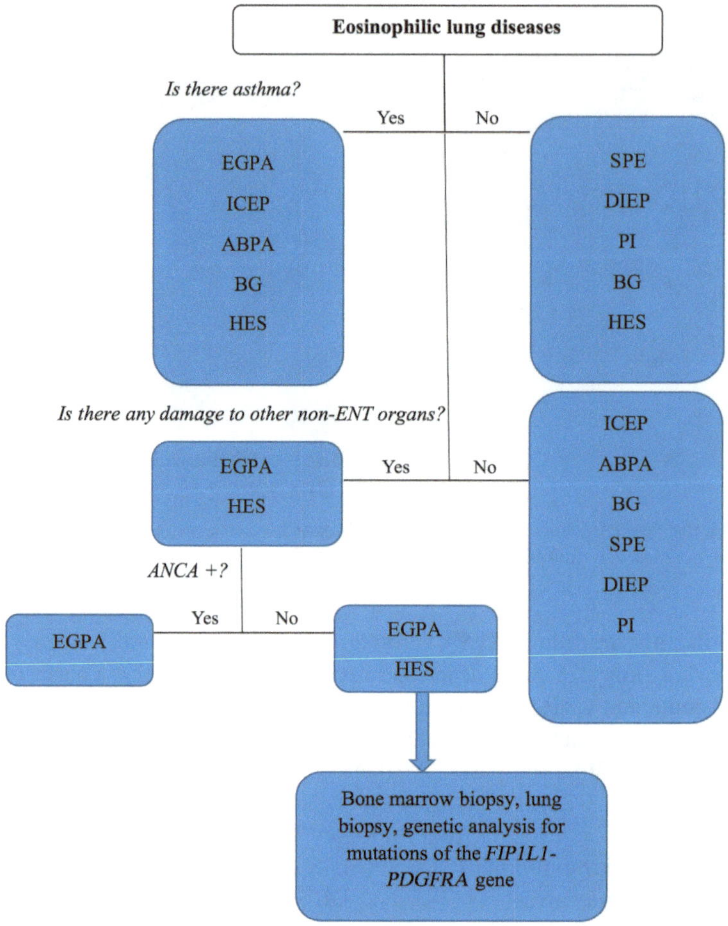

FIG. 7.3.9 Diagnostic algorithm of eosinophilic pulmonary diseases.

pathology does not completely rule out EGPA for which onset can start from lung lesions. However, an isolated process in the lungs at HES is actually not possible. An important test confirming EGPA is an increased serum p-ANCA titer, which is not characteristic of other EPD. With a negative result of this analysis (in 60% of cases), it is necessary to consider a combination of other signs, and in the case of differentiation with HES, biopsy of bone marrow or lungs or genetic testing for key HES mutations is required.

It should always be remembered that eosinophilia of the blood and asthma can be an atopic background for a disease that has a different (noneosinophilic) genesis of lung lesions. For example, cryptogenic or secondary organized pneumonia, hypersensitivity pneumonitis, and infectious lung lesions can develop in patients with asthma and eosinophilia. In such cases, to clarify the genesis of lung changes, BAL should be performed, and if possible, histological material should be obtained. Elevated levels of eosinophils in BAL fluid will indicate EPD.

References

[1] Jeong YJ, Kim KI, Seo IJ, Lee CH, Lee KN, Kim KN, et al. Eosinophilic lung diseases: a clinical, radiologic, and pathologic overview. Radiographics 2007;27(3):617–37.

[2] Cottin V, Cordier JF. Eosinophilic lung diseases. Immunol Allergy Clin North Am 2012;32(4):557–86.

[3] Cottin V, Cordier JF. Eosinophilic lung diseases. In: Murray & Nadel's textbook of respiratory medicine. 6th ed. Philadelphia: Elsevier Saunders; 2016. p. 1221–42.

[4] Allen J, Wert M. Eosinophilic pneumonias. J Allergy Clin Immunol Pract 2018;6(5):1455–61.

[5] Taylor MR, Keane CT, O'Connor P, Mulvihill E, Holland C. The expanded spectrum of toxocaral disease. Lancet 1988;1(8587):692–5.

[6] Siddiqui AA, Berk SL. Diagnosis of *Strongyloides stercoralis* infection. Clin Infect Dis 2001;33(7):1040–7.

[7] Khoury P, Bochner BS. Consultation for elevated blood eosinophils: clinical presentations, high value diagnostic tests, and treatment options. J Allergy Clin Immunol Pract 2018;6(5):1446–53.

[8] Johkoh T, Müller NL, Akira M, Ichikado K, Suga M, Ando M, et al. Eosinophilic lung diseases: diagnostic accuracy of thin-section CT in 111 patients. Radiology 2000;216(3):773–80.

[9] Webb WR, Muller NL, Naidich DP. High-resolution CT of the lung. 4th ed. Philadelphia: Lippincott Williams and Wilkins; 2009. p. 347–9.

[10] Mayo JR, Muller NL, Road J, et al. Chronic eosinophilic pneumonia: CT findings in six cases. AJR Am J Roentgenol 1989;153:727–30.

[11] Müller NL, Staples CA, Miller RR. Bronchiolitis obliterans organizing pneumonia: CT features in 14 patients. AJR Am J Roentgenol 1990;154(5):983–7.

[12] Arakawa H, Kurihara Y, Niimi H, Nakajima Y, Johkoh T, Nakamura H. Bronchiolitis obliterans with organizing pneumonia versus chronic eosinophilic pneumonia: high-resolution CT findings in 81 patients. AJR Am J Roentgenol 2001;176(4):1053–8.

[13] Mehrian P, Doroudinia A, Rashti A, Aloosh O, Dorudinia A. High-resolution computed tomography findings in chronic eosinophilic vs. cryptogenic organising pneumonia. Int J Tuberc Lung Dis 2017;21(11):1181–6.

[14] Kim SJ, Lee KS, Ryu YH, Yoon YC, Choe KO, Kim TS, et al. Reversed halo sign on high-resolution CT of cryptogenic organizing pneumonia: diagnostic implications. AJR Am J Roentgenol 2003;180(5):1251–4.

[15] Cordier JF, Cottin V. Eosinophilic pneumonias. In: Schwarz MI, King Jr TE, editors. Interstitial lung disease. 5th ed. Shelton, CT: People's Medical Publishing House-USA; 2011. p. 833–93.

[16] Oliver JW, Deol I, Morgan DL, Tonk VS. Chronic eosinophilic leukemia and hypereosinophilic syndromes. Proposal for classification, literature review, and report of a case with a unique chromosomal abnormality. Cancer Genet Cytogenet 1998;107(2):111–7.

[17] Valent P, Klion AD, Horny HP, Roufosse F, Gotlib J, Weller PF, et al. Contemporary consensus proposal on criteria and classification of eosinophilic disorders and related syndromes. J Allergy Clin Immunol 2012;130(3):607–12. e9.

[18] Weller PF, Bubley GJ. The idiopathic hypereosinophilic syndrome. Blood 1994;83(10):2759–79.

[19] Zimmermann N, Wikenheiser-Brokamp KA. Hypereosinophilic syndrome in the differential diagnosis of pulmonary infiltrates with eosinophilia. Ann Allergy Asthma Immunol 2018;121(2):179–85.

[20] Dulohery MM, Patel RR, Schneider F, Ryu JH. Lung involvement in hypereosinophilic syndromes. Respir Med 2011;105(1):114–21.

[21] Slabbynck H, Impens N, Naegels S, Dewaele M, Schandevyl W. Idiopathic hypereosinophilic syndrome-related pulmonary involvement diagnosed by bronchoalveolar lavage. Chest 1992;101(4):1178–80.

[22] Schnabel A, Csernok E, Braun J, Gross WL. Inflammatory cells and cellular activation in the lower respiratory tract in Churg-Strauss syndrome. Thorax 1999;54(9):771–8.

[23] Agarwal R, Chakrabarti A, Shah A, Gupta D, Meis JF, Guleria R, et al. Allergic bronchopulmonary aspergillosis: review of literature and proposal of new diagnostic and classification criteria. Clin Exp Allergy 2013;43(8):850–73.

[24] Ward S, Heyneman LE, Flint JD, Leung AN, Kazerooni EA, Müller NL. Bronchocentric granulomatosis: computed tomographic findings in five patients. Clin Radiol 2000;55(4):296–300.

[25] Bartal C, Sagy I, Barski L. Drug-induced eosinophilic pneumonia: a review of 196 case reports. Medicine (Baltimore) 2018;97(4):e9688.

Chapter 8

Lung disease related to connective tissue diseases

Alexander Averyanov[a,b], Evgeniya Kogan[c], Victor Lesnyak[d], Olesya Danilevskaya[e]

[a]Clinical Department, Pulmonology Research Institute under FMBA of Russia, Moscow, Russia, [b]Pulmonary Division, Federal Research Clinical Center under FMBA of Russia, Moscow, Russia, [c]Anatomic Pathology Department, Sechenov University, Moscow, Russia, [d]Radiology Department, Federal Research Clinical Center under FMBA of Russia, Moscow, Russia, [e]Endoscopy Department, Pulmonology Research Institute under FMBA of Russia, Moscow, Russia

Chapter 8.1

Respiratory involvement in rheumatoid arthritis

Rheumatoid arthritis (RA) is one of the most common connective tissue diseases (CTD) and affects approximately 1% of the adult population. Its prevalence is three times higher in females than in males [1] and usually begins between the ages of 20–50 years. RA is a progressive systemic autoimmune disorder usually manifesting as erosive synovitis characterized mainly by the involvement of the small joints and internal organs. Extraarticular manifestations are often noted and include subcutaneous nodules, vasculitis, mononeuritis, pericarditis, and episcleritis [2]. RA associated with interstitial lung disease (RA-ILD) and pleural abnormalities of varying severity are found in approximately 25% of patients with RA [3]. According to the National Center for Health Statistics (the United States), the signs of RA-ILD at autopsy are found in 6.6% of deceased patients with RA [4]. It is predicted that lungs are involved in up to 20% of lethal outcomes in patients with RA [5]. Airways' abnormalities are even more common (40%–60%), albeit with less clinical significance [6].

The risk factors and pathogenesis of pulmonary manifestations of RA are not well understood. Older age, male (gender), smoking, genetic predisposition associated with the DRB1 allele of human leukocyte antigen, and increased anticyclic citrullinated peptide antibody (ACPA) levels are the most widely recognized predisposing factors of RA-ILD [7]. ACPAs are formed in response to citrullination, which is the conversion of arginine to citrulline, a powerful trigger of immune response [8].

ACPA is not only one of the important pathogenic factors of synovial damage in RA but also a potential independent factor for the development of ILD. Patients with elevated ACPA titers might have interstitial and bronchial abnormalities in the absence of signs of RA or other CTDs [9]. During follow-up, only a small number of these patients develop articular inflammation. Strong correlations between ACPA levels and severity of interstitial lung lesions in patients with RA are well established [10].

Interstitial and pleural lesions result from a cascade of autoimmune responses to primary damage to epithelial cells. Paulin et al. [11] proposed a novel hypothesis, according to which, in patients with RA-ILD having a pattern of nonspecific interstitial pneumonia, the primary immune response to the production of citrullinated peptides occurs in the synovial membranes of the joints and that the lungs are secondarily affected due to the presence of cross antigens. According to another proposed mechanism of RA-ILD, which is similar to that considered for idiopathic pulmonary fibrosis (IPF), the chronic primary microinjury to the respiratory epithelium, such as that occurring with tobacco smoke, leads to the aberrant activation of myofibroblasts and epithelial cells and citrullination, followed by the development of ACPAs and secondary joint damage. Concurrently, usual interstitial pneumonia becomes a morphological manifestation of RA-ILD [11].

The pathogenesis of bronchial abnormalities, especially bronchiectasis, is associated with persistent respiratory tract infections that cause citrullination in the bronchial tree and immune response in genetically predisposed patients [12].

Difficult to Diagnose Rare Diffuse Lung Disease. https://doi.org/10.1016/B978-0-12-815375-8.00008-X

Morphology

Lung lesions in RA can be divided into following four main groups: parenchymal, pleural, airways, and vascular. Some authors also distinguish infectious and drug-induced lesions and AA amyloidosis-associated with RA, albeit as a secondary disease [13, 14].

The parenchymal pattern includes ILD; nodules; and Caplan syndrome, a rare variant of rheumatoid pneumoconiosis. Rheumatoid nodules or granulomas are the tissue response to lung injury observed only in RA. The size of the nodules ranges from a few millimeters to 5–7 cm; they can be single or multiple and can be located in the subpleural or paraseptal areas [15].

Microscopic examination of the central section of the nodules reveals a necrotic zone surrounded by a layer of large histiocytes and epithelioid cells; multinucleated cells such as foreign bodies and Langhans cells can also be found (Figs. 8.1.1–8.1.3) [13]. Rheumatoid nodules can be observed as cavitary forms, inoculated with fungal flora, and be the background for the development of pulmonary adenocarcinoma. Differential diagnosis should include staining for mycobacteria and fungi.

Rheumatoid nodules may have deposits of coal and silica dust, which are observed in Caplan syndrome, which is found among workers in foundry, marble, asbestos, and other industries related to dust exposure [16].

ILD is the pulmonary manifestation with the most significant clinical presentation and prognosis in patients with RA. Most often, ILD represents as usual interstitial pneumonia (41%–61%) or nonspecific interstitial pneumonia (11%–33%) (Fig. 8.1.4), with corresponding manifestations on chest high-resolution computed tomography (HRCT) [17, 18]. Organizing pneumonia (Fig. 8.1.5), acute interstitial pneumonia, and lymphoid interstitial pneumonia are less common. The rarest presentation of pulmonary disorders associated with RA is pulmonary amyloidosis. In approximately 6% of the patients, several histological forms of ILD (overlap syndrome) can be found simultaneously [19].

Pleural lesions in RA can be a consequence of autoimmune inflammation; however, they can also occur as infectious complications. Microscopic examination reveals nonspecific inflammatory abnormalities with lymphohistiocytic infiltration of the pleura (Figs. 8.1.6 and 8.1.7) with or without the formation of fibrinous exudate and the formation of lymphoid follicles. The airway pathologies in RA include bronchiectasis and nonspecific chronic, obliterating, and follicular bronchiolitis (Figs. 8.1.8 and 8.1.9). Vascular involvement can be diverse, including vasculitis, capillaritis (Fig. 8.1.10), pulmonary hypertension, and plexiform arteriopathy.

Clinical presentation

The clinical presentation of RA-ILD is not specific. The main manifestations are nonproductive cough and dyspnea, which, however, may not be obvious due to the limited mobility of patients with lower-limb arthritis [20]. Patients with RA-

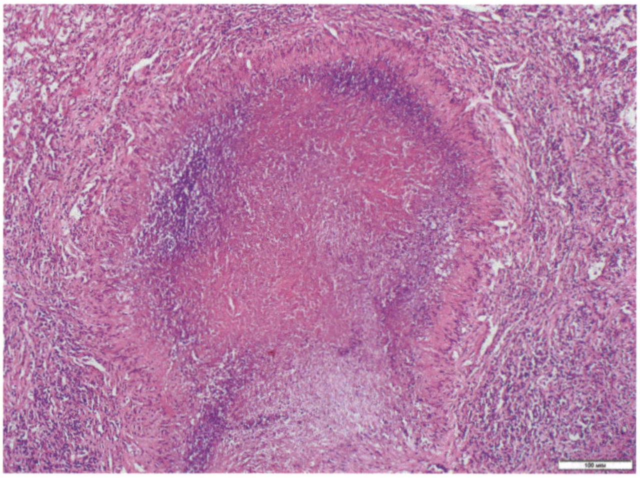

FIG. 8.1.1 Rheumatoid arthritis. Rheumatoid nodule with central necrosis with surrounding palisading epithelioid, histiocytic, and multinucleated Langhans cells and lymphocytes. Hematoxylin and eosin (H&E) staining, 100×.

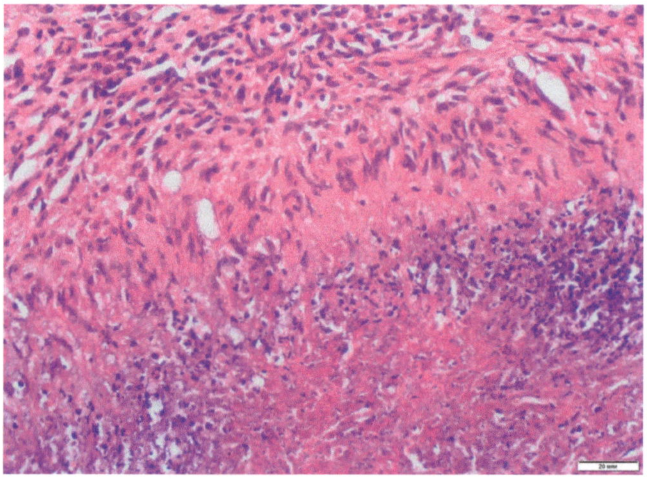

FIG. 8.1.2 Rheumatoid arthritis. The same preparation. Rheumatoid nodule with central necrosis surrounding palisading epithelioid, histiocytic, and multinucleate Langhans cells and lymphocytes. H&E stain, 400×.

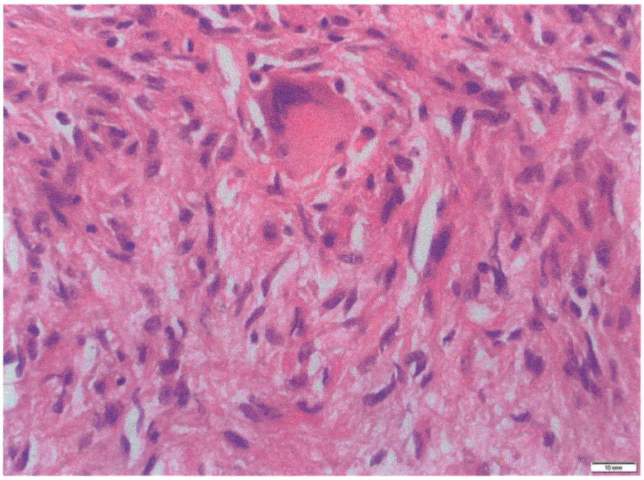

FIG. 8.1.3 Rheumatoid arthritis. Rheumatoid nodule with a multinucleated Langhans cell and lymphocytes. H&E stain, 600×.

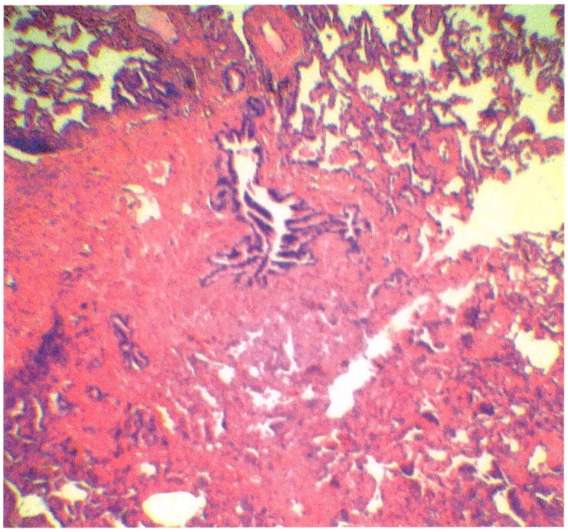

FIG. 8.1.4 Rheumatoid arthritis. Interstitial fibrosis and inflammatory lymphohistiocytic infiltration with obliterating bronchiolitis. H&E stain, 100×.

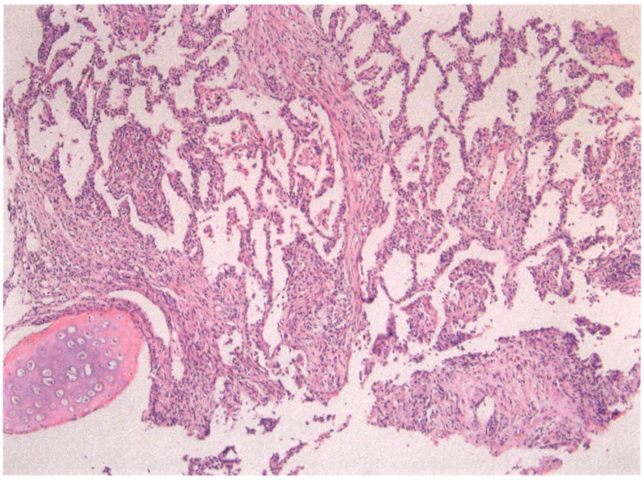

FIG. 8.1.5 Rheumatoid arthritis. Interstitial fibrosis and inflammatory lymphohistiocytic infiltration with foci of carnification. H&E stain, 100×.

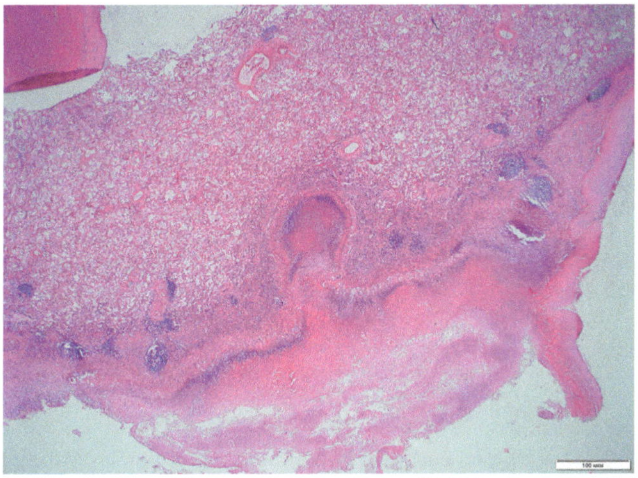

FIG. 8.1.6 Rheumatoid arthritis. Chronic pleurisy with lymphoid follicles, rheumatoid nodule located subpleurally, and fibrinous exudate on the surface. H&E stain, 2×.

ILD, contrary to those with idiopathic interstitial pneumonia, may not have obvious clinical symptoms for long time periods despite pronounced X-ray abnormalities. Nevertheless, Velcro crackles are auscultated in many patients [20], whereas clubbing is a rarer finding than idiopathic fibrotic lung diseases [21].

Patients with RA-ILD and those with IPF can tolerate exacerbations, with a sharp deterioration in symptoms of respiratory failure and appearance of signs of diffuse alveolar damage on chest HRCT [22]. Patients with RA, with bronchiectasis, especially those treated with biological drugs, are at increased risk of lower respiratory tract infections and mucopurulent bronchitis [23]. Involvement of the pleura, especially fibrinous pleurisy, leads to pleural pain. Rheumatoid nodules in the lungs are usually asymptomatic.

Functional tests can usually detect moderate restrictive defects and serve to assess the severity of functional disorders and monitor the course of RA-ILD. In addition, low carbon monoxide diffusion capacity (DLCO) and forced vital capacity are predictors of increased mortality risk in these patients [24, 25].

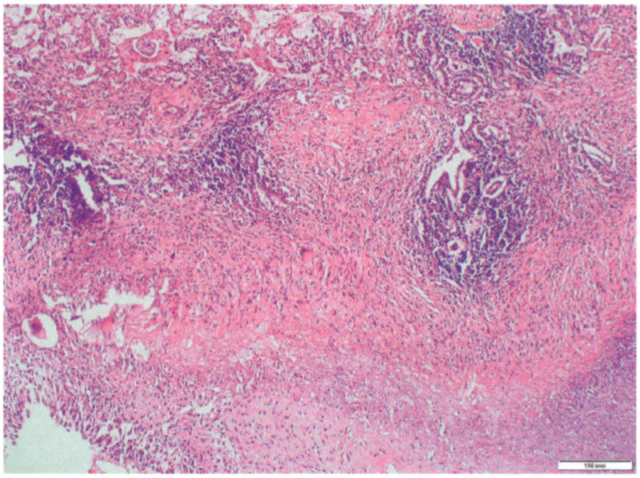

FIG. 8.1.7 Rheumatoid arthritis. Chronic pleurisy with lymphoid follicles. H&E stain, 100×.

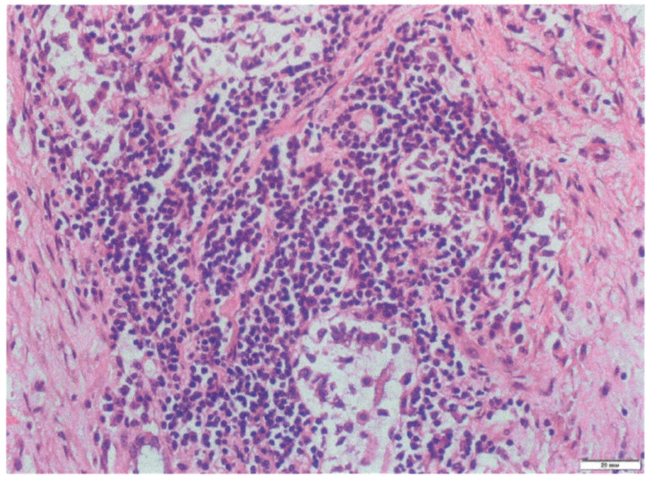

FIG. 8.1.8 Rheumatoid arthritis. Lymphocytic bronchiolitis. H&E stain, 400×.

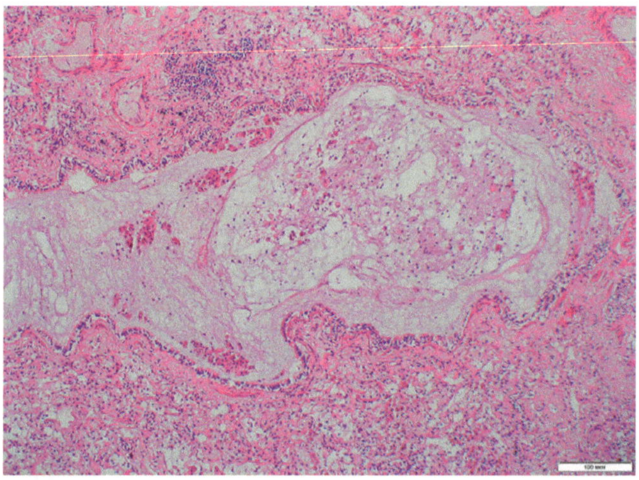

FIG. 8.1.9 Rheumatoid arthritis. Bronchiectasis with mucus in the lumen. H&E stain, 100×.

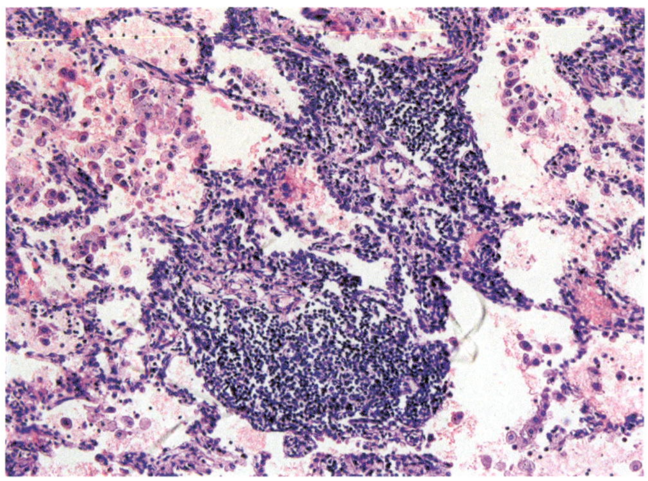

FIG. 8.1.10 Rheumatoid arthritis. Destructive vasculitis of small arteries with lymphoid infiltrates. H&E stain, 400×.

High-resolution computed tomography

RA-ILD

The characteristics of HRCT of RA-ILD closely overlap with HRCT signs of corresponding histological variants of idiopathic forms (see the relevant chapters). In summary the following are the most common HRCT findings (Figs. 8.1.11–8.1.17) [26]: reticular abnormalities (thickening of the interlobular and intralobular septa, traction bronchiectasis, and bronchiolectasis); honeycombing; ground-glass opacity (GGO); peripheral and subpleural distribution of GGO and fibrosis, mainly in the posterior basal fields; thickening of the visceral pleura or pleural effusion; bronchiectasis without interstitial fibrosis; large rheumatoid nodules; consolidation; and thin-walled cysts. The last two HRCT findings are rare, predominantly found in organizing and lymphocytic interstitial pneumonia.

Specific HRCT signs of the RA-ILD subtypes are summarized in Table 8.1.1.

As with other CTDs, interstitial lung lesions may precede the onset of the joint involvement in approximately 16% of the patients or can develop simultaneously [21, 27].

In general the usual interstitial pneumonia (UIP) variant, which is associated with worse prognosis than the nonspecific interstitial pneumonia variant [28], is nonetheless carries a more favorable prognosis compared with IPF in patients without RA [29, 30].

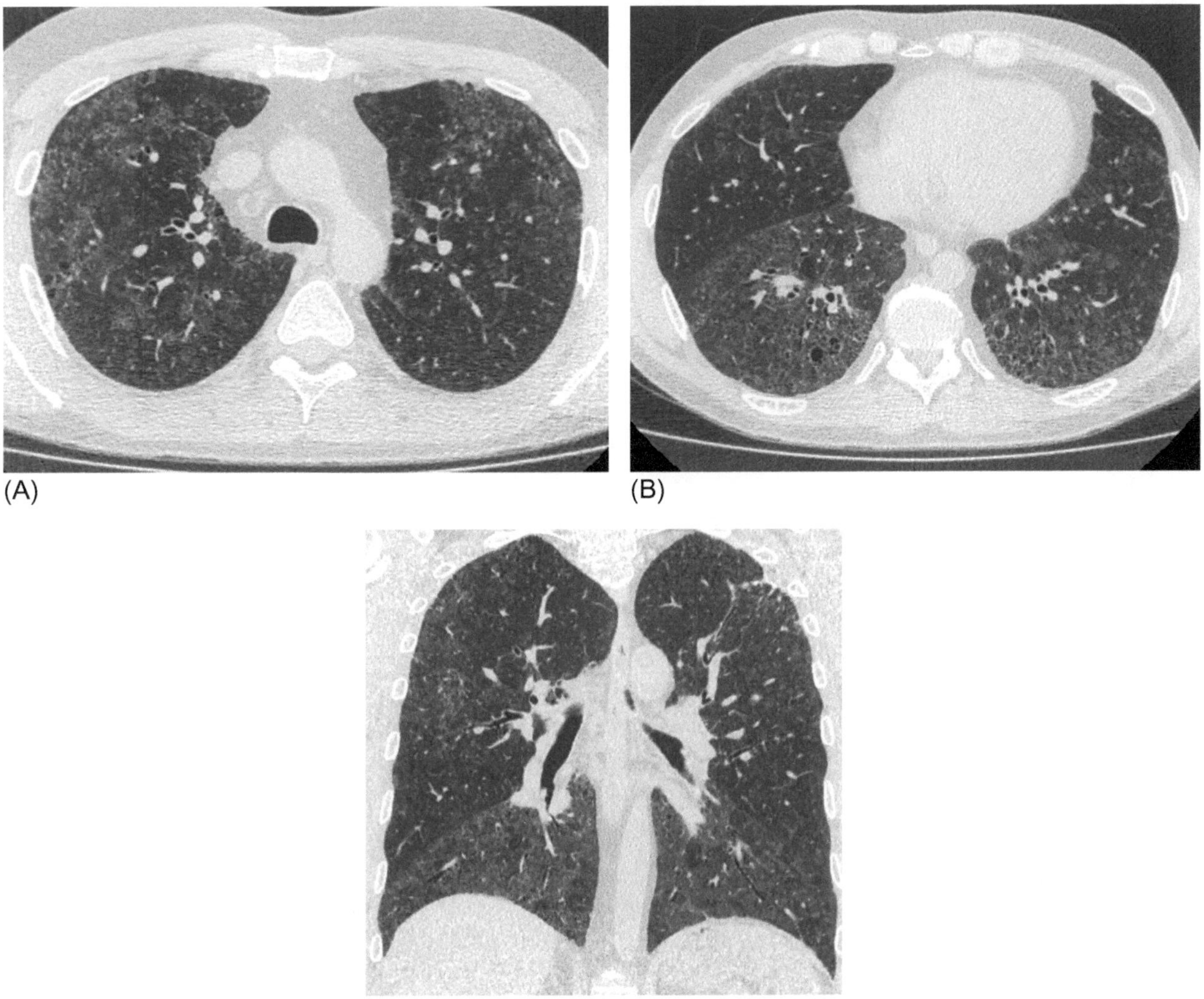

(A) (B) (C)

FIG. 8.1.11 Rheumatoid arthritis-related interstitial lung disease. Bilateral symmetrical areas of ground-glass opacity, moderate reticular abnormalities (A, B). Multiple, small, ill-defined centriacinar nodules and single, small subpleural cysts (A). Traction bronchiectasis associated with the ground-glass opacity (B). Abnormalities are most pronounced in the lower lobes (C). Computed tomography pattern is consistent with nonspecific interstitial pneumonia.

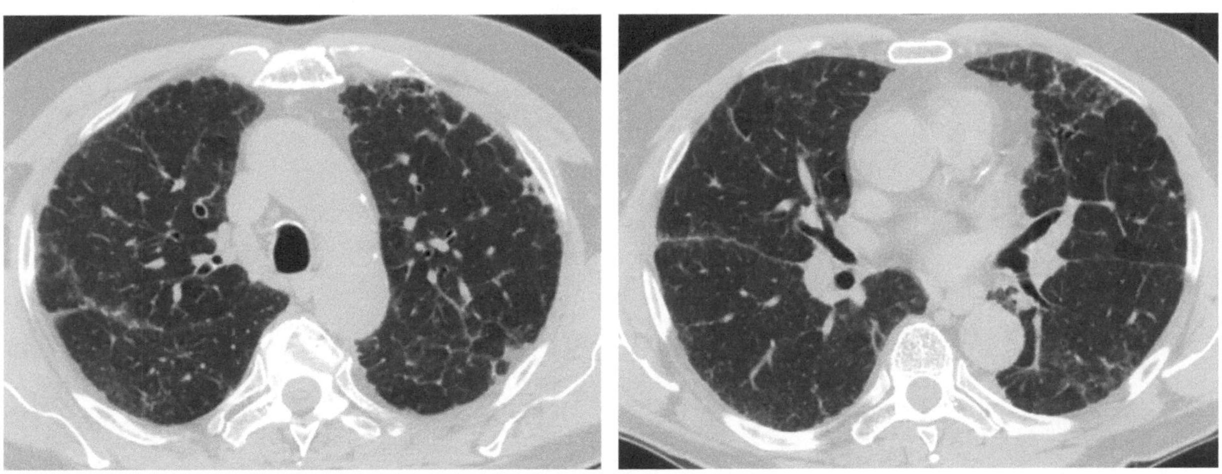

(A) (B)

FIG. 8.1.12 Rheumatoid arthritis-related interstitial lung disease. Pronounced reticular abnormalities, foci of subpleural fibrosis, patchy areas of ground-glass opacity (A, B). Traction bronchiectasis in the lingular segment of the left lung (B). Computed tomography pattern corresponds to probable usual interstitial pneumonia.

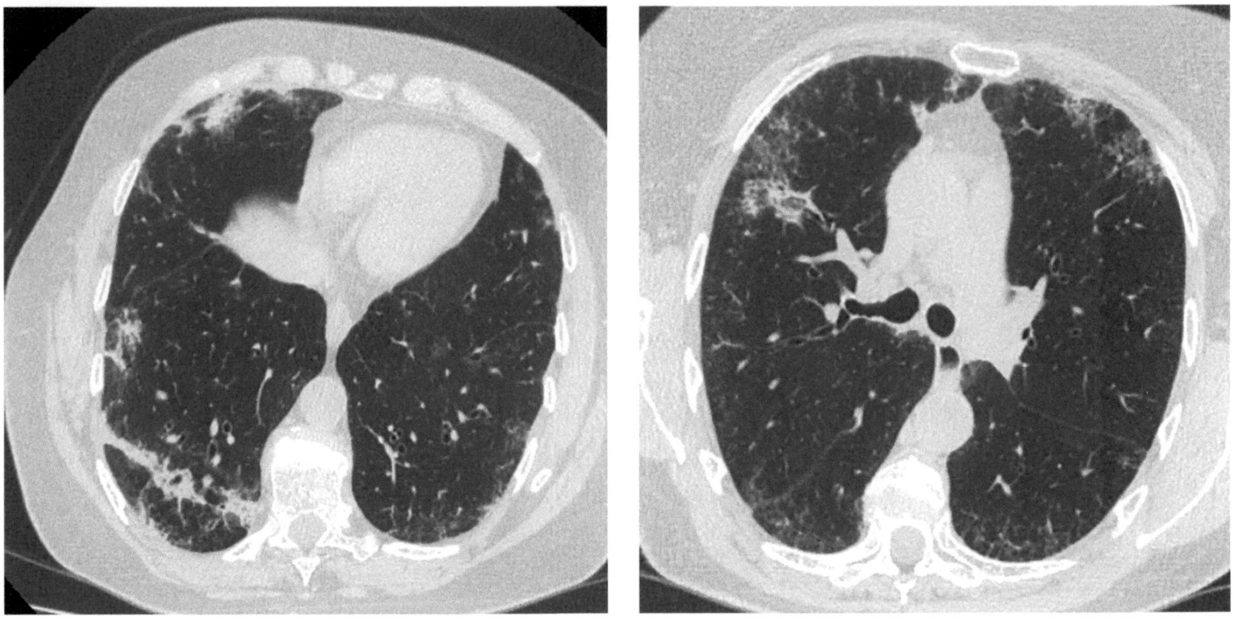

(A) (B)

FIG. 8.1.13 Rheumatoid arthritis-related interstitial lung disease. Bilateral asymmetrical areas of ground-glass opacity and consolidation with subpleural and peribronchovascular distribution. Moderately pronounced reticular abnormalities (A, B) and poorly differentiated centriacinar nodules (A). Single bronchiectasis in the lower lobes (B).

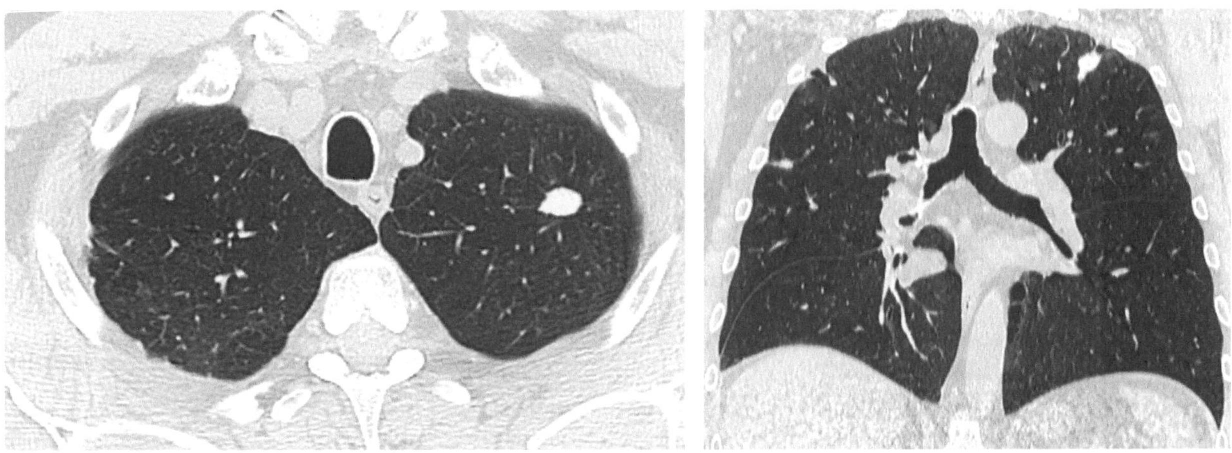

(A) (B)

FIG. 8.1.14 Rheumatoid arthritis. Rheumatoid nodule in the upper left lobe with sharp contours (A). In coronal reconstruction, several dense, irregularly shaped nodules associated with the visceral pleura are visualized in the upper lobes (B).

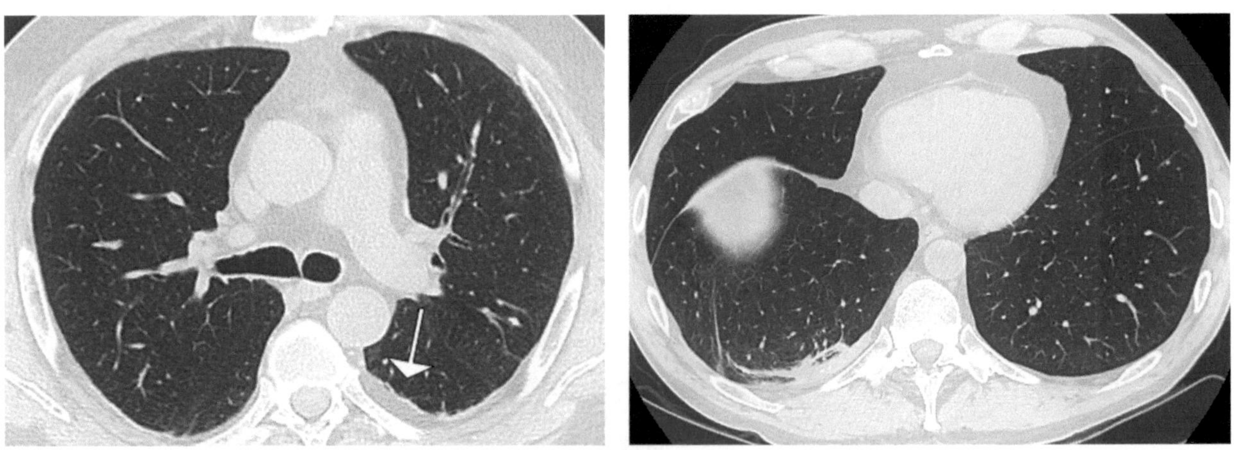

(A) (B)

FIG. 8.1.15 Rheumatoid arthritis. Pleural abnormalities. Local thickening of the visceral pleura in the left lung *(arrow)* (A). Thickening of the visceral pleura on the right with subpleural linear fibrosis in the lower right lobe (B).

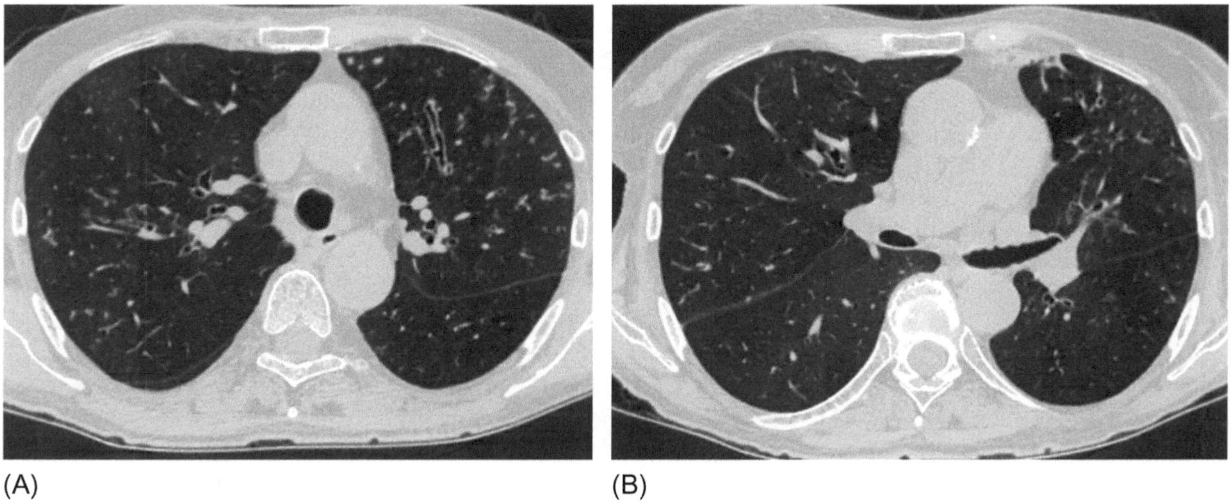

FIG. 8.1.16 Rheumatoid arthritis. Bronchial abnormalities. Bronhiectasis, mainly in the left upper lobe. Multiple small nodules, the tree-in-bud phenomenon (A and B).

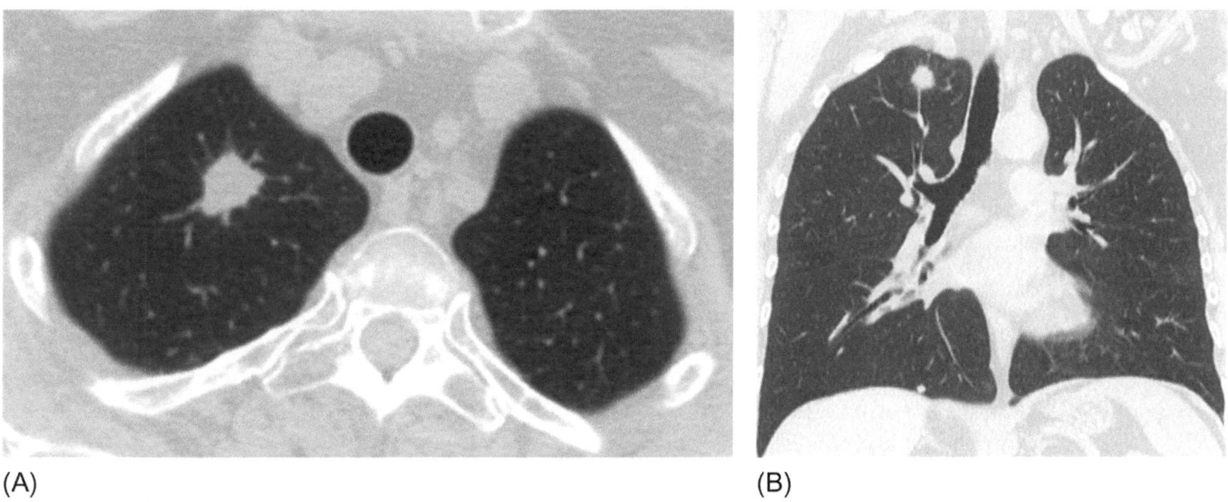

FIG. 8.1.17 Rheumatoid arthritis. Rheumatoid nodule in the upper right lobe with irregular contours, spicularity (A, B), retraction of the visceral pleura (B). It is indistinguishable from tumor by computed tomography findings.

TABLE 8.1.1 High-resolution computed tomography findings of interstitial lung disease variants of rheumatoid arthritis

RA-ILD variant	HRCT characteristics
Usual interstitial pneumonia	Subpleural and basal honeycombing, reticular abnormalities, traction bronchiectasis
Nonspecific interstitial pneumonia	Reticular abnormalities, areas of GGO, mild honeycombing, subpleural sparing
Organizing pneumonia	Patchy areas of consolidation alternating with GGO, which exhibit subpleural and peribronchovascular distribution. Often, there is a direct and reversed halo sign
Lymphoid pneumonia	GGO, thin-walled cysts, reticular abnormalities
Amyloidosis	Nodular thickening of the interlobular and intralobular septa, subpleural areas of consolidation with calcification, intrathoracic lymphadenopathy

GGO, ground-glass opacity; *HRCT*, high-resolution computed tomography; *ILD*, interstitial lung disease; *RA*, rheumatoid arthritis.

Nodular lesions

Nodules in the lungs can be found in up to 20% of the patients with RA, and they are usually accompanied with periarticular rheumatoid nodules with a long-term RA seropositivity; the nodules develop in males more frequently [31]. The nodules, which range in size from several millimeters to several centimeters, are usually located subpleurally (Fig. 8.1.14) and may have necrotic cavities [32]. The number and size of nodules do not depend on the severity of RA and, as a rule, do not manifest clinically, except for patients with pneumothorax due to pleural rupture [15].

The nodules are very difficult to differentiate radiographically from metastatic or primary tumor lesions of the lungs, since they may have similar characteristics such as spicularity and retraction of the adjacent pleura.

Caplan syndrome

In a 1952 study of approximately 14,000 coal miners, Caplan found that those who suffered from RA had more frequent lung lesions with an atypical radiological characteristic that was distinct from simple silicosis [33]. Caplan syndrome, that is, rheumatoid pneumoconiosis, is characterized by single or multiple nodular or nodal formations, which are up to 5 cm in diameter. Smaller lesions may be surrounded by GGOs, while larger ones may have calcification and cavities [34].

The nodules in Caplan syndrome tend to localize in the upper lobes. Histologically the signs of pneumoconiosis and rheumatoid nodules are simultaneously present in Caplan syndrome and are observed as a zone of central necrosis surrounded by an infiltrating layer of lymphocytes and macrophages with pigment inclusions. In contrast to silicosis a more rapid appearance and growth of nodules is observed in Caplan syndrome [26].

Pleural lesions

Some pleural abnormalities are found in 50% of patients with RA at the time of autopsy; however, the pleural involvement often does not present clinically or is not clinically significant while the patient is alive [35]. The primary clinical and radiological pleural signs are pleural effusion and thickening of the visceral pleura, respectively (Fig. 8.1.15) [36]. Pleural involvement in RA is found more frequently in males than females. Pleural effusion, an infrequent finding, is observed in approximately 5% of the patients with active RA [37]. Pleurisy is usually unilateral, has a moderate or small volume, and is often concomitant with exudative pericarditis [38]. Approximately 40% of the patients with rheumatoid pleural effusion have signs of RA-ILD [26]. Effusion usually disappears within a few weeks by treatment with systemic or topical steroids [37].

Since pleural effusion in RA is not a typical complication, it is necessary to rule out other etiologies. Cytology of the pleural fluid in RA is not specific. However, biochemical analysis can be utilized for more specific findings of RA. Rheumatoid exudate is characterized by a low glucose level of less than 50 mg/dL, a high lactate dehydrogenase level (more than two times higher than the upper limit of the norm), and a low pH level of <7.30 [36]. A high rheumatoid factor (RF) titer ($>1:320$) in the pleural effusion fluid reflects serum values of RF, but this is not a proof of RA as the cause of effusion [39]. With the recurrent nature of pleurisy, video-assisted thoracoscopic surgery with pleural biopsy may be required in patients with a controlled course of RA. Thickening of the visceral pleura can be present in RA-ILD or as a separate sign and indicates involvement of the pleura in granulomatous inflammatory response, identical to that seen in rheumatoid nodules [35].

Lower airway abnormalities

The primary pathological abnormalities of the lower airway in RA are bronchiectasis, obliterative bronchiolitis, and follicular bronchiolitis (Fig. 8.1.16). Bronchiectasis that is not associated with ILD occurs in approximately 20% of patients with RA. The cause of bronchiectasis is not quite clear, and its frequency is not associated with the severity of RA but depends on the level of autoantibodies [40].

A more frequent association with the heterozygosity of the cystic fibrosis transmembrane conductance regulator gene was found in patients with RA and bronchiectasis, compared with patients with bronchiectasis in the absence of RA [41]. Repeated lower airway infections and smoking are other risk factors for bronchiectasis. However, the specific mechanisms of bronchiectasis in adult patients with RA are not completely understood, although infectious factors probably play a significant role in the pathogenesis inducing citrullination in the epithelium of the respiratory tract [12, 42].

The presence of bronchiectasis in RA increases the risk of death by approximately 7.3 times compared with the general population, by 5 times compared with patients with RA without lung lesion, and by 2.4% compared with patients with bronchiectasis without RA [43].

Obliterative (constrictive) bronchiolitis (OB) is a rarer variant involving terminal lower airways, which manifests by progressive dyspnea and the presence of mosaic zones of hypoventilation and hyperperfusion on CT and air traps that are detectable more easily during expiration. OB is more commonly associated with penicillamine intake but can also occur without a history of its use. OB usually occurs in combination with bronchiectasis [26].

Follicular bronchiolitis (FB) is also a rare RA-associated pulmonary pathology. The underlying pathology is lymphoid follicular hyperplasia in the terminal bronchioles, which results in their narrowing or obliteration. On HRCT, FB is manifested by disseminated centrilobular nodules and the tree-in-bud phenomenon. FB can occur concomitantly with lymphoid interstitial pneumonia [35].

Vascular lesions

Pulmonary hypertension occurs in approximately 20% of patients with rheumatoid lung and more often depends on the degree of involvement of the lung parenchyma in RA-ILD.

Pulmonary vasculitis in RA is a rare involvement of the small lung vessels during the systemic disease. However, rheumatoid pulmonary vasculitis can determine the prognosis in a patient and is sometimes the first manifestation of RA [44].

Differential diagnosis

Differential diagnosis of rheumatoid lung depends on the lesion pattern. Nodular manifestations must be differentiated from peripheral tumors, whereas the differential diagnosis of RA-ILD should include drug-induced pneumonitis, hypersensitivity pneumonitis, and opportunistic infections; pleural effusion associated with RA always requires consideration of other possible causes.

Without histological verification, definitive diagnosis of a single rheumatoid nodule from a lung tumor is very difficult (Figs. 8.1.17 and 8.1.18). One of the reference signs that can be used in diagnosis is the nodule size. Nodules of up to 5 mm in diameter are benign in 99% of the cases. The probability of malignancy is 6%–28% for nodules with a diameter of 5–10 mm and 64%–82% for those with a diameter above 2 cm [45]. The absence of significant growth and signs of invasion into the surrounding parenchyma lends support for a benign process, only after an observation period. The duration of observation is subject to debate; however, the evaluation period has been determined based on the doubling in the size of a neoplasm since 1950s. If this interval is within the range of 1–18 months, the probability of lung carcinoma is deemed as high [46]. The lack of growth of dense formations within 2 years virtually rules out malignancy potential and the need for further control [47]. The precise recommendations for the management of patients with focal lung formations can be found in the Fleischner Society 2017 Guidelines for Management of Incidentally Detected Pulmonary Nodules in Adults (see Chapter 10.2) [48]. Positron emission tomography can help differentiate rheumatoid nodules from malignant tumors due to lower accumulation of ^{18}F-fluorodeoxyglucose in benign lesions [49].

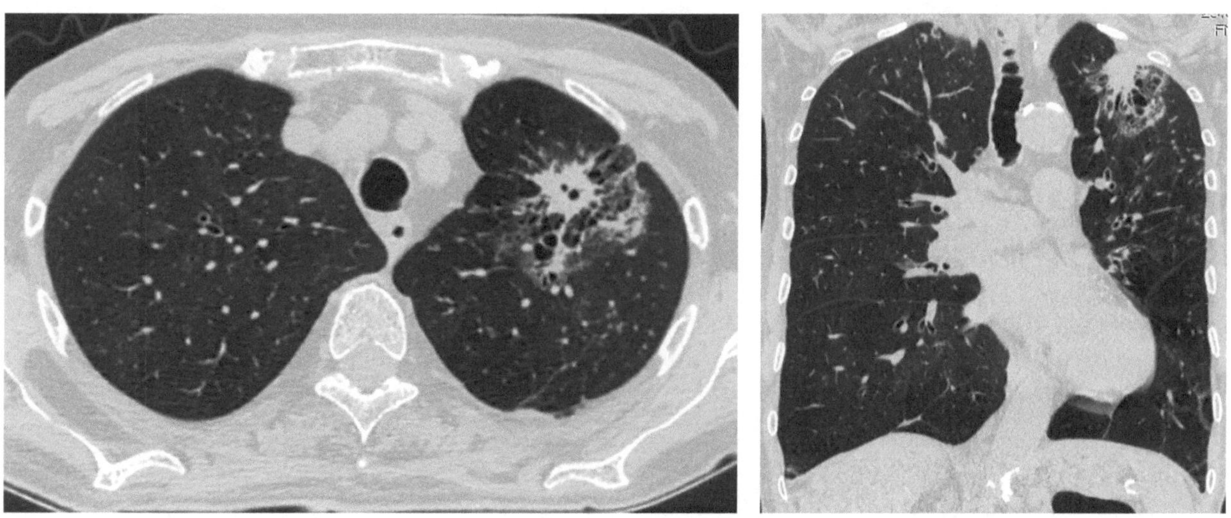

(A) (B)

FIG. 8.1.18 Adenocarcinoma in a 79-year-old patient with rheumatoid arthritis. Nodule with irregular, spicular contours, retraction of the visceral pleura, surrounded by ground-glass opacity, and multiple cysts (A, B). The air-bubble sign in the center of the nodule (A).

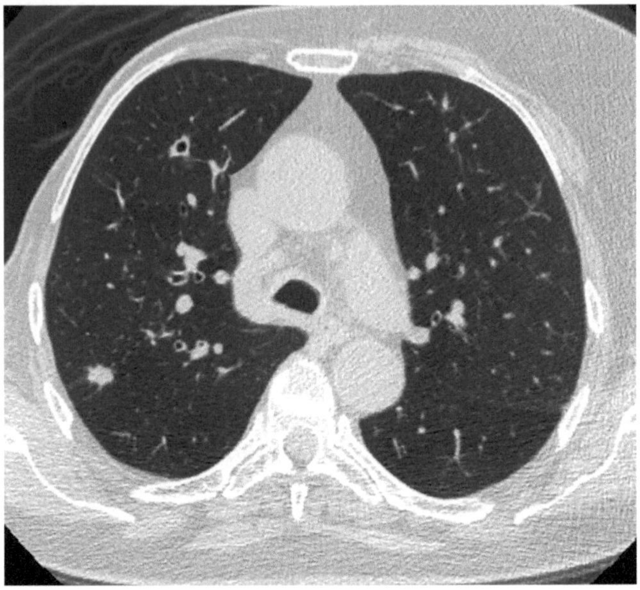

FIG. 8.1.19 Granulomatosis with polyangiitis. Dense nodule with irregular contours in the right upper lobe. In the anterior segment a small thick-walled cavity is visualized, which is a result of destructive vasculitis.

In some cases with less pronounced articular involvement, kidney lesions, and signs of necrotic nodules in the lung parenchyma, differential diagnosis from granulomatosis with polyangiitis (GPA) might be necessary (Fig. 8.1.19). In such cases the focus should be on signs of hemorrhagic vasculitis of the upper respiratory tract and high serum antineutrophil cytoplasmic antibodies (c-ANCA) titers. Normal ANCA levels render the diagnosis of GPA unlikely.

RA-ILD has to be differentiated from interstitial pulmonary lesions of another origin (Table 8.1.2). Drug-induced pneumonitis, especially that resulting from methotrexate treatment, a first-line drug for RA, develops usually 1.5–2 years after the onset of therapy in 0.6%–11% of patients receiving the treatment [50]. Patients with initially lower DLCO have the greatest risk for the development of drug-induced pneumonitis, by approximately 10 times [51]. Some studies recommend that DLCO should be determined before initiating methotrexate treatment and other drugs should be considered in patients with a DLCO below 70% [52]. Histologically, methotrexate-mediated lung lesions can be categorized as methotrexate-induced pneumonitis (MIP), diffuse alveolar damage, and hypersensitivity pneumonitis (HP). Organizing pneumonia is also a rare variant of MIP (Fig. 8.1.20) [53].

HRCT findings of MIP are nonspecific and correspond to histological pattern. In patients with HP, patchy areas of GGO with lobular air trapping and centrilobular nodules predominate. Acute interstitial pneumonia is characterized by extensive GGO and consolidation areas. In patients with organizing pneumonia, multiple consolidation foci usually alternate with patches of GGO. Treatment with other immunosuppressants and biological drugs, such as leflunomide, azathioprine, sulfasalazine, and tumor necrosis factor-alpha inhibitors, can also lead to the development of drug-induced interstitial lung disease (DILD) or exacerbation of the existing RA-ILD (Fig. 8.1.21) [54]. However, in general, the frequency of DILD in the patients treated with biologics is not significantly different than those receiving traditional therapy [55]. Many experts believe that lung biopsy is not necessary for the definitive diagnosis of DILD

TABLE 8.1.2 Differential diagnosis of lung lesions in rheumatoid arthritis

	RA-ILD	Drug-induced pneumonitis	*Pneumocystis jirovecii* pneumonia
History	RA	Intake of cytostatic and biological drugs, especially methotrexate	Intake of immunosuppressants
Clinical presentation	Slowly progressing dyspnea	Progressive dyspnea; acute course with rapidly progressive respiratory failure is possible	Febrile fever, gradually worsening respiratory failure over several weeks
Laboratory analysis	Moderately accelerated ESR, CRP correlated with RA activity	Increased ESR	Leukopenia, lymphopenia, high ESR, moderately elevated CRP
BAL	Moderate neutrophilia, moderate lymphocytosis	Lymphocytosis >30%	Neutrophilia, PCR positive for *P. jirovecii*
CT findings	Subpleural and basal honeycombing, GGO, reticular abnormalities, traction bronchiectasis	GGO opacity, lobular areas of air trapping, intralobular nodules, consolidation in cases of acute alveolar damage	Patchy areas of GGO and nodules in the upper lobes; in progressing cases, diffuse GGO with air trappings, consolidation, cysts in later stage

BAL, bronchoalveolar lavage; *CPR*, C-reactive protein; *CT*, computed tomography; *ESR*, erythrocyte sedimentation rate; *GGO*, ground-glass opacity; *ILD*, interstitial lung disease; *PCR*, polymerase chain reaction; *RA*, rheumatoid arthritis.

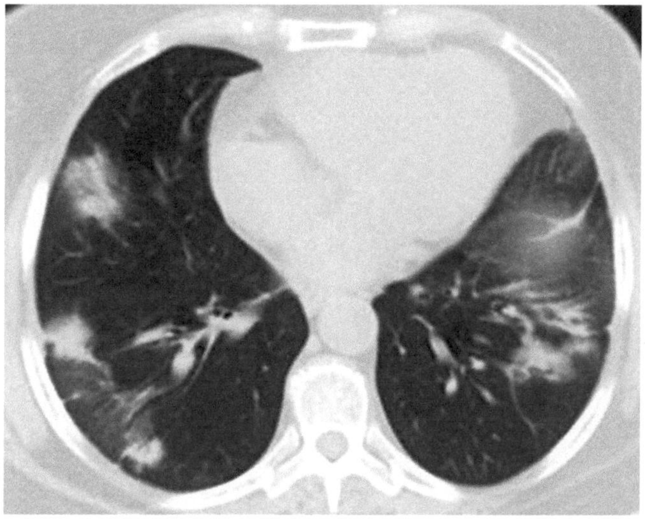

FIG. 8.1.20 Methotrexate-induced pneumonitis in a patient with rheumatoid arthritis. Bilateral areas of ground-glass opacity and consolidation with subpleural and peribronchovascular distribution. Pattern of organizing pneumonia.

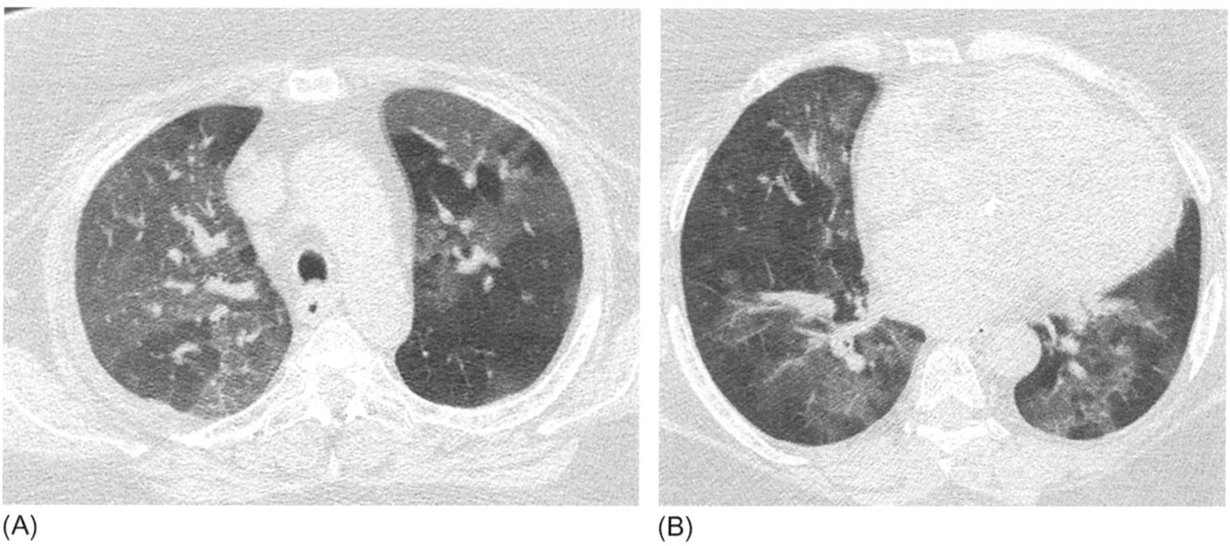

FIG. 8.1.21 Leflunomide-induced pneumonitis in a patient with rheumatoid arthritis. Diffuse ground-glass opacities with air trapping (hypersensitivity pneumonitis pattern) (A, B).

and that HRCT and bronchoalveolar lavage analysis are sufficient, which exclude lung infections and reveal significant lymphocytosis (>30%), respectively [56].

Secondary immunodeficiency, which increases the risk of opportunistic infections, often develops with RA treatment. The risk of reactivating latent tuberculosis in patients with RA receiving biological drugs is well known [57].

Pneumocystis jirovecii pneumonia (PP) is another life-threatening complication of cytostatic and steroid therapy that can be very aggressive similar to MIP and in cases is characterized by diffuse alveolar damage; PP can also be fatal [58]. In addition to the well-known increased risk for PP in patients treated with methotrexate and systemic steroids, new biological drugs, including rituximab, infliximab, etanercept, adalimumab, tocilizumab, and abatacept, can also result in the development of severe PP with a high mortality rate (10%–22%), although the incidence of this complication is not high (<0.5%) [59].

The combination of fever, progressive dyspnea, and extensive areas of GGO on HRCT that are more pronounced in the upper fields always requires consideration of PP in the differential diagnosis (Fig. 8.1.22). Visualization of *Pneumocystis jirovecii* in induced sputum or positive polymerase chain reaction of the bronchoalveolar lavage fluid is important in diagnostic tests that should always be used in patients with suspicious PP.

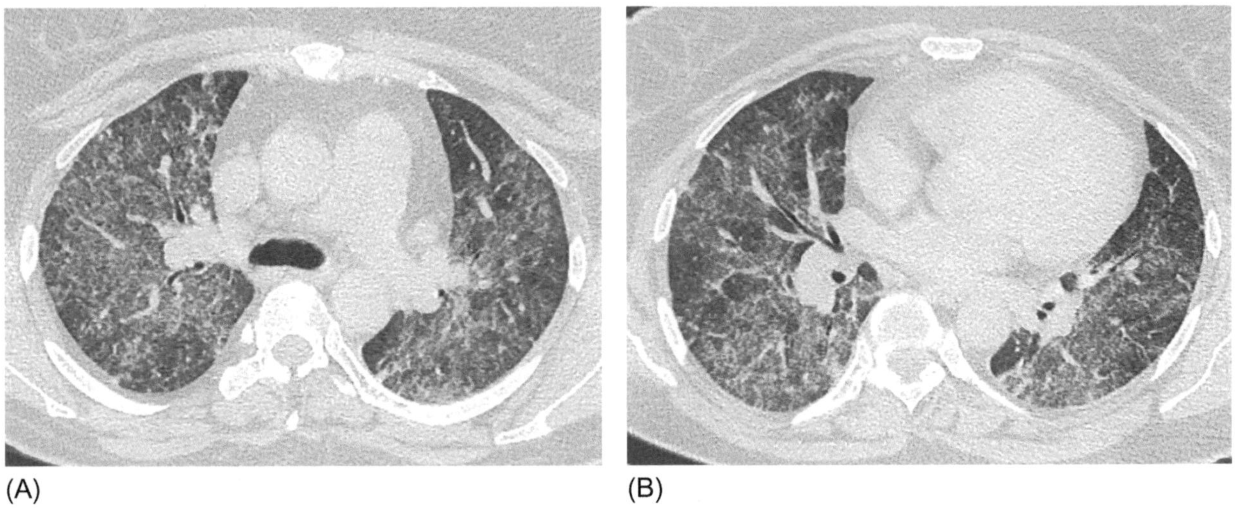

FIG. 8.1.22 Severe *Pneumocystis jirovecii* pneumonia in a patient taking methotrexate. Diffuse bilateral areas of ground-glass opacity with single unaffected hyperinflated lobules. Subpleural sparing (A, B). Findings are similar to those of hypersensitivity pneumonitis.

Treatment

The treatment strategies for RA-ILD are based on the severity of lung lesions and the rate of disease progression. The approaches depend on the three courses of disease progression that can occur in patients with RA-ILD as proposed by Chan et al. [19].

Group 1 includes patients with asymptomatic disease in which lung function remains stable. For these patients, no additional treatment is required other than the baseline therapy for RA. However, methotrexate and antitumor necrosis factor-alpha drugs should be used with care.

Group 2 includes patients with clinical manifestations of interstitial pneumonia, who have a slow increase in symptoms and restrictive abnormalities. In addition to the first-line RA therapy, these patients should receive mycophenolate mofetil (2 g/day) with or without *N*-acetylcysteine (1800 mg/day). In the absence of an effect and in the presence of an HRCT pattern of UIP, new antifibrotic drugs (pirfenidone and nintedanib), which are currently evaluated in clinical trials, might be recommended in this group of patients [23, 60].

Group 3 includes patients with RA with rapidly progressive RA-ILD. Six cycles of pulse therapy with methylprednisolone (10 mg/kg) with cyclophosphamide (15 mg/kg) at intervals of 3–4 weeks are recommended in these patients. In case of improvement, these drugs can be replaced by mycophenolate mofetil with *N*-acetylcysteine.

All patients with clinically significant RA-ILD (groups 2 and 3) should avoid the use of methotrexate. Rituximab can be considered for treatment of RA-ILD in patients with severe articular involvement and severe inflammatory interstitial lung abnormalities [23].

For patients with exacerbation of RA-ILD, oral prednisone (1 mg/kg/day) with or without an immunosuppressant is recommended [60].

The prognosis of pulmonary disorders associated with RA is determined by the histological form: the prognosis is worst for UIP, with an average survival rate of 8.27 years for definite UIP in RA [28]. Other risk factors of mortality in RA-ILD are age above 64.7 years, history of smoking, low forced vital capacity (<58.7% of predicted), and DLCO (<38%) as well as a rapid reduction in forced viral capacity or DLCO [25].

References

[1] Alamanos Y, Drosos AA. Epidemiology of adult rheumatoid arthritis. Autoimmun Rev 2005;4(3):130–6.

[2] Grassi W, De Angelis R, Lamanna G, Cervini C. The clinical features of rheumatoid arthritis. Eur J Radiol 1998;27(Suppl. 1):S18–24.

[3] Dawson JK, Fewins HE, Desmond J, Lynch MP, Graham DR. Fibrosing alveolitis in patients with rheumatoid arthritis as assessed by high resolution computed tomography, chest radiography, and pulmonary function tests. Thorax 2001;56(8):622–7.

[4] Olson AL, Swigris JJ, Sprunger DB, Fischer A, Fernandez-Perez ER, Solomon J, et al. Rheumatoid arthritis-interstitial lung disease-associated mortality. Am J Respir Crit Care Med 2011;183(3):372–8.

[5] Brown KK. Rheumatoid lung disease. Proc Am Thorac Soc 2007;4(5):443–8.

[6] Yunt ZX, Solomon JJ. Lung disease in rheumatoid arthritis. Rheum Dis Clin North Am 2015;41(2):225–36.
[7] Johnson C. Recent advances in the pathogenesis, prediction, and management of rheumatoid arthritis-associated interstitial lung disease. Curr Opin Rheumatol 2017;29(3):254–9.
[8] McInnes IB, Schett GN. The pathogenesis of rheumatoid arthritis. N Engl J Med 2011;365(23):2205–19.
[9] Fischer A, Solomon JJ, du Bois RM, Deane KD, Lynch DA, Olson AL, et al. Lung disease with anti-CCP antibodies but not rheumatoid arthritis or connective tissue disease. Respir Med 2012;106(7):1040–7.
[10] Giles JT, Danoff SK, Sokolove J, Wagner CA, Winchester R, Pappas DA, et al. Association of fine specificity and repertoire expansion of anticitrullinated peptide antibodies with rheumatoid arthritis associated interstitial lung disease. Ann Rheum Dis 2014;73(8):1487–94.
[11] Paulin F, Doyle TJ, Fletcher EA, Ascherman DP, Rosas IO. Rheumatoid arthritis-associated interstitial lung disease and idiopathic pulmonary fibrosis: shared mechanistic and phenotypic traits suggest overlapping disease mechanisms. Rev Invest Clin 2015;67(5):280–6.
[12] Quirke AM, Perry E, Cartwright A, Kelly C, De Soyza A, Eggleton P, et al. Bronchiectasis is a model for chronic bacterial infection inducing autoimmunity in rheumatoid arthritis. Arthritis Rheumatol 2015;67(9):2335–42.
[13] Corte T, Du Bois R, Wells A. Connective tissue diseases. In: Murray & Nadel's textbook of respiratory medicine. 6th ed. Philadelphia, PA: Elsevier Saunders; 2016. p. 1167–87.
[14] Guinee DG, Travis WD. The lung in connective tissue disease. In: Hasleton P, Flieder DB, editors. Spencer's pathology of the lung. 6th ed. New York: Cambridge University Press; 2013. p. 804–46.
[15] Alunno A, Gerli R, Giacomelli R, Carubbi F. Clinical, epidemiological, and histopathological features of respiratory involvement in rheumatoid arthritis. Biomed Res Int 2017;2017:7915340.
[16] Schreiber J, Koschel D, Kekow J, Waldburg N, Goette A, Merget R. Rheumatoid pneumoconiosis (Caplan's syndrome). Eur J Intern Med 2010;21(3):168–72.
[17] Assayag D, Elicker BM, Urbania TH, Colby TV, Kang BH, Ryu JH, et al. Rheumatoid arthritis-associated interstitial lung disease: radiologic identification of usual interstitial pneumonia pattern. Radiology 2014;270(2):583–8.
[18] Kim EJ, Collard HR, King Jr TE. Rheumatoid arthritis-associated interstitial lung disease: the relevance of histopathologic and radiographic pattern. Chest 2009;136(5):1397–405.
[19] Chan E, Chapman K, Kelly C. Interstitial lung disease in rheumatoid arthritis: a review. Arthtitis Res 2013;3(Topical Reviews Series 7):1–9.
[20] Spagnolo P, Lee JS, Sverzellati N, Rossi G, Cottin V. The lung in rheumatoid arthritis—focus on interstitial lung disease. Arthritis Rheumatol 2018;70(10):1544–54.
[21] Lee HK, Kim DS, Yoo B, Seo JB, Rho JY, Colby TV, et al. Histopathologic pattern and clinical features of rheumatoid arthritis-associated interstitial lung disease. Chest 2005;127(6):2019–27.
[22] Suda T, Kaida Y, Nakamura Y, Enomoto N, Fujisawa T, Imokawa S, et al. Acute exacerbation of interstitial pneumonia associated with collagen vascular diseases. Respir Med 2009;103(6):846–53.
[23] Doyle TJ, Dellaripa PF. Lung manifestations in the rheumatic diseases. Chest 2017;152(6):1283–95.
[24] Assayag D, Lubin M, Lee JS, King TE, Collard HR, Ryerson CJ. Predictors of mortality in rheumatoid arthritis-related interstitial lung disease. Respirology 2014;19(4):493–500.
[25] Solomon JJ, Chung JH, Cosgrove GP, Demoruelle MK, Fernandez-Perez ER, Fischer A, et al. Predictors of mortality in rheumatoid arthritis-associated interstitial lung disease. Eur Respir J 2016;47(2):588–96.
[26] Webb RW, Higgins CB. Thoracic imaging: pulmonary and cardiovascular radiology. 2nd ed. Philadelphia, PA: Lippincott Williams & Wilkins; 2011.
[27] Henriet AC, Diot E, Marchand-Adam S, de Muret A, Favelle O, Crestani B, et al. Organising pneumonia can be the inaugural manifestation in connective tissue diseases, including Sjögren's syndrome. Eur Respir Rev 2010;19(116):161–3.
[28] Yunt ZX, Chung JH, Hobbs S, Fernandez-Perez ER, Olson AL, Huie TJ, et al. High resolution computed tomography pattern of usual interstitial pneumonia in rheumatoid arthritis-associated interstitial lung disease: Relationship to survival. Respir Med 2017;126:100–4.
[29] Kim EJ, Elicker BM, Maldonado F, Webb WR, Ryu JH, Van Uden JH, et al. Usual interstitial pneumonia in rheumatoid arthritis-associated interstitial lung disease. Eur Respir J 2010;35(6):1322–8.
[30] Park JH, Kim DS, Park IN, Jang SJ, Kitaichi M, Nicholson AG, et al. Prognosis of fibrotic interstitial pneumonia: idiopathic versus collagen vascular disease-related subtypes. Am J Respir Crit Care Med 2007;175(7):705–11.
[31] Franquet T. High-resolution CT of lung disease related to collagen vascular disease. Radiol Clin North Am 2001;39(6):1171–87.
[32] Sidhu HS, Bhatnagar G, Bhogal P, Riordan R. Imaging features of the pleuropulmonary manifestations of rheumatoid arthritis: pearls and pitfalls. J Clin Imaging Sci 2011;1:32.
[33] Caplan A. Certain unusual radiological appearances in the chest of coal-miners suffering from rheumatoid arthritis. Thorax 1953;8(1):29–37.
[34] Capitani E, Schweller M, da Silva CM, Metze K, Cerqueira EM, Bértolo MB. Rheumatoid pneumoconiosis (Caplan's syndrome) with a classical presentation. J Bras Pneumol 2009;35(9):942–6.
[35] Massey H, Darby M, Edey A. Thoracic complications of rheumatoid disease. Clin Radiol 2013;68(3):293–301.
[36] Corcoran JP, Ahmad M, Mukherjee R, Redmond KC. Pleuro-pulmonary complications of rheumatoid arthritis. Respir Care 2014;59(4):e55–9.
[37] Balbir-Gurman A, Yigla M, Nahir AM, Braun-Moscovici Y. Rheumatoid pleural effusion. Semin Arthritis Rheum 2006;35(6):368–78.
[38] Mayberry JP, Primack SL, Muller NL. Thoracic manifestations of systemic autoimmune diseases: radiographic and high-resolution CT findings. Radiographics 2000;20(6):1623–35.
[39] Hassan T, Al-Alawi M, Chotirmall SH, McElvaney NG. Pleural fluid analysis: standstill or a work in progress. Pulm Med 2012;2012:716235.
[40] Demoruelle MK, Weisman MH, Simonian PL, Lynch DA, Sachs PB, Pedraza IF, et al. Brief report: airways abnormalities and rheumatoid arthritis-related autoantibodies in subjects without arthritis: early injury or initiating site of autoimmunity? Arthritis Rheum 2012;64(6):1756–61.

[41] Puechal X, Bienvenu T, Genin E, Berthelot JM, Sibilia J, Gaudin P, et al. Mutations of the cystic fibrosis gene in patients with bronchiectasis associated with rheumatoid arthritis. Ann Rheum Dis 2011;70(4):653–9.
[42] Wilczynska MM, Condliffe AM, McKeon DJ. Coexistence of bronchiectasis and rheumatoid arthritis: revisited. Respir Care 2013;58(4):694–701.
[43] Ong HK, Lee AL, Hill CJ, Holland AE, Denehy L. Effects of pulmonary rehabilitation in bronchiectasis: a retrospective study. Chron Respir Dis 2011;8(1):21–30.
[44] Tourin O, de la Torre Carazo S, Smith DR, Fischer A. Pulmonary vasculitis as the first manifestation of rheumatoid arthritis. Respir Med Case Rep 2013;8:40–2.
[45] Wahidi MM, Govert JA, Goudar RK, Gould MK, McCrory DC, American College of Chest Physicians. Evidence for the treatment of patients with pulmonary nodules: when is it lung cancer?: ACCP evidence-based clinical practice guidelines (2nd edition). Chest 2007;132(3 Suppl.):94S–107S.
[46] Khan A, Al-Jahdali H, Irion K, Arabi M, Koteyar SS. Solitary pulmonary nodule: a diagnostic algorithm in the light of current imaging technique. Avicenna J Med 2011;1(2):39–51.
[47] Gould M, Donington J, Lynch W, Mazzone PJ, Midthun DE, Naidich DP, et al. Evaluation of individuals with pulmonary nodules: when is it lung cancer? Diagnosis and management of lung cancer, 3rd ed: American College of Chest Physicians evidence-based clinical practice guidelines. Chest 2013;143(5 Suppl.):e93S–e120S.
[48] MacMahon H, Naidich DP, Goo JM, Lee KS, Leung ANC, Mayo JR, et al. Guidelines for management of incidental pulmonary nodules detected on CT images: from the Fleischner Society 2017. Radiology 2017;284(1):228–43.
[49] Gupta P, Ponzo F, Kramer EL. Fluorodeoxyglucose (FDG) uptake in pulmonary rheumatoid nodules. Clin Rheumatol 2005;24(4):402–5.
[50] Lake F, Proudman S. Rheumatoid arthritis and lung disease: from mechanisms to a practical approach. Semin Respir Crit Care Med 2014;35(2):222–38.
[51] Saravanan V, Kelly CA. Reducing the risk of methotrexate pneumonitis in rheumatoid arthritis. Rheumatology (Oxford) 2004;43(2):143–7.
[52] Hamblin MJ, Horton MR. Rheumatoid arthritis-associated interstitial lung disease: diagnostic dilemma. Pulm Med 2011;2011:872120.
[53] Imokawa S, Colby TV, Leslie KO, Helmers RA. Methotrexate pneumonitis: review of the literature and histopathological findings in nine patients. Eur Respir J 2000;15(2):373–81.
[54] Roubille C, Haraoui B. Interstitial lung diseases induced or exacerbated by DMARDS and biologic agents in rheumatoid arthritis: a systematic literature review. Semin Arthritis Rheum 2014;43(5):613–26.
[55] Dixon WG, Hyrich KL, Watson KD, Lunt M, BSRBR Control Centre Consortium, Symmons DP, British Society for Rheumatology Biologics Register. Influence of anti-TNF therapy on mortality in patients with rheumatoid arthritis-associated interstitial lung disease: results from the British Society for Rheumatology Biologics Register. Ann Rheum Dis 2010;69(6):1086–91.
[56] Lateef O, Shakoor N, Balk RA. Methotrexate pulmonary toxicity. Expert Opin Drug Saf 2005;4(4):723–30.
[57] Cantini F, Nannini C, Niccoli L, Petrone L, Ippolito G, Goletti D. Risk of tuberculosis reactivation in patients with rheumatoid arthritis, ankylosing spondylitis, and psoriatic arthritis receiving non-anti-TNF-targeted biologics. Mediators Inflamm 2017;2017:8909834.
[58] Yale SH, Limper AH. Pneumocystis carinii pneumonia in patients without acquired immunodeficiency syndrome: associated illnesses and prior corticosteroid therapy. Mayo Clin Proc 1996;71(1):5–13.
[59] Mori S, Sugimoto M. *Pneumocystis jirovecii* pneumonia in rheumatoid arthritis patients: risks and prophylaxis recommendations. Clin Med Insights Circ Respir Pulm Med 2015;9(Suppl. 1):29–40.
[60] Ha YJ, Lee YJ, Kang EH. Lung involvements in rheumatic diseases: update on the epidemiology, pathogenesis, clinical features, and treatment. Biomed Res Int 2018;2018:6930297.

Chapter 8.2

Systemic sclerosis-related interstitial lung disease

Systemic sclerosis (SSC) is a connective tissue disease (CTD), characterized by three pathogenetic signs, namely, angiopathy of small vessels, the production of autoantibodies, and fibroblast dysfunction, leading to an increased deposition of extracellular matrix, primarily collagen [1].

Women tend to suffer more often than men, with the highest incidence occurring in the second half of life (45–65 years) [2]. There are three subsets of SSC depending on the degree of skin involvement [3]:

- Diffuse cutaneous SSC is characterized by changes in the skin that spread to the extremities for a year, including above the elbow and knee joints, trunk, or face; early development of Raynaud's phenomenon and visceral pathology (interstitial lung disease (ILD); lesion of the gastrointestinal tract, myocardium, and kidneys; a decrease in the number of capillaries; and the development of avascular areas as detected by nail fold capillaroscopy); and detection of antibodies to topoisomerase-1 (Scl-70).
- Limited cutaneous SSC is characterized by skin fibrosis that is only limited to the distal parts of the extremities (arms below the elbow joints) and the face; quite often, Raynaud's phenomenon precedes other manifestations of the disease for a long time; pulmonary damage, gastrointestinal tract, telangiectasia develop much later, and the dilatation of the nail fold capillaries without pronounced avascular areas as detected by nail fold capillaroscopy.
- Visceral SSC is characterized by the absence of skin lesions; the disease manifests itself with Raynaud's phenomenon, ILD, sometimes the development of acute kidney injury, lesion of the heart and gastrointestinal tract, and detection of antinuclear antibodies (ANAs) (Scl-70, anticentromere antibodies).

In 1980 the American College of Rheumatology (ACR) proposed the "major" and "minor" criteria for the diagnosis of SSC. However, over the past 30 years, the knowledge of SSC has significantly increased, the nail fold capillaroscopy method was widely introduced, and the immunologic tests that have become one of the essential markers of the disease were developed. As a result, in 2013, ACR and the European League Against Rheumatism (EULAR) revised the diagnostic criteria, and a new approach to the diagnosis of SSC was proposed [4] (Table 8.2.1).

Pulmonary arterial hypertension (PAH) and ILD are the two main manifestations of respiratory system lesion in SSC, which largely determine the prognosis in patients. During autopsy, ILD is found in 90% of those who died of SSC, and in a lifetime, signs of interstitial lesions are manifested in approximately 65% of patients [5, 6]. SSC-associated ILD (SSC-ILD)

TABLE 8.2.1 Criteria for the diagnosis of SSC (extract from [4])

Criteria	Addition	Points
Skin thickening of fingers of both hands extending proximal to the metacarpophalangeal joints (*sufficient criterion*)		9
Skin thickening of fingers (*only count the highest score*)	Puffy fingers Sclerodactyly of fingers (thickening extends proximal to the metacarpophalangeal joints)	2 4
Fingertip lesions (*only count the highest score*)	Digital tip ulcers Fingertip pitting scars	2 3
Telangiectasia		2
Abnormal nail fold capillaries in capillaroscopy		2
Pulmonary arterial hypertension (PAH) and/or interstitial lung disease (maximum score is 2)	PAH ILD	2
Raynaud's phenomenon		3
Scleroderma-related antibodies (maximum score is 3)	Anticentromere Antitopoisomerase I Anti-RNA polymerase III	3

The diagnosis of SSC is recognized if the patient has a set of signs of 9 or more points.

most often develops in the first 5 years from the time the diagnosis of SSC was established; hence the patients' respiratory system should be monitored especially carefully during this period [7].

It is hypothesized that the disease is caused by a recurrent microinjury of the alveolar epithelium and endothelium [8], as confirmed by an increase in serum alveolar damage markers (surfactant protein D and KL-6) [9]. The risks of further development of inflammation and fibrosis are determined by genetic and epigenetic factors. The epithelial cell apoptosis triggers a pathway of immune responses involving TGF-β, thrombin, and others, which activate lymphocyte and fibroblastic reactions from hyperproduction and the deposition of collagen and other extracellular matrix proteins through signaling pathways and cause pulmonary fibrosis in these patients [8,10]. In recent years the phenomenon of epithelial-mesenchymal transition as a mechanism of fibroproliferation has also been actively discussed [11].

Morphology

Nonspecific interstitial pneumonia (NSIP) is the most frequent ILD variant in SSC (Fig. 8.2.1), accounting for approximately 76% of cases, while usual interstitial pneumonia (UIP) is the rarer variant (~11%) (Fig. 8.2.2) [12]. It should be noted that both histopathologic types of interstitial pneumonia can be found in biopsy samples of the same patient, even when studying several biopsies within the same anatomical unit of the lung [13]. Based on the result, honeycombing develops in both UIP and NSIP (Fig. 8.2.3). Organizing pneumonia and desquamative interstitial pneumonia are possible but rare morphological patterns [14].

Fibrosing interstitial pneumonia with worsening SSC, that is, UIP and NSIP, is difficult to differentiate from idiopathic variants of this pulmonary pathology. In rare cases, lung cancer is possible (Fig. 8.2.4). In SSC, vascular involvement is manifested by the concentric narrowing of pulmonary arterioles due to the thickening of the intima by fibromyxoid tissue and hypertrophy of the media (Fig. 8.2.5). Most often, 50% of cases that have limited SSC phenotype (also known as the CREST syndrome: calcifications, Raynaud's phenomenon, esophageal hypomotility, sclerodactyly, and telangiectasia) are more likely to develop PAH [15]. In other cases the incidence of PAH in SSC is lower (10%–33%), which generally correlates with the degree of pulmonary interstitial fibrosis [16]. Chronic pleurisy is quite rare (approximately 7% of cases) and is characterized by fibroplastic thickening and patchy lymphoplasmacytic pleural infiltration (Fig. 8.2.6) [17].

Clinical presentation

The clinical presentation of SSC-ILD is nonspecific and is generally already noticeable at the full-scale stages of the disease. Patients often have physical restrictions due to muscle and joint lesions that lead to the development of dyspnea only with significant involvement of the lung parenchyma in the pulmonary damage. However, even at the beginning of the development of the interstitial process, up to 50% of patients exhibited decreased lung volume and pulmonary diffusion capacity, with the absence of significant clinical symptoms [18]. The second most common symptom is dry cough. However, as the disease progresses, the cough becomes productive (up to 46% of cases) [119]. Typical Velcro crackles are usually auscultated in the basal regions even before the onset of severe dyspnea. The development of pulmonary hypertension can lead to an

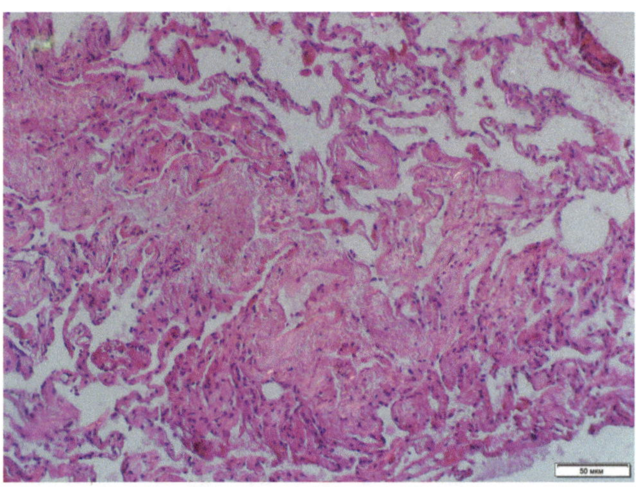

FIG. 8.2.1 SSC. NSIP, a fibroplastic variant with pronounced sclerosis of the alveolar septa. Hematoxylin and eosin (H&E) staining, 100×.

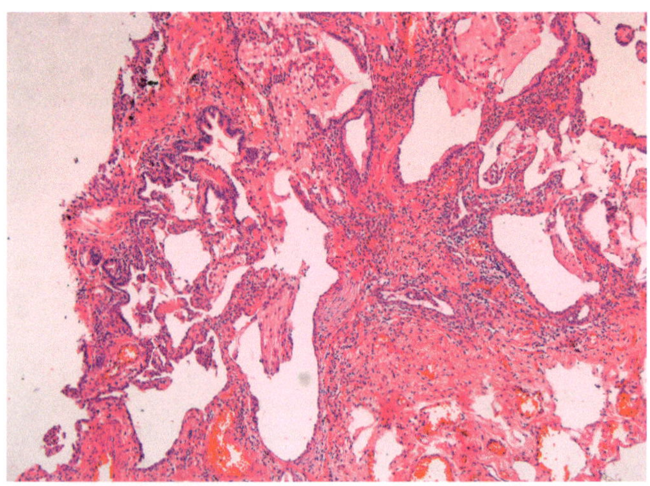

FIG. 8.2.2 SSC. UIP. H&E stain, 100×.

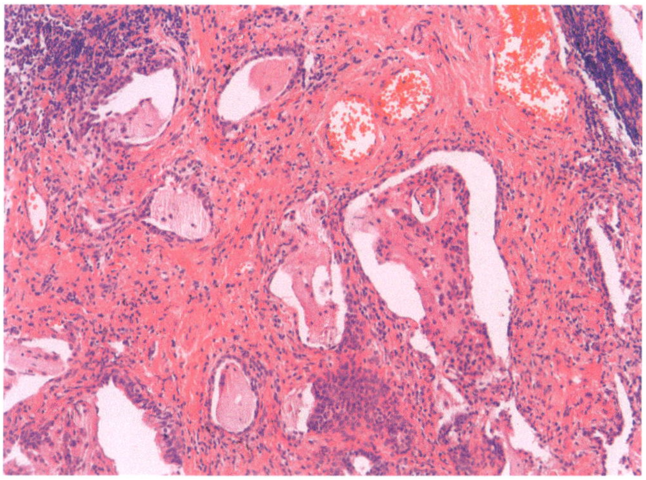

FIG. 8.2.3 SSC. Honeycombing in the outcome of UIP. H&E stain, 100×.

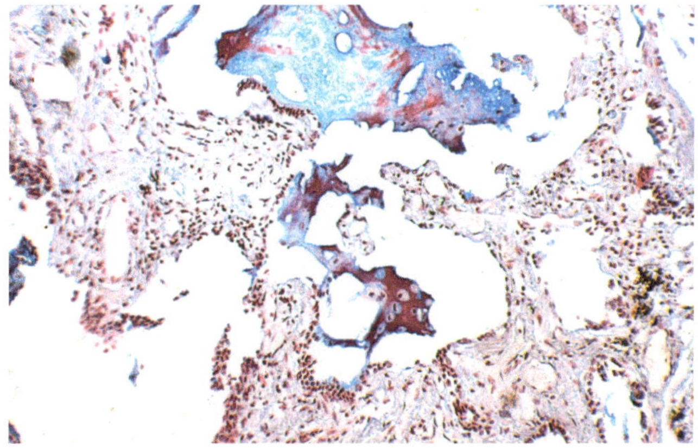

FIG. 8.2.4 SSC. Squamous cell lung cancer in the honeycombing. Kreyberg staining, 100×.

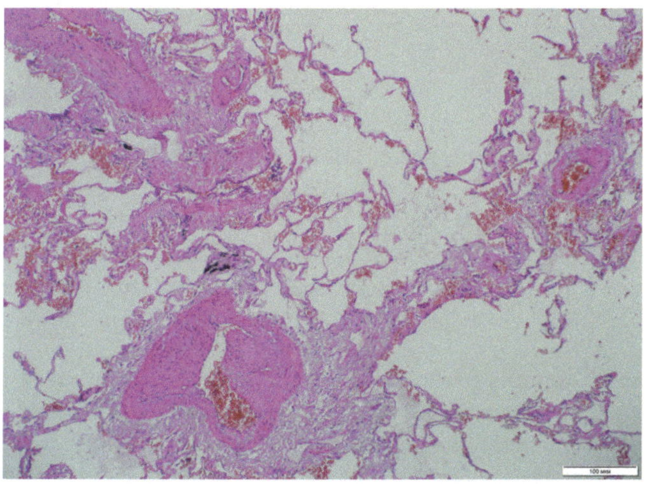

FIG. 8.2.5 SSC associated with pulmonary hypertension. Arteries with thickened wall and narrowed lumen. H&E stain, 100×.

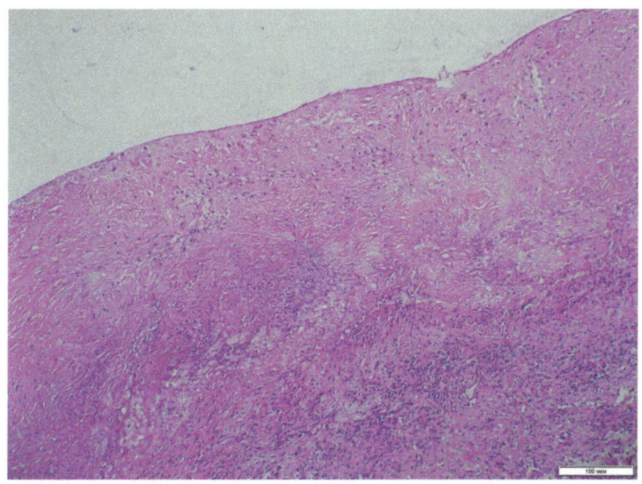

FIG. 8.2.6 SSC. Fibroplastic chronic pleurisy with pronounced tissue sclerosis. H&E stain, 100×.

earlier appearance of dyspnea. Clubbing sign is extremely rare, unlike in patients with idiopathic pulmonary fibrosis (IPF) [20]. Upon the initial medical examination of patients, respiratory symptoms are often assessed, while the characteristic skin manifestations of SSC are being overlooked. Often the disease is diagnosed only with the involvement of the lungs. Therefore, when treating patients with ILD, the specialist should always pay attention to obvious visual manifestations of SSC, such as changes in the skin of fingers (Fig. 8.2.7), telangiectasia, and alterations in capillaroscopy (Fig. 8.2.8), and take into consideration the history of Raynaud's phenomenon. The latter fact is especially important if SSC starts in the lungs and in case of the absence of skin alterations.

For quite some time, it has been believed that after the first few years of SSC-ILD active progression, it can stabilize and emerge on a functional plateau [21]. Recently published observations of 171 patients with SSC-ILD for the period 1998–2016 demonstrated that there are several disease phenotypes, including rapidly progressing forms, which create

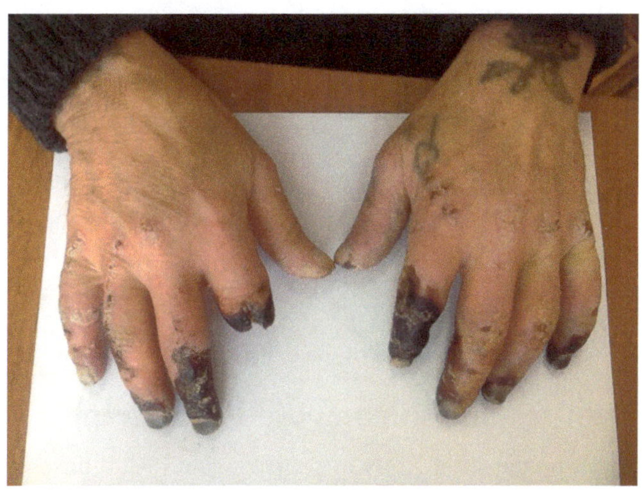

FIG. 8.2.7 Changes in the fingers that are critical manifestations of SSC, namely, swelling, induration, hyperpigmentation of the skin, and digital necrosis. *(Case courtesy of Dr. P.I. Novikov, Sechenov First Moscow State Medical University, Moscow, Russia.)*

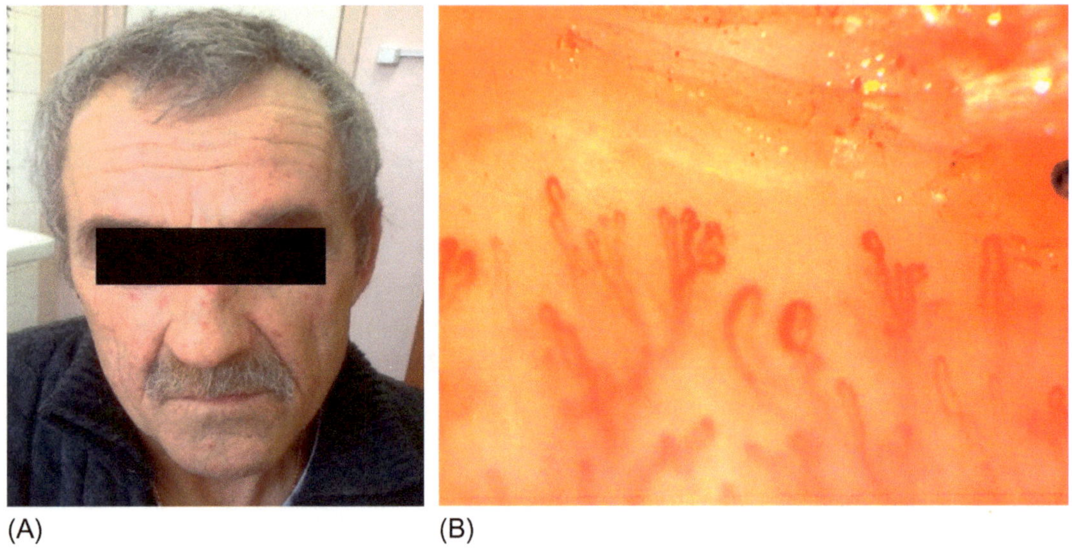

FIG. 8.2.8 Skin manifestations of SSC in the form of telangiectasia on the facial skin (A), capillaroscopic picture—capillary dilatation with SSC (B). *(Case courtesy of Dr. P.I. Novikov, Sechenov First Moscow State Medical University, Moscow, Russia.)*

the illusion of stabilization in the total cohort of the remaining patients while quitting the general statistical population (due to death) [22]. The authors believe that the identification of disease phenotypes with different rates of progression, which do not change with time, will help personalize the approach to therapy, thus avoiding excessive immunosuppression, especially since the effect of therapy on the survival rate of patients with SSC-ILD has not been proved [22].

Functional tests, as with other ILDs, include the measurement of lung volumes and lung diffusion capacity in which indexes are reduced in proportion to the severity of lung parenchyma lesions. However, in patients with radiologically confirmed SSC-ILD, the forced volume vital capacity in 31% of cases is more than 80%, while in 43% of cases, the diffusing capacity of the lungs for carbon monoxide (DLCO) exceeds 60% of pred., which indicates the possibility of an interstitial process in the lungs without significant functional changes [23].

In SSC-ILD, there is a need to assess the pressure in the pulmonary artery as an important pathogenetic and a prognostic factor of the disease. The control of cardiorespiratory function in patients with SSC should be performed every 3–6 months for the timely detection of involvement in the pathological process of the lungs and pulmonary artery [7]. There is evidence that the prevalence of the interstitial process in the lungs directly correlates with the level of nitric oxide in exhaled air [24].

In regard to prognosis, not only the basic level of FVC and DLCO but also the rate of their decline is important, which has a great predictive value [25].

Laboratory diagnostics

The diagnosis of SSC should currently be confirmed by the presence of specific ANAs, which are revealed in 90%–100% of patients [20]. Without them, even in the presence of clinical signs, another disease is more likely, for example, scleroderma-like syndromes, such as drug-induced disease or diseases caused by chemical and dusty environmental factors. It must be taken into account that up to 30% of healthy people have positive, although low, ANA titers (1:40). Antibodies (IgG) to centromere B and antibodies to topoisomerase I (Scl-70) are most commonly detected. Antibodies to RNA polymerases I, II, and III are much less common, as are the antibodies PM-Scl to TH-ribonucleoprotein and U_1-ribonucleoprotein. The detection of Scl-70 antibodies is a high-risk factor for the development and progression of interstitial pulmonary lesions in patients with diffuse SSC [13], whereas anticentromere antibodies are rarely associated with pulmonary disease [26]. The fecal calprotectin level is significantly higher in patients with SSC-ILD than in the general SSC group, which, along with the severity and duration of the disease, older age, and skin score, determines the negative prognosis [27].

High-resolution computed tomography

Based on the chest X-ray of patients with SSC, diffuse interstitial changes are observed in 25%–40% of cases; however, radiography, even digital, cannot be recommended for the assessment of SSC-ILD due to its lack of sensitivity [19]. HRCT pattern usually corresponds to the histological variant of the interstitial lung lesion. Since SSC-ILD is most often based on NSIP, its characteristics dominate in the high-resolution computed tomography (HRCT) (Fig. 8.2.9) [19]. UIP is a more rare form with a set of specific signs, such as honeycombing and reticular abnormalities (Fig. 8.2.10). Finally, HRCT detected other conditions associated with extrapulmonary manifestations of the disease, such as esophageal dilatation and the involvement of the pulmonary artery as a result of pulmonary hypertension in patients with SSC (Fig. 8.2.11) [16].

In general, HRCT findings in SSC-ILD include the following [16, 19]:

- Ground-glass opacity (GGO)
- Reticular pattern (thickening of the inter- and intralobular septa)
- Subpleural and posterior basal distribution of the lesion
- Bronchial abnormalities (thickening of the bronchial walls, traction bronchiectasis, and bronchiolectasis)
- Honeycombing
- Irregular pleural thickening
- Esophageal dilatation
- Pericardial changes
- Expansion of the pulmonary artery

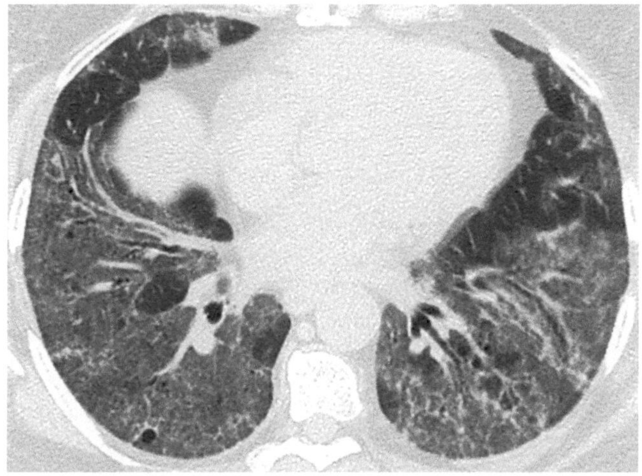

FIG. 8.2.9 SSC. Pattern of NSIP. Bilateral diffuse areas of GGO with thickened intralobular septa, traction bronchiectasis, and bronchiolectasis.

GGO is the most common sign found in 66%–93% patients with SSC-ILD, and it may be the only manifestation of the disease in 7%–22% of patients [28, 29]. For quite some time a number of researchers associated the presence of GGO sign with the activity of the inflammatory process and its prevalence over fibrotic changes, which suggests the reversibility of changes in the treatment process. Contrary to the assumption that the presence of GGO is a good prognostic factor for the response to treatment, Shah et al. showed that with long-term follow-up, only two of patients with SSC and GGO exhibited significant improvement, while none of the patients exhibited a positive HRCT dynamics, where disease manifested itself as GGO only. The authors concluded that this sign in patients with SSC-ILD is a marker of poor prognosis [28].

Based on own experience, in a study of 169 patients with limited and diffuse SSC forms, when 82% of patients exhibited ILD signs, the GGO phenomenon was not an isolated symptom in any case and was always combined with linear and reticular abnormalities [30].

The studies conducted revealed that signs of pulmonary fibrosis in patients with SSC are less common than GGO, but nevertheless, they are found in more than half of patients with ILD [19, 31]. According to data from this study, linear and reticular changes in the lung parenchyma occur in all patients with SSC-ILD in which a third of them are the only manifestations of ILD. Thus the intralobular septal thickening is noted in 98%; this is the earliest sign of the

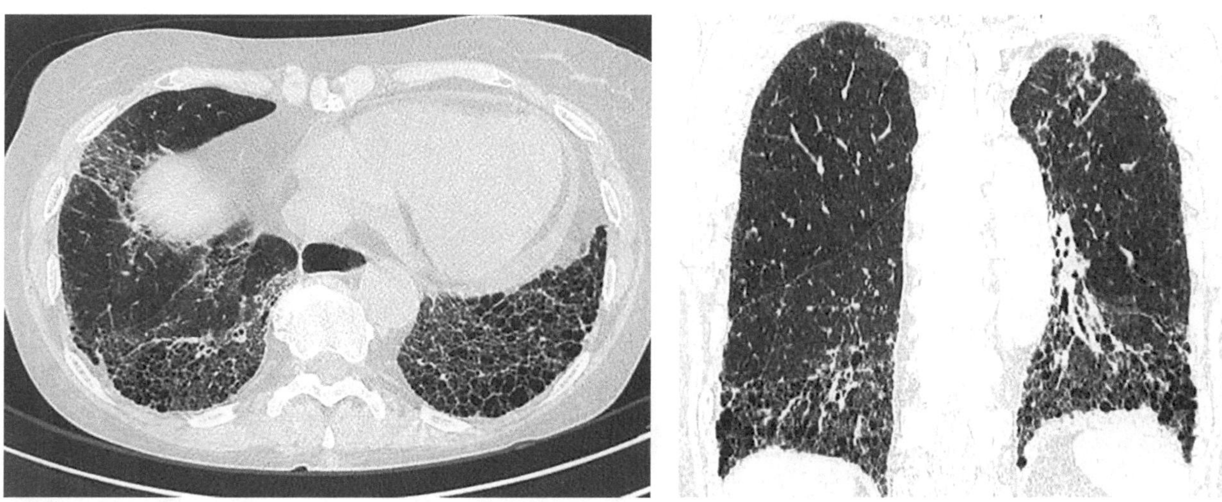

(A) (B)

FIG. 8.2.10 SSC. Pattern of probable UIP. Honeycombing, thickened intralobular septa, traction bronchiolectasis, and dilated esophagus. Area of GGO in the right lung shrinkage areas (A). Basal and subpleural pattern of distribution of abnormalities (B).

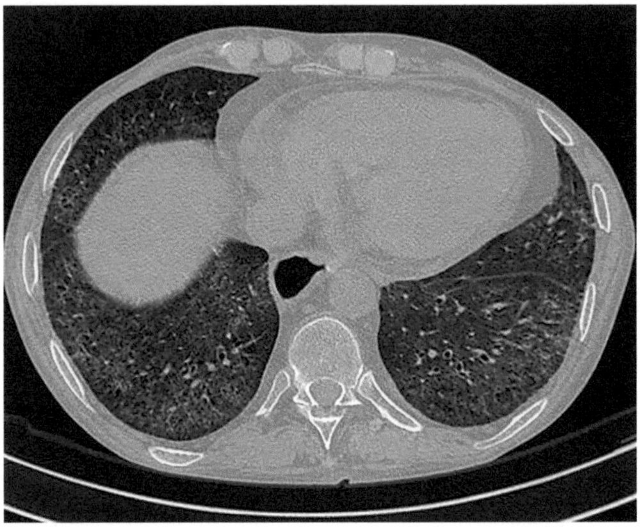

FIG. 8.2.11 SSC. Dilatation of the lower esophagus. Diffuse GGO, reticular abnormalities, bronchiolectasis, and subpleural sparing are noted in the lower zones of the lungs.

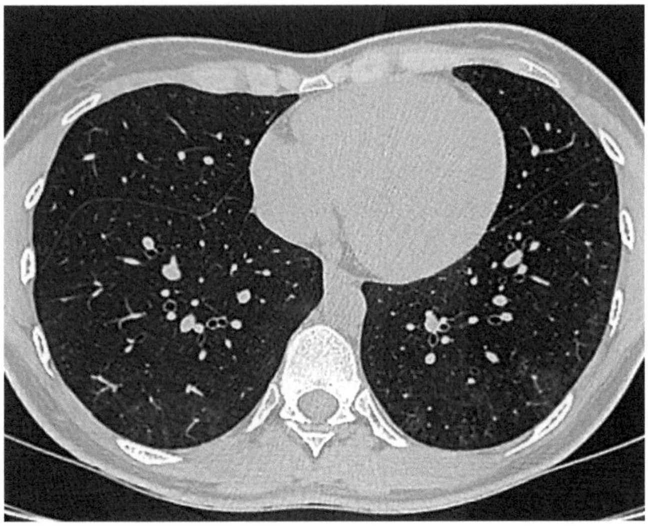

FIG. 8.2.12 SSC. ILD initial manifestations in the form of a mild subpleural GGO associated with irregular intralobular septal thickening.

presence of interstitial fibrosis, which requires a careful analysis of HRCT (Fig. 8.2.12). Also the thickening of the interlobular septa is observed in 77% of cases, which is not an isolated sign and is usually combined with intralobular thickenings, GGO, and bronchiolectasis (Fig. 8.2.13). Changes in the bronchi are characteristic signs of lung involvement in SSC, which occur in 51% of patients. Bronchiectasis has a traction mechanism and is not noted as the only sign of the disease. Bronchiolectasis is detected in the peripheral regions of the lungs with consistent and obvious interstitial alterations in the peribronchiolar parenchyma and is usually irreversible upon follow-up (Fig. 8.2.14). An additional sign is the inflammatory-fibrous thickening of bronchovascular bundles, which is found in one-third of patients (Fig. 8.2.15).

The most severe manifestation of pulmonary fibrosis is honeycombing, which can accompany both NSIP and UIP and occurs in 17%–59% of patients with SSC-ILD [19, 30, 31]. In addition, subpleural cysts are found in some patients (up to 17%) [32]. The development of honeycombing in one-third of cases was observed (Fig. 8.2.10). Honeycombing patches were detected five times more often in patients with diffuse involvement of the pulmonary parenchyma in the ILD.

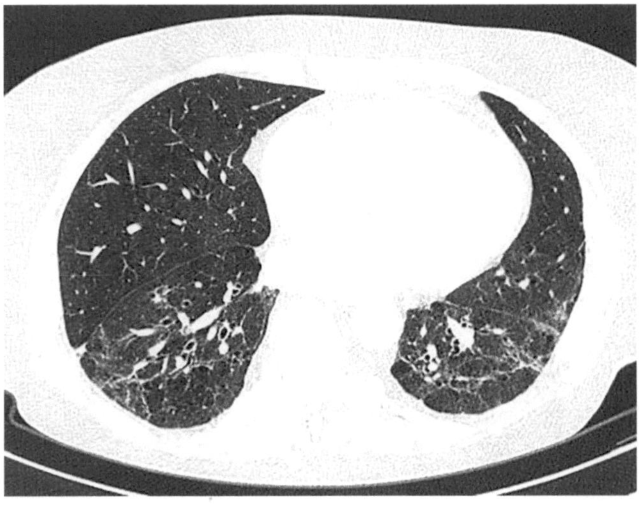

FIG. 8.2.13 SSC. Areas of GGO in the lower lobes associated with irregular septal thickening, traction bronchiectasis, and bronchiolectasis.

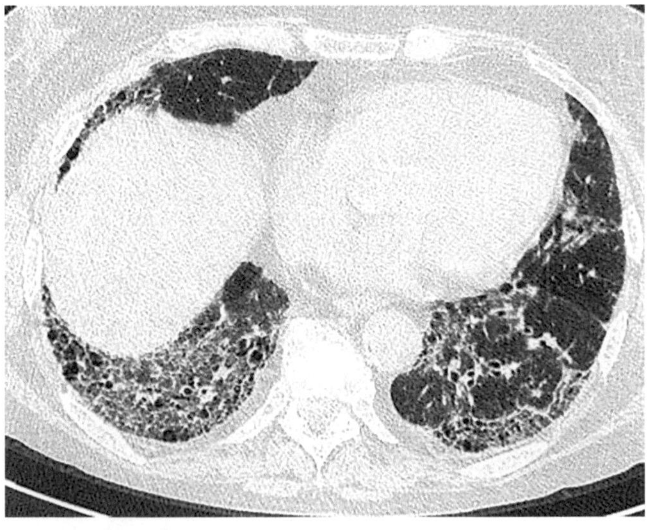

FIG. 8.2.14 SSC. Diffuse areas of GGO with the lumens of the dilated bronchi and bronchioles inside.

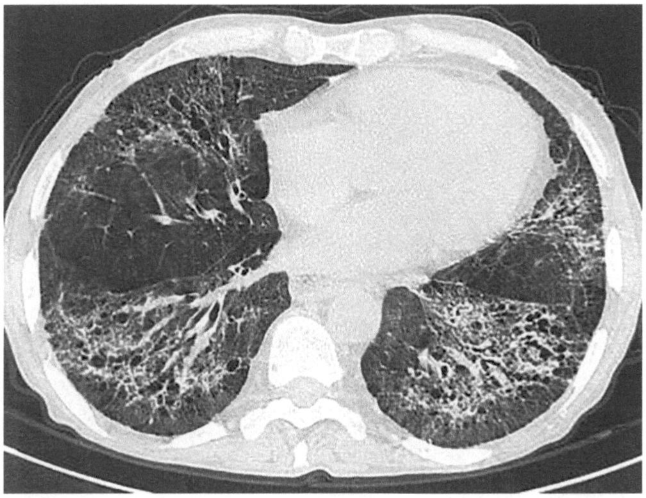

FIG. 8.2.15 SSC. Traction bronchiectasis with bronchovascular thickening

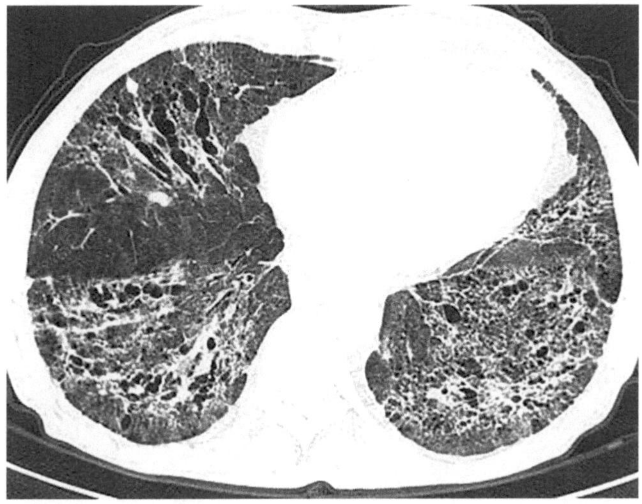

FIG. 8.2.16 SSC. Multiple cyst-like bronchiectasis and bronchiolectasis forming a pattern of intraparenchymal honeycombing.

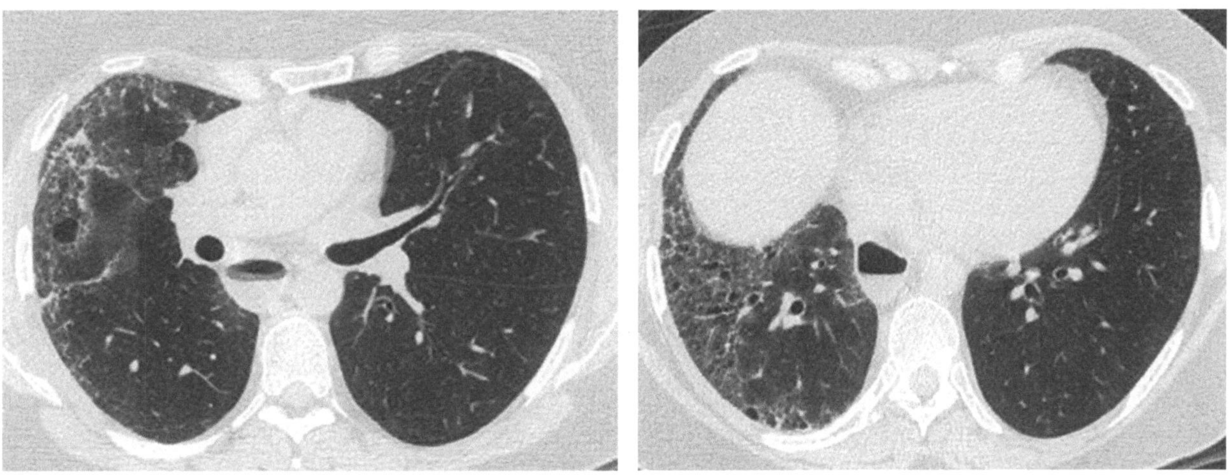

(A) (B)

FIG. 8.2.17 SSC. Unilateral ILD. Pronounced subpleural reticular abnormalities, GGO, and honeycombing in the right lung. Left lung has no signs of lesion (A and B).

Multiple small cysts are very often associated with traction bronchiectasis and bronchiolectasis and are located not only in subpleural but also in intraparenchymal areas (Fig. 8.2.16).

In our practice a unique case of unilateral pulmonary fibrosis was observed in a patient with SSC with a fully preserved parenchyma of another lung for several years (Fig. 8.2.17) [33].

The prevalence of pulmonary fibrosis has direct correlations with the development of PAH and negative prognosis [34]. In a study by Goh et al. [35], the involvement of more than 20% of lung parenchyma in the fibrotic process was found to be associated with rapid progression and poor prognosis, while the involvement of less than 20% was found to be more favorable, with the possibility of process stabilization.

Pleural lesions in the SSC included small nodules and pseudoplaques connected with the pleura, with focal or diffuse thickening of the visceral pleura. According to different authors, their frequency varies significantly. Remy-Jardin et al. found small subpleural nodules in 89% of patients and pleural thickening in one-third of patients [31]. In a study by Goldin et al. [19], certain pleural alterations are described only in 3.8% of patients with SSC. Our own data indicate the involvement of pleura in the pulmonary lesion in more than half of cases. Changes in the visceral pleura are observed mainly in the lower lobes, which is manifested by its local thickening (Fig. 8.2.18). Pleural effusion is not typical for SSC, and we have never noted the presence of fluid in pleural cavities [30].

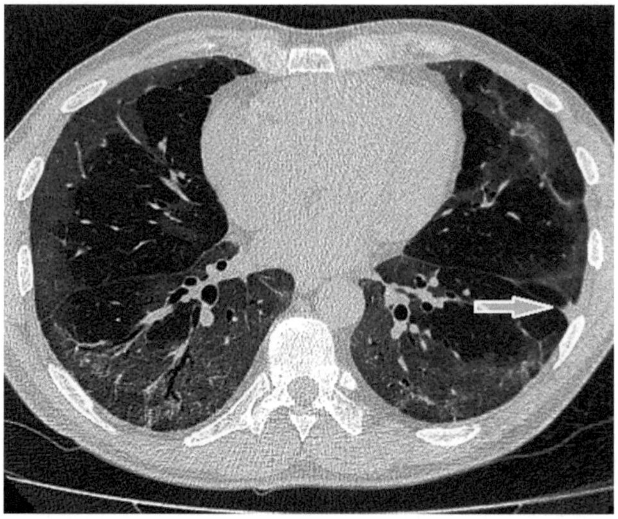

FIG. 8.2.18 SSC. Bilateral areas of GGO associated with subpleural sparing and mild reticular changes. Thickened visceral pleura on the left *(arrow)*.

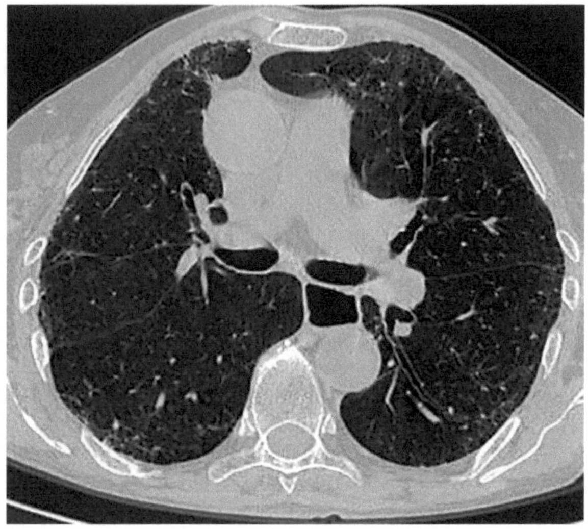

FIG. 8.2.19 SSC. Esophagus dilatation and subpleural reticular abnormalities.

According to the analyzed literature, pericardial alterations in patients with SSC are found quite often. Thus these lesions were detected by Fisher et al. [36] in 59% of patients, in the form of pericardial effusion, thickening of pericardial layers, and anterior pericardial pocket. The authors believe that abnormalities in the pericardium in general indicate an increased level of pressure in the pulmonary artery and effusion may be due to direct diffusion of fluid from it or of veins entering into the right atrium.

Esophageal dilatation is a significant and specific sign of lesion in the mediastinum SSC, which is found in a large number of patients with SSC-ILD (40%–80%) [37, 38]. The criterion for esophageal dilatation includes dilatation by more than 9 mm, although the average size is approximately 23 mm in diameter, while in severe cases the CT pattern may resemble achalasia (the lumen contains fluid or food debris) (Figs. 8.2.11 and 8.2.19). Esophageal dilatation is a significant differential sign of SSC that is rare found in other CTD [39].

Until recently, only esophageal dilatation was considered a specific HRCT sign that helped to differentiate SSC-ILD from IPF, idiopathic NSIP, and hypersensitivity pneumonitis. However, two publications in 2018 proposed additional criteria specific to SSC-ILD. These include the so-called "four corners" sign (FCS) [40], "straight-edge" sign (SES), and the "exuberant honeycombing" sign (EHS) [41]. FCS suggests an isolated involvement in the fibrotic and inflammatory process of the upper anterolateral zones (at the level from the top of the aortic arch to the carina) and "posterior corners" (posterosuperior lower lobes areas between the carina and the inferior pulmonary veins) (Fig. 8.2.20). SES implies a clear

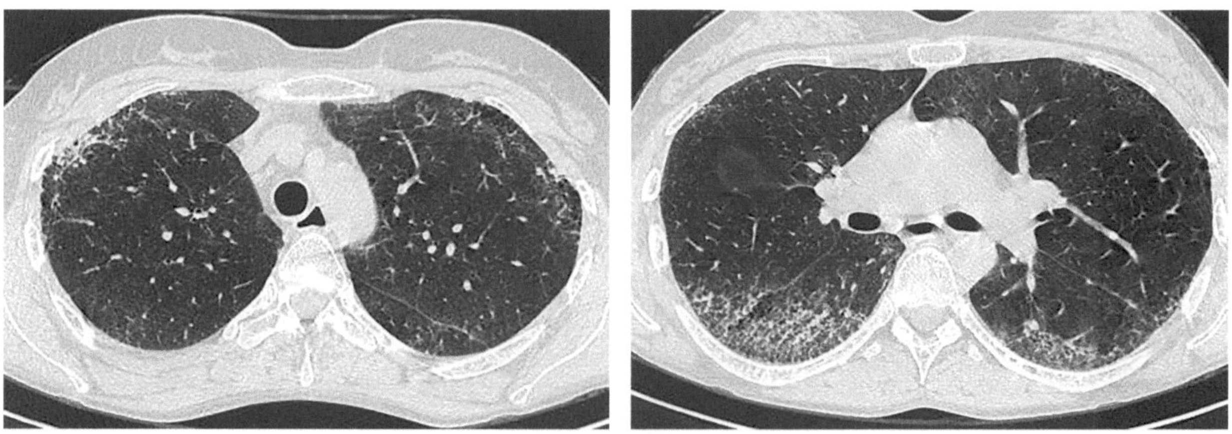

(A) (B)

FIG. 8.2.20 SSC. Isolated reticular abnormalities and honeycombing in symmetrical anterolateral angles of upper lobes (A). GGO and traction bronchiolectasis in posterolateral zones of low lobes (B). "Four corners" sign.

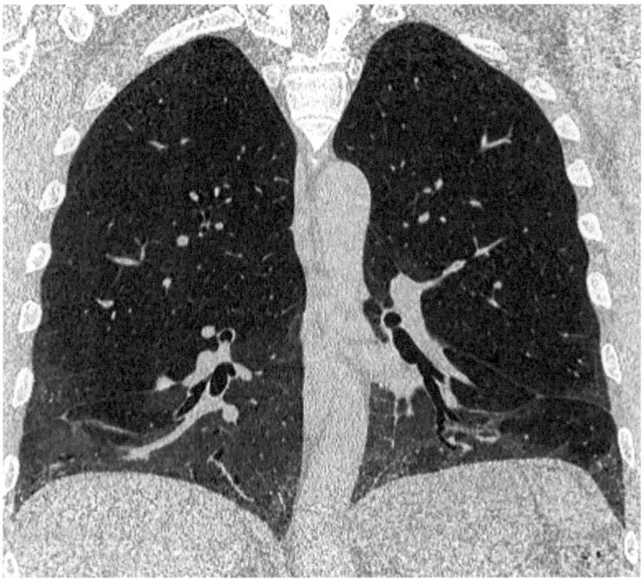

FIG. 8.2.21 SSC. Coronal slice shows sharply demarcated diffuse GGO areas in the lower zones from unaffected upper zones. "Straightedge" sign.

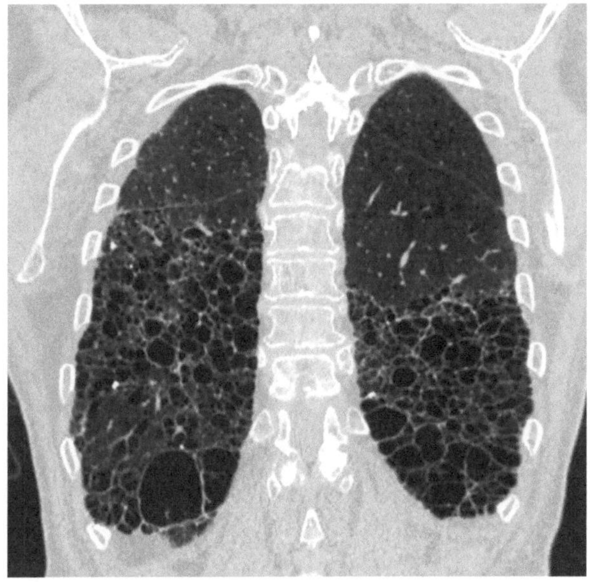

FIG. 8.2.22 SSC. Coronal reconstruction shows multiple bilateral cystic transformation of lung parenchyma. Upper lobes are not affected. "Exuberant honeycombing" sign.

isolation of areas of interstitial damage from a healthy lung (Fig. 8.2.21). EHS is a cyst-like lesion that differs from a typical honeycombing by the abundance of cysts and their diffuse filling of the affected area (Fig. 8.2.22).

An expansion of the pulmonary artery trunk by more than 29 mm is a sign of pulmonary hypertension and is revealed in approximately 12% of patients with SSC [30]. However, pulmonary hypertension is the most important prognostic factor. Its presence determines 50% mortality over 3 years in patients with SSC [42]. The absence of a visible expansion of the pulmonary artery does not rule out pulmonary hypertension.

An enlarged lymph node of the mediastinum is a nonspecific sign that accompanies many ILDs. According to Bhalla et al. [37], it occurs in approximately 60% of cases in patients with SSC-ILD. Our research does not support this phenomenon, as out of 169 patients examined, intrathoracic lymphadenopathy was observed only in 8 (5%) cases [30].

Foci and areas of consolidation are not typical for SSC-ILD and appear in organizing pneumonia. Atypical manifestations can also include the appearance of small nodules that are either a consequence of lymphocytic bronchiolitis or the addition of a pulmonary infection. In our opinion the occurrence of focal changes in the lungs in SSC requires a bronchoalveolar lavage (BAL) with bacteriologic and polymerase chain reaction analysis for respiratory pathogens.

The emergence of a single nodule with fibrous alterations may indicate the development of a malignant neoplasm, the risk of which increases by more than six times in patients with SSC-ILD [43]. The combination of SSC-ILD and lung cancer was observed in our experience only in one case (Fig. 8.2.23) [30].

BAL can be useful both to rule out possible alternative diseases (respiratory infections) and to evaluate the activity of the inflammatory interstitial process in the lungs in SSC. In accordance with the criteria of the European Respiratory and American Thoracic Societies, BAL fluid of a healthy person should contain no more than 3% neutrophils, 2% eosinophils, and 15% lymphocytes, with alveolar macrophages constituting the cell mass [44]. In SSC-ILD, alterations in BAL cytosis are found in 38%–100% of cases. Main abnormalities include the increase in the number of neutrophils and eosinophils [45]. An increase in the level of granulocytes in BAL is associated with a worse prognosis and a decrease in the survival rate in these patients [12]. There have been attempts to differentiate the histological forms of interstitial pneumonia in SSC through BAL fluid cellular analysis; however, no significant differences were found in most studies, with the exception of two studies where eosinophilia was more frequent with NSIP than with UIP pattern [12, 46]. A number of authors believe that both transbronchial and open lung biopsy in patients with SSC and ILD do not provide additional diagnostic information, except in cases of atypical manifestations or suspected concomitant process, for example, tumor [20, 47].

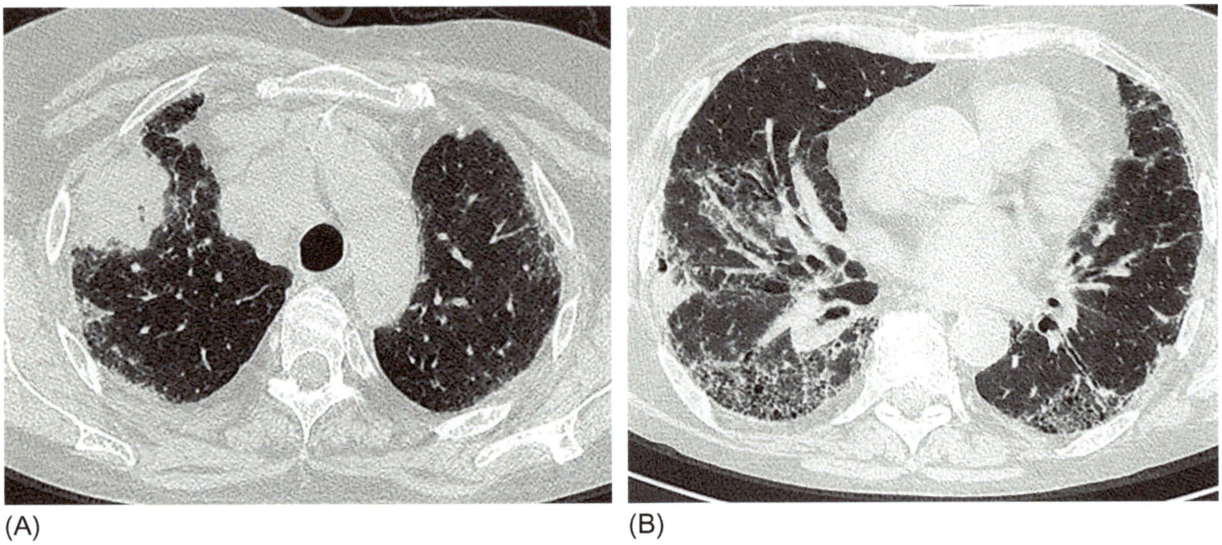

FIG. 8.2.23 SSC. Squamous cell carcinoma in the right upper lobe (A). Reticular changes and honeycombing in the lower zones (B).

Probe-based confocal laser endomicroscopy (pCLE)

In our practice, pCLE was performed in two patients with SSC-ILD with the examination of 42 bronchopulmonary areas through the analysis of 207 informative images. The endomicroscopic image in SSC is close to that in IPF and is characterized by the lack of visualization of normal alveolar structures in the affected pulmonary segments. Instead, brightly fluorescent elements of the connective tissue are determined, which "fuse" the alveolar lumens (Fig. 8.2.24A). Such areas alternate with parts of the preserved structure of the respiratory system distal compartment (Fig. 8.2.24B). In the lumen of the terminal bronchioles (Fig. 8.2.25A) and further in separate alveolar sacs, mobile, small, rounded structures with bright fluorescence were observed (Fig. 8.2.25B). The thickening of the interalveolar septa was observed in the areas of the most pronounced reticular alterations, according to the HRCT data (Fig. 8.2.26). Moreover, there were occasionally single giant cells with a pronounced autofluorescence (Fig. 8.2.27).

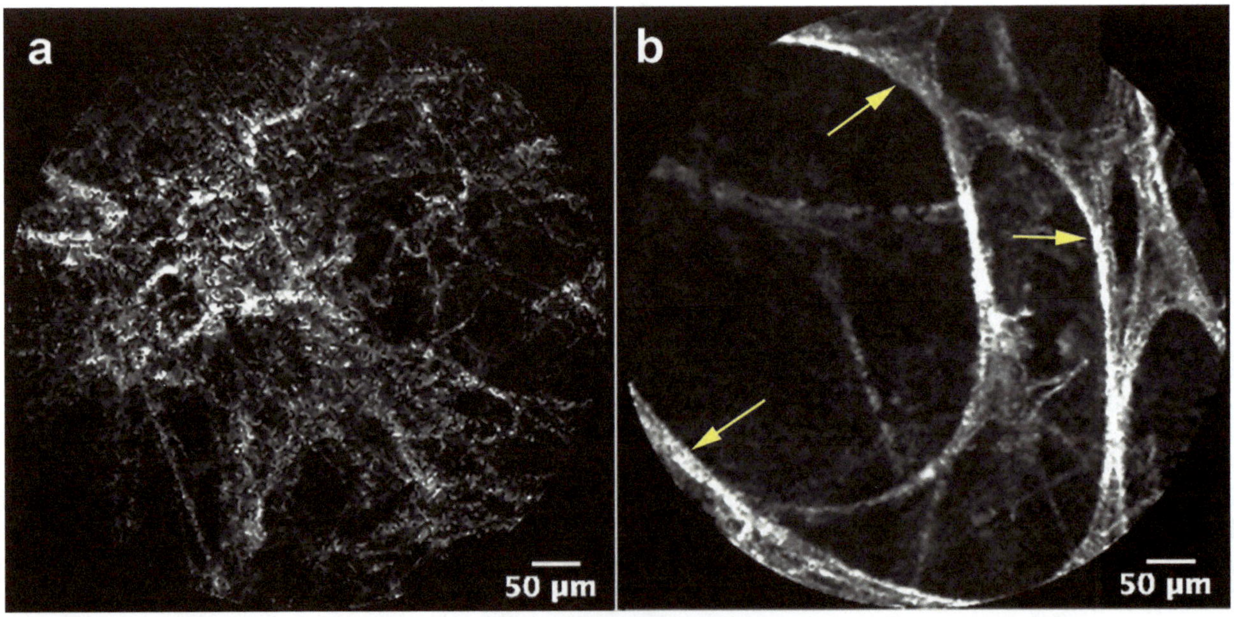

FIG. 8.2.24 pCLE pattern in a 42-year-old patient with unilateral pulmonary lesion in SSC. There is almost a total lesion of the basal parts of the right lung (A) that represents bright fluorescent randomly located connective tissue elements against which the alveoli are practically not differentiated, while the normal alveolar structure is preserved in the left lung (B). *Yellow arrows* indicate the elastic fibers of the alveolar walls, having a regular rounded shape.

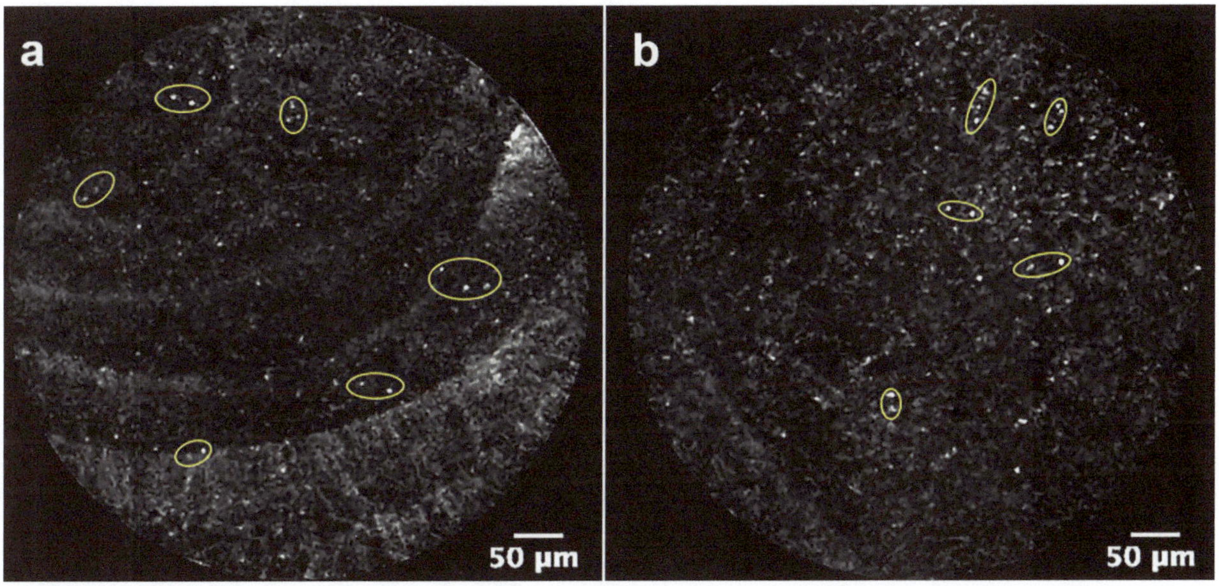

FIG. 8.2.25 pCLE of distal airways in a patient with SSC. (A) The image of a distal bronchiole with typical pronounced cross striation in a patient with SSC. (B) Viscous secretions in the alveolar lumen in one of the bronchopulmonary segments. Both endomicroscopic images represent diffusely scattered luminous rounded formations sized 3–5 µm, the largest of which are encircled with *yellow ellipses*.

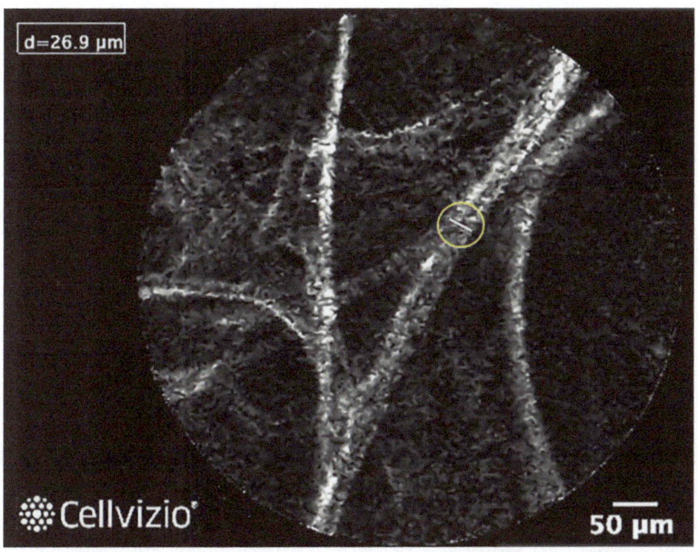

FIG. 8.2.26 Alveoloscopic image of a patient with SSC. The interalveolar septa are thickened by more than two times as compared with the normal pCLE pattern. A septum thickness of 26.9 µm (marked with an *ellipse*) was obtained upon measurement.

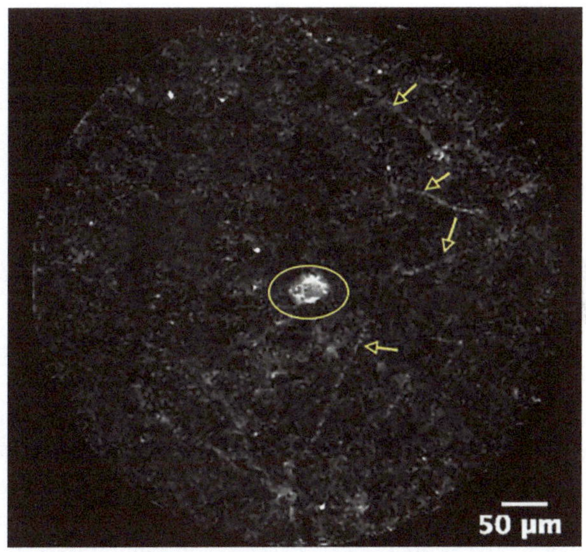

FIG. 8.2.27 Alveoloscopic image of a patient with SSC. There is a difficulty visualizing separately thinned, flabby, and partially destroyed interalveolar septa (*arrows*) due to the presence of viscous secretion in the alveolar passages lumen. A single fluorescent structure of 39.4µm (marked with an ellipse) is determined in the center of the field of vision.

Differential diagnosis

The differential diagnosis of pulmonary manifestations of SSC includes those diseases that manifest similar clinical and radiological patterns but are either idiopathic or have an exogenous trigger. Certainly, vivid clinical manifestations of SSC with confirmation of the presence of specific autoantibodies leave no doubt in SSC-ILD; however, in establishing the diagnosis at the onset process or at the ANA terminal titers, a number of additional signs have to be analyzed in order to lean toward the true disease (Table 8.2.2).

TABLE 8.2.2 Differential signs of SSC-ILD with nonrheumatic interstitial lung lesions

	NSIP	IPF	HP	SSC	DILD
Anamnesis	Arthralgic syndrome is common. The onset of the disease is at the age of 40–50 years	Onset of the disease is at the age of over 50 years, prolonged smoking	Exposure with potential allergen	Raynaud's phenomenon	The intake of bleomycin, amiodarone, cyclophosphamide, and methotrexate
Aspects of the clinical picture	More often nonsmoking women	More often men with a clubbing symptom	Intensification of dyspnea, coughing after contact with an allergen	More often women older than 40 years. Sclerodactyly, dry, thickened skin of fingers	The appearance of respiratory symptoms usually coincides in time with drug intake
CT signs	GGO, moderate reticular changes, symmetrical subpleural areas of the preserved parenchyma. Predominantly basal location	Honeycombing with predominant subpleural and basal location. Severe reticular changes	GGO, lobular areas of air traps. Intralobular nodules, subpleural sparing. Uniform distribution with the capture of the upper lobes	Identical to NSIP. Esophageal dilatation, expansion of the pulmonary artery. Four-corners sign, straight-edge sign, and exuberant honeycombing	Identical to NSIP. Areas of consolidation as a manifestation of organizing pneumonia are often noted. Upper-lobe localization may prevail

BOOP, bronchiolitis obliterans organizing pneumonia; *DILD*, drug-induced interstitial lung disease; *HP*, hypersensitivity pneumonitis; *IPF*, idiopathic pulmonary fibrosis; *NSIP*, nonspecific interstitial pneumonia; *SSC*, systemic scleroderma.

In all patients with signs of interstitial inflammation and fibrosis in the lungs, CTDs should be ruled out, including SSC, as well as hypersensitivity pneumonitis and drug-induced pneumonitis; therefore history taking and filling out special questionnaires to identify possible exogenous factors are of paramount importance. The overwhelming majority of patients with SSC have Raynaud's phenomenon manifestations for several years (sometimes decades), preceding the clinical and radiological symptoms.

The identification of this phenomenon in a patient with interstitial lung lesions indicates with a high probability of connective tissue disease or interstitial pneumonia with autoimmune features.

Treatment

Immunosuppressants are the first-line drugs for the treatment of SSC-ILD. Cyclophosphamide (CF) has the greatest evidence of efficacy in the treatment of this condition since 1993. Its efficacy in improving pulmonary function was demonstrated in 93% of patients in combination with low doses of systemic glucocorticosteroids (GCS) [48]. More recent studies have demonstrated the effect of CF not only on pulmonary function but also on the survival rate of patients with SSC-ILD [49]. However, in randomized, placebo-controlled studies, CF demonstrated an effect only during the first year of treatment; in the longer term, its results did not differ from placebo [50]. CF is currently the drug of choice for the treatment of SSC-ILD, as recommended by EULAR. It is prescribed intravenously at doses of 500–750 mg/m^2 per month or orally at doses of 1–2 mg/kg/day, depending on the efficacy and tolerability of the drug [51, 52]. According to the metaanalysis the method of administering CF (oral or intravenous) does not significantly affect the level of change in the indexes of functional pulmonary tests [52]. The duration of the CF course should be at least 6 months. Compared with azathioprine, CF exhibited advantages in influencing functional parameters [53]. Predictors of poor response to intravenous pulse therapy with CF include high serum CRP levels, KL-6, surfactant protein D (SP-D) (>2000 U/mL), and low baseline DLCO values. On the contrary the decrease in SP-D levels (>200 ng/mL) after the first intravenous administration of CF argues for the efficiency of continued therapy [54].

EULAR experts recommend the use of cyclosporine in a dose of 1–2 mg/kg/day orally in case of rapidly progressive and initially severe forms of SSC-ILD [55].

Mycophenolate mofetil (MM) is a cytostatic drug that is actively used in a number of autoimmune diseases and post-solid organ transplantation, and it has also proved to be highly effective in blocking SSC-ILD [56]. In the largest study of Tashkin et al. [57], MM demonstrated a better safety and tolerability profile than CF, although it did not exceed its efficiency. MM can currently be recommended as an alternative to CF [58].

Systemic GCSs are not recommended as independent drugs for the treatment of SSC-ILD but only as a supplement in small doses (up to 15 mg/day) to CF in severe and progressive forms of the disease [59].

Rituximab has been the most studied among the new biological drugs for the treatment of systemic autoimmune diseases. In minor studies, it revealed good efficacy in improving pulmonary function, including the diffusion capacity of the lungs with a good safety profile [60]. This drug may be recommended in case of intolerance or contraindications to CF. With the ineffectiveness of MM, CF, and rituximab, tocilizumab as rescue therapy can be prescribed, which demonstrated a managing effect on the disease progression with a satisfactory safety profile in case series in most patients with refractory SSC-ILD [61].

In randomized studies, other antifibrotic agents, such as D-penicillamine and interferons alpha and gamma, have not proved their effectiveness in SSC-associated interstitial pulmonary fibrosis [62, 63].

Proton pump inhibitors are prescribed to patients who experienced frequent damage to the esophagus and severe gastroesophageal reflux [40].

Do all patients with SSC-ILD need the treatment? Experts of the European Respiratory Society believe that cytostatic therapy is indicated only for patients who have an extensive interstitial process or the risks of its progression [64]. Prevalence is estimated by the volume of involvement in the interstitial lesion of more than 30% of lung tissue or if the patient has a FVC lower than 70% of predicted values. The risks of progression include high levels of antibodies to topoisomerase I, male gender, and lung damage at the beginning of the systemic process development. Such patients should take pulse therapy with CF at a dose of 0.5–2 g/m^2 once a month for 6–18 months. In the future, upon reaching remission, it can be changed to MM (2 g/day) or azathioprine 1–2 mg/kg/day, and in case no effect is observed, the possibility of administering cyclosporine or rituximab should be considered [55, 64]. Hematopoietic stem-cell transplantation may be considered in individual patients with refractory and progressive SSC-ILD but with great care and only in specialized centers with experience in such treatment [55].

New antifibrotic drugs, such as pirfenidone and nintedanib, included in the recommendations for the treatment of IPF are certainly candidates for the treatment of at least patients exhibiting SSC with the pattern of UIP. The first pilot study that included five patients with SSC-ILD performed in Japan, demonstrated the efficiency of pirfenidone in improving lung function in all patients [65]. The first large randomized trial SENSCIS, in 576 patient with SSC-ILD have shown significant reduction in annual rate of FVC decline in the group took nintedanib 150 mg twice a day (52,4 ml per year) compared with placebo (93,3 ml per year) [66].

With the progression of SSC-ILD, patients may undergo lung transplantation. Earlier studies demonstrated a worse prognosis for survival rate after surgery, but recent reports indicated comparable results of transplantation in patients with SSC and nonassociated with SSC fibrotic lung disease [67]. Thus the presence of a SSC and its complications, such as esophageal dilatation, should not be considered a contraindication for transplantation [68].

The prognosis of the disease depends on the response to treatment and the development of such complications as pulmonary hypertension and lung cancer. The average survival rate of patients with SSC-ILD in one of the latest studies amounted to 11.2 years from the time of diagnosis establishment [22].

Factors of poor prognosis for SSC-ILD included male gender; elderly age; more than 20% prevalence of interstitial lesions on HRCT; the X-ray pattern of UIP; low FVC and DLCO, as well as the rate of their decrease; and the presence of pulmonary hypertension [69–71].

References

[1] Wollheim FA. Classification of systemic sclerosis: visions and reality. Rheumatology (Oxford) 2005;44(10):1212–6.

[2] Mayes MD, Lacey Jr JV, Beebe-Dimmer J, Gillespie BW, Cooper B, Laing TJ, et al. Prevalence, incidence, survival, and disease characteristics of systemic sclerosis in a large US population. Arthritis Rheum 2003;48(8):2246–55.

[3] LeRoy EC, Black C, Fleischmajer R, Jablonska S, Krieg T, Medsger Jr TA, et al. Scleroderma (systemic sclerosis): classification, subsets and pathogenesis. J Rheumatol 1988;15(2):202–5.

[4] van den Hoogen F, Khanna D, Fransen J, Johnson SR, Baron M, Tyndall A, et al. 2013 Classification criteria for systemic sclerosis: an American college of rheumatology/European league against rheumatism collaborative initiative. Arthritis Rheum 2013;65(11):2737–47.

[5] Varga J. Systemic sclerosis: an update. Bull NYU Hosp Jt Dis 2008;66(3):198–202.

[6] Steele R, Hudson M, Lo E, Baron M, Canadian Scleroderma Research Group. Clinical decision rule to predict the presence of interstitial lung disease in systemic sclerosis. Arthritis Care Res (Hoboken) 2012;64(4):519–24.

[7] Schoenfeld SR, Castelino FV. Interstitial lung disease in scleroderma. Rheum Dis Clin North Am 2015;41(2):237–48.

[8] Akter T, Silver RM, Bogatkevich GS. Recent advances in understanding the pathogenesis of scleroderma-interstitial lung disease. Curr Rheumatol Rep 2014;16(4):411.

[9] Hant FN, Ludwicka-Bradley A, Wang HJ, et al. Surfactant protein D and KL-6 as serum biomarkers of interstitial lung disease in patients with scleroderma. J Rheumatol 2009;36(4):773–80.

[10] Tamby MC, Chanseaud Y, Guillevin L, Mouthon L. New insights into the pathogenesis of systemic sclerosis. Autoimmun Rev 2003;2(3):152–7.

[11] Jimenez SA, Piera-Velazquez S. Endothelial to mesenchymal transition (EndoMT) in the pathogenesis of systemic sclerosis-associated pulmonary fibrosis and pulmonary arterial hypertension. Myth or reality? Matrix Biol 2016;51:26–36.

[12] Bouros D, Wells AU, Nicholson AG, Colby TV, Polychronopoulos V, Pantelidis P, et al. Histopathologic subsets of fibrosing alveolitis in patients with systemic sclerosis and their relationship to outcome. Am J Respir Crit Care Med 2002;165(12):1581–6.

[13] Flaherty KR, Towes GB, Travis WD, Colby TV, Kazerooni EA, Gross BH, et al. Clinical significance of histologic classification of idiopathic interstitial pneumonia. Eur Repir J 2002;19(2):275–83.

[14] Kim EA, Lee KS, Johkoh T, Kim TS, Suh GY, Kwon OJ, et al. Interstitial lung diseases associated with collagen vascular diseases: radiologic and histopathologic findings. Radiographics 2002;22(Spec issue):S151–65.

[15] Yousem SA. The pulmonary pathologic manifestations of the CREST syndrome. Hum Pathol 1990;21(5):467–74.

[16] Capobianco J, Grimberg A, Thompson BM, Antunes VB, Jasinowodolinski D, Meirelles GS. Thoracic manifestations of collagen vascular diseases. Radiographics 2012;32(1):33–50.

[17] Urisman A, Jones K. Pulmonary pathology in connective tissue disease. Semin Respir Crit Care Med 2014;35(2):201–12.

[18] Steen VD, Medsger Jr TA. Severe organ involvement in systemic sclerosis with diffuse scleroderma. Arthritis Rheum 2000;43(11):2437–44.

[19] Goldin JG, Lynch DA, Strollo DC, Suh RD, Schraufnagel DE, Clements PJ, et al. High-resolution CT scan findings in patients with symptomatic scleroderma-related interstitial lung disease. Chest 2008;134(2):358–67.

[20] Corte T, Du Bois R, Wells A. Connective tissue diseases. In: Broaddus VC, Mason RJ, Ernst JD, MD KTE Jr., Lazarus SC, Murray JF, et al., editors. Murray and Nadel's textbook of respiratory medicine. 6th ed. Elsevier; 2016. p. 1165–87.

[21] Moore OA, Proudman SM, Goh N, Corte TJ, Rouse H, Hennessy O, et al. Quantifying change in pulmonary function as a prognostic marker in systemic sclerosis related interstitial lung disease. Clin exp rheumatol 2015;33(4 Suppl. 91):S111–6.

[22] Guler SA, Winstone TA, Murphy D, Hague C, Soon J, Sulaiman N, et al. Does systemic sclerosis-associated interstitial lung disease burn out? Specific phenotypes of disease progression. Ann Am Thorac Soc 2018;15(12):1427–33.

[23] Showalter K, Hoffmann A, Rouleau G, Aaby D, Lee J, Richardson C, et al. Performance of forced vital capacity and lung diffusion cutpoints for associated radiographic interstitial lung disease in systemic sclerosis. J Rheumatol 2018;45(11):1572–6.

[24] Tiev KP, Le-Dong NN, Duong-Quy S, Hua-Huy T, Cabane J, Dinh-Xuan AT. Exhaled nitric oxide, but not serum nitrite and nitrate, is a marker of interstitial lung disease in systemic sclerosis. Nitric Oxide 2009;20(3):200–6.

[25] Volkmann ER, Tashkin DP, Sim M, Li N, Goldmuntz E, Keyes-Elstein L, et al. Short-term progression of interstitial lung disease in systemic sclerosis predicts long-term survival in two independent clinical trial cohorts. Ann Rheum Dis 2019;78(1):122–30.

[26] Steen VD. Autoantibodies in systemic sclerosis. Semin Arthritis Rheum 2005;35(1):35–42.

[27] Caimmi C, Bertoldo E, Venturini A, Caramaschi P, Frulloni L, Ciccocioppo R, et al. Relationship between increased fecal calprotectin levels and interstitial lung disease in systemic sclerosis. J Rheumatol 2018. pii: jrheum.171445.

[28] Shah RM, Jimenez S, Wechsler R. Significance of ground glass opacity on HRCT in long-term follow-up of patients with systemic sclerosis. J Thorac Imaging 2007;22(2):120–4.

[29] Pandey A, Wilcox P, O'Brien J, Brown EJ, Leipsic J. Significance of various pulmonary and extrapulmonary abnormalities on HRCT of the chest in scleroderma lung. Indian J Radiol Imaging 2013;23(4):304–7.

[30] Ananyeva LP, Teplova LV, Lesnyak VN, Koneva OA, Ovsyannikova OB, Starovoitova MN, et al. Interstitial lung injury in systemic scleroderma from high-resolution computer tomography data. Sci Pract Rheumatol 2011;2:30–9 [in Russian].

[31] Remy-Jardin M, Remy J, Wallaert B, Bataille D, Hatron PY. Pulmonary involvement in progressive systemic sclerosis: sequential evaluation with CT, pulmonary function tests, and bronchoalveolar lavage. Radiology 1993;188(2):499–506.

[32] Schurawitzki H, Stiglbauer R, Graninger W, Herold C, Pölzleitner D, Burghuber OC, et al. Interstitial lung disease in progressive systemic sclerosis: high-resolution CT versus radiography. Radiology 1990;176(3):755–9.

[33] Lesnyak VN, Danilevskaya OV, Averyanov AV, Ananyeva LP. Unilateral pulmonary fibrosis and systemic sclerosis. Am J Respir Crit Care Med 2014;190(9):1067–8.

[34] Pandey AK, Wilcox P, Mayo JR, Sin D, Moss R, Ellis J, et al. Predictors of pulmonary hypertension on high-resolution computed tomography of the chest in systemic sclerosis: a retrospective analysis. Can Assoc Radiol J 2010;61(5):291–6.

[35] Goh NS, Desai SR, Veeraraghavan S, Hansell DM, Copley SJ, Maher TM, et al. Interstitial lung disease in systemic sclerosis: a simple staging system. Am J Respir Crit Care Med 2008;177(11):1248–54.

[36] Fischer A, Misumi S, Curran-Everett D, Meehan RT, Ulrich SK, Swigris JJ, et al. Pericardial abnormalities predict the presence of echocardiographically defined pulmonary artery hypertension in systemic sclerosis related interstitial lung disease. Chest 2007;131(4):988–92.

[37] Bhalla M, Silver RM, Sheperd JA, McLoud TC. Chest CT in patients with scleroderma: prevalence of asymptomatic esophageal dilatation and mediastinal lymphadenopathy. AJR Am J Roentgenol 1993;161(2):269–72.

[38] Vonk MC, Van Die CE, Snoeren MM, et al. Oesophageal dilatation on high-resolution computed tomography scan of the lungs as a sign of Scleroderma. Ann Rheum Dis 2008;67(9):1317–21.

[39] Bussone G, Mouthon L. Interstitial lung disease in systemic sclerosis. Autoimmun Rev 2011;10(5):248–55.

[40] Walkoff L, White DB, Chung JH, Asante D, Cox CW. The four corners sign: a specific imaging feature in differentiating systemic sclerosis-related interstitial lung disease from idiopathic pulmonary fibrosis. J Thorac Imaging 2018;33(3):197–203.

[41] Chung JH, Cox CW, Montner SM, Adegunsoye A, Oldham JM, Husain AN, et al. CT features of the usual interstitial pneumonia pattern: differentiating connective tissue disease-associated interstitial lung disease from idiopathic pulmonary fibrosis. AJR Am J Roentgenol 2018;210(2):307–13.

[42] Chaisson NF, Hassoun PM. Systemic sclerosis-associated pulmonary arterial hypertension. Chest 2013;144(4):1346–56.
[43] Pearson JE, Silman AJ. Risk of cancer in patients with scleroderma. Ann Rheum Dis 2003;62(8):697–9.
[44] Goldstein RA, Rohatgi PK, Bergofsky EH, Block ER, Daniele RP, Dantzker DR, et al. Clinical role of bronchoalveolar lavage in adults with pulmonary disease. Am Rev Respir Dis 1990;142(2):481–6.
[45] Kowal-Bielecka O, Kowal K, Highland KB, Silver RM. Bronchoalveolar lavage fluid in scleroderma interstitial lung disease: technical aspects and clinical correlations: review of the literature. Semin Arthritis Rheum 2010;40(1):73–88.
[46] Kim DS, Yoo B, Lee JS, Kim EK, Lim CM, Lee SD, et al. The major histopathologic pattern of pulmonary fibrosis in scleroderma is nonspecific interstitial pneumonia. Sarcoidosis Vasc Diffuse Lung Dis 2002;19(2):121–7.
[47] Volkmann ER, Tashkin DP. Treatment of systemic sclerosis-related interstitial lung disease: a review of existing and emerging therapies. Ann Am Thorac Soc 2016;13(11):2045–56.
[48] Silver RM, Warrick JH, Kinsella MB, Staudt LS, Baumann MH, Strange C. Cyclophosphamide and low-dose prednisone therapy in patients with systemic sclerosis (scleroderma) with interstitial lung disease. J Rheumatol 1993;20(5):838–44.
[49] White B, Moore WC, Wigley FM, Xiao HQ, Wise RA. Cyclophosphamide is associated with pulmonary function and survival benefit in patients with scleroderma and alveolitis. Ann Intern Med 2000;132(12):947–54.
[50] Tashkin DP, Elashoff R, Clements PJ, Roth MD, Furst DE, Silver RM, et al. Effects of 1-year treatment with cyclophosphamide on outcomes at 2 years in scleroderma lung disease. Am J Respir Crit Care Med 2007;176(10):1026–34.
[51] Hoyles RK, Ellis RW, Wellsbury J, Lees B, Newlands P, Goh NS, et al. A multicenter, prospective, randomized, double-blind, placebo-controlled trial of corticosteroids and intravenous cyclophosphamide followed oral azathioprine for the treatment of pulmonary fibrosis in scleroderma. Arthritis Rheum 2006;54(12):3962–70.
[52] Tzelepsis GE, Plastiras SC, Karadimitrakis SP, Vlachoyiannopoulos PG. Determinants of pulmonary function improvement in patients with scleroderma and interstitial lung disease. Clin Exp Rheumatol 2007;25(5):734–9.
[53] Nadashkevich O, Davis P, Fritzler M, Kovalenko W. A randomized unblinded trial of cyclophosphamide versus azathioprine in the treatment of systemic sclerosis. Clin Rheumatol 2006;25(2):205–12.
[54] Sumida H, Asano Y, Tamaki Z, Aozasa N, Taniguchi T, Toyama T. Prediction of therapeutic response before and during i.v. cyclophosphamide pulse therapy for interstitial lung disease in systemic sclerosis: a longitudinal observational study. J Dermatol 2018;45(12):1425–33.
[55] Kowal-Bielecka O, Fransen J, Avouac J, Becker M, Kulak A, Allanore Y, et al. Update of EULAR recommendations for the treatment of systemic sclerosis. Ann Rheum Dis 2017;76(8):1327–39.
[56] Gerbino AJ, Goss CH, Molitor JA. Effect of mycophenolate mofetil on pulmonary function in scleroderma-associated interstitial lung disease. Chest 2008;133(2):455–60.
[57] Tashkin DP, Roth MD, Clements PJ, Furst DE, Khanna D, Kleerup EC, et al. Mycophenolate mofetil versus oral cyclophosphamide in scleroderma-related interstitial lung disease (SLS II): a randomised controlled, double-blind, parallel group trial. Lancet Respir Med 2016;4(9):708–19.
[58] Ueda T, Sakagami T, Kikuchi T, Takada T. Mycophenolate mofetil as a therapeutic agent for interstitial lung diseases in systemic sclerosis. Respir Investig 2018;56(1):14–20.
[59] Bérezné A, Valeyre D, Ranque B, Guillevin L, Mouthon L. Interstitial lung disease associated with systemic sclerosis: what is the evidence for efficacy of cyclophosphamide? Ann NY Acad Sci 2007;1110:271–84.
[60] Daoussis D, Liossis SN, Tsmandas AC, Kalogeropoulou C, Paliogianni F, Sirinian C, et al. Effect of long-term treatment with rituximab on pulmonary function and skin fibrosis in patients with diffuse systemic sclerosis. Clin Exp Rheum 2012;30(2 Suppl. 71):17–22.
[61] Narváez J, LLuch J, Alegre Sancho JJ, Molina-Molina M, Nolla JM, Castellví I. Effectiveness and safety of tocilizumab for the treatment of refractory systemic sclerosis associated interstitial lung disease: a case series. Ann Rheum Dis 2018. pii: annrheumdis-2018-214449.
[62] Black CM, Silman AJ, Herrick AI, Denton CP, Wilson H, Newman J, et al. Interferon-alpha does not improve outcome at one year in patients with diffuse cutaneous scleroderma: results of a randomized, double-blind, placebo-controlled trial. Arthritis Rheum 1999;42(2):299–305.
[63] Clements PJ, Furst DE, Wong WK, Mayes M, White B, Wigley F, et al. High-dose versus low-dose D-penicillamine in early diffuse systemic sclerosis: analysis of a two-year, double-blind, randomized, controlled clinical trial. Arthritis Rheum 1999;42(6):1194–203.
[64] Cappelli S, Guiducci S, Randone S, Cerinic M. Immunosuppression for interstitial lung disease in systemic sclerosis. Eur Respir Rev 2013;22(129):236–43.
[65] Miura Y, Saito T, Fujita K, Tsunoda Y, Tanaka T, Takoi H, et al. Clinical experience with pirfenidone in five patients with scleroderma-related interstitial lung disease. Sarcoidosis Vasc Diffuse Lung Dis 2014;31(3):235–8.
[66] Distler O, Highland KB, Gahlemann M, Azuma A, Fischer A, Mayes MD, et al. Nintedanib for systemic sclerosis–associated interstitial lung disease. N Engl J Med 2019 Jun 27;380(26):2518–28.
[67] Jablonski R, Dematte J, Bhorade S. Lung transplantation in scleroderma: recent advances and lessons. Curr Opin Rheumatol 2018;30(6):562–9.
[68] Shah RJ, Boin F. Lung transplantation in patients with systemic sclerosis. Curr Rheumatol Rep 2017;19(5):23.
[69] Johnson SR, Swiston JR, Granton JT. Prognostic factors for survival in scleroderma associated pulmonary arterial hypertension. J Rheumatol 2008;35(8):1584–90.
[70] Hax V, Bredemeier M, Didonet Moro AL, Pavan TR, Vieira MV, Pitrez EH, et al. Clinical algorithms for the diagnosis and prognosis of interstitial lung disease in systemicsclerosis. Semin Arthritis Rheum 2017;47(2):228–34.
[71] Panopoulos S, Bournia VK, Konstantonis G, Fragiadaki K, Sfikakis PP, Tektonidou MG. Predictors of morbidity and mortality in early systemic sclerosis: long-term follow-up data from a single-centre inception cohort. Autoimmun Rev 2018;17(8):816–20.

Chapter 8.3

Lung disease in polymyositis/dermatomyositis

Polymyositis and dermatomyositis (PM/DM) belong to the group of idiopathic autoimmune inflammatory myopathies with frequent involvement of the skin and lungs in the systemic inflammatory process [1]. They usually are considered together due to the common pathogenesis and clinical presentation, except for skin lesions that are more typical for dermatomyositis. The diseases are rare; their prevalence is approximately 10 cases in one million people [2]. The two peaks of incidence are in childhood (age 10–15 years) and at age 35–65 years. Women are affected more often than men [3]. Due to the low prevalence and a limited number of cases, data on the frequency of lung lesions in PM/DM vary significantly, from 20% to 65% [4–6]. Lung involvement usually determines the severity of the course and prognosis.

The pathogenesis of interstitial lung disease (ILD) associated with idiopathic autoimmune myopathies (ILD-MP), as with other pulmonary manifestations of connective tissue diseases (CTD), has not been studied sufficiently. Possible triggers can be viruses and drugs (statins, immune checkpoint inhibitor, and malignant tumors) [7]. Several viruses (influenza virus, echovirus and coxsackievirus virus, human T-cell lymphotropic virus type 1, and parvovirus B19) have an affinity for muscle tissue and the endothelium and epithelium of lung structures, which makes them potential triggers [8].

A history of chronic inflammatory diseases of the upper and lower airways also is associated with the development of idiopathic inflammatory myopathies [9]. Apparently, primary damage to the endothelium and the emergence of cross antigens in the lung tissue lead to a primary inflammatory response, which can enter the fibroproliferative phase resulting in pulmonary fibrosis [8].

Morphology

The interstitial process in the lungs is represented histologically by nonspecific interstitial pneumonia (NSIP) in more than half of the cases, usual interstitial pneumonia (UIP), or bronchiolitis obliterans with organizing pneumonia (BOOP) (Fig. 8.3.1). Lymphoid interstitial pneumonia and diffuse alveolar damage (DAD) are much less common [10]. In addition, rare cases of acute fibrinous, aspiration, and organizing pneumonia have been described, as well as diffuse alveolar hemorrhages, pulmonary vasculitis and capillaritis, and plexogenic arteriopathy, as well as inflammatory pseudotumor development [11].

Mixed histopathologic findings often are revealed during biopsy from different parts of the lungs [12]. Morphological patterns of lung damage are indistinguishable from idiopathic ones. Morphological changes in the lungs may precede the expanded clinical presentation of PM/DM [13]. Thus histopathologic pulmonary abnormalities are not critical to the diagnosis of ILD associated with PM/DM (ILD-PM/DM) and usually require multidisciplinary discussion.

Clinical presentation and diagnosis

PM/DM usually is diagnosed on the basis of the following criteria [14]: (1) weakness in the proximal muscles of the limbs with possible involvement of other striated muscles, manifested by dysphagia and dyspnea; (2) characteristic changes in skeletal muscles in biopsy, including degeneration, regeneration, necrosis, and interstitial mononuclear infiltrates; (3) increased levels of enzymes released in the inflammation of skeletal muscles (creatine phosphokinase, aspartate aminotransferase, alanine aminotransferase, and/or lactate dehydrogenase); (4) characteristic changes in electromyogram; and (5) dermatologic symptoms, including purple eyelids with periorbital edema; exfoliative erythematous dermatitis, especially on the opisthenars in the area of the metacarpophalangeal and upper interphalangeal joints (Gottron sign) (Fig. 8.3.2); and erythematous rashes in the elbows, knees, medial ankles, face, neck, and upper torso (décolleté area).

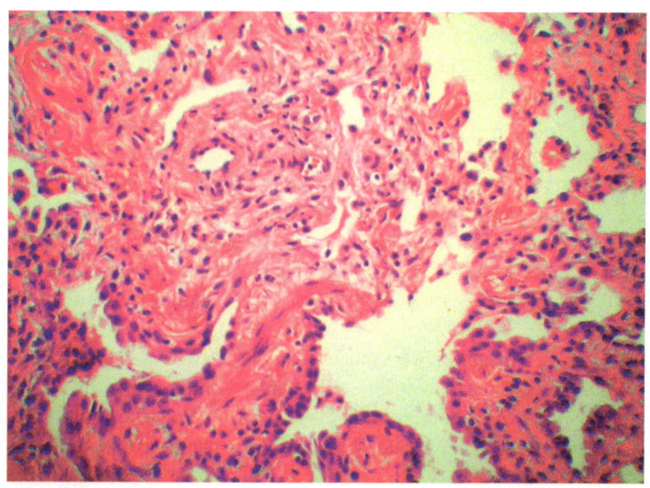

FIG. 8.3.1 Polymyositis. Hematoxylin and eosin staining, 400×.

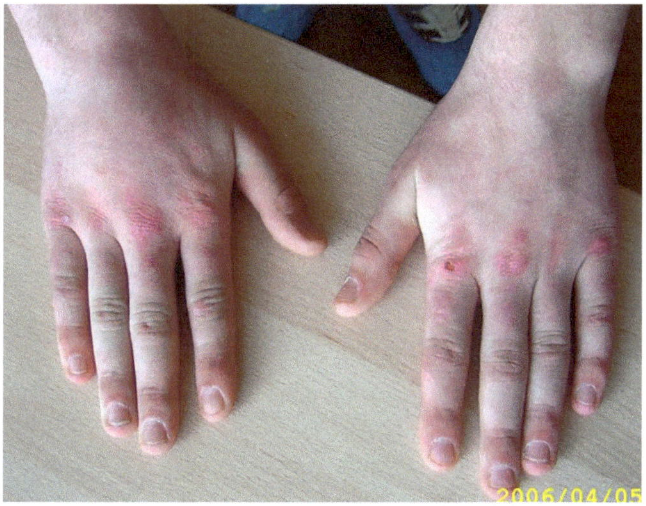

FIG. 8.3.2 Polymyositis/dermatomyositis. Gottron sign.

A definite diagnosis of PM/DM is valid in the presence of the first four criteria. Therefore a skeletal muscle biopsy and electromyography are required for the confirmation of PM/DM diagnosis.

Lungs are affected more often in PM/DM patients with anti-Jo1 syndrome, assuming positive titers of anti-aminoacyl-tRNA synthetase (anti-ARS) antibodies (anti-Jo-1, anti-PL-7, anti-PL-12, anti-EJ, anti-OJ, anti-KS, anti-Zo, and anti-Ha) and ≥1 symptoms, such as myositis, arthritis, ILD, fever, Reynaud's phenomenon or erythema, hyperkeratosis, and crunch in the interphalangeal joints (mechanic's hands) [15, 16]. The highest incidence of ILD-PM/DM development occurs in patients with increased anti-PL-12 autoantibody (up to 90%), anti-KS (up to 100%), and anti-OJ (up to 100%) levels [17, 18].

A few patients with PM/DM with lung damage may have no signs of myopathy [19]. In this case, such a common test as serum creatine phosphokinase (CPK) cannot be used as a basis for diagnosis, since CPK level is correlated strongly with the degree of skeletal muscle damage. This subtype of PM/DM is called amyopathic dermatomyositis (AD) in which ILD often has a rapidly progressing course and is resistant to corticosteroids and immunosuppressive drugs [20]. The disease has typical dermatomyositis skin manifestations in the absence of skeletal muscle damage [21]. The 6-month mortality rate in these patients reaches 55% [22]. AD is strongly associated with the presence of serum antibodies to the melanoma differentiation-associated gene 5 (anti-*MDA5*) [23].

In approximately 20% of patients with pulmonary involvement, interstitial pneumonia develops earlier than the extrapulmonary symptoms of PM/DM, which complicates early diagnosis significantly [14]. In anti-Jo1 syndrome the onset of the disease with lung lesions reaches 50% of cases [24].

Symptoms of pulmonary involvement are the same as for other interstitial diseases and include cough, slowly progressive dyspnea in physical exertion, and general fatigue [14]. The clubbing sign can be noted rarely. Dyspnea in patients with PM/DM may be due to a simultaneous lesion of the diaphragm; therefore the severity of ILD does not always correlate with the degree of respiratory failure [25]. Because of the weakness of the pharyngeal muscles in patients with PM/DM, aspiration pneumonia often occurs, with the development of inflammation [3]. Pulmonary infections can occur not only due to aspiration but also due to immunosuppressive treatment. In addition, significant risks for urinary tract infections, catheter-related bloodstream infections, and sepsis, which often cause death, are described in patients with PM/DM who receive the treatment [26–28].

The causative agents of pulmonary infections are more often opportunistic pathogens such as *P. jirovecii* (especially in patients not receiving cotrimoxazole prophylaxis), *Candida* and *Aspergillus* spp., cytomegalovirus, and tuberculosis and nontuberculosis mycobacterium, whereas gram-negative enterobacteria predominate in aspiration pneumonia [26].

It should be noted that PM/DM can be a manifestation of paraneoplastic syndrome with a typical clinical presentation and the appearance of specific antibodies. In this case the radical treatment of a tumor leads to the elimination of antibodies and clinical symptoms, including the reverse of interstitial lung damage [29].

Functional pulmonary tests usually reveal a restrictive pattern and decreased diffusing capacity for carbon monoxide (DLco). Unlike other ILDs, due to the weakness of the respiratory muscles in patients with PM/DM, the maximal inspiratory pressure and maximal expiratory pressure are reduced, and residual volume is increased without decreased FEV1/FVC ratio [25].

Bronchoalveolar lavage (BAL) in ILD-PM/DM is not the main diagnostic tool. In a rapidly progressive disease, neutrophilia and mild eosinophilia are detected in cytological counts [30]. In patients with anti-Jo1 syndrome, an increased level of lymphocytes (20%–50%) and eosinophils (up to 10%) is revealed [16]. BAL is an important method to rule out pulmonary infections and lower airway malignancies. Transbronchial lung biopsy usually is not effective in diagnosing ILD; it should be performed in cases of suspected malignancy and granulomatosis [25].

High-resolution computed tomography

The high-resolution computer tomography (HRCT) characteristics of the pulmonary lesion in PM/DM most often correspond to the NSIP, UIP, and BOOP radiological patterns and more rarely to DAD and bronchial and bronchiolar abnormalities. On the basis of our own 78 patients with PM/DM, 46 of whom had changes in the lungs, we conducted the analysis of the primary radiological signs (Table 8.3.1) [31].

TABLE 8.3.1 HRCT signs in patients with polymyositis/dermatomyositis (own data)

CT signs	Incidence $N = 46$
Thickening of the intralobular septa	43 (93.5%)
Thickening of the bronchial walls	37 (80.4%)
Thickening of interlobular septa	33 (71.7%)
Nodules	2 (4.3%)
Ground-glass opacity	40 (87%)
Consolidation	8 (17.4%)
Traction bronchiectasis	19 (41.3%)
Honeycombing	1 (2.1%)
Cysts	1 (2.1%)
Subpleural linear bands	24 (52.1%)
Mediastinal lymphadenopathy	3 (6.5%)

The most common signs in PM/DM were reticular abnormalities, ground-glass opacity (GGO), thickening of the bronchial walls, and subpleural linear attenuation, whereas consolidation was much less common, which corresponded to the pattern of organizing pneumonia (Figs. 8.3.3–8.3.5) [13].

Due to the small number of cases, the data in the studies on evaluation of HRCT signs of ILD associated with PM/DM vary greatly. The frequency was 45%–100% for GGO, 20%–100% for consolidation, and 0%–27% for honeycombing [32–34].

Hypotension and dilatation of the upper and middle esophagus are considered specific signs of PM/DM and may be combined with parenchymal abnormalities or can be an independent sign (Fig. 8.3.6) [35].

In cases of anti-aminoacyl-tRNA synthetase antibody-positive interstitial pneumonia, the most frequent findings are GGO (100%), traction bronchiectasis (90%), consolidation (75%), and reticular signs (75%) with a 100% distribution of abnormalities in the lower areas, whereas the honeycombing and peribronchovascular distribution of changes were not found in any patient [16]. At that the prevailing HRCT pattern was NSIP (85%), whereas the remaining cases showed BOOP [16]. A frequent finding that accompanies the BOOP pattern in PM/DM is the reversed halo sign [35]. A minor pleural effusion is a possible finding in PM/DM, which occurs in up to 10% of cases (Fig. 8.3.5) [16]. The presence of areas of consolidation and GGO in the lower lung lobes is associated with a worse 90 day survival rate compared with the pattern of only GGO or reticular abnormalities [36].

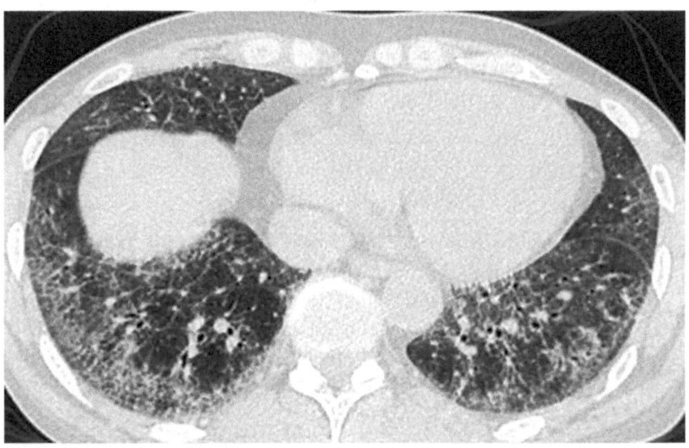

FIG. 8.3.3 Polymyositis/dermatomyositis. Bilateral subpleural areas of ground-glass opacity in the lower segments, associated with thickened intralobular septa. Pattern of nonspecific interstitial pneumonia.

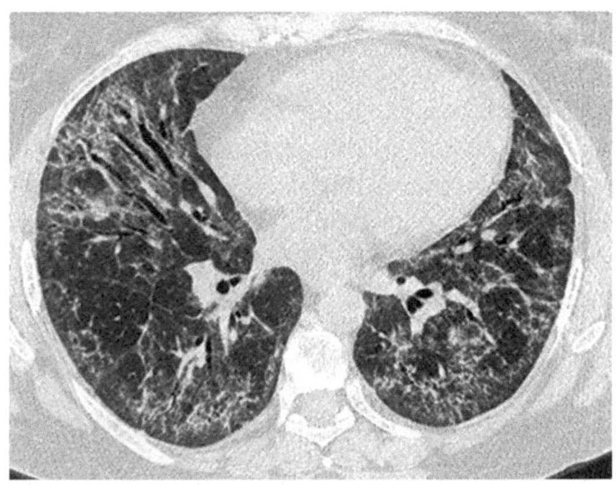

FIG. 8.3.4 Polymyositis/dermatomyositis. Diffuse bilateral areas of ground-glass opacity associated with severe reticular abnormalities and traction bronchiectasis. Subpleural sparing. Fibrous nonspecific interstitial pneumonia pattern.

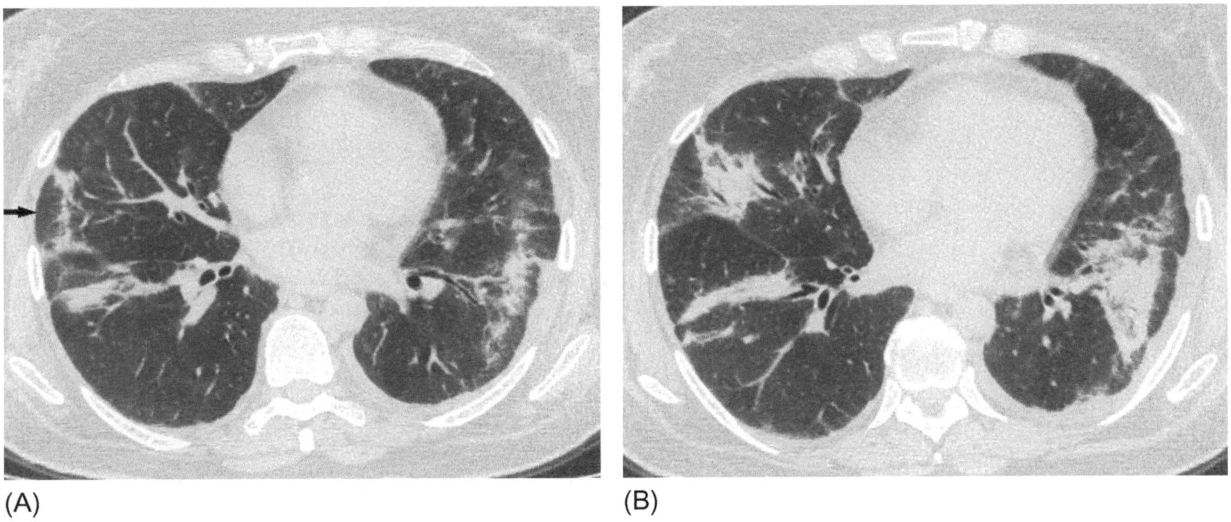

FIG. 8.3.5 Polymyositis/dermatomyositis. Bilateral consolidation areas with an adjacent ground-glass opacity. The abnormalities have a peribroncho-vascular and subpleural distribution (A, B). In the right lung, there is an atoll sign *(arrow)* (A). In the left lung, there is a small left-sided hydrothorax (B). Pattern of bronchiolitis obliterans with organizing pneumonia.

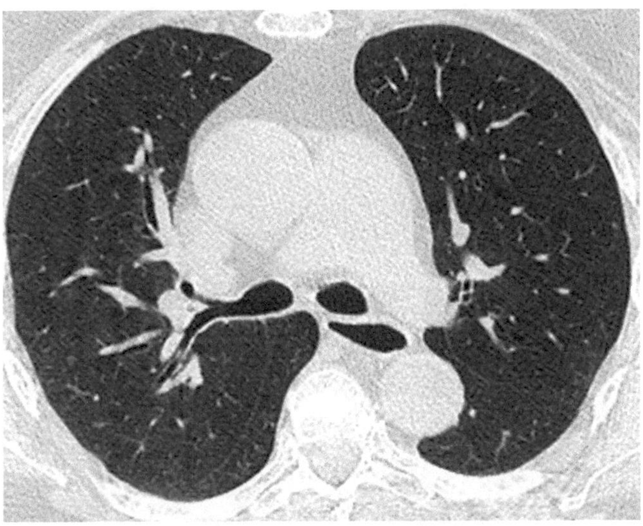

FIG. 8.3.6 Polymyositis/dermatomyositis. Dilatation of the middle third of the esophagus. There are no visible abnormalities of the pulmonary parenchyma.

Under the influence of treatment, such phenomena as GGO and consolidation disappear or are minimized in respondents to therapy or increase with the emergence of reticular abnormalities in refractory forms of the disease (Fig. 8.3.7). Massive areas of consolidation typical for DAD, in the case of patient survival, are transformed into interstitial fibrosis (Fig. 8.3.8) [33]. However, Ingegnoli et al. [37], with 12 months of follow-up of ILD-PM/DM patients treated with steroids and immunosuppressants, noted visual stabilization of the disease only in 26.6% of cases, and in other cases, the progression of radiological signs was registered.

Differential diagnosis

Because PM/DM can be a manifestation of malignant paraneoplastic syndrome, patients should be examined carefully to rule out the primary tumor.

In case of an ILD, its primary or secondary origin always should be differentiated. Since both nonspecific and usual interstitial and organizing pneumonia can be idiopathic and associated with other diseases and conditions, including those that are drug-induced, a careful analysis of the circumstances preceding changes in the lungs is required, especially since the biopsy in this case most likely will not be able to answer the question about the origin of interstitial pneumonia. Questions about contact with potential aeroallergens (especially birds), the professional environment, drug intake, chronic diseases of other organs, and immunodeficiency must be asked when taking a history of a patient with interstitial changes in the lungs. It should be noted that idiopathic NSIP often is accompanied by moderate myalgia, which requires ruling out PM/DM. However, as already mentioned, since respiratory symptoms can occur earlier than muscle lesions, making a final diagnosis is very difficult in this case, but certainly, in addition to CPK serum level, anti-aminoacyl-tRNA synthetase antibodies should be determined. Respiratory infections, aspiration-related and opportunistic, can change significantly the clinical and radiological pattern of the underlying disease (Fig. 8.3.9). Increase in dyspnea, the emergence of new areas of GGO, consolidation, nodules on HRCT, and inflammatory changes in the blood require a differential diagnosis between the exacerbation of interstitial pneumonia, drug-induced pneumonitis, and the pulmonary infection. BAL with PCR and cultural examination usually is required in such cases to detect a possible pathogen. In the cytological formula of a BAL

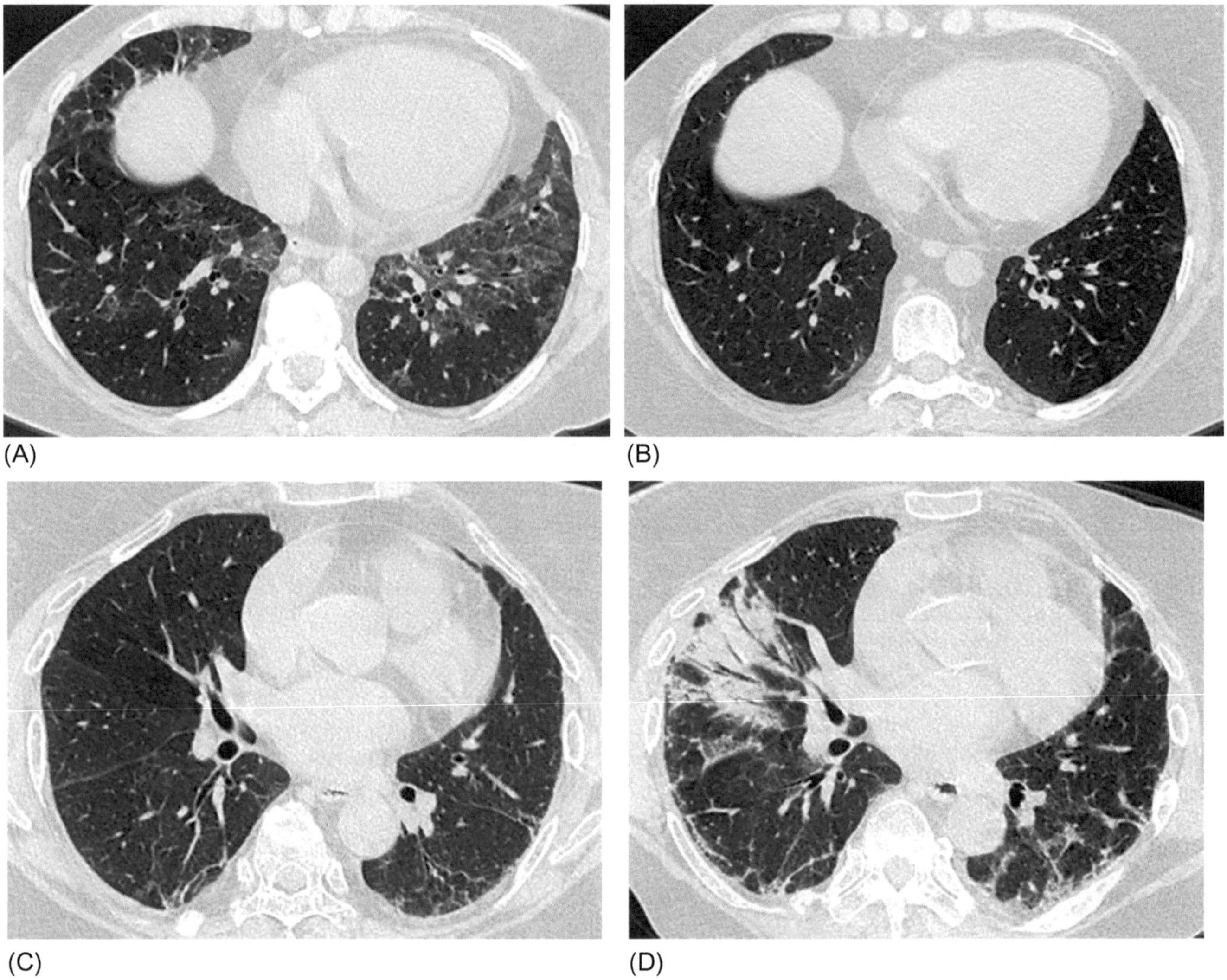

FIG. 8.3.7 Changes in time of HRCT images in treatment of two patients with polymyositis/dermatomyositis. Bilateral patchy areas of ground-glass opacity, associated with thickening of interlobular septa (A), are initially visualized. After 3 months of treatment, significant reduction of signs is noted (B). The initially mild reticular abnormalities and ground-glass opacity in the lower areas (C). After 6 months, despite therapy with corticosteroids and cyclophosphamide, there is a significant increase in reticular changes, the emergence of massive peribronchovascular areas of consolidation in the right lung (D).

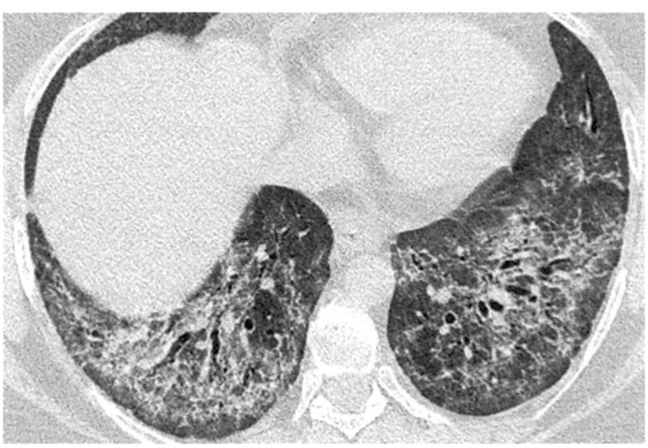

FIG. 8.3.8 Amyopathic dermatomyositis. After diffuse alveolar damage, diffuse interstitial fibrosis with traction bronchiectasis developed in the lower lobes.

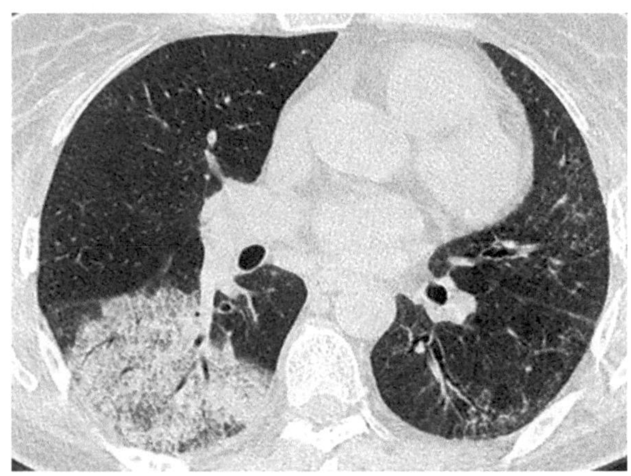

FIG. 8.3.9 Right-sided community-acquired pneumonia in a patient with polymyositis/dermatomyositis. On the right, in the lower lobe, there is a massive area of heterogeneous consolidation, an air bronchogram sign. On the left, there are moderately expressed subpleural reticular abnormalities and traction bronchiectases.

fluid, with exacerbation of interstitial pneumonia, neutrophilia usually is found, whereas with drug-induced pulmonary disease, lymphocytosis of more than 30% is revealed.

Treatment

Therapy of ILD-MP usually is started with high doses of systemic corticosteroids (0.75–1 mg/kg per day of prednisone per os) for at least 2–3 months and a long, slow reduction of the dose to the maintenance dose. In case of aggressive forms, especially those accompanied by severe lesions of the esophagus and lungs, intravenous pulse therapy with methylprednisolone at 1000 mg/day is possible during the first 3 days [38]. Nawata et al. [39] showed that the effect of corticosteroids was higher in patients with a higher blood creatine phosphokinase level. In case of refractory disease, inability to achieve remission of PM/DM for 3 months with adequate doses of corticosteroids, second-line therapy with immunosuppressive drugs is indicated [40]. In small studies, positive results were obtained for cyclophosphamide, azathioprine, cyclosporine, mycophenolate mofetil, and tacrolimus [41–44]. Rituximab and intravenous immunoglobulin demonstrated the effect in clinical reports in case of aggressive forms [45, 46]. With rapidly progressive AD associated with ILD and severe respiratory failure, Huang et al. [47] also showed the efficiency of extracorporeal membrane oxygenation combined with double filtration plasmapheresis.

Hozumi et al. [48] reported that combined systemic glucocorticosteroids (GCS) with calcineurin inhibitors compared with GCS monotherapy improved the 2-year progression-free survival rate in patients with anti-aminoacyl-tRNA synthetase antibody-positive PM/DM. According to Sugiyama et al. [28], such a combination when using prednisone at 0.5 mg/kg body weight compared with prednisone monotherapy at 1 mg/kg or intravenous cyclophosphamide was the most effective and safe. Some researchers believe that the presence of serum anti-MDA5 antibodies requires more active immunosuppressive therapy from the onset of disease in patients with ILD-PM/DM [49].

In general, experts note that most cases of ILD-PM/DM, especially with early detection, respond to treatment or are at least stabilized, but usually, there is no clear association between the reduction of symptoms of myositis and dynamics of pulmonary abnormalities [14]. Yamasaki et al. [50], with 27 years of follow-up of 197 patients with idiopathic inflammatory myopathies, found that their 1, 5, and 10 year survival rates were 85%, 75%, and 67%, respectively.

Factors of negative prognosis for ILD-PM/DM include male sex [51], old age [52], the presence of serum anti-MDA5 antibodies [53], upper-lobe lesions, and hypocapnia [28]. In contrast, anti-aminoacyl-tRNA synthetase antibody-positive patients with ILD have the best response to steroid and immunosuppressive therapy, as well as a good survival prognosis [54, 55]. Patients with AD and primary dermatomyositis with pulmonary lesion have the highest mortality rates in the first 6 months after diagnosis [50]. However, in general, the 5-year survival rate in ILD-MP amounts to 50%–87% [38].

References

[1] Plotz PH, Dalakas M, Leff RL, Love LA, Miller FW, Cronin ME. Current concepts in the idiopathic inflammatory myopathies: polymyositis, dermatomyositis and related disorders. Ann Intern Med 1989;111(2):143–57.

[2] Olson AL, Brown KK, Fischer A. Connective tissue disease-associated lung disease. Immunol Allergy Clin North Am 2012;32(4):513–36.

[3] Schwarz MI. The lung in polymyositis. Clin Chest Med 1998;19(4):701–12.

[4] Labirua A, Lundberg IE. Interstitial lung disease and idiopathic inflammatory myopathies: progress and pitfalls. Curr Opin Rheumatol 2010;22(6):633–8.

[5] Selva-O'Callaghan A, Labrador-Horrillo M, Muñoz-Gall X, Martínez-Gomez X, Majó-Masferrer J, Solans-Laque R, et al. Polymyositis/dermatomyositis-associated lung disease: analysis of a series of 81 patients. Lupus 2005;14(7):534–42.

[6] Connors GR, Christopher-Stine L, Oddis CV, Danoff SK. Interstitial lung disease associated with the idiopathic inflammatory myopathies: what progress has been made in the past 35 years? Chest 2010;138(6):1464–74.

[7] Adler BL, Christopher-Stine L. Triggers of inflammatory myopathy: insights into pathogenesis. Discov Med 2018;25(136):75–83.

[8] Brown KK. Rheumatoid lung disease. Proc Am Thorac Soc 2007;4(5):443–8.

[9] Svensson J, Holmqvist M, Lundberg IE, Arkema EV. Infections and respiratory tract disease as risk factors for idiopathic inflammatory myopathies: a population-based case-control study. Ann Rheum Dis 2017;76(11):1803–8.

[10] Douglas WW, Tazelaar HD, Hartman TE, Hartman RP, Decker PA, Schroeder DR, et al. Polymyositis-dermatomyositis associated interstitial lung disease. Am J Respir Crit Care Med 2001;164(7):1182–5.

[11] Urisman A, Jones KD. Pulmonary pathology in connective tissue disease. Semin Respir Crit Care Med 2014;35(2):201–12.

[12] Tansey D, Wells AU, Colby TV, Ip S, Nikolakoupolou A, du Bois RM, et al. Variations in histological patterns of interstitial pneumonia between connective tissue disorders and their relationship to prognosis. Histopathology 2004;44(6):585–96.

[13] Priyangika SM, Karunarathna WG, Liyanage I, Gunawardana M, Udumalgala S, Rosa C, et al. Organizing pneumonia as the first manifestation of anti-synthetase syndrome. BMC Res Notes 2016;9:290.
[14] Vij R, Strek M. Diagnosis and Treatment of connective tissue disease-associated interstitial lung disease. Chest 2013;143(3):814–24.
[15] Koreeda Y, Higashimoto I, Yamamoto M, Takahashi M, Kaji K, Fujimoto M, et al. Clinical and pathological findings of interstitial lung disease patients with anti-aminoacyl-tRNA synthetase autoantibodies. Intern Med 2010;49(5):361–9.
[16] Yura H, Sakamoto N, Satoh M, Ishimoto H, Hanaka T, Ito C, et al. Clinical characteristics of patients with anti-aminoacyl-tRNA synthetase antibody positive idiopathic interstitial pneumonia. Respir Med 2017;132:189–94.
[17] Kalluri M, Sahn SA, Oddis CV, Gharib SL, Christopher-Stine L, Danoff SK, et al. Clinical profile of anti-PL-12 autoantibody. Cohort study and review of the literature. Chest 2009;135(6):1550–6.
[18] Hamaguchi Y, Fujimoto M, Matsushita T, Kaji K, Komura K, Hasegawa M, et al. Common and distinct clinical features in adult patients with anti-aminoacyl-tRNA synthetase antibodies: heterogeneity within the syndrome. PLoS One 2013;8(4):e60442.
[19] Mukae H, Ishimoto H, Sakamoto N, Hara S, Kakugawa T, Nakayama S, et al. Clinical differences between interstitial lung disease associated with clinically amyopathic dermatomyositis and classic dermatomyositis. Chest 2009;136(5):1341–7.
[20] Ye S, Chen XX, Lu XY, Wu MF, Deng Y, Huang WQ, et al. Adult clinically amyopathic dermatomyositis with rapid progressive interstitial lung disease: a retrospective cohort study. Clin Rheumatol 2007;26(10):1647–54.
[21] Bailey EE, Fiorentino DF. Amyopathic dermatomyositis: definitions, diagnosis, and management. Curr Rheumatol Rep 2014;16(12):465.
[22] Sun Y, Liu Y, Yan B, Shi G. Interstitial lung disease in clinically amyopathic dermatomyositis (CADM) patients: a retrospective study of 41 Chinese Han patients. Rheumatol Int 2013;33(5):1295–302.
[23] Sato S, Hoshino K, Satoh T, Fujita T, Kawakami Y, Fujita T, et al. RNA helicase encoded by melanoma differentiation-associated gene 5 is a major autoantigen in patients with clinically amyopathic dermatomyositis: association with rapidly progressive interstitial lung disease. Arthritis Rheum 2009;60(7):2193–200.
[24] Mimori T, Nakashima R, Hosono Y. Interstitial lung disease in myositis: clinical subsets, biomarkers, and treatment. Curr Rheumatol Rep 2012;14(3):264–74.
[25] Fathi M, Lundberg I, Tornling G. Pulmonary complications of polymyositis and dermatomyositis. Semin Respir Crit Care Med 2007;28(4):451–8.
[26] Murray SG, Schmajuk G, Trupin L, Lawson E, Cascino M, Barton J, et al. A population-based study of infection-related hospital mortality in patients with dermatomyositis/polymyositis. Arthritis Care Res (Hoboken) 2015;67(5):673–80.
[27] Marie I, Menard JF, Hachulla E, Cherin P, Benveniste O, Tiev K, et al. Infectious complications in polymyositis and dermatomyositis: a series of 279 patients. Semin Arthritis Rheum 2011;41(1):48–60.
[28] Sugiyama Y, Yoshimi R, Tamura M, Takeno M, Kunishita Y, Kishimoto D, et al. The predictive prognostic factors for polymyositis/dermatomyositis-associated interstitial lung disease. Arthritis Res Ther 2018;20(1):7.
[29] Rozelle A, Trieu S, Chung L. Malignancy in the setting of the anti-synthetase syndrome. J Clin Rheumatol 2008;14(5):285–8.
[30] Schnabel A, Reuter M, Biederer J, Richter C, Gross WL. Interstitial lung disease in polymyositis and dermatomyositis: clinical course and response to treatment. Semin Arthritis Rheum 2003;32(5):273–84.
[31] Lesnyak V, Averyanov A, Antelava O, Ananyeva L. CT evaluation of pulmonary abnormalities in polymyositis and dermatomyositis patients. Am J Respir Crit Care Med 2013;187:A2925.
[32] Ikezoe J, Johkoh T, Kohno N, Takeuchi N, Ichikado K, Nakamura H. High-resolution CT findings of lung disease in patients with polymyositis and dermatomyositis. J Thorac Imag 1996;11(4):250–9.
[33] Bonnefoy O, Ferretti G, Calaque O, Coulomb M, Begueret H, Beylot-Barry M, et al. Serial chest CT findings in interstitial lung disease associated with polymyositis-dermatomyositis. Eur J Radiol 2004;49(3):235–44.
[34] Cottin V, Thivolet-Bejui F, Reynaud-Gaubert M, Cadranel J, Delaval P, Ternamian PJ, et al. Interstitial lung disease in amyopathic dermatomyositis, dermatomyositis and polymyositis. Eur Respir J 2003;22(2):245–50.
[35] Capobianco J, Grimberg A, Thompson BM, Antunes VB, Jasinowodolinski D, Meirelles GS. Thoracic manifestations of collagen vascular diseases. Radiographics 2012;32(1):33–50.
[36] Tanizawa K, Handa T, Nakashima R, Kubo T, Hosono Y, Aihara K, et al. The prognostic value of HRCT in myositis-associated interstitial lung disease. Respir Med 2013;107(5):745–52.
[37] Ingegnoli F, Lubatti C, Ingegnoli A. Interstitial lung disease outcomes by high-resolution computed tomography (HRCT) in Anti-Jo1 antibody-positive polymyositis patients: a single centre study and review of the literature. Autoimmun Rev 2012;11(5):335–40.
[38] Wells AU, Hirani N. Interstitial lung disease guideline. Thorax 2008;63(Suppl. V):v1–v58.
[39] Nawata Y, Kurasawa K, Takabayashi K, Miike S, Watanabe N, Hiraguri M, et al. Corticosteroid resistant interstitial pneumonitis in dermatomyositis/polymyositis: prediction and treatment with cyclosporine. J Rheumatol 1999;26(7):1527–33.
[40] Findlay AR, Goyal NA, Mozaffar T. An overview of polymyositis and dermatomyositis. Muscle Nerve 2015;51(5):638–56.
[41] Morganroth PA, Kreider ME, Werth VP. Mycophenolate mofetil for interstitial lung disease in dermatomyositis. Arthritis Care Res 2010;62:1496–501.
[42] Wilkes MR, Sereika SM, Fertig N, Lucas MR, Oddis CV. Treatment of antisynthetase-associated interstitial lung disease with tacrolimus. Arthritis Rheum 2005;52(8):2439–46.
[43] Cavagna L, Caporali R, Abdi-Ali L, Dore R, Meloni F, Montecucco C. Cyclosporine in anti-jo1-positive patients with corticosteroid-refractory interstitial lung disease. J Rheumatol 2013;40(4):484–92.
[44] Hallowell RW, Danoff SK. Interstitial lung disease associated with the idiopathic inflammatory myopathies and the antisynthetase syndrome: recent advances. Curr Opin Rheumatol 2014;26(6):684–9.

[45] Bakewell CJ, Raghu G. Polymyositis associated with severe interstitial lung disease: remission after three doses of IV immunoglobulin. Chest 2011;139(2):441–3.

[46] Vandenbroucke E, Grutters JC, Altenburg J, Boersma WG, ter Borg EJ, van den Bosch JM. Rituximab in life threatening antisynthetase syndrome. Rheumatol Int 2009;29(12):1499–502.

[47] Huang J, Liu C, Zhu R, Su Y, Lin J, Lu J, et al. Combined usage of extracorporeal membrane oxygenation and double filtration plasmapheresis in amyopathic dermatomyositis patient with severe interstitial lung disease. A case report. Medicine (Baltimore) 2018;97(22):e10946.

[48] Hozumi H, Fujisawa T, Nakashima R, Yasui H, Suzuki Y, Kono M. Efficacy of glucocorticoids and calcineurin inhibitors for anti-aminoacyl-tRNA synthetase antibody-positive polymyositis/dermatomyositis-associated interstitial lung disease: a propensity score-matched analysis. J Rheumatol 2019. pii: jrheum.180778.

[49] Nakashima R, Hosono Y, Mimori T. Clinical significance and new detection system of autoantibodies in myositis with interstitial lung disease. Lupus 2016;25(8):925–33.

[50] Yamasaki Y, Yamada H, Ohkubo M, Yamasaki M, Azuma K, Ogawa H, et al. Long-term survival and associated risk factors in patients with adult-onset idiopathic inflammatory myopathies and amyopathic dermatomyositis: experience in a single institute in Japan. J Rheumatol 2011;38(8):1636–43.

[51] Schiopu E, Phillips K, MacDonald PM, Crofford LJ, Somers EC. Predictors of survival in a cohort of patients with polymyositis and dermatomyositis: effect of corticosteroids, methotrexate and azathioprine. Arthritis Res Ther 2012;14(1):R22.

[52] Marie I, Hatron PY, Levesque H, Hachulla E, Hellot MF, Michon-Pasturel U, et al. Influence of age on characteristics of polymyositis and dermatomyositis in adults. Medicine (Baltimore) 1999;78(3):139–47.

[53] Gono T, Kawaguchi Y, Satoh T, Kuwana M, Katsumata Y, Takagi K, et al. Clinical manifestation and prognostic factor in anti-melanoma differentiation-associated gene 5 antibody-associated interstitial lung disease as a complication of dermatomyositis. Rheumatology (Oxford) 2010;49(9):1713–9.

[54] Hozumi H, Fujisawa T, Nakashima R, Johkoh T, Sumikawa H, Murakami A, et al. Comprehensive assessment of myositis-specific autoantibodies in polymyositis/dermatomyositis-associated interstitial lung disease. Respir Med 2016;121:91–9.

[55] Spath M, Schroder M, Schlotter-Weigel B, Walter MC, Hautmann H, Leinsinger G, et al. The long-term outcome of anti-Jo-1-positive inflammatory myopathies. J Neurol 2004;251(7):859–64.

Chapter 8.4

Lung disease in systemic lupus erythematosus

Systemic lupus erythematosus (SLE) is an autoimmune disease characterized by the production of a wide range of autoantibodies and inflammatory lesions of many organs and systems [1]. This indistinct definition is due to the diversity of morphological substrates and clinical manifestations of SLE. SLE affects mainly teenagers and women; the ratio of affected men and women is 1:13 [2]. In different regions the prevalence of SLE varies significantly; for Western Europe, this figure is less than 50 per 100,000 population, while in the United States, it exceeds 240 per 100,000 population [3]. The age of most patients at diagnosis ranges from 15 to 45 years [1].

The etiology and pathogenesis of SLE are unclear. A correlation with the polymorphism of a number of genes, such as 1q23, 2q35-37, 6p21-11, and 12q24, exists [4]. Excessive exposure to sunlight and Epstein-Barr virus carriage can be disease triggers. A number of drugs, such as procainamide, hydralazine, sulfanilamides, interferon-alpha, and antibodies to tumor necrosis factor-alpha, can induce a lupus-like syndrome that is reversible upon cessation of the drug load [5]. Apparently, apoptosis of skin cells and neutrophils result in a critical mass of nucleoproteins and nucleic acids that get presented to dendritic cells and trigger an interferon-I-dependent mechanism of autoantibody formation [6]. The autoantibodies can be detected long before the clinical manifestations of the disease [7].

Morphology

Pulmonary involvement in SLE is most often manifested as serous, hemorrhagic, or fibrinous pleuritis, as lung parenchyma lesions in acute lupus pneumonitis, diffuse alveolar damage with hyaline membranes, pulmonary vasculitis with hemorrhages, nonspecific interstitial pneumonia, organizing pneumonia, usual interstitial pneumonia, follicular bronchiolitis, and lymphoid interstitial pneumonia [8] (Figs. 8.4.1–8.4.3). Vascular lesions manifest as antiphospholipid syndrome, pulmonary embolism (PE), and pulmonary hypertension. In the presence of Libman-Sacks endocarditis, chronic venous congestion occurs in the lungs, with the development of brown induration of the lungs. In addition, patients with SLE may develop shrinking lung syndrome caused by bilateral paralysis of the phrenic nerve or diaphragm dysfunction.

Clinical presentation

The clinical presentation of the disease depends on the lesions in affected organs and systems and can be very diverse. Table 8.4.1 shows the frequency of organ involvement in the autoimmune process and the symptoms of SLE [9].

In this table the respiratory manifestations of SLE are presented only as pleuritis and alveolar hemorrhages, but the pathological spectrum of respiratory lesions is actually wider and includes the following variants (Table 8.4.2):

According to the Spanish Rheumatology Society Lupus Registry, 996 (31%) of 3215 patients with SLE had pleuropulmonary manifestations among which pleural diseases (21%), lupus pneumonitis (3.6%), pulmonary thromboembolism

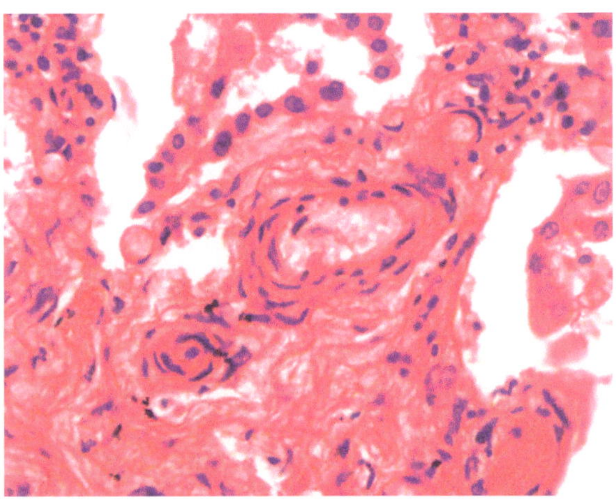

FIG. 8.4.1 Systemic lupus erythematosus. Productive vasculitis of small arteries of the lung. Hematoxylin and eosin (H&E) staining, 400×.

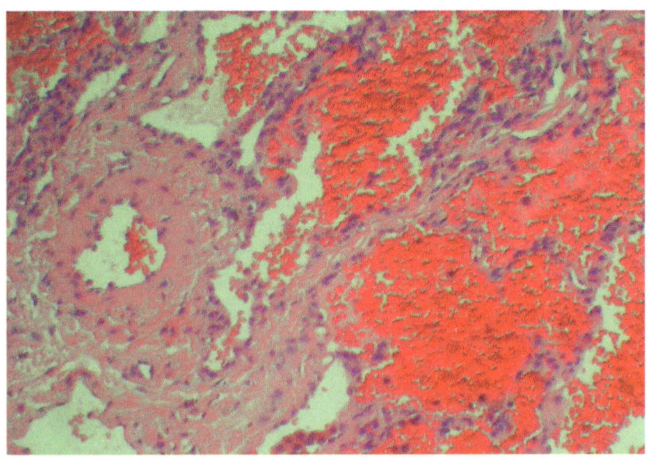

FIG. 8.4.2 Systemic lupus erythematosus. Intra-alveolar hemorrhage. H&E staining, 400×.

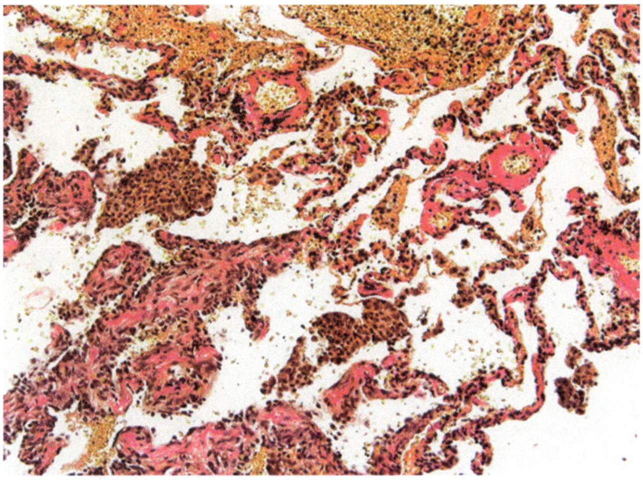

FIG. 8.4.3 Systemic lupus erythematosus. Pulmonary hemosiderosis. H&E staining, 100×.

(2.9%), pulmonary hypertension (2.4%), diffuse interstitial lung disease (2%), alveolar hemorrhage (0.8%), and shrinking lung syndrome (0.8%) were the most frequent [11].

The upper airways are affected by SLE less often than the lower respiratory tract; however, up to 48% of patients have certain manifestations of lupus inflammation, such as nasal and laryngeal ulceration, rhinitis, pruritus and dryness in the nose, laryngitis, ulceration of the laryngeal mucosa, edema and vocal cord paralysis, and cartilaginous lesions [12, 13]. Variants with life-threatening larynx stenosis due to necrotizing vasculitis have been described [14].

The lower airway changes in the form of thickening of the bronchial walls and bronchiectasis are not always clinically pronounced [15]. However, presentations include forms of constrictive bronchiolitis with a significant airflow limitation, signs of mosaic hyperinflation, and bronchiectasis (Fig. 8.4.4) that can determine the quality of life and prognosis of patients with SLE [16]. Fibrinous pleuritis or pleural effusion is the most common manifestation of lupus serositis and in 5% of cases may be the first sign of the disease (Fig. 8.4.5) [17]. Clinically significant pleural effusions occur in approximately half of SLE patients. However, pleural lesions are revealed in the vast majority of autopsies of affected individuals, and exudate in the pleural cavity is found in two-thirds of patients [18]. The predictors of pleural effusion for SLE are fever, Reynaud's phenomenon, serum anti-DNA antibodies, and infarction or resection of the bowel [19]. The main symptoms of pleural diseases include chest pain, unproductive cough, and dyspnea on physical exertion. Pleural effusion in patients with SLE is not always associated with the disease activity and may be due to PE or other related diseases (viz., heart failure, tuberculosis, pneumonia, malignant diseases, or hypoalbuminemia) [20].

Chronic interstitial lung disease associated with SLE (ILD-SLE) develops in 2%–15% of patients [11, 21]. ILD-SLE is more often found in patients with increased titers of anti-Ro (SSA) antibodies [22]. Since this type of antibody is regarded

TABLE 8.4.1 Frequency of symptoms in SLE

Symptom	Frequency (%)
Skin manifestations	88
Arthritis/arthralgia	76
Neuropsychic symptoms	66
Pleuritis/pericarditis	63
Anemia	57
Reynaud's phenomenon	44
Vasculitis	43
Atherosclerosis	37
Nephritis	31
Thrombocytopenia	30
Sensory motor neuropathy	28
Valve lesion	18
Alveolar hemorrhages	12
Pancreatitis	10
Myositis	5
Myocarditis	5

(Extracted from Goldman's Cecil medicine. 24th ed. Philadelphia, PA: Saunders Elsevier; 2011. 2704 p.)

TABLE 8.4.2 Variants of respiratory system lesions in SLE

Pleura lesion	Dry pleuritis Exudative pleuritis
Parenchymal lesions	Acute lupus pneumonitis Acute respiratory distress syndrome Diffuse alveolar hemorrhages Chronic interstitial pneumonia Diaphragmatic dysfunction/shrinking lung syndrome
Vascular lesions	Acute reversible hypoxemia Pulmonary artery thromboembolism Pulmonary arterial hypertension
Airway lesions	Small airway obstruction (constrictive bronchiolitis) Lesions of the upper airways
Drug-induced lesions	Nonspecific interstitial pneumonia, organizing pneumonia, usual interstitial pneumonia, acute interstitial pneumonia
Infectious lesions	Pneumonia caused by opportunistic pathogens (pneumocystis, fungi, cytomegalovirus, and mycobacteria)

(Extract from Kamen DL, Strange C. Pulmonary manifestations of systemic lupus erythematosus. Clin Chest Med 2010;31(3):479–88 with supplements.)

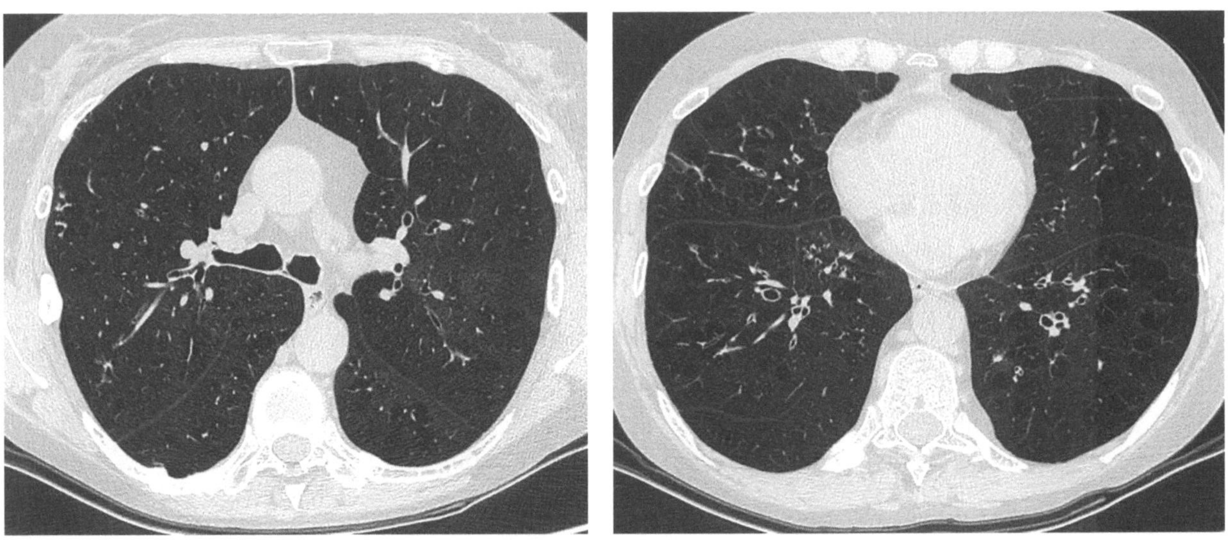

(A) (B)

FIG. 8.4.4 Constrictive bronchiolitis in a nonsmoking 47-year-old woman, with a 29-year history of systemic lupus erythematosus. Diffuse reduced attenuation, lobular areas of air trapping and emphysema, and multiple small bronchiectasis. Areas of linear fibrosis after pneumonia are also visible (A, B).

as a marker for Sjögren syndrome, the development of ILD is probably the result of the overlap of SLE with secondary Sjögren syndrome [23]. Nonspecific interstitial pneumonia (Fig. 8.4.6) is the most frequent histopathologic substrate of ILD-SLE [24]. The risk factors for ILD development are higher in patients with SLE who have higher activity and duration of the disease, hypocomplementemia, and high anti-dsDNA antibody levels [25, 26].

The clinical presentation and examination results are traditional for chronic interstitial pneumonia and include progressive dyspnea, dry cough, Velcro crackles, a restrictive functional pattern, and a decrease in diffusion capacity for carbon monoxide (DLCO) [21].

Acute lupus pneumonia (pneumonitis) is rare, but one of the most dangerous pulmonary manifestations of SLE. Its course corresponds to the type of acute interstitial pneumonia with diffuse alveolar damage, sometimes with signs of

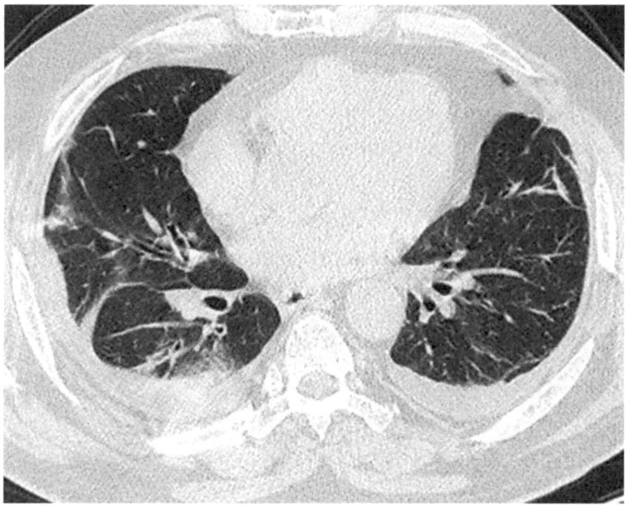

FIG. 8.4.5 Systemic lupus erythematosus. Bilateral small pleural effusion. Moderate reticular abnormalities and linear opacities.

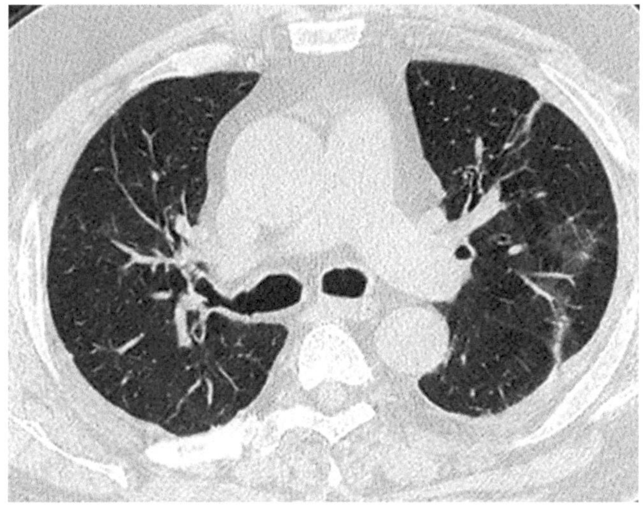

FIG. 8.4.6 Systemic lupus erythematosus. Bilateral patchy areas of ground-glass opacity. Irregular reticular abnormalities.

capillaritis [18]. Clinically the most prominent syndrome is an acute respiratory failure with high fever and the presence of bilateral massive consolidation areas and diffuse or patchy ground-glass opacity (GGO) on chest high-resolution computed tomography (HRCT) images. The mortality in these cases reaches 50% [27].

Diffuse alveolar hemorrhage (DAH) is also a potentially life-threatening condition in patients with SLE; it usually develops in patients with a long history of disease and is often associated with lupus nephritis, hypocomplementemia, and the presence of antiphospholipid antibodies [28, 29]. DAH clinically manifests as progressive dyspnea correlating with the severity of the parenchymal hemorrhagic lesion, cough, and hemoptysis. However, the latter symptom is not obligate, and its absence does not rule out DAH. Usually a rapid fall of erythrocyte, hemoglobin, and hematocrit levels is found, and bilateral symmetrical diffuse or patchy areas of GGO associated with thickened interlobular septa with a perihilar distribution of abnormalities are usually detected on HRCT (Fig. 8.4.7). Inside the GGOs the emergence of foci of consolidation is possible. When the DAH diagnosis is unclear, bronchoalveolar lavage can reveal the presence of red blood cells and siderophages in the BAL fluid as evidence of alveolar hemorrhages. A biopsy is rarely required to confirm the diagnosis. Histologically, signs of capillaritis or bland hemorrhage are always found [30].

Shrinking lung syndrome is a consequence of the weakness of the diaphragm due to myopathy of the diaphragm or neuropathy of the phrenic nerve [31]. It is manifested by dyspnea during physical exertion and radiographically detected

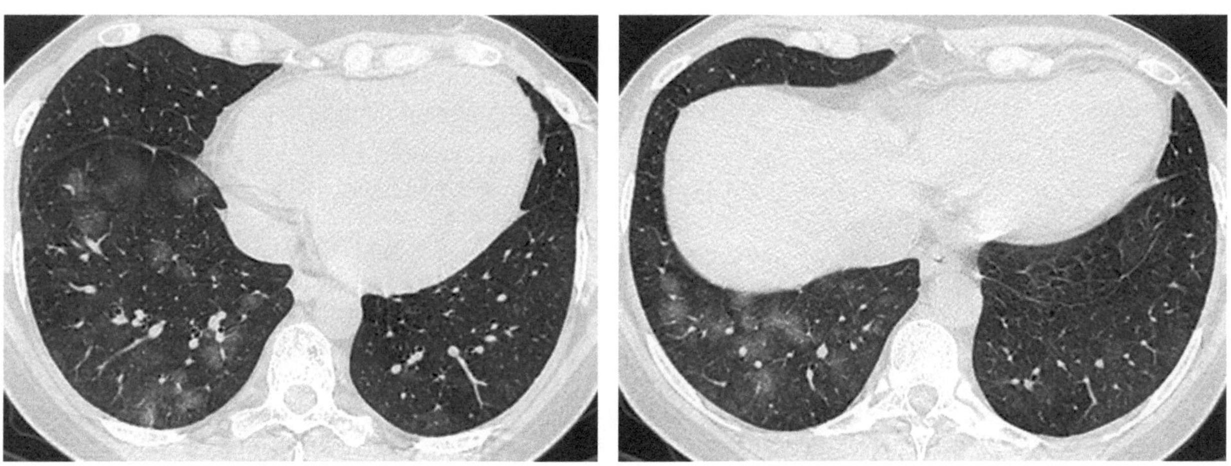

(A) (B)

FIG. 8.4.7 Alveolar hemorrhages in a patient with systemic lupus erythematosus. Spotted areas of ground-glass opacity, predominantly in the right lung, associated with slight thickening of the interlobular and intralobular septa. Subpleural areas remain unaffected (A, B).

by an elevated diaphragm (often bilateral). The restrictive pattern in the respiratory function study is not accompanied by a decrease in lung diffusion capacity (D_{LCO}/V_A) [31].

Patients with SLE and antiphospholipid syndrome (approximately 1/3 of patients) have a high risk of PE, peripheral thrombosis, pulmonary microthrombosis, alveolar hemorrhages, and hemolytic uremic syndrome. Dyspnea and hemoptysis usually are prominent in the clinical presentation of PE. CT angiography (in the absence of renal failure) is an important tool to confirm or exclude the presence of thrombotic masses in the pulmonary artery [10].

Finally a possible pulmonary complication of SLE, apparently due to leukocyte aggregation in the pulmonary capillaries, is acute reversible hypoxemia, which manifests as acute onset respiratory failure in the absence of visible parenchymal changes in the lungs on HRCT and gets resolved as a result of steroid therapy and respiratory support for 72 h [32].

Apparently the age of SLE onset is important for the development of pulmonary pathology. In a study by Medlin et al., late-onset patients (over 50 years of age) had nearly threefold increased odds of developing ILD, compared with the odds of those with the early onset of SLE [33].

Diagnosis

The diagnosis of SLE is established in the presence of 4 of the 11 criteria [34]:

1. Butterfly-shaped lupus erythema on the face without the involvement of the nasolabial triangle
2. Discoid rash represented by elevated erythematous nodules with closely adherent keratosis squamosa and follicular clots (old nodules may have atrophic scars)
3. Photosensitization (appearance of skin rashes after ultraviolet insolation)
4. Ulcerous defects of the mouth or nasopharynx mucosa (usually painful)
5. Nonerosive arthritis involving two or more joints
6. Serositis (pleural effusion or pericarditis)
7. Kidney lesion:
   - proteinuria of more than 0.5 g/day or more than 3%
   - increased cytosis of urine sediment-erythrocyturia (hemoglobinuria), casts
8. Neurological disorders (with the exception of drug-induced or metabolically induced disorders):
   - seizures
   - psychosis
9. Hematologic disorders:
   - hemolytic anemia with reticulocytosis
   - leukopenia (<4000 in 1 mm^3), recorded at least two times
   - lymphopenia (<1500 in 1 mm^3), recorded at least two times
   - thrombocytopenia (<100,000 in 1 mm^3) in the absence of a drug-induced condition
10. Immunologic disorders:
    - antibodies to native DNA in elevated titer
    - the presence of antibodies to the Sm nuclear antigen
    - antiphospholipid antibodies (IgG or IgM antibodies to cardiolipin; positive test result for lupus anticoagulant and anti-beta-2 glycoprotein IgM or IgG or false-positive test result for syphilis)
11. Increased titer of antinuclear antibodies (with the exception of drug-induced lupus syndrome)

The diagnosis of SLE-associated lung disease is established based on the clinical presentation and instrumental examination results, namely, HRCT, studies of pleural fluid, bronchoalveolar lavage, and functional tests.

The emergence of a pleural effusion in a patient with an established diagnosis of SLE in the absence of concomitant changes in the lung parenchyma usually does not require analysis of the pleural fluid. However, in the case of an unspecified underlying disease, suspected paraneoplastic lupus syndrome, or the presence of other possible causes of effusion, cytological and biochemical analysis of the pleural fluid should be performed. Lupus pleuritis is characterized by serous or serous-hemorrhagic effusion with exudate characteristics but with normal or slightly reduced glucose level and pH and normal lactate dehydrogenase (LDH) concentration, which distinguishes it from the exudate in rheumatoid arthritis, which is characterized by a decrease in glucose and pH, but with an increase in LDH [17, 35]. However, pleural effusion in the form of transudate is also possible. An increase in the level of antinuclear antibodies in the pleural fluid used is considered evidence of lupoid pleuritis. However, studies have shown that this immunologic marker is neither specific nor sensitive for diagnosing SLE [36]. BAL is indicated for the differential diagnosis of pulmonary infections with diffuse alveolar damage, namely, acute lupus pneumonia (ALP) or alveolar hemorrhages. The prevalence of neutrophils in the composition of the

BAL fluid in combination with erythrocytosis is typical for ALP, while DAH is manifested only in the hemorrhagic nature of the BAL [37].

The respiratory function and DLCO are important for evaluating and monitoring the restrictive pattern and limiting the gas exchange surface that accompanies many lung lesions.

Computed tomography

Since the variants of lung damage in SLE are diverse, the set of radiological signs is wide and depends on specific morphological patterns (Figs. 8.4.4–8.4.7 and Table 8.4.3).

In contrast to other connective tissue diseases in which interstitial pneumonias dominate as clinically significant respiratory manifestations, in SLE, pleural effusion, DAH, and PE occur most often.

Differential diagnosis

In this chapter, we will discuss the differential signs of acute lung parenchymal lesions in SLE that we consider to be the most challenging.

Table 8.4.4 presents the main criteria for diagnosis (without morphological findings).

In patients with SLE who receive steroid therapy, with the appearance of new infiltrates in the lungs, PE and pneumonia caused by opportunistic pathogens should always be considered in the differential range. Pulmonary infections are the most common infectious complications of SLE (approximately 38% of cases) and often become the cause of death [38].

Moreover, opportunistic infections can cover on the primary pathological substrate, for example, ILD, complicating the diagnosis (Fig. 8.4.8). We recommend all patients with SLE and primary identified abnormalities in the lungs to perform BAL not only for the purpose of cytological evaluation of the lavage fluid but also for PCR analysis for opportunistic pathogens such as *P. jirovecii*, mold, and yeast infection (*Aspergillus* spp., *Cryptococcus* spp., and *Mucor* spp.) detected most frequently in these patients [39, 40].

TABLE 8.4.3 Dominant CT findings as a manifestation of noninfectious lung lesions in SLE

Radiological symptom	Possible substrate
Ground-glass opacity	Acute lupus pneumonia Alveolar hemorrhages Nonspecific interstitial pneumonia Organizing pneumonia Secondary Sjögren syndrome
Consolidation	Acute lupus pneumonia Alveolar hemorrhages Organizing pneumonia Pulmonary embolism
Signs of interstitial fibrosis (irregular thickening of the interlobular and intralobular septa, bronchiectases, and bronchiolectases)	Nonspecific interstitial pneumonia Usual interstitial pneumonia Secondary Sjögren syndrome
Honeycombing	Usual interstitial pneumonia
Upper-lobe localization of abnormalities	Alveolar hemorrhages
Cysts	Secondary Sjögren syndrome
Predominantly basal localization	Acute lupus pneumonia Nonspecific and usual interstitial pneumonia
Subpleural localization of abnormalities	Pulmonary embolism Organizing pneumonia Nonspecific and usual interstitial pneumonia
Perihilar distribution of abnormalities	Diffuse alveolar hemorrhages
Pleural effusion	Lupus exudative pleuritis

TABLE 8.4.4 Differential signs of primary and infectious lung lesions of acute course in patients with SLE

	ALP	DAH	PE	BP	PP
Onset	Acute, subacute	Acute, subacute	Acute	Acute	Subacute
Prevailing clinical symptoms	Dyspnea, high fever	Dyspnea, hemoptysis, anemia	Dyspnea, hemoptysis, chest pain	Fever, cough	High fever, dyspnea
Procalcitonin nanogram/milliliter	<0.5	<0.5	<0.5	>1	<1
D-dimer	Normal or slightly increased	Moderately increased	Significantly increased	Increased in severe course	Normal
BAL	Neutrophils >40%, erythrocytes	Erythrocytosis, siderophages	Erythrocytosis	Neutrophilia, bacteria	Neutrophilia + PCR for *P. jirovecii*
CT signs	Bilateral extensive consolidation (more in the basal fields), ground-glass opacity, thickening of the interlobular septa	Areas of consolidation and ground-glass opacity, thickening of intralobular and interlobular septa, perihilar pattern of distribution of the abnormalities	Subpleural consolidation areas. In CT angiography, filling defects of contrast-enhanced vessels	More often unilateral areas of consolidation, turning into ground-glass opacity, nodules	Multiple patches of ground-glass opacity (more in the upper lobes). Air traps. Thickening of interlobular septa. Centrilobular nodules
Effect of steroids	++	–	–	–	+
Effect of antibiotics	–	–	–	+++	–

ALP, acute lupus pneumonia; *BP*, bacterial pneumonia; *DAH*, diffuse alveolar hemorrhages; *PE*, pulmonary embolism; *PP*, *Pneumocystis jirovecii* pneumonia. –, Nontypical; +, possible; ++, frequent; +++, typical.

Mycoplasma pneumoniae, *Streptococcus pneumoniae*, *Acinetobacter baumannii*, *Mycobacterium tuberculosis*, and nontuberculous mycobacteria are also described as causative agents of severe infectious lung lesions in patients with SLE [38, 41].

Since ALP can imitate bacterial or fungal pneumonia [42], the detection of pathogens and neutrophilia in BAL fluid should be considered critical.

Thus *P. jirovecii* can be a trigger for diffuse alveolar damage without the development of pneumocystis pneumonia [43].

Treatment

The choice of treatment for respiratory pathology in patients with SLE depends on the form of the lesion. Thus exudative pleuritis usually responds to treatment with nonsteroidal antiinflammatory drugs or with average doses of corticosteroids (20–40 mg/day) [10, 24].

In cases of ALP, high doses of oral prednisone (1–1.5 mg/kg) are indicated, and in the absence of an effect for 72 h, pulse therapy with methylprednisolone (1000 mg/day for 3 days) is prescribed [44].

In refractory cases the administration of cyclophosphamide (0.5–1 g/m^2 for 6 months, 1–2 times per month) is recommended, as well as mycophenolate mofetil, intravenous immunoglobulin, and plasma exchange [45, 46].

The diagnosis of DAH determines the combination therapy with high doses of glucocorticosteroids and cyclophosphamide in the earlier dosages since steroid monotherapy is associated with a high level of mortality [47]. Plasma exchange can also be effective in such patients [48]. Rituximab can also be considered as a reserve therapy for cases of recurrence of life-threatening DAH [49, 50].

Approaches to the treatment of ILD-SLE are not fully developed. Systemic glucocorticosteroids (0.5–1 mg/kg of body weight per day) are usually considered as the first line of treatment, but some authors, based on the similar morphological pattern with systemic sclerosis, have suggested the immediate start of a combined therapy with medium doses of prednisolone and cyclophosphamide or azathioprine [10].

The diaphragmatic dysfunction usually responds to steroid therapy, but immunosuppressive drugs are also considered in cases of inefficiency of the initial treatment, and the number of reports of the successful treatment with rituximab has increased [51–53].

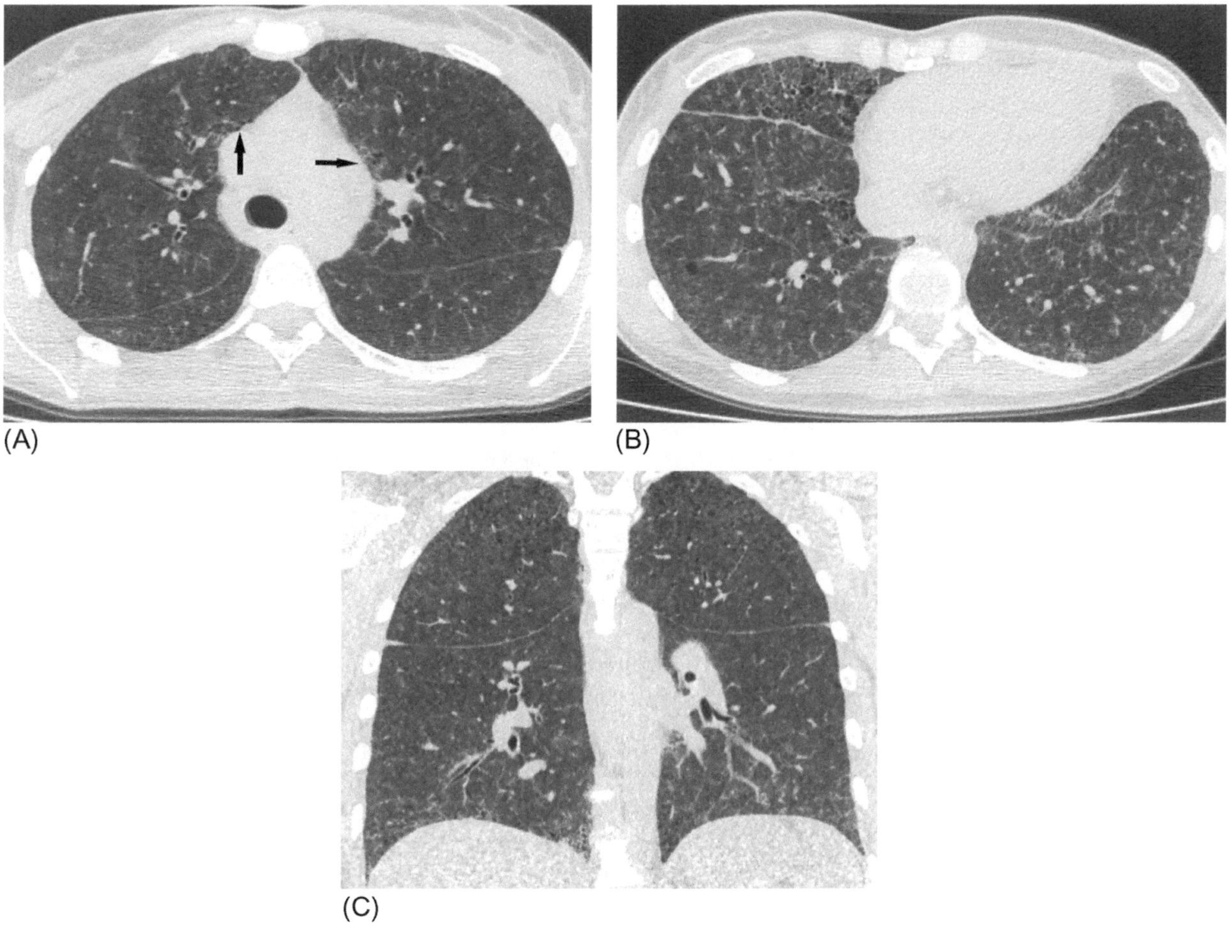

FIG. 8.4.8 *Pneumocystis jirovecii* pneumonia in a 30-year-old patient with a 16-year history of systemic lupus erythematosus and interstitial lung disease. Multiple centrilobular poorly differentiated nodules, subpleural small cysts *(arrows)*, and mild reticular abnormalities (A). Multiple patchy areas of ground-glass opacity, centrilobular nodules, and moderate reticular abnormalities; subpleural honeycombing in the lower and middle lobes and a single air-filled cyst on the right (B). Coronal reconstruction revealing predominantly lower field lesions (C).

Theophylline can be used as an additional drug that improves muscle function [54].

In patients with antiphospholipid syndrome and SLE, aspirin therapy is recommended for the prevention of thromboembolic complications. In patients with PE a lifelong administration of anticoagulants is indicated [10].

Acute reversible hypoxemia is usually resolved within 3 days under the influence of monotherapy with steroids or their combination with aspirin [55].

References

[1] Alarcon GS, Friedman AW, Straaton KV, Moulds JM, Lisse J, Bastian HM, et al. Systemic lupus erythematosus in three ethnic groups: III. A comparison of characteristics early in the natural history of the LUMINA cohort. LUpus in MInority populations: NAture vs. Nurture. Lupus 1999;8(3):197–209.

[2] Danchenko N, Satia J, Anthony M. Epidemiology of systematic lupus erythematosus: a comparison of worldwide disease burden. Lupus 2006;15(5):308–18.

[3] Rees F, Doherty M, Grainge MJ, Lanyon P, Zhang W. The worldwide incidence and prevalence of systemic lupus erythematosus: a systematic review of epidemiological studies. Rheumatology (Oxford) 2017;56(11):1945–61.

[4] Ghodke-Puranik Y, Niewold TB. Immunogenetics of systemic lupus erythematosus: a comprehensive review. J Autoimmun 2015;64:125–36.

[5] Kamen D. Environmental influences on systemic lupus erythematosus expression. Rheum Dis Clin North Am 2014;40(3):401–12.

[6] Ali A, Sayyed Z, Ameer MA, Arif AW, Kiran F, Iftikhar A, et al. Systemic lupus erythematosus: an overview of the disease pathology and its management. Cureus 2018;10(9):e3288.

[7] Tsokos GC, Lo MS, Costa Reis P, Sullivan KE. New insights into the immunopathogenesis of systemic lupus erythematosus. Nat Rev Rheumatol 2016;12(12):716–30.

[8] Vivero M, Padera RF. Histopathology of lung disease in the connective tissue diseases. Rheum Dis Clin North Am 2015;41(2):197–211.

[9] Goldman's Cecil medicine. 24th ed. Philadelphia, PA: Saunders Elsevier; 2011. 2704 p.

[10] Kamen DL, Strange C. Pulmonary manifestations of systemic lupus erythematosus. Clin Chest Med 2010;31(3):479–88.

[11] Narváez J, Borrell H, Sánchez-Alonso F, Rúa-Figueroa I, López-Longo FJ, Galindo-Izquierdo M, et al. Primary respiratory disease in patients with systemic lupus erythematosus: data from the Spanish rheumatology society lupus registry (RELESSER) cohort. Arthritis Res Ther 2018;20(1):280.

[12] Teitel AD, MacKenzie CR, Stern R, Paget SA. Laryngeal involvement in systemic lupus erythematosus. Semin Arthritis Rheum 1992;22(3):203–14.

[13] Kusyairi KA, Gendeh BS, Sakthiswary R, Shaharir SS, Haizlene AH, Yusof KH. The spectrum of nasal involvement in systemic lupus erythematosus and its association with the disease activity. Lupus 2016;25(5):520–4.

[14] Langford CA, Van Waes C. Upper airway obstruction in the rheumatic diseases. Rheum Dis Clin North Am 1997;23(2):345–63.

[15] Keane MP, Lynch 3rd JP. Pleuropulmonary manifestations of systemic lupus erythematosus. Thorax 2000;55(2):159–66.

[16] Kawahata K, Yamaguchi M, Kanda H, Komiya A, Tanaka R, Dohi M, et al. Severe airflow limitation in two patients with systemic lupus erythematosus: effect of inhalation of anticholinergics. Mod Rheumatol 2008;18(1):52–6.

[17] Swigris JJ, Fischer A, Gillis J, Meehan RT, Brown KK. Pulmonary and thrombotic manifestations of systemic lupus erythematosus. Chest 2008;133(1):271–80.

[18] Olson AL, Brown KK, Fischer A. Connective tissue disease-associated lung disorders. Immunol Allergy Clin North Am 2012;32(4):513–36.

[19] Ryu S, Fu W, Petri MA. Associates and predictors of pleurisy or pericarditis in SLE. Lupus Sci Med 2017;4(1):e000221.

[20] Palavutitotai N, Velazquez ML, Highland KB. Pulmonary manifestations of systemic lupus erythematosus and Sjögren's syndrome. Curr Opin Rheumatol 2018;30(5):449–64.

[21] Mathai SC, Danoff SK. Management of interstitial lung disease associated with connective tissue disease. BMJ 2016;352:h6819.

[22] Xu D, Tian X, Zhang W, Zhang X, Liu B, Zhang F. Sjogren's syndrome-onset lupus patients have distinctive clinical manifestations and benign prognosis: a case-control study. Lupus 2010;19(2):197–200.

[23] Pasoto SG, Adriano de Oliveira Martins V, Bonfa E. Sjögren's syndrome and systemic lupus erythematosus: links and risks. Open Access Rheumatol 2019;11:33–45.

[24] Lopez Velazquez M, Highland KB. Pulmonary manifestations of systemic lupus erythematosus and Sjögren's syndrome. Curr Opin Rheumatol 2018;30(5):449–64.

[25] Memet B, Ginzler EM. Pulmonary manifestations of systemic lupus erythematosus. Semin Respir Crit Care Med 2007;28(4):441–50.

[26] Alamoudi OS, Attar SM. Pulmonary manifestations in systemic lupus erythematosus: association with disease activity. Respirology 2015;20(3):474–80.

[27] Matthay RA, Schwarz MI, Petty TL, Stanford RE, Gupta RC, Sahn SA, et al. Pulmonary manifestations of systemic lupus erythematosus: review of twelve cases of acute lupus pneumonitis. Medicine (Baltimore) 1975;54(5):397–409.

[28] Santos-Ocampo AS, Mandell BF, Fessler BJ. Alveolar hemorrhage in systemic lupus erythematosus: presentation and management. Chest 2000;118(4):1083–90.

[29] Zuily S, Domingues V, Suty-Selton C, Eschwège V, Bertoletti L, Chaouat A, et al. Antiphospholipid antibodies can identify lupus patients at risk of pulmonary hypertension: a systematic review and meta-analysis. Autoimmun Rev 2017;16(6):576–86.

[30] Martınez-Martınez MU, Abud-Mendoza C. Diffuse alveolar hemorrhage in patients with systemic lupus erythematosus. Clinical manifestations, treatment, and prognosis. Rheumatol Clin 2014;10(4):248–53.

[31] Hardy K, Herry I, Attali V, Cadranel J, Similowski T. Bilateral phrenic paralysis in a patient with systemic lupus erythematosus. Chest 2001;119(4):1274–7.

[32] Abramson SB, Dobro J, Eberle MA, Benton M, Reibman J, Epstein H, et al. Acute reversible hypoxemia in systemic lupus erythematosus. Ann Intern Med 1991;114(11):941–7.

[33] Medlin JL, Hansen KE, McCoy SS, Bartels CM. Pulmonary manifestations in late versus early systemic lupus erythematosus: a systematic review and meta-analysis. Semin Arthritis Rheum 2018;48(2):198–204.

[34] Petri M, Orbai A-M, Alarcón GS, Gordon C, Merrill JT, Fortin PR, et al. Derivation and validation of the Systemic Lupus International Collaborating Clinics classification criteria for systemic lupus erythematosus. Arthritis Rheum 2012;64(8):2677–86.

[35] Palavutitotai N, Buppajarntham T, Katchamart W. Etiologies and outcomes of pleural effusions in patients with systemic lupus erythematosus. J Clin Rheumatol 2014;20(8):418–21.

[36] Murray and Nadel's textbook of respiratory medicine. 5th ed. Philadelphia: Saunders/Elsevier; 2010. p. 1719–63.

[37] Sişmanlar Eyüboğlu T, Aslan AT, Özdemir Y, Gezgin Yıldırım D, Buyan N, Boyunağa Ö. Isolated acute lupus pneumonitis as the initial presentation of systemic lupus erythematosus in an 8-year-old girl. Auto Immun Highlights 2018;9(1):4.

[38] Teh CL, Wan SA, Ling GR. Severe infections in systemic lupus erythematosus: disease pattern and predictors of infection-related mortality. Clin Rheumatol 2018;37(8):2081–6.

[39] Quadrelli SA, Alvarez C, Arce SC, Paz L, Sarano J, Sobrino EM, et al. Pulmonary involvement of systemic lupus erythematosus: analysis of 90 necropsies. Lupus 2009;18(12):1053–60.

[40] Lao M, Wang X, Ding M, Yang Z, Chen H, Liang L, et al. Invasive fungal disease in patients with systemic lupus erythematosus from Southern China: a retrospective study. Lupus 2019;28(1):77–85.

[41] Shaharir SS, Sulaiman Sahari N, Mohamed Fuad Z, Zukiman WZHW, Mohd Yusof NH, Sulong A, et al. Non-tuberculous mycobacterium bacteraemia in a pregnant systemic lupus erythematosus (SLE) patient: a case review and pooled case analysis. Clin Rheumatol 2018;37(3):837–47.

[42] Jain V, Aziz M, Banoub MG, Neuman JT, Sidlow R. Pulmonary systemic lupus erythematosus mimicking a pneumonia in a postpartum female. Case Rep Rheumatol 2018;2018:5379192.

[43] Martínez-Rísquez MT, Friaza V, de la Horra C, Martín-Juan J, Calderón EJ, Medrano FJ. Pneumocystis jirovecii infection in patients with acute interstitial pneumonia. Rev Clin Esp 2018;218(8):417–20.

[44] Mira-Avendano IC, Abril A. Pulmonary manifestations of Sjogren syndrome, systemic lupus erythematosus, and mixed connective tissue disease. Rheum Dis Clin North Am 2015;41(2):263–77.

[45] Chen MC, Wu YL, Lee KL, Lai KS, Chung CL. Lupus pneumonitis presenting with high titre of anti-Ro antibody. Respirol Case Rep 2017;6(1):e00280.

[46] Swigris JJ, Olson AL, Fischer A, Lynch DA, Cosgrove GP, Frankel SK, et al. Mycophenolate mofetil is safe, well tolerated, and preserves lung function in patients with connective tissue disease related interstitial lung disease. Chest 2006;130(1):30–6.

[47] Abud-Mendoza C, Diaz-Jouanen E, Alarcon-Segovia D. Fatal pulmonary hemorrhage in systemic lupus erythematosus. Occurrence without hemoptysis. J Rheumatol 1985;12(3):558–61.

[48] Ednalino C, Yip J, Carsons SE. Systematic review of diffuse alveolar hemorrhage in systemic lupus erythematosus: focus on outcome and therapy. J Clin Rheumatol 2015;21(6):305–10.

[49] Martínez-Martínez MU, Abud-Mendoza C. Diffuse alveolar hemorrhage in patients with systemic lupus erythematosus. Clinical manifestations, treatment, and prognosis. Reumatol Clin 2014;10(4):248–53.

[50] Tse JR, Schwab KE, McMahon M, Simon W. Rituximab: an emerging treatment for recurrent diffuse alveolar hemorrhage in systemic lupus erythematosus. Lupus 2015;24(7):756–9.

[51] Soubrier M, Dubost JJ, Piette JC, Urosevic Z, Rami S, Oualid T, et al. Shrinking lung syndrome in systemic lupus erythematosus. A report of three cases. Rev Rhum Engl Ed 1995;62(5):395–8.

[52] Benham H, Garske L, Vecchio P, Eckert BW. Successful treatment of shrinking lung syndrome with rituximab in a patient with systemic lupus erythematosus. J Clin Rheumatol 2010;16(2):68–70.

[53] Goswami RP, Mondal S, Lahiri D, Basu K, Das S, Ghosh P, et al. Shrinking lung syndrome in systemic lupus erythematosus successfully treated with rituximab. QJM 2016;109(9):617–8.

[54] Van Veen S, Peeters AJ, Sterk PJ, Breedveld FC. The "shrinking lung syndrome" in SLE, treatment with theophylline. Clin Rheumatol 1993;12(4):462–5.

[55] Martinez-Taboada VM, Blanco R, Armona J, Fernandez-Sueiro JL, Rodriguez-Valverde V. Acute reversible hypoxemia in systemic lupus erythematosus: a new syndrome or an index of disease activity? Lupus 1995;4(4):259–62.

Chapter 8.5

Pulmonary disease in Sjögren syndrome

Sjögren syndrome (SS) is a chronic inflammatory autoimmune disease characterized by lymphocytic infiltration of exocrine glands, primarily lacrimal and salivary, resulting in their impaired function and ocular and oral dryness [1]. Primary SS is distinguished when it is an independent nosological form, also called Sjögren disease. Secondary SS occurs in other connective tissue diseases (CTDs), most often in rheumatoid arthritis, systemic sclerosis, and systemic lupus erythematosus [2]. SS is one of the most common CTDs, with a prevalence of 0.5% of the population, and women are affected approximately nine times more often than men [3].

The **pathogenesis** of SS is unclear, but the autoimmune mechanisms for the development of the disease are beyond discussion. Apparently, in the presence of a genetic predisposition, a primary lesion of the salivary gland epithelial cells by viruses (Epstein-Barr virus, cytomegalovirus, human herpes virus type 8, and human T-lymphotropic virus type 1) leads to recruitment of immune cells, such as dendritic cells, NK cells, and T cells, with release of cytokines and chemokines. These include IFN γ, which contributes to the IFN signature and acts as a link between innate and adaptive immunity. The IFN signature mediates lymphocytic infiltration and lymphocyte activation and stimulates the production of B-cell activating factor (BAFF). BAFF with IFN γ promotes autoantibody production by B cells, with the subsequent development of autoimmune glandular epithelitis with secretory dysfunction [4]. The formation of immune complexes usually determines the extraglandular areas of the lesion in SS, including the lungs and mediastinum [5].

The histopathologic pattern of the lungs in SS is diverse and is presented as a lesion of the airways (follicular lymphocytic bronchiolitis and constrictive bronchiolitis), various interstitial lung diseases (ILD), and the development of primary malignant lymphoma of the lungs [6]. The ILDs, namely, nonspecific interstitial pneumonia (NSIP) and lymphoid interstitial pneumonia (LIP), are the most common, and usual interstitial pneumonia (UIP), organizing pneumonia (OP), diffuse alveolar damage, and amyloidosis are less common presentations (Figs. 8.5.1 and 8.5.2). NSIP and OP overlap is often found [7].

Clinical presentation

Approximately 80% of SS patients show signs of sicca syndrome, such as xerophthalmia (dry eye) and xerostomia (dry mouth), due to reduced production of saliva and conjunctival fluid [8].

As a result, dryness, a sandy-gritty sensation, itching in the eyelids, oral dryness, pain when swallowing, hoarseness, progressive caries, and rhagades also occur, as well as recurring parotitis or gingivitis [6]. The remaining 20% of patients present in the so-called atypical fashion in which sicca symptoms are minimal or absent [8]. Approximately a third of SS patients have a history of Raynaud's phenomenon, and arthralgia, myalgia, and arthritis of the small joints of the hands are frequent symptoms but without erosive synovitis, which distinguishes these cases from rheumatoid arthritis [9]. One

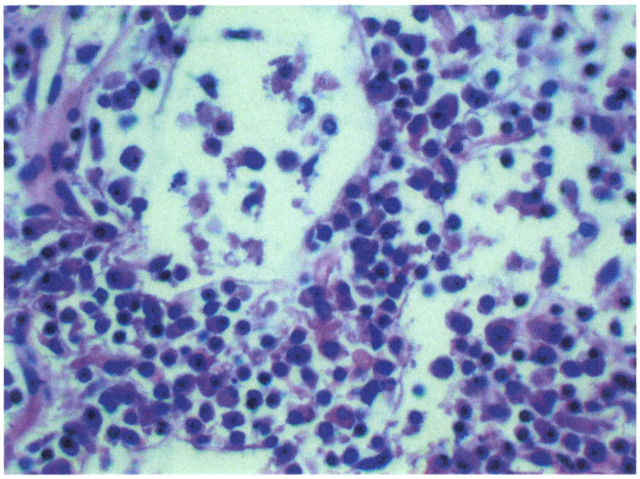

FIG. 8.5.1 SS. Lymphocytic interstitial pneumonia. Hematoxylin and eosin (H&E) staining, 600×.

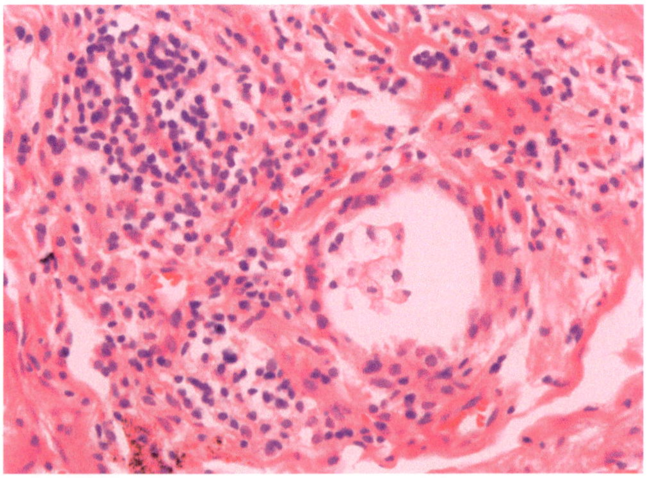

FIG. 8.5.2 SS. Usual interstitial pneumonia with hyperplasia of the lymphoid tissue. H&E stain, 200×.

frequent extraglandular symptom is chronic fatigue, which is detected in 80% of patients [10]. Rarer manifestations include demyelinating disease, neuropathy, depression, and constipation [8, 10]. Most patients demonstrate submandibular and cervical lymphadenopathy, without obvious soreness of lymph nodes or edema of salivary glands. In a small number of patients, hypergammaglobulinemic purpura and petechial rash are detected, usually on the shins, but found less often on the hips and abdomen, with hyperpigmentation after rash elimination. Involvement of the respiratory organs in the autoimmune process is a typical situation in SS, which is found in up to 75% of patients [11, 12]. The main clinical and radiological pulmonary patterns are bronchial (bronchiolar) lesions, interstitial lung disease (ILD), and lymphoproliferative disorders [13]. The greatest clinical significance that often determines SS prognosis is the development of ILD [13]. ILD patients are older and more often have Raynaud's phenomenon and gastrointestinal tract involvement [14].

The common signs of ILD in SS (ILD-SS) are unproductive cough and dyspnea on physical exertion. Chronic unproductive cough is the most frequent pulmonary symptom in SS, is often troublesome, and reduces the quality of life. It is also associated with dryness in the trachea and frequent development of concomitant gastroesophageal reflux and is often found in patients without ILD [15]. In some patients, crackles may be heard in the posterior basal areas of the lungs, but in general, this phenomenon occurs in a smaller proportion of patients [16]. Clubbing is not typical for ILD-SS, as it is found in approximately 6% of cases [17].

In 25%–60% of patients, ILD precedes the full-scale presentation of SS; therefore the definition of anti-SSA and anti-SSB antibodies, as well as other autoimmune markers, is an important diagnostic tool in patients with unclear interstitial pulmonary abnormalities [14, 18]. An asymptomatic onset of ILD-SS is noted in 24% of all cases [14].

Diagnosis

Diagnosis of SS is made on the basis of a characteristic clinical presentation supplemented with confirmation of a systemic autoimmune inflammatory process.

General laboratory changes often include hypergammaglobulinemia and high ESR. In addition, increased levels of rheumatoid factor, antinuclear antibodies, and circulating immune complexes are often determined. In some patients, anemia and lymphopenia develop. The key autoimmune markers of SS are circulating antibodies against the ribonucleoprotein complexes Ro/SSA and La/SSB [6]. However, the presence of anti-SSB only in the serum without anti-SSA antibodies is not associated with SS and thus cannot confirm the diagnosis alone [19].

Since 2002, six criteria have been used to establish the diagnosis of SS [9]:

1. Symptoms in the eyes (at least one)
   - **(A)** Daily persistent dryness in the eyes for more than 3 months
   - **(B)** Recurrent sandy-gritty sensation or sensation of foreign body in the eyes
   - **(C)** The need for rewetting drops for eyes (at least three times a day)
2. Symptoms in the oral cavity (at least one)
   - **(A)** Daily sensation of dry mouth for more than 3 months
   - **(B)** Recurrent or persistent swelling of the salivary glands
   - **(C)** Frequent need for drinking to swallow dry food
3. Positive ocular tests
   - **(A)** Schirmer's test—wetting less than 5 mm of a filter paper placed in the conjunctival sac under the lower eyelid for 5 min
   - **(B)** A score of 4 or more according to the Bijsterveld system
4. Histopathology
   - **(A)** Biopsy of the salivary glands demonstrates lymphocytic sialoadenitis
5. Symptoms in the salivary glands (at least one)
   - **(A)** Decreased salivation.
   - **(B)** Parotid sialography shows diffuse sialectasia.
   - **(C)** Pathological changes in salivary sialography.
6. Positive antibody titers to Ro/SSA and/or La/SSB

Diagnosis of SS is considered proved in the presence of at least four out of six criteria, and one must be related to histopathology (No. 4) or antibodies (No. 6).

SS diagnosis is also valid in case of three of four objective criteria, namely, the ocular tests (No. 3), histopathology (No. 4), salivary symptoms (No. 5), and autoantibodies (No. 6) [9].

However, in 2016, the American College of Rheumatology and the European League Against Rheumatism reviewed these criteria and proposed a more simple approach to the diagnostics of primary SS based on the following symptoms [20]:

1. Labial salivary gland with focal lymphocytic sialoadenitis and focus score of >1 foci/4 mm^2 (3 points)
2. Anti-Ro positive (3 points)
3. Ocular staining score >5 in at least one eye (1 point)
4. Schirmer's test <5 mm/5 min in at least one eye (1 point)
5. Unstimulated whole saliva flow rate <0.1 mL/min (1 point)

An individual with a score of ≥4 is classified as primary SS [20].

A comparative analysis of Billings and Hadavand revealed that the new criteria for 2016 are more sensitive but less specific than the previous criteria of 2002 [21]. Despite apparent clinical and serological signs, the diagnosis of SS is usually made a little less than 4 years after the onset of symptoms [8].

Assessment of lung damage in SS includes traditional lung functional tests with the obligatory study of lung diffusion capacity for carbon monoxide (DLCO). In cases of bronchiolitis an obstructive pattern is usually detected, and in ILD-SS, both a restrictive pattern and a decrease in DLCO and an obstructive and mixed pattern are usually noted [18].

Bronchoalveolar lavage (BAL) usually reveals a mild increase in lymphocytes (>20%), neutrophils (>4%), and eosinophils (>3%) in patients with signs of interstitial pneumonia [16, 22]. The value of the analysis of BAL fluid is primarily in the differential diagnosis of ILD from pulmonary infections and malignancies. A lung biopsy is usually not required in ILD-SS patients as it increases the risk of inducing an exacerbation or other complications and does not provide additional diagnostic information that could aid in the disease management of patients. The exception would be suspected amyloidosis or lymphoma, when the final diagnosis is not possible without histopathologic evidence.

Radiological signs

On normal chest radiographs, certain abnormalities in SS are detected in no more than 35% of cases and include a reticular or reticulonodular pattern with predominant involvement of the basal regions [23].

High-resolution computed tomography (HRCT) as a more sensitive method, reveals the features of a pulmonary lesion much more frequently and more diversely. In general the following variants are typical for chest pathology associated with SS [13]:

- Respiratory lesion in the form of follicular or constrictive bronchiolitis
- Interstitial lung disease
- Lymphoproliferative syndrome
- Cystic syndrome
- Mediastinal syndrome

A lesion of the large and small bronchi and bronchioles occurs in 48%–68% of SS patients [12, 24]. It is based on the lymphoplasmacytic infiltration of the respiratory tract and peribronchiolar space, detected on HRCT in the form of thickening of the bronchial and bronchiolar walls that are traced to the subpleural areas and centrilobular nodules. Bronchiolar obstruction can lead to the formation of hyperinflation areas (air traps) creating a pattern of mosaic attenuation, which better manifests during expiration [25]. In the case of severe constrictive bronchiolitis, bronchiectasis can be formed.

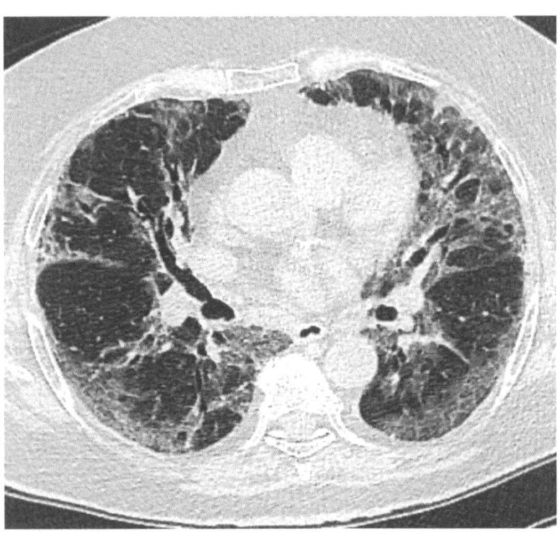

FIG. 8.5.3 Interstitial lung disease associated with Sjögren syndrome. Bilateral patchy areas of ground-glass opacity, predominantly with subpleural distribution and reticular abnormalities. Pattern of nonspecific interstitial pneumonia.

Interstitial lung disease develops much less frequently than bronchial abnormalities in SS patients (3.4%–17%) [14, 18, 26]. Most often, ILD-SS is represented by NSIP, which is the dominant variant (up to 61% of cases); UIP is found in up to 25% of cases, and OP is revealed in up to 20% of cases, with respective CT signs characteristic of each (Figs. 8.5.3 and 8.5.4) [14, 16, 18]. A study by Manfredi et al. showed HRCT abnormalities presenting as a definite or possible UIP pattern in 12 of 13 ILD-SS patients [26]. LIP, considered specific for SS and thus distinguishing it from ILD in other CTDs by the presence of cysts, is actually only detected in 0.9%–17% of patients with ILD-SS (Fig. 8.5.5) [25, 27]. Interestingly the UIP and LIP patterns are associated with higher titers of antinuclear antibodies (>1:320), and NSIP and OP develop more often at lower titers (>1:40) [18].

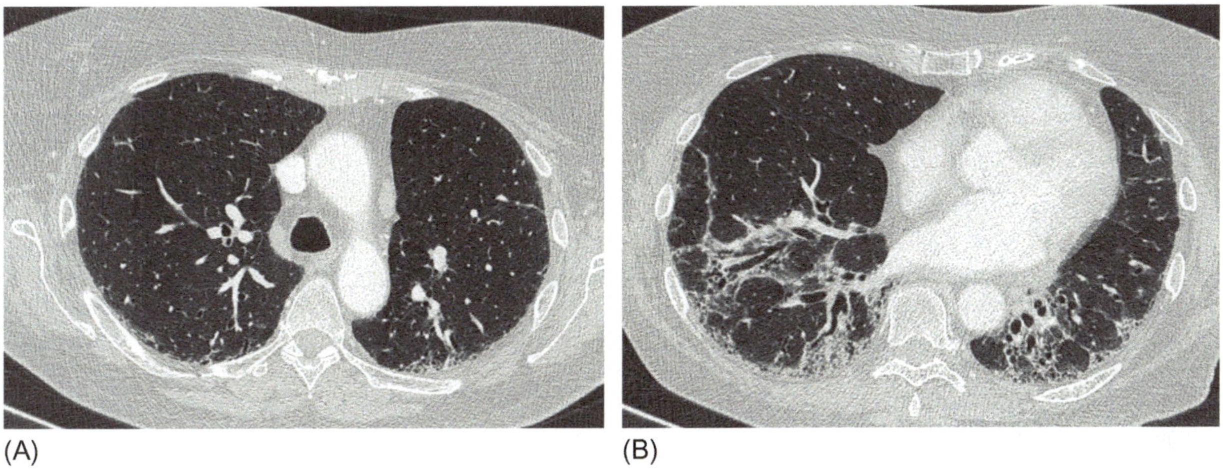

FIG. 8.5.4 Interstitial lung disease associated with Sjögren syndrome. Subpleural honeycombing is more pronounced in the basal regions (A and B). In the lower lobes, severe reticular abnormalities, traction bronchiectasis, and linear attenuations are visualized. Pattern of probable usual interstitial pneumonia. *(Case courtesy of Prof. V.I. Vasilyev, V.A. Nasonova Research Institute of Rheumatology, Moscow, Russia.)*

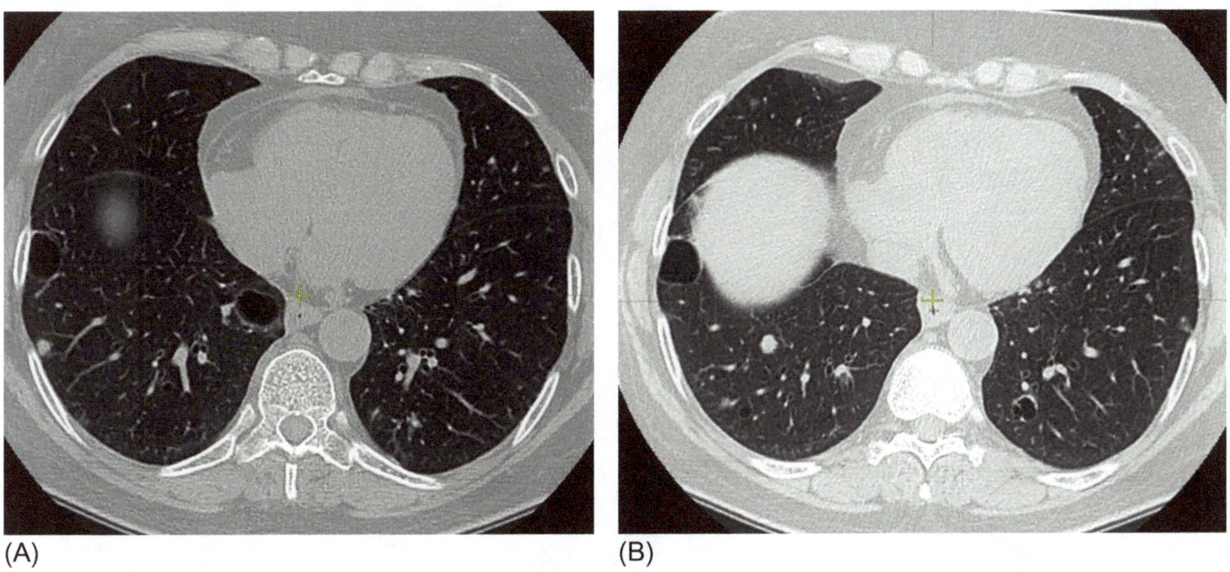

FIG. 8.5.5 Interstitial lung disease associated with Sjögren syndrome. Thin-walled cysts from 4 to 25 mm in diameter in the lower pulmonary lobes. Separate nodules and mild subpleural reticular changes are noted (A and B). Pattern of lymphoid interstitial pneumonia. *(Case courtesy of Prof. V.I. Vasilyev, V.A. Nasonova Research Institute of Rheumatology, Moscow, Russia.)*

Summarizing the HRCT manifestations of interstitial pneumonia, a set of signs traditional for ILD can be considered, namely, ground-glass opacity (GGO), reticular pattern, subpleural linear fibrosis, traction bronchiectasis, and honeycombing [23, 28]. The abnormalities have a predominantly peripheral and basal distribution. The peribronchovascular or subpleural areas of consolidation, alternating, OP surrounded by GGO as a manifestation of OP, are less common (up to 20%), and thin-walled cysts (a specific sign of LIP) are noted in 7%–17% of patients [23, 28, 29].

The largest study in the Chinese population revealed that the most frequent HRCT signs of ILD-SS were the reticular pattern (92.7%), GGO (87.4%), vascular bundle thickening (82%), and pleural effusion (62.1%), whereas the remaining abnormalities, namely, consolidation (34.5%), nodules (31.2%), fibrotic band (31.1%), bronchiectasis (30.1%), cysts (25.2%), and honeycombing (20.4%), were much less common [17]. However, Roca et al. [14] determined that the most frequent HRCT signs of ILD-SS were linear opacities found in 75% of patients, but GGO was revealed only in 45% of cases.

GGO is a typical sign of NSIP, LIP, and OP and is found in most ILD-SS patients. The reticular pattern is a characteristic of NSIP and UIP, and honeycombing is typical for UIP and fibrotic forms of NSIP. Consolidation areas represent a characteristic sign of OP but can also appear in bacterial pneumonia, which, however, have a more acute course than OP.

Lymphoproliferative syndrome includes tumor and nontumor lesions. Nontumor manifestations of lymphoproliferative processes include diffuse lymphoid hyperplasia (DLH), nodular lymphoid hyperplasia (NLH), and amyloidosis [13]. NLH and DLH are characterized by nodular changes due to lymphocyte infiltration of the lymphatic drainage pathways, namely, the interlobular septa, vascular bundles, or subpleural spaces [13]. In DLH, they can visually resemble granulomatous perilymphatic dissemination in sarcoidosis and lymphogenous metastases. NLH and DLH are often combined with LIP but can also be independent manifestations of SS. Amyloid deposits in SS may accompany both the primary forms of the disease (amyloid A) and amyloid-producing lymphoma and plasmacytoma (amyloid AL). In primary SS, amyloidosis is represented by nodules and areas of consolidation distributed along the lymphatic vessels, namely, the interlobular septa and visceral pleura, often with calcification of amyloid masses and the formation of cysts (Fig. 8.5.6) [30]. Lymphomas, especially MALT lymphomas, are tumor manifestation of SS.

On HRCT, lymphomas are represented by solitary or several foci or extensive areas of consolidation with a sign of air bronchogram surrounded by GGO (halo sign) [31].

Such areas can be distinguished from organizing pneumonia with contrast enhancement as lymphomas usually actively accumulate a contrast agent [32] (Fig. 8.5.7).

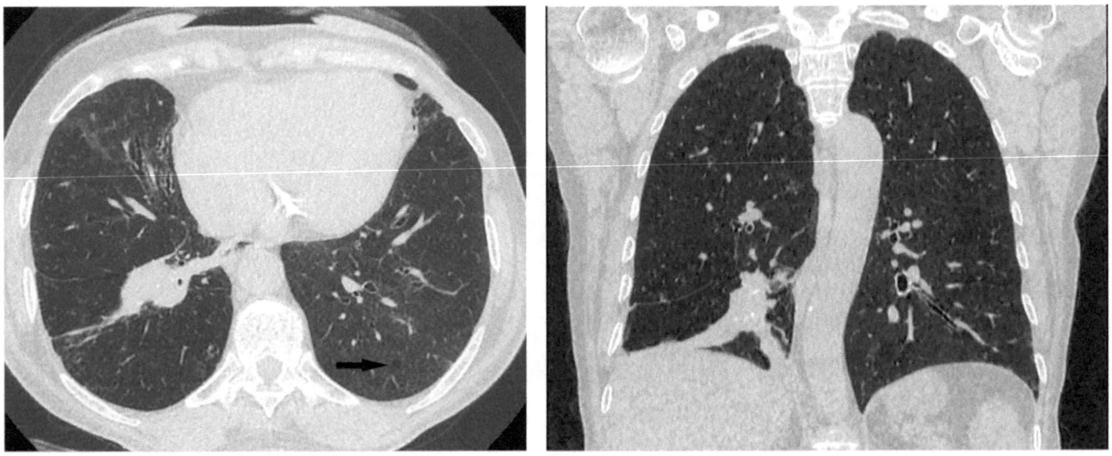

(A) (B)

FIG. 8.5.6 Amyloidosis proved by surgical lung biopsy in a patient with Sjögren syndrome. An area of consolidation in the right lung, traction bronchiectasis, and ground-glass opacity foci in the middle lobe can be seen. On the left, there is a single small cyst *(arrow)* (A). The coronal section shows that amyloid masses that propagate along the bronchial-vascular bundles contain calcification inside (B). *(Case courtesy of Prof. V.I. Vasilyev, V.A. Nasonova Research Institute of Rheumatology, Moscow, Russia.)*

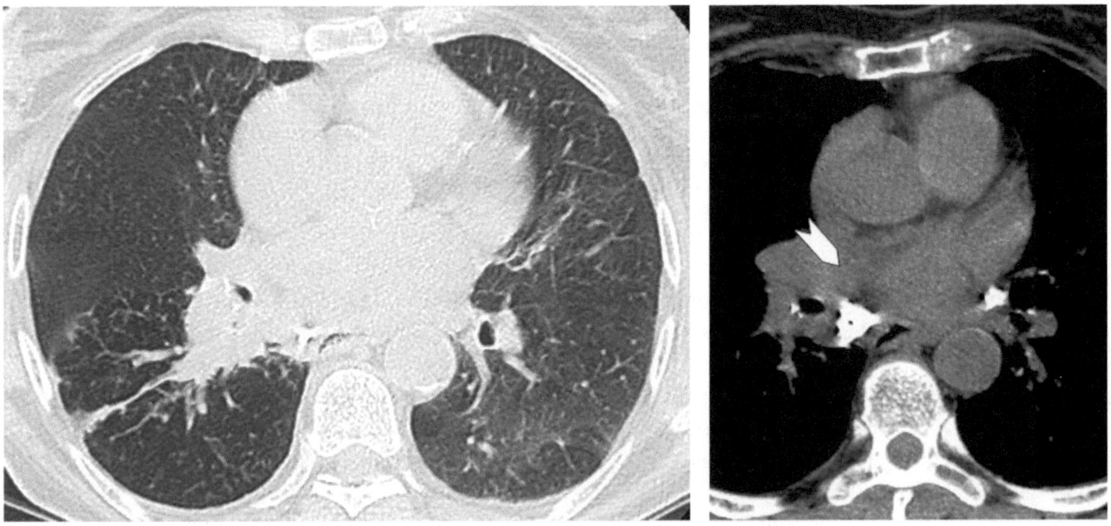

(A) (B)

FIG. 8.5.7 Sjögren syndrome. Right lung lymphoma. Massive peribronchovascular infiltration in the right lung associated with a dense cord with the visceral pleura (A). Mass of lymphoma in the mediastinum *(arrow)* (B).

Cystic syndrome is a specific pattern of lung damage in SS, although it is far from ubiquitous. LIP, lymphocytic bronchiolitis, amyloidosis, and DLH all contribute to the formation of cysts. Cystic abnormalities are less characteristic for other morphological variants of lung damage in SS. The formation of cysts is determined by two main mechanisms, namely, a valve mechanism, when air traps appear because of lymphoid infiltration of the bronchioles, and a destructive mechanism due to damage to the alveolar walls with lymphocytic proteases [13]. Cysts are usually thin walled and 5–30mm in diameter and are located in the lower fields, either subpleurally or around the vascular-bronchial bundles (Fig. 8.5.5) [33].

Mediastinal syndrome manifests with lymphadenopathy, lymphoid hyperplasia of the thymus, and its cystic transformation. Systemic lymphoplasmacytic infiltration is a characteristic manifestation of SS and, among others, is observed in the mediastinal lymph nodes. In CT analysis of SS patients, attention should always be paid to the anterior mediastinum so as not to miss the changes in the thymus. The emergence of a solid component may indicate malignancy [13].

Differential diagnosis

Taking into account the diversity of variants for lung lesions in SS, the differential diagnosis is performed depending on the leading radiological pattern. Differential diagnosis of NSIP, OP, and LIP is described in detail in the relevant sections of Chapter 2.

ILD-SS most often has to be differentiated from opportunistic infections that can manifest during steroid intake and immunosuppressive therapy. Pneumocystis or cytomegalovirus pneumonia, fungal lesions, mycobacterioses, and tuberculosis are those latent infections that should be considered as possible causes of diffuse interstitial abnormalities in the lungs. It is especially difficult to differentiate the acute exacerbation of ILD-SS from an infectious disease, since both clinical and radiological manifestations can be very similar, namely, fever, dyspnea, and inflammatory changes in the blood. The emergence of new areas of GGO and consolidation are a characteristic of both of these conditions.

Lymphoproliferative syndrome in the form of DLH requires a differential diagnosis with lung sarcoidosis and lymphogenous metastases. The overlap of pulmonary sarcoidosis and SS is described in 1%–2% of SS patients [34]. Primarily, such association is diagnosed simultaneously in 60% of cases [35]. Amyloidosis of the lungs in SS is found in approximately 6% of cases, most often in the form of nodular (AL) amyloidosis [16, 35]. In addition, dense nodules, consolidation foci, and masses are usually found in tissues with a perilymphatic distribution, often with signs of calcification and intrathoracic lymphadenopathy. Such an HRCT pattern requires a differential diagnosis with sarcoidosis and pulmonary lymphoma; the final diagnosis can only be established histopathologically.

The cystic pattern is a hallmark of lymphoid interstitial pneumonia and requires a differential diagnosis with other conditions that have cysts as a radiological sign, which is quite difficult with a diffuse cystic lesion (Fig. 8.5.8; see Chapters 2.7 and 9.4).

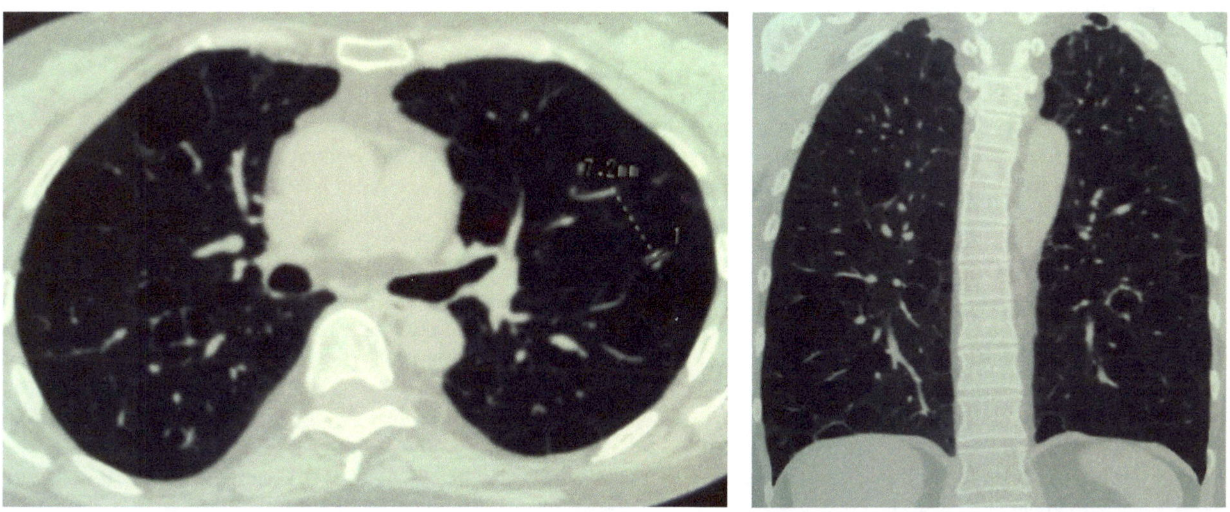

(A) (B)

FIG. 8.5.8 Sjögren syndrome. Multiple bilateral thin-walled cysts, mainly of medium and large size (A). The coronal section shows the distribution of cysts predominantly in the lower and middle areas (B). *(Case courtesy of Prof. V.I. Vasilyev, V.A. Nasonova Research Institute of Rheumatology, Moscow, Russia.)*

Treatment

Treatment of ILD-SS is started with oral systemic steroids (prednisone 0.5–1 mg/kg/day), with gradual dose reduction in 8–12 weeks to reach a maintenance therapy dose of 10 mg/day [28]. This is usually sufficient to at least stabilize the process, and more often, there is an obvious positive tendency toward functional disorder of the lungs. In the face of life-threatening complications or insufficiency of steroid efficacy, azathioprine may be added or used as a monotherapy in a dose of 1–2 mg/kg/day [36]. Apparently the HRCT or morphological pattern of ILD determines the success of corticosteroid therapy. Roca et al. [14] reported on an improved response to systemic steroids of SS patients with the NSIP, OP, and LIP patterns in contrast to the steroid resistance observed in UIP. Cyclophosphamide and hydroxychloroquines also showed activity in limited studies [37]. Rituximab can be considered a rescue drug in case of inefficiency of steroids and immunosuppressants and progression of ILD-SS [8, 38]. Several reports have been published on the successful treatment of ILD-SS with tocilizumab, tacrolimus, and abatacept [39].

Yamano et al. studied patients with ILD associated with CTD, including four SS patients, and deployed a therapeutic regimen that included intravenous (i.v.) methylprednisolone (1000 mg i.v. 3 days a week for 2 weeks), followed by low-dose prednisolone (10 mg/day) and tacrolimus. The researchers noted significantly improved forced vital capacity, DLCO, 6-min walk distance, lowest oxygen saturation on pulse oximetry, MMRC, and SGRQ with satisfactory tolerability and the absence of significant side effects [40].

For patients with dry, severe cough that accompanies various lung diseases, regular inhalations of hypertonic (3% or 7%) saline solution or secretory stimulants such as pilocarpine and cevimeline are recommended [35].

In general the prognosis for ILD-SS patients is rather favorable, with the 5-year survival rate reaching 84% [8]. However, approximately 38% of ILD-SS patients have an impaired pulmonary function despite treatment [14]. According to Roca et al. [14] the factors associated with the progression of ILD-SS are older age, esophageal involvement, and UIP pattern on HRCT. The main factor affecting mortality is the development of non-Hodgkin lymphomas, which occurs in 5%–10% of SS patients [41].

References

[1] Ferro F, Marcucci E, Orlandi M, Baldini C, Bartoloni-Bocci E. One year in review 2017: primary Sjogren's syndrome. Clin Exp Rheumatol 2017;35(2):179–91.

[2] Mavragani CP. Mechanisms and new strategies for Sjögren's syndrome. Annu Rev Med 2017;68:331–43.

[3] Peri Y, Agmon-Levin N, Theodor E, Shoenfeld Y. Sjögren's syndrome: the old and the new. Best Pract Res Clin Rheumatol 2012;26(1):105–17.

[4] Sandhya P, Kurien BT, Danda D, Scofield RH. Update on pathogenesis of Sjogren's syndrome. Curr Rheumatol Rev 2017;13(1):5–22.

[5] Voulgarelis M, Tzioufas AG. Current aspects of pathogenesis in Sjogren's syndrome. Ther Adv Musculoskelet Dis 2010;2(6):325–34.

[6] Mavragani CP, Moutsopoulos HM. Sjögren's syndrome. Annu Rev Pathol 2014;9:273–85.

[7] Vivero M, Padera RF. Histopathology of lung disease in the connective tissue diseases. Rheum Dis Clin North Am 2015;41(2):197–211.

[8] Vivino FB. Sjogren's syndrome: clinical aspects. Clin Immunol 2017;182:48–54.

[9] Vitali C, Bombardieri S, Jonsson R, Moutsopoulos HM, Alexander EL, Carsons SE, et al. Classification criteria for Sjögren's syndrome: a revised version of the European criteria proposed by the American-European Consensus Group. Ann Rheum Dis 2002;61(6):554–8.

[10] Lackner A, Ficjan A, Stradner MH, Hermann J, Unger J, Stamm T, et al. It's more than dryness and fatigue: the patient perspective on health-related quality of life in primary Sjögren's syndrome—a qualitative study. PLoS One 2017;12(2):e0172056.

[11] Palm O, Garen T, Berge Enger T, Jensen JL, Lund MB, Aaløkken TM, et al. Clinical pulmonary involvement in primary Sjögren's syndrome: prevalence, quality of life and mortality—a retrospective study based on registry data. Rheumatology (Oxford) 2013;52(1):173–9.

[12] Koyama M, Johkoh T, Honda O, Mihara N, Kozuka T, Tomiyama N, et al. Pulmonary involvement in primary Sjögren's syndrome: spectrum of pulmonary abnormalities and computed tomography findings in 60 patients. J Thorac Imaging 2001;16(4):290–6.

[13] Egashira R, Kondo T, Hirai T, Kamochi N, Yakushiji M, Yamasaki F, et al. CT findings of thoracic manifestations of primary Sjögren syndrome: radiologic-pathologic correlation. Radiographics 2013;33(7):1933–49.

[14] Roca F, Dominique S, Schmidt J, Smail A, Duhaut P, Lévesque H, et al. Interstitial lung disease in primary Sjögren's syndrome. Autoimmun Rev 2017;16(1):48–54.

[15] Flament T, Bigot A, Chaigne B, Henique H, Diot E, Marchand-Adam S. Pulmonary manifestations of Sjogren's syndrome. Eur Respir Rev 2016;25(140):110–23.

[16] Ito I, Nagai S, Kitaichi M, Nicholson AG, Johkoh T, Noma S, et al. Pulmonary manifestations of primary Sjogren's syndrome: a clinical, radiologic, and pathologic study. Am J Respir Crit Care Med 2005;171(6):632–8.

[17] Dong X, Zhou J, Guo X, Li Y, Xu Y, Fu Q, et al. A retrospective analysis of distinguishing features of chest HRCT and clinical manifestation in primary Sjögren's syndrome-related interstitial lung disease in a Chinese population. Clin Rheumatol 2018;37(11):2981–8.

[18] Reina D, Roig Vilaseca D, Torrente-Segarra V, Cerdà D, Castellví I, Díaz Torné C, et al. Sjögren's syndrome-associated interstitial lung disease: a multicenter study. Reumatol Clin 2016;12(4):201–5.

[19] Baer A, DeMarco M, Shiboski S, Lam M, Challacombe S, Daniels T, et al. The SSB-positive/SSA-negative antibody profile is not associated with key phenotypic features of Sjogren's syndrome. Ann Rheum Dis 2015;74(8):1557–61.
[20] Shiboski CH, Shiboski SC, Seror R, Criswell LA, Labetoulle M, Lietman TM, et al. 2016 American College of Rheumatology/European League Against Rheumatism classification criteria for primary Sjögren's syndrome: a consensus and data-driven methodology involving three international patient cohorts. Ann Rheum Dis 2017;76(1):9–16.
[21] Billings M, Amin Hadavand M, Alevizos I. Comparative analysis of the 2016 ACR-EULAR and the 2002 AECG classification criteria for Sjögren's syndrome: findings from the B GNIH cohort. Oral Dis 2018;24(1–2):184–90.
[22] Salaffi F, Manganelli P, Carotti M, Baldelli S, Blasetti P, Subiaco S, et al. A longitudinal study of pulmonary involvement in primary Sjögren's syndrome: relation between alveolitis and subsequent lung changes on high-resolution computed tomography. Br J Rheumatol 1998;37(3):263–9.
[23] Webb WR, Muller NL, Naidich DP. High-resolution CT of the lung. 4th ed. Philadelphia: Lippincott Williams and Wilkins; 2009. p. 236–8.
[24] Franquet T, Giménez A, Monill JM, Díaz C, Geli C. Primary Sjögren's syndrome and associated lung disease: CT findings in 50 patients. AJR Am J Roentgenol 1997;169(3):655–8.
[25] Howling SJ, Hansell DM, Wells AU, Nicholson AG, Flint JD, Müller NL. Follicular bronchiolitis: thin-section CT and histologic findings. Radiology 1999;212(3):637–42.
[26] Manfredi A, Sebastiani M, Cerri S, Cassone G, Bellini P, Casa GD, et al. Prevalence and characterization of non-sicca onset primary Sjögren syndrome with interstitial lung involvement. Clin Rheumatol 2017;36(6):1261–8.
[27] Parambil JG, Myers JL, Lindell RM, Matteson EL, Ryu JH. Interstitial lung disease in primary Sjögren syndrome. Chest 2006;130(5):1489–95.
[28] Vij R, Strek M. Diagnosis and treatment of connective tissue disease-associated interstitial lung disease. Chest 2013;143(3):814–24.
[29] Verschakelen JA. The role of high-resolution computed tomography in the work-up of interstitial lung disease. Curr Opin Pulm Med 2010;16(5):503–10.
[30] Jeong YJ, Lee KS, Chung MP, Han J, Chung MJ, Kim KI, et al. Amyloidosis and lymphoproliferative disease in Sjögren syndrome: thin-section computed tomography findings and histopathologic comparisons. J Comput Assist Tomogr 2004;28(6):776–81.
[31] Bae YA, Lee KS, Han J, Ko YH, Kim BT, Chung MJ, et al. Marginal zone B-cell lymphoma of bronchus-associated lymphoid tissue: imaging findings in 21 patients. Chest 2008;133(2):433–40.
[32] Shah RM, Friedman AC. CT angiogram sign: incidence and significance in lobar consolidations evaluated by contrast-enhanced CT. AJR Am J Roentgenol 1998;170(3):719–21.
[33] Lynch D. Lung disease related to collagen vascular disease. J Thorac Imaging 2009;24(4):299–309.
[34] Ramos-Casals M, Font J, Garcia-Carrasco M, Brito MP, Rosas J, Calvo-Alen J, et al. Primary Sjogren syndrome: hematologic patterns of disease expression. Medicine (Baltimore) 2002;81(4):281–92.
[35] Lopez Velazquez M, Highland KB. Pulmonary manifestations of systemic lupus erythematosus and Sjögren's syndrome. Curr Opin Rheumatol 2018;30(5):449–64.
[36] Deheinzelin D, Capelozzi VL, Kairalla RA, Barbas Filho JV, Saldiva PH, de Carvalho CR. Interstitial lung disease in primary Sjögren's syndrome. Clinical-pathological evaluation and response to treatment. Am J Respir Crit Care Med 1996;154(3 Pt 1):794–9.
[37] Kokosi M, Riemer EC, Highland KB. Pulmonary involvement in Sjögren syndrome. Clin Chest Med 2010;31(3):489–500.
[38] Seror R, Sordet C, Guillevin L, Hachulla E, Masson C, Ittah M, et al. Tolerance and efficacy of rituximab and changes in serum B cell biomarkers in patients with systemic complications of primary Sjögren's syndrome. Ann Rheum Dis 2007;66(3):351–7.
[39] Thompson G, Mclean-Tooke A, Wrobel J, Lavender M, Lucas M. Sjögren syndrome with associated lymphocytic interstitial pneumonia successfully treated with tacrolimus and abatacept as an alternative to rituximab. Chest 2018;153(3):e41–3.
[40] Yamano Y, Taniguchi H, Kondoh Y, Ando M, Kataoka K, Furukawa T, et al. Multidimensional improvement in connective tissue disease-associated interstitial lung disease: two courses of pulse dose methylprednisolone followed by low-dose prednisone and tacrolimus. Respirology 2018;23(11):1041–8.
[41] Voulgarelis M, Ziakas PD, Papageorgiou A, Baimpa E, Tzioufas AG, Moutsopoulos HM. Prognosis and outcome of non-Hodgkin lymphoma in primary Sjogren syndrome. Medicine (Baltimore) 2012;91(1):1–9.

Chapter 9

Diffuse cystic lung disease

Alexander Averyanov[a,b], Evgeniya Kogan[c], Victor Lesnyak[d], Olesya Danilevskaya[e], Igor E. Stepanyan[f]

[a]*Clinical Department, Pulmonology Research Institute under FMBA of Russia, Moscow, Russia,* [b]*Pulmonary Division, Federal Research Clinical Center under FMBA of Russia, Moscow, Russia,* [c]*Anatomic Pathology Department, Sechenov University, Moscow, Russia,* [d]*Radiology Department, Federal Research Clinical Center under FMBA of Russia, Moscow, Russia,* [e]*Endoscopy Department, Pulmonology Research Institute under FMBA of Russia, Moscow, Russia,* [f]*Central TB Research Institute, Moscow, Russia*

Diffuse cystic lung diseases are a group of heterogeneous diseases in which the leading HRCT and morphological feature are the formation of multiple cysts in the lungs. Cysts are cavities, usually filled with air, clearly delimited from the pulmonary parenchyma by a thin (<2 mm) wall [1]. It is assumed that the leading mechanisms of cyst formation are the check-valve obstruction of the distal airways (follicular bronchiolitis and metastases), ischemic injury of terminal bronchioles with their subsequent dilatation (amyloidosis and lymphocytic pneumonia), and local activation of endogenous proteolytic enzymes such as matrix metalloproteinase and podoplanin with the decrease in antiprotease protection (lymphangioleiomyomatosis (LAM), Langerhans cell histiocytosis (LCH), and Birt-Hogg-Dubé syndrome (BHDS)) [2–4].

In 2015 the experts of American Thoracic Society proposed the following classification scheme for diffuse cystic lung diseases [5] (Table 9.1).

Most of the diseases listed according to this classification are rare or orphan. In many of them, cystic lung disease is not ubiquitous. In patients with LAM and in most cases of LCH and BHDS, the formation of cysts in lung tissue is an indispensable HRCT morphological pattern that is often the basis for the diagnosis.

TABLE 9.1 Classification of diffuse cystic lung diseases [5] (with the permission of American Thoracic Society)

Group	Diseases
Neoplastic	LAM, LCH, non–Langerhans cell histiocytoses, including Erdheim–Chester disease, other primary or metastatic lesions (adenocarcinomas, sarcomas, and leiomyomas)
Genetically determined	Birt-Hogg-Dubé syndrome, Proteus syndrome, Ehlers-Danlos syndrome, Marfan syndrome, neurofibromatosis, congenital bronchopulmonary malformation
Associated with lymphoproliferative disorders	Follicular bronchiolitis, lymphocytic interstitial pneumonia, Sjögren syndrome, amyloidosis, light-chains deposition disease
Infectious	*Pneumocystis jirovecii* pneumonia, staphylococcal infection, recurrent respiratory papillomatosis, paragonimiasis, coccidioidomycosis
Associated with interstitial lung disease	Hypersensitivity pneumonitis, desquamate interstitial pneumonia
Smoking-related	LCH, desquamate interstitial pneumonia, respiratory bronchiolitis
Other/miscellaneous	Posttraumatic pseudocysts, fire-eater's lungs, hyper-IgE syndrome
Cyst mimicking diseases	Pulmonary emphysema, honeycomb lung, bronchiectasis

LAM, lymphangioleiomyomatosis; LCH, Langerhans cell histiocytosis.

Difficult to Diagnose Rare Diffuse Lung Disease. https://doi.org/10.1016/B978-0-12-815375-8.00009-1

References

[1] Hansell DM, Bankier AA, MacMahon H, McLoud TC, Müller NL, Remy J. Fleischner Society: glossary of terms for thoracic imaging. Radiology 2008;246(3):697–722.

[2] Masuzawa M, Mikami T, Numata Y, Tokuyama W, Masuzawa M, Murakumo Y, et al. Association of D2-40 and MMP-1 expression with cyst formation in lung metastatic lesions of cutaneous angiosarcoma on the scalp: immunohistochemical analysis of 23 autopsy cases. Hum Pathol 2013;44(12):2751–9.

[3] Zhe X, Yang Y, Jakkaraju S, Schuger L. Tissue inhibitor of metalloproteinase-3 downregulation in lymphangioleiomyomatosis: potential consequence of abnormal serum response factor expression. Am J Respir Cell Mol Biol 2003;28(4):504–11.

[4] Furuya M, Tanaka R, Koga S, Yatabe Y, Gotoda H, Takagi S, et al. Pulmonary cysts of Birt-Hogg-Dubé syndrome: a clinicopathologic and immunohistochemical study of 9 families. Am J Surg Pathol 2012;36(4):589–600.

[5] Gupta N, Vassallo R, Wikenheiser-Brokamp KA, McCormack FX. Diffuse cystic lung disease. Part I. Am J Respir Crit Care Med 2015;191(12):1354–66.

Chapter 9.1

Pulmonary Langerhans cell histiocytosis

Langerhans cell histiocytosis (LCH) is a rare disease characterized by infiltration of specific dendritic cells (CD1a+) called Langerhans cells into the lung parenchyma [1]. LCH is a variety of histiocytosis, which are a group of diseases caused by the proliferation of monocyte-macrophage and dendritic cell lines. Histiocytoses combine a variety of pathological processes—from fatal neoplasms to single benign focal granulomas. In 1997 the International Society of Histiocytosis proposed a three-group classification scheme for these disorders [1]: Group I is LCH; group II is non-Langerhans cell monocyte-macrophage proliferating histiocytoses (Erdheim-Chester disease and Rozai-Dorfman disease); group III is malignant histiocytoses. Lung involvement is most common in LCH patients.

A disease characterized by local formation of granulomas from Langerhans cells in bones was first described in 1913 by N.I. Taratynov. In 1921 A. Hand, based on the observations of A. Schuller and H.A. Christian and his own experience, described a disease that manifested as hepatosplenomegaly, lymphadenopathy, focal bone lesions, polyuria, and exophthalmos. In 1924 E. Letterer described a fatal disease in a child, manifested by a combination of fever, bilateral otitis media, hepatosplenomegaly, and generalized lymphadenopathy. In 1933 A. Siwe described a 16-month-old girl who presented with fever, hepatosplenomegaly, peripheral blood neutrophilia, and a destructive fibula lesion. The term "histiocytosis" was first used in 1944 by M. Lichtenstein and N. L. Jaffe. In 1953 they suggested combining Hand-Schuller-Christian disease, Abt-Letterer-Siwe syndrome, and eosinophilic granuloma under the general name "histiocytosis X" [2].

The prevalence of LCH is not accurately established, but among patients with diffuse parenchymal lung diseases, it occurs in approximately 4%–5% of cases [3]. Young people (20–40 years old) are more commonly affected [4]. Epidemiological data on the gender distribution of patients are contradictory. Some reported a prevalence of males [5], while others suggested a higher incidence among females [6]. Possible differences are explained by the different prevalence rates of smoking during that time period among American and European females.

The etiology of LCH is not established, but rather is classified as a group of smoking-related interstitial lung diseases, along with desquamate interstitial pneumonia and respiratory bronchiolitis with interstitial lung disease. Some authors also include acute eosinophilic pneumonia in this classification [7]. Indeed, up to 95% of patients with a confirmed diagnosis of LCH are active or former smokers or otherwise have regular contact with tobacco smoke [6, 7]. In cases of multisystem lesions in children, smoking is unlikely to be the only risk factor for LCH; however, smoking certainly worsens existing lung damage. In some patients, smoking cessation reverses pathological changes in the lungs [8]. Nevertheless, high prevalence of smoking habit in contrast with the rarity of LCH gives the reason to suppose that smoking is rather a risk factor for LCH in genetically predisposed persons than the cause of the disease.

Langerhans cells are a subpopulation of CD1a+ dendritic cells that are formed from CD34+ bone marrow stem cells. Their role is to induce the initial antigen-specific immune response. Contact with cigarette smoke launches a chain of local inflammatory reactions and the release of cytokines that in turn recruit Langerhans cells, inducing their accumulation in lung tissue [9].

Langerhans cells are agents of inflammation and fibrosis through the expression of various factors including metalloproteinases and inflammatory mediators. Due to the production of chemoattractants such as IL-4, IL-5, and granulocyte-macrophage colony-stimulating factor, Langerhans cells attract a large number of eosinophilic granulocytes into granulomas [4]. In most patients with systemic LCH and 28% of patients with isolated pulmonary LCH, a proto-oncogenic $BRAF^{V600E}$ mutation is found that is natural for many tumor diseases [10, 11]. The clinical significance of this mutation has not been sufficiently studied. In general, this is a younger group of patients, with a higher risk of disease recurrence. Oftentimes children with $BRAF^{V600E}$ mutation will not respond to first-line therapies (vinblastine and corticosteroids), although this does not affect overall survival [12, 13].

Morphology

LCH is characterized by hyperplasia of Langerhans cells and the development of granulomatosis, interstitial fibrosis, and multiple cavitations [14, 6] (Figs. 9.1.1–9.1.4).

At an early granulomatous stage, clusters of large histiocytes and eosinophils are found in the lungs, resembling granulomas, and located in perivascular and peribronchiolar regions. These cells range in size from a few mm to 15 mm [6]. In this case, bronchiole clearance is sharply narrowed, and in the surrounding lung tissue, there is a picture of microcystic cavities. Granulomatous infiltrates also spread to the small branches of the pulmonary artery located near the bronchioles, contributing to productive vasculitis with hemosiderosis foci in certain vessels [15]. A characteristic morphological sign of such histiocytes is the expression of surface markers of Langerhans cells, revealed either under electron microscopy (Birbeck granules) or during immunohistochemical studies (CD1A [Figs. 9.1.5–9.1.6], langerin, S-100, and HLA-DR) [16]. According to our data, eosinophils in histiocytic infiltrates are found only in approximately half of cases.

The late (fibrous) stage is characterized by the presence of large fibrous nodes located centrilobularly. Cavitation is often detected in the center of the nodes secondary to vasculitis causing ischemic tissue damage. In adjacent lung tissue, multiple cysts develop, as well as peribronchiolar fibrosis, up to the area of honeycombing [14].

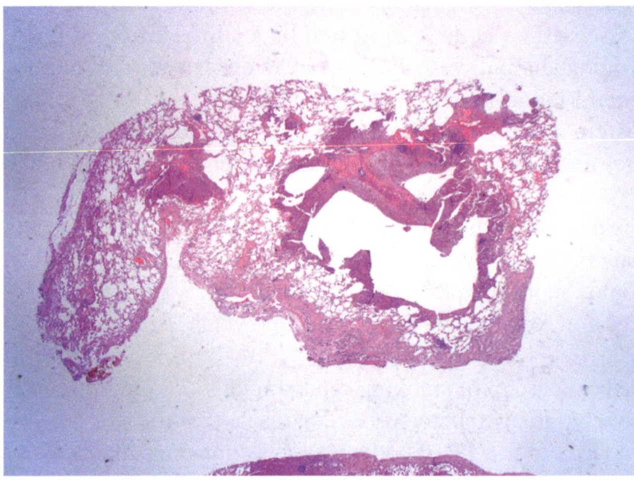

FIG. 9.1.1 LCH. Focal lesion around the bronchioles. The beginning of the formation of cystic cavities. Hematoxylin and eosin (H&E) stain, ×1.25.

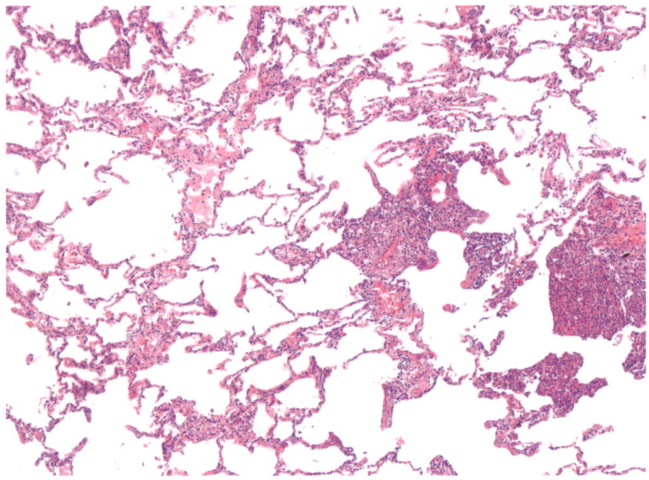

FIG. 9.1.2 LCH. Focal lesion around the bronchioles. Perifocal and obstructive pulmonary emphysema. H&E stain, ×100.

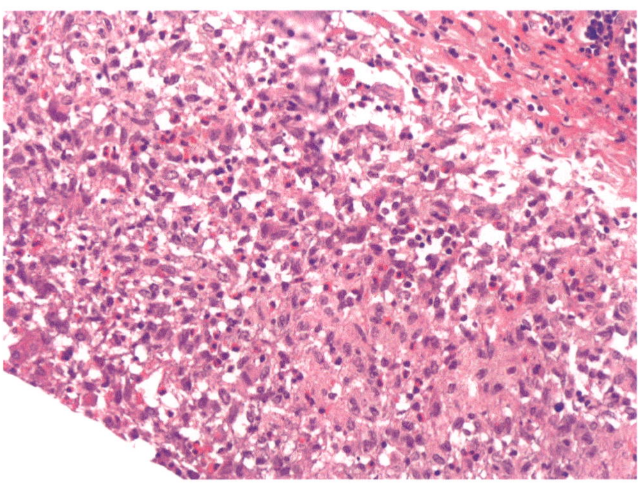

FIG. 9.1.3 LCH. Langerhans cells in the composition of the pulmonary infiltrate into the walls of the bronchiole. H&E stain, ×400.

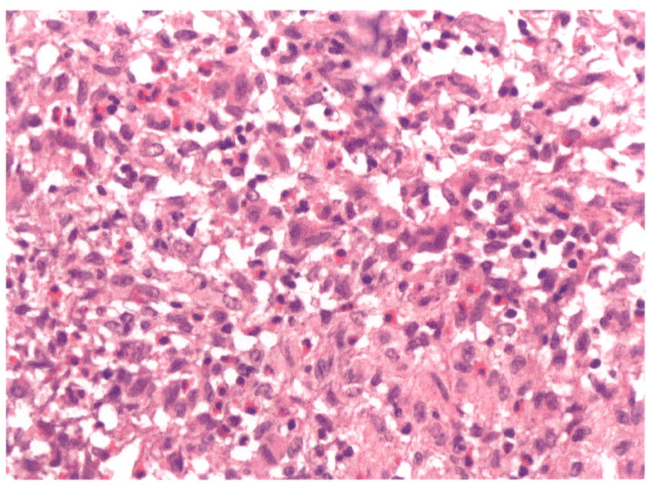

FIG. 9.1.4 LCH. Langerhans cells in the composition of the pulmonary infiltrate into the walls of the bronchiole. H&E stain, ×600.

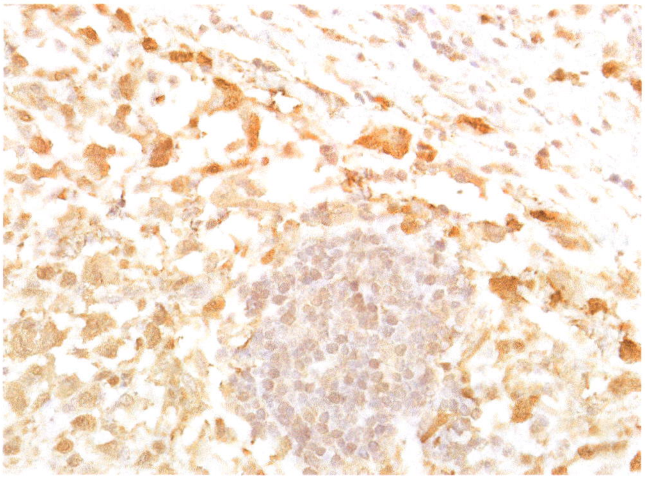

FIG. 9.1.5 CD1a in Langerhans cells in the pulmonary infiltrate into the bronchiole wall and lymphoid follicle that does not contain a marker. Immunohistochemical reaction with diaminobenzidin, ×600.

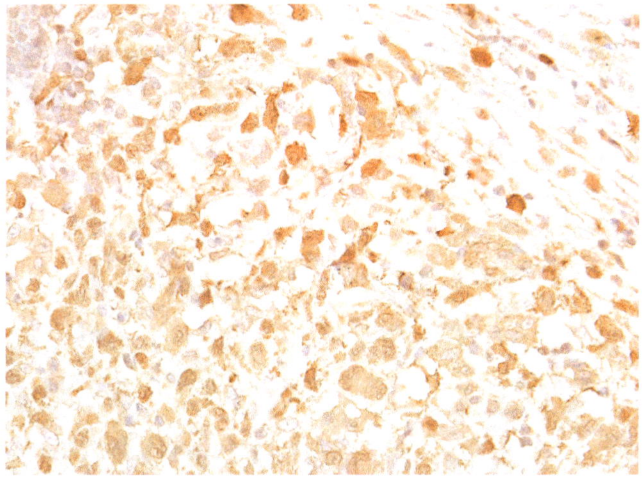

FIG. 9.1.6 CD1a in Langerhans cells in the pulmonary infiltrate. Immunohistochemical reaction with diaminobenzidin, ×600.

The development of bronchial obstruction in LCH can be explained by two groups of factors. First, airflow limitation may be related to the localization of granulomas and infiltrates in the walls of bronchioles. Second the overwhelming majority of LCH patients exhibit smokers' chronic bronchitis, leading to generalized bronchial obstruction.

In patients with LCH, granulomas are localized in the walls of bronchioles. This specific localization occurs because Langerhans cells that form the basis of the granulomatous process in patients with LCH are predominantly located in the epithelial layer of bronchioles. The association of smoking and LCH is also not accidental. Nicotine may trigger autoimmune processes and induce the proliferation of Langerhans cells [17, 18].

Differential diagnosis is based on lung biopsy data of the monoorganic LCH form and should be performed in patients with chronic interstitial pneumonia and pulmonary eosinophilia. In difficult cases, diagnosis should be based on histological and immunohistochemical data, such as positive expression of CD1a S-100 protein cells in LCH. LCH can develop simultaneously with lung carcinomas and *Pneumocystis jirovecii* pneumonia [19].

Differential diagnosis with non-Langerhans cellular and monocyte-macrophage proliferating histiocytoses (Erdheim-Chester disease and Rosai-Dorfman disease) is based on the characteristics of clinical multisystemic manifestations. In rare lung lesions, histiocytic proliferates express markers of monocytic and macrophage cells, such as CD68+, and in some cases even S-100, but do not contain CD1A and langerin [20, 21].

In malignant histiocytoses such as Langerhans cell sarcoma and malignant histiocytosarcoma, expressions of S-100, CD1A, and langerin are preserved in tumor cells, but there are all signs of cellular atypism with high mitotic activity. These conditions are highly invasive and characterized by infiltrating growth and metastatic lesions [22].

Clinical presentation

Symptoms of LCH can be divided into general, respiratory, and extrapulmonary. General symptoms (malaise, sweating, fatigue subfebrile body temperature and weight loss) are observed in approximately 20% of patients. Respiratory symptoms include cough, shortness of breath, chest pain (infrequently), and occasional hemoptysis, whereas extrapulmonary symptoms include pain, polyuria, and polydipsia due to lesions of the pituitary gland, flat bones, skin, soft tissues, peripheral lymph nodes, etc. Extrapulmonary lesions occur in 10%–15% of patients with LCH [23]. Approximately 15% of patients develop spontaneous pneumothorax [24], and in 25% of patients the disease is asymptomatic until the time of detection [4, 25]. Physical examination usually does not reveal significant changes. Skin cyanosis is observed in cases of severe respiratory failure. On auscultation, it is often not possible to discover specific symptoms, although some patients exhibit rales due to the concomitant occurrence of smoker bronchitis. LCH patients exhibit higher incidence of pulmonary hypertension than patients with other interstitial lung diseases due to arterial vessel damage [26].

Laboratory analysis. Changes in blood tests are nonspecific, or absent.

Pulmonary function testing usually reveals airflow limitation or mixed restrictive and obstructive changes in pulmonary ventilation. In 60%–90% of patients, lung diffusion capacity is impaired [27]. In half of patients, obstructive disorders tend to progress. This progression is more dramatic than that observed in patients with COPD and can involve the small bronchi in the underlying pathological process, in contrast to smoker bronchitis. A 2-year follow-up of treated LCH patients showed that 60% of patients exhibited declines in pulmonary function and only 20% had improvement in FEV_1 and DLCO [28].

Radiological imaging

Chest X-rays usually reveal bilateral disseminated changes in the form of small nodules and/or enhancement of the pulmonary pattern, mainly in the upper and middle lung sections (Fig. 9.1.7); small cavities and bullae of various sizes can be seen.

HRCT is the most important method for achieving accurate diagnosis.

The main HRCT signs of LCH (Fig. 9.1.8) include [29, 30]

- multiple small nodules that are 1–5 mm in diameter,
- thin-walled or thick-walled cysts,
- a predominant distribution of pathological changes in the upper and middle zones of the lungs,
- spared costophrenic angles,
- moderate reticular changes around nodules and cysts.

HRCT changes reflect the pathomorphological dynamics of LCH. The first findings are usually multiple small nodules, rarely exceeding 5 mm in size, located in centrilobular or peribronchiolar areas. At the beginning of the process, the nodules are homogeneous, often with irregular and sharp contours (Figs. 9.1.9A and 9.1.10). As the disease develops, they can enlarge. In the center of nodules greater than 5 mm in diameter, lucent foci often appear. These reflect the formation of cysts or zones of dilated bronchioles when they are surrounded by granulomas (Fig. 9.1.9B) [31]. The nodules occur in 80%–90% of patients with LCH and may be the only HRCT sign in those with disease durations of up to 6 months [31]. At this stage the onset of LCH can be very difficult to differentiate radiologically from sarcoidosis, disseminated tuberculosis, or lymphogenous metastases in the lungs. A study found that 47% of 51 LCH patients exhibited nodules with diameters that did not exceed 3 mm, and in 45% of patients the diameter ranged from 3 to 1 cm [32]. As the disease progresses the number of nodules decreases, first replaced by thick-walled small and then thin-walled larger cysts (Fig. 9.1.11) [33].

Cyst appearance is an important CT sign of LCH, occurring in 90%–100% of patients [30, 31]. Cysts with a wall thickness of 2 mm or less are classified as thin-walled and the remaining regarded as thick-walled [32]. Since there are few CT studies of patients with LCH and existing studies include patients with varying disease durations, the frequency of occurrence of thick-walled and thin-walled cysts differs accordingly. Brauner [31] found thick-walled cavities in 39% of patients; Grenier [32] found them in 53%, and Kim [30] observed them in 82%. Thin-walled cysts are a more frequent finding (>80% of cases, according to all authors). Often the cysts have irregular, bizarre forms and are grouped together (Fig. 9.1.12). At earlier stages, their size does not exceed 10 mm in diameter, but over time, especially with the addition of mechanical factors (valve mechanisms), a bullous transformation can occur, and the cysts become difficult to distinguish from diffuse emphysema (Fig. 9.1.12) [34]. In some patients, if smoking cessation is successful, the cysts can completely disappear, with restoration of the normal structure of the parenchyma [30]. However, in general, the evolution of nodules into thick-walled and then thin-walled cysts with subsequent generalized cystic degeneration of lung tissue is quite typical and described by many authors. At the final stage of cystic lung disease, it is not easy to differentiate LCH from LAM and diffuse emphysema (Figs. 9.1.12 and 9.1.13).

Moderate reticular changes in the form of thickening of intralobular septa and linear fibrosis are observed in half of the advanced stage patients, but are not specific criteria for radiographic diagnosis. Ground-glass opacities are possible, but not obligatory, feature (Fig. 9.1.14) that usually manifest around the zones of greatest damage [33].

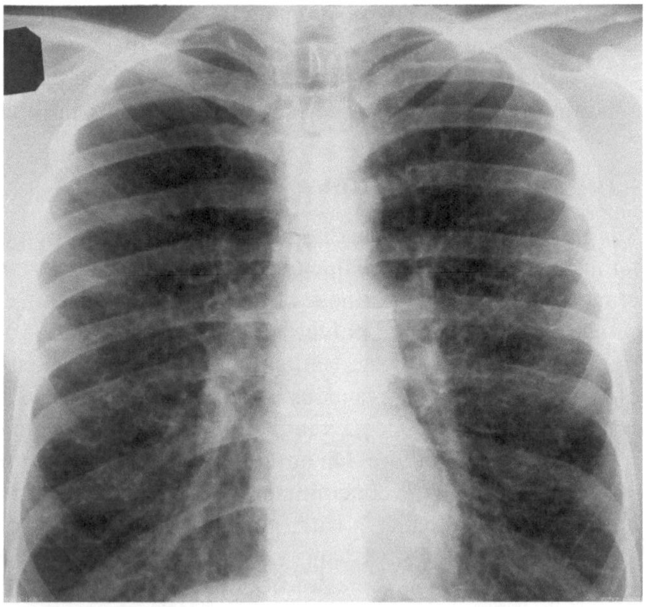

FIG. 9.1.7 Radiographic image in a patient with LCH. Diffuse reticular abnormalities in both lungs. Some small thin-walled cavities are visible.

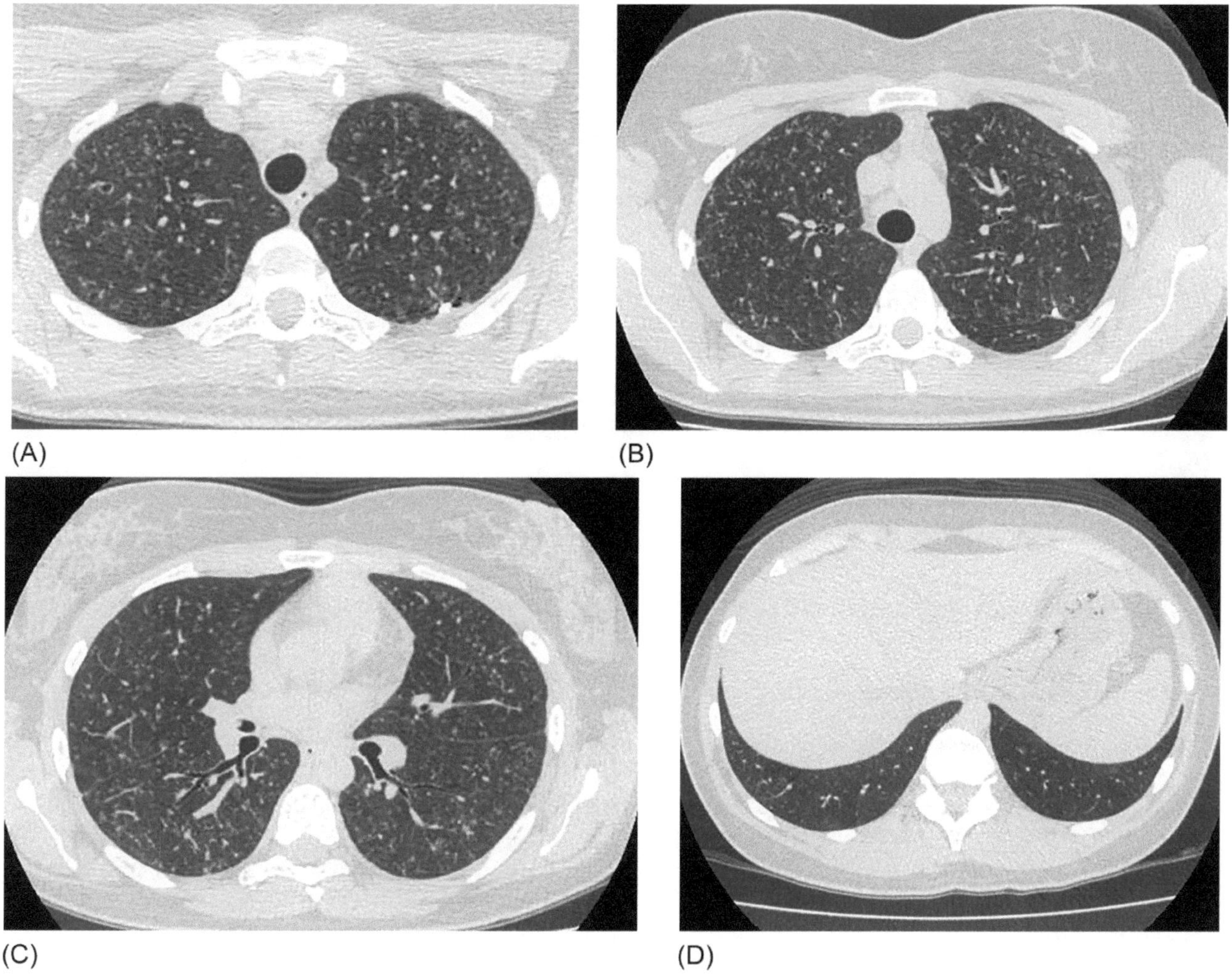

FIG. 9.1.8 Early manifestations of LCH. Multiple small centrilobular nodules, randomly distributed in the lung tissue. Single thin-walled cavities in the upper lobes (arrows). Mild thickening of intralobular septa (A–C). The pulmonary tissue of the costophrenic angles is typically not involved in the pathological processes (D).

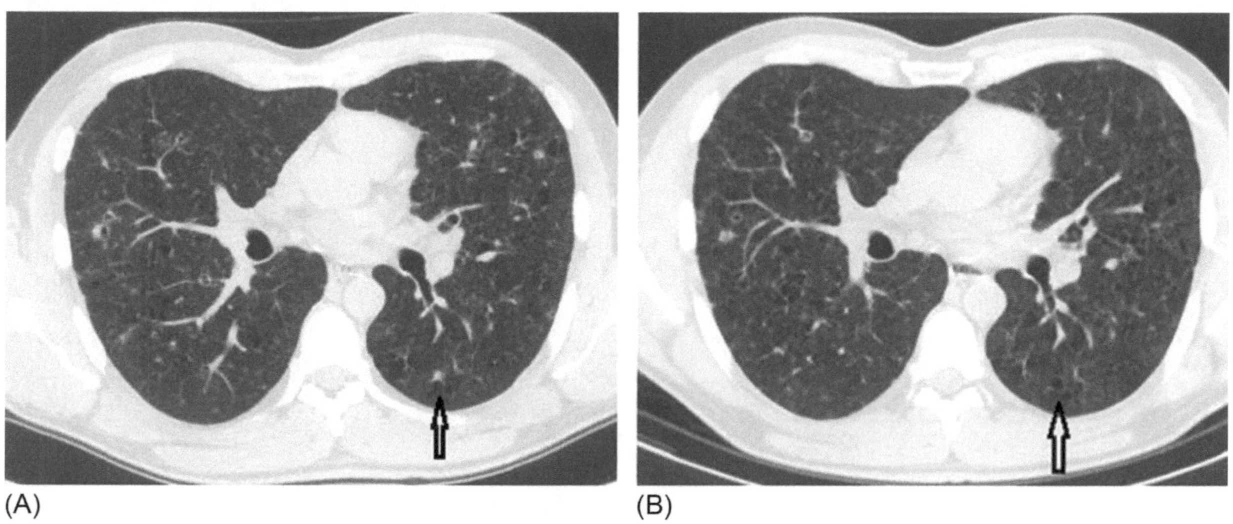

FIG. 9.1.9 LCH. (A) Multiple interstitial nodules and small thin-walled cavities. (B) HRCT scan dynamics over six months: the cavity (arrows) is visualized at the site of the focus.

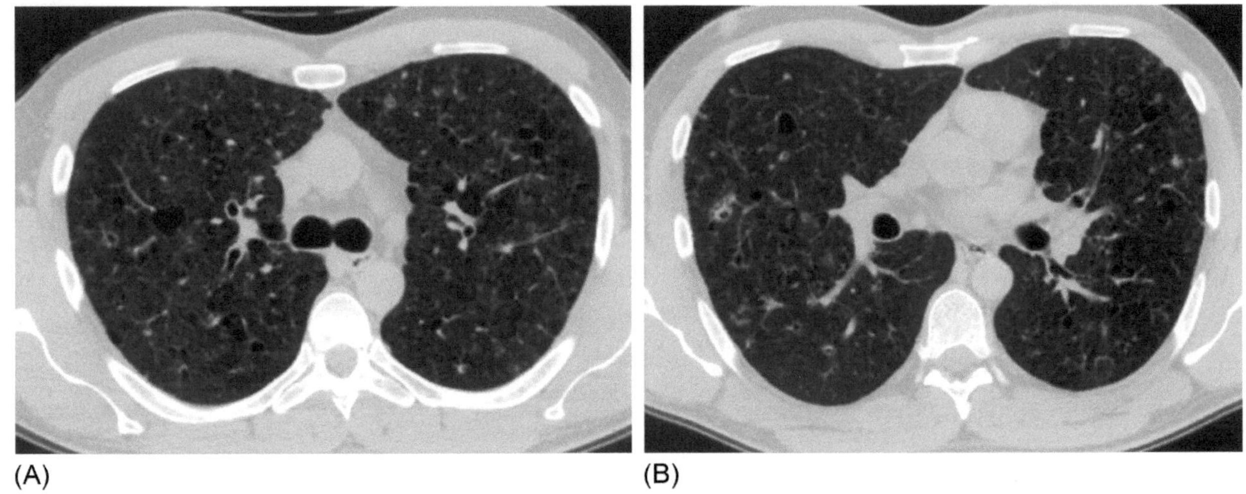

FIG. 9.1.10 LCH. Axial scans at the level of the carina (A) and pulmonary arteries (B). Single small homogeneous nodules. Multiple thin-walled cavities sized from 3 to 15 mm. Nodules in the stage of cavity formation.

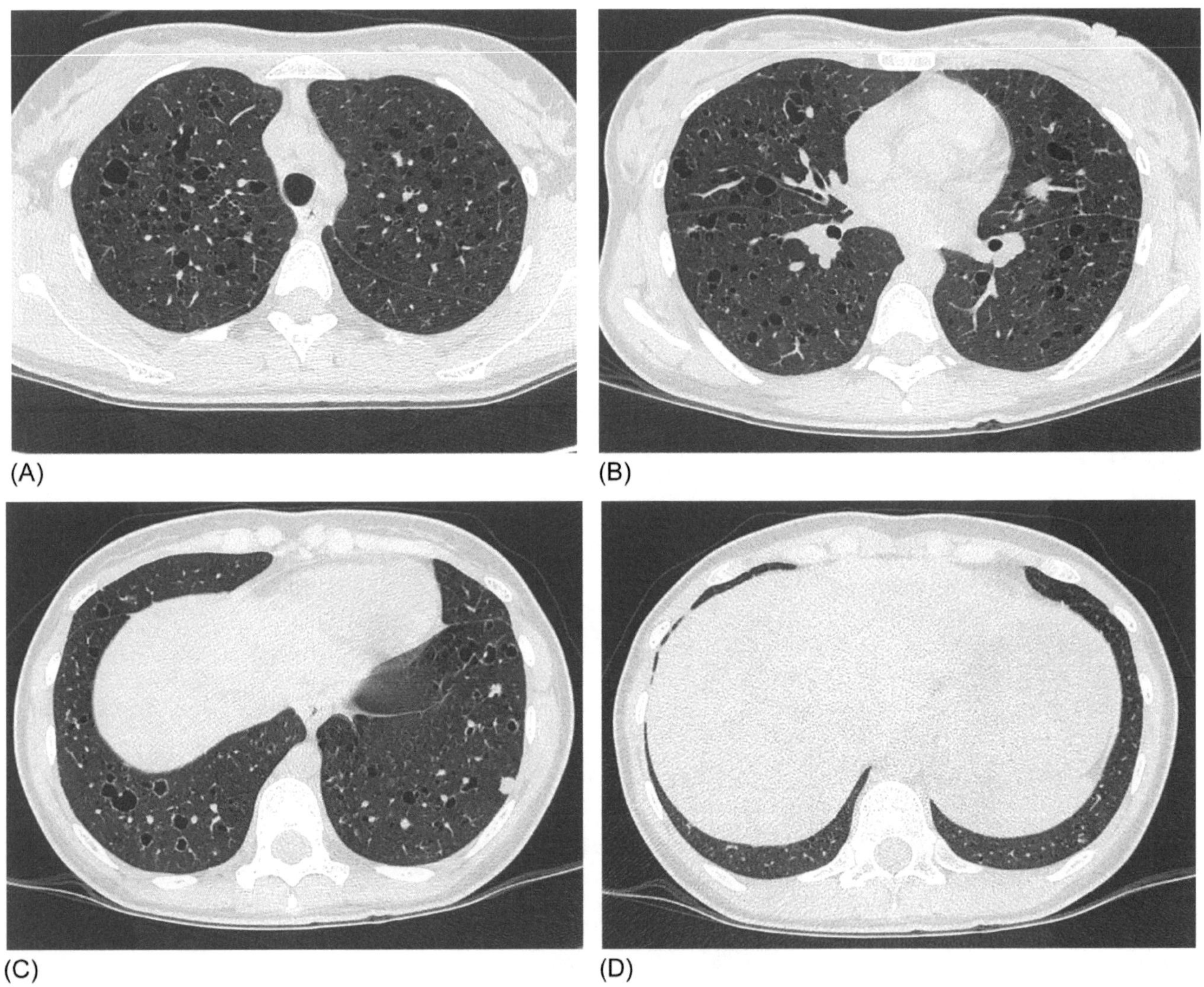

FIG. 9.1.11 The classical pattern in the full-scaled stage of LCH: multiple thin- and thick-walled cysts, mainly in the upper parts of the lungs (A, B). Their number decreases toward the basal segments (C). Minimal changes in the costophrenic zones (D). Single noncavitary nodules.

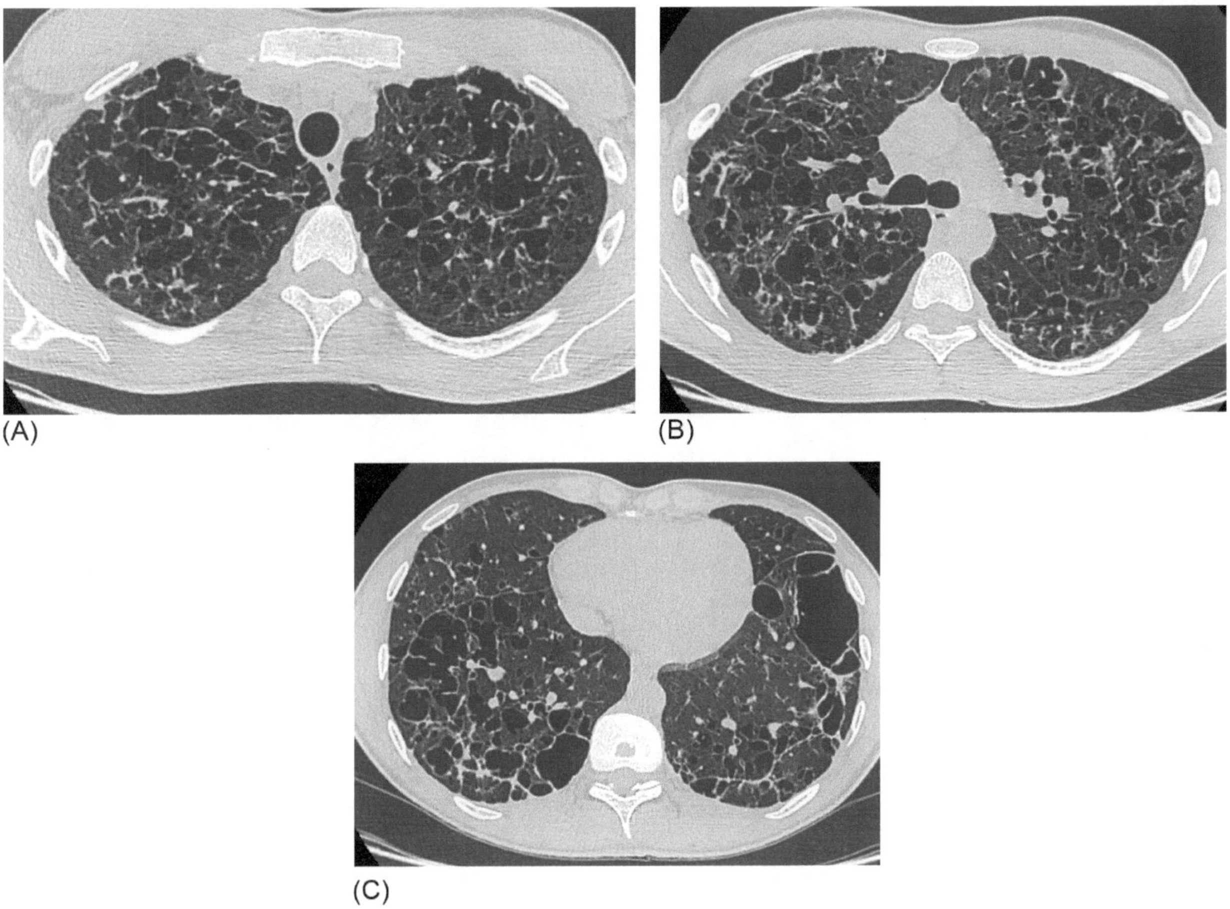

FIG. 9.1.12 Late-stage LCH: Extensive cystic replacement of lung parenchyma. Large cavities form secondary to the confluence of smaller adjacent cavities. These cavities often exhibit an elongated, irregular shape and are indistinguishable from diffuse emphysema (A, B); some cavitary formations correspond to the borders of the pulmonary lobules (B); in the costophrenic angles, changes are minimal (C).

It should be noted that the distinctive localization of pathological changes to the upper and middle parts of the lungs, with sparing of basal sections (especially in the costophrenic angles), affects only adult patients. In children the same pattern is distributed evenly on CT, leaving no lesion-free areas [35].

Patients with LCH often develop pulmonary vascular lesions, leading to severe pulmonary hypertension (Fig. 9.1.15). The disease course is often very malignant, which determines the prognosis of the disease, and does not always correlate with the severity of pulmonary functional disorders. Patients with severe LCH, in addition to arterial involvement, are also predisposed to pulmonary vein occlusions [36].

To clarify extrapulmonary LCH lesions (in the presence of symptoms), bone radiography or MRI of the brain is performed (with signs of diabetes insipidus). Positron emission tomography with ^{18}F-fludeoxyglucose can reveal systemic lesion foci in addition in 50% of patients with LCH and can be used to monitor the efficacy of therapy for extrapulmonary manifestations of the disease, especially within the bones. However, for primary diagnosis of LCH, this method does not have sufficient specificity, although it has a high sensitivity. [37].

Bronchoalveolar lavage and lung biopsy

Bronchoscopy should be performed in patients with suspected LCH to analyze BAL fluid and for pulmonary biopsy. Cytological analysis of the BAL can reveal an increased neutrophils and eosinophils, with a normal level of lymphocytes. Also, with LCH, an increased number of alveolar macrophages are frequently detected, including hyperpigmented varieties, which is believed to reflect a history of smoking [38]. Increases in CD1a+ cells within the BAL that are greater than 5% are revealed in one-quarter of LCH patients [39]. Antibodies to langerin (CD207) are potentially useful markers for diagnosing LCH via BAL. Smetana indicated significantly greater amounts of these antibodies in patients with LCH, compared with those with sarcoidosis or idiopathic pulmonary fibrosis [40]. In our opinion, analysis of BAL fluid in

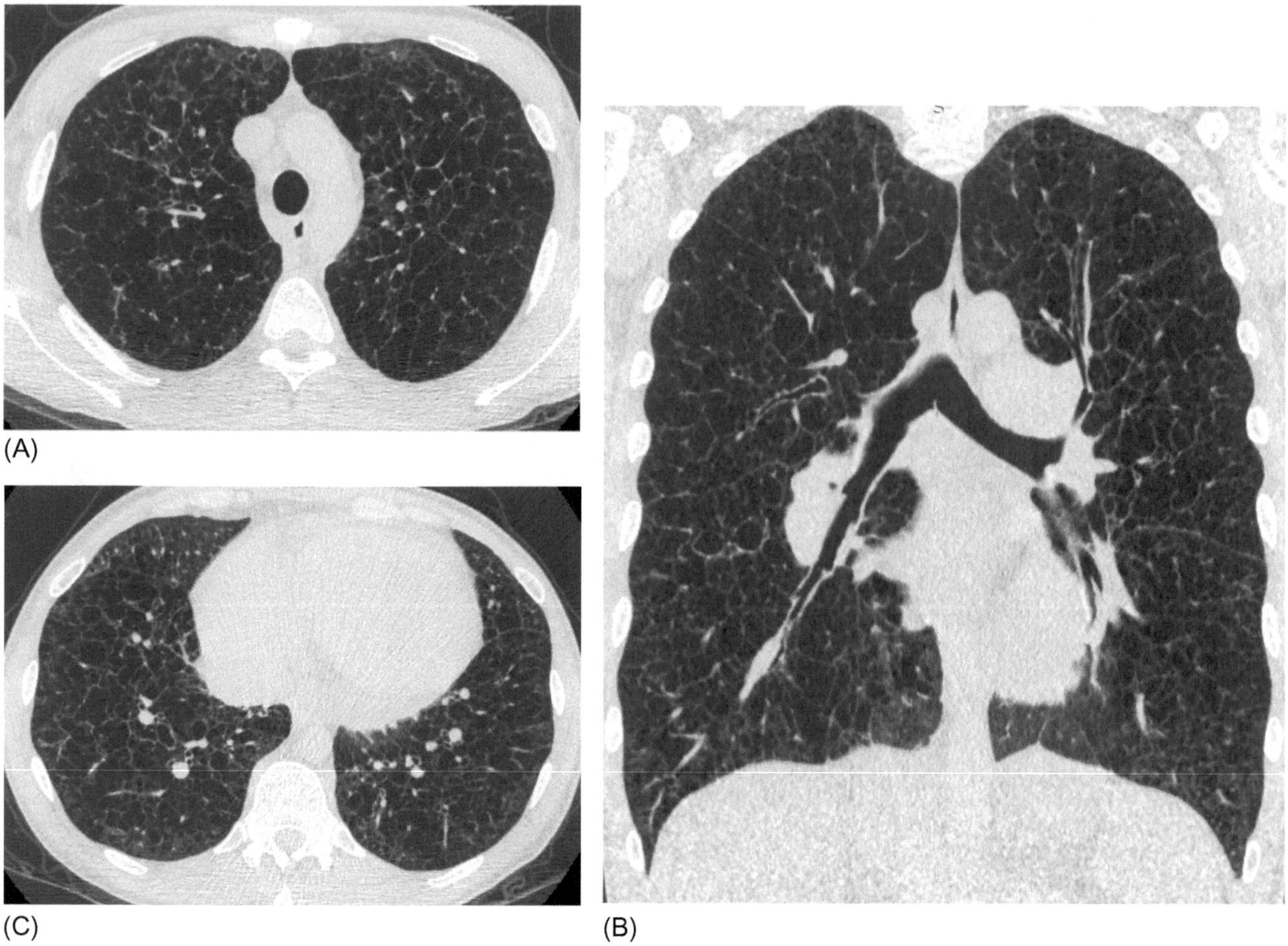

FIG. 9.1.13 The 25 years old patient with terminal-stage LCH. Diffuse cystic transformation of lung tissue (A–C).

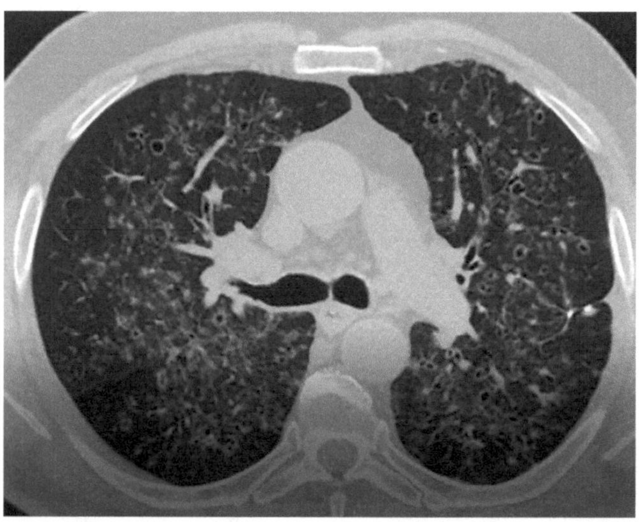

FIG. 9.1.14 LCH. Multiple, predominantly thick-walled small cavities; intralobular nodules; and diffuse areas of ground-glass opacity between nodular and cystic changes.

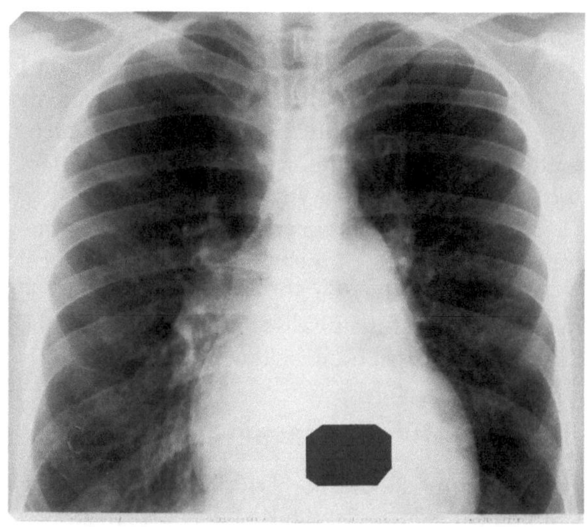

FIG. 9.1.15 X-ray image of the chest in a patient with LCH. Note the heart configuration changes, due to the development of severe pulmonary hypertension. Specifically, bulging of the arch of the right atrium, the vascular arch due to the superior vena cava, the pulmonary artery trunk, and the left ventricle arch due to its displacement by the right ventricle.

patients with suspected LCH should also include a PCR study for tuberculosis and nontuberculosis mycobacteria, as well as *Pneumocystis jirovecii* pneumonia to exclude these diseases from the differential LCH series.

The diagnostic value of transbronchial lung biopsy for LCH is rather low (10%–40%), unlike LAM, where in most cases the morphological material is sufficient for diagnosis [41]. Biopsy with video-assisted thoracoscopy (VAT) is a potential alternative to transbronchial lung biopsy for patients with LCH. However, the recent introduction of cryobiopsy technology has proved highly efficacious, permitting diagnosticians to obtain a sufficient amount of diagnostic material. At the same time the incidence of pneumothorax (6.8%) was comparable with that observed with transbronchial forceps lung biopsy [42].

Probe-based confocal laser endomicroscopy

In two patients with LCH, 35 bronchopulmonary regions were studied, 175 informative images were obtained, and 1223 measurements were performed. The elastic frame could be preserved (Fig. 9.1.16E), or pathological changes in the structure of the acinus (Fig. 9.1.16A–D and F) were observed. In 31 out of 35 regions, the presence of alveolar macrophages in the lumen of the alveoli (smoker's cells and activated macrophages), including gigantic ones, was established (Fig. 9.1.16E and F). The number of cells ranged from single (one point) to occupying the entire field of view (five points). The elastic fibers of the interalveolar septa in 16 out of 35 regions looked "flabby" and thinned (Fig. 9.1.16A). The presence of cystic changes was noted in 5 of the 35 regions (Fig. 9.1.16C). In addition, in 2 out of 35 surveyed regions, single large (up to 232.2 μm) structureless brightly fluorescent complexes were recorded, similar to those in alveolar proteinosis (Fig. 9.1.16D).

Diagnosis

The diagnosis of LCH is established on the basis of characteristic clinical manifestations such as slow progressive dyspnea, spontaneous pneumothorax, diabetes insipidus in young active smokers, changes in the lungs as evidenced on HRCT, or extrapulmonary manifestations of the disease such as flat bone defects. Diagnosis often requires confirmation by cryobiopsy or VAT and subsequent histological and immunohistochemical (for CD1a, langerin, and protein S100) biopsy studies. However, in cases with a classic clinical and radiological presentation, where there is no doubt about the diagnosis, morphological verification is not obligatory [43]. LCH at the stage of isolated nodular changes must be differentiated from lung sarcoidosis, tuberculosis, silicosis and other pneumoconiosis, metastatic lesions, and sometimes HP. If cysts are observed in the lungs, a differential diagnosis should be made with other diffuse cystic and cystic-like lung diseases (LAM, Birt-Hogg-Dubé syndrome, infectious lesions, etc.), as described in Chapter 9.4.

Treatment and prognosis

Smoking cessation is a primary treatment for patients with LCH. Despite the fact that not all patients will exhibit spontaneous remission, the obvious relationship between smoking and disease cannot be ignored. The decision to initiate drug therapy depends on symptom severity, signs of multisystemic lesions, and speed of disease progression. If respiratory failure is minimal, signs of systemicity are absent, no pulmonary function impairment is observed over 3–6 months, the patient quits smoking, and dynamic observation (the "wait and see" approach) is possible, given the hope for stability or regression of pathological manifestations. Predictors of favorable and unfavorable LCH courses (with the exception of pulmonary hypertension) have not been determined to date [43]. Systemic glucocorticosteroids were considered first-line drugs in patients with progressive LCH, at a starting dose of 0.5–1 mg/kg of body weight per day and followed by a slow dose reduction [44]. However, the indications for this approach should be carefully considered, since the effect is not observed in all patients. For more severe cases of lung damage, a combination of corticosteroids and immunosuppressants (cyclophosphamide, vinblastine, and methotrexate) is used. There was renewed hope for helping patients with severe cystic lung lesions following dissemination of the first results from a study that used a purine nucleoside analogue of cladribine. In a report on the treatment of three patients with subcutaneous administration of the drug at a dose of 0.1 mg/kg/day for 5 days of each month, Gwenael Lorillon et al. noted that, in all cases, there was a significant improvement in clinical status and pulmonary function, including diffusion capacity of the lungs, decreased zones of cystic abnormalities [45]. Another phase 2 study combined cladribine and cytarabine in patients with multisystemic life-threatening LCH who were refractory to glucocorticosteroids and vinblastine therapy. The authors showed a positive effect in 92% of cases and a 5-year survival of 85% but with a large number of toxic side effects [46].

In patients with the BRAF mutation, targeted therapy with a BRAF vemurafenib inhibitor appears promising. However, no randomized control trial data exist; rather we have only descriptions of individual cases of successful treatment, mainly in those with non-Langerhans histiocytosis or a combination thereof [47].

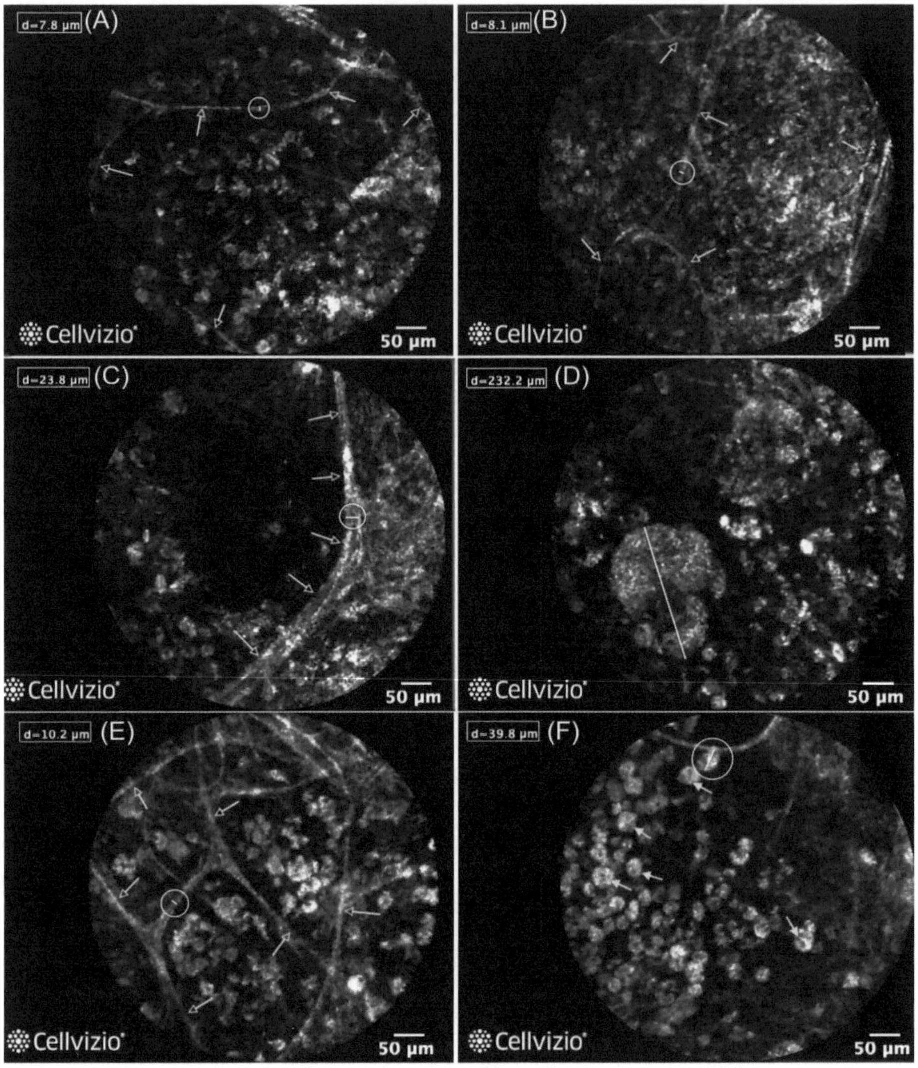

FIG. 9.1.16 pCLE pattern in patients with LCH. (A) The elastic fibers of the alveolar walls *(arrows)* are thinned, and their tension is reduced. Consequently the rounded shape of the alveolar cavities is lost. (B) Complete filling of the lumen of the alveoli with cellular and noncellular fluorescent elements (five points) and the interalveolar septa *(arrows)* are thinned. Such an endomicroscopic picture may correspond to the granuloma. (C) Due to a significant increase in the size of the alveolar cavities, their structure is only partially visualized (*arrows* indicate the thickened wall of the bulla). The diameter of the field of view is $600 \times 500\,\mu m$ or $0.28\,mm^2$; it is impossible to establish the exact size of the cavity during alveoscopy, since it occupies several fields of vision. (D) In the lumen of the alveoli on the background of big amount of alveolar macrophages, structureless irregularly shaped conglomerates with high autofluorescence are visible. (E) Normal acini structure (interalveolar septa indicated by *arrows*) with high amount of alveolar macrophages. (F) Almost half of the field of view is occupied by alveolar macrophages, many of which are giant (shown by *arrows*). Due to the large number of cellular elements, the alveolar structures are nearly impossible to visualize. Measurement areas are marked with *ellipses*.

Severe pulmonary hypertension usually requires the prescription of long time oxygen therapy and inhibitors of phosphodiesterase (sildenafil) and/or endothelin receptor blockers [43]. In cases of bronchial obstruction, bronchodilators are indicated. Patients with confirmed LCH, generalized changes in the lungs, and progressive respiratory failure are candidates for lung transplantation.

In general the LCH course is difficult to predict. A 6-year follow-up of 45 patients with LCH revealed that 27% of them died or required lung transplantation [5]. The average life expectancy after diagnosis was 13 years, according to the follow-up of 102 patients, and a third of patients developed respiratory failure [23].

In adults, a relatively favorable LCH course is possible. Cases of long-term stabilization and spontaneous regression are described. The main causes of death in patients with LCH are respiratory failure, pulmonary hypertension, and tumor diseases (hemoblastoses, lymphomas, and lung cancer). The markers of unfavorable prognosis in LCH patients are pulmonary hypertension, recurrent spontaneous pneumothorax, diabetes insipidus, childhood, and long-term use of steroid drugs, although these factors reflect the severity of clinical manifestations, rather than serve as reliable early predictors [7, 48].

References

[1] Favara BE, Feller AC, Pauli M, Jaffe ES, Weiss LM, Arico M, et al. Contemporary classification of histiocytic disorders. The WHO Committee on Histiocytic/Reticulum Cell Proliferations. Reclassification Working Group of the Histiocyte Society. Med Pediatr Oncol 1997;29(3):157–66.

[2] Komp DM. Historical perspectives of Langerhans cell histiocytosis. Hematol Oncol Clin North Am 1987;1(1):9–21.

[3] Gaensler EA, Carrington CB. Open biopsy for chronic diffuse infiltrative lung disease: clinical, roentgenographic, and physiological correlations in 502 patients. Ann Thorac Surg 1980;30(5):411–26.

[4] Tazi A, Soler P, Hance AJ. Adult pulmonary Langerhans cell histiocytosis. Thorax 2000;55(5):405–16.

[5] Delobbe A, Durieu J, Duhamel A, Wallaert B. Determinants of survival in pulmonary Langerhans' cell granulomatosis (histiocytosis X). Eur Respir J 1996;9(10):2002–6.

[6] Travis WD, Borok Z, Roum JH, Zhang J, Feuerstein I, Ferrans VJ, et al. Pulmonary Langerhans cell granulomatosis (histiocytosis X). A clinicopathologic study of 48 cases. Am J Surg Pathol 1993;17(10):971–86.

[7] Vassalo R, Ryu JH. Smoking-related interstitial lung diseases. Clin Chest Med 2012;33(1):165–78.

[8] Mogulkoc N, Veral A, Bishop PW, Bayindir U, Pickering CA, Egan JJ. Pulmonary Langerhans' cell histiocytosis: radiologic resolution following smoking cessation. Chest 1999;115(5):1452–5.

[9] Casolaro MA, Bernaudin JF, Saltini C, Ferrans VJ, Crystal RG. Accumulation of Langerhans' cells on the epithelial surface of the lower respiratory tract in normal subjects in association with cigarette smoking. Am Rev Respir Dis 1988;137(2):406–11.

[10] Sahm F, Capper D, Preusser M, Meyer J, Stenzinger A, Lasitschka F, et al. BRAFV600E mutant protein is expressed in cells of variable maturation in Langerhans cell histiocytosis. Blood 2012;120(12):e28–34.

[11] Hervier B, Haroche J, Arnaud L, Charlotte F, Donadieu J, Néel A, et al. Association of both Langerhans cell histiocytosis and Erdheim-Chester disease linked to the BRAFV600E mutation. Blood 2014;124(7):1119–26.

[12] Berres ML, Lim KP, Peters T, Price J, Takizawa H, Salmon H, et al. BRAF-V600E expression in precursor versus differentiated dendritic cells defines clinically distinct LCH risk groups. J Exp Med 2015;212(2):281.

[13] Héritier S, Emile JF, Barkaoui MA, Thomas C, Fraitag S, Boudjemaa S, et al. BRAF mutation correlates with high-risk Langerhans cell histiocytosis and increased resistance to first-line therapy. J Clin Oncol 2016;34(25):3023–30.

[14] Colby TV, Lombard C. Histiocytosis X in the lung. Hum Pathol 1983;14(10):847–56.

[15] Soler P, Kambouchner M, Valeyre D, Hance AJ. Pulmonary Langerhans cell granulomatosis (histiocytosis X). Annu Rev Med 1992;43:105–15.

[16] Tazi A, Bonay M, Grandsaigne M, Battesti JP, Hance AJ, Soler P. Surface phenotype of Langerhans cells and lymphocytes in granulomatous lesions from patients with pulmonary histiocytosis X. Am Rev Respir Dis 1993;147(6 Pt 1):1531–6.

[17] Yang SR, Chida AS, Bauter MR, Shafiq N, Seweryniak K, Maggirwar SB, et al. Cigarette smoke induces proinflammatory cytokine release by activation of NF-kappaB and posttranslational modifications of histone deacetylase in macrophages. Am J Physiol Lung Cell Mol Physiol 2006;291(1):L46–57.

[18] Caux C, Dezutter-Dambuyant C, Schmitt D, Banchereau J. GM-CSF and TNF-alpha cooperate in the generation of dendritic Langerhans cells. Nature 1992;360(6401):258–61.

[19] Ph H, Flieder DB, editors. Spencer's pathology of the lung. 6th ed. New York: Cambridge University Press; 2016.

[20] Chung JH, Park MS, Shin DH, Choe KO, Kim SK, Chang J, et al. Pulmonary involvement with Erdhiem-Chester disease. Respirology 2005;10(3):389–92.

[21] Ben Ghorbti I, Naffati H, Khanfir M, Kchir MN, Mrad K, Ben Romdhane K, et al. Disseminated form of Rossai-Dorfmann disease. A case report. Res Med Interne 2005;26(5):415–9.

[22] Pileri SA, Grogan TM, Harris NL, Banks P, Campo E, Chan JK, et al. Tumours of hisitiocytes and accessory dendritic cells: an immunohistochemical approach to classification from the international Lymphoma Study Group based on 61 cases. Histopathology 2002;41(1):1–29.

[23] Vassallo R, Ryu JH, Schroeder DR, Decker PA, Limper AH. Clinical outcomes of pulmonary Langerhans' cell histiocytosis in adults. N Engl J Med 2002;346(7):484–90.

[24] Lacronique J, Roth C, Battesti JP, Basset F, Chretien J. Chest radiological features of pulmonary histiocytosis X: a report based on 50 adult cases. Thorax 1982;37(2):104–9.

[25] Vassallo R, Ryu JH, Colby TV, Hartman T, Limper AH. Pulmonary Langerhans'-cell histiocytosis. N Engl J Med 2000;342(26):1969–78.

[26] Fartoukh M, Humbert M, Capron F, Maître S, Parent F, Le Gall C, et al. Severe pulmonary hypertension in histiocytosis X. Am J Respir Crit Care Med 2000;161(1):216–23.

[27] Crausman RS, Jennings CA, Tuder RM, Ackerson LM, Irvin CG, King Jr TE. Pulmonary histiocytosis X: pulmonary function and exercise pathophysiology. Am J Respir Crit Care Med 1996;153(1):426–35.

[28] Tazi A, Marc K, Dominique S, de Bazelaire C, Crestani B, Chinet T, et al. Serial CT and lung function testing in pulmonary Langerhans' cell histiocytosis. Eur Respir J 2012;40(4):905–12.

[29] Brauner MW, Grenier P, Tijani K, Battesti JP, Valeyre D. Pulmonary Langerhans cell histiocytosis: evolution of lesions on CT scans. Radiology 1997;204(2):497–502.

[30] Kim HJ, Lee KS, Johkoh T, Tomiyama N, Lee HY, Han J, et al. Pulmonary Langerhans cell histiocytosis in adults: high-resolution CT-pathology comparisons and evolutional changes at CT. Eur Radiol 2011;21(7):1406–15.

[31] Brauner MW, Grenier P, Mouelhi MM, et al. Pulmonary histiocytosis X: evaluation with high-resolution CT. Radiology 1989;172(1):255–8.

[32] Grenier P, Valeyre D, Cluzel P, Brauner MW, Lenoir S, Chastang C. Chronic diffuse interstitial lung disease: diagnostic value of chest radiography and high-resolution CT. Radiology 1991;179(1):123–32.

[33] Webb WR, Muller NL, Naidich DP. High-resolution CT of the lung. 4th ed. Philadelphia: Lippincott Williams and Wilkins; 2009. p. 368–81.

[34] Nair A, Hansell DM. High-resolution computed tomography features of smoking-related interstitial lung disease. Semin Ultrasound CT MR 2014;35(1):59–71.

[35] Seely JM, Salahudeen Sr. S, Cadaval-Goncalves AT, Jamieson DH, Dennie CJ, Matzinger FR, et al. Pulmonary Langerhans cell histiocytosis: a comparative study of computed tomography in children and adults. J Thorac Imaging 2012;27(1):65–70.

[36] Hamada K, Teramoto S, Narita N, Yamada E, Teramoto K, Kobzik L. Pulmonary veno-occlusive disease in pulmonary Langerhans cell granulomatosis. Eur Respir J 2000;15(2):421–3.

[37] Obert J, Vercellino L, van Der Gucht A, de Margerie-Mellon C, Bugnet E, Chevret S, et al. ^{18}F-fluorodeoxyglucose positron emission tomography-computed tomography in the management of adult multisystem Langerhans cell histiocytosis. Eur J Nucl Med Mol Imaging 2017;44(4):598–610.

[38] Hance AJ, Basset F, Saumon G, Danel C, Valeyre D, Battesti JP, et al. Smoking and interstitial lung disease. The effect of cigarette smoking on the incidence of pulmonary histiocytosis X and sarcoidosis. Ann N Y Acad Sci 1986;465:643–56.

[39] Torre O, Harari S. The diagnosis of cystic lung diseases: a role for bronchoalveolar lavage and transbronchial biopsy? Respir Med 2010;104(Suppl 1): S81–5.

[40] Smetana K, Mericka O, Saeland S, Homolka J, Brabec J, Gabius HJ. Diagnostic relevance of Langherin detection in cells from bronchoalveolar lavage of patients with pulmonary Langerhans cell, sarcoidosis and idiopathic pulmonary fibrosis. Virchows Arch 2004;444(2):171–4.

[41] Harari S, Torre O, Cassandro R, Taveira-DaSilva AM, Moss J. Bronchoscopic diagnosis of Langerhans cell histiocytosis and lymphangioleiomyomatosis. Respir Med 2012;106(9):1286–92.

[42] Dhooria S, Sehgal IS, Aggarwal AN, Behera D, Agarwal R. Diagnostic yield and safety of cryoprobe transbronchial lung biopsy in diffuse parenchymal lung diseases: systematic review and meta-analysis. Respir Care 2016;61(5):700–12.

[43] Suri HS, Yi ES, Nowakowski GS, Vassallo R. Pulmonary langerhans cell histiocytosis. Orphanet J Rare Dis 2012;7:16.

[44] Schönfeld N, Frank W, Wenig S, Uhrmeister P, Allica E, Preussler H, et al. Clinical and radiologic features, lung function and therapeutic results in pulmonary histiocytosis X. Respiration 1993;60(1):38–44.

[45] Lorillon G, Bergeron A, Detourmignies L. Cladribine is effective against cystic pulmonary Langerhans cell histiocytosis. Am J Respir Crit Care Med 2012;186(9):930–2.

[46] Donadieu J, Bernard F, van Noesel M, Barkaoui M, Bardet O, Mura R, et al. Cladribine and cytarabine in refractory multisystem Langerhans cell histiocytosis: results of an international phase 2 study. Blood 2015;126(12):1415–23.

[47] Váradi Z, Bánusz R, Csomor J, Kállay K, Varga E, Kertész G, et al. Effective BRAF inhibitor vemurafenib therapy in a 2-year-old patient with sequentially diagnosed Langerhans cell histiocytosis and Erdheim-Chester disease. OncoTargets Ther 2017;10:521–6.

[48] Chaowalit N, Pellikka PA, Decker PA, Aubry MC, Krowka MJ, Ryu JH, et al. Echocardiographic and clinical characteristics of pulmonary hypertension complicating pulmonary Langerhans cell histiocytosis. Mayo Clin Proc 2004;79(10):1269–75.

Chapter 9.2

Lymphangioleiomyomatosis

Lymphangioleiomyomatosis (LAM) is a rare, systemic neoplastic disease that is associated with cystic lung transformation, chylous fluid accumulations, and abdominal tumors, including angiomyolipomas and lymphangioleiomyomas [1]. LAM is characterized by overgrowth of perivascular epithelioid cells (PEC) resembling smooth muscle cells. According to its WHO classification, LAM belongs to the group of PEC tumors [2]. In the vast majority of cases, LAM develops in women of childbearing age.

LAM occurs sporadically and in association with tuberous sclerosis complex (TSC) [1]. TSC is an autosomal dominant disease characterized by angiofibromatosis of the face (Fig. 9.2.1), nontraumatic periungual fibroids, hypopigmented spots (more than three), "shagreen skin," multiple retinal hamartomas, cortical tubers, subependymal nodes, giant cell astrocytomas, multiple or single rhabdomyomas of the heart, multiple renal angiomyolipomas, and directly LAM ("large" criteria) [3].

LAM occurs in approximately 26%–34% of patients with TSC, and cases of this pathology are described not only in women but also in men and male children with TSC [4, 5].

The mechanisms of LAM have not been adequately studied; however, it is known that various mutations of TSC suppressor genes (TSC-1 and TSC-2) are found in patients suffering from both TSC-associated and sporadic LAM [6]. The presence of such a genetic defect is associated with abnormal growth of PEC and their abnormal response to female sex hormones. Estrogens regulate gene transcription and also act as a stimulus for the proliferation and migration of smooth muscle cells to other organs and tissues. This is the second most important etiopathogenetic factor of LAM.

The cytokine activity of PEC plays an important role in LAM pathogenesis. The increased level of serum response factor induces the expression of types 2 and 14 matrix metalloproteinases (MMPs; usually overexpressed in LAM). LAM cells cleave to various components of the extracellular matrix and penetrate into organs and tissues because of overexpression of MMPs [7]. Thus LAM is a multicentric tumorlike process caused by a genetic defect, which determines the abnormal response of (predominantly) smooth muscle cells to female sex hormones. LAM cells can exhibit uncontrolled growth and expression of hormonal receptors and proteases and are prone to metastatic behavior [8]. The proliferation of LAM cells predominantly occurs in the lungs and lymphatic system, which determines the clinical manifestations of this disease. In 80% of cases, progesterone and estrogen receptors are found on the surface of LAM cells, confirming the disease's hormonal dependence. Moreover the occurrence of LAM in women of predominantly childbearing age, provocation of LAM manifestations due to estrogen treatment, onset during menstruation or pregnancy, and mostly favorable course after the onset of menopause are evidences supporting the hormone-dependent nature of this disease [9].

The origin of PEC has not yet been established, but is associated with migrating mesenchymal stem cells [3, 10–12]. PEC can be found in the blood, urine, and pleural effusion of patients with LAM. This is most likely due to the ability of the stem cells to invade blood vessels and spread through the bloodstream and lymph circulation. Like tumors, LAM can recur after lung transplantation, while retaining the existing gene mutations of primary LAM.

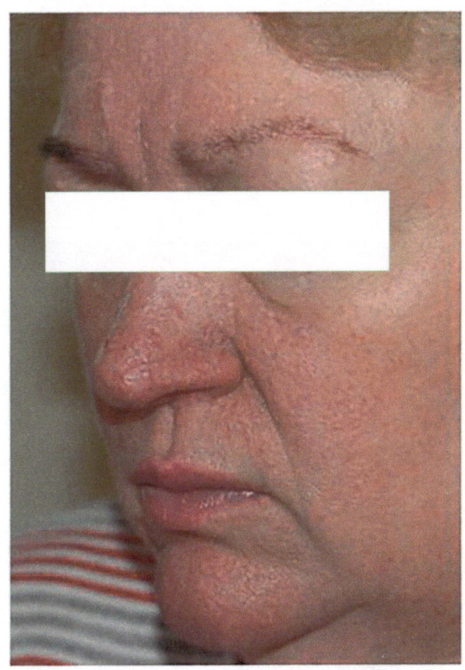

FIG. 9.2.1 Multiple small angiofibromas of the face including the forehead, eyelid, and nose in a patient with TSC.

Morphology

At macroscopic examination of the lungs with LAM, the following changes are observed [13]: bilateral lesions; significant density of the lungs; multiple small (approximately 0.5–1.5 cm in diameter), whitish, fluid-filled nodules located in the subpleural space; the presence of large air cavities in some parts of the lungs; hyperplasia of the lymph nodes; often hemo- and chylothorax due to the destruction of the walls of the blood and lymphatic vessels and rupture of the subpleural cysts; and the chylous effusion in the cavities of the pleura, pericardium, and peritoneum.

Microscopy reveals a number of pathological changes such as multifocal proliferation of immature smooth muscle and PEC of LAM cells along and around lymphatic and blood vessels and bronchioles [13, 14]; as a result of such proliferation, obstruction of lymph drainage and distal airways and venous occlusions develop. These in turn cause the formation of multiple small cysts, interstitial hemorrhages, and hemosiderosis [15, 16] (Figs. 9.2.2–9.2.4). MMP-1 is likely involved in cyst formation [2, 9], and it is intensively synthesized by epithelioid cells [17]. Bronchopulmonary, paratracheal, mediastinal, and retroperitoneal lymph nodes may be involved in the pathological process, where not only the hyperplasia of the lymphoid tissue is observed but also the accumulations of PEC, which in some cases precede lung damage.

At immunohistochemical examination the expression of smooth muscle actin is found in LAM cells (Fig. 9.2.5) and muscle myosin, melan-A, HMB-45, calponin, tyrosinase and microphthalmia of transcription factor, and the absence of vimentin expression (Fig. 9.2.6) [16, 18, 19]. In addition, a system of insulin-like growth factors and binding proteins has great significance in myoid proliferating tissues. Activation of factors indicative of an increase in the proliferative potential of myoid cells, for example, PCNA, bcl-2, c-myc, and a family of insulin-like growth factors, was noted [14, 20, 21]. In some cases, hormonal dependence of LAM is due to high expression of progesterone and estrogen receptors in myoid-like cells (Figs. 9.2.7–9.2.9) [18, 19].

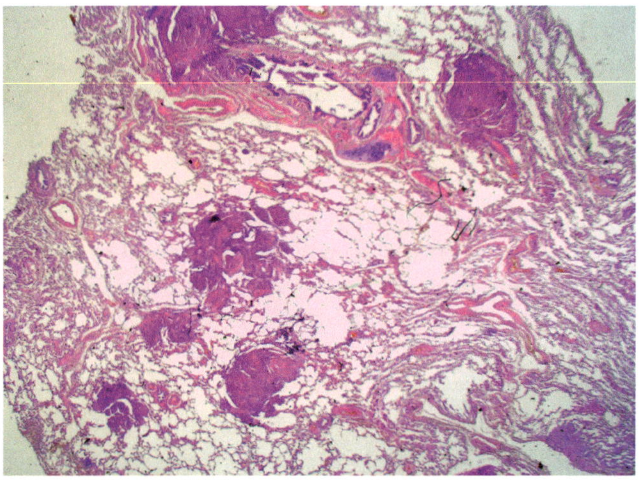

FIG. 9.2.2 LAM. Focal solid round and microcystic structures of the honeycombing. Hematoxylin and eosin (H&E) stain, 1.25×.

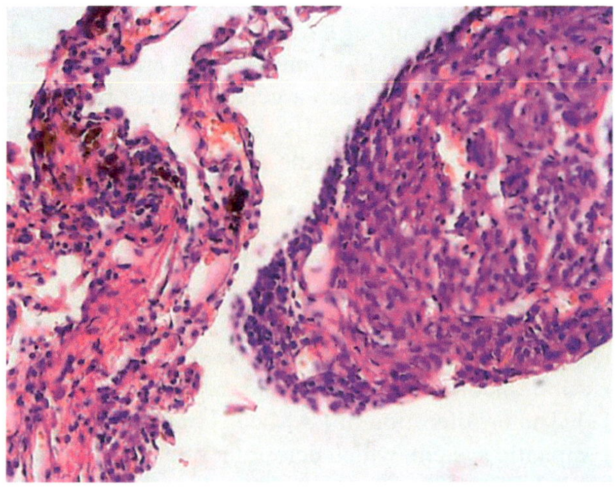

FIG. 9.2.3 LAM. Cyst with polypoid formation from smooth muscle cells. H&E stain, 400×.

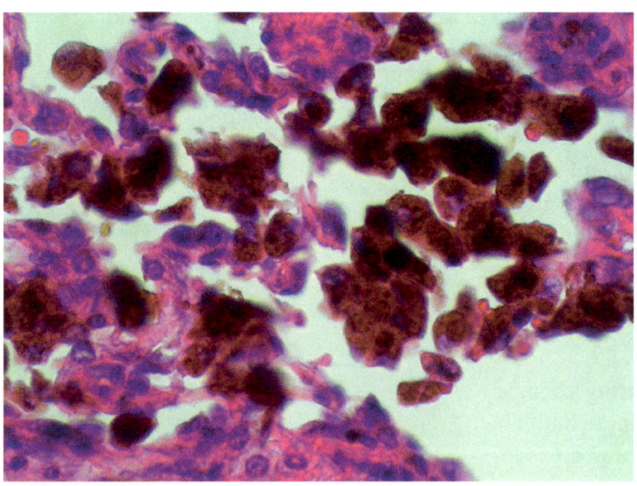

FIG. 9.2.4 LAM. Hemosiderosis of the lung tissue. H&E stain, 600×.

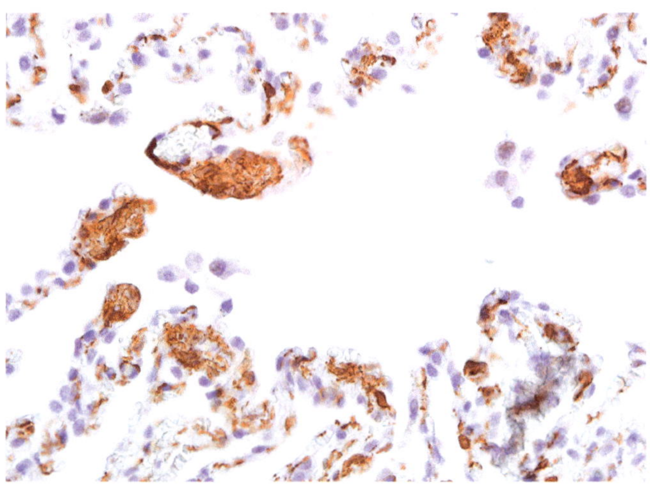

FIG. 9.2.5 LAM. Cyst with polypoid formation from smooth muscle cells. Immunohistochemical study: smooth muscle actin (SMA)-positive leiomyocytes in polypoid structures, 400×.

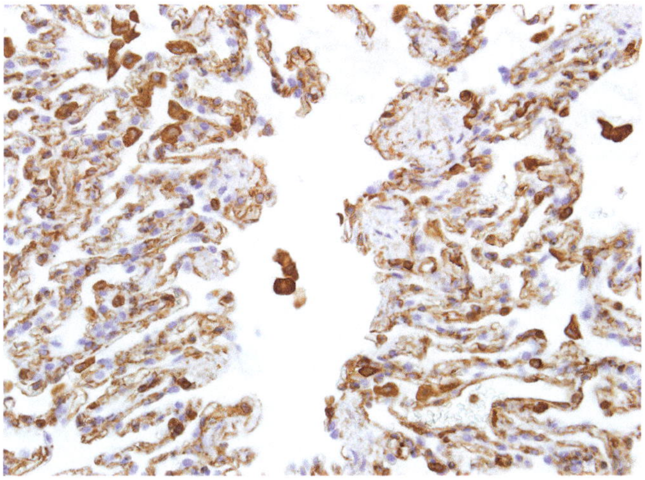

FIG. 9.2.6 LAM. Cyst with polypoid formation from smooth muscle cells. Immunohistochemical study: Vimentin (VIM) negative leiomyocytes in polypoid structures, 400×.

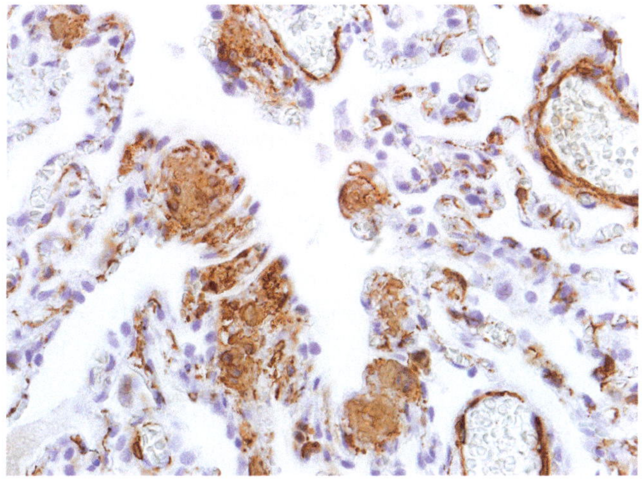

FIG. 9.2.7 LAM. Cyst with polypoid formation from smooth muscle cells. Immunohistochemical study: HMB45-positive leiomyocytes in polypoid structures, 400×.

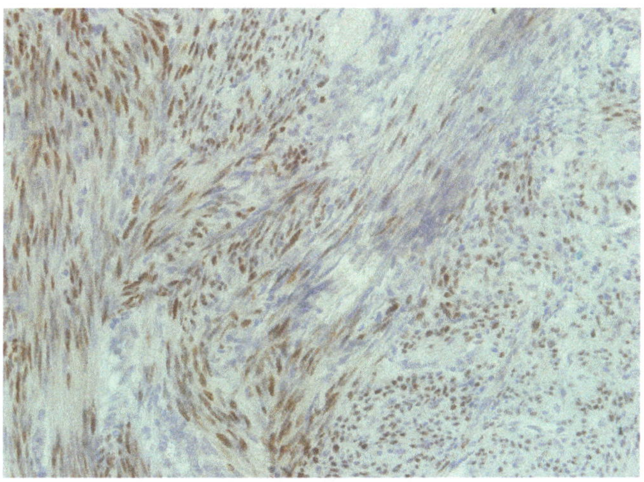

FIG. 9.2.8 LAM. Hyperplasia of smooth muscle cells. Immunohistochemical study: progesterone receptors (PR)-positive leiomyocytes in polypoid structures, 200×.

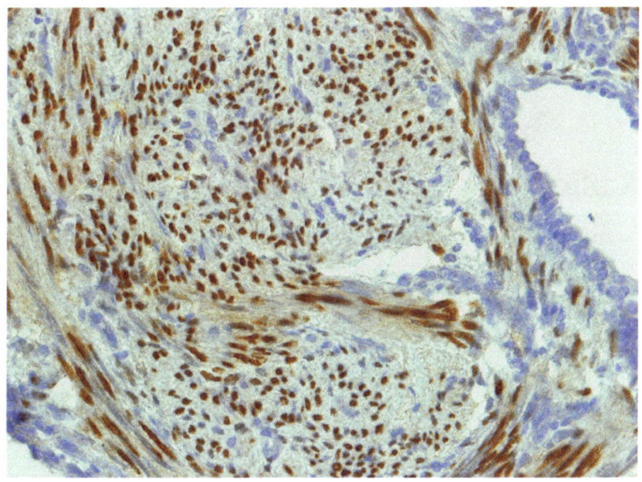

FIG. 9.2.9 LAM. Hyperplasia of smooth muscle cells. Immunohistochemical study: estrogen receptors (ER)-positive leiomyocytes in polypoid structures, 400×.

When examining the biopsy material of patients with LAM, great difficulties may arise in differential diagnosis with processes in the lungs that are accompanied by smooth muscle and fibroblastic proliferation, as well as the formation of cysts. These include usual interstitial pneumonia (UIP), benign metastasizing leiomyoma, leiomyosarcoma, LCH, and obstructive pulmonary emphysema.

During differential diagnosis with UIP, the nature of myoid cell proliferation and localization requires attention. UIP is characterized by the formation of myofibroblastic foci that are localized in the bronchiole-alveolar junction zone, followed by the development of the honeycombing and adenomatosis.

In benign metastasizing leiomyoma, myoid proliferating tissues have the appearance of tumor nodes from smooth muscle cells, there are no cysts in the lung, and a female patient has leiomyoma in the uterus. During IHC testing, no HMB-45 expression is observed.

In lung leiomyosarcoma, myoid proliferating tissue of atypical cells forms a tumor node; there are no cysts in the lung. Immunohistochemically there is no expression of HMB-45.

Obstructive pulmonary emphysema have thin walls and do not have polypoid lesions of myoid cells. IHC reveals no expression of HMB-45.

Clinical presentation

LAM can be asymptomatic over a long time; the disease often starts with spontaneous pneumothorax, usually recurrent, or angiomyolipoma of the kidney.

The main characteristic of symptomatic LAM include the following [22]:

- Dyspnea (87% of cases).
- Recurrent spontaneous pneumothorax (65%).
- Cough (51%).
- Hemoptysis (22%).
- Chest pain (34%) of various genesis, intensifying with breathing.
- Recurrent chylothorax (28%). It is characteristic that the development of pneumo- and chylothorax often coincides with menstruation.
- Chylopericardium and chylous ascites.
- Pulmonary hypertension.

Biomarker

In 2016 the American Thoracic Society and the Japanese Respiratory Society issued a recommendation pertaining to patients with LAM HRCT pattern of lung lesions but without additional extrapulmonary manifestations of the disease. The recommendation was to use the vascular endothelial growth factor-D (VEGF-D) as a serum LAM biomarker. Its level equal to or above 800 pg/mL provides 100% specificity of the LAM diagnosis, almost excluding other cystic diseases. A lower level of VEGF-D (in the presence of cystic changes in the lungs) does not exclude the diagnosis of LAM and requires additional verification methods [23].

Radiological imaging

Standard X-ray examination is able to detect changes only in the late stages of the disease; in 26% of patients the radiological pattern may appear normal (Fig. 9.2.10) [24].

The typical HRCT LAM pattern includes multiple small cysts, usually 2–5 mm in diameter, some of which reach 25–30 mm in size. Cysts are rounded, do not merge with one another, and are located in the unchanged parenchyma (Fig. 9.2.11).

The size of the cysts usually correlates with the degree of damage to the lung tissue [25]. Nevertheless, a total small cystic lesion is possible (Fig. 9.2.12). In advanced forms, almost all the lung is replaced by cysts to the point where it is impossible to evaluate the free parenchyma. In such cases, it is very difficult to differentiate LAM from the pulmonary emphysema and the terminal stage of LCH (Fig. 9.2.13).

The thickness of the cyst walls does not exceed 2 mm. In more than half the cases, the distribution of cysts is uniform; some authors describe a tendency for cysts to localize in the upper zones in 10%–39% of patients (Fig. 9.2.14) [25, 26]. A subpleural distribution resembling a "honeycomb lung" is also possible (Fig. 9.2.15).

HRCT findings in patients with LAM typically include the presence of a limited number (from 2 to 10) of cysts, with the earlier described characteristics.

Despite the fact that LAM does not affect the pleura, many patients experience certain pleural symptoms associated with surgical aid in case of pneumo- or chylothorax. Calcification of pleura and local pleural thickening occurs in 24% and 13% of LAM cases, respectively, after pleurodesis, sometimes mimicking pleural tumors [27].

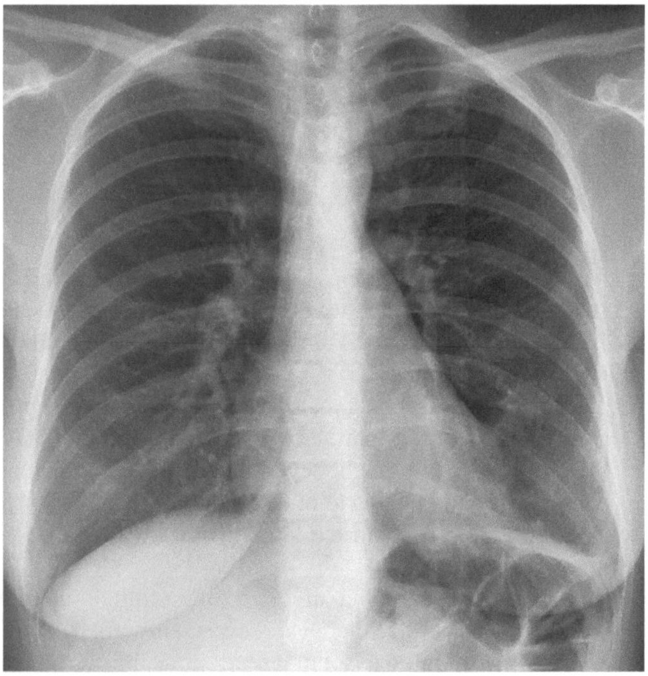

FIG. 9.2.10 LAM. Chest radiograph appears normal.

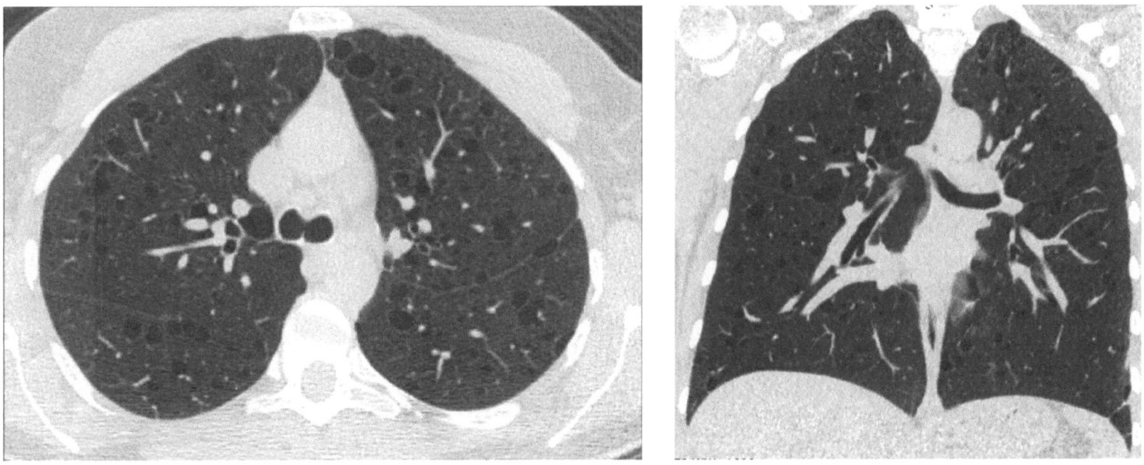

(A) (B)

FIG. 9.2.11 LAM. Axial (A) and coronal (B) scans of the same patient with a normal chest X-ray image. Multiple diffusely distributed thin-walled cysts of small and medium sizes from 2 to 15 mm in diameter inside a normal lung parenchyma.

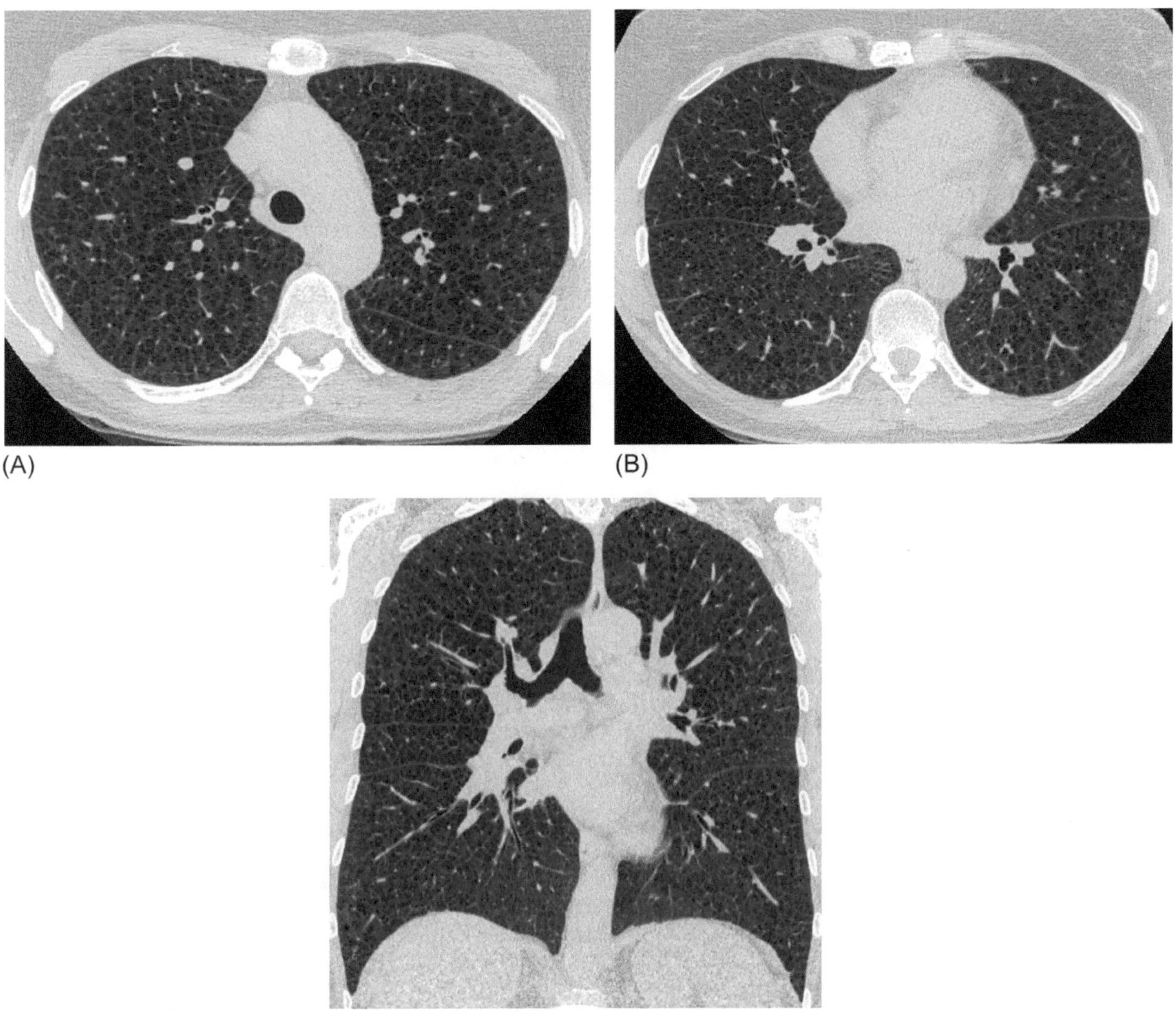

(A) (B) (C)

FIG. 9.2.12 LAM. Diffuse small cystic lung disease (A–C). Cysts do not exceed 5 mm in diameter.

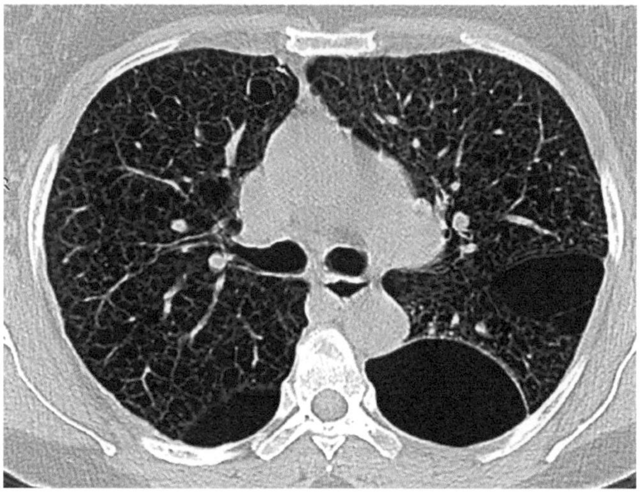

FIG. 9.2.13 Terminal stage of LAM. Diffuse cystic-bullous lung transformation appears similar to diffuse pulmonary emphysema.

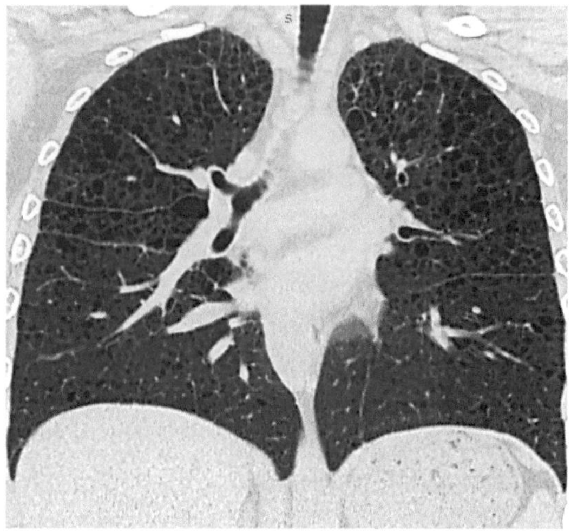

FIG. 9.2.14 LAM: preferential upper-lobe cyst distribution.

Hilar, retrocrural, and mediastinal lymph node enlargement is an infrequent finding on HRCT. Abbott et al. found this sign in 1 of 18 patients with LAM [25]. Earlier studies found enlarged intrathoracic lymph nodes in up to 40% of patients [28]. Lymphadenopathy is detected more frequently in the retroperitoneal space, compared with the mediastinum. Sometimes, when combined with angiomyolipoma of the kidney, lymphadenopathy is mistaken for metastatic lesions (Fig. 9.2.16) [29].

In most patients, if LAM associated with TSC in addition to cysts, multiple small nodules are identified, usually a few millimeters in diameter. These nodules reflect micronodular pneumocytic hyperplasia as hamartoma-like proliferation of type II alveolocytes (Fig. 9.2.17). Some patients may have only nodules, without cystic changes [30].

Pulmonary function tests are not critical for the primary diagnosis of LAM, but are important for monitoring the disease course. The most common disorders detected in patients include decreased diffusion lung capacity (82%–97%) and an obstructive pattern (57%) [31, 32]. Hyperinflation is detected up to 6% of patients, and 25% exhibit positive bronchodilation tests. This is the basis for the prescription of bronchodilators [32].

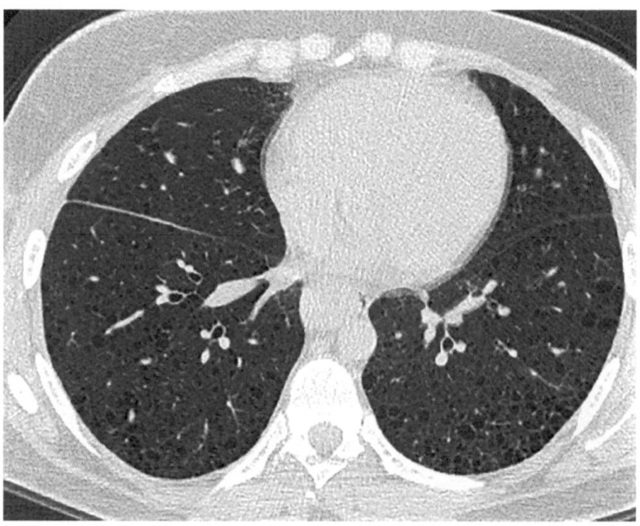

FIG. 9.2.15 Both intraparenchymal and subpleural cyst distribution in a LAM patient, resembling honeycombing. In contrast to the true "honeycombing" observed in fibrotic processes, the cyst walls are much thinner.

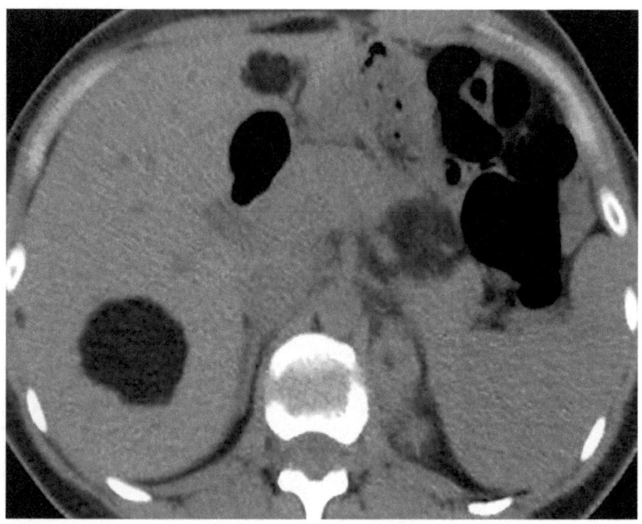

FIG. 9.2.16 LAM. A package of lymph nodes in the retroperitoneal space on the left. Angiolipoma in the right and left lobes of the liver.

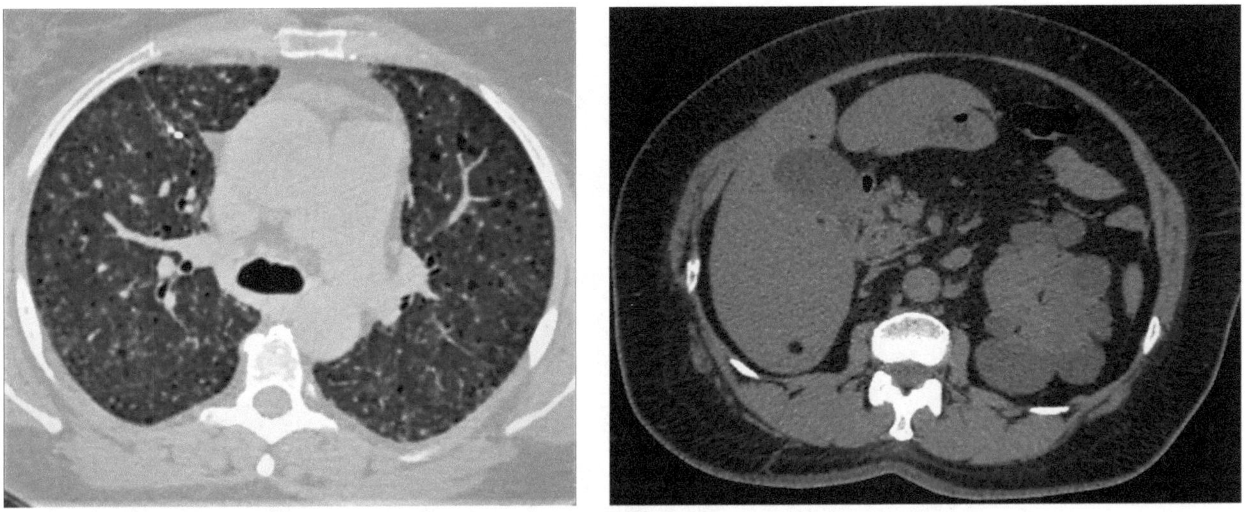

(A) (B)

FIG. 9.2.17 Tuberous sclerosis complex in a 52-year-old female. Numerous small cavities with wall thicknesses of up to 1.5 mm and multiple polymorphic nodules up to 5 mm in diameter (A). Multiple small angiolipomas of the liver and kidney cysts (B).

Probe-based confocal laser endomicroscopy

In two female patients with LAM, 33 bronchopulmonary regions were studied, 165 informative images were obtained, and 1152 measurements were performed. Among the characteristic pCLE features in 16 of 33 regions, most of the interalveolar septa were thinned, and one of them was thickened (Fig. 9.2.18A). An increase in the diameter of individual alveolar structures (Fig. 9.2.18B) was observed in 12 (36.4%) regions. This likely corresponded to the cystic transformation of the pulmonary parenchyma. A viscous secret in the lumen with the small brightly fluorescent elements (Fig. 9.2.18C) was present in 14 (42.4%) regions. The dystelectasis signs (Fig. 9.2.18D) were noticed in 17 (51.5%) regions. Cellular elements in the lumen of the alveoli were not met. Visualization during pCLE was the worst in comparison with all studied nosological forms, most likely due to the large amount of intraluminal contents of a specific composition.

A differential diagnosis for LAM should be performed in patients with pulmonary emphysema, when the thinning of the alveolar septa and an increase in the diameter of the alveolar structures are noted.

Diagnosis

In 2017 modified diagnostic criteria for LAM were proposed [33].

Diagnosis is accurate if there is a corresponding clinical history and radiological signs (according to HRCT) plus at least one of the following phenomena:

- *Diagnosis of tuberous sclerosis complex*
- *Kidney angiomyolipoma*
- *Increased level of VEGF-D > 800 pg/mL*
- *Chylous effusion—pleural or ascite, confirmed by tap and biochemical analysis of the fluid*
- *Lymphangioleiomyoma*
- *Demonstration of LAM cells or LAM cell clusters on cytological examination of effusions or lymph nodes—histopathologic confirmation of LAM by lung biopsy or biopsy of retroperitoneal or pelvic masses*

HRCT data should not be the only confirmation of the diagnosis; additional signs are required. If these signs are not present, transbronchial forceps lung biopsy (TBLB) should be carried out to obtain histological material. This is due to the rather high diagnostic value of TBLB that enables to verify the diagnosis in up to 60% of patients, with a total complication rate of approximately 14%, and 6% of them are pneumothoraces [34, 35].

Regarding differential diagnosis, there are a number of diseases manifested by the formation of cysts or bullae in the lungs. These include LCH, Birt-Hogg-Dubé syndrome (BHDS), lymphocytic interstitial pneumonia (LIP), cystic metastases in the lungs, bullous and centriacinar pulmonary emphysema, and several others (see Chapter 9.4).

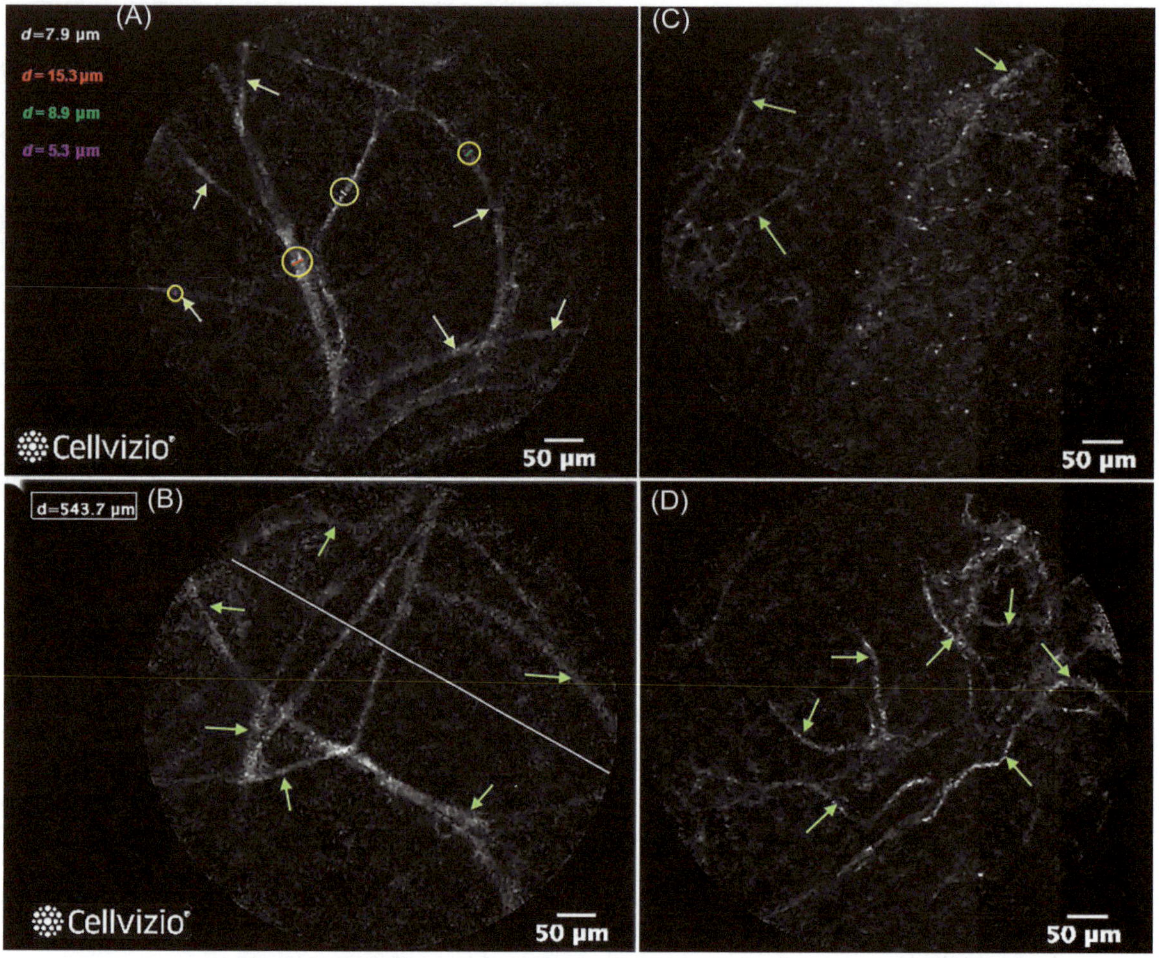

FIG. 9.2.18 pCLE pattern in patients with LAM. The wave length is 488 nm, the field of vision is 600×500 μm or 0.28 mm^2. (A) Most of the elastic fibers are thinned to varying degrees; one of them is thickened. Measurement zones are shown in ellipses. (B) Alveolar structures are visualized, some of which have abnormally large diameters. (C) On the background of individual alveolar cavities with dystelectasis phenomena, almost the entire field of vision is occupied by a viscous secret with small particles (3–4 μm), which have high autofluorescent activity. (D) Dystelectasis signs are noted with a decrease in the airiness of the alveolar structures and the loss of the correct round shape of the alveolar sacs. Elastic fibers are indicated by *arrows*.

Treatment and prognosis

Attempts of antiestrogen therapy for LAM, with tamoxifen, progesterone, and bilateral ovariectomy, performed no better than placebo during well-organized clinical studies, although they led to separate positive effects [36]. Effective and conservative LAM treatments have recently become available [22]. A new and promising LAM therapy includes the use of m-TOR inhibitors with antiproliferative properties, previously used as immunosuppressants.

Based on the results of the first clinical studies, sirolimus stabilized FEV1 and forced volume vital capacity, improved quality of life, reduced the chylous effusion, and reduced the size of renal angiomyolipomas in patients with LAM, both sporadic and TSC-associated [37–39]. More recent studies using reduced doses of sirolimus (1 mg/day) on a fairly large population of patients with LAM showed significant superiority over placebo in patients with FEV1 <70% for functional parameters, quality of life, and VEGF-D levels. At the same time, there was no increase in serious adverse events in female patients taking sirolimus [40]. In 2015 the FDA approved sirolimus for LAM treatment in the United States, and in 2016 ATS/JRS included it in the LAM management guidelines [23]. The main indication for prescribed sirolimus is a confirmed diagnosis of LAM with a decrease in FEV1 <70% of predicted. In addition, in patients with higher FEV1, but with a rapid decline (>90 mL/year), sirolimus therapy may also be prescribed, despite the lack of appropriate studies. In cases of chylous pleural effusion or ascites, it is also necessary to consider the possibility of treatment with sirolimus before initiating drainage procedures. Drug therapy effects usually emerge a few months following initiation and disappear after drug withdrawal [23].

The second agent from the group of m-TOR inhibitors is the rapamycin analogue, everolimus [41]. Patients with LAM who are treated with everolimus exhibit decreased size of angiolipoma and DLCO stabilization, compared with placebo [42].

From 2008 to 2015, at least six randomized, multicenter, placebo-controlled clinical trials of the efficacy and safety of m-TOR inhibitors were conducted on approximately 300 patients with sporadic and TSC-associated LAM [41].

The choice of the dose is determined by the nature of the disease. In patients with TSC-associated LAM, extrapulmonary tumor manifestations predominate. These tumors increase the chance of severe complications and can be treated with higher doses of m-TOR inhibitors. Intake of 2 mg of sirolimus produces its plasma concentrations of 5–15 ng/mL [37]. The effective dose of everolimus TSC-LAM is 10 mg [42]. In patients with sporadic LAM, there is a possibility of effective use of m-TOR inhibitors at significantly lower doses [43].

m-TOR inhibitors have a wide range of adverse effects, including infections (viral, bacterial, and fungal), complications from immunosuppression, cytopenia due to hematopoietic suppression, gastrointestinal disorders, hepato- and nephrotoxicity, dyslipidemia, drug-induced pneumonitis, and slow wound healing. Careful monitoring for intolerance is necessary, and depending on their severity, corrective therapy may be indicated. Corrective therapy involves reducing or temporarily withdrawing drug dosage. The fetotoxicity of m-TOR inhibitors justifies strict adherence to contraception during the treatment period [37].

Currently the use of m-TOR inhibitors is a therapy of choice for patients with LAM, both sporadic and TSC-associated.

There are other potential conservative treatments for LAM including doxycycline inhibiting the activity of matrix metalloproteinases, antiestrogen preparations, progesterone, tyrosine kinase inhibitors (imatinib and nintedanib), and anti-VEGF-D antibodies [30, 44]. Although positive effects were observed in individual studies and clinical cases, none of these drugs have proved superior to placebo and consequently are not recommended for use in patients with LAM.

Patients with LAM can exhibit resting oxygen saturation levels of 90%. Symptomatic treatment includes the prescription of bronchodilators that relieve dyspnea and home oxygen therapy.

The last stage of treatment of LAM is lung transplantation; however, in this case, recurrence of the disease in the transplanted lung is not excluded [45].

In patients with LAM, prognosis depends on the disease course. Although there are rapidly progressing forms, the 10-year survival rate is generally 80%–90%, and the average life expectancy from the onset of symptoms is approximately 30 years [30]. The disease usually has a slowly progressive nature, but after the onset of menopause the rate of disease development often decreases [46].

References

[1] Johnson SR, Cordier JF, Lazor R, Cottin V, Costabel U, Harari S, et al. European Respiratory Society guidelines for the diagnosis and management of lymphangioleiomyomatosis. Eur Respir J 2010;35(1):14–26.

[2] Hasleton P, Flieder DB, editors. Spencer's Pathology of the Lung. 6th ed. New York: Cambridge University Press; 2013.

[3] Moss J, Avila NA, Barnes PM, Litzenberger RA, Bechtle J, Brooks PG, et al. Prevalence and clinical characteristics of lymphangioleiomyomatosis (LAM) in patients with tuberous sclerosis complex. Am J Respir Crit Care Med 2001;164(4):669–71.

[4] Mboyo A, Flurin V, Foulet-Roge A, Bah G, Orain I, Weil D. Conservative treatment of a mesenteric lymphangiomyomatosis in an 11-year-old girl with a long follow-up period. J Pediatr Surg 2004;39(10):1586–9.

[5] Miyake M, Tateishi U, Maeda T, Kusumoto M, Satake M, Arai Y, et al. Pulmonary lymphangioleiomyomatosis in a male patient with tuberous sclerosis complex. Radiat Med 2005;23(7):525–7.

[6] Carsillo T, Astrinidis A, Henske EP. Mutations in the tuberous sclerosis complex gene TSC2 are a cause of sporadic pulmonary lymphangioleiomyomatosis. Proc Natl Acad Sci U S A 2000;97(11):6085–90.

[7] Henry MT, McMahol K, Mackarel AJ, Prikk K, Sorsa T, Maisi P, et al. Matrix metalloproteinases and inhibitor of metalloproteinase-1 in sarcoidosis and IPF. Eur Respir J 2002;20(5):1220–7.

[8] Crooks D, Pacheco-Rodriguez G, DeCastro R, McCoy Jr JP, Wang JA, Kumaki F, et al. Molecular and genetic analysis of disseminated neoplastic cells in lymphangioleiomyomatosis. Proc Natl Acad Sci U S A 2004;101(50):17462–7.

[9] Sinclar W, Wright JL, Churg A. Lymphangioleiomyomatosis presenting in postmenopausal women. Thorax 1985;40(6):475–6.

[10] Hancock E, Osborne J. Lymphangioleiomyomatosis: a review of the literature. Respir Med 2002;96(1):1–6.

[11] Chorianopoulos D, Stratakos G. Lymphangioleiomyomatosis and tuberous sclerosis complex. Lung 2008;186(4):197–207.

[12] Crino PB, Nathanson KL, Henske EP. The tuberous sclerosis complex. N Engl J Med 2006;355(13):1345–56.

[13] Corrin B, Liebow AA, Friedman PJ. Pulmonary lymphangioleiomyomatosis: a review. Am J Pathol 1975;79(2):348–82.

[14] Ferrans VJ, Yu ZX, Nelson WK, Valencia JC, Tatsuguchi A, Avila NA, et al. Lymphangioleiomyomatosis (LAM). A review of clinical and morphological features. J Nippon Med Sch 2000;67(5):311–29.

[15] Carrington CB, Cugell DW, Gaensler EA, Marks A, Redding RA, Schaaf JT, et al. Lymphangioleiomyomatosis. Physiologic-pathologic-radiologic correlations. Am Rev Respir Dis 1977;116(6):977–95.

[16] Matsui K, Beasley MB, Nelson WK, Barnes PM, Bechtle J, Falk R, et al. Prognostic significance of pulmonary lymphangioleiomyomatosis histologic score. Am J Surg Pathol 2001;25(4):479–84.

[17] Matsui K, Takeda K, Yu Z-X, Travis WD, Moss J, Ferrans VJ. Role for activation of matrix metalloproteinases in the pathogenesis of pulmonary lymphangioleiomyomatosis. Arch Pathol Lab Med 2000;124(2):267–75.

[18] Matsumoto Y, Horiba K, Usuki J, Chu SC, Ferrans VJ, Moss J. Markers of cell proliferation and expression of melanosomal antigen in lymphangioleiomyomatosis. Am J Respir Cell Mol Biol 1999;21(3):327–36.

[19] Grzegorek I, Lenze D, Chabowski M, Janczak D, Szolkowska M, Langfort R, et al. Immunohistochemical evaluation of pulmonary lymphangioleiomyomatosis. Anticancer Res 2015;35(6):3353–60.

[20] Fetsch PA, Fetsch JF, Marincola FM, Travis W, Batts KP, Abati A. Comparison of melanoma antigen recognized by T cells (MART-1) to HMB-45: additional evidence to support a common lineage for angiomyolipoma, lymphangioleiomyomatosis, and clear cell sugar tumor. Mod Pathol 1998;11(8):699–703.

[21] Usuki J, Horida K, Chu SC, Moss J, Ferrans VJ. Lymphangioleiomyomatosis analysis of proteins of the Bcl-2 family in pulmonary lymphangioleiomyomatosis: association of Bcl-2 expression with hormone receptor status. Arch Pathol Lab Med 1998;122(10):895–902.

[22] Harari S, Torre O, Moss J. Lymphangioleiomyomatosis: what do we know and what are we looking for? Eur Respir Rev 2011;20(119):34–44.

[23] McCormack FX, Gupta N, Finlay GR, Lisa R, et al. Official American Thoracic Society/Japanese Respiratory Society Clinical Practice Guidelines: lymphangioleiomyomatosis diagnosis and management. Am J Respir Crit Care Med 2016;194(6):748–61.

[24] Chu SC, Horiba K, Usuki J, Avila NA, Chen CC, Travis WD, et al. Comprehensive evaluation of 35 patients with lymphangioleiomyomatosis. Chest 1999;115(4):1041–52.

[25] Abbott GF, Rosado-de-Christianson ML, Frazier AA, Franks TJ, Pugatch RD, Galvin JR. From the archives of the AFIP: lymphangioleiomyomatosis: radiologic-pathologic correlation. Radiographics 2005;25(3):803–28.

[26] Lim KE, Tsai YH, Hsu Y, Hsu WC. Pulmonary lymphangioleiomyomatosis: high-resolution CT findings in 11 patients and compared with the literature. Clin Imaging 2004;28(1):1–5.

[27] Avila NA, Dwyer AJ, Rabel A, DeCastro RM, Moss J. CT of pleural abnormalities in lymphangioleiomyomatosis and comparison of pleural findings after different types of pleurodesis. AJR Am J Roentgenol 2006;186(4):1007–12.

[28] Sherrier RH, Chiles C, Roggli V. Pulmonary lymphangioleiomyomatosis: CT findings. AJR Am J Roentgenol 1989;153(5):937–40.

[29] Avila NA, Dwyer AJ, Moss J. Imaging features of lymphangioleiomyomatosis: diagnostic pitfalls. AJR Am J Roentgenol 2011;196(4):982–6.

[30] Webb WR, Muller NL, Naidich DP. High-resolution CT of the lung. 5th ed. Philadelphia: Lippincott Williams and Wilkins; 2015. p. 500–10.

[31] Mason RJ, Murray JF, Broaddus VC, Nadel JA. Murray & Nadel's textbook of respiratory medicine. 6th ed. vol. 2. USA: Elsevier Inc; 2015. p. 1243–74.

[32] Ryu JH, Moss J, Beck GJ, Lee JC, Brown KK, Chapman JT, et al. The NHLBI lymphangioleiomyomatosis registry: characteristics of 230 patients at enrollment. Am J Respir Crit Care Med 2006;173(1):105–11.

[33] Gupta N, Finlay GA, Kotloff RM, Strange C, Wilson KC, Young LR, et al. Lymphangioleiomyomatosis diagnosis and management: high-resolution chest computed tomography, transbronchial lung biopsy, and pleural disease management. An Official American Thoracic Society/Japanese Respiratory Society Clinical Practice Guideline. Am J Respir Crit Care Med 2017;196(10):1337–48.

[34] Harari S, Torre O, Cassandro R, Taveira-DaSilva AM, Moss J. Bronchoscopic diagnosis of Langerhans cell histiocytosis and lymphangioleiomyomatosis. Respir Med 2012;106(9):1286–92.

[35] Meraj R, Wikenheiser-Brokamp KA, Young LR, Byrnes S, McCormack FX. Utility of transbronchial biopsy in the diagnosis of lymphangioleiomyomatosis. Front Med 2012;6(4):395–405.

[36] Casanova A, Ancochea J. Lymphangioleiomyomatosis: new therapeutic approaches. Arch Bronconeumol 2011;47(12):579–80.

[37] Davies DM, de Vries PJ, Johnson SR, McCartney DL, Cox JA, Serra AL, et al. Sirolimus therapy for angiomyolipoma in tuberous sclerosis and sporadic lymphangioleiomyomatosis: a phase 2 trial. Clin Cancer Res 2011;17(12):4071–81.

[38] McCormack FX, Inoue Y, Moss J, Singer LG, Strange C, Nakata K, et al. Efficacy and safety of sirolimus in lymphangioleiomyomatosis. N Engl J Med 2011;364(17):1595–606.

[39] Taveira-DaSilva AM, Hathaway O, Stylianou M, Moss J. Changes in lung function and chylous effusions in patients with lymphangioleiomyomatosis treated with sirolimus. Ann Intern Med 2011;154(12):797–805.

[40] Ando K, Kurihara M, Kataoka H, Ueyama M, Togo S, Sato T, et al. The efficacy and safety of low-dose sirolimus for treatment of lymphangioleiomyomatosis. Respir Investig 2013;51(3):175–83.

[41] Yates DH. mTOR treatment in lymphangioleiomyomatosis: the role of everolimus. Expert Rev Respir Med 2016;10(3):249–60.

[42] Bissler JJ, Kingswood JC, Radzikowska E, Zonnenberg BA, Frost M, Belousova E, et al. Everolimus for angiomyolipoma associated with tuberous sclerosis complex or sporadic lymphangioleiomyomatosis (EXIST-2): a multicentre, randomised, double-blind, placebo-controlled trial. Lancet 2013;381(9869):817–24.

[43] Goldberg H, Harari S, Cottin V, Rosas IO, Peters E, Biswal S, et al. Everolimus for the treatment of lymphangioleiomyomatosis: a phase II study. Eur Respir J 2015;46(3):783–94.
[44] Mavroudi M, Zarogoulidis P, Katsikogiannis N, Tsakiridis K, Huang H, Sakkas A, et al. Lymphangioleiomyomatosis: current and future. J Thorac Dis 2013;5(1):74–9.
[45] Karbowniczek M, Astrinidis A, Balsara BR, Testa JR, Lium JH, Colby TV, et al. Recurrent lymphangioleiomyomatosis after transplantation: genetic analyses reveal a metastatic mechanism. Am J Respir Crit Care Med 2003;167(7):976–82.
[46] Harknett EC, Chang WY, Byrnes S, Johnson J, Lazor R, Cohen MM, et al. Use of variability in national and regional data to estimate the prevalence of lymphangioleiomyomatosis. QJM 2011;104(11):971–9.

Chapter 9.3

Birt-Hogg-Dubé syndrome

Birt-Hogg-Dubé syndrome (BHDS) is a rare autosomal dominant disease characterized by the development of fibrofolliculomas, renal tumors, and the formation of pulmonary cysts. BHDS is often accompanied by recurrent pneumothorax [1]. BHDS is caused by mutation of the gene located in the XVII chromosome that codes for the synthesis of folliculin (FLCN). Folliculin is a tumor suppressor that implements its effects through m-TOR pathways [2, 3].

Clinical manifestations of BHDS include cutaneous, pulmonary, and renal symptoms. Skin lesion occurs in 90% of patients in European and American populations; however, studies in East Asia show a significantly lower prevalence (approximately 20% of cases) [4, 5]. Typical cutaneous manifestations include fibrofolliculomas, trichodiscomas, and perifollicular fibromas (cutaneous hamartomas) that are mainly localized to the face, neck, and upper body. Fibrofolliculomas are benign tumors of the hair follicles that look like multiple whitish papules, usually appearing in patients older than age 20 [1]. Fibrofolliculomas should be differentiated from angiofibromas in TSC. The latter are smaller and contain telangiectasias [6]. Some researchers believe that angiofibroma is clinically and histologically indistinguishable from fibrofolliculoma [7].

Kidney damage in BHDS includes the development of cancer and cysts. The incidence of renal cancer in these patients is 29%–34%, exceeding the frequency in the general population by 6.9 times [8, 9]. The most common histological variants of renal cancer are hybrid oncocytic tumors composed of features of chromophobe renal cell carcinoma. Renal oncocytoma (50%) and chromophobic (34%) and clear cell (9%) cancers [10, 11] are less common. Cystic changes in the kidneys are not uncommon in patients with BHDS, but the frequency of their development in this pathology has not been studied [1].

The lungs are involved in 80%–85% of patients [1, 12].

Bilateral asymmetrical thin-walled cysts of different shapes and sizes are typical. Small cysts are usually round, while larger cysts often have lenticular, oval, or irregular shape, sometimes with internal septa. Most cysts are located in the lower and middle sections of the lung, and approximately 40% are subpleural [13]. The total number of cysts is usually small. Toro et al. estimated an average of approximately 16 in one patient. However, multicystic variants of the lesion (up to 166 cysts) are also possible [14]. Proximal pulmonary veins or arteries may abut or be located inside the cysts [15]. A characteristic feature of BHDS-related cysts is their stable size over many years. This is different than most other cystic lung diseases, where increases in the number and volume of cystic lesions are usually noted (Fig. 9.3.1) [16].

Cyst formation in BHDS likely results from a violation of intercellular adhesion, which damages epithelial cells and may mechanically expand the alveolar space [17]. This, the so-called stretch hypothesis, suggests that cysts appear in areas that experience the greatest changes in the alveolar volume during breathing or ventilation (i.e., in the lower lobes and subpleural areas, with weakness of the supporting pleural "anchor points") [18]. Spontaneous pneumothorax is most often the first clinical manifestation of lung damage with BHDS, and its development often outstrips the emergence of cutaneous signs of the disease [19]. In general the risk of pneumothorax in these patients is increased 50 times, and it is higher at a younger age (up to 40 years) [20]. According to Gupta et al., 76% of 104 patients with BHDS experienced at least one spontaneous pneumothorax, and the average number of cases of pneumothorax was 3.6 [12].

The diagnosis of BHDS is usually based on the major and minor criteria proposed in 2009 (Table 9.3.1) [1].

Later, Schmidt and Linehan modified the criteria that allow for definitive disease diagnosis if there is a mutation in the folliculin gene (Table 9.3.2) [21].

However, 7%–9% of patients with no known FLCN gene mutations exhibit a complete set of major and minor BHDS diagnosis criteria. This is apparently explained by the incompleteness of the known spectrum of possible genetic disorders related to this disease [6].

Patients with BHDS are primarily managed by follow-up and control of the development of possible neoplasms, primarily of the kidneys. In the absence of lung symptoms, control of the cysts is not required, since the likelihood of growth is low [22]. Patients should be warned about an increased risk of pneumothorax, especially during activities associated with sharp differences in atmospheric pressure (diving and high altitude climbing). When recurrent pneumothorax develops, pleurodesis, bullectomy, or pleural covering is usually performed [23].

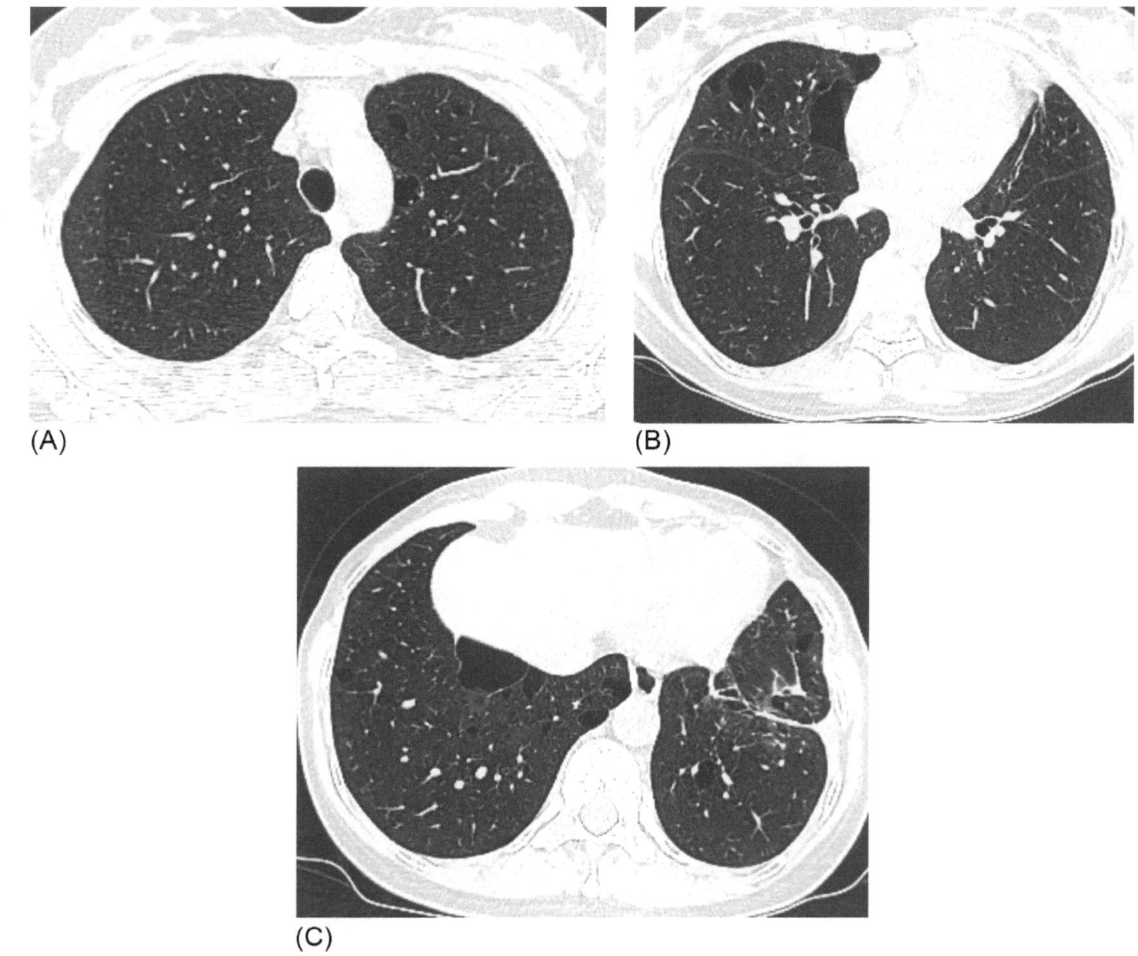

FIG. 9.3.1 Birt-Hogg-Dubé syndrome in 43-year-old woman. Small round and oval thin-walled cysts in the left upper lobe (A). A few subpleural cysts with irregular shapes are in the right middle lobe (B). Several cysts with oval and irregular shapes with subpleural predominance are visible in the basal segments. Fibrotic band in the left lower lobe is a consequence of surgery (C). *(Case courtesy of Prof. S.N. Avdeev, Sechenov First Moscow State Medical University, Moscow, Russia.)*

TABLE 9.3.1 Diagnostic criteria for BHDS [1]

Major criteria	≥Five fibrofolliculomas/trichodiscomas, at least one is confirmed histologically Pathogenic FLCN germline mutation
Minor criteria	Multiple cysts in the lungs: bilateral basally located cysts with no other cause with or without pneumothorax Renal cancer: at age 50 or younger, multifocal or bilateral cancers or renal cancer with mixed chromophobic and oncocytic histological pattern

The diagnosis is accurate in the presence of one major or two minor criteria.

TABLE 9.3.2 Modified BHDS criteria [21]

Major criteria (high likelihood of BHDS)	• Two cutaneous papules clinically compatible with fibrofolliculoma/trichodiscoma with at least one histologically confirmed fibrofolliculoma; adult onset
Minor criteria (suspicious for BHDS)	• Multiple bilateral basilar lung cysts with or without spontaneous pneumothorax (<40 years of age) especially with family history of these lung manifestations • Bilateral multifocal chromophobe or hybrid oncocytic renal tumors especially if there is a family history of such manifestations • Combinations of cutaneous, lung, or renal manifestations known to be associated with BHDS in an individual or his family
Definitive diagnosis	• Positive germline FLCN mutation test

References

[1] Menko FH, van Steensel MA, Giraud S, Friis-Hansen L, Richard S, Ungari S, et al. Birt-Hogg-Dube syndrome: diagnosis and management. Lancet Oncol 2009;10(12):1199–206.

[2] Hasumi Y, Baba M, Ajima R, Hasumi H, Valera VA, Klein ME, et al. Homozygous loss of BHD causes early embryonic lethality and kidney tumor development with activation of mTORC1 and mTORC2. Proc Natl Acad Sci U S A 2009;106(44):18722–7.

[3] Kahnoski K, Khoo SK, Nassif NT, Chen J, Lobo GP, Segelov E, et al. Alterations of the Birt-Hogg-Dube gene (BHD) in sporadic colorectal tumours. J Med Genet 2003;40(7):511–5.

[4] Toro JR, Wei MH, Glenn GM, Weinreich M, Toure O, Vocke C, et al. BHD mutations, clinical and molecular genetic investigations of Birt-Hogg-Dube syndrome: a new series of 50 families and a review of published reports. J Med Genet 2008;45(6):321–31.

[5] Park HJ, Park CH, Lee SE, Lee GD, Byun MK, Lee S, et al. Birt-Hogg-Dube syndrome prospectively detected by review of chest computed tomography scans. PLoS ONE 2017;12(2):e0170713.

[6] Tong Y, Schneider JA, Coda AB, Hata TR, Cohen PR. Birt-Hogg-Dubé syndrome: a review of dermatological manifestations and other symptoms. Am J Clin Dermatol 2018;19(1):87–101.

[7] Toro JR. Birt-Hogg-Dubé syndrome. In: Pagon RA, Adam MP, Ardinger HH, Wallace SE, Amemiya A, Bean LJ, et al., editors. GeneReviews. Seattle: University of Washington; 2006.

[8] Schmidt LS, Nickerson ML, Warren MB, Glenn GM, Toro JR, Merino MJ, et al. Germline BHD-mutation spectrum and phenotype analysis of a large cohort of families with Birt-Hogg-Dubé syndrome. Am J Hum Genet 2005;76(6):1023–33.

[9] Zbar B, Alvord WG, Glenn G, Turner M, Pavlovich CP, Schmidt L, et al. Risk of renal and colonic neoplasms and spontaneous pneumothorax in the Birt-Hogg-Dubé syndrome. Cancer Epidemiol Biomark Prev 2002;11(4):393–400.

[10] Pavlovich CP, Grubb RL, Hurley K, Glenn GM, Toro J, Schmidt LS, et al. Evaluation and management of renal tumors in the Birt-Hogg-Dubé syndrome. J Urol 2005;173(5):1482–6.

[11] Benusiglio PR, Giraud S, Deveaux S, Méjean A, Correas JM, Joly D, et al. French National Cancer Institute Inherited Predisposition to Kidney Cancer Network. Renal cell tumour characteristics in patients with the Birt-Hogg-Dubé cancer susceptibility syndrome: a retrospective, multicentre study. Orphanet J Rare Dis 2014;9:163.

[12] Gupta N, Kopras EJ, Henske EP, James LE, El-Chemaly S, Veeraraghavan S, et al. Spontaneous pneumothoraces in patients with Birt-Hogg-Dubé syndrome. Ann Am Thorac Soc 2017;14(5):706–13.

[13] Tobino K, Gunji Y, Kurihara M, Kunogi M, Koike K, Tomiyama N, et al. Characteristics of pulmonary cysts in Birt-Hogg-Dubé syndrome: thin-section CT findings of the chest in 12 patients. Eur J Radiol 2011;77(3):403–9.

[14] Toro JR, Pautler SE, Stewart L, Glenn GM, Weinreich M, Toure O, et al. Lung cysts, spontaneous pneumothorax, and genetic associations in 89 families with Birt-Hogg-Dubé syndrome. Am J Respir Crit Care Med 2007;175(10):1044–53.

[15] Lee JE, Cha YK, Kim JS, Choi JH. Birt-Hogg-Dubé syndrome: characteristic CT findings differentiating it from other diffuse cystic lung diseases. Diagn Interv Radiol 2017;23(5):354–9.

[16] Ayo DS, Aughenbaugh GL, Yi ES, Hand JL, Ryu JH. Cystic lung disease in Birt-Hogg-Dubé syndrome. Chest 2007;132(2):679–84.

[17] Gupta N, Sunwoo BY, Kotloff RM. Birt-Hogg-Dubé syndrome. Clin Chest Med 2016;37(3):475–86.

[18] Kennedy JC, Khabibullin D, Henske EP. Mechanisms of pulmonary cyst pathogenesis in Birt-Hogg-Dube syndrome: The stretch hypothesis. Semin Cell Dev Biol 2016;52:47–52.

[19] Johannesma PC, van den Borne BEEM, Gille JJP, Nagelkerke AF, van Waesberghe JTM, Paul MA, et al. Spontaneous pneumothorax as indicator for Birt-Hogg-Dubé syndrome in paediatric patients. BMC Pediatr 2014;14:171.

[20] Zbar B, Alvord WG, Glenn G, Turner M, Pavlovich CP, Schmidt L, et al. Risk of renal and colonic neoplasms and spontaneous pneumothorax in the Birt-Hogg-Dubé syndrome. Cancer Epidemiol Biomark Prev 2002;11(4):393–400.

[21] Schmidt LS, Linehan WM. Clinical features, genetics and potential therapeutic approaches for Birt-Hogg-Dubé syndrome. Expert Opin Orphan Drugs 2015;3(1):15–29.

[22] Jensen DK, Villumsen A, Skytte AB, Madsen MG, Sommerlund M, Bendstrup E. Birt-Hogg-Dubé syndrome: a case report and a review of the literature. Eur Clin Respir J 2017;4(1):1292378.

[23] Takegahara K, Yoshino N, Usuda J. A case of recurrent pneumothorax associated with Birt-Hogg-Dubé syndrome treated with bilateral simultaneous surgery and total pleural covering. Ann Thorac Cardiovasc Surg 2017;23(6):309–12.

Chapter 9.4

Differential diagnosis of cystic diseases

The diagnosis of diffuse cystic lung diseases does not cause significant difficulties, but in practice, there are many more cases requiring a systemic approach and application of differential diagnosis algorithms:

- *The first question that should answered when meeting a patient with multiple cysts in the lungs is "are they really cysts, or we dealing with mimicking cysts lesions?"*

The cyst is traditionally defined as an air-filled lucency or low-attenuating area bordered by a thin wall (usually <2 mm) with a well-defined interface with normal lung tissue [1]. Cysts must be distinguished from pulmonary emphysema (centriacinar, bullous, and paraseptal), cavitary lesions in the lungs due to destruction of the parenchyma, cystic bronchiectasis, or honeycombing (Table 9.4.1) (Figs. 9.4.1–9.4.5).

Bullous pulmonary emphysema has characteristic features that distinguish it from true cysts. Bullae (areas of emphysema greater than 1 cm in diameter, surrounded by a thin wall) are usually asymmetrical, subpleural, and limited in number and may have internal septa (Fig. 9.4.1). Bullous pulmonary emphysema can be accompanied by paraseptal, centriacinar, or panacinar emphysema and other signs of COPD such as thickening of the walls of the bronchi, peribronchial fibrosis, and air traps [2].

In contrast to cysts, for centriacinar emphysema, there is a lack of a clear wall with preservation inside or around the periphery areas of intralobular artery (Fig. 9.4.2). In patients with congenital connective tissue dysplasia (Marfan syndrome and Ehlers-Danlos syndrome), the round foci of centriacinar emphysema that resemble cysts are not accompanied by a change in the bronchial tree and air trapping (Fig. 9.4.3). The cavities of destruction are usually formed in the thickness of the inflammatory or tumor infiltration and have a much thicker walls than the cysts (>4 mm) (Fig. 9.4.4) [3]. Diseases that lead to the formation of cavitary lesions in the lungs are usually associated with pronounced clinical and laboratory symptoms such as fever, hemoptysis, and increases in inflammatory markers in the blood analyses.

Honeycombing is a manifestation of alveolar disruption that features dilatation and fibrous thickening of alveolar walls and respiratory bronchioles. Honeycombing looks like several overlapping rows of small subpleural cysts with a thickness of 1–3 mm and a size of several millimeters to 1 centimeter. Honeycombing is predominantly subpleural and frequently

TABLE 9.4.1 Cystiform lung lesions

	Characteristics	Diseases/conditions
Bullae	Rounded, more than 1 cm in diameter air cavities with a thin (<1 mm) wall, often combined with centriacinar or paraseptal emphysema	COPD, Ehlers-Danlos syndrome, Marfan syndrome, alpha-1-antitrypsin deficiency
Centriacinar emphysema	Enlarged airspace distal to terminal bronchioles. Has no visible wall. Often with the preserved intralobular artery in the form of a central point on the HRCT	
Paraseptal emphysema	Small (up to 1 cm) cavities abut to or surrounded by the visceral pleura. Primarily in the upper lobes	Smoking, COPD, young tall lean people. The most common cause of spontaneous pneumothorax
Cavitary destruction of lung parenchyma	Nodal or focal formations inside the zones of consolidation. Thick-walled (>4 mm) and can contain fluid	Destructive pneumonia (bacterial, pneumocystic, fungal, and parasitic), septic metastases, lung tumors, granulomatosis with polyangiitis, rheumatoid nodes
Honeycombing	Subpleural small-sized cysts clustered in several rows (3–10 mm in diameter)	Idiopathic pulmonary fibrosis, NSIP chronic HP, lung lesions in systemic connective tissue diseases
Cystic bronchiectases	Sacciform dilations of the proximal bronchi with an artery abut to the wall, with a wall thickness of more than 1 mm	Cystic fibrosis, tracheobronchomegaly, Williams-Campbell syndrome

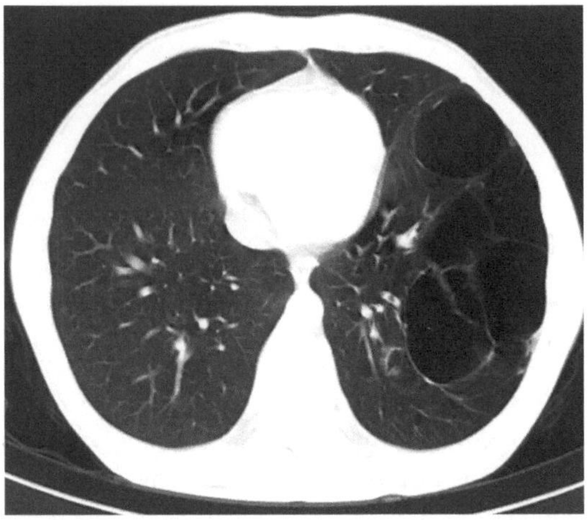

FIG. 9.4.1 Bullous emphysema in COPD patient. Multiple bullae of the left lung. Internal septa can be seen inside several bullae.

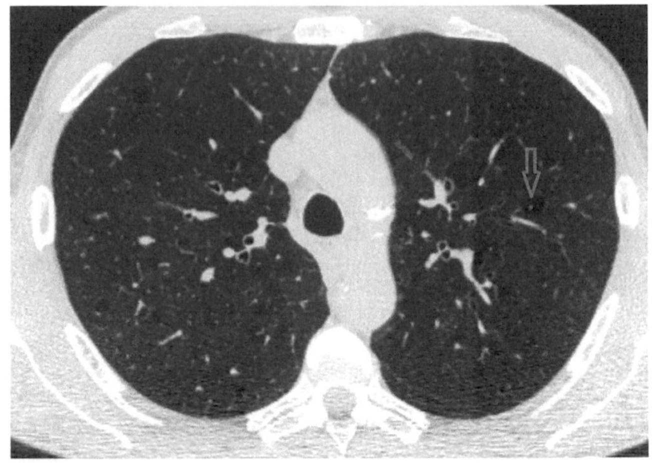

FIG. 9.4.2 Centrilobular emphysema in a smoking patient. There are no visible walls in the areas of lucency. In some of them, preserved intralobular vessels *(gray arrow)* are seen.

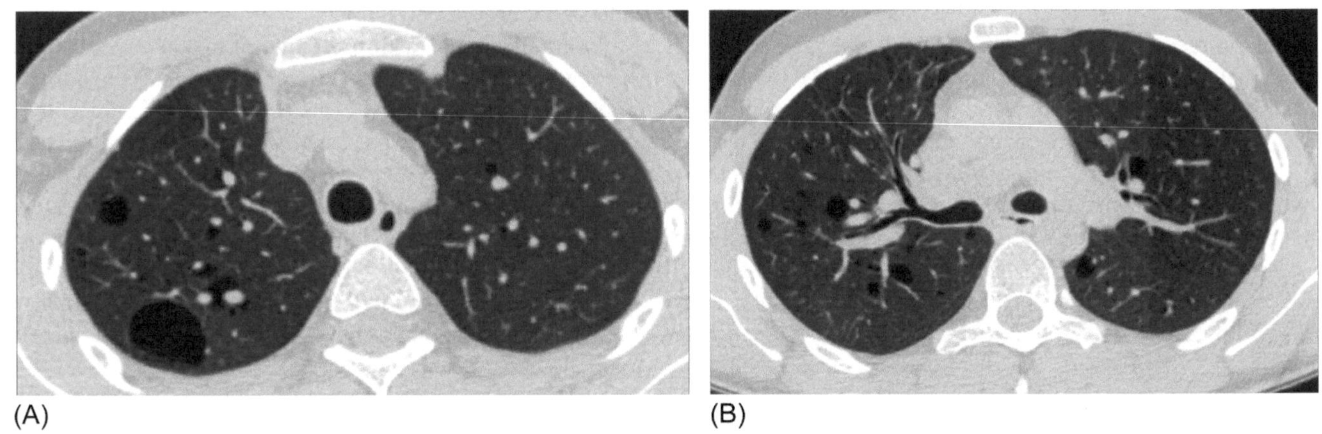

FIG. 9.4.3 Centriacinar and bullous emphysema in a patient with Ehlers-Danlos syndrome. In the thickness of the parenchyma, rounded areas of hyperlucency are revealed; most of them are adjacent to the branches of the pulmonary artery. Some of the cavities do not have a clearly differentiated wall. The surrounding parenchyma of the lungs is normal (A and B).

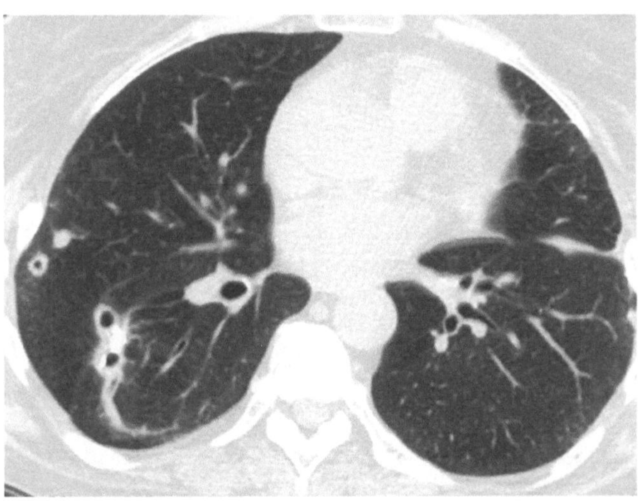

FIG. 9.4.4 Granulomatosis with polyangiitis. Nodal infiltrates with destruction in the tissue of right lung and formation of thick-walled cavities.

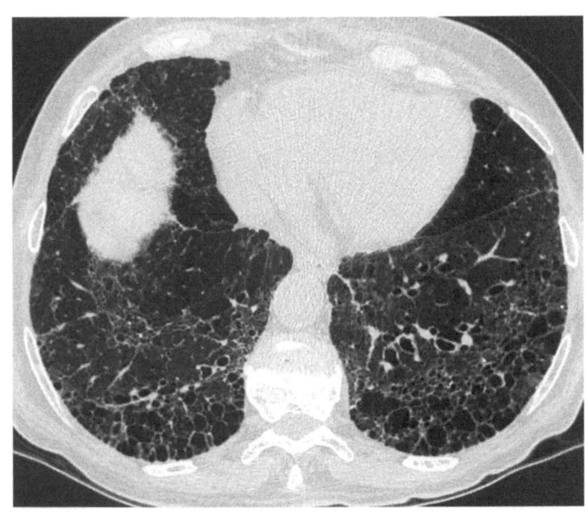

FIG. 9.4.5 Honeycombing in a patient with idiopathic pulmonary fibrosis. Unlike the typical honeycomb pattern, there are multiple large cysts with relatively thin walls.

affects the basal parts of the lungs and is usually surrounded by pronounced reticular changes. In addition, the cyst sizes never change depending on the phase of breathing [4]. Difficulties of the differential diagnosis can arise when confronted with atypical forms of honeycombing with large cystic transformation (Fig. 9.4.5).

It can be difficult to estimate the true thickness of the cyst wall because the compression of the surrounding parenchyma makes it appear thicker. At the same time the cavities of destruction can have thin walls, lending to an appearance that is practically indistinguishable from true cysts (Fig. 9.4.6).

Interlobular interstitium with the vasculature surrounding centrilobular emphysema with impaired collateral ventilation sometimes creates the impression of a thin wall, like in a cyst (Fig. 9.4.3A). Infection of the cyst can also thicken its wall. In any case, cavitary lesions with wall thicknesses of 2–3 mm should be considered as possible cysts. Cystic bronchiectasis

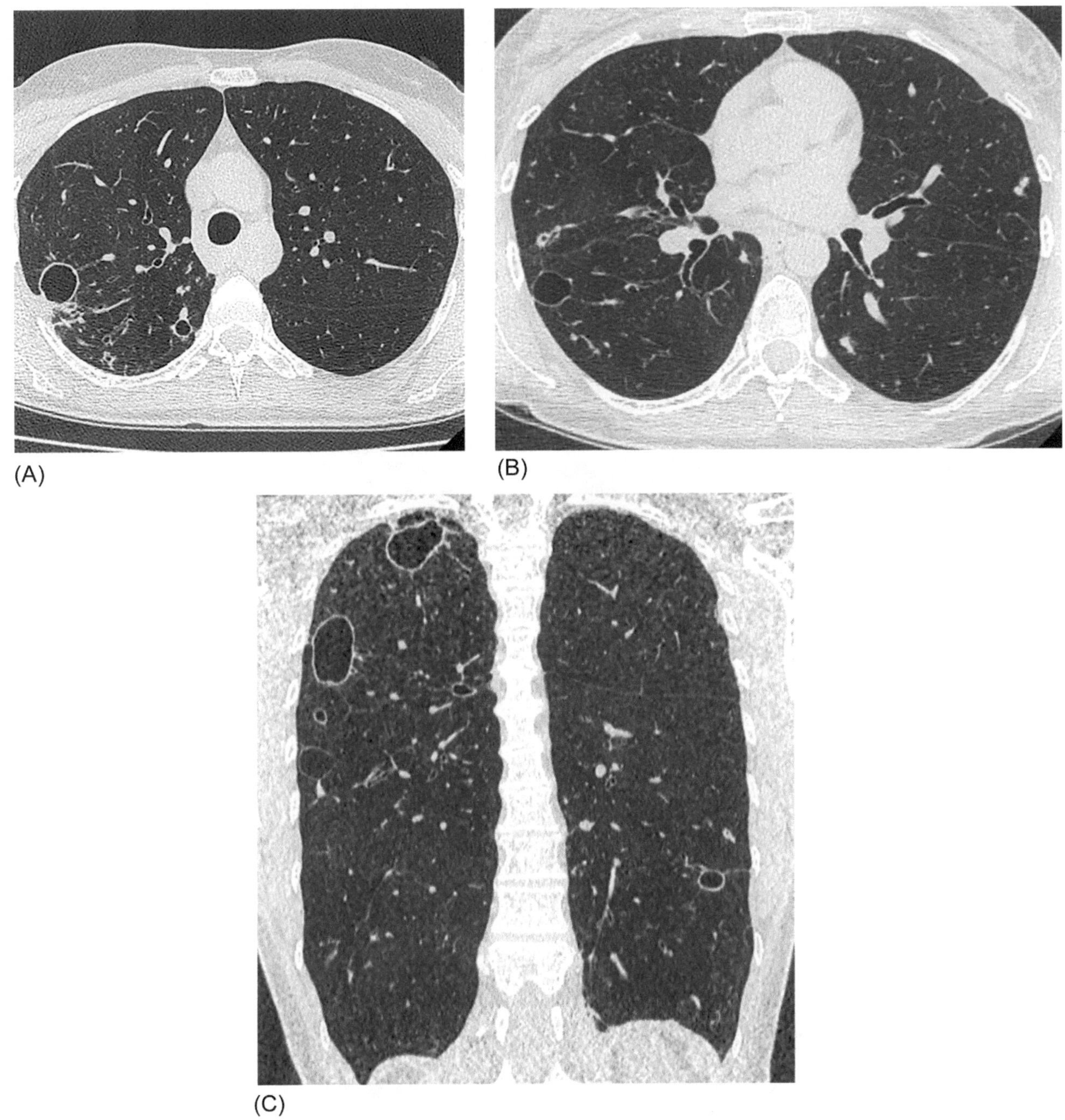

FIG. 9.4.6 Mycobacteriosis of the lungs caused by *Mycobacterium avium* complex. Multiple rounded cavities with a wall thickness of 1–2 mm. Around some of cavities, infiltration of the parenchyma (A) is visible. Single, randomly scattered, and dense nodules are revealed. Within one nodule a central lucency (B) is traced. Most of cystic lesions are distributed in the upper lobes (C).

usually exhibits this wall thickness, a concomitant branch of the bronchial artery is always proximal, and there are signs of a prolonged inflammatory process in the bronchial tree and peribronchial fibrosis (Fig. 9.4.7).

The next question to be considered for a differential diagnosis is the characteristics of observable cysts. Cysts may be monomorphic rounded or polymorphic, polygonal, diffusely distributed, or concentrated in certain zones (Table 9.4.2). It is also necessary to take into account the size of cystic formations.

Patients with cysts plus other HRCT phenomena should be carefully considered during differential diagnosis. The most common findings observed in combination with cystic changes are centrilobular nodules, ground-glass opacities, consolidation, and enlargement of the intrathoracic lymph nodes (Table 9.4.3).

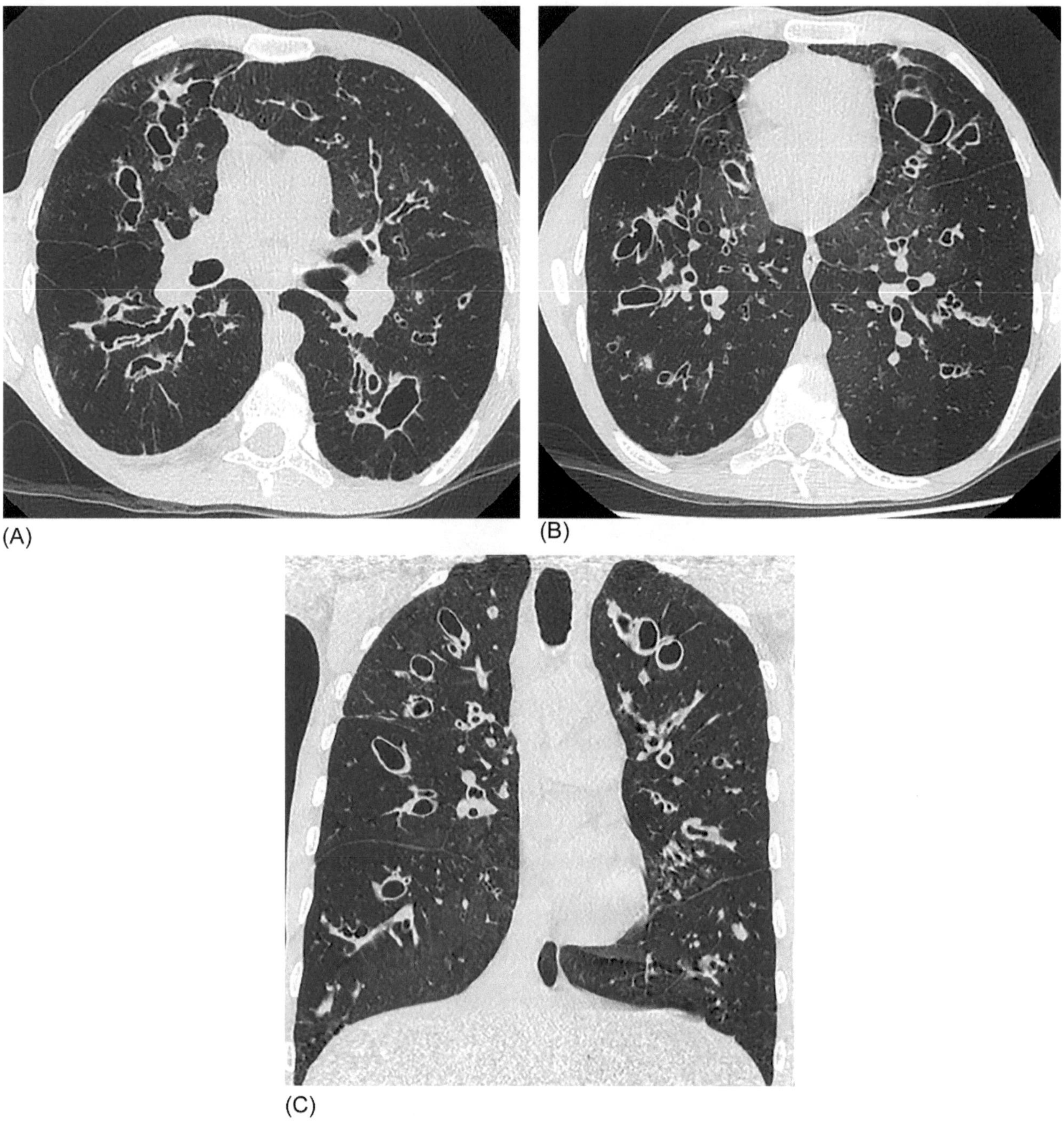

FIG. 9.4.7 Cystic bronchiectasis in a patient with cystic fibrosis. Multiple pockets with a relatively thick wall (up to 3 mm) and peribronchial fibrosis (A); mosaic perfusion and separate nodules, probably of infectious origin are visible (B, C).

TABLE 9.4.2 Special aspects of cysts in various diffuse cystic lung diseases

	Monomorphic	Polymorphic	Diffuse/casual	Mostly upper lobes	Mostly lower lobes	Diameter >10 mm
LAM	+	–	+	–	–	±
LCH	–	+	+	+	–	–
BHDS	–	+	+	–	+	++
LP/FB	+	±	±	–	+	+
CM	+		+	–	+	±
Infectious	–	+	+	+[a]	–	±
AM, LCDD	–	+	+	–	+	++

[a] *Characteristic of pneumocystis jirovecii pneumonia.*
LAM, lymphangioleiomyomatosis; LCH, Langerhans cellular histiocytosis; BHDS, Birt-Hogg-Dubé syndrome; LP, lymphocytic pneumonia; FB, follicular bronchiolitis; CM, cystic metastases; AM, amyloidosis; LCDD, light-chain deposition disease; ++, typical sign; +, often present; ±, possible sign, –, noncharacteristic sign.

TABLE 9.4.3 HRCT findings in diffuse cystic lung diseases

	Intralobular nodules	GGO	Consolidation	Mediastinal lymphadenopathy	Reticular signs
LAM	–	–	–	±	–
LCH	++	+	–	–	+
BHDS	–	–	–	–	–
LP	+	++	–	±	++
CM	+	+	+	+	–
Infectious	+	++	++	+	+
AM, LCDD	++	+	+	++	+

GGO, ground-glass opacity; LAM, lymphangioleiomyomatosis; LCH, Langerhans cell histiocytosis; BHDS, Birt-Hogg-Dubé syndrome; LP, lymphocytic pneumonia; FB, follicular bronchiolitis; CM, cystic metastases; AM, amyloidosis; LCDD, light-chain deposition disease; ++, typical sign; +, often present; ±, possible sign; –, noncharacteristic sign.

When analyzing cases with a combination of small nodular and cystic lung lesions, most typical for LCH, the following questions should be answered:

1. *What type of nodules does the patient exhibit—interstitial or acinar (localized in airspace)*? Interstitial nodules have well-defined margins and a high density. In contrast, acinar nodules can exhibit ill-defined contours and attenuation close to GGO and less than peripheral vessels [5]. For LCH, amyloidosis, and the light-chain deposition disease (LCDD), interstitial nodules are typical. Acinar nodules exhibiting filling of alveoli and bronchioles by the cell exudative substrate (Fig. 9.4.8) and are revealed in patients with pneumonia, including viral and pneumocystic; HP; bronchiolitis; organizing pneumonia; and vasculitis.
2. *What is the predominant distribution of nodular changes—perilymphatic, centrilobular, or random?*

The perilymphatic pattern presupposes a peribronchovascular layout of nodules. Around the interlobular septa and visceral pleura, this distribution is typical of sarcoidosis, pneumonoconiosis, lymphogenous metastases, amyloidosis of the lungs, and LCDD. Usually there is a blotching of the distribution including concentration in some segments and sparse distribution in other segments.

Random distribution is usually associated with diffuse, hematogenically disseminated processes, such as miliary pulmonary tuberculosis and hematogenous metastases, but occasionally can occur in patients with sarcoidosis. These changes are usually bilateral and symmetrical and are found in most parts of the lungs [6].

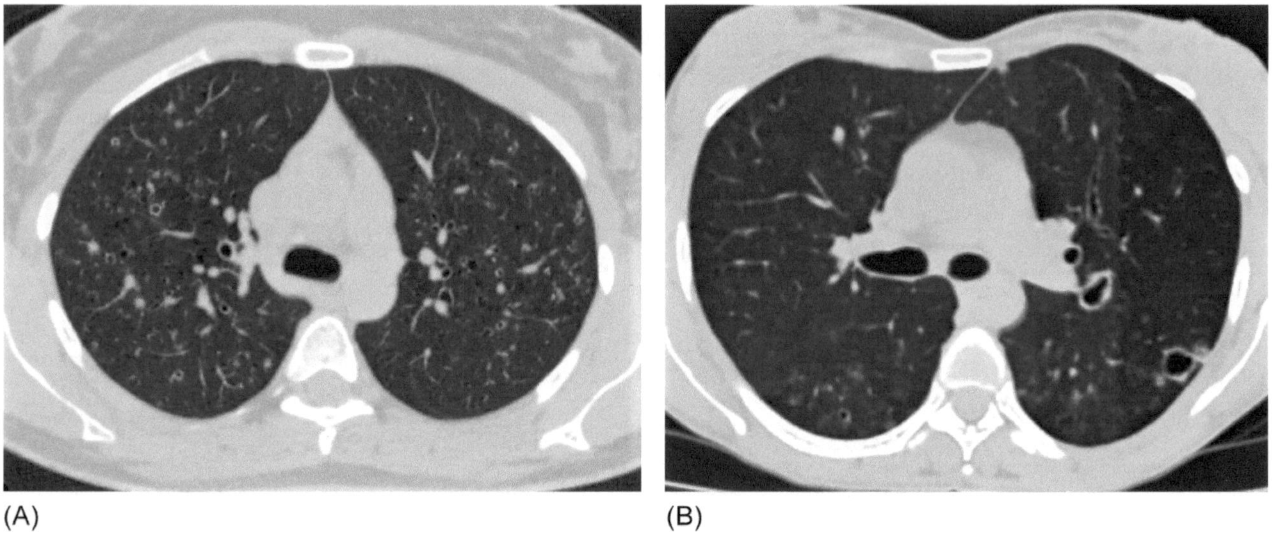

FIG. 9.4.8 Interstitial nodules in a patient with LCH (A). Dense, with well-defined margins, some with lucent centers (B). Multiple intralobular acinar nodules in a patient with pneumocystic pneumonia. The nodules have ill-defined margins and a ground-glass attenuation.

Intralobular nodules may occur in a variety of diseases, accompanied by damage to acinus and interstitial tissue (idiopathic and secondary interstitial pneumonia; HP; bronchiolitis; infectious diseases, including HIV infection; etc.). However, almost all of these diseases are manifested by centrilobular nodules with ill-defined margins and ground-glass attenuation (either they are mixed—interstitial and centrilobular). In LCH the nodules often but not always have sharp contours and a high attenuation. Similar characteristics of nodules are seen in sarcoidosis but usually in association with perilymphatic distribution, which is an important sign that differentiates sarcoidosis from LCH.

The next question is there intrathoracic lymphadenopathy?

Enlargement of intrathoracic lymph nodes is an important feature of diseases with a cystic pattern such as amyloidosis, LCDD, and (occasionally) LAM, but is not among the diagnostic features of LCH and BHDS.

Another question—does the patient have a "tree in bud" phenomenon?

This radiological sign reflects the presence of dilated bronchioles filled with exudate and usually accompanies heterogeneous bronchiolitis [5]. In most cases, this phenomenon indicates an endobronchial generalization of the infectious process. This is seen in cases of tuberculosis, nontuberculous mycobacteriosis of the lungs, fungal invasions, bronchiectasis, etc. The "tree in bud" sign suggests the exclusion of LCH as a diagnosis.

Clinical and laboratory manifestations are always taken into account when analyzing a patient with a combination of small focal and cystic lung lesions. These include acute or subacute onset, signs of active inflammation, and immunodeficiency and suggest infectious lung damage. Young and middle age, active smoking, polydipsia, and polyuria are clinical markers of LCH. A history of mieloma, chronic inflammatory disease and renal failure suggests possible amyloidosis or LCDD.

Using only HRCT, it is difficult to differentiate LCH from TSC. Both conditions exhibit multiple small nodules, cysts, and moderate reticular changes (Fig. 9.4.9). TSC is characterized by thin-walled cysts without deformity and a tendency to fuse. Here extrapulmonary manifestations such as facial angiofibromatosis; retinal hamartomas; tumor lesions of the brain, kidneys, and liver; and family history are decisive factors for the diagnosis. Diagnosis confirmation requires genetic testing for mutations of the TSC-1 and TSC-2 genes.

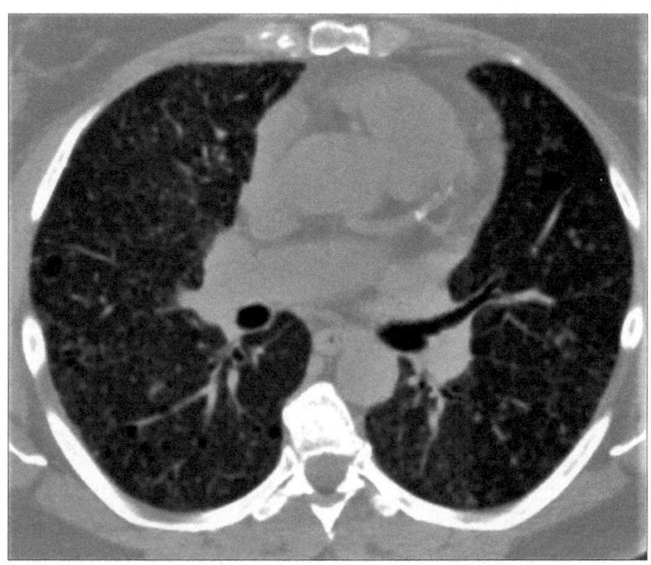

FIG. 9.4.9 Lesion of lungs with TSC. Multiple small nodules, thin-walled cysts of small and medium size, thickening of interlobular septa.

The GGO occurs in a variety of lung diseases, is completely uncharacteristic of LAM and BHDS, and is noted in most patients with LIP and infectious cysts. In some diseases, accompanied by a pronounced GGO (HP and desquamative interstitial pneumonia), single cysts can develop, but they never predominate on general CT. Cystic metastases in the lungs may initially look like small GGO foci.

The combination of cystic formations within sites of consolidation most often indicates an infectious process in the lungs (*pneumocystis jirovecii* pneumonia, mycobacteriosis, and tuberculosis) or deposition lung diseases (amyloidosis and LCDD) (Figs. 9.4.10 and 9.4.11).

Reticular changes in combination with multiple cysts usually occur with LIP, amyloidosis of the lungs, and infections and in the late stages of LCH, but are not characteristic of LAM and BHDS.

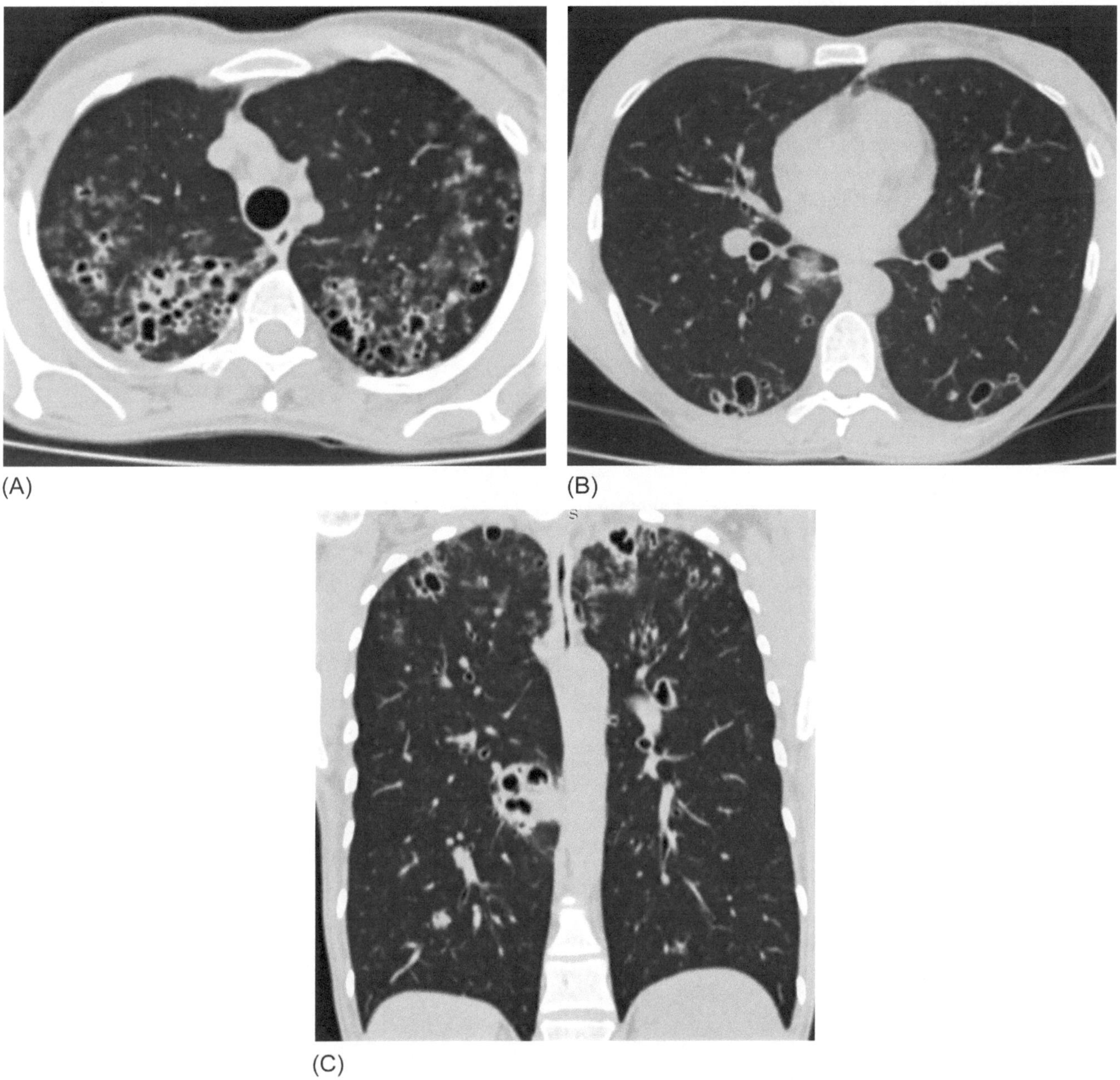

FIG. 9.4.10 *Pneumocystis jirovecii* pneumonia in a patient with HIV not receiving antiretroviral therapy. Multiple thick-walled small cavities in the upper lobes; the space between them is filled with high attenuation and GGO (A). In the lower lobes, there are single cavities and foci of GGO (B). On the coronal reconstruction upper lobe-predominant abnormalities are visible (C).

While performing a differential diagnosis of diffuse cystic lung diseases, a number of other conditions are considered, usually in addition to the big three (LAM, LCH, and BHDS). Differential diagnoses are conditions where cysts can be the only HRCT manifestation of the disease or dominate the general CT presentation. These can include cystic metastases in the lungs, LIP, lung damage with amyloidosis, and LCDD (Table 9.4.4).

Lymphocytic interstitial pneumonia (LIP) may accompany some autoimmune diseases, such as Sjögren syndrome, systemic lupus erythematosus, primary biliary cirrhosis, autoimmune thyroiditis, and HIV infection, but sometimes occurs as an idiopathic form. In contrast to LAM, cysts with LIP are typically larger than 10 mm in diameter and limited in number, exhibit a lower lobe distribution, and are accompanied by HRCT signs of interstitial inflammation such as a GGO, thickening of interlobular septa, and intralobular nodules [7] (Fig. 9.4.12).

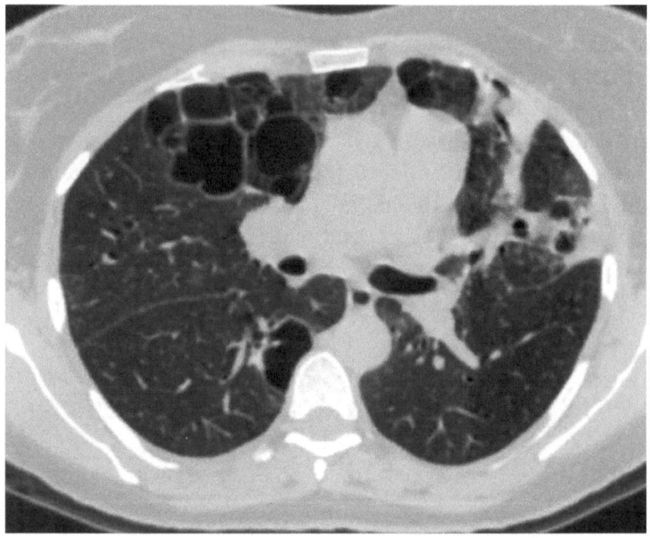

FIG. 9.4.11 LCDD. Multiple cysts with a wall thickness of 1–2 mm are in the right lung. Peribronchovascular consolidation with associated thick-wall cysts and separate thin-walled cysts are visible in the left lung.

A number of malignant tumors such as leiomyosarcoma, epithelioid-cell sarcoma, and endometrial stromal sarcoma can metastasize to the lungs and exhibit cyst group similar to that observed with LAM (Fig. 9.4.13). The cysts are thin-walled and of different sizes. If limited in number, they are usually located in the basal regions of the lungs [7]. We found a casuistic case of a primary multiple cystic pulmonary adenocarcinoma that originally presented with thin-walled cysts (Fig. 9.4.14).

Cyst formation is a form of lung damage seen in patients with amyloidosis. In addition to cystic transformation, there are usually other HRCT findings of the disease such as multiple nodules up to 15 mm in diameter, sometimes with

TABLE 9.4.4 Differential signs of diffuse cystic lung lesions

	CT signs	Additional characteristics
Lymphangioleiomyomatosis	Multiple symmetrical thin-walled cysts of 2–5 mm in diameter in unchanged parenchyma. Additional multiple small nodules in TSC	Women of childbearing age Chylothorax Renal angiomyolipoma Lymphangioleiomyoma of the uterus Retroperitoneal lymphadenopathy VEGF-D in serum > 800 pg/mL
Pulmonary Langerhans cell histiocytosis	Small thin- and thick-walled deformed cysts up to 10 mm, mainly in the upper lobes, interstitial nodules of 1–5 mm, costophrenic angles spared	Heavy smokers, young and middle age
Birt-Hogg-Dubé syndrome	A limited number of thin-walled cysts located in the middle basal areas. No progression with dynamic observation	Family history of pneumothorax, multiple papular eruptions on the face, renal cancer
Lymphocytic interstitial pneumonia	Large single cysts, mainly in the basal parts. GGO, thickening of interlobular septa, centriacinar nodules	Frequent association with autoimmune, lymphoproliferative diseases
Cystic metastases in the lungs	Single cysts of different sizes in the middle and basal parts of the lungs. Foci of GGO or nodules are possible	Malignant abdominal and pelvic tumors. High level of CA-125 Symptoms of cancer
Amyloidosis	Cysts of various sizes in combination with nodules, consolidation and intrathoracic lymphadenopathy, calcification within the zones of parenchymal infiltration, changes lean toward the basal divisions	Sjögren syndrome, MALT lymphoma, renal failure
Light-chain deposition disease	Cysts of medium and large size, contain vessels in the wall of the cyst, dense nodules with perilymphatic distribution, intrathoracic lymphadenopathy	Macroglobulinemia, myeloma disease, increased level of gamma globulins in the blood, renal failure

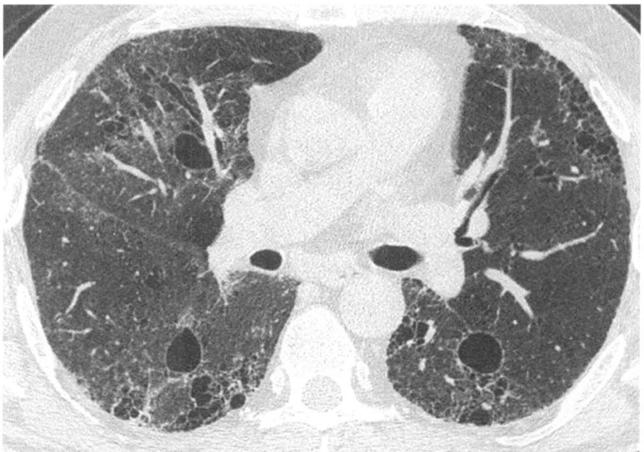

FIG. 9.4.12 HRCT of a patient with lymphoid interstitial pneumonia. Diffuse zones of GGO associated with thin-walled large cysts, reticular abnormalities, honeycombing, and traction bronchiectasis are in the lower regions.

cavitation. Additionally foci of consolidation are observed, inside of which dense calcification and intrathoracic lymphadenopathy are often found [8]. However, the forms of the disease in which the cystic pattern predominates are also described [9]. Cysts in cases of amyloidosis are thin-walled, rounded, and tend to accumulate in lower regions of the lungs. Their size can vary from 6 to 20 mm in diameter [10]. Most often, cysts are formed with amyloidosis associated with Sjögren syndrome and with lymphocytic interstitial pneumonia. A confirmation of amyloidosis of the lungs is the detection of amyloid masses stained with Congo red via polarization microscope or fluorescence microscopy.

The distinguishing feature of amyloid masses in the lungs is their low accumulation of (18) F-2-deoxyglucose on PET-CT scanning. This helps differentiate them from tumors and sarcoidosis [10].

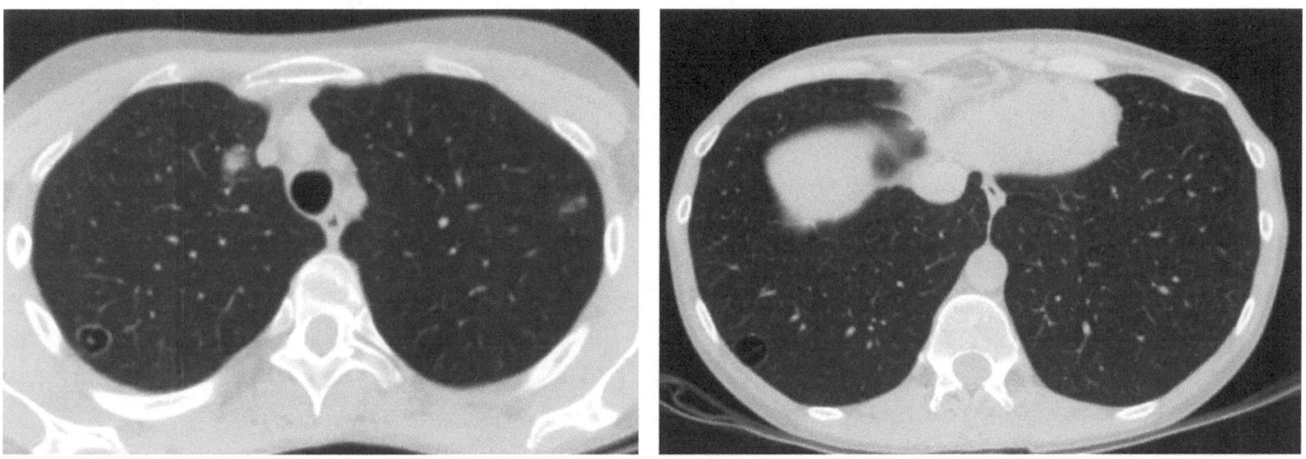

(A) (B)

FIG. 9.4.13 Hematogenic cystic metastases of ovarian leiomyosarcoma in the lungs. Single thin-walled cysts; inside a preserved lobular artery is visible. (A, B). Single nodules with irregular contours are in the upper lobes (A)

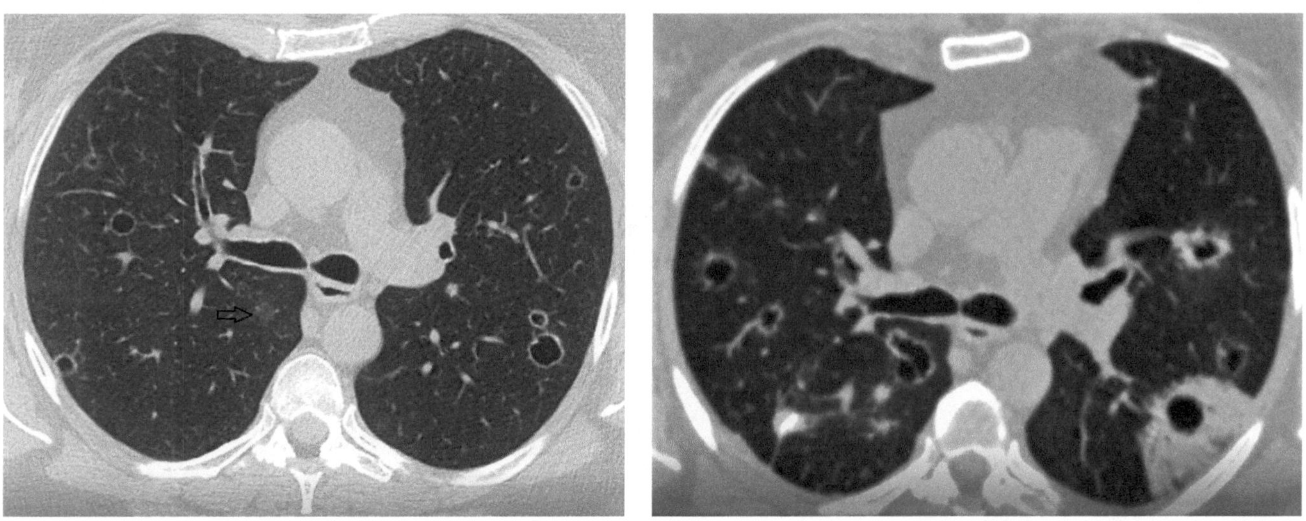

(A) (B)

FIG. 9.4.14 Cystic pulmonary adenocarcinoma. Randomly distributed cavities with wall thicknesses of 1–2 mm. In the right lung the focal area of GGO *(arrow)* (A) is detected. After nine months the number of cavities increased and their walls thickened and deformed, with the emergence of perifocal zones of consolidation and calcification (B).

Light-chain deposition disease (LCDD) is a systemic disease in which hyperproduction and the deposition of light-chain immunoglobulin occurs in various tissues and organs. Two-thirds of patients with multiple myeloma exhibit LCDD. LCDD is also associated with Waldenstrom macroglobulinemia and lymphoma [11].

Despite a pathogenesis similar to pulmonary amyloidosis, LCDD is associated with granular, and not fibrillary, deposition of amorphous eosinophilic material in the walls of the alveoli, vessels, and respiratory tract. Unlike amyloids, it does not have a characteristic apple-green birefringence luminescence in the polarization light during Congo red staining [12].

In LCDD the kidneys are affected more often than the lungs. Patients can exhibit nephrotic syndrome and kidney failure, as well as heart and liver failure. Isolated forms of lung disease are also described.

Typical manifestations of the pulmonary process include focal lesions in the lungs and multiple cysts. According to S. Sheard et al. the cysts vary in size from 3 to 35 mm, averaging 10 mm. The cysts themselves usually have a regular and round shape [13]. Cysts are usually bilateral, randomly distributed throughout the pulmonary parenchyma, and larger in the lower lobes [12]. The peculiarity of the cysts is the presence of their wall or the circumflexion of the cysts with pulmonary vessels (Fig. 9.4.15) [13].

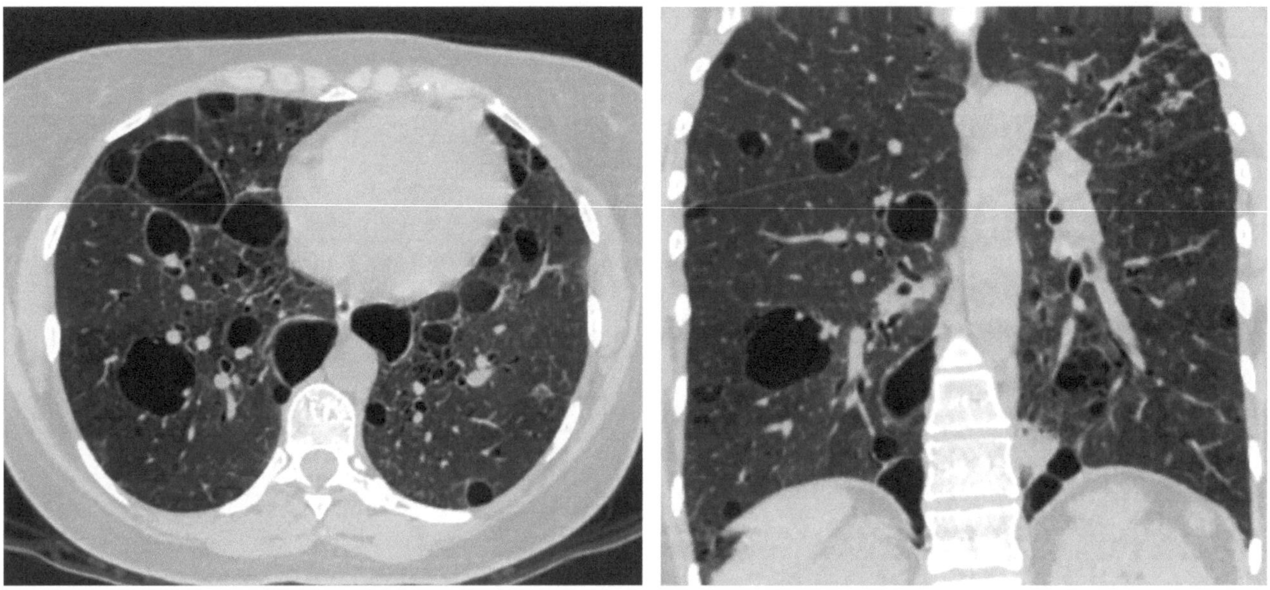

(A) (B)

FIG. 9.4.15 Light-chain deposition disease of the 51-year-old woman. Multiple cysts, in the lower lobes of predominantly large size, contain vessels in the wall (A, B). Several dense nodules are in the upper left lobe; area of consolidation is in the left low lobe.

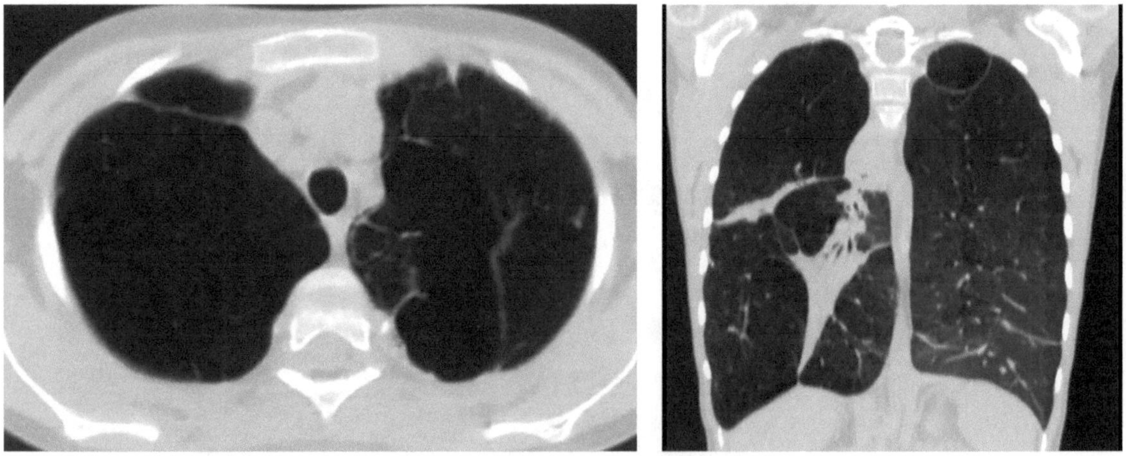

(A) (B)

FIG. 9.4.16 Proteus syndrome in 14-year-old girl. Giant bulla with irregular contour in the upper left lobe (A). Coronal reconstruction shows diffuse emphysema and fibritic masses in the right lung (B).

Nodules in LCDD are usually multiple, ranging in size from a few millimeters to 2 cm, and have a perilymphatic pattern of distribution [13]. The appearance of zones of consolidation in individual patients is described [12]. Intrathoracic lymphadenopathy is possible, but not necessary finding in the LCDD. Primary diagnostics of LCDD includes the study of gamma globulins in serum. At their increased level, further diagnosis is made by serum and urine protein electrophoresis (PEL) with immunofixation (IFE), to demonstrate the presence of monoclonal protein [14]. A number of rare diseases, such as Proteus syndrome, Ehlers-Danlos syndrome, neurofibromatosis, and Marfan syndrome, may manifest with pulmonary emphysema or intraparenchymal bullae, which are difficult to distinguish from true cysts (see Figs. 9.4.3 and 9.4.16).

Despite the fact that CT scans of diffuse cystic lung diseases are well-studied, even experienced specialists cannot always identify the disease. M. Koyama et al. reported that the correct diagnosis was made by two independent qualified radiologists in only 72% of patients with LCH and LAM and in 81% of patients with LIP. In contrast, interstitial pneumonia was accurately diagnosed in 100% of patients [15].

References

[1] Hansell DM, Bankier AA, MacMahon H, McLoud TC, Muller NL, Remy J. Fleischner Society: glossary of terms for thoracic imaging. Radiology 2008;246(3):697–722.

[2] COPDGene CT Workshop Group, Barr RG, Berkowitz EA, Bigazzi F, Bode F, Bon J, et al. A combined pulmonary-radiology workshop for visual evaluation of COPD: study design, chest CT findings and concordance with quantitative evaluation. COPD 2012;9(2):151–9.

[3] Ryu JH, Swensen SJ. Cystic and cavitary lung diseases: focal and diffuse. Mayo Clin Proc 2003;78(6):744–52.

[4] Aquino SL, Webb WR, Zaloudek CJ, Stern EJ. Lung cysts associated with honeycombing: change in size on expiratory CT scans. AJR Am J Roentgenol 1994;162(3):583–4.

[5] Webb WR, Muller NL, Naidich DP. High-resolution CT of the lung. 4th ed. Philadelphia: Lippincott Williams and Wilkins; 2009. p. 368–81.

[6] McLoud TC, Boiselle PM. Thoracic radiology the requisites. Philadelphia, PA: Mosby Elsevier; 2010.

[7] Seaman DM, Meyer CA, Gilman MD, McCormack FX. Diffuse cystic lung disease at high-resolution CT. AJR Am J Roentgenol 2011;196(6):1305–11.

[8] Czeyda-Pommersheim F, Hwang M, Chen SS. Amyloidosis: modern cross-sectional imaging. Radiographics 2015;35(5):1381–92.

[9] Chew KM, Clarke MJ, Dubey N, Seet JE. Nodular pulmonary amyloidosis with unusual, widespread lung cysts. Singap Med J 2013;54(5):e97–9.

[10] Baqir M, Kluka EM, Aubry MC, Hartman TE, Yi ES, Bauer PR, et al. Amyloid-associated cystic lung disease in primary Sjögren's syndrome. Respir Med 2013;107(4):616–21.

[11] Rho L, Qiu L, Strauchen JA, Gordon RE, Teirstein AS. Pulmonary manifestations of light chain deposition disease. Respirology 2009;14(5):767–70.

[12] Colombat M, Stern M, Groussard O, Droz D, Brauner M, Valeyre D, et al. Pulmonary cystic disorder related to light chain deposition disease. Am J Respir Crit Care Med 2006;173(7):777–80.

[13] Sheard S, Nicholson AG, Edmunds L, Wotherspoon AC, Hansell DM. Pulmonary light-chain deposition disease: CT and pathology findings in nine patients. Clin Radiol 2015;70(5):515–22.

[14] Boppana S, Sacher RA. Light-chain deposition disease. Medscape reference, http://emedicine.medscape.com/article/202585-overview; June 17, 2014.

[15] Koyama M, Johkoh T, Honda O, Tsubamoto M, Kozuka T, Tomiyama N, et al. Chronic cystic lung disease: diagnostic accuracy of high-resolution CT in 92 patients. AJR Am J Roentgenol 2003;180(3):827–35.

Chapter 10

Tumors that mimic diffuse parenchymal lung disease

Oleg Pikin[a], Evgeniya Kogan[b], Victor Lesnyak[c], Alexander Averyanov[d,e]

[a]Thoracic Surgery Department, Hertzen Research Institute of Oncology, Moscow, Russia, [b]Anatomic Pathology Department, Sechenov University, Moscow, Russia, [c]Radiology Department, Federal Research Clinical Center under FMBA of Russia, Moscow, Russia, [d]Clinical Department, Pulmonology Research Institute under FMBA of Russia, Moscow, Russia, [e]Pulmonary Division, Federal Research Clinical Center under FMBA of Russia, Moscow, Russia

Chapter 10.1

Primary malignant lung tumors

Chapter 10.1.1

Pulmonary adenocarcinoma

Lung cancer is one of the most common malignant tumors, and in most countries of the world, it is the oncological disease with the highest mortality [1, 2]. The main method for detecting and staging lung cancer is through X-ray diagnostics. Lung cancers are divided into two main groups: small-cell lung cancer and non-small cell lung cancer (NSCLC). NSCLC is characterized by various radiographic and clinical manifestations and is most often associated with smoking [3]. Pulmonary adenocarcinoma, a subtype of NSCLC, is observed especially in women and nonsmokers, and its incidence rate has shown a substantial increase during the last 70 years [4]. This is the histological subtype of lung cancer that causes the greatest difficulties in diagnosis, especially in differentiating it from diffuse nontumor lung diseases. On high-resolution computed tomography (HRCT), adenocarcinoma can appear as a solid mass, a consolidation focus, or ground-glass opacity in one lung, but it can often be bilateral, taking the form of multiple infiltration zones of varying opacities or interstitial changes; this can cause difficulties in differential diagnosis with inflammatory lung diseases, and it is often referred to as "masquerader" [5] (Fig. 10.1.1.1).

These radiological changes in lung tissue are most frequently observed in a specific subtype of pulmonary adenocarcinoma previously called bronchioloalveolar carcinoma (BAC). The changes characteristic of BAC were first described more than 130 years ago, although the first detailed characterization of BAC as a special subtype of high-grade pulmonary adenocarcinoma was presented by Leibow in 1960 [6]. Recently, views about the pathomorphology of pulmonary adenocarcinoma have undergone significant changes, as reflected in the new classification of the European Respiratory Society, the American Thoracic Society, the International Association for the Study of Lung Cancer in 2011, and the World Health Organization in 2015 (Table 10.1.1.1) [7, 8].

Almost all the histological variants of pulmonary adenocarcinoma in computed tomography can be represented by diffuse or interstitial abnormalities in lung tissue. This can hinder differential diagnostics with conditions such as infectious lung lesions, organizing pneumonia, granulomatosis with polyangiitis, and chronic eosinophilic pneumonia.

Difficult to Diagnose Rare Diffuse Lung Disease. https://doi.org/10.1016/B978-0-12-815375-8.00010-8

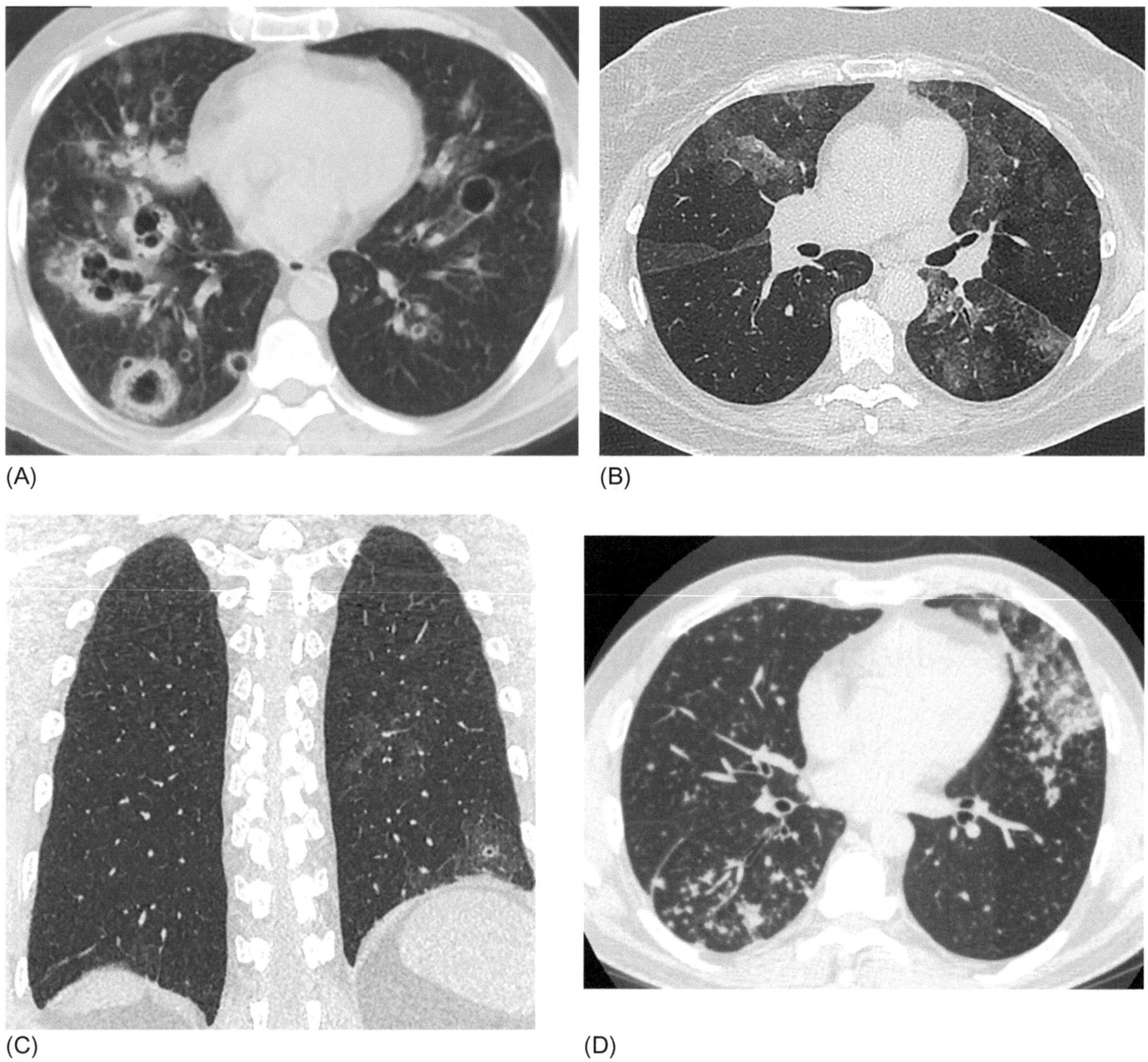

FIG. 10.1.1.1 Diversity of CT of semiotics of pulmonary adenocarcinoma: (A) Bilateral thick- and thin-walled lung cysts of different sizes, randomly distributed foci of consolidation, ground-glass opacity, and solid nodules. (B) Bilateral, extensive ground-glass opacity (GGO) zones, sharply demarcated by the interlobar pleura, associated thickened interlobular septa are visible ("crazy paving"). It resembles a picture of alveolar proteinosis. (C) Multisegmental areas of light GGO in the left lung; there is a small cavity formation inside one such area in the lower lobe—the air-bubble sign. (D) Multiple scattered nodules with a perilymphatic distribution, sometimes coalescing into larger foci. Patchy areas of GGO in the left lung. A similar pattern can be observed with pulmonary sarcoidosis.

Morphology

The main feature of adenocarcinomas that mimic diffuse nontumor lung diseases is their characteristic growth along the anatomical structures, preserving the alveolar structure. Tumor cells grow as a monolayer on the alveolar septum, which serves as a platform for them; this phenomenon has been described as "lepidic growth," because the tumor cells resemble butterflies sitting on the septum. Atypical adenomatous hyperplasia (AAH), adenocarcinoma in situ, and minimally invasive adenocarcinoma are characterized by preservation of the lung alveolar structure and the absence of stromal invasion with desmoplasia, unlike the other adenocarcinomas (Fig. 10.1.1.7–10.1.1.10). Apparently, it is this feature of adenocarcinoma from among those of the former BAC that is associated with the CT phenomena also observed in interstitial lung diseases, namely, widespread areas of ground-glass opacity, thickening of the intralobular and interlobular septa, and "crazy paving."

TABLE 10.1.1.1 IASLC/ATS/ERS classification of lung adenocarcinoma in resection specimens [7]

Classification
Preinvasive lesions
Atypical adenomatous hyperplasia (Fig. 10.1.1.2)
Adenocarcinoma in situ (≤3 cm formerly BAC) (Fig. 10.1.1.3)
Nonmucinous
Mucinous (Fig. 10.1.1.4)
Mixed mucinous/nonmucinous
Minimally invasive adenocarcinoma (≤3-cm lepidic predominant tumor with ≤5-mm invasion)
Nonmucinous
Mucinous (Fig. 10.1.1.4)
Mixed mucinous/nonmucinous
Invasive adenocarcinoma
Lepidic predominant (formerly nonmucinous BAC pattern, with >5-mm invasion) (Fig. 10.1.1.5)
Acinar predominant
Papillary predominant
Micropapillary predominant
Solid predominant with mucin production
Variants of invasive adenocarcinoma
Invasive mucinous adenocarcinoma (formerly mucinous BAC) (Fig. 10.1.1.6)
Colloid
Fetal (low and high grade)
Enteric

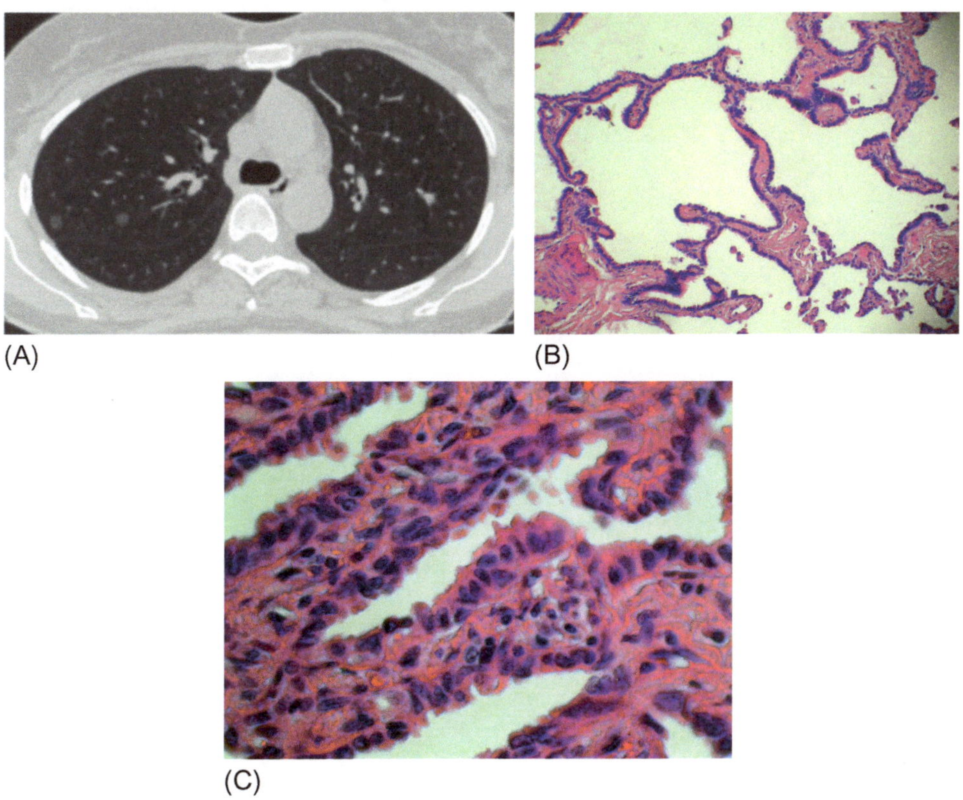

FIG. 10.1.1.2 Atypical adenomatous hyperplasia. (A) Two 3-mm foci of ground-glass opacity in the upper right lobe. (B) This nodular lesion consists of atypical pneumocytes proliferating along preexisting alveolar walls without features of invasive growth; hematoxylin and eosin staining, 100×. (C) Focus of slightly atypical proliferating pneumocytes with enlarged hyperchromatic and pleomorphic nuclear; few cells show multinucleation; hematoxylin and eosin staining, 400×.

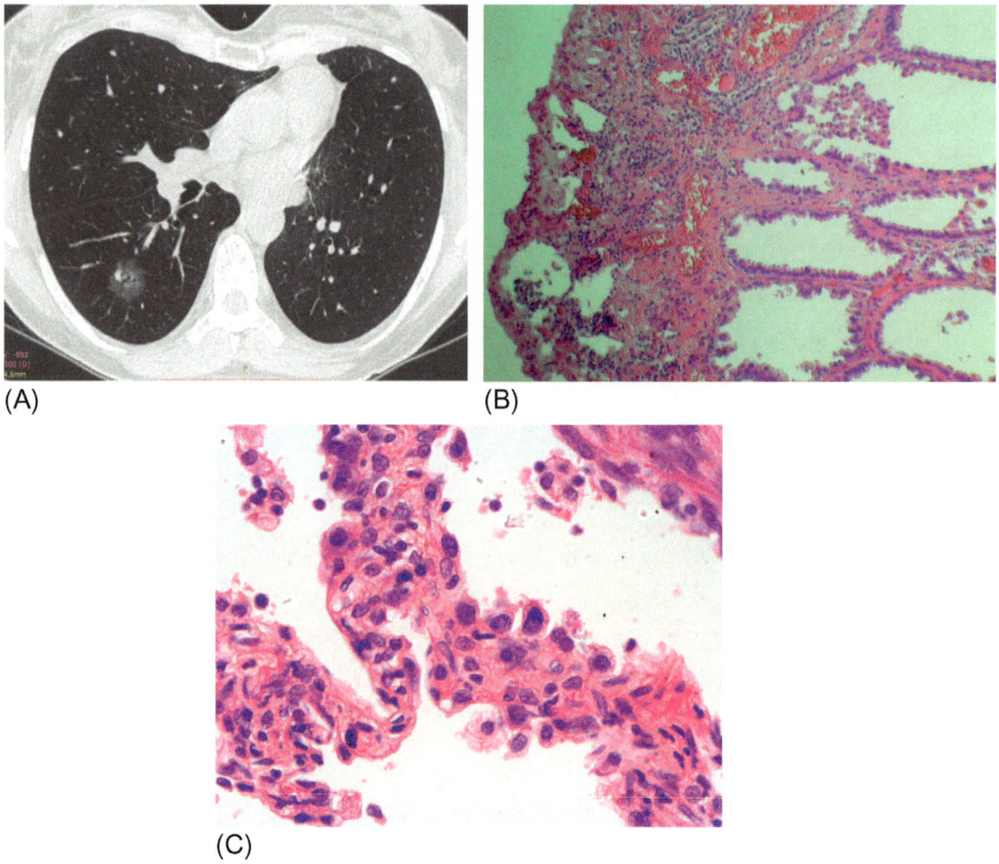

FIG. 10.1.1.3 Nonmucinous adenocarcinoma in situ. (A) 20-mm focus of GGO with a small solid component in the lower lobe of the right lung. (B) This circumscribed nonmucinous tumor grows purely with a lepidic pattern. No foci of invasion or scarring are observed; hematoxylin and eosin staining, 100×. (C) The tumor shows atypical pneumocytes proliferating along the slightly thickened but preserved alveolar walls; hematoxylin and eosin staining, 400×.

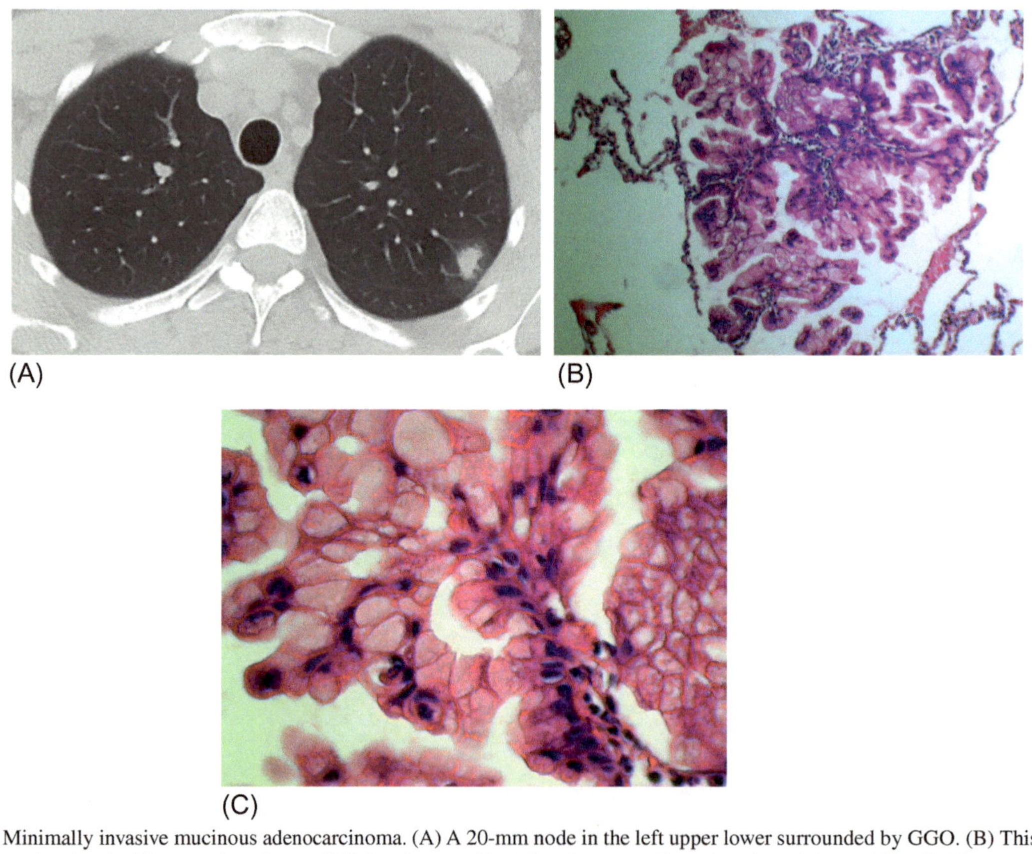

FIG. 10.1.1.4 Minimally invasive mucinous adenocarcinoma. (A) A 20-mm node in the left upper lower surrounded by GGO. (B) This mucinous MIA consists of a tumor showing lepidic growth and a small (0.5-cm) area of invasion; hematoxylin and eosin staining, ×100. (C) The tumor cells consist of mucinous columnar cells growing mostly in a lepidic pattern along the surface of the alveolar walls; hematoxylin and eosin staining, ×400.

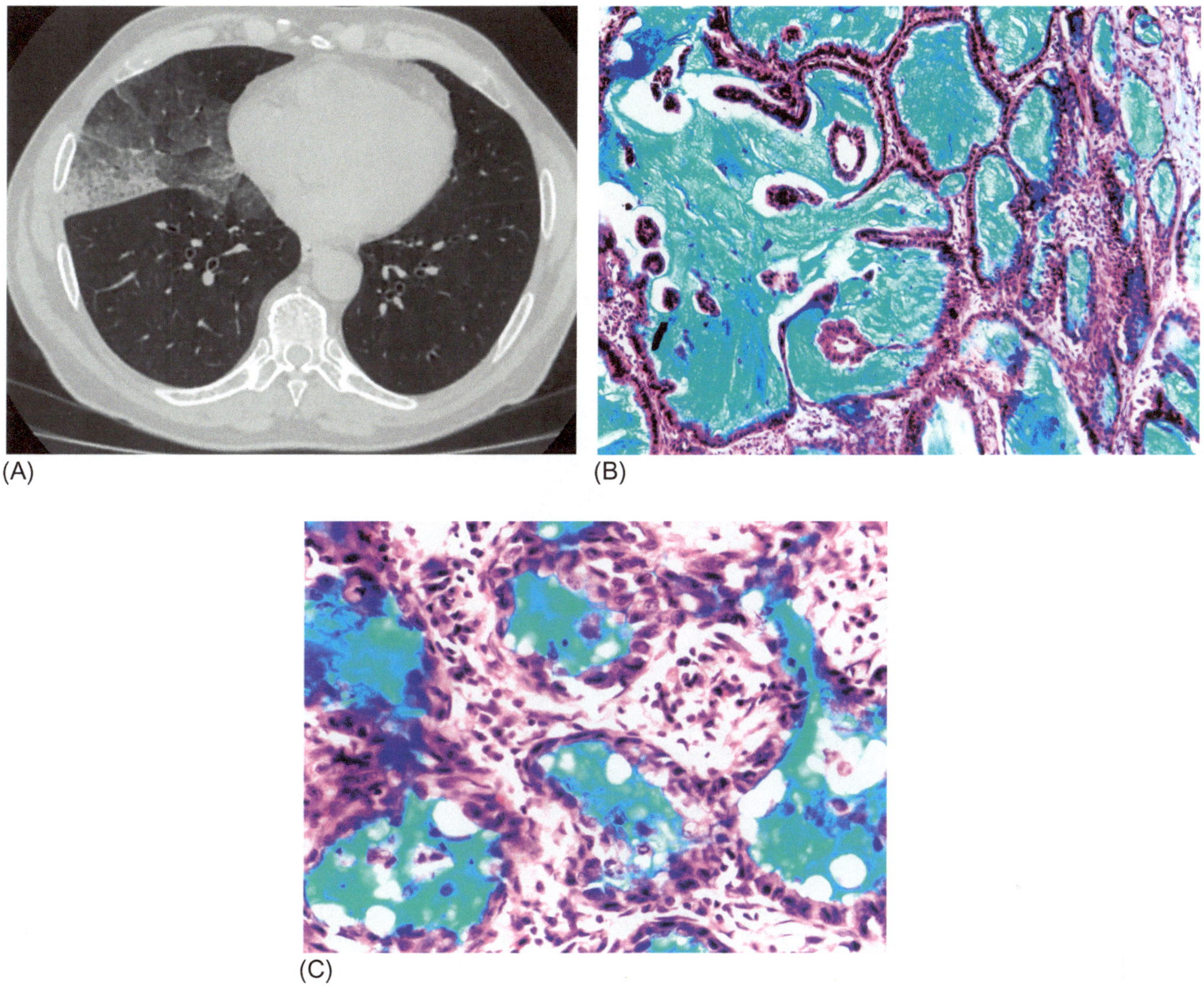

FIG. 10.1.1.5 Invasive adenocarcinoma with lepidic growth. (A) Diffuse area of GGO with thickened intra- and interlobular septa in the right middle lobe. (B) This area of invasive mucinous adenocarcinoma demonstrates a pure lepidic growth. The tumor consists of columnar cells filled with abundant mucin in the apical cytoplasm and shows small basal-oriented nuclei; alcian blue staining, 100×. (C) Nevertheless, elsewhere this tumor demonstrated invasion associated with desmoplastic stroma and an acinar pattern; alcian blue staining, 400×.

Immunohistochemically nonmucinous variants of adenocarcinomas express cytokeratin CK-7 and thyroid transcription factor (TTF-1), but they do not express cytokeratin CK-20 [9].

Mucinous variants of adenocarcinoma in situ and minimally invasive adenocarcinoma usually express cytokeratins CK-7 and CK-20 but not the thyroid transcription factor TTF-1 [7, 10].

Both mucinous and nonmucinous tumors can cause concomitant changes in alveolar septa, such as pronounced lymphoplasmacytic infiltration, amyloid deposits, osteocartilaginous metaplasia, and fibrosis [7, 8].

Clinical presentation

If the localized form of a pulmonary adenocarcinoma that belongs to the former BAC category is virtually asymptomatic or manifests only as a cough, then the diffuse version often resembles bacterial pneumonia, with the only difference being that the symptoms do not appear acutely but increase over weeks; the leading complaints are progressive dyspnea and a productive cough, sometimes sufficient to be classed as bronchorrhea. However, the significance of this symptom for the diagnosis of mucinous adenocarcinomas is obviously exaggerated [11], although, with extensive bilateral and lobar lung lesions, a patient can discharge up to a liter of viscous mucilaginous sputum per day. Some patients have hemoptysis. Elevated body temperature may be present. Most patients exhibit general symptoms, such as weakness, weight loss, and loss of appetite [12].

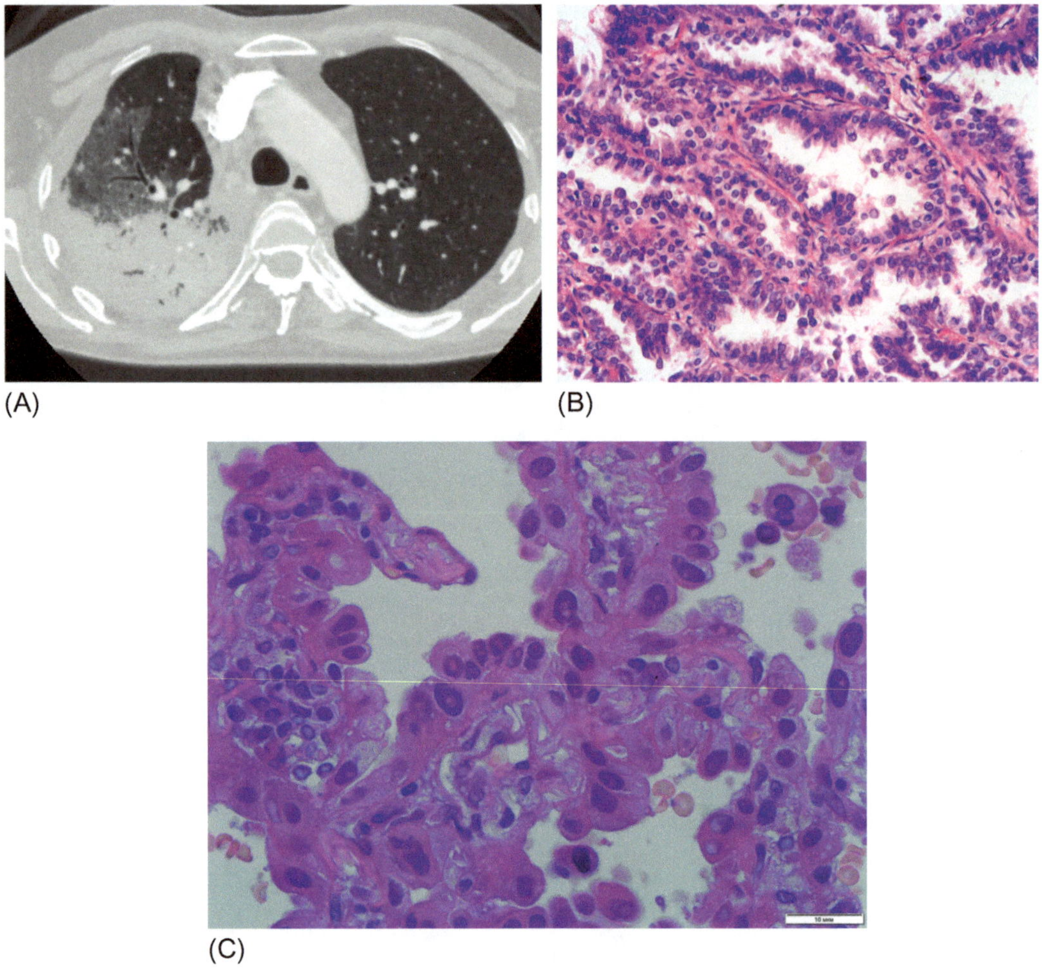

FIG. 10.1.1.6 CT and histological patterns of invasive mucinous adenocarcinoma. (A) An extensive zone of consolidation in the right upper lobe. A GGO area with thickened interlobular septa reflects a tumor invasion. (B) Lepidic predominant pattern with mostly lepidic growth. The lepidic pattern consists of a proliferation of type II pneumocytes and Clara cells along the surface of the alveolar walls; hematoxylin and eosin staining, 200×. (C) The lepidic pattern consists of a proliferation of type II pneumocytes along the surface of the alveolar walls; hematoxylin and eosin staining, 600×.

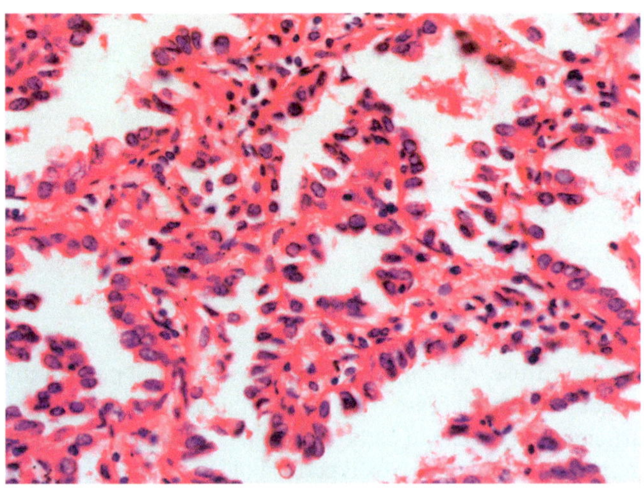

FIG. 10.1.1.7 Lepidic predominant adenocarcinoma consisting of atypical cuboidal and cylindrical cells proliferating along the surface of the alveolar walls; hematoxylin and eosin staining, 400×.

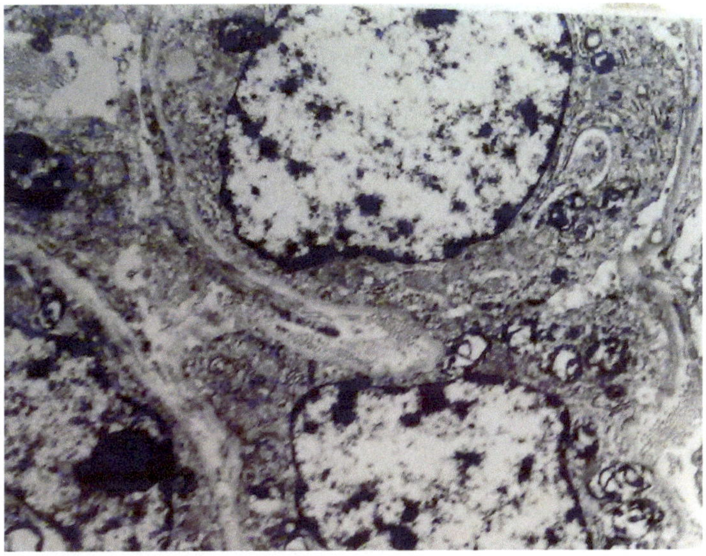

FIG. 10.1.1.8 Electron microscopy of cancer cells with lamellar granules in the cytoplasm (surfactant containing).

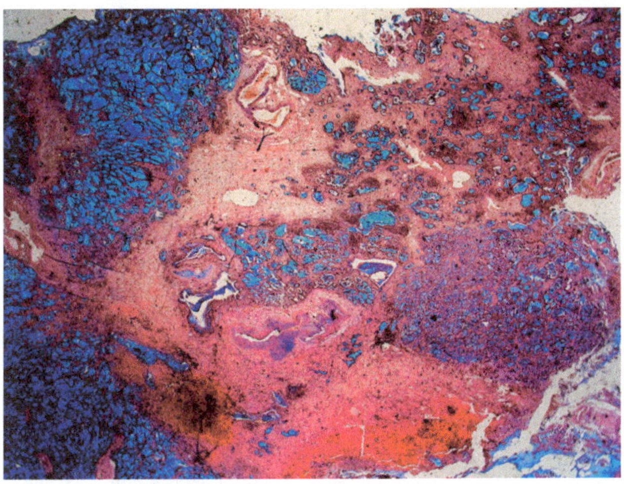

FIG. 10.1.1.9 Invasive mucinous adenocarcinoma with background fibrosis ("cancer in the scar"). Mucin in the lumens of glandular structures; alcian blue staining, 2.5×.

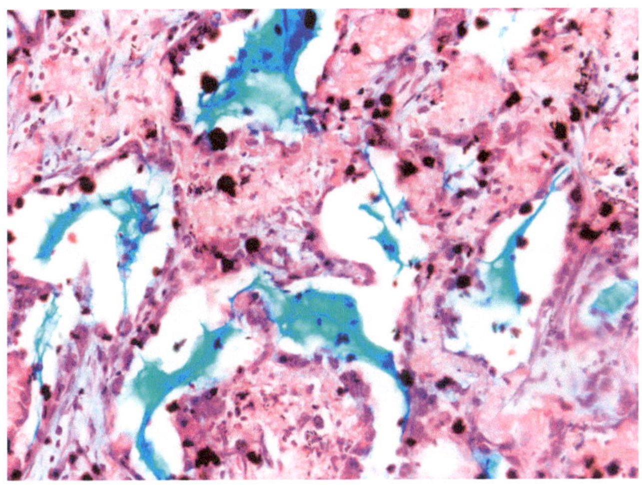

FIG. 10.1.1.10 Invasive mucinous adenocarcinoma with background fibrosis ("cancer in the scar"). Mucin in the lumens of glandular structures; alcian blue staining, 400×.

Radiological findings

The forms of adenocarcinoma previously classified as BAC present a diverse radiological picture, but three main variants of symptomatology can be distinguished [13]:

1. Ground-glass opacity or a nonsolid type of lesion
2. A mixed type, with the simultaneous presence of zones of lesion of different opacities
3. A solid variant, represented only by the consolidation site or sites

Characteristic radiographic signs of the solitary nodal form of adenocarcinoma are streaking toward the lung root and the parietal pleura, as well as the cellular structure of the tumor node, against which background small bronchial lumens can be seen. On HRCT, contrast enhancement reveals the vessels inside the infiltration zone, known as the "angiogram sign" [14].

Each histological variant of the tumor manifests differently on HRCT. So, in the study of Lee et al. [15], all patients with solitary nonmucinous minimally invasive adenocarcinoma showed only ground-glass opacity zones, whereas the mucinous variant exhibited both isolated areas of consolidation and these combined with ground-glass opacity. If, in the peripheral zone around the central consolidation, there is increased attenuation of the pulmonary tissue with ground-glass opacity (the halo sign) (Fig. 10.1.1.11), this indicates the presence of an invasive tumor component [12]. The traditional concept of a malignant lung tumor as a steadily progressing process is not always true for adenocarcinomas in situ. Monitoring patients with areas of ground-glass opacity in the lung tissue, subsequently identified as BAC, for more than 10 years has demonstrated that a CT image can remain stable for several years and can even show a reduction in the lesion area, albeit with simultaneous opacification of the central zone [16]. Unlike other forms of lung cancer that manifest as a solitary focus, invasive adenocarcinomas can appear as multiple sites of lung parenchyma opacities with different radiographic characteristics (Fig. 10.1.1.12), with primary multiple forms accounting for 22% of all adenocarcinomas [17].

The radiographic pattern of adenocarcinomas is not limited to ground-glass opacity and consolidation. Patsios et al. [14] identified the following additional signs: "air bubbles," appearing as small rounded radiolucent areas with sharp margins within the solid mass lesion (Fig. 10.1.1.13); the appearance of pseudocavities (sometimes multiple) in the consolidation zones (Fig. 10.1.1.14); and "crazy paving," revealing the accumulation of glycoprotein secretion in the alveoli and thickening of the interlobular septa (Fig. 10.1.1.15). These signs, along with traditional characteristics (ground-glass opacity consolidation and a combination of these), can present as single findings or in combination with other signs; this makes the HRCT diagnosis of adenocarcinoma rather unspecific, because it assumes a wide differential range of diseases that manifest with similar radiological findings. An atypical radiological variant of multifocal adenocarcinoma is a cystic form that shows infiltrative growth only in the late stages of the disease (Fig. 10.1.1.16).

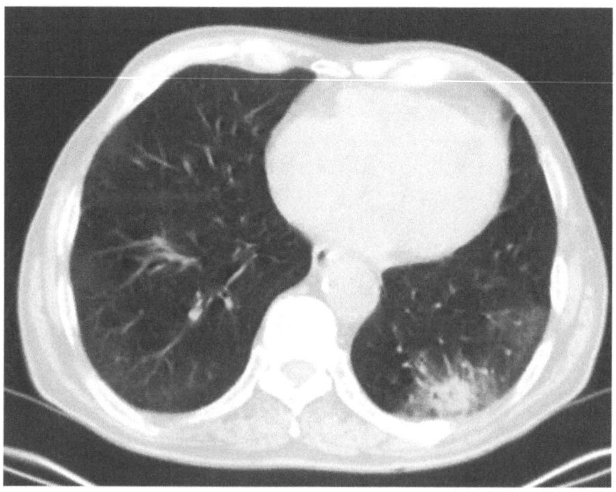

FIG. 10.1.1.11 Adenocarcinoma. Tumor node surrounded by a zone of ground-glass opacity (a halo sign). The air-bubble sign and spiculation are visible.

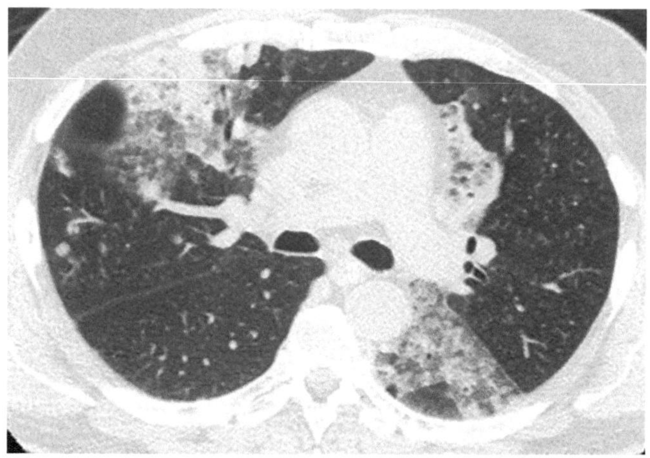

FIG. 10.1.1.12 A variety of CT signs of the multiple adenocarcinoma. There are bilateral zones of consolidation and ground-glass opacity, which are clearly delimited by the interlobar pleura in the left lung. Separately located solid nodules are visible.

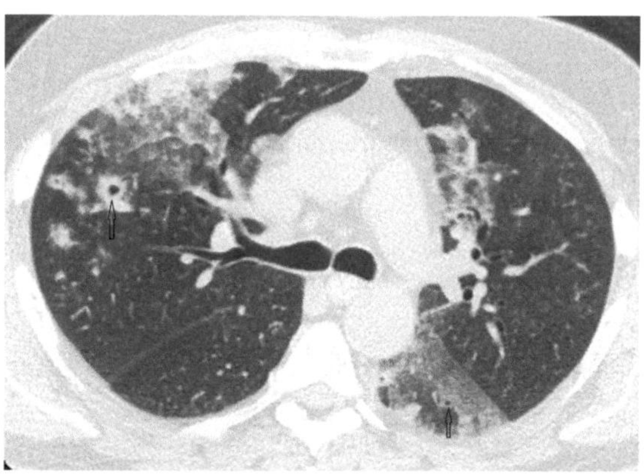

FIG. 10.1.1.13 Mucinous adenocarcinoma. Air-bubble sign *(arrows)*.

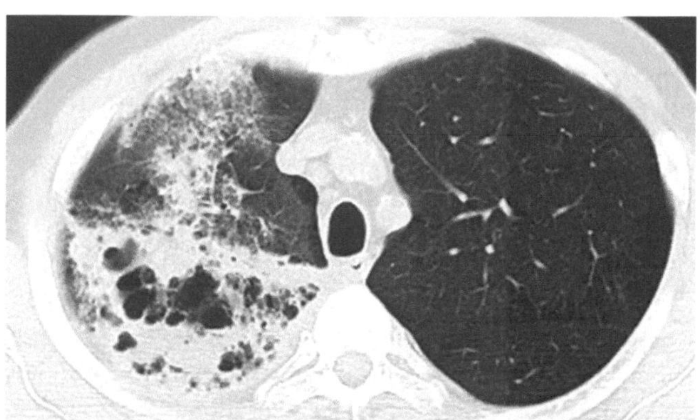

FIG. 10.1.1.14 Invasive mucinous adenocarcinoma. Multiple pseudocavities inside the consolidation zone, which are a consequence of the tumor growth.

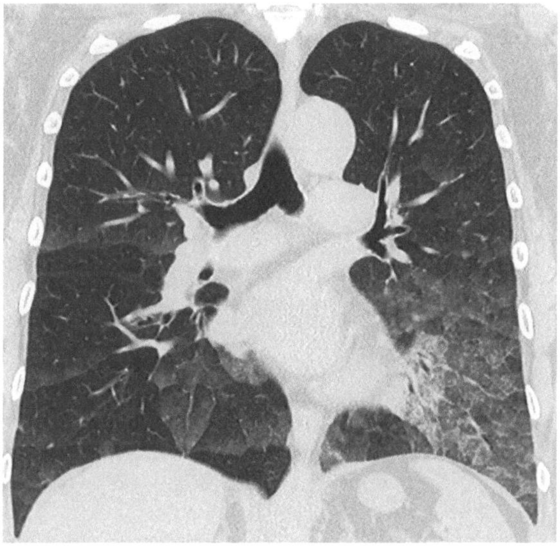

FIG. 10.1.1.15 Invasive nonmucinous adenocarcinoma. Bilateral areas of ground-glass opacity with thickened interlobular septa (crazy paving), mainly in the left lower lobe.

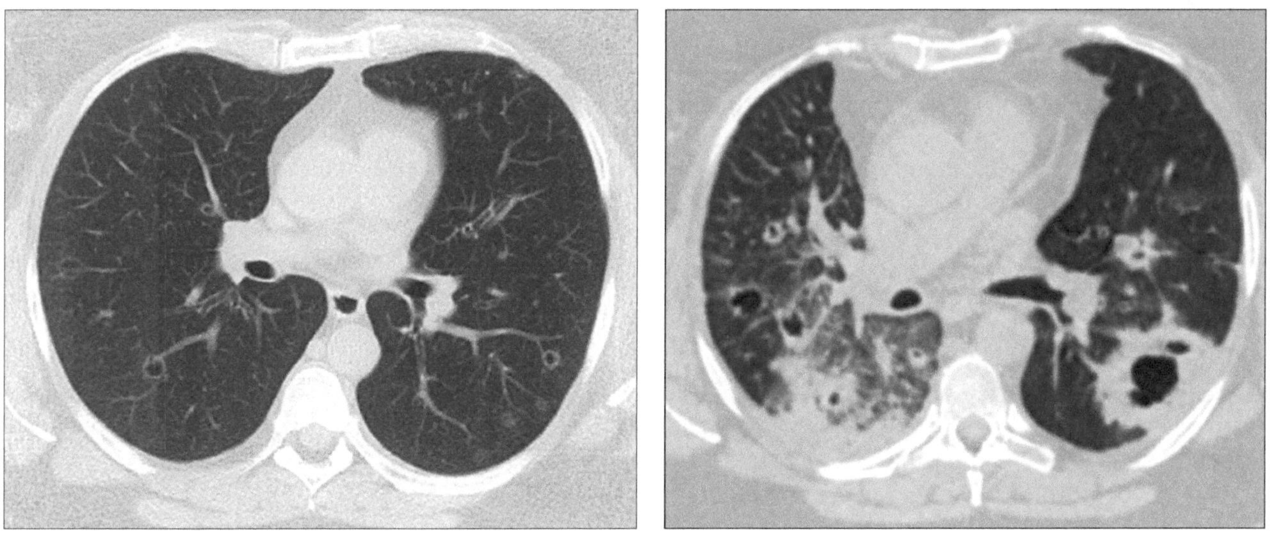

(A) (B)

FIG. 10.1.1.16 Invasive adenocarcinoma with nonusual cystic HRCT pattern. (A) Separately located thin-walled cysts connected to the vessels. In the left lower lobe, small ill-defined nodules with a lucent center are also visible. (B) The same patient after 18 months. Bilateral zones of GGO and consolidation with cavities in the left lung. There are several rounded thick walled cavities from 3 to 20mm in diameter.

Differential diagnosis

Areas of ground-glass opacity less than 10 mm in diameter are most often signs of AAH, a localized moderately atypical zone with the proliferation of type II alveolocytes or Clara cells that cover the alveolar walls, possibly spreading to the respiratory bronchioles (Fig. 10.1.1.2).

In the past, AAH was considered to be a premalignancy disease; in the modern classification, it refers to the early noninvasive form, which may precede or appear concomitantly with primary and metastatic pulmonary adenocarcinoma (Fig. 10.1.1.17) [18].

It is extremely difficult to differentiate adenocarcinomas in the former BAC or AAH by HRCT alone without an evaluation of the dynamics. Large adenocarcinomas prove malignant, as do adenocarcinomas where the lesion area increases in size during the follow-up period or where there is an air bronchogram in the ground-glass opacity zone or the presence of

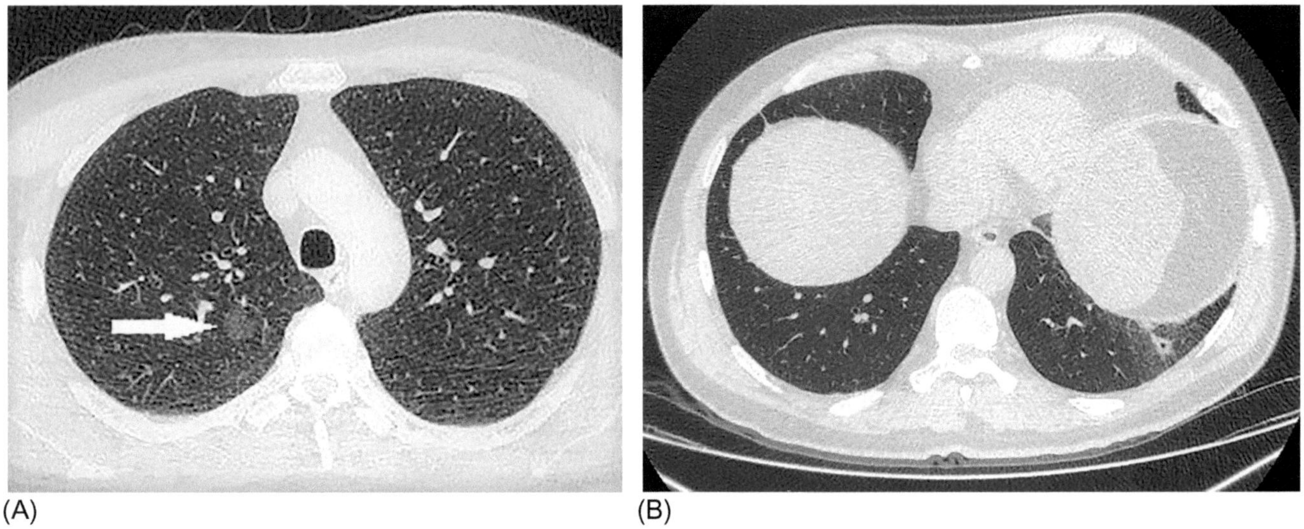

FIG. 10.1.1.17 (A) Atypical adenomatous hyperplasia. The rounded focus of ground-glass opacity of 15 mm in diameter *(arrow)* in the upper right lobe. (B) Adenocarcinoma. The same patient. Thick-walled small cavity surrounded by GGO in the left lower lobe.

consolidation sites. In a comparative study of the CT imaging of patients with AAH and BAC, Oda et al. [19] demonstrated highly significant differences in two main criteria: sphericity of the opacity zone, which was characteristic of AAH; and visualization of an air bubble in the ground-glass opacity that was observed seven times more often in BAC. It is believed that air bronchograms are caused by alveolar collapse or fibrosis in the area of the tumor, resulting in stretching of the bronchial walls. In the same study, it was found that the diameter was more than 10 mm in 47% of patients with AAH, whereas the diameter was less than 10 mm in only 14% of patients with BAC.

In clinical practice, the differential diagnosis of adenocarcinomas with unresolved bacterial pneumonia is important, but in most cases the X-ray and CT semiotics are nonspecific. The most specific signs of pulmonary adenocarcinoma are vacuole-like air bubbles; those of pneumonia show local thickening of the pleura and thickening of the bronchial wall [20].

Table 10.1.1.2 presents the frequency of various CT signs in patients with adenocarcinoma and pneumonia. In patients with pulmonary emphysema, these signs may appear together with pneumonic infiltration (Fig. 10.1.1.18).

Ground-glass opacity air bronchograms and CT angiograms are not pathognomonic for adenocarcinoma because they are equally common in unresolved pneumonia. Inflammatory signs such as the presence of minor pleural effusion and local hypertrophy of extrapleural adipose tissue are observed in pneumonia, and local pleural retraction is often noted in patients with adenocarcinoma, but these differences are not reliable (Fig. 10.1.1.19) [20–22].

TABLE 10.1.1.2 Differential and diagnostic CT criteria for pulmonary adenocarcinoma and unresolved pneumonia

CT sign	Adenocarcinoma	Pneumonia
Ground-glass opacity	++	++
Air bronchogram	++	++
Air bubble	+++	+
Thickening of the pleura	+	+++
Pleural retraction	+++	+
Pleural effusion	++	+
Thickening of the bronchus wall	+	+++
Local hypertrophy of extrapleural fat tissue	–	+
CT angiography	++	++

–, not observed; +, occurs rarely; ++, occurs with similar frequency; +++, occurs often.

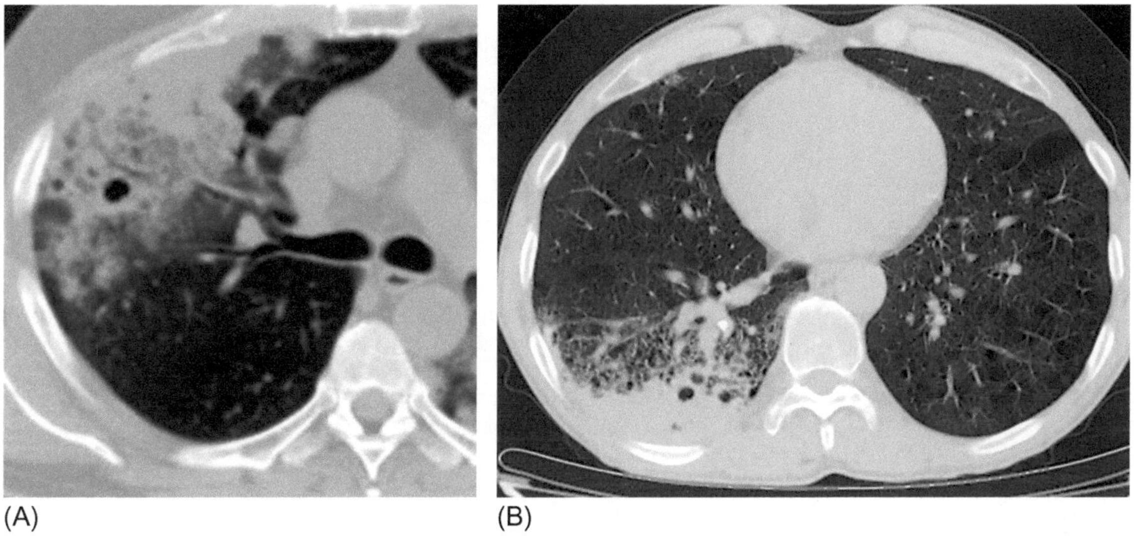

FIG. 10.1.1.18 (A) Mucinous invasive adenocarcinoma. Ground-glass opacity, area of consolidation, and an air-bubble sign are in the right upper lobe. (B) Bacterial pneumonia caused by *Pseudomonas aeruginosa* in a COPD patient. In the consolidation zone, small air cavities free from infiltration are visible—these are areas of emphysema. The visceral pleura in the adjacent area is thickened.

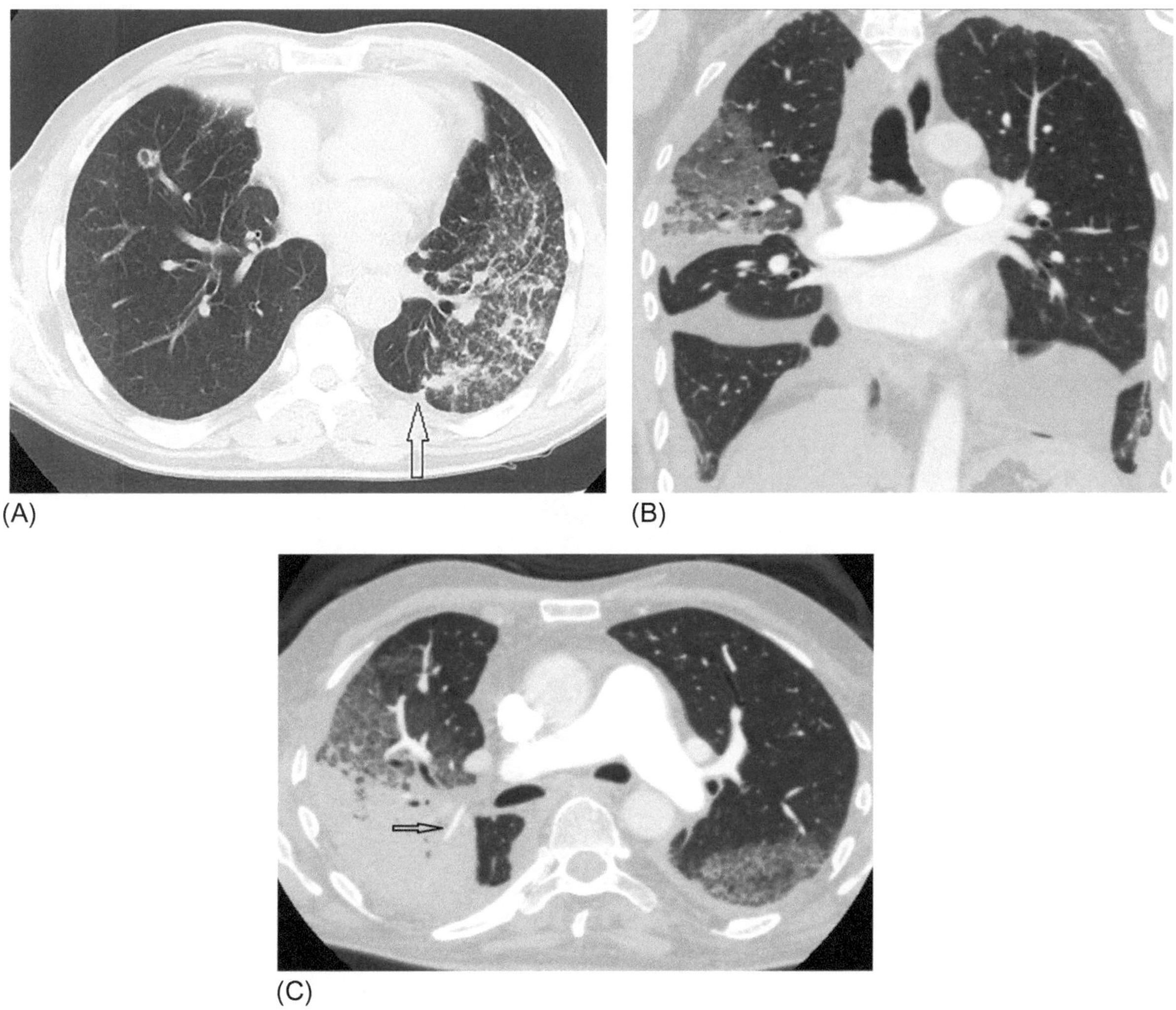

FIG. 10.1.1.19 Additional CT signs of lung adenocarcinoma: (A) visceral pleural retraction *(arrow)*. (B) Pleural effusion. (C) CT angiography sign *(arrow)*.

TABLE 10.1.1.3 Clinical criteria for pulmonary adenocarcinoma and unresolved pneumonia

Clinical characteristics	Adenocarcinoma (lepidic growth)	Unresolved pneumonia
Acute onset	No	Yes
Mucous, foamy sputum in large volume	Yes (in 5%–10% of patients with the mucinous form)	No
Changes in laboratory indicators	No or increased ESR	Increased level of leukocytes, CRP, and procalcitonin
Stable X-ray presentation	Yes (for months and sometimes years)	Usually, there is a tendency for the form of infiltration to change, with signs of fibrosis around the consolidation zone over the long term
Clinical response to antibiotic therapy	No	Yes
Presence of tumor cells in sputum or bronchial lavage	Yes	No
Conformation of diagnosis with transthoracic biopsy	Yes (in about 80% of cases)	No

In addition to the radiographic findings, it is necessary in the differential diagnostic series to take into account both the clinical manifestations of the disease and the anamnestic data (Table 10.1.1.3). Bacterial pneumonia is evidenced by its acute onset, more pronounced fatigue and febrile syndromes, purulent sputum, pronounced inflammatory changes in the blood test, and high levels of CRP and procalcitonin.

The algorithm for the differential diagnosis of pulmonary adenocarcinoma and unresolved pneumonia is presented in Fig. 10.1.1.20.

HRCT findings similar to that of pulmonary adenocarcinoma can be observed with an organizing pneumonia, manifesting as an inflammatory response of the lung tissue in systemic connective tissue diseases, organ transplantation, drug-induced and other lesions, or an idiopathic disease (a cryptogenic organizing pneumonia). The typical CT signs of organizing pneumonia are unilateral or (more often) bilateral patchy zones of inhomogeneous consolidation, located subpleurally or peribronchially, with bronchial wall thickening and dilatation in abnormal areas (Fig. 10.1.1.21) [23].

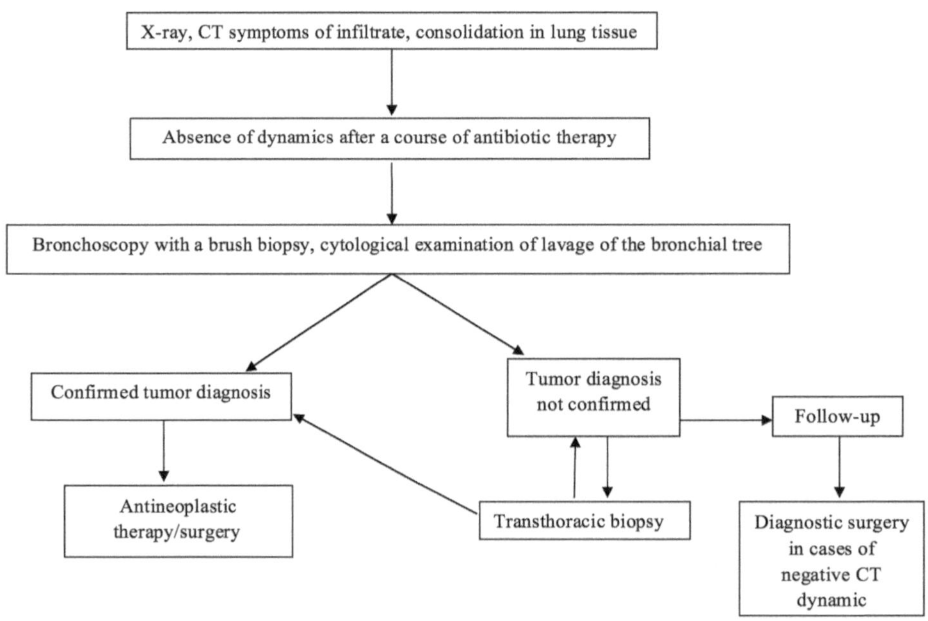

FIG. 10.1.1.20 Algorithm for the differential diagnosis of adenocarcinoma and unresolved pneumonia.

It is typical to find a high density of infiltrates, close to that found in the liver. In addition to zones of consolidation, there may be zones of ground-glass opacity. Unlike pulmonary adenocarcinoma, which is characterized by the halo sign, organizing pneumonia can present with a sickle-shaped zone of consolidation around the area of ground-glass opacity or unchanged lung tissue (reversed halo sign) (Fig. 10.1.1.21). Organizing pneumonia is characterized by changes in the size, configuration, and density of infiltrates, either spontaneously or influenced by the treatment [24]. The greatest difficulties arise in the differential diagnosis of a solitary type of organizing pneumonia and a lung tumor; the pneumonia can occur asymptomatically, and it can have radiographic characteristics resembling those of a malignant neoplasm (Fig. 10.1.1.22). The most common localization of solitary forms of organizing pneumonia is in the upper lobes of the lungs [23]. In the absence or technical impossibility of biopsy, the pathognomonic sign of the inflammatory nature of the changes is their positive dynamics under the influence of steroid therapy. Sometimes, pulmonary adenocarcinomas have to be differentiated not only from organizing pneumonia but also from other interstitial pneumonias (such as nonspecific, desquamative interstitial pneumonia and idiopathic pulmonary fibrosis), as well as rheumatoid nodules and granulomatosis [25]. For nonspecific interstitial pneumonia, reticular changes are typical, as well as the presence of generalized areas of ground-glass opacity and traction bronchiectasis (Fig. 10.1.1.23). Foci of consolidation may be present, but they are of limited size and are combined with interstitial fibrosis and bronchiectasis [26]. Desquamative interstitial pneumonia is characterized by symmetrical bilateral zones of ground-glass opacity in the lower regions and moderate reticular changes, but not by the presence of consolidation or other features natural for adenocarcinoma. Nevertheless, adenocarcinomas can exhibit histological findings such as tumor cells desquamated into alveoli, simulating macrophages; this can result in an erroneous morphological diagnosis of desquamative pneumonia, especially when there is only a limited volume of histological material [27]. Very rarely, lepidic growth with thickening of the alveolar walls can appear as honeycomb-like changes in the lung parenchyma, which can incorrectly be interpreted as pulmonary fibrosis [5].

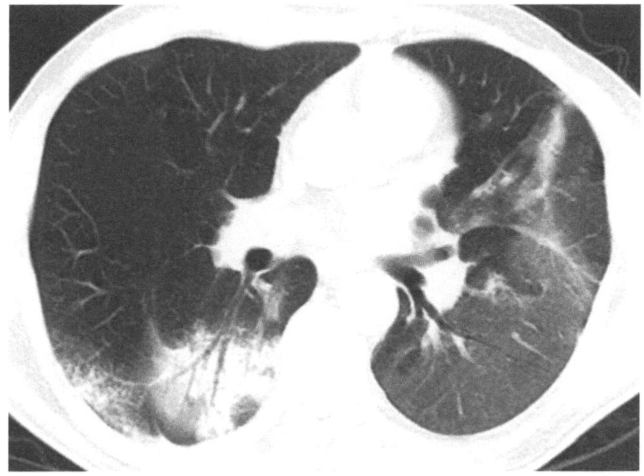

FIG. 10.1.1.21 Cryptogenic organizing pneumonia. Bilateral areas of GGO, consolidation with an air-bronchogram and atoll sign in the right lung.

In some cases, granulomatosis with polyangiitis (GP) can also present CT findings similar to those of pulmonary adenocarcinoma (Fig. 10.1.1.24).

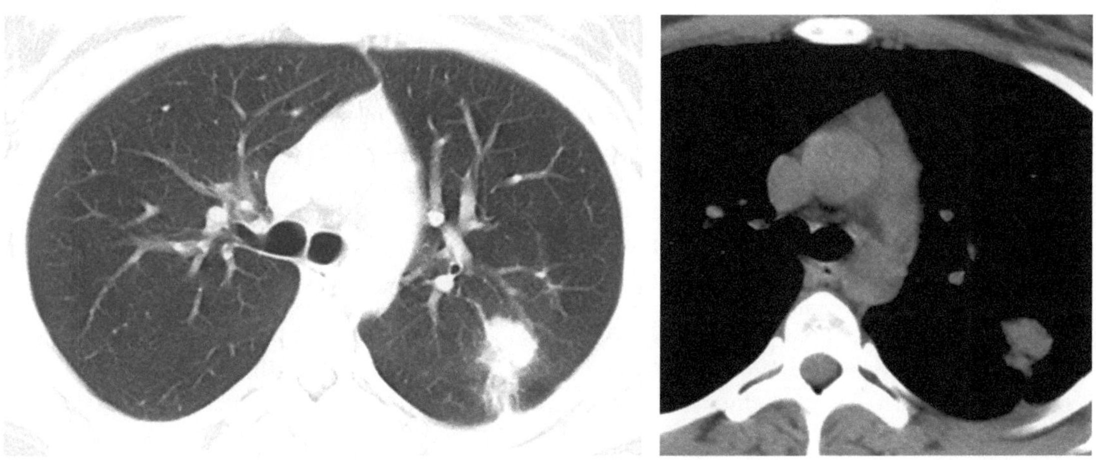

(A) (B)

FIG. 10.1.1.22 The solitary form of cryptogenic organizing pneumonia. The focus of consolidation surrounded by GGO with visceral pleura retraction is visible. The pattern resembles peripheral lung cancer.

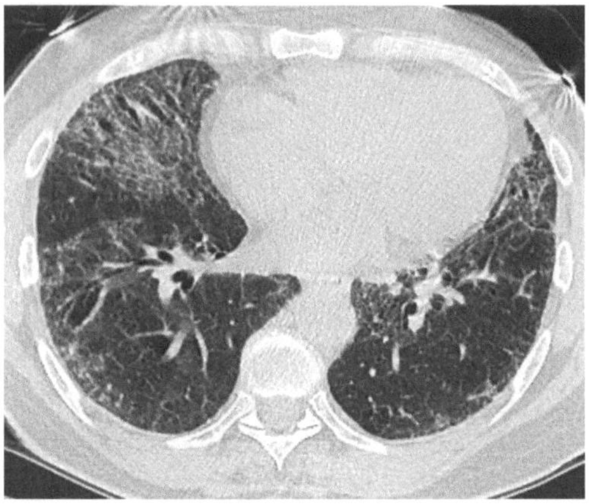

FIG. 10.1.1.23 Nonspecific interstitial pneumonia (NSIP). Bilateral areas of GGO with traction bronchiectasis and reticular abnormalities are visible.

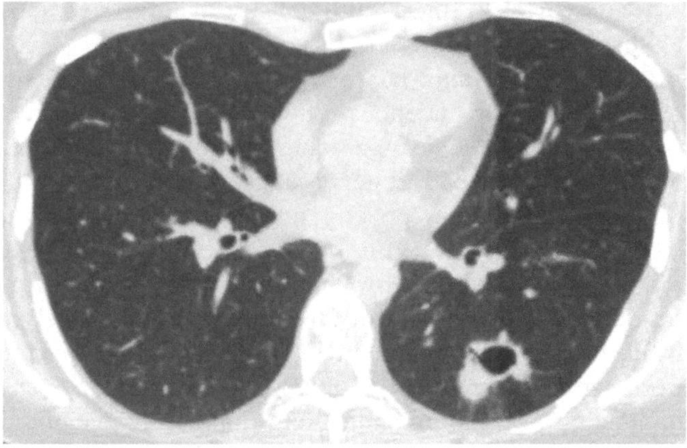

FIG. 10.1.1.24 Granulomatosis with polyangiitis. Focus of consolidation with irregular shape, spicularity contours, and internal cavity in the left lower lobe. Findings are very similar to adenocarcinoma pattern.

These may be areas of consolidation, which can be limited or large (>10 cm in diameter). The larger the size of the high-density zone, the more likely it is to have the appearance of cavities [28]. Very often, the consolidated areas are surrounded by a shadow of ground-glass opacity (the halo sign); however, unlike with adenocarcinoma, the opposite finding is also possible, with the ground-glass opacity surrounded by a band of consolidated tissue, the so-called reversed halo or atoll sign. This is generally infrequent, and it can require an additional differential diagnosis, such as in cases of organizing pneumonia, eosinophilic pneumonia, and drug-induced pneumonitis [29, 30]. Alveolar hemorrhages often occur in cases of GP; these manifest on CT as areas of ground-glass opacity with clear outlines, usually not affecting the subpleural zones, and they can be limited areas or involve subtotal spreading. In the differential diagnosis, additional clinical and laboratory characteristics of GP are more helpful. As a rule, lung damage is preceded by sinusitis resistant to the usual treatment, with ulceration of the nasal mucosa; up to 80% of patients have renal vessels lesions, manifested by hematuria [28]. An important diagnostic marker of GP is the presence of antibodies specific to proteinase-3 neutrophil cytoplasm (c-ANCA) in blood serum, which are detected in approximately 90% of cases [31].

It is sometimes necessary to differentiate adenocarcinoma of the lung, characterized by radiological "crazy paving," from other diseases that exhibit the same pattern, including alveolar proteinosis and lipoid pneumonia. The characteristic radiographic signs of alveolar proteinosis are "geographic" zones that are clearly distinguished from healthy lung tissue, with the characteristic "crazy paving." Importantly, areas of consolidation are rarely observed in alveolar proteinosis, and lymphadenopathy, the halo sign, and "air bubbles" are virtually never seen [32] (Fig. 10.1.1.25). Exogenous lipoid pneumonia usually develops in patients who have chronically aspirated or inhaled oil substances, so a careful history enables suspicion of this disease.

The differential diagnostic range of pulmonary adenocarcinoma with diffuse and interstitial lung diseases is presented in the Table 10.1.1.4.

The detection of tumor cells in sputum or bronchoalveolar lavage fluid or histological confirmation is decisive in the diagnosis of pulmonary adenocarcinoma. In diffuse forms, the diagnosis can almost always be verified with bronchoscopy, but limited and localized forms are much more difficult to diagnose.

When a malignant tumor this confirmed, surgical treatment is indicated for a patient functionally fit for surgery. Currently, the optimal volume of surgery is considered to be lobectomy with mediastinal lymphadenectomy. The prognosis following treatment depends mainly on the histological subtype of pulmonary adenocarcinoma (Table 10.1.1.5).

If surgery is contraindicated, radiation therapy or stereotactic ablative radiation therapy is usually performed; their effectiveness approaches that of surgery at the first stage of the disease [33, 34]. For multiple lung lesions, drug treatment is administered. It is essential to undertake morphological verification of the diagnosis by determining the molecular and genetic characteristics of the tumor. Mutations of EGFR, ALK, ROS-1, KRAS, TTF-1, and PDL-1 are of clinical importance: their presence or absence predetermines the sensitivity of the tumor to the targeted therapy and immunotherapy [35–37].

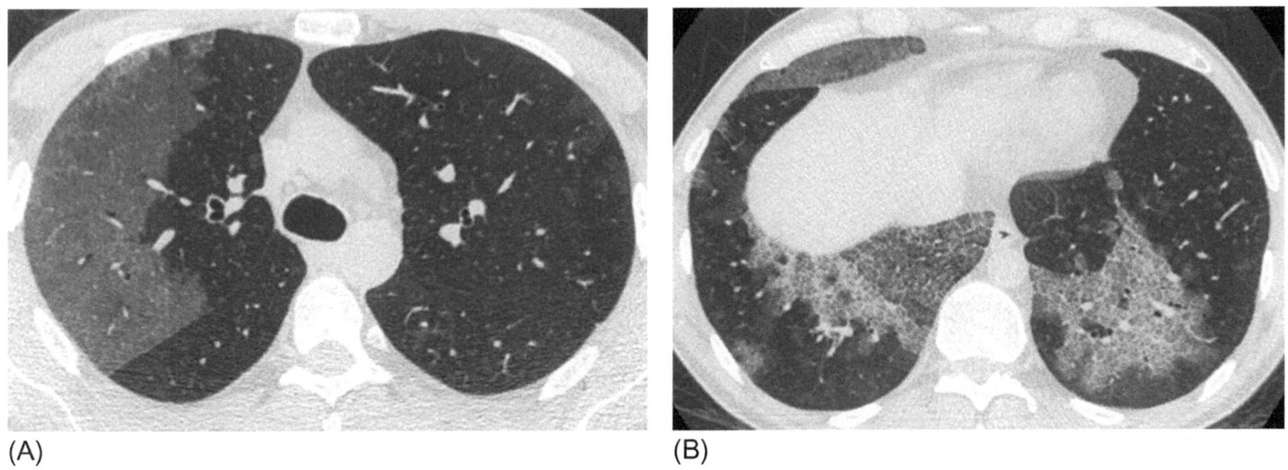

FIG. 10.1.1.25 Pulmonary alveolar proteinosis. Bilateral patchy areas of GGO sharply bordered from normal lung parenchyma (A, B). In the lower lobes GGO associated with a thickening of interlobular and intralobular septa—crazy-paving sign (B).

TABLE 10.1.1.4 Differential diagnostic criteria of pulmonary adenocarcinoma and interstitial diseases

	HRCT signs	Additional symptoms
Adenocarcinoma	Bilateral consolidation zones surrounded by ground-glass opacity, "air bubbles," pseudocavities, and intrathoracic lymphadenopathy	Bronchorrhea can be observed with the mucinous form. Symptoms of fatigue and hemoptysis
Atypical adenomatous hyperplasia	Local spherical focus of ground-glass opacity up to 10 mm in diameter. There can be multiple foci	Absence of clinical symptoms. It can present concomitantly with a tumor lesion
Organizing pneumonia	As a rule, bilateral subpleural areas of consolidation, which can change the configuration and size, areas of ground-glass opacity. Subpleural and peribronchovascular distribution of abnormalities. The reversed halo sign	Subacute onset, fever, and fatigue, resembling bacterial pneumonia. Good response to corticosteroid therapy. It can develop together with chronic autoimmune diseases, or it can be drug-induced
Nonspecific interstitial pneumonia	Reticular changes, ground-glass opacity interstitial fibrosis, and bronchiectasis. Zones of consolidation are limited. Honeycombing in long-term cases	It often develops in patients with systemic diseases and is accompanied by arthralgia
Granulomatosis with polyangiitis	Subpleural zones of consolidation of various sizes, often with cavitation in large nodes. Halo or reversed halo sign (the atoll sign). Linear scars. Pleural effusion is possible	As a rule, it is combined with lesions of the upper respiratory tract (such as sinusitis) and kidneys
Chronic eosinophilic pneumonia	Extensive zones of subpleural consolidation, mainly in the upper lobes. Ground-glass opacity is usually present. Migration of infiltrates	Eosinophilia of peripheral blood and bronchoalveolar lavage. A fast response to systemic steroids. Asthma symptoms
Alveolar proteinosis	Clearly demarcated "geographic" areas of ground-glass opacity with thickened interlobular septa, creating a crazy paving pattern	Often found in middle-aged male smokers. Bronchoalveolar lavage shows a cloudy liquid containing a large amount of protein, which produces a white precipitate upon sedimentation

TABLE 10.1.1.5 Prognostic criteria for pulmonary adenocarcinoma according to the histotype

Morphological characteristics	Histotype	Prognosis
Noninvasive adenocarcinoma		
Only lepidic growth (<3 cm)	Adenocarcinoma in situ	Good
Predominantly lepidic growth, ≤3 cm, invasion <5 mm	Adenocarcinoma with minimal invasion	Good
Invasive adenocarcinoma		
Lepidic	Predominantly lepidic growth	Intermediate
Papillary	Predominantly papillary	Intermediate
Acinar	Predominantly acinar	Intermediate
Solid	Predominantly solid	Poor
Micropapillary	Mainly micropapillary	Poor
Mucinous	Invasive mucinous adenocarcinoma	Poor

References

[1] Torre LA, Siegel RL, Jemal A. Lung cancer statistics. Adv Exp Med Biol 2016;893:1–19.

[2] Dubey AK, Gupta U, Jain S. Epidemiology of lung cancer and approaches for its prediction: a systematic review and analysis. Chin J Cancer 2016;35(1):71.

[3] Hassanein M, Callison JC, Callaway-Lane C, Aldrich MC, Grogan EL, Massion PP. The state of molecular biomarkers for the early detection of lung cancer. Cancer Prev Res (Phila) 2012;5(8):992–1006.

[4] Lee PN, Forey BA, Coombs KJ, Lipowicz PJ, Appleton S. Time trends in never smokers in the relative frequency of the different histological types of lung cancer, in particular adenocarcinoma. Regul Toxicol Pharmacol 2016;74:12–22.

[5] Mehić B, Duranović Rayan L, Bilalović N, Dohranović Tafro D, Pilav I. Lung adenocarcinoma mimicking pulmonary fibrosis-a case report. BMC Cancer 2016;16(1):729.

[6] Liebow AA. Bronchiolo-alveolar carcinoma. Adv Intern Med 1960;10:329–58.

[7] Travis WD, Brambilla E, Noguchi M, Nicholson AG, Geisinger KR, Yatabe Y, et al. International Association for the Study of Lung Cancer/American Thoracic Society/European Respiratory Society International Multidisciplinary Classification of Lung Adenocarcinoma. J Thorac Oncol 2011;6(2):244–85.

[8] Travis WD, Brambilla E, Nicholson AG, Yatabe Y, Austin JHM, Beasley MB, et al. WHO Panel. The 2015 World Health Organization classification of lung tumors: impact of genetic, clinical and radiologic advances since the 2004 classification. J Thorac Oncol 2015;10(9):1243–60.

[9] Travis WD, Brambilla E, Müller-Hermelink HK, Harris CC, editors. Pathology and genetics of tumours of the lung, pleura, thymus and heart. World Health Organization classification of tumours. Lyon, France: IARC Press; 2004.

[10] Goldstraw P, Crowley J, Chansky K, Giroux DJ, Groome PA, Rami-Porta R. The IASLC Lung Cancer Staging Project: proposals for the revision of the TNM stage groupings in the forthcoming (seventh) edition of the TNM Classification of malignant tumours. J Thorac Oncol 2007;2(8):706–14.

[11] Lee KS, Kim Y, Han J, Ko EJ, Park CK, Primack SL. Bronchioloalveolar carcinoma: clinical, histopathologic, and radiologic findings. Radiographics 1997;17(6):1345–57.

[12] Dirican N, Baysak A, Cok G, Goksel T, Aysan T. Clinical characteristics of patients with bronchioloalveolar carcinoma: a retrospective study of 44 cases. Asian Pac J Cancer Prev 2013;14(7):4365–8.

[13] Travis WD, Garg K, Franklin WA, Wistuba II, Sabloff B, Noguchi M, et al. Evolving concepts in the pathology and computed tomography imaging of lung adenocarcinoma and bronchioloalveolar carcinoma. J Clin Oncol 2005 May 10;23(14):3279–87.

[14] Patsios D, Roberts HC, Paul NS, Chung T, Herman SJ, Pereira A, et al. Pictorial review of the many faces of bronchioloalveolar cell carcinoma. Br J Radiol 2007;80(960):1015–23.

[15] Lee HY, Lee KS, Han J, Kim BT, Cho YS, Shim YM, et al. Mucinous versus nonmucinous solitary pulmonary nodular bronchioloalveolar carcinoma: CT and FDG PET findings and pathologic comparisons. Lung Cancer 2009;65(2):170–5.

[16] Min JH, Lee HY, Lee KS, Han J, Park K, Ahn MJ, et al. Stepwise evolution from a focal pure pulmonary ground-glass opacity nodule into an invasive lung adenocarcinoma: an observation for more than 10 years. Lung Cancer 2010;69(1):123–6.

[17] Read WL, Page NC, Tierney RM, Piccirillo JF, Govindan R. The epidemiology of bronchioloalveolar carcinoma over the past two decades: Analysis of the SEER database. Lung Cancer 2004;45(2):137–42.

[18] Park CM, Goo JM, Lee HJ, Lee CH, Kim HC, Chung DH, et al. CT findings of atypical adenomatous hyperplasia in the lung. Korean J Radiol 2006;7(2):80–6.

[19] Oda S, Awai K, Liu D, Nakaura T, Yanaga Y, Nomori H, et al. Ground-glass opacities on thin-section helical CT: differentiation between bronchioloalveolar carcinoma and atypical adenomatous hyperplasia. AJR Am J Roentgenol 2008;190(5):1363–8.

[20] Kim TH, Kim SJ, Ryu YH, Chung SY, Seo JS, Kim YJ, et al. Differential CT features of infectious pneumonia versus bronchioloalveolar carcinoma (BAC) mimicking pneumonia. Eur Radiol 2006;16(8):1763–8.

[21] Mir E, Sareen R, Kulshreshtha R, Shah A. Bronchioloalveolar cell carcinoma presenting as a "non-resolving consolidation" for two years. Pneumonol Alergol Pol 2015;83(3):208–11.

[22] Akira M, Atagi S, Kawahara M, Luchi K, Jonkoh T. High-resolution CT findings of diffuse bronchioloalveolar carcinoma in 38 patients. AJR Am J Roentgenol 1999;173(6):1623–9.

[23] Cottin V, Cordier JF. Cryptogenic organizing pneumonia. Semin Respir Crit Care Med 2012;33(5):462–75.

[24] Maldonado F, Daniels CE, Hoffman EA, Yi ES, Ryu JH. Focal organizing pneumonia on surgical lung biopsy: causes, clinicoradiologic features, and outcomes. Chest 2007;132(5):1579–83.

[25] Thompson WH. Bronchioloalveolar carcinoma masquerading as pneumonia. Respir Care 2004;49(11):1349–53.

[26] Kligerman SJ, Groshong S, Brown KK, Lynch DA. Nonspecific interstitial pneumonia: radiologic, clinical, and pathologic considerations. Radiographics 2009;29(1):73–87.

[27] Raparia K, Ketterer J, Dalurzo ML, Chang YH, Colby TV, Leslie KO. Lung tumors masquerading as desquamative interstitial pneumonia (DIP): report of 7 cases and review of the literature. Am J Surg Pathol 2014;38(7):921–4.

[28] Martinez F, Chung JH, Digumarthy SR, Kanne JP, Abbott GF, Shepard JA, et al. Common and uncommon manifestations of Wegener granulomatosis at chest CT: radiologic-pathologic correlation. Radiographics 2012;32(1):51–69.

[29] Ananthakrishnan L, Sharma N, Kanne JP. Wegener's granulomatosis in the chest: high-resolution CT findings. AJR Am J Roentgenol 2009;192(3):676–82.

[30] Beom JW, Lee JH. Case of invasive mucinous adenocarcinoma mimicking chronic eosinophilic pneumonia. Thorac Cancer 2014;5(2):179–83.

[31] Gaffo AL. Diagnostic approach to ANCA-associated vasculitides. Rheum Dis Clin N Am 2010;36(3):491–506.

[32] Ishii H, Trapnell B, Tazawa R, Inoue Y, Akira M, Kogure Y, et al. Comparative study of high-resolution CT findings between autoimmune and secondary pulmonary alveolar proteinosis. Chest 2009;136(5):1348–55.

[33] Franks KN, Jain P, Snee MP. Stereotactic ablative body radiotherapy for lung cancer. Clin Oncol (R Coll Radiol) 2015;27(5):280–9.

[34] Shultz DB, Diehn M, Loo BW. To SABR or not to SABR? Indications and contraindications for stereotactic ablative radiotherapy in the treatment of early-stage, oligometastatic, or oligoprogressive non-small cell lung cancer. Semin Radiat Oncol 2015;25(2):78–86.

[35] Wakelee H, Kelly K, Edelman MJ. 50 Years of progress in the systemic therapy of non-small cell lung cancer. Am Soc Clin Oncol Educ Book 2014;177–89.

[36] Domingues D, Turner A, Silva MD, Marques DS, Mellidez JC, Wannesson L, et al. Immunotherapy and lung cancer: current developments and novel targeted therapies. Immunotherapy 2014;6(11):1221–35.

[37] Anagnostou VK, Brahmer JR. Cancer immunotherapy: a future paradigm shift in the treatment of non-small cell lung cancer. Clin Cancer Res 2015;21(5):976–84.

Chapter 10.1.2

Primary lung lymphoma

Primary lung lymphoma is a rare disease that manifests as local or diffuse lesions of lung tissue. It is extremely difficult to diagnose. Lymphomas account for less than 1% of all primary malignant lung tumors, and among patients with lymphomas, a primary lesion of the lung is diagnosed in only 0.34% of cases [1, 2]. According to the accepted criteria, the diagnosis of primary lung lymphoma is confirmed when there are no signs of extrathoracic disease within 3 months after the diagnosis is established [3].

Lymphoma of the lung presents as a clonal lymphoid proliferation characterized by a lesion in one or both lungs (in the parenchyma or bronchus or both). A primary lung lymphoma is most likely to be a high-grade small-cell B-cell lymphoma or a lymphoma with mucosa-associated lymphoid tissue (MALT lymphoma) or bronchial-associated lymphoid tissue (BALT lymphoma). Less frequently it is low-grade large-cell B-cell lymphoma and lymphomatoid granulomatosis. MALT lymphomas account for about 58%–87% of primary lung lymphomas, whereas high-grade B-cell lymphomas account for 11%–19%; most often, this lymphoma histotype is diagnosed in immunocompromised patients with HIV infection or after organ transplantation, and it is often concomitant with the Epstein-Barr virus [4]. Lymphomatoid granulomatosis is extremely rare, with fewer than 1000 observations reported in the literature; it presents a variable clinical picture and course, from spontaneous regression to a fatal outcome [5].

The clinical presentation in patients with primary lung lymphoma includes symptoms such as cough (50%–53%), dyspnea (21%–39%), chest pain (17%–21%), hemoptysis (6%–10%), fever (about 16%), and weakness (11%), with 22%–32% of patients asymptomatic [6, 7]. According to Dong [8], clinical symptoms in primary and secondary lymphomas are virtually the same. Laboratory tests often show anemia and increased ESR and LDG levels [9]. Up to 30% of patients exhibit increased monoclonal secretion of IgG or IgM [10].

Diagnosis and differential diagnosis

Computed tomography

CT findings for primary lung lymphoma are nonspecific. Usually, there are bilateral, multiple lesions of the lung tissue; this is observed in about 60%–79% of cases. Other findings include pulmonary tissue consolidation masses more than 3 cm in diameter (in 33%–88% of cases), consolidation areas less than 3 cm (39%–62%), bronchiectasis in the consolidation zone (up to 58%), and intrathoracic lymphadenopathy (28%–29%) [6, 11] (Fig. 10.1.2.1). Air bronchograms are often

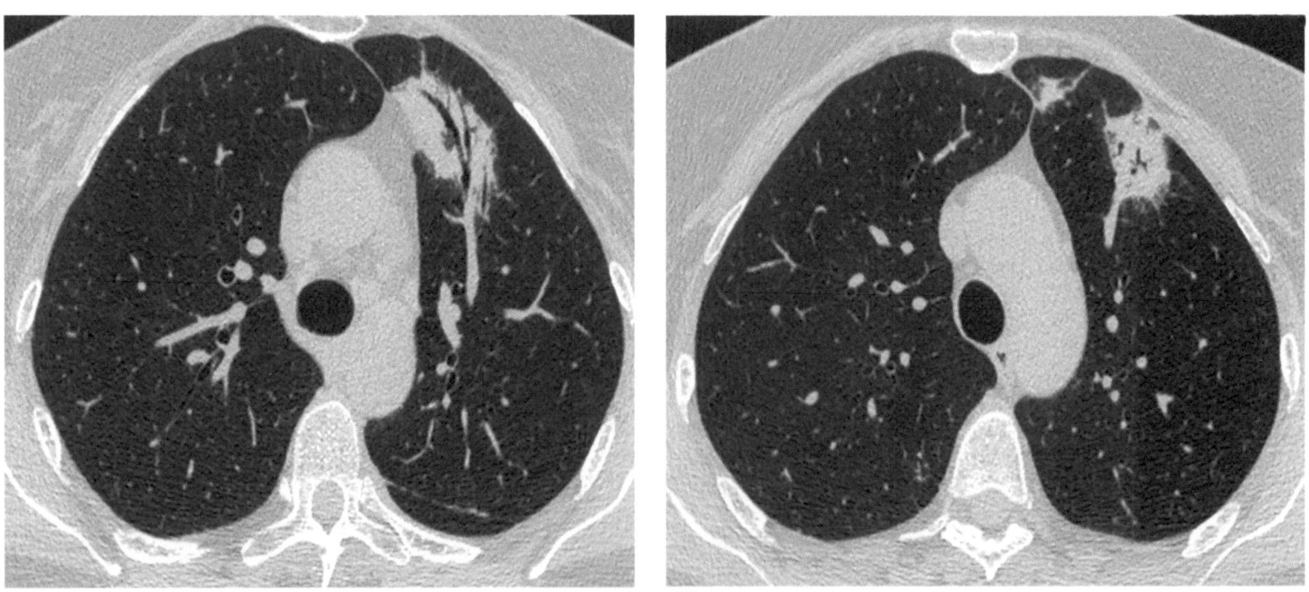

(A) (B)

FIG. 10.1.2.1 Primary pulmonary lymphoma. Focal area of consolidation containing air bronchogram with sharp (A) and flame-shaped (B) contours.

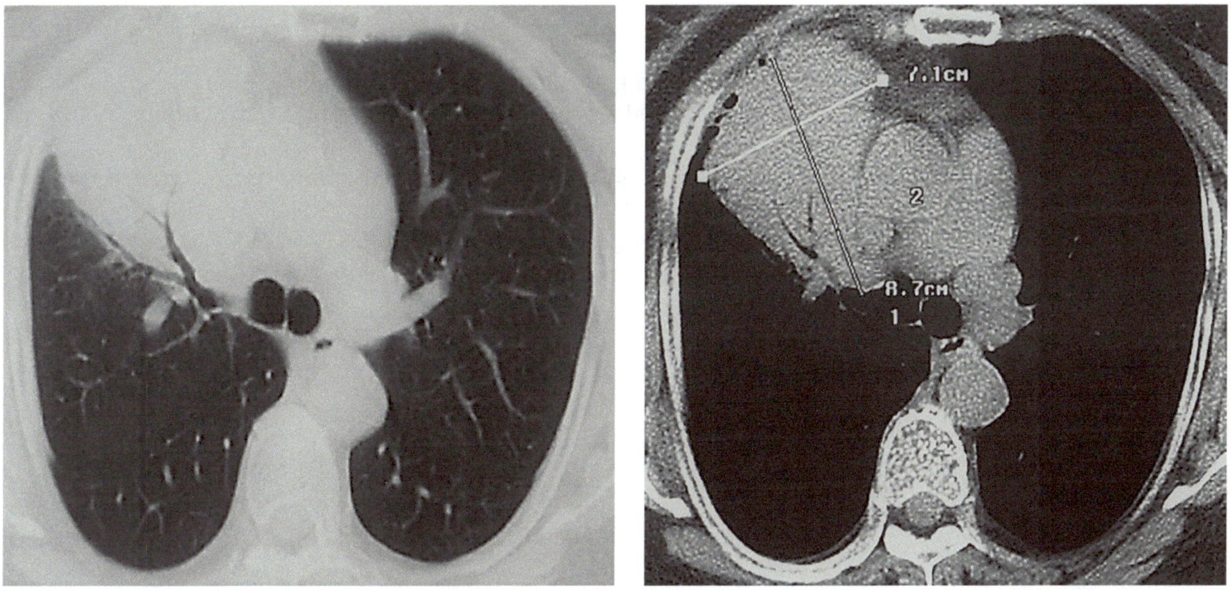

(A) (B)

FIG. 10.1.2.2 Primary pulmonary lymphoma. Consolidated mass with air bronchogram in the right middle lobe.

detected (in up to 88% of cases) because the tumor develops peribronchially without destroying the bronchial tree [11] (Fig. 10.1.2.2). Approximately one-fifth of patients have pleural effusions, and a few patients exhibit cavities in the consolidation areas [7]. Some researchers consider the halo sign (a shadow of ground-glass opacity around the consolidation zone) in combination with the air bronchogram to be the most characteristic finding for primary lymphomas, along with pathological changes that remains stable for several weeks and even months, and the predominant localization of the pathological process in the right middle lobe [7, 10] (Fig. 10.1.2.2). Involvement of the peripheral parts of the lungs, extensive consolidation zones (larger than 3 cm) and cavities are more common in primary lymphomas than in secondary lymphomas, whereas intrathoracic lymphadenopathy is less frequently observed [8].

Bronchoalveolar lavage has limited benefit for the diagnosis of lymphomas, although it is required to rule out other diseases, primarily pulmonary infections. Nevertheless, studies have suggested analysis of the bronchoalveolar lavage fluid to determine the immunoglobulin heavy chain (*IgH*) and *MALT1* gene rearrangements as genetic markers of the disease [12, 13]. Morphological verification of the diagnosis usually requires invasive diagnostic methods. Material for histological examination can be obtained through thick-needle biopsy under the supervision of ultrasound or CT. However, this is informative in only 10.0% of cases; in the remaining 90.0% of cases, it is insufficient to provide a clear morphological diagnosis, resulting in the need to use surgical diagnostic methods, such as thoracoscopy or thoracotomy with the lung resection. The majority of authors have reported that the material obtained with transbronchial biopsy is insufficient to establish a correct diagnosis [2, 5, 14]. In a study of 17 patients with primary lung lymphoma, Eynden et al. [2] reported that none of the patients were accurately diagnosed before surgery. Thoracotomy was performed for 16 of the patients and thoracoscopy for one. In a planned morphological study, B-cell lymphoma of low-grade malignancy was diagnosed in 14 patients (83%), B-cell lymphoma of high-grade malignancy was diagnosed in two patients (12%), and lymphoid granulomatosis was registered in one patient (6%) [2].

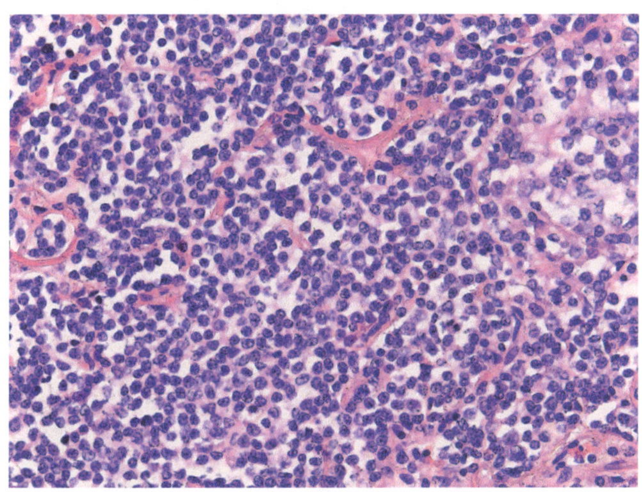

FIG. 10.1.2.3 Pulmonary lymphoma from the cells of the marginal zone; small cells with irregularly shaped nuclei. Hematoxylin and eosin staining, 400×.

The cornerstone for the diagnosis of primary lung lymphoma is a morphological study in which diffuse proliferation of small or large atypical lymphoid cells is detected (Fig. 10.1.2.3). The lymphoma histotype is usually confirmed by immunohistochemical examination showing positive expression of the markers CD-20 and CD-79, which are characteristic for B-cell lymphoma (Fig. 10.1.2.4).

Primary lung lymphoma usually has to be differentiated from organizing pneumonia, lung tumors, sarcoidosis, Kaposi sarcoma, and with granulomatosis with poliangiitis, it should also be differentiated from infiltrative pulmonary tuberculosis and invasive aspergillosis (Fig. 10.1.2.5–10.1.2.7) [5].

The differential diagnostic range for lung lymphomas, adenocarcinoma, and infiltrative lung diseases is presented in Table 10.1.2.1.

In summary, if the clinical picture of the disease does not correspond to the findings of the X-ray study, thoracoscopic or open lung biopsy is indicated to obtain material for a complete histological and immunohistochemical study.

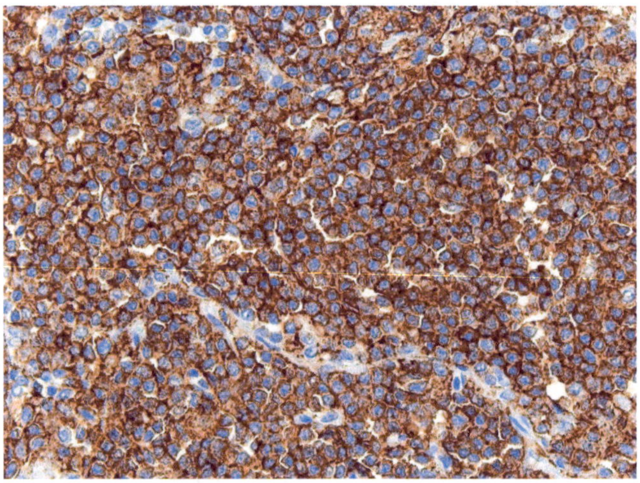

FIG. 10.1.2.4 Pulmonary lymphoma from cells in the marginal zone. Immunohistochemical study: CD-20 positive lymphoid cells 400×.

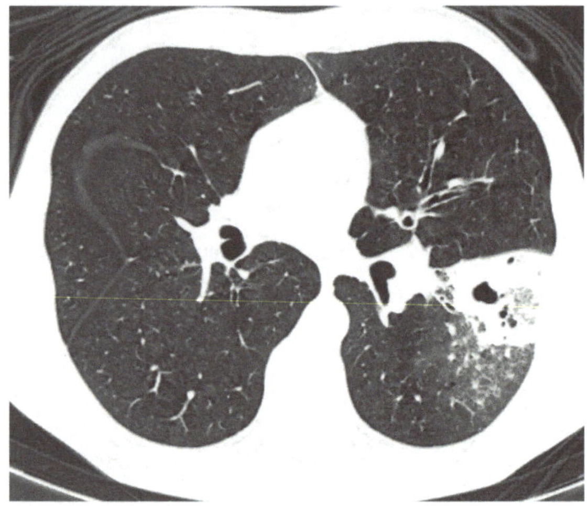

FIG. 10.1.2.5 Infiltrative tuberculosis of the low left lobe. Large mass containing cavities surrounded by ground-glass opacity and ill-defined nodules is visible.

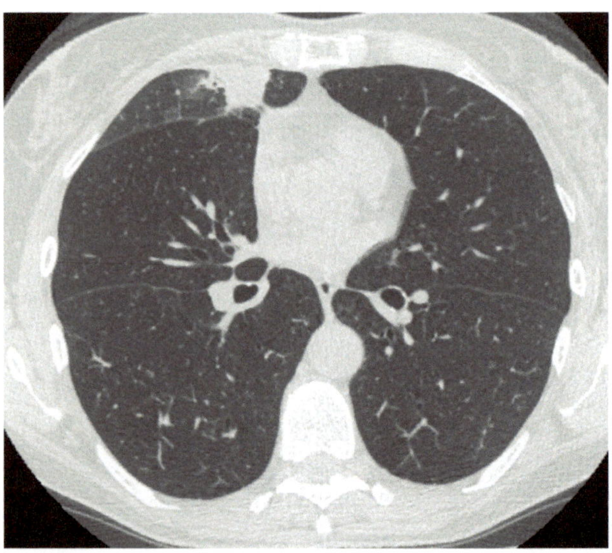

FIG. 10.1.2.6 Cryptogenic organizing pneumonia is proven by surgical biopsy. Subpleural consolidation area surrounded by light shadow of ground-glass opacity.

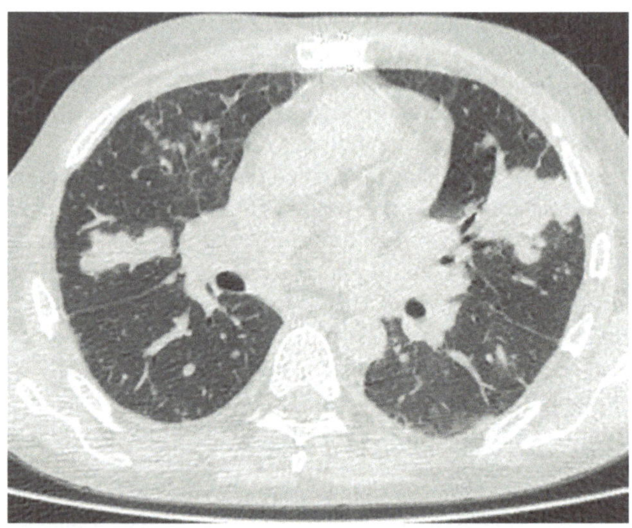

FIG. 10.1.2.7 Kaposi sarcoma in AIDS patient. HRCT shows bilateral tuberous masses with peribronchovascular distribution and multiple nodules. *(Case courtesy of Prof. S.N. Avdeev, Sechenov First Moscow State Medical University, Moscow, Russia).*

TABLE 10.1.2.1 Differential diagnostic range for primary lymphoma of the lung

Sign	Primary lung lymphoma	Adenocarcinoma	Granulomatosis with polyangiitis	Organizing pneumonia	Pulmonary tuberculosis
History	Gradual onset; associated with immunodeficiency and organ transplantation	Indication of the presence of malignant tumors in relatives	Rhinitis, sinusitis, blood-tinged discharge from the nose	Onset usually after a respiratory infection	Contacts with patients with tuberculosis; immunodeficiency
Clinical presentation	Shortness of breath, nonproductive cough, prolonged fever	Weight loss, nonproductive cough, dyspnea, possible foamy sputum	Cough, hemoptysis, dyspnea, chest pain	Fever, dyspnea with physical exertion, nonproductive cough	Fever, weight loss, sweating, cough, hemoptysis
c-ANCA	–	–	+	–	–
BAL	Most often not informative	Tumor cells	Lymphocytosis in the period of low activity, CD4/CD8 >3, neutrophilia in the high-activity phase	Foamy macrophages, eosinophilia in 2%–25%, cases, lymphocytosis >25%	Moderate lymphocytosis, neutrophilia CD4/CD8 <2; BK+
CT signs	Multiple lesions, areas of consolidation, air bronchogram	Solitary or multiple poorly defined foci; a combination of a solid component and ground-glass opacity air-bubble sign, air bronchogram	Nodal infiltrates of various sizes up to 10 cm in diameter, often with cavities inside large nodes; the halo sign; pleural effusion is possible; thickening of the walls of the trachea and large bronchi	Areas of bilateral subpleural consolidation without cavities, changes in size and configuration, bordering areas of ground-glass opacity; a reversed halo sign that mimics the cavities in the consolidation zone	Infiltrates with cavities, mainly localized in the upper lobe; signs of bronchogenic seeding, such as small nodules around the consolidation zones; lymphadenopathy
Diagnostic value of transthoracic biopsy	Informative in no more than 10% of cases	Informative in approximately 80% of cases (identifying tumor cells)	Necrotic polymorphocellular granulomas; vasculitis; foci of "geographic" necrosis	Limited value	Granulomas with the presence of caseous necrosis

References

[1] Tao H, Nakata M, Saeki H, Kurita A, Takashima S. Unsuspected primary pulmonary malignant lymphoma. Jpn J Thorac Cardiovasc Surg 2002;50(12):533–6.

[2] Vanden Eynden F, Fadel E, de Perrot M, de Montpreville V, Mussot S, Dartevelle P. Role of surgery in the treatment of primary pulmonary B-cell lymphoma. Ann Thorac Surg 2007;83(1):236–40.

[3] Cardenas-Garcia J, Talwar A, Shah R, Fein A. Update in primary pulmonary lymphomas. Curr Opin Pulm Med 2015;21(4):333–7.

[4] Cadranel J, Wislez M, Antoine M. Primary pulmonary lymphoma. Eur Respir J 2002;20(3):750–62.

[5] Matsumoto T, Otsuka K, Funayma Y, Imai Y, Tomii K. Primary pulmonary lymphoma mimicking a refractory lung abscess: A case report. Oncol Lett 2015;9(4):1575–8.

[6] Graham BB, Mathisen DJ, Mark EJ, Takvorian RW. Primary pulmonary lymphoma. Ann Thorac Surg 2005;80(4):1248–53.

[7] Yao D, Zhang L, Wu PL, Gu XL, Chen YF, Wang LX, et al. Clinical and misdiagnosed analysis of primary pulmonary lymphoma: a retrospective study. BMC Cancer 2018;18(1):281.

[8] Dong Y, Zeng M, Zhang B, Han L, Liu E, Lian Z, et al. Significance of imaging and clinical features in the differentiation between primary and secondary pulmonary lymphoma. Oncol Lett 2017;14(5):6224–30.

[9] Yu H, Chen G, Zhang R, Jin X. Primary intravascular large B-cell lymphoma of lung: a report of one case and review. Diagn Pathol 2012;7:70.

[10] Cottin V, Cordier JF, Richeldi L, editors. Orphan lung diseases: a clinical guide to rare lung disease. London: Springer-Verlag; 2015.

[11] King LJ, Padley SP, Wotherspoon AC, Nicholson AG. Pulmonary MALT lymphoma: imaging findings in 24 cases. Eur Radiol 2000;10(12):1932–8.

[12] Nicholson AG, Wotherspoon AC, Diss TC, Butcher DN, Sheppard MN, Isaacson PG, et al. Pulmonary B-cell non-Hodgkin's lymphomas. The value of immunohistochemistry and gene analysis in diagnosis. Histopathology 1995;26(5):395–403.

[13] Kido T, Yatera K, Noguchi S, Sakurai Y, Nagata S, Kozaki M, et al. *D*etection of MALT1 gene rearrangements in BAL fluid cells for the diagnosis of pulmonary mucosa-associated lymphoid tissue lymphoma. Chest 2012;141(1):176–82.

[14] Miyahara N, Eda R, Umemori Y, Murakami T, Kunichika N, Makihata K, et al. Pulmonary lymphoma of large B-cell type mimicking Wegener's granulomatosis. Intern Med 2001;40(8):786–90.

Chapter 10.2

Metastatic lung disease

Metastases in the lungs are detected in 6%–30% of patients with tumors of any site, either during a primary examination or at follow-up after the treatment of malignant neoplasms. According to autopsy data, the most common lung metastases are associated with chorion carcinoma, melanoma, osteogenic sarcoma, kidney cancer, and testicular tumors. 65%–93% of patients with pulmonary metastases have also extrathoracic metastases [1–3].

Often, single metastases have to be differentiated from metachronous peripheral primary malignant tumors, benign tumors, and focal lesions with a nontumor lung genesis, such as a rheumatoid node, solitary organizing pneumonia, tuberculosis, or a fungal infection [4]. In general, multiple metastases detected at follow-up in a patient with a history of tumor do not present any difficulties for diagnosis, especially when there are extrathoracic manifestations of the disease. In typical cases, metastases are rounded or less often oval in shape, sharply marginated with a random peripheral distribution. Usually, they are localized in the middle and lower parts of the lungs (Fig. 10.2.1). Less often, they are lobulated, irregular with spicular, tuberous, or ill-defined contours (Fig. 10.2.2) [5].

Friedmann et al. [6] noted that metastases up to 2 cm in diameter usually have a rounded shape with clear borders, whereas large nodes are often polymorphic and irregular. The structure of metastases is usually homogeneous. They often have uniform density, but they can be partially calcified, as is the case, for example, for metastases of osteogenic sarcoma and chondrosarcoma; breast cancer; and cancers of thyroid gland, ovaries, and colon [5]. Nodules of up to 5 mm are usually benign, but the likelihood of malignancy increases significantly with enlarged size [7, 8]. Minor metastases are usually located between the centrilobular pulmonary arteries and the interlobular septum, whereas the larger metastases often constrict the branches of the pulmonary vessels [9]. The number of metastases can vary from a single focus to massive dissemination, and it does not help in differentiating benign and malignant processes [5].

An increase in intrapulmonary lymph nodes may resemble a metastatic lung disease. In an analysis of focal lesions of the lungs of up to 1 cm in diameter, Yokomise et al. [10] found in increased intrapulmonary lymph nodes in 46.2% of cases but metastases in only 7.7%. In a comparison of CT signs of intrapulmonary lymph nodes and pulmonary metastases, Hyodo et al. [11] observed that the lymph nodes were characterized by a significantly smaller distance from the pleura (a mean distance of 4.6 vs. 16 mm) and by the presence of a series of linear densities extending from the nodules (Fig. 10.2.3), which were demonstrated histologically to be ectatic lymphoid or blood vessels. Unlike primary lung tumors, lymph nodes are almost always visualized below the level of the carina. Finally, the borders of the lymph nodes are always sharp, whereas tumors often have ill-defined borders [12]. Most lymph nodes are characterized by a round or oval shape; more

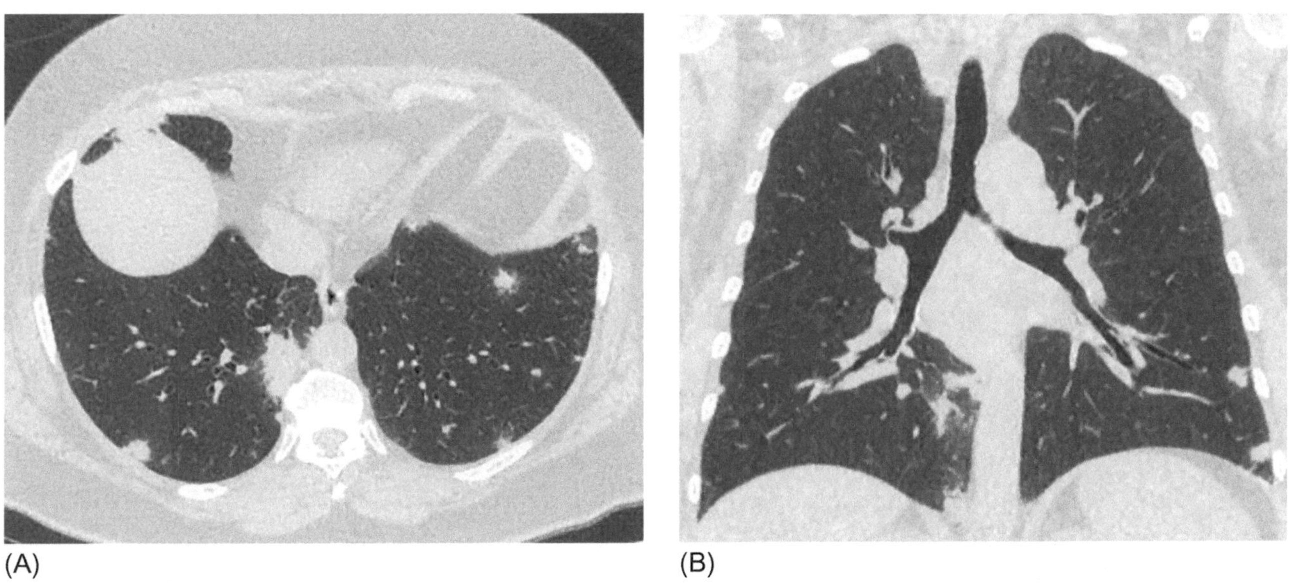

FIG. 10.2.1 Multiple metastases of chorion carcinoma in the lungs. Bilateral nodules mainly round in shape, some of them with sharp contours; others are surrounded by a ground-glass opacity shadow (A). Nodules localized subpleurally in the lower parts (B).

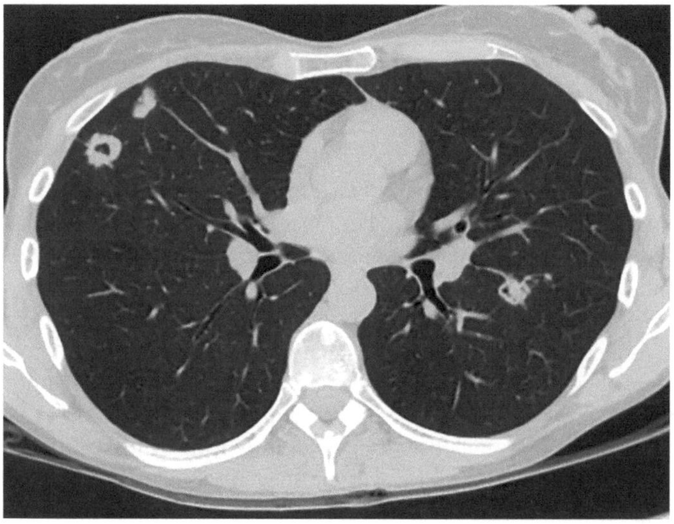

FIG. 10.2.2 Metastases of colon cancer in the lungs. Bilateral nodular opacities with spiculated contours. Nodules are distributed under visceral pleura in the low lobes.

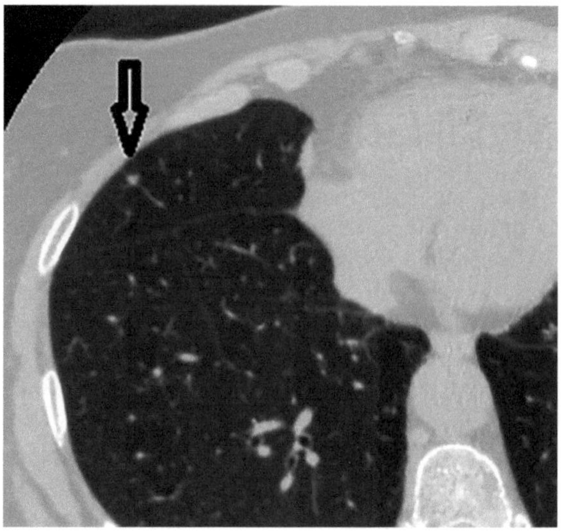

FIG. 10.2.3 Intrapulmonary lymph node. The subpleural polygon-shaped focus with sharp contours and adjacent linear attenuation *(arrow)*.

rarely it is angular, as is the case for tumor lesions [13]. A nontumor increase in intrapulmonary lymph nodes is usually associated with blockage of the lymphatic outflow associated with the dust factor [14].

Patients with focal lung lesions are managed according to an assessment of the risk factors for malignancy and the size and density characteristics of the identified lesions. The latest (2017) recommendations of the Fleischner Society are presented in Table 10.2.1 [15].

TABLE 10.2.1 Fleischner Society 2017 guidelines for management of incidentally detected pulmonary nodules in adults [15] (with permission of Fleischner Society)

A: Solid nodules[a]				
		Size		
Nodule Type	**<6 mm (<100 mm³)**	**6–8 mm (100–250 mm³)**	**>8 mm (>250 mm³)**	**Comments**
Single				
Low risk[b]	No routine follow-up	CT at 6–12 months, then consider CT at 18–24 months	Consider CT at 3 months, PET/CT, or tissue sampling	Nodules <6 mm do not require routine follow-up in low-risk patients (recommendation 1A)
High risk[b]	Optional CT at 12 months	CT at 6–12 months, then CT at 18–24 months	Consider CT at 3 months, PET/CT, or tissue sampling	Certain patients at high risk with suspicious nodule morphology, upper-lobe location, or both may warrant 12-month follow-up (recommendation 1A)
Multiple				
Low risk[b]	No routine follow-up	CT at 3–6 months, then consider CT at 18–24 months	CT at 3–6 months, then consider CT at 18–24 months	Use most suspicious nodule as guide to management. Follow-up intervals may vary according to size and risk (recommendation 2A)
High risk[b]	Optional CT at 12 months	CT at 3–6 months, then at 18–24 months	CT at 3–6 months, then at 18–24 months	Use most suspicious nodule as guide to management. Follow-up intervals may vary according to size and risk (recommendation 2A)

TABLE 10.2.1 Fleischner Society 2017 guidelines for management of incidentally detected pulmonary nodules in adults [15] (with permission of Fleischner Society)—cont'd

B: Subsolid nodules[a]			
	Size		
Nodule Type	<6 mm (<100 mm^3)	>6 mm (>100 mm^3)	Comments
Single			
Ground glass	No routine follow-up	CT at 6–12 months to confirm persistence, then CT every 2 years until 5 years	In certain suspicious nodules <6 mm, consider follow-up at 2 and 4 years. If solid component(s) or growth develops, consider resection (recommendations 3A and 4A)
Part solid	No routine follow-up	CT at 3–6 months to confirm persistence. If unchanged and solid component remains, 6 mm, annual CT should be performed for 5 years	In practice, part-solid nodules cannot be defined as such until >6 mm, and nodules < 6 mm do not usually require follow-up. Persistent part-solid nodules with solid components >6 mm should be considered highly suspicious (recommendations 4A-4C)
Multiple	CT at 3–6 months. If stable, consider CT at 2 and 4 years	CT at 3–6 months. Subsequent management based on the most suspicious nodule (s)	Multiple <6 mm pure ground-glass nodules are usually benign, but consider follow-up in selected patients at high risk at 2 and 4 years (recommendation 5A)

[a]*Dimensions are average of long and short axes, rounded to the nearest millimeter.*
[b]*Consider all relevant risk factors.*
These recommendations do not apply to lung cancer screening, patients with immunosuppression, or patients with known primary cancer.

To assess the risk of malignancy of foci in the lungs, it is proposed to use the logarithmic formula from the clinical recommendations of the American College of Chest Physicians of 2013, which provides an assessment based on six factors of estimated risk of malignancy [16]:

- Elderly age (with the risk increasing by 4% for each year of life)
- Smoking (current or past)
- A history of cancer of extrapulmonary localization, >5 years before nodules detection
- The size of the nodule
- Spiculation
- Upper-lobe localization

$$\text{Probability of malignancy} = e^x / \left(1 + e^x\right),$$

where e is the irrational constant of the natural logarithm, equal to approximately 2.718, and x is derived from the set of risk factors corrected by the following coefficients:

$$x = -6.8272 + (0.0391 \times \text{age in years}) + (0.7917 \times \text{smoker}) + (1.3388 \times \text{cancer}) \\ + (0.1274 \times \text{diameter in mm}) + (1.0407 \times \text{radiance}) + (0.7838 \times \text{upper-lobe localization}).$$

Age and diameter have values equal to their values in years and millimeters, respectively, and smoker, radiance, and upper-lobe localization are estimated in a qualitative way, using the value 1 if there is a sign and 0 if this factor is absent [16].

The probability of malignancy <5%, calculated according to the formula, corresponds to a low risk; if it is more than 65%, it corresponds to high risk; in cases where it is 5%–65%, this corresponds to intermediate risk [16].

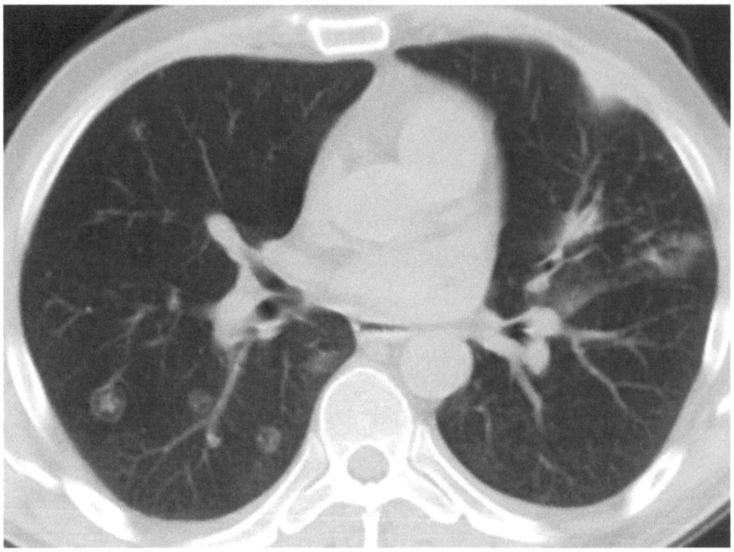

FIG. 10.2.4 Multiple cystic metastases of kidney cancer in the lungs. Bilateral cavitary nodules containing irregular opacities are visible.

In about 8% of patients with lung metastasis, the CT scan of the lung metastatic lesion is atypical, hindering diagnosis and resulting in errors [17]. The cases most difficult to diagnose are those with cyst-shaped metastases accompanied by CT scans of interstitial lesions, and those that are embologenic. Most primary cystic lung tumors (up to 82%) are squamous cell carcinoma, with adenocarcinoma and large-cell carcinoma accounting for most of the remaining malignant cystic lesions (11%) (Fig. 10.2.4) [18].

The primary tumor in cystic lung metastases is most often a squamous carcinoma of the colon, ovaries, uterus, stomach, pancreas, or other organs [19]. Various types of sarcomas, especially angiosarcoma and leiomyosarcoma, can be accompanied by metastatic spreading to the lungs, with the formation of cysts and often the development of spontaneous pneumothorax [20]. Benign metastatic leiomyoma can manifest as multiple nodular and cystic metastases in the lungs several years after the removal of the uterine leiomyoma; it is more often observed in premenopausal women [21].

Honda et al. [19] analyzed the characteristics of cysts in malignant and benign (primarily infectious) processes in the lungs and concluded that irregularity of the external and internal borders was characteristic of tumor cavities, as was more frequent identification of notches and spiculation (Fig. 10.2.5).

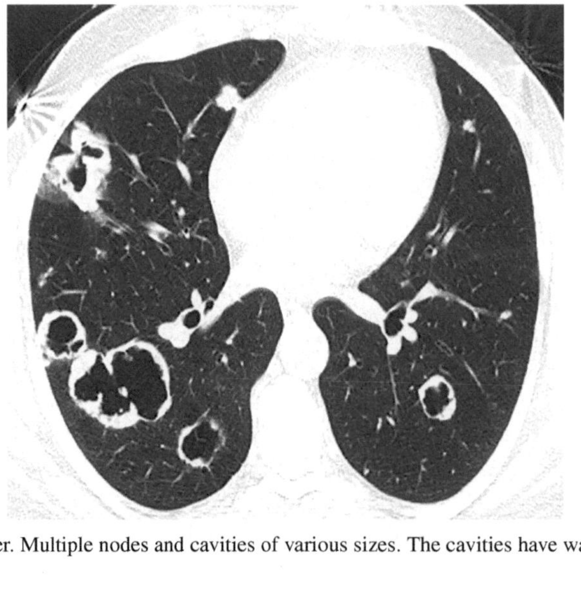

FIG. 10.2.5 Metastases of cervical cancer. Multiple nodes and cavities of various sizes. The cavities have walls from 1 mm to 1 cm of thickness, sharp external and notched internal contours.

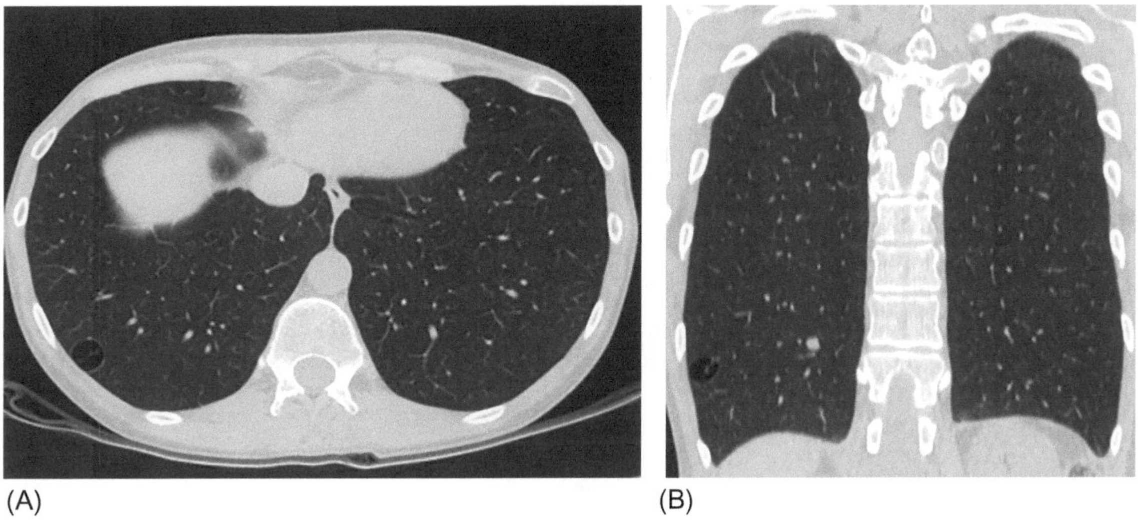

FIG. 10.2.6 Metastases of gastric cancer in the lungs. (A) A single, subpleurally located thin-walled cyst in the right lower lobe. (B) There is also a nodular metastasis of 8 mm on the coronal section.

Sometimes the so-called "Cheerio" sign, first described in adenocarcinomas, is identified; this includes multiple small nodules a few millimeters in diameter, with cavities inside so that they resemble the ring-shaped breakfast cereal. This sign can also be observed with Langerhans cell histiocytosis [20]. Single thin-walled metastases are the most difficult to differentiate because they can be indistinguishable from pulmonary cysts; additional signs are only observed with progression, such as consolidation, focal shadows, and lymphadenopathy (Fig. 10.2.6).

Tumor metastasis in the lungs that have masks of interstitial nonneoplastic diseases mainly present with adenocarcinoma (in variants related to the former BAC), which can generally be differentiated with by multiple foci of organizing pneumonia, pulmonary infections, pulmonary sarcoidosis, and GP (see Chapter 10.1.1) [22–24].

A rarer variant of lung metastases that mimics diffuse nonneoplastic parenchymal processes is lymphangitic carcinomatosis (LCM), which accounts for 6%–8% of all metastatic lung lesions. The most common primary tumors are cancers of the breast (33%), stomach (29%), and pancreas (17%) [25]. In a study by Prakash [26], 83% of cases of LCM were a consequence of lung cancer. In most cases, LCM is based on the hematogenous microembolization of pulmonary vessels with tumor cells, followed by their spread to the lymphatic vessels; however, it is also possible for there to be a primarily lymphogenous pathway of retrograde dissemination through the axillary lymph nodes [27].

A review of 20 publications [28] showed that LCM was more common in women (accounting for 81% of cases) and that the mean age of patients was 53 years. The dominant symptom was progressive dyspnea that lasted for several weeks (in 69% of cases), and 12.5% of patients reported the acute onset of respiratory failure [28]. Rare but possible symptoms of the disease include hemoptysis, fever, and weight loss. In more than half of cases, blood tests showed an increase in ESR [29]. In 58% of patients, the primary chest X-ray or CT revealed minimal deviations or the absence of pathological changes [28]. The 21-day survival after diagnosis was less than 50% [28]. The main CT signs of LCM are reticular changes, sometimes associated with a nodular lesion [30]. An early description of CT signs of the disease reported thickening of bronchovascular bundles (in 95% of cases) and interlobular septa (85%). There were nodules in 20% of cases and pleural effusion in 40% [31] (Fig. 10.2.7). About 31% of patients exhibited ground-glass opacity [29], usually located in the areas with

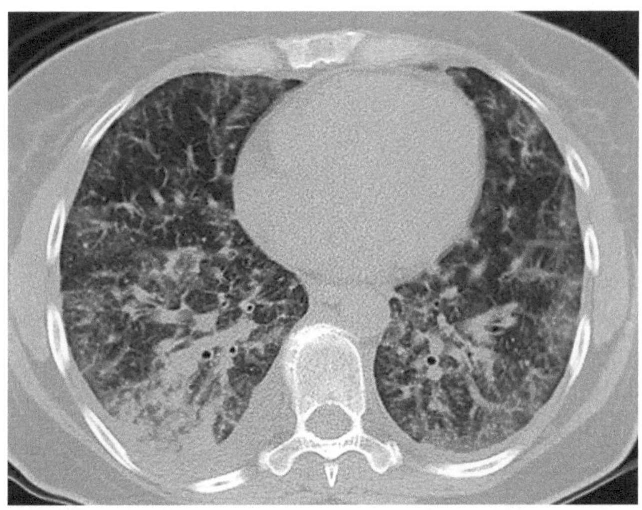

FIG. 10.2.7 Lymphangitic carcinomatosis in gastric cancer. Bilateral thickening of bronchovascular bundles, multiple small nodules, and subpleural zones of ground-glass opacity. Right-sided pleural effusion bordering with a site of heterogeneous consolidation in the lower lobe of the right lung.

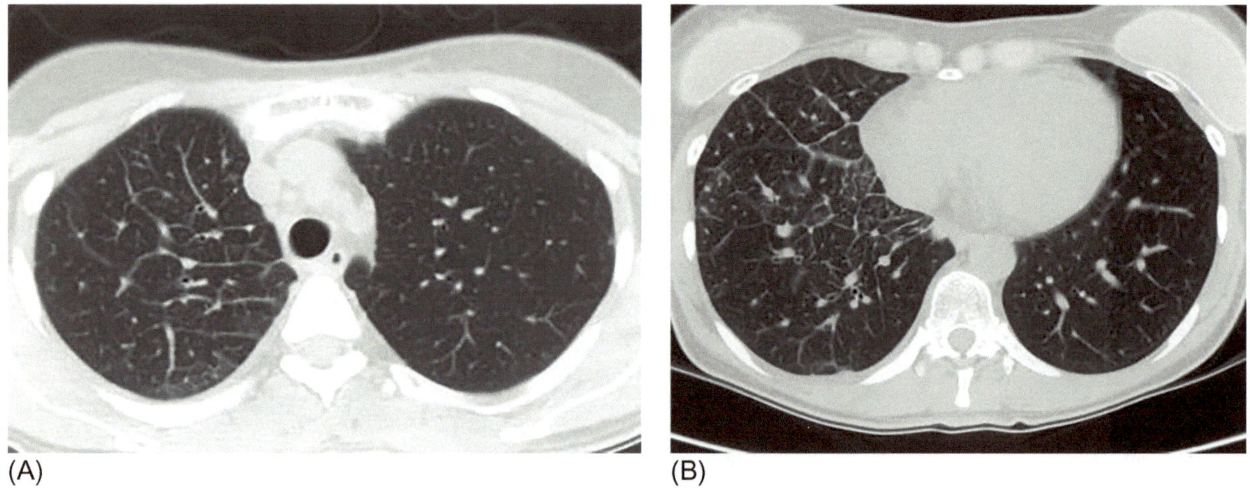

FIG. 10.2.8 Lymphogenous carcinomatosis in breast cancer. Thickened interlobular septa and vascular structures in the right lung. Separate small nodules are also visualized (A and B).

the most severe reticular changes, which can manifest as the "crazy paving" pattern (Fig. 10.2.8). The findings were diffuse and unilateral in about half the cases, unilateral and focal in another 37% of cases, and bilateral in 14% of patients; however, in this study, most of the patients had lung cancer as the primary tumor [27]. Despite quite substantial reticular changes, the architecture of the lung tissue is generally not distorted; the pulmonary lobules retain their structure and are not subjected to fibrotic changes, as observed in interstitial lung diseases [31]. Lymphadenopathy of the axillary and intrathoracic lymph nodes is observed in 38%–94% of patients [27, 29].

The differential diagnosis of metastatic lung disease includes disorders that manifest similar findings, including thickening of the peribronchovascular interstitium and interlobular septa, such as congestive heart failure, pulmonary infections including pneumocystis pneumonia, interstitial pneumonia, and also diseases manifested by perilymphatic nodular pattern combined with intrathoracic lymphadenopathy (sarcoidosis, pneumoconiosis, or amyloidosis). In patients with a progressive oncological disease exhibiting characteristic CT signs, the diagnosis of LCM does not usually require verification. A more balanced approach should be used for patients with a history of treated tumor and new nodules in the lungs. The enticement to consider new symptoms in the context of a known disease can lead to the wrong diagnosis. As an example, multiple small nodules in the lungs of a patient who underwent an operation for spindle cell carcinoma of the stomach were identified as metastases, but a morphological study of biopsy material obtained during video-assisted thoracoscopy resulted in a diagnosis of sarcoidosis (Fig. 10.2.9).

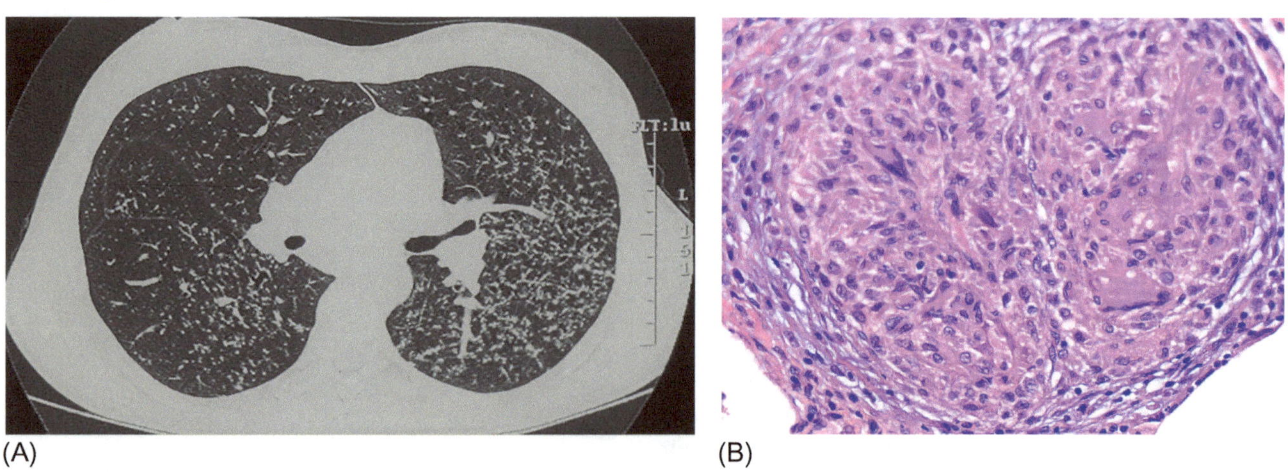

FIG. 10.2.9 Sarcoidosis in a patient who underwent resection of the stomach for spindle cell carcinoma. Multiple small nodules with a perilymphatic pattern and intrathoracic lymphadenopathy. The abnormalities were primarily treated as metastases (A). The diagnosis was established with video-assisted thoracoscopy biopsy. A histological picture of the lung tissue of the same patient. Hematoxylin-eosin staining, ×400 magnification (B).

Drug-induced pneumonitis from tumor chemotherapy should also be considered in a differential range in such patients. In most cases, bronchoscopy with cytological and PCR analysis of the bronchoalveolar lavage fluid allows tumor cells to be detected and infections to be ruled out, but when there is doubt, a transbronchial biopsy of the lungs is required, which is usually highly informative for LCM. PET-CT is also possible where the diagnosis is in doubt or there are difficulties with histological confirmation. The regions affected by LCM are characterized by a significantly higher uptake of ^{18}F-FDG, which has a sensitivity for LCM of 86% and a specificity of 100% [27].

Tumor embolism of the pulmonary artery is a rare condition that can mimic thromboembolism and some diffuse parenchymal lung diseases. Microscopic pulmonary tumor embolism is the most difficult to diagnose. This pathology develops in patients with various forms of cancer (most often of the mammary gland, stomach, lung, or liver) and manifests as rapidly progressing hypoxemic respiratory failure, often with no visible changes in radiographs and CT scans (both native and with contrast enhancement) [32, 33]. This explains why a diagnosis is often only made on autopsy. In general, tumor microembolisms of pulmonary vessels are found in 2%–4% of autopsies of cancer patients, but only half of these patients demonstrated symptoms of increasing pulmonary hypertension while alive [34]. Another problem of accurate diagnosis is the long period after the so-called successful treatment of the primary tumor and the absence of signs of recurrence [35].

Typically, it takes several days to several weeks from the onset of dyspnea to the death of the patient [33]. The disease involves the multiple embolism of small pulmonary vessels by tumor cells without penetrating the parenchyma, often with microthrombosis in situ. This is concomitant with pulmonary hypertension in up to half of cases and with acute right ventricular failure in 15%–20% of cases [36, 37]. The most effective method for diagnosing microscopic pulmonary tumor embolism is a pulmonary biopsy; however, because of the absence of visible signs of the parenchymal process, it is rarely performed. Ventilation-perfusion lung scanning should be considered as a diagnostic method. This often reveals a characteristic "segmental contour" pattern, a break in the zones of perfusion distal to the segmental level, which creates a segmental map pattern [36]. However, a similar finding can be observed in other diffuse microvascular lesions.

Intravascular metastases of chorion carcinoma can mimic thromboembolism of the pulmonary artery and diffuse infiltrative lung diseases. Chorion carcinoma is one of the variants of gestational trophoblastic disease, which is most often detected in women during their reproductive period, usually after delivery or abortion. In 80% the cases, chorion carcinoma metastasizes to the lungs, sometimes with the rapid development of respiratory failure [38]. Because of its high degree of malignancy, when the chorion carcinoma enters the pulmonary artery, it rapidly invades the vessels and spreads to the lung tissue, with the formation of large infiltrates and masses, of which some tend to cavitate (Fig. 10.2.10). Similar CT findings can lead to erroneous primary diagnoses of bacterial pneumonia, organizing pneumonia, or pulmonary thromboembolism. Cases of spontaneous regression of trophoblastic masses have been described; this can further hinder the correct interpretation of pulmonary symptoms [39]. The first clinical symptoms of chorion carcinoma metastasis are usually dyspnea and hemoptysis [40, 41]. An inherent extrapulmonary symptom is amenorrhea, and the laboratory confirmation is a high level of chorionic gonadotropin.

The differential signs for several metastatic lung tumors are presented in the Table 10.2.2.

In summary, the differential diagnosis of a suspected metastatic lung disease is not always simple, both because of the tumor history and because of the atypical forms of the secondary pulmonary lesions.

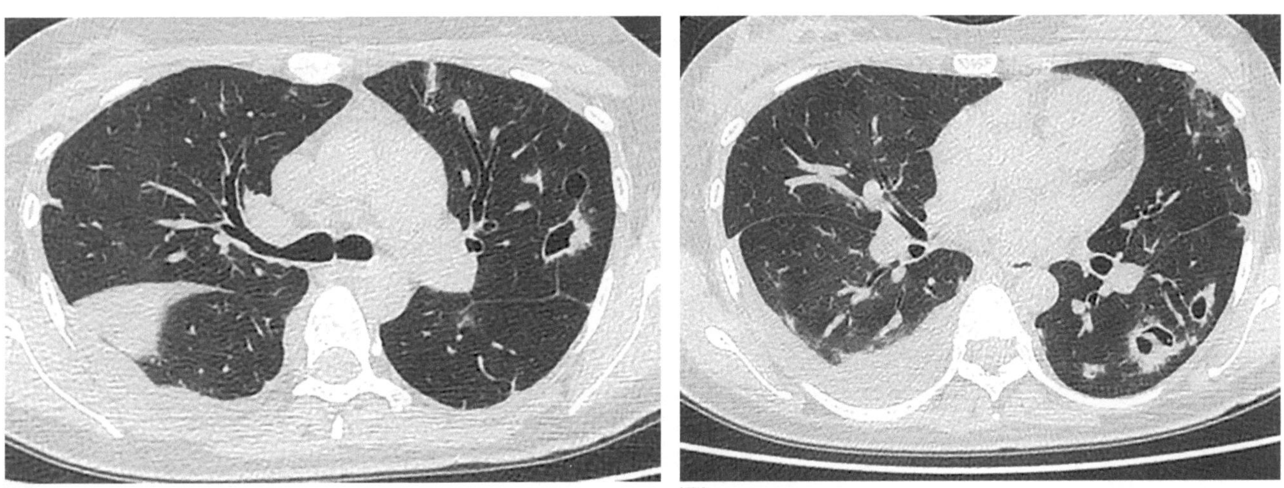

(A) (B)

FIG. 10.2.10 Embologenic metastases of chorion carcinoma in the lungs. A variety of CT signs—nodules, thick- and thin-walled cavities with irregular shape, right-sided pleural effusion (A, B). The picture was initially diagnosed as a septic lung injury.

TABLE 10.2.2 Differential signs for various metastatic lung lesions

	Nodular metastases	Lymphangitic carcinomatosis	Chorion carcinoma	Microscopic tumor cell embolism of the pulmonary artery
Localization of the primary tumor	Breast, kidney, testis, osteoblastic sarcoma, melanoma	Breast, stomach, prostate, pancreas, lung, sarcoma	Uterus, ovaries; a primary pulmonary form is possible	Breast, stomach, lung, liver
Symptoms	Often absent; with multiple metastases, there is weight loss and weakness	Increasing dyspnea	Increasing dyspnea, hemoptysis	Rapidly increasing dyspnea
CT signs	A spherical or oval nodules with sharp borders; large metastases often have spicularity and rough contours; cavities within large nodes are possible	Thickening of the interlobular septa and vascular-bronchial bundles; pleural effusion is common; a normal CT image is possible	Nodules and masses; the emergence of cavities is possible; contrasting filling defects in the pulmonary artery	Most often absent, although there may be mosaic perfusion
Diagnostic methods	Surgical biopsy	Transbronchial biopsy, bronchoalveolar lavage, PET-CT	Chorionic gonadotropin in blood serum	Biopsy, ventilation-perfusion scan
Differential range	Pulmonary infections, including septic metastases, organizing pneumonia, sarcoidosis, enlarged intrapulmonary lymph nodes, rheumatoid nodes	Congestive heart failure, pulmonary infections, sarcoidosis, interstitial pneumonia, drug-induced pneumonitis	Pulmonary thromboembolism, organizing pneumonia; pulmonary infections	Pulmonary thromboembolism, pulmonary hypertension; heart failure

References

[1] Pastorino U, Buyse M, Friedel G, Ginsberg RJ, Girard P, Goldstraw P, et al. Long-term results of lung metastasectomy: prognostic analyses based on 5206 cases. The international registry of lung metastases. J Thorac Cardiovasc Surg 1997;113(1):37–49.

[2] Sternberg DI, Sonett JR. Surgical therapy of lung metastases. Semin Oncol 2007;34(3):186–96.

[3] Pfannschmidt J, Dienemann H, Hoffmann H. Surgical resection of pulmonary metastases from colorectal cancer: a systematic review of published series. Ann Thorac Surg 2007;84(1):324–38.

[4] Seemann MD, Seemann O, Luboldt W, Bonél H, Dienemann H, Staebler A. Differentiation of malignant from benign solitary pulmonary lesions using chest radiography, spiral CT and HRCT. Lung Cancer 2000;29(2):105–24.

[5] Davis SD. CT evaluation for pulmonary metastases in patients with extrathoracic malignancy. Radiology 1991;180(1):1–12.

[6] Friedmann G, Bohndorf K, Krüger J. Radiology of pulmonary metastases: cornparison of imaging techniques with operalive findings. Thorac Cardiovasc Surg 1986;34:Spec No. 2: 120–4.

[7] Johnson Jr H, Fantone J, Flye MW. Histologic evaluation of the nodules resected in the treatment of pulmonary metastatic disease. J Surg Oncol 1982;21(1):1–4.

[8] Flye MW, Woltering G, Rosenberg SA. Aggressive pulmonary resection for metastatic osteogenic and soft tissue sarcomas. Ann Thorac Surg 1984;37(2):123–7.

[9] Murata K, Takahashi M, Mori M, Kawaguchi N, Furukawa A, Ohnaka Y, et al. Pulmonary metastatic nodules: CT-pathologic correlation. Radiology 1992;182(2):331–5.

[10] Yokomise H, Mizuno H, Ike O, Wada H, Hitomi S, Itoh H. Importance of intrapulmonary lymph nodes in the differential diagnosis of small pulmonary nodular shadows. Chest 1998;113(3):703–6.

[11] Hyodo T, Kanazawa S, Dendo S, Kobayashi K, Hayashi H, Kouno Y, et al. Intrapulmonary lymph nodes: thin-section CT findings, pathological findings, and CT differential diagnosis from pulmonary metastatic nodules. Acta Med Okayama 2004;58(5):235–40.

[12] Takenaka M, Uramoto H, Shimokawa H, So T, Hanagiri T, Aoki T, et al. Discriminative features of thin-slice computed tomography for peripheral intrapulmonary lymph nodes. Asian J Surg 2013;36(2):69–73.

[13] Matsuki M, Noma S, Kuroda Y, Oida K, Shindo T, Kobashi Y. Thin-section CT features of intrapulmonary lymph nodes. J Comput Assist Tomogr 2001;25(5):753–6.

[14] Bankoff MS, McEniff NJ, Bhadelia RA, Garcia-Moliner M, Daly BD. Prevalence of pathologically proven intrapulmonary lymph nodes and their appearance on CT. AJR Am J Roentgenol 1996;167(3):629–30.

[15] MacMahon H, Naidich DP, Goo JM, Lee KS, Leung ANC, Mayo JR, et al. Guidelines for Management of incidental Pulmonary nodules Detected on CT images: From the Fleischner Society 2017. Radiology 2017;284(1):228–43.

[16] Gould M, Donington J, Lynch W, et al. Evaluation of individuals with pulmonary nodules: when is it lung cancer? Diagnosis and management of lung cancer, 3rd ed: American College of Chest Physicians evidence-based clinical practice guidelines. Chest 2013;143(5 Suppl):e93S–e120S.

[17] Seo B, Im JG, Goo JM, Chung MJ, Kim MY. Atypical pulmonary metastases: spectrum of radiologic findings. Radiographics 2001;21(2):403–17.

[18] Chaudhuri MR. Primary pulmonary cavitating carcinomas. Thorax 1973;28(3):354–66.

[19] Honda O, Tsubamoto M, Inoue A, Johkoh T, Tomiyama N, Hamada S, et al. Pulmonary cavitary nodules on computed tomography: differentiation of malignancy and benignancy. J Comput Assist Tomogr 2007;31(6):943–9.

[20] Cordier J.-F, editor. Orphan lung diseases. Eur Respir Monogr 2011;54:64–5.

[21] Loukeri AA, Pantazopoulos IN, Tringidou R, Giampoudakis P, Valaskatzi A, Loukeri PA, et al. Benign metastasizing leiomyoma presenting as cavitating lung nodules. Respir Care 2014;59(7):e94–7.

[22] Chander K, Feldman L, Mahajan R. Spontaneous regression of lung metastases: possible BOOP connection? Chest 1999;115(2):601–2.

[23] Rolston KV, Rodriguez S, Dholakia N, Whimbey E, Raad I. Pulmonary infections mimicking cancer: a retrospective, three-year review. Support Care Cancer 1997;5(2):90–3.

[24] Tolaney SM, Colson YL, Gill RR, Schulte S, Duggan MM, Shulman LN, et al. Sarcoidosis mimicking metastatic breast cancer. Clin Breast Cancer 2007;7(10):804–10.

[25] Bruce DM, Heys SD, Eremin O. Lymphangitis carcinomatosa: a literature review. J R Coll Surg Edinb 1996;41(1):7–13.

[26] Prakash P, Kalra MK, Sharma A, Shepard JA, Digumarthy SR. FDG PET/CT in assessment of pulmonary lymphangitic carcinomatosis. AJR Am J Roentgenol 2010;194(1):231–6.

[27] Masson RG, Krikorian J, Lukl P, et al. Pulmonary microvascular cytology in the diagnosis of lymphangitic carcinomatosis. N Engl J Med 1989;321:71–6.

[28] Charest M, Armanious A. Prognostic implication of the lymphangitic carcinomatosis pattern on perfusion lung scan. Can Assoc Radiol J 2012;63(4):294–303.

[29] Grenier P, Chevret S, Beigelman C, Brauner MW, Chastang C, Valeyre D. Chronic diffuse infiltrative lung disease: determination of the diagnostic value of clinical data, chest radiography, and CT and Bayesian analysis. Radiology 1994;191(2):383–90.

[30] Webb WR, Müller NL, Naidich DP. High-resolution CT of the lung. 5th ed. Philadelphia: Lippincott Williams and Wilkins; 2015. p. 280–5.

[31] Johkoh T, Ikezoe J, Tomiyama N, Nagareda T, Kohno N, Takeuchi N, et al. CT findings in lymphangitic carcinomatosis of the lung: correlation with histologic findings and pulmonary function tests. AJR Am J Roentgenol 1992;158(6):1217–22.

[32] Chatkin JM, Fritscher LG, Fiterman J, Fritscher CC, da Silva VD. Microscopic pulmonary neoplastic emboli: report of a case with respiratory failure but normal imaging. Prim Care Respir J 2007;16(2):115–7.

[33] Vlenterie M, Desar IM, van Herpen CM, Tol J. Fatal microscopic pulmonary tumour embolisms in patients with breast cancer: necessary knowledge for future medical practice. Neth J Med 2014;72(1):28–31.

[34] Abati A, Landucci D, Danner RL, Solomon D. Diagnosis of pulmonary microvascular metastases by cytologic evaluation of pulmonary artery catheter-derived blood specimens. Hum Pathol 1994;25(3):257–62.

[35] McCabe JM, Bhave PD, McGlothlin D, Teerlink JR. Running from her past: a case of rapidly progressive dyspnea on exertion. Circulation 2011;124(21):2355–61.

[36] Roberts KE, Hamele-Bena D, Saqi A, Stein CA, Cole RP. Pulmonary tumor embolism: a review of the literature. Am J Med 2003;115(3):228–32.

[37] Kridel R, Myit S, Pache JC, Gaspoz JM. Pulmonary tumor embolism: a rare cause of acute right heart failure with elevated D-dimers. J Thorac Oncol 2008;3(12):1482–3.

[38] Berkowitz RS, Goldstein DP. Chorionic tumors. N Engl J Med 1996;335(23):1740–8.

[39] Niimi K, Yamamoto E, Nishino K, Fujiwara S, Ino K, Kikkawa F. Spontaneous regression of gestational trophoblastic neoplasia. Gynecol Oncol Rep 2017;21:98–100.

[40] Zhang W, Liu B, Wu J, Sun B. Hemoptysis as primary manifestation in three women with choriocarcinoma with pulmonarymetastasis: a case series. J Med Case Rep 2017;11(1):110.

[41] Ibi T, Hirai K, Bessho R, Kawamoto M, Koizumi K, Shimizu K. Choriocarcinoma of the lung: report of a case. Gen Thorac Cardiovasc Surg 2012;60(6):377–80.

Chapter 11

Drug-induced pulmonary diseases

Alexander Averyanov[a,b], Evgeniya Kogan[c], Victor Lesnyak[d]

[a]Clinical Department, Pulmonology Research Institute under FMBA of Russia, Moscow, Russia, [b]Pulmonary Division, Federal Research Clinical Center under FMBA of Russia, Moscow, Russia, [c]Anatomic Pathology Department, Sechenov University, Moscow, Russia, [d]Radiology Department, Federal Research Clinical Center under FMBA of Russia, Moscow, Russia

The lungs, like other organs, can be targets for direct or indirect lesions of drug-related diseases. Toxic reactions to drugs occur in 2%–5% of hospitalized patients [1], and pulmonary pathology in drug-induced disorders accounts for less than 5% of all registered adverse events [2]. Therefore the relatively low rate of drug-related lesions of the lungs contributes to the insufficient knowledge on drug-induced pulmonary disease (DIPD) among specialists. According to www.pneumotox.com, among the more than 1400 drugs, procedures, and substances that can cause respiratory problems, there are more than 350 drugs that can cause drug-induced interstitial lung disease (ILD), with the list constantly updated [3]. For example, sarcoid reaction to pirfenidone, which is a relative new drug for the treatment of idiopathic pulmonary fibrosis (IPF), was described for the first time in a recent report [4].

There are four primary mechanisms of drug-induced damage to the lung tissue [5]:

1. Oxidative stress: Certain drugs lead to the formation of a large number of free radicals and singlet oxygen, which can damage the epithelium and endothelium. An example is nitrofurans.
2. Direct cytotoxic damage to the lung parenchyma: Immunosuppressants exert direct cytotoxic parenchymal injury. Destructive effects on the DNA of epithelial cells are also possible, such as that seen with bleomycin.
3. Accumulation of phospholipids in macrophages and alveolocytes, leading to their accumulation in the distal respiratory tract. This mechanism underlies lung lesion observed in amiodarone lung injury and exogenous lipoid pneumonia.
4. Immune-mediated reactions: This mechanism can involve a wide range of drugs, such as sulfanilamides, hydralazine, monoclonal antibodies to tumor necrosis factor alpha, and other biological drugs.

Main morphological presentations of drug-induced lesions of the lungs are the following [6]: diffuse alveolar damage (DAD) and acute respiratory distress syndrome; bronchiolitis obliterans organizing pneumonia (BOOP); constrictive bronchiolitis; eosinophilic pneumonia; hypersensitivity pneumonitis (HP); nonspecific interstitial pneumonia (NSIP); diffuse alveolar hemorrhage; pulmonary vasculitis; and granulomatous reactions, including sarcoid-like reactions.

Each of these morphological forms has specific clinical and radiological presentations, which are already described in respective chapters. Awareness of the specific aspects of drug-induced lesions facilitates the establishment of diagnosis significantly (Table 11.1). Among the most common causes of DIPD identified in the large studies to date, anticancer agents (23%–51%), antirheumatic medications (6%–72%), antibiotics (6%–26%), nonsteroidal anti-inflammatory drugs (0%–23%), psychiatric drugs (0%–9%), and antiarrhythmic drugs (0%–9%) are the most prominent [3].

The succeeding texts are the detailed descriptions of the best-known drug-induced lesions of the lungs.

Amiodarone-induced lung disease (AILD)

Despite more than 40-year history of use, amiodarone remains one of the most effective and widely prescribed drugs for treatment of heart rhythm disorders, especially atrial fibrillation and ventricular arrhythmias. However, due to its cumulative properties, amiodarone often has adverse effects, approximately 15% during the first year of treatment and up to 50% with longer use [7]. The most common target organs for amiodarone accumulation and development of drug-induced lesion are the eyes (corneal deposits and photophobia), thyroid gland (hypo- and hyperthyroidism), liver (drug-induced hepatitis and dyspeptic disorders), skin (photosensitization), and nervous system (peripheral neuropathy) [8]. The lungs are affected less frequently, comprising 4%–6% of all complications; however, they are associated with clinically the most significant adverse, even lethal, outcomes [5].

Difficult to Diagnose Rare Diffuse Lung Disease. https://doi.org/10.1016/B978-0-12-815375-8.00011-X

TABLE 11.1 The most common morphological and radiological manifestations of drug-induced lung lesions

Morphological form	Drug	Computed tomography findings
Diffuse alveolar damage, ARDS	Amiodarone Bleomycin Busulfan Gefitinib Imatinib Cocaine Melphalan Mitomycin Monoclonal antibodies Gold salts Cyclophosphamide Cytosine arabinoside	Bilateral massive consolidation areas, mainly in subpleural and basal fields. Ground-glass opacity can surround these areas or can exhibit a patchy distribution
BOOP	Amiodarone Bleomycin Busulfan Gefitinib Interferon Methotrexate Nitrofurans Penicillamine Gold salts Sulfasalazine Cyclophosphamide	Subpleural and peribronchovascular consolidation areas surrounded by or alternating with ground-glass opacities
Constrictive bronchiolitis	Lomustine Penicillamine Gold salts Sulfasalazine	Thickening of the bronchial walls, tree-in-bud appearance, mosaic perfusion, air traps
Hypersensitivity pneumonitis	Amitriptyline Leflunomide Methotrexate Paclitaxel Fluoxetine Cyclophosphamide	Diffuse, ill-defined intralobular nodules; ground-glass opacities; lobulated hyperinflation
Nonspecific interstitial pneumonia	Amiodarone Bleomycin Methotrexate Carmustine Penicillamine Sulfasalazine Chlorambucil	Ground-glass opacities, reticular abnormalities, traction bronchiectasis, honeycombing
Eosinophilic pneumonia	Imatinib Daptomycin Mesalamine Minocycline Methotrexate Nitrofurans Penicillamine Sulfasalazine NSAIDs	Areas of consolidation and ground-glass opacities, primarily subpleural and in upper lobes; configuration and localization vary
Alveolar hemorrhage, pulmonary vasculitis[a]	Anticoagulants Amphotericin B Cocaine Cyclosporine Penicillamine Propylthiouracil Rituximab	Bilateral patchy areas of ground-glass opacity with parahilar distribution; multiple, often confluent centrilobular nodules; no damage in subpleural and apical sites

TABLE 11.1 The most common morphological and radiological manifestations of drug-induced lung lesions—cont'd

Morphological form	Drug	Computed tomography findings
Sarcoid-like and other granulomatous reactions	Nivolumab Pembrolizumab Vemurafenib Etanercept Infliximab Interferon alpha Methotrexate Seasonal Influenza Vaccine	Increased intrathoracic lymph nodes, well-defined nodules with peribronchovascular distribution

[a]*An extended list of drugs that cause alveolar hemorrhage is presented in the chapter 4.3.*
ARDS, acute respiratory distress syndrome; BOOP, bronchiolitis obliterans organizing pneumonia; NSAID, nonsteroidal antiinflammatory drug.
(Modified from Roden AC, Camus P. Iatrogenic pulmonary lesions. Semin Diagn Pathol 2018;35(4):260-71; Webb RW, Higgins CB Thoracic imaging: pulmonary and cardiovascular radiology. 2nd ed. Philadelphia, PA: Lippincott Williams & Wilkins; 2011.)

The proposed leading mechanism for the development of amiodarone-associated pulmonary lesions includes the accumulation of amiodarone-containing phospholipid complexes in alveolar macrophages and type II alveolocytes [9]. In turn, suppression of phospholipase by amiodarone leads to the accumulation of phospholipids [10]. One study showed that the risk factors for amiodarone lung were duration of the intake and age over 60 years [11].

Earlier work also revealed a link between the cumulative dose and the underlying amiodarone-associated lung disease [12]. However, cases with acute amiodarone-induced lung injury following intravenous or oral administration during the first weeks of treatment are well known [13–15]. In general, amiodarone-induced lung disease manifests during the first 2 years of administration, and the daily dose is at least 400 mg in majority of the patients [16]. However, there are numerous cases where AILD develops in patients administered 200-mg or less amiodarone daily [17]. Thus there is no clear relationship between the dose and frequency of drug treatment and the severity of amiodarone-induced lung disease.

Generally the morphological variants of amiodarone-induced lung lesions are organizing pneumonia (Fig. 11.1) and NSIP (Fig. 11.2), and the proportion of DAD accounts for no more than 10% of all pulmonary manifestations of drug-induced pneumonitis (Fig. 11.3). Other variants including lymphocytic pneumonia, follicular bronchiolitis, diffuse lymphoid hyperplasia, and eosinophilic pneumonia are also described as manifestations of amiodarone toxicity [18].

The presence of type II pneumocytes and alveolar macrophages with finely vacuolated cytoplasm in biopsy samples reflects a history of amiodarone administration but does not indicate pulmonary toxicity [19].

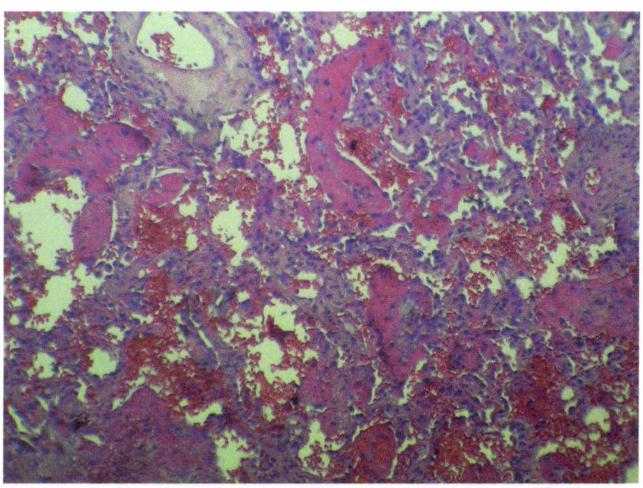

FIG. 11.1 Amiodarone-induced pneumonitis. Diffuse alveolar damage, organization stage. Diffuse sclerosis, edema, and lymphohistiocytic infiltration of alveolar septa with hyaline membranes. Hematoxylin and eosin (H&E) staining, 200×.

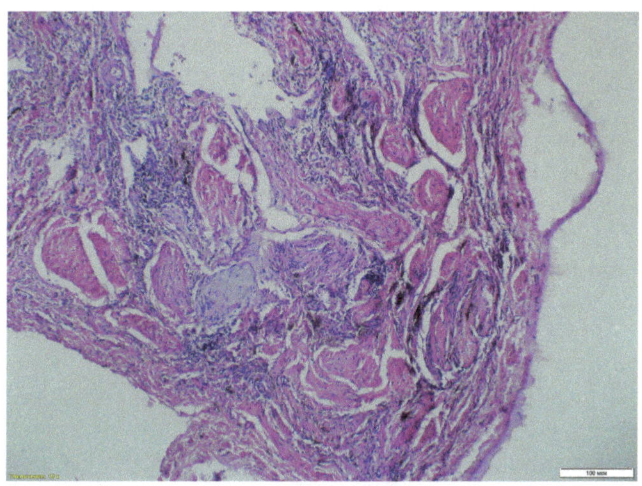

FIG. 11.2 Amiodarone-induced pneumonitis. Organizing pneumonia. Granulation tissue fills the lumen of the alveoli and bronchioles. Sclerosis, edema, and lymphohistiocytic infiltration of alveolar septa. H&E staining, 100×.

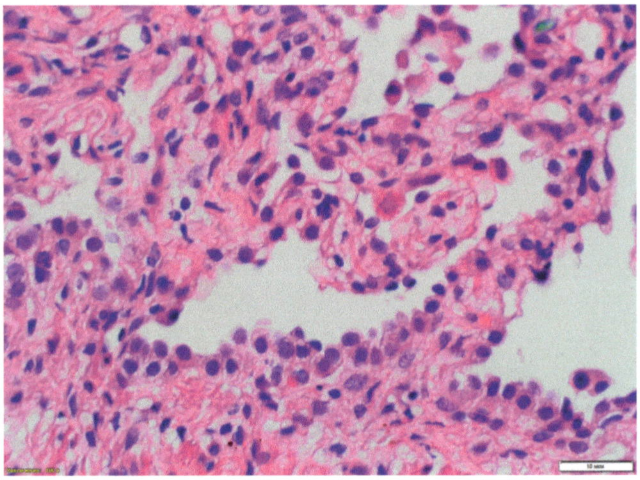

FIG. 11.3 Amiodarone-induced pneumonitis, nonspecific interstitial pneumonia. Diffuse sclerosis, edema, and lymphohistiocytic infiltration of alveolar septa and small vessels. H&E stain, ×400.

Clinical presentation

Clinically the majority of patients with AILD exhibit subacute or a gradually progressing dyspnea, dry cough, and fever. A small number of patients may have pleural pain. Auscultation usually fails to detect any distinctive sounds, although it is sometimes possible to reveal Velcro crackles in posterior basal fields in some patients with nonspecific interstitial pneumonia pattern [5]. In cases of diffuse alveolar damage, the symptoms of acute respiratory failure with the development of respiratory distress syndrome are dominant. In the case of amiodarone-induced BOOP, clinical symptoms may resemble bacterial pneumonia (BP); however, this variant of AILD follows a subacute course.

Diagnosis

In patients with a chronic course, functional tests usually reveal a moderate restrictive pattern. In one study of patients with a newly diagnosed amiodarone-induced lung injury, the average FVC was 74%, and carbon monoxide diffusing capacity (DLCO) was 45% of the normal value [20]. Moderate leukocytosis and increased erythrocyte sedimentation rate (ESR) and C-reactive protein can be detected in patients with amiodarone lung [5]. However, these changes are nonspecific and generally reflect severity of the interstitial inflammation.

Bronchoalveolar lavage (BAL) is critical for the definite diagnosis of amiodarone lung. First, BAL aids in ruling out several diseases with similar clinical and radiological patterns, such as eosinophilic pneumonia, hypersensitivity pneumonitis, and pneumonia due to bacteria and opportunistic pathogens. Second a decrease in the total number of macrophages in BAL fluid is a sensitive, albeit not very specific, sign of toxic amiodarone alveolitis [21]. However, macrophage number, which comprises approximately 66% of all BAL fluid cells, is significantly lower in patients compared with healthy individuals and those who are taking amiodarone but do not exhibit pulmonary symptoms who have approximately 85% macrophages in the BAL fluid [21]. Foamy macrophages in the BAL fluid are also found frequently in patients with drug-induced disease and those receiving amiodarone without complications [22]. Nevertheless, the absence of foamy macrophages aids in ruling out the diagnosis of amiodarone pneumonitis [5].

High resolution computed tomography

Characteristic high-resolution computer tomography (HRCT) findings of the amiodarone lung depend on the morphological pattern. In BOOP due to amiodarone, the presence of bilateral areas of consolidation, located more often subpleurally and peribronchovascularly, with patchy areas of ground-glass opacity (GGO) (Fig. 11.4) is typical [23].

With treatment or spontaneously the configuration and localization of these sites can change, which makes them similar to eosinophilic infiltrates. Density of the consolidated areas is usually higher than that of the soft tissues, accounting for 82–174 Hounsfield units (HU), which is considered a specific sign of amiodarone infiltrates due to the presence of the iodide component [23, 24].

In the NSIP-type amiodarone lung, HRCT scans usually reveal areas of GGO, reticular abnormalities, and sites of linear fibrosis; subpleural areas of honeycombing and traction bronchiectasis may also be found (Fig. 11.5). Changes are similar to those that occur in idiopathic nonspecific interstitial pneumonia (Fig. 11.6).

Finally, DAD manifests as bilateral massive consolidation zones involving the lower parts, which captures more of the overlying parenchyma as it progresses [23]. Pleural effusion, which is often observed in patients, is not typical for amiodarone toxicity itself but due to concomitant heart failure that occurs in most patients with severe arrhythmias [25].

Usually, in patients with suspicious AILD, density assessment of the liver and spleen parenchyma is recommended, as they are almost always higher than normal. However, this sign is often observed in amiodarone-treated patients without pulmonary manifestations (Fig. 11.7) [25].

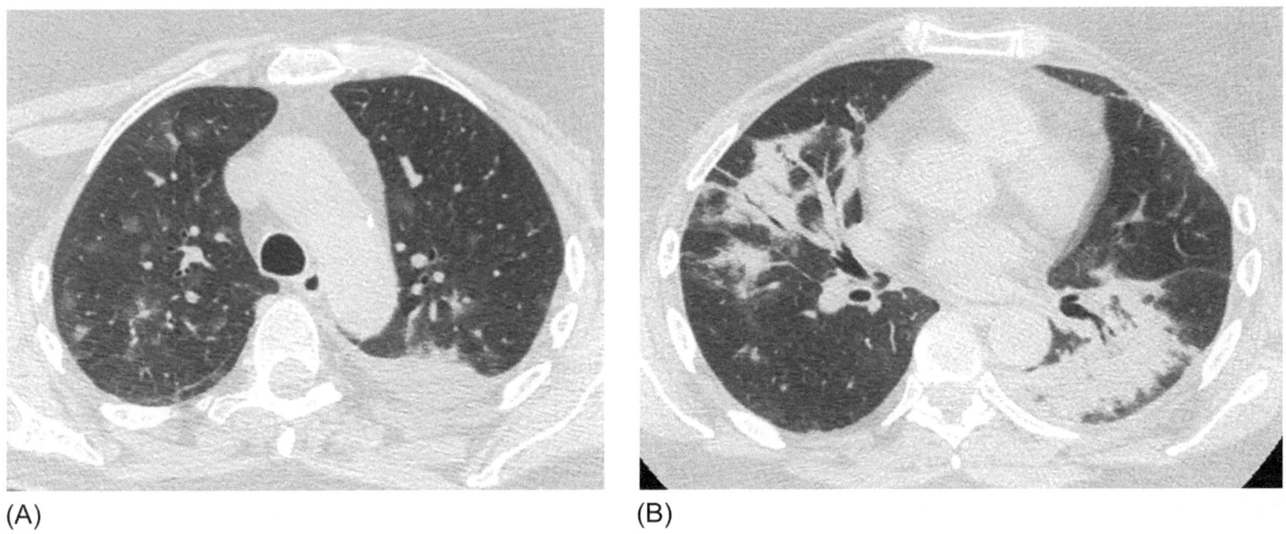

FIG. 11.4 Amiodarone-induced pneumonitis. Bilateral patchy areas of ground-glass opacity, with unevenly thickened interlobular septa (A). Foci of ground-glass opacity, fragmented reticular opacities, and subpleural linear fibrosis (B). Abnormalities are present in both the upper and lower lung fields.

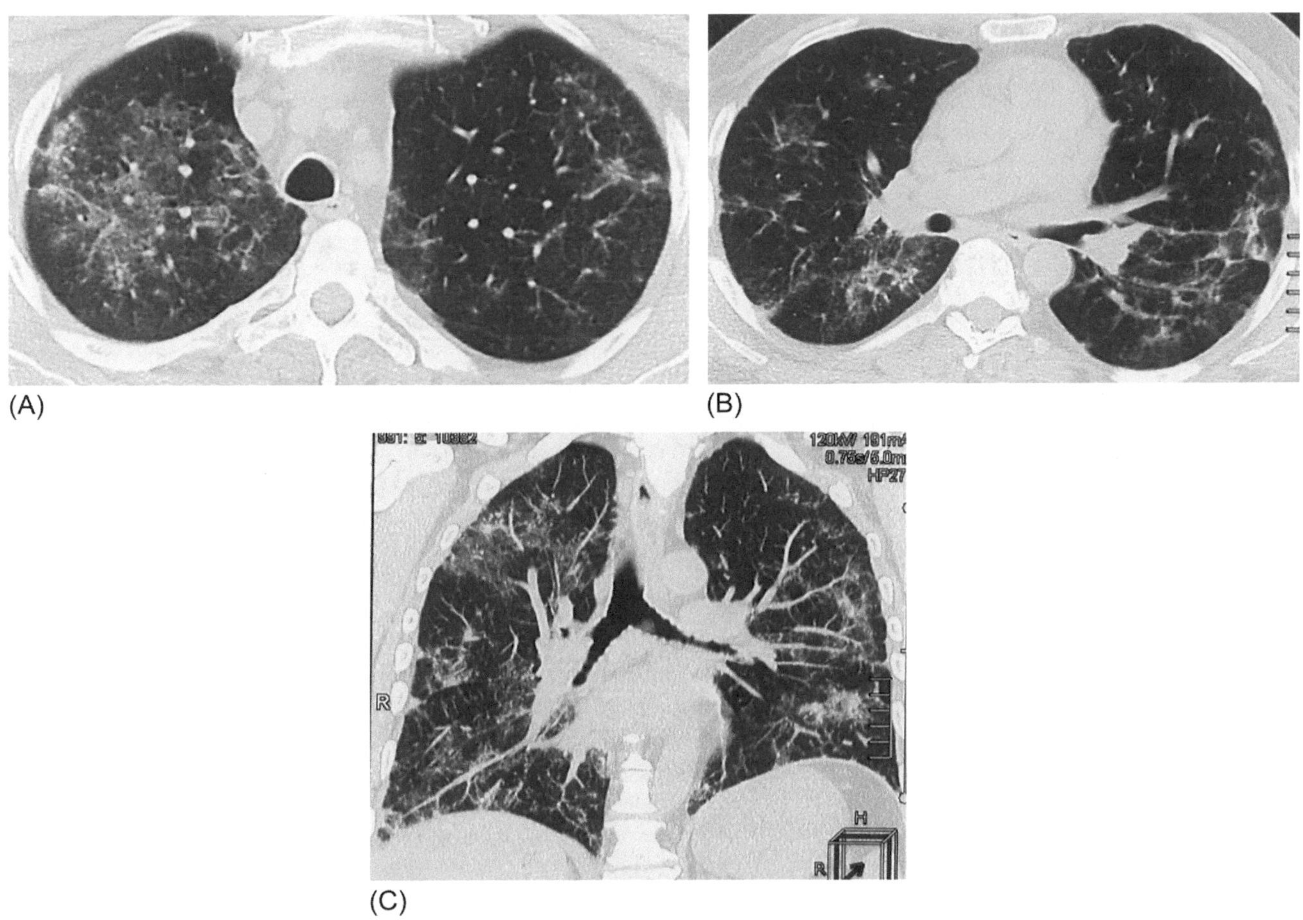

FIG. 11.5 Amiodarone-induced pneumonitis. Randomly distributed patchy areas of ground-glass opacity and reticular abnormalities are visible (A–C).

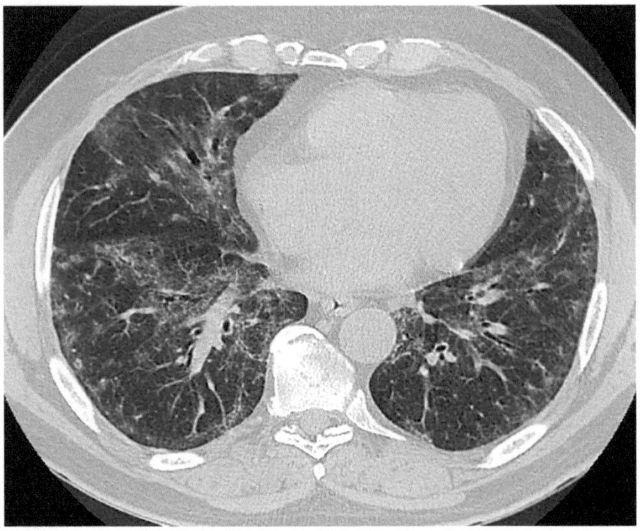

FIG. 11.6 Idiopathic nonspecific interstitial pneumonia. Diffuse areas of ground-glass opacity and fine reticular abnormalities in the lower lobes. Subpleural sparing *(arrows)*.

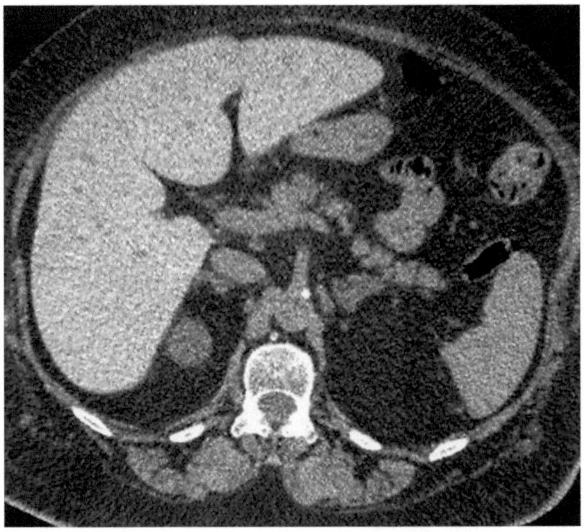

FIG. 11.7 Abnormality of the abdominal parenchymal organs with prolonged intake of amiodarone. Density of the liver parenchyma is evenly increased and reaches 85 HU, the spleen tissue density is approximately 75 HU. The density of these tissues is visibly higher than that of the spinal muscles.

Morphological evaluation was a widely used approach to verify the diagnosis of AILD; however, after the development of acute respiratory distress syndrome (ARDS) in a significant number of patients in the postoperative period (up to 50%) [26], surgical approaches to obtain biopsy material are recommended only in cases of emergency. The association of ARDS with thoracic interventions in patients currently taking amiodarone or with a history of amiodarone treatment is attributed to the additional damaging effect of high oxygen concentration on the respiratory tract during intraoperative invasive lung ventilation [5].

Differential diagnosis

The differential diagnosis of amiodarone-induced lung lesions should include infectious diseases, ILD of other origins, and pulmonary manifestations of cardiac rhythm disorders and heart failure.

Elderly people suffering from chronic diseases who are treated with amiodarone are at high risk for bacterial pneumonia (BP). Comorbidities such as diabetes mellitus and history of stroke render them even more susceptible to pulmonary infections. Moreover the course of pneumonia in these patients can be obscure in the absence of overt fever or pronounced cough. In such cases, it is very difficult to differentiate BP from BOOP, as the clinical and radiological presentations can be very similar. However, a more acute onset, purulent sputum, frequently unilateral presentation, and a limited number of consolidation areas are characteristics of BP. Peripheral blood leukocytosis is usually not present and can also be observed in BOOP. In the first days of disease, procalcitonin level may help in diagnosis, as it increases with bacterial infections but remains normal in BOOP. BP can be detected by auscultation more frequently than BOOP for which bronchial obstruction (a frequent finding concomitant with BP) is generally not a characteristic, and pneumonic crackles are absent in more than half of the patients. High degree of virulent microorganisms from sputum also indicates the infectious nature of the pulmonary lesion. Finally, response to adequate antibiotic therapy in BP is obvious. Conversely, whereas there may be some improvement due to inexplicable reasons in BOOP, the disease is not completely resolved.

ILDs, such as hypersensitivity pneumonitis and IPF, as well as eosinophilic lesions, may be considered in the differential diagnosis of AILD. Their differential characteristics are detailed in the relevant sections. It is especially difficult to differentiate atypical or initial forms of IPF, which develop in older individuals with cardiac problems and those who take amiodarone (Fig. 11.8). The lack of improvement with steroids and pronounced subpleural zones of honeycombing may indicate IPF. As for cryptogenic OP and NSIP, that can be diagnostic alternatives to their drug-induced analogue, in our opinion, it is unlikely that their presence in patients receiving amiodarone can be considered to indicate independent disease. It is possible that OP and NSIP may only be due to a secondary reaction of the lung tissue to exogenous factors or endogenous processes, and the idiopathic nature is explained only by the inability to establish these relationships with existing diagnostic tools.

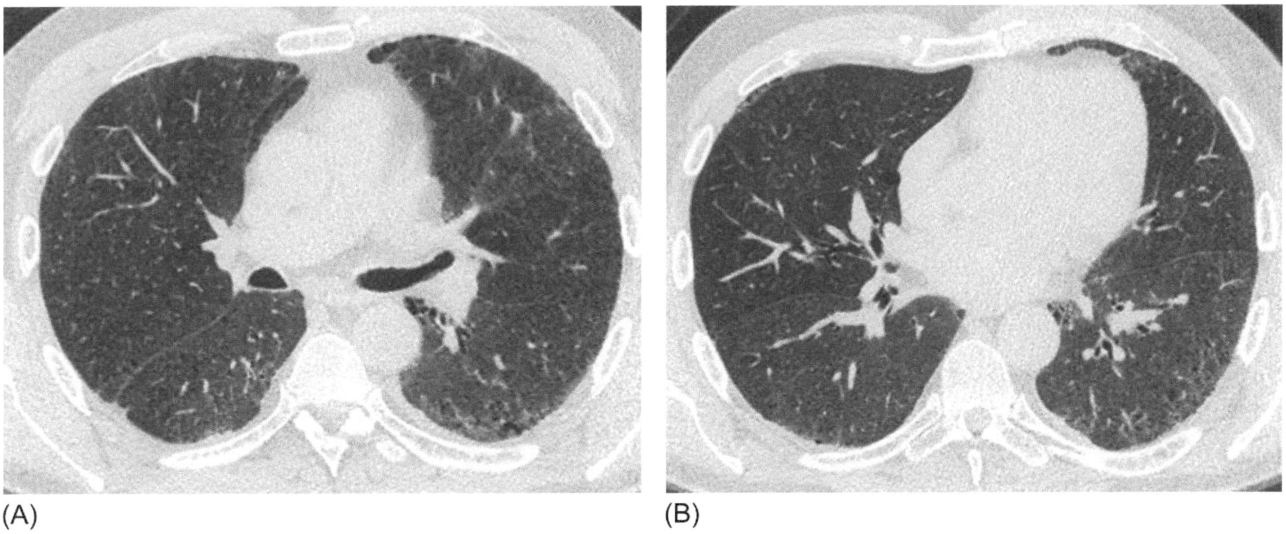

FIG. 11.8 Initial manifestations of idiopathic pulmonary fibrosis, confirmed by video-assisted thoracoscopy, in a patient treated with amiodarone. Mild subpleural reticular abnormalities, bronchioloectases, and single small cysts located below the visceral pleura (A and B).

Heart failure and pulmonary thromboembolism (PTE), which can occur in these patients, should always be in the differential range of AILD. The greatest difficulties arise if a patient taking amiodarone has both interstitial changes in the lungs and thromboembolism (Fig. 11.9). Repeated thromboembolism of small pulmonary artery branches may not have a vivid clinical pattern of acute respiratory failure and may manifest as either recurrent hemoptysis or slowly increasing dyspnea during the development of pulmonary hypertension. Additionally, subpleural consolidation due to pulmonary infarction, sometimes resembling BOOP, can occur in the lungs. In patients with new PTE, there is usually an increase in D-dimer levels in the blood, which is however not sensitive in long-term infarction. Furthermore, pulmonary hypertension is not an evidence of PTE, as it often occurs in patients with congestive heart failure. In that case, CT angiography of the pulmonary artery is necessary, which clarifies the diagnosis in most patients.

Pulmonary edema can be observed in patients with prolonged congestive heart failure or stenosis of the pulmonary veins, which can sometimes complicate radio-frequency ablation. Signs of interstitial pulmonary edema on chest radiographs include the appearance of Kerley lines, indicating thickening of the interlobar fissures, peribronchial edema, and blurring of pulmonary vessels. CT of the chest reveals thickening of the interlobular septa, thickening of the peribronchovascular interstitium, and GGO with perihilar distribution [24]. It is the last characteristic that can be confusing for specialists, as it can be observed in interstitial inflammatory processes. GGO with interstitial pulmonary edema is usually bilateral, diffuse, and somewhat symmetrical and tends to occur in the central zones, whereas subpleural spaces remain spared, which reflect the preserved lymphatic outflow in the periphery of the lungs (Fig. 11.10). However, patchy and subpleural areas of GGO can also occur in interstitial pulmonary edema [24]. In cases where the origin of the changes in the lungs is unclear, clinical presentation should be taken in consideration as well. A patient with interstitial edema will have upright position, due to the dramatically worsening breathlessness in the horizontal position, whereas a patient with interstitial pneumonia does not deteriorate as quickly when lying down, although they tend to occupy a semisitting position. A significantly elevated plasma level of a brain natriuretic peptide in patients with left ventricular failure is also helpful in the diagnosis, so are the findings

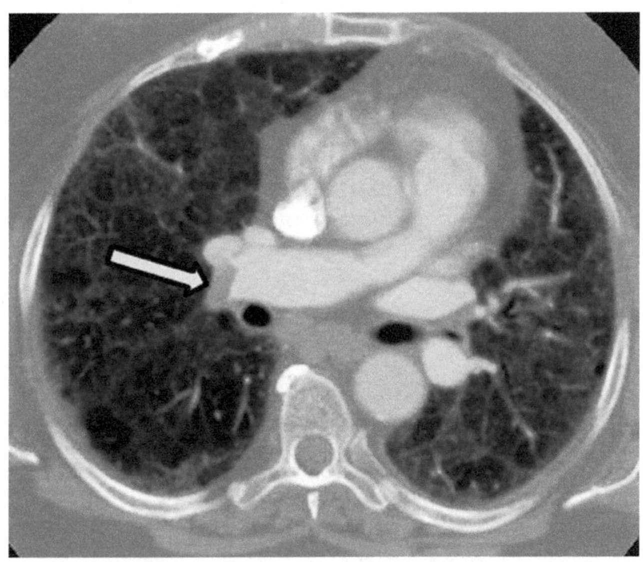

FIG. 11.9 Thromboembolism of the right inferior pulmonary artery *(arrow)* in a patient with idiopathic pulmonary fibrosis who was treated with amiodarone due to concomitant arrhythmia.

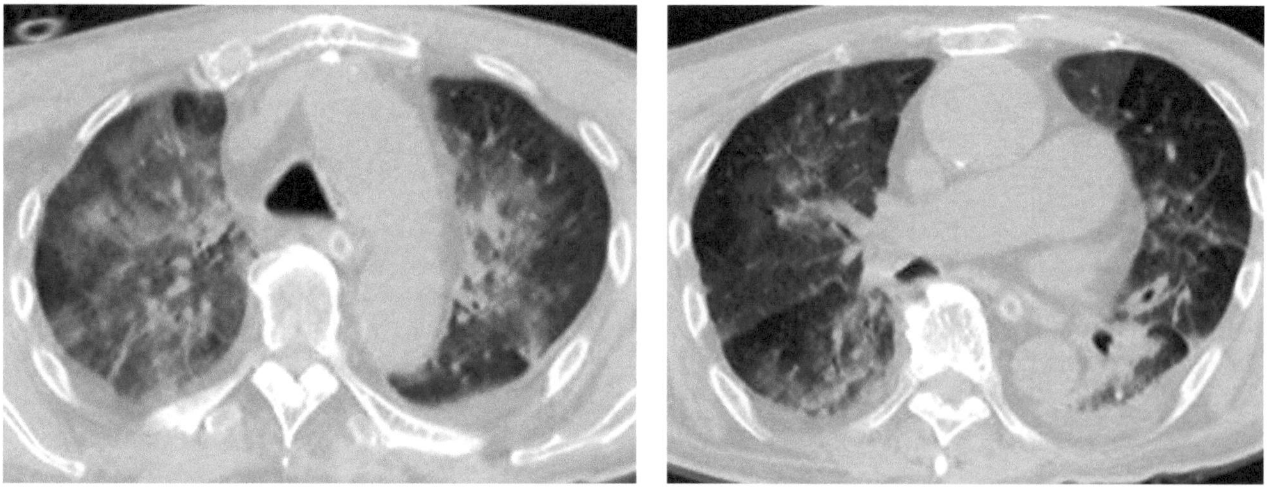

(A) (B)

FIG. 11.10 Pulmonary edema in a patient treated with amiodarone. Extensive areas of ground-glass opacities of varying severity with perihilar distribution. Thickening of interlobular septa and peribronchial interstitium (A and B). Bilateral hydrothorax (B).

of echo and doppler cardiography to determine cardiac ejection fraction, although the latter approach cannot be used to rule out heart failure, because many conditions including bradyarrhythmias and tachycardias may be accompanied by a normal ejection fraction.

The diagnosis of pulmonary vein stenosis is more complicated. Radio-frequency catheter ablation of the pulmonary veins should precede pulmonary vein stenosis, which occurs in 1%–3% of patients undergoing the procedure. Clinically, pulmonary vein stenosis manifests as progressive dyspnea and cough, often with mucous sputum and hemoptysis in one-third of the patients. Computed tomography (CT) angiography, magnetic resonance imaging, and endocardial ultrasound are used for diagnosis [27].

Treatment and prognosis

There are no clear clinical guidelines for the management of patients with AILD. One unambiguous approach is discontinuation of the drug. Further steps should be determined based on the specific clinical situation. In patients who are asymptomatic or with mild symptoms and limited lesions, a waiting period with regularly scheduled functional and CT studies is possible; however, there are no defined timelines or frequencies. We perform the first control evaluation at 1 month after the diagnosis. In improving or stable patients, further examination is performed 3 months later, and the interval is extended to 6 months. In patients with severe respiratory failure or massive pulmonary tissue involvement, systemic corticosteroids are administered at a prednisolone dose of 40–60 mg/day [6]. Due to the long elimination half-life of amiodarone, the treatment course must be a minimum of 2 months, followed by a dose reduction, and the total treatment duration is 6 months or more [5]. If it is not possible to discontinue amiodarone due to the absence of alternative drugs, supportive treatment with systemic steroids can be considered, as reported in numerous cases [28]. Although a clear clinical improvement is observed with steroid treatment and amiodarone discontinuation in most cases, structural changes in the lungs are not complete reversible. In a study of 15 patients with amiodarone-induced lung disease who were followed for 8–36 months, Mankikian et al. reported that none of the patients achieved complete recovery by pulmonary CT scan and DLCO, although the FVC reached reference values [20].

Methotrexate-induced pneumonitis

Methotrexate is an antagonist of folic acid that is used as immunosuppressant for the treatment of a wide range of cancers and inflammatory diseases such as rheumatoid arthritis, the Crohn disease, ulcerative colitis, and psoriasis. According to the Japanese registry of adverse drug reactions including 3341 drugs, methotrexate is the leading drug causing ILD [29]. Between 60% and 93% of patients receiving methotrexate are believed to have at least one adverse drug reaction during treatment [30]. The skin, the gastrointestinal tract, and the central nervous system are the more frequently affected sites, whereas the lungs are less frequently involved in methotrexate-induced adverse events (2%–11.6% of all treated cases), but the course of pulmonary involvement can be life-threatening [31, 32]. In a study published 20 years ago, the mortality rate of methotrexate-induced lung injury reached 15%–20% [33]. Nevertheless, one of the most recent metaanalyses based on 22 studies with more than 8500 participants did find an increased risk of death from methotrexate-induced lung disease (MILD) in patients with rheumatoid arthritis, although the moderate risk of pulmonary lesions due to both MILD and infectious complications was confirmed [34].

Most cases of MILD occur in the first 6 months of treatment; however, early forms of drug-induced pneumonitis 3 days after treatment initiation and late reactions a few weeks after withdrawal were described [33]. Therefore the frequency and severity of pulmonary involvement are not directly dose-dependent [35, 36].

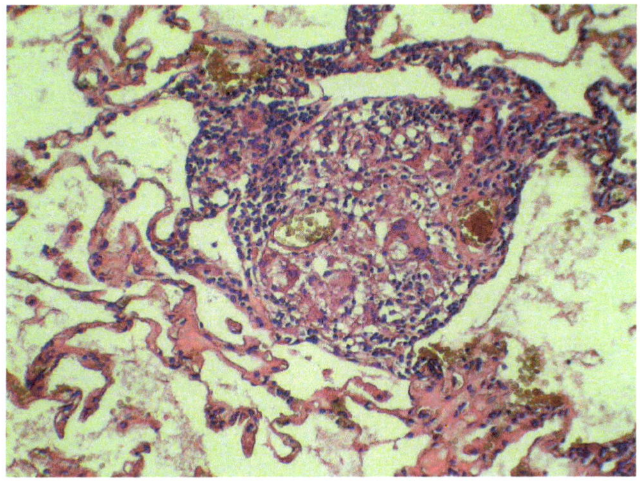

FIG. 11.11 Sarcoid-like reaction with the formation of granulomas in a patient treated with methotrexate. H&E stain, ×200.

Risk factors for methotrexate-associated lung damage are the following [37]: age over 60 years; hypoalbuminemia; diabetes mellitus; administration of high doses of methotrexate; the presence of an underlying pulmonary disease; reduced lung function prior to the initiation of methotrexate therapy; and intake of drugs that prevent binding of methotrexate to plasma proteins, such as aspirin, sulfonamides, penicillin, barbiturates, and nonsteroidal antiinflammatory drugs.

Histological changes in methotrexate-mediated lung damage include lymphocytic alveolitis (71%), organizing pneumonia (10%), hyaline membranes (8%), and the formation of granulomas without signs of necrosis (35%) (Fig. 11.11) [35]. In general the disease course can be categorized into acute, with diffuse alveolar damage, and subacute, with hypersensitivity pneumonitis or organizing pneumonia patterns [35]. Accordingly the clinical presentation is characterized by the rate of increase and the level of severity of dyspnea, which is present in all patients with methotrexate-induced lung injury. Additionally the acute presentation is usually accompanied by fever, chills, and severe dry cough. The progression of respiratory failure often requires introduction of invasive lung ventilation. Conversely the subacute presentation is less aggressive. In most cases subtle crackles in the posterior basal zone can be auscultated in addition to the traditional symptoms of dyspnea and unproductive cough.

Diagnosis

In general, blood tests reveal eosinophilia and moderate leukocytosis in half of the cases [5]. Lymphopenia should suggest the possibility of opportunistic infections that may be developing, which should be differentiated from MILD in all cases.

BAL is a useful tool to both rule out infections and diagnose subacute forms of drug-induced pneumonitis that is characterized by high lymphocytosis of more than 30% in the BAL fluid [38]. The absence of BAL lymphocytosis immediately puts the diagnosis of MILD in doubt, except for acute forms and in patients who have received corticosteroids previously. However, patients with neutrophilia in the BAL fluid were described, although this phenomenon is likely an exception [39]. An increase in the $CD4^+/CD8^+$ T lymphocyte ratio is also typical, but the wide variations in this marker hinders its utility as an ideal marker to differentiate MILD from other conditions such as ILD associated with rheumatoid arthritis [38, 40].

High-resolution computed tomography

HRCT characteristics are nonspecific and are determined by the histological variants [23]. The most common sign is GGO, reflecting interstitial lymphocytic infiltration. GGO is found in all patients, but consolidation zones with DAD appear very quickly. Organizing pneumonia, a form of MILD, usually manifests as bilateral subpleural areas of consolidation, surrounded by or alternating with patches of GGOs (Fig. 11.12) [41]. In general the posterior basal distribution of consolidation is prevalent. In patients with the hypersensitivity pneumonitis pattern, GGO, areas of lobular hyperinflation, and ill-defined intralobular nodules or linear opacities can often be identified (Fig. 11.13). In some cases an increase in intrathoracic lymph nodes and nodular changes in the lungs may mimic sarcoidosis [5]. After withdrawal of methotrexate, radiological abnormalities usually regress; however, in approximately 10% of the patients, traces of fibrosis and even honeycombing may be observed [24].

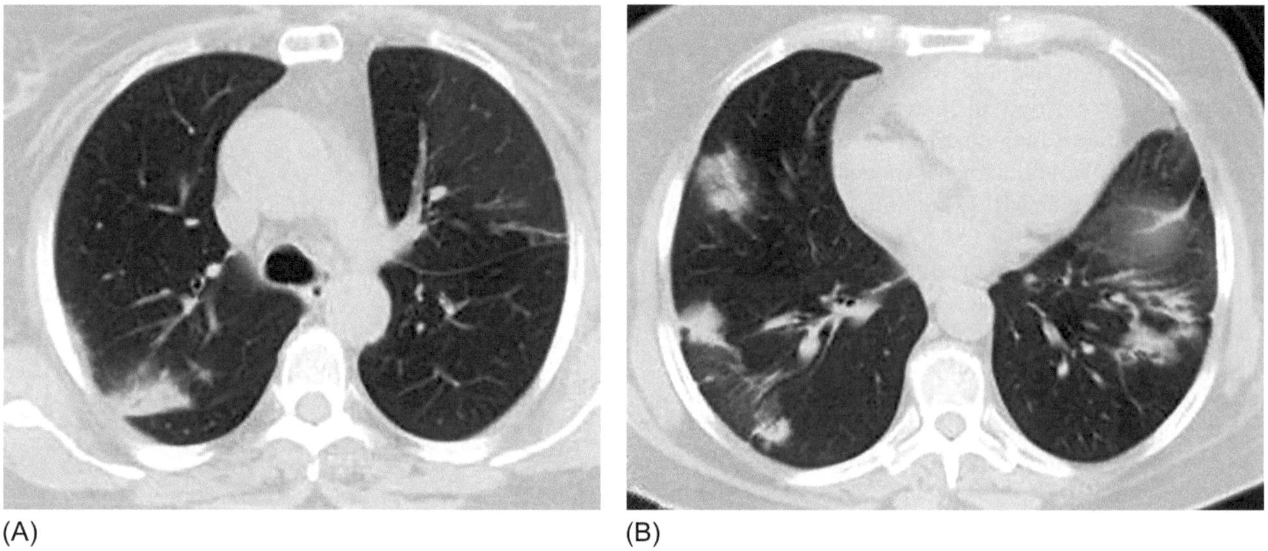

FIG. 11.12 Methotrexate-induced pneumonitis. Pattern of organizing pneumonia. Bilateral subpleural and peribronchovascular zones of consolidation and ground-glass opacities (A, B).

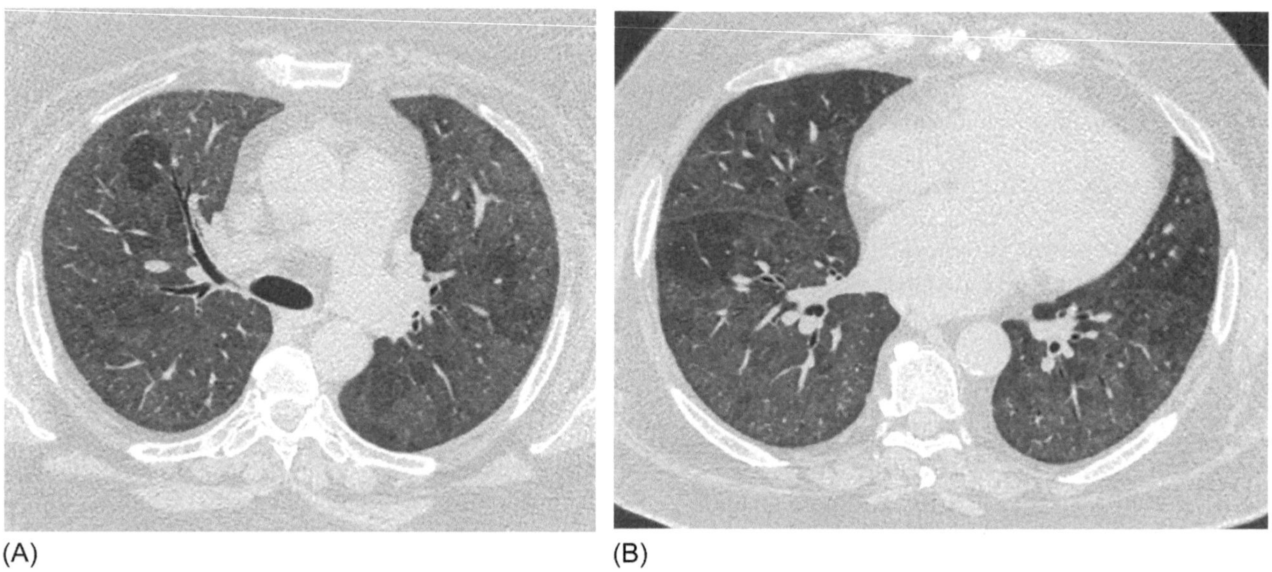

FIG. 11.13 Methotrexate-induced pneumonitis. Pattern of hypersensitivity pneumonitis. Diffuse areas of ground-glass opacities, distributed throughout the lung parenchyma, inside which lobular areas of the hyperinflation are visible. Subpleural sparing (A, B).

In 1987 Searles and McKendry developed the diagnostic criteria for MILD (Table 11.2). The definitive diagnosis of MILD requires fulfillment of a combination of first or second major criteria with third major criteria and at least with three minor ones [42].

Differential diagnosis

Differentiation of new infiltrative changes in the lungs of patients treated with methotrexate is not an easy task. The diagnostic range should encompass the spectrum from pneumonia caused by opportunistic or traditional pathogens to primary lung lesions in autoimmune diseases such as rheumatoid arthritis and inflammatory bowel disease. The most important signs for the differential diagnosis of MILD are presented in Table 11.3.

Differentiation of MILD from ILD caused by an underlying disease, such as rheumatoid arthritis and the Crohn disease, is not possible based on CT and histological data (Fig. 11.14). An important differential diagnostic sign is lymphocytosis in the

TABLE 11.2 Criteria for the diagnosis of methotrexate-induced pneumonitis

Major criteria	1. Histologically confirmed hypersensitivity pneumonitis without signs of infectious lesion 2. Radiologically confirmed alveolar or interstitial infiltration 3. Negative bacteriological blood and sputum culture
Minor criteria	1. Dyspnea up to 8 weeks 2. Unproductive cough 3. O_2 saturation < 90% in atmospheric air 4. DLCO < 70% of predicted value 5. Blood leukocytes < 15×10^9/L

DLCO, carbon monoxide diffusing capacity.

BAL fluid, which exceeds 30% (average 57.7% ± 24.4%) in most patients with MILD [38]. *Pneumocystis jirovecii* pneumonia (PP) has many clinical and radiological signs common with MILD (Fig. 11.15). Tokuda et al. [43] comparing the clinical and laboratory characteristics of patients with MILD and PP with rheumatoid arthritis (PP-RA) in patients treated with methotrexate found that the levels of C-reactive protein and the sialo-carbohydrate glycoprotein KL-6, a constituent of alveolar mucin, were significantly different between the two groups, with average C-reactive protein levels of 116 and 86 mg/l and average KL-6 levels of 814 and 1204 units/ml in the MILD and PP-RA groups, respectively. In addition, total IgG levels were higher in the MILD group (1551 vs. 1056 mg/dL in the PP-RA group). Additionally, half of the patients with rheumatoid arthritis had a history of interstitial lung lesions. All other clinical and laboratory signs, including the rate of symptom development that occurred over several days, frequency and intensity of fever, leukocyte and lymphocyte counts, lactate dehydrogenase level, and albumin, did not differ significantly between the two groups [43].

Although that study by Tokuda et al. did not compare the cytological findings of BAL, other studies indicate that severe lymphocytosis is a characteristic of methotrexate-associated lung lesions, whereas moderate lymphocytosis and neutrophilia are usually observed in PP. In addition, in cases where it was possible to analyze lymphocyte subpopulations, a decrease in the $CD4^+/CD8^+$ T lymphocyte ratio is typical for PP, as opposed to an increase observed frequently in MILD [40, 44].

The HRCT characteristics of MILD and PP-RA are also very similar, and Tokuda et al. comment that definitive differentiation of MILD and PP-RA by HRCT is impossible (Fig. 11.15). The definitive diagnosis should be based on the presence of *P. jirovecii* in BAL fluid. In cases with negative results for *Pneumocystis*, MILD should be considered as the diagnosis [43]. Lung biopsy is not considered mandatory study for the definitive diagnosis and is required only if there is no improvement with treatment and sometimes in cases of life-threatening conditions [37].

Treatment and prognosis

As with other drug-induced lesions of the lungs, discontinuation of methotrexate usually leads to clinical improvement and a decrease in BAL lymphocytosis within a few days, whereas radiological changes persist for up to several weeks [38, 45]. In severe cases with acute respiratory failure and need for respiratory support, high doses of systemic steroids may be required. Usually a starting dose of prednisone at 1 mg/kg body weight with subsequent reduction is reported to be effective. As a rule the duration of steroid therapy should not be too long. The disappearance of the signs of interstitial disease, such as GGO and consolidation, is a criterion for steroid discontinuation. The presence of irreversible local fibrosis that persists in approximately 10% of the patients after the completion of drug-induced disease is not an indication for continued steroid therapy. Long-term monitoring after MILD is not recommended [46]. Subsequent prescription of methotrexate should be avoided, as this may cause a new, more severe reemergence of MILD [47].

Lung damage by anticancer drugs

Anticancer drugs are the most common causes of DIPD according to several studies [48, 49]. This large group includes both traditional chemotherapeutic drugs (bleomycin, gemcitabine, and docetaxel) and newer ones including mammalian target of rapamycin inhibitors (sirolimus, temsirolimus, and everolimus), epidermal growth factor receptor-tyrosine kinase inhibitors (gefitinib and erlotinib), and immune checkpoint inhibitors (nivolumab, pembrolizumab, avelumab, and durvalumab). In general the frequencies of pulmonary toxicity due to these drugs vary, ranging from 1% to 3.6% [3] with a maximum prevalence in everolimus (up to 10.4%) and bleomycin (up to 46%) [50, 51].

TABLE 11.3 Differential diagnosis of methotrexate-induced pneumonitis

	MILD	ILD associated with underlying disease	Bacterial pneumonia	*Pneumocystis jirovecii* pneumonia
Clinical presentation	Progressive dyspnea when prolonged use of methotrexate	Slowly increasing dyspnea	Acute onset, cough with purulent sputum, fever	Fever, rapidly progressive respiratory failure
Laboratory data	Increased ESR, leukocytosis $< 15 \times 10^9$/L, procalcitonin < 0.5 ng/mL	Moderately accelerated ESR, correlated with activity of the underlying disease, procalcitonin < 0.5 ng/ml	Leukocytosis, accelerated ESR, CRP usually >100 mg/L, procalcitonin > 1 ng/mL	Normal or moderately increased leukocyte count, lymphopenia, procalcitonin < 1 ng/mL, CRP before onset of invasive lung ventilation usually below 100 mg/L
BAL	Significant lymphocytosis > 30%	Moderate neutrophilia, moderate lymphocytosis, increase in $CD4^+/CD8^+$ in the Crohn disease	Severe neutrophilia, the presence of microorganisms microscopically	Moderate neutrophilia, lymphocytosis, reduced $CD4^+/CD8^+$, PCR positive for *Pneumocystis jirovecii*
CT signs	GGO, intralobular nodules, lobular air trapping, bilateral subpleural foci of consolidation (with BOOP), extensive bilateral consolidation with diffuse alveolar damage	GGO with basal and subpleural distribution, reticular abnormalities, traction bronchiectasis, subpleural and basal honeycombing	Consolidation, frequently one site, the lack of reticular changes or fibrosis	Diffuse or patchy GGO, air trapping, consolidation in later stage, cysts. The changes start and are most pronounced in upper lobes

BAL, bronchoalveolar lavage; CRP, C-reactive protein; CT, computed tomography; ESR, erythrocyte sedimentation rate; GGO, ground-glass opacity; ILD, interstitial lung disease; BOOP, bronchiolitis obliterans organizing pneumonia; PCR, polymerase chain reaction.

Bleomycin-induced interstitial lung disease (BIILD) is the best studied of this type of complication. Bleomycin is directly toxic to epithelial and endothelial cells in the lungs, especially on type II pneumocytes, causing apoptosis and subsequent activation of alveolar macrophages and fibroblasts [52]. Bleomycin injury to the respiratory tract also leads to the activation of T lymphocytes with increased expression of proinflammatory cytokines and impaired surfactant function, which altogether leads to the development of interstitial inflammation and pulmonary fibrosis [53]. Risk factors of BIILD are age of over 70 years; high cumulative dose (usually higher than 400 IU); impaired renal function; and, to a lesser extent, smoking, intake of granulocyte colony-stimulating factor, and inhalation of high oxygen concentrations during anesthesia [51]. Despite the dose-dependent incidence of bleomycin-induced lung damage, severe cases with exposure to bleomycin doses of less than 100 IU were also described [54]. The average time from the onset of administration to the development of pulmonary symptoms is 4.2 months [55].

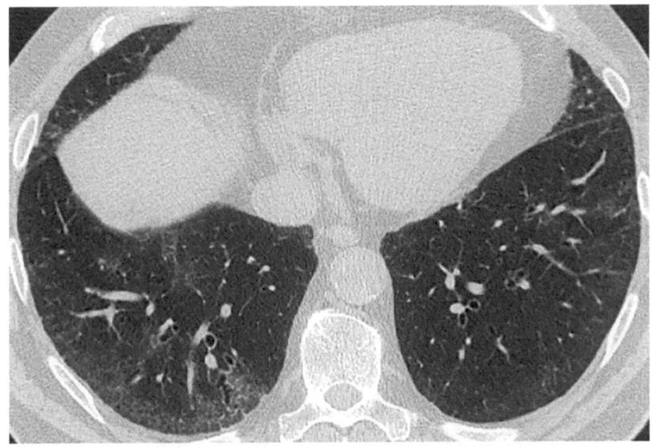

FIG. 11.14 Nonspecific interstitial pneumonia, confirmed histologically, in a patient with rheumatoid arthritis, who was treated with methotrexate. Subpleural areas of ground-glass opacity in the lower lobes, more pronounced on the right, with slightly thickened intralobular septa and onset of traction bronchiectasis. Bronchoalveolar lavage fluid contained 7% lymphocytes.

The clinical and morphological presentation of BIILD includes DAD, NSIP, and organizing pneumonia [23]. The rarer variants are hypersensitivity pneumonitis and usual interstitial pneumonia [6, 56].

Symptoms of BIILD are nonspecific and depend on the morphological variant of the lung lesion. In DAD, progressive respiratory failure rapidly develops with bilateral massive zones of GGO with or without consolidation. This critical presentation requires differential diagnosis from other similar conditions, namely, infections, acute eosinophilic pneumonia, and acute interstitial pneumonia. To rule out the first two causes as the etiology, BAL fluid examination with cell counts and polymerase chain reaction for *P. jirovecii*, cytomegalovirus, and *Haemophilus influenzae* are necessary. Lung biopsy does not always reveal the cause of DAD. The morphological signs are frequently nonspecific and include hyaline membranes and organizing pneumonia, although it is possible to detect intranuclear tubular structures in type II pneumocytes typical for bleomycin- or busulfan-induced damage by ultrastructural analysis in some cases [19]. Thus only a history of bleomycin use with the exclusion of other causes of DAD is necessary for the definitive diagnosis of BBD.

In other variants the disease course of BIILD course often includes a gradually advancing dyspnea. Up to 39% of the patients are asymptomatic and can only be detected by CT evaluation (Figs. 11.16 and 11.17) [3].

The diagnosis of BIILD and drug-induced pneumonitis caused by other immunosuppressive agents is based on identification of the chronological relationship between drug treatment and appearance of interstitial changes in the lungs together

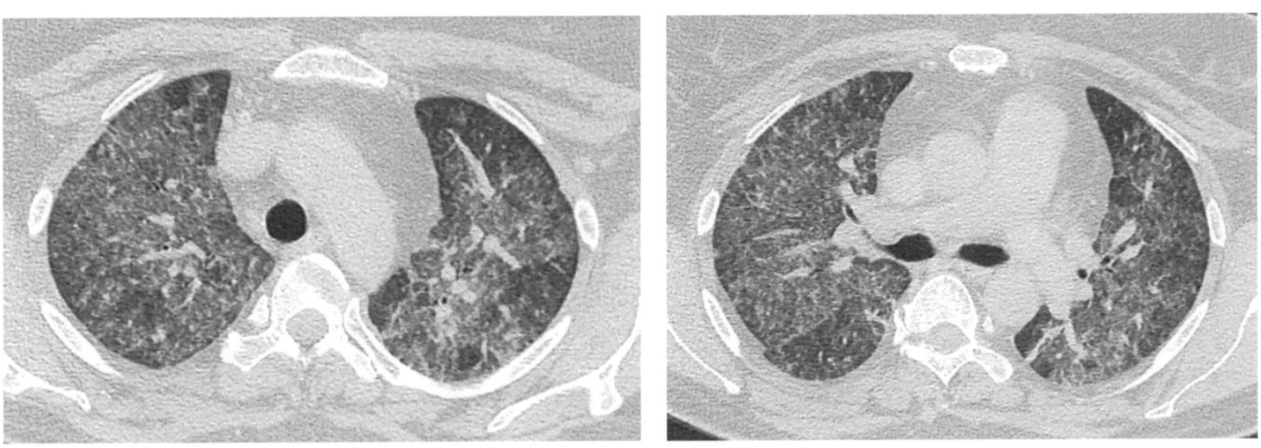

(A) (B)

FIG. 11.15 *Pneumocystis jirovecii* pneumonia in a patient with rheumatoid arthritis, who was treated with methotrexate. Diffuse ground-glass opacities associated with thickened intralobular and interlobular septa, lobular air trapping, and subpleural sparing (A, B). Findings closely resemble the hypersensitivity pneumonitis pattern.

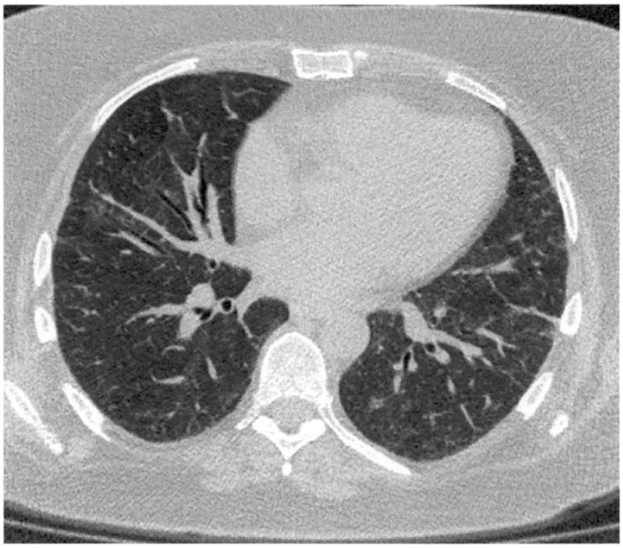

FIG. 11.16 Bleomycin-induced pneumonitis without clinical manifestations. Diffuse zones of mildly increased attenuation, which is pronounced peribronchovascularly. Mild reticular subpleural abnormalities and linear opacities are visible.

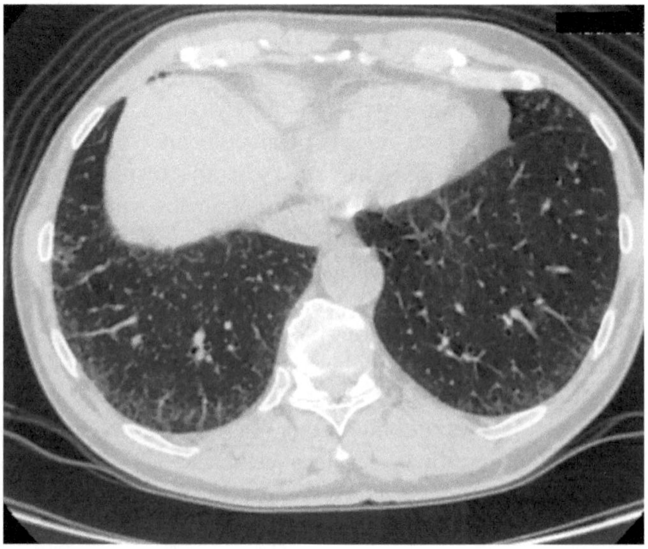

FIG. 11.17 Bleomycin-induced pneumonitis without clinical manifestations. Subpleural layer of ground-glass opacities with thickened intralobular septa.

with the exclusion of other possible causes of diffuse parenchymal disease, which include primarily opportunistic infections, most often *Pneumocystis* spp. For example, in a study by Solazzo et al. [57], in 12 of the 26 patients who received everolimus and developed HRCT signs of interstitial lung disease, the cause was pulmonary infection. In 5 (43%) of the 12 patients, the etiology was *P. jirovecii* despite the prophylactic administration of trimethoprim-sulfamethoxazole, and drug-induced pneumonitis was noted in the remaining 14 patients.

Thus BAL is an important diagnostic tool that should be used to rule out diseases similar to DIPD. Due to nonspecific morphological patterns [58], the value of lung biopsy appears to be lower in DIPD than in other interstitial diseases. Similar to BAL analysis, lung biopsy should be necessary in patients with doubtful diagnosis of DIPD after a thorough assessment of the benefits and risks. Absence of specific disease symptoms generally makes the diagnosis of DIPD only more or less likely, based on the history of drug treatment and data related to the pulmonary toxicity of the specific drug on www.pneumotox.com or other validated sources.

Treatment options are available only for the most frequent forms of DIPD: AILD and MILD. Nonetheless, in most cases, suspicious DIPD requires the withdrawal of the candidate drug, while improvement of the disease confirms the diagnosis. However, in the absence of alternative drugs and the presence of progression risk of the underlying disease, continuation of the drug treatment is permissible in combination with additional systemic corticosteroids [3]. In cases of life-threatening conditions due to DIPD, such as diffuse alveolar damage, we suggest that the culpable drug should always be withdrawn. Systemic steroids are the primary option for the treatment of DIPD; however, common approaches regarding the dose, duration, and rules for dose reduction remain unclear [3, 6]. An acute onset and severe lung lesions, especially DAD, are indicators of unfavorable DIPD outcomes [3].

References

[1] Classen DC, Pestotnik SL, Evans RS, Burke JP. Computerized surveillance of adverse drug events in hospital patients. JAMA 1991;266(20):2847–51.

[2] Nebeker JR, Barach P, Samore MH. Clarifying adverse drug events: a clinician's guide to terminology, documentation, and reporting. Ann Intern Med 2004;140(10):795–801.

[3] Skeoch S, Weatherley N, Swift AJ, Oldroyd A, Johns C, Hayton C, et al. Drug-induced interstitial lung disease: a systematic review. J Clin Med 2018;7(10). pii: E356.

[4] Kolaitis NA, Kukreja J, Jones KD, Hays SR, Leard LE. Pirfenidone-induced sarcoid-like reaction: a novel complication. Chest 2018;154(4):e89–92.

[5] Mason RJ, Broaddus VC, Martin TR, King Jr TE, Schraufnagel DE, Murray JF, et al, editors. Drug induced pulmonary disease. In: Murray and Nadel's textbook of respiratory medicine. 5th ed. Philadelphia, PA: Saunders Elsevier; 2010. p. 1691–718.

[6] Roden AC, Camus P. Iatrogenic pulmonary lesions. Semin Diagn Pathol 2018;35(4):260–71.

[7] Jafari-Fesharaki M, Scheinmann MM. Adverse effects of amiodarone. Pacing Clin Electrophysiol 1998;21(1) Pt 1. 108–20.

[8] Goldschlager N, Epstein AE, Naccarelli GV, Olshansky B, Singh B, Collard HR, et al. Practice Guidelines Sub-committee, North American Society of Pacing and Electrophysiology (HRS). A practical guide for clinicians who treat patients with amiodarone: 2007. Heart Rhythm 2007;4(9):1250–9.
[9] Kennedy JI, Myers JL, Plumb VJ, Fulder JD. Amiodarone pulmonary toxicity. Clinical, radiologic, and pathologic correlations. Arch Intern Med 1987;147(1):50–5.
[10] Halliwell WH. Cationic amphiphilic drug-induced phospholipidosis. Toxicol Pathol 1997;25(1):53–60.
[11] Ernawati DK, Stafford L, Hughes JD. Amiodarone-induced pulmonary toxicity. Br J Clin Pharmacol 2008;66(1):82–7.
[12] Wolkove N, Baltzan M. Amiodarone pulmonary toxicity. Can Respir J 2009;16(2):43–8.
[13] Olshansky B, Sami M, Rubin A, Kostis J, Shorofsky S, Slee A, et al. Use of amiodarone for atrial fibrillation in patients with preexisting pulmonary disease in the AFFIRM study. Am J Cardiol 2005;95(3):404–5.
[14] Ashrafian H, Davey P. Is amiodarone an underrecognized cause of acute respiratory failure in the ICU? Chest 2001;120(1):275–82.
[15] Kharabsheh S, Abendroth CS, Kozak M. Fatal pulmonary toxicity occurring within two weeks of initiation of amiodarone. Am J Cardiol 2002;89(7):896–8.
[16] Schwaiblmair M, Berghaus T, Haeckel T. Amiodarone-induced pulmonary toxicity: an under-recognized and severe adverse effect? Clin Res Cardiol 2010;99(11):693–700.
[17] Yamada Y, Shiga T, Matsuda N, Hagiwara N, Kasanuki H. Incidence and predictors of pulmonary toxicity in japanese patients receiving low-dose amiodarone. Circ J 2007;71(10):1610–6.
[18] Larsen BT, Vaszar LT, Colby TV, Tazelaar HD. Lymphoid hyperplasia and eosinophilic pneumonia as histologic manifestations of amiodarone-induced lung toxicity. Am J Surg Pathol 2012;36(4):509–16.
[19] Leslie K, Wick M. Practical pulmonary pathology: a diagnostic approach. 2nd ed. London, UK: Elsevier Health Sciences; 2011. p. 120–31.
[20] Mankikian J, Favelle O, Guillon A, Guilleminault L, Cormier B, Jonville-Béra AP, et al. Initial characteristics and outcome of hospitalized patients with amiodarone pulmonary toxicity. Respir Med 2014;108(4):638–46.
[21] Ohar JA, Jackson F, Dettenmeier PA, Bedrossian CW, Tricomi SM, Evans RG. Bronchoalveolar lavage cell count and differential are not reliable indicators of amiodarone-induced pneumonitis. Chest 1992;102(4):999–1004.
[22] Myers JL, Kenedy JI, Plumb VJ. Amiodarone lung: pathologic findings in clinically toxic patients. Hum Pathol 1987;18(4):349–54.
[23] Erasmus JJ, McAdams HP, Rossi SE. High-resolution CT of drug-induced lung disease. Radiol Clin N Am 2002;40(1):61–72.
[24] Webb RW, Higgins CB. Thoracic imaging: pulmonary and cardiovascular radiology. 2nd ed. Philadelphia, PA: Lippincott Williams & Wilkins; 2011.
[25] Kuhlman JE, Teigen C, Ren H, Hruban RH, Hutchins GH, Fishman EK. Amiodarone pulmonary toxicity: CT findings in symptomatic patients. Radiology 1990;177(1):121–5.
[26] Greenspon AJ, Kidwell GA, Hurley W, Mannion J. Amiodarone-related postoperative adult respiratory distress syndrome. Circulation 1991;84(5 Suppl. III). III-407–III-415.
[27] Holmes Jr. DR, Monahan KH, Packer D. Pulmonary vein stenosis complicating ablation for atrial fibrillation: clinical spectrum and interventional considerations. JACC Cardiovasc Interv 2009;2(4):267–76.
[28] Zaher C, Hamer A, Peter T, Mandel W. Low-dose steroid therapy for prophylaxis of amiodarone-induced pulmonary infiltrates. N Engl J Med 1983;308(13):779.
[29] Chisaki Y, Aoji S, Yano Y. Analysis of adverse drug reaction risk in elderly patients using the japanese adverse drug event report (JADER) Database. Biol Pharm Bull 2017;40(6):824–9.
[30] Goodman TA, Polisson RP. Methotrexate: adverse reactions and major toxicities. Rheum Dis Clin N Am 1994;20(2):513–28.
[31] Borchers AT, Keen CL, Cheema GS, Gershwin ME. The use of methotrexate in rheumatoid arthritis. Semin Arthritis Rheum 2004;34(1):465–83.
[32] Kinder AJ, Hassell AB, Brand J, Brownfield A, Grove M, Shadforth MF. The treatment of inflammatory arthritis with methotrexate in clinical practice: treatment duration and incidence of adverse drug reactions. Rheumatology (Oxford) 2005;44(1):61–6.
[33] Kremer JM, Alarcon GS, Weinblatt ME, Kaymakcian MV, Macaluso M, Cannon GW, et al. Clinical, laboratory, radiographic, and histopathologic features of methotrexate-associated lung injury in patients with rheumatoid arthritis: a multicenter study with literature review. Arthritis Rheum 1997;40(10):1829–37.
[34] Conway R, Low C, Coughlan RJ, O'Donnell MJ, Carey JJ. Methotrexate and lung disease in rheumatoid arthritis: a meta-analysis of randomized controlled trials. Arthritis Rheumatol 2014;66(4):803–12.
[35] Imokawa S, Colby TV, Leslie KO, Helmers RA. Methotrexate pneumonitis: review of the literature and histopathological findings in nine patients. Eur Respir J 2000;15(2):373–81.
[36] Elsasser S, Dalquen P, Soler M, Perruchoud AP. Methotrexate-induced pneumonitis: appearance four weeks after discontinuation of treatment. Am Rev Respir Dis 1989;140(4):1089–92.
[37] Lateef O, Shakoor N, Balk RA. Methotrexate pulmonary toxicity. Expert Opin Drug Saf 2005;4(4):723–30.
[38] Fuhrman C, Parrot A, Wislez M, Prigent H, Boussaud V, Bernaudin JF, et al. Spectrum of CD4 to CD8 T-cell ratios in lymphocytic alveolitis associated with methotrexate-induced pneumonitis. Am J Respir Crit Care Med 2001;164(7):1186–91.
[39] Yamakawa H, Yoshida M, Takagi M, Kuwano K. Late-onset methotrexate-induced pneumonitis with neutrophilia in bronchoalveolar lavage fluid. BMJ Case Rep 2014;2014. pii: bcr2014206123.
[40] Schnabel A, Richter C, Bauerfeind S, Gross WL. Bronchoalveolar lavage cell profile in methotrexate induced pneumonitis. Thorax 1997;52(4):377–9.
[41] Muller NL, Silva CS, editors. High-yield imaging: chest. 1st ed. Philadelphia, PA: Saunders Ilsivier; 2010. p. 538–9.
[42] Searles G, McKendry RJ. Methotrexate pneumonitis in rheumatoid arthritis: potential risk factors. Four case reports and a review of the literature. J Rheumatol 1987;14(6):1164–71.

[43] Tokuda H, Sakai F, Yamada H, Johkoh T, Imamura A, Dohi M, et al. Clinical and radiological features of Pneumocystis pneumonia in patients with rheumatoid arthritis, in comparison with methotrexate pneumonitis and Pneumocystis pneumonia in acquired immunodeficiency syndrome: a multicenter study. Intern Med 2008;47(10):915–23.

[44] Tamai K, Tachikawa R, Tomii K, Nagata K, Otsuka K, Nakagawa A, et al. Prognostic value of bronchoalveolar lavage in patients with non-HIV pneumocystis pneumonia. Intern Med 2014;53(11):1113–7.

[45] Cooper Jr JA, White DA, Matthay RA. Drug-induced pulmonary disease: cytotoxic drugs. Part 1. Am Rev Respir Dis 1986;133(2):321–40.

[46] Cottin V, Tebib J, Massonnet B, Souquet PJ, Bernard JP. Pulmonary function in patients receiving long-term low-dose methotrexate. Chest 1996;109(4):933–8.

[47] Kremer JM, Phelps CT. Long-term prospective study of the use of methotrexate in the treatment of rheumatoid arthritis. Update after a mean of 90 months. Arthritis Rheum 1992;35(2):138–45.

[48] Piciucchi S, Romagnoli M, Chilosi M, Bigliazzi C, Dubini A, Beomonte ZB, et al. Prospective evaluation of drug-induced lung toxicity with high-resolution CT and transbronchial biopsy. Radiol Med 2011;116(2):246–63.

[49] Tamura M, Saraya T, Fujiwara M, Hiraoka S, Yokoyama T, Yano K, et al. High-resolution computed tomography findings for patients with drug-induced pulmonary toxicity, with special reference to hypersensitivity pneumonitis-like patterns in gemcitabine-induced cases. Oncologist 2013;18(4):454–9.

[50] Iacovelli R, Palazzo A, Mezi S, Morano F, Naso G, Cortesi E. Incidence and risk of pulmonary toxicity in patients treated with mTOR inhibitors for malignancy. A meta-analysis of published trials. Acta Oncol 2012;51(7):873–9.

[51] Froudarakis M, Hatzimichael E, Kyriazopoulou L, Lagos K, Pappas P, Tzakos AG, et al. Revisiting bleomycin from pathophysiology to safe clinical use. Crit Rev Oncol Hematol 2013;87(1):90–100.

[52] Hay J, Shahzeidi S, Laurent G. Mechanisms of bleomycin-induced lung damage. Arch Toxicol 1991;65(2):81–94.

[53] Schmidt R, Ruppert C, Markart P, Lubke N, Ermert L, Weissmann N, et al. Changes in pulmonary surfactant function and composition in bleomycin-induced pneumonitis and fibrosis. Toxicol Appl Pharmacol 2004;195(2):218–31.

[54] McLeod BF, Lawrence HJ, Smith DW, Vogt PJ, Gandara DR. Fatal bleomycin toxicity from a low cumulative dose in a patient with renal insufficiency. Cancer 1987;60(11):2617–20.

[55] O'Sullivan JM, Huddart RA, Norman AR, Nicholls J, Dearnaley DP, Horwich A. Predicting the risk of bleomycin lung toxicity in patients with germ-cell tumours. Ann Oncol 2003;14(1):91–6.

[56] Cleverley JR, Screaton NJ, Hiorns MP, Flint JD, Müller NL. Drug-induced lung disease: high-resolution CT and histological findings. Clin Radiol 2002;57(4):292–9.

[57] Solazzo A, Botta C, Nava F, Baisi A, Bonucchi D, Cappelli G. Interstitial lung disease after kidney transplantation and the role of everolimus. Transplant Proc 2016;48(2):349–51.

[58] Romagnoli M, Bigliazzi C, Casoni G, Chilosi M, Carloni A, Dubini A, et al. The role of transbronchial lung biopsy for the diagnosis of diffuse drug-induced lung disease: a case series of 44 patients. Sarcoidosis Vasc Diffuse Lung Dis 2008;25(1):36–45.

Index

Note: Page numbers followed by *f* indicate figures and *t* indicate tables.

M

N

O

P